AF323458

Taming the Forces

between

Quarks and Gluons

— Calorons Out of The Box —

Scientific Papers by
Pierre van Baal

Taming the Forces
between
Qvarks and Gluons
— Calorons Out of The Box —

Scientific Papers by
Pierre van Baal

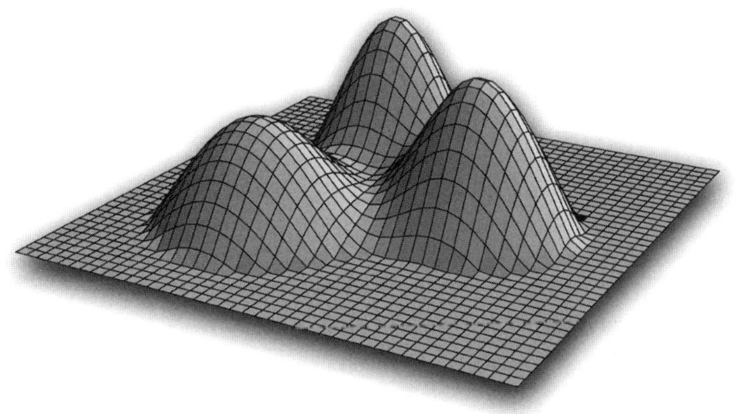

edited by

Gerard 't Hooft
Utrecht University, The Netherlands

Chris P. Korthals Altes
NIKHEF, The Netherlands & CNRS, France

World Scientific

NEW JERSEY · LONDON · SINGAPORE · BEIJING · SHANGHAI · HONG KONG · TAIPEI · CHENNAI

Published by

World Scientific Publishing Co. Pte. Ltd.

5 Toh Tuck Link, Singapore 596224

USA office: 27 Warren Street, Suite 401-402, Hackensack, NJ 07601

UK office: 57 Shelton Street, Covent Garden, London WC2H 9HE

British Library Cataloguing-in-Publication Data
A catalogue record for this book is available from the British Library.

The editors and publisher would like to thank the following publishers of the journals and books for their assistance and permission to include the selected reprints found in this volume:

American Physical Society (*Phys. Rev. Letts.*, *Phys. Rev. D*)
Elsevier (*Annals of Physics, Nucl. Phys. B, Nucl. Phys. B (Proc. Suppl.), Phys. Lett. B*)
Institute of Physics, Jagiellonian University (*Acta Phys. Polon. B*)
Koninklijke Nederlandse Akademie van Wetenschappen (*Proc. K. Ned. Akad. Wet.*)
China Science Publishing & Media Ltd. (*Proc. of ISATQP-Shanxi*)
SISSA and IOP Publishing (*JHEP*)
Springer Science+Business Media (*Comm. Math. Phys.*)

TAMING THE FORCES BETWEEN QUARKS AND GLUONS — CALORONS OUT OF THE BOX
Scientific Papers by Pierre van Baal

ISBN   978-981-4447-86-7

Printed in Singapore by World Scientific Printers.

CONTENTS

PREFACE

It was still in the 1980s that visitors of Disney world could choose to make a roller coaster ride into the "World of the Small". Riding through the intestines of mice, passing worms and insects, the visitor raced to blood cells, DNA molecules and then single atoms were visited. Finally the tour ended at a blinking light. "This is the nucleus of an atom", the narrator said, "and nobody knows what's in there". That was the end of the ride.

But already in 1980, a lot more was known about what's inside the nucleus of an atom. After the neutron was discovered by Chadwick, in 1932, it was soon realized that atomic nuclei are filled with protons and neutrons. For a long time, protons and neutrons were regarded as elementary objects, until numerous observations of new particles revealed that protons and neutrons must have a substructure. In 1964, Gell-Mann presented his theory that these particles are made of more elementary constituents that he called quarks. Then, in the 1970s a lot more became clear. There is a unique force that controls the behavior of quarks: they are kept together in grouplets of three, or alternatively, bound to their antiparticles, the anti-quarks. The force that is responsible for this, now called "quantum chromodynamics", turned out to be be closely related to other subnuclear forces that had been tamed at around the same time: the electric, magnetic and weak forces. However, the force between quarks is very strong. The force-carrying particles, the gluons, are roaming around in the protons and neutrons as well, and they may cast these into new excited states, states that vibrate or rotate faster than the more stable protons and neutrons themselves.

All these exciting things were not told to the visitor of Disney world, but maybe this was because, indeed, a lot remained unknown as well. Handling and understanding the mathematics of the behavior of quarks and gluons was far from easy. These forces act a bit like the forces in waves of an ocean: if there are just tiny ripples, these spread in an orderly manner, passing right through one another without any noticeable disturbance, but when the waves become high, that's quite something else. They roll over and by doing this, become a lot less predictable.

It was October 1980, now over thirty years ago, that Pierre van Baal began his thesis work on this problem, at Utrecht University. His starting point was to look at what happens if you confine the quarks and gluons inside a mathematical box. We cannot make real boxes in nuclear systems, but we can study the mathematics of what you would get. This mathematics becomes more manageable, just like what it would be like to study ocean waves if we would replace the ocean by a small pond. Even in a pond, given enough wind, waves can become high and irregular. Mathematically, a pond becomes more like an ocean if you replace the sides of the pond by periodic boundary conditions. Understanding such a system, may be an important first step towards understanding the real thing. In July 1984, Pierre graduated, but he realized that his work was not yet finished. Periodic boundary conditions allow for all sorts of interesting modifications, and each of these generate different structures inside the pond.

Pierre went to the University of New York in Stony Brook, and then to CERN, Geneva, where he continued his research.

Behavior of water waves depends on the temperature of the water, and indeed, also quarks and gluons behave differently when their temperature is modified. But, unlike water in a pond, quarks and gluons at a finite temperature behave as if their box is *four* dimensional. New kinds of structures become relevant. Pierre concentrated on structures he called "calorons". Again, we are then far away from the perturbative regime, that is, the regime where everything takes the shape of simple ripples.

Physicists do have other ways of studying the theory of quarks and gluons. One can try to let a supercomputer do the job, but then also, one has to consider these particles to be confined to a box. It now became very interesting to check to what extent the computer simulations concur with Pierre's theoretical findings. They do, but both groups of scientists can learn a lot from each other.

Pierre became full professor in Leiden twenty years ago, in 1992, and he continued to make important contributions and discoveries in this field, until his career was abruptly disturbed when, one day, end of July 2005, he was found missing at his institute. Knowing that not showing up without notice was not his habit, his colleagues became worried, and with the help of police broke into his apartment, where he was found unconscious. Pierre had suffered from a dangerous stroke in the brain. Partly paralyzed, he began a slow and demanding struggle to regain his health. He has now improved a lot, not in the least because of his strong will, and makes considerable efforts to resume his scientific work.

In the mean time we decided that the moment has come to make the balance of his discoveries and contributions. Several of our friends and colleagues inquired whether a compilation of Pierre van Baal's work could be made available. Here we have made a selection of his most important and influential papers, enlightened by his own comments. The work is intended for advanced researchers, postdoctoral workers as well as graduate students. It is for the benefit of the latter that we include in the introduction a review of how QCD came into being, and its salient features. The chronological order chosen is also the order this work can be best presented.

We thank Ana Achúcarro, Dmitri Diakonov as well as Tony Gonzalez-Arroyo for their strong support in this project. Shortly before this volume came to completion, we learned the sad news that Dmitry (or Mitya, as friends and colleagues called him) would not live to see it; he passed away last December 26.

Gerard 't Hooft
Chris P. Korthals Altes,
editors.

INTRODUCTION

"Und läg er nur noch immer in dem Grase!
In jeden Quark begräbt er seine Nase."

Mephisto in the Prolog im Himmel of Faust I
Johann Wolfgang von Goethe

I. WHAT THIS COLLECTION OF REPRINTS IS ABOUT

Mephisto's wisecrack about humans and the human condition was inspired by their quest for knowledge. His words are a worthy accompaniment of the effort described in this book: to understand the strong force that keeps the building stones of matter, the quarks, tightly bound inside protons and guarantees their stability. This, in a nutshell, is what this reprint collection is about.

Indeed the theory of quantum chromo dynamics (QCD) is of Faustian pretensions. Our learned Doctor would have objected against its mystical appearance:

QCD assumes the existence of quarks but it denies their detectability, indeed,
a Faustian conundrum!

The reader may well ask: *where on Earth did physicists dig up such a theory?*

Seasoned scientists, of course, know the answer: *This is the theory that works!* And, once we understand the nature of strong and weak forces a bit better, we realize that this theory is not so strange at all. It is one of the mathematically sound quantum field theories that is appropriate for systems of the strongly interacting subatomic particles: *mesons* and *baryons*.

This book is to commemorate the profound contributions made to this theory by Prof. Pierre van Baal. When Pierre started his PhD work on this new theory, not yet much was understood about the internal structure of mesons and baryons. The theory claims that forces due to a direct generalization of the concept of electro-magnetism are responsible for the peculiar behavior of the mysterious constituents of these particles, called quarks and gluons. Gradually, we understood more and more.

We begin this work with a general, less technical Introduction, putting the general development of the strong interaction theory in a historical perspective, and illustrating how prescient Pierre's work has been. The technical matters start in section V of this introduction and continue in the reprint section of the book. Especially the mathematically inclined reader is advised to go to Pierre's original articles in this volume.

A. A short history of Quantum Chromo Dynamics

In the early sixties the world of particle physics had been confronted with quite large differences in time scales. On the one hand there are stable particles, such as the proton, electron and their anti particles, and of course the photon. On the other hand the neutron can live outside the nucleus for not much more than fifteen minutes. It decays into a proton, an electron, and an elusive particle called the neutrino. Time scales ranging from microseconds to minutes, or even many years, are all typical for the weak interactions; these time scales vary a lot as they depend strongly on the energy available for a reaction to take place.

Experiments, such as the ones at Brookhaven's synchrotron, unveiled a world of mesons, and resonances of mesons colliding with protons. A meson like the neutral pion decays in 8.4×10^{-17} sec. into two photons. This time scale is typical for decays that require the electro magnetic force, the force that produces photons.

When a particle enters into an excited mode, called *resonance*, this state is much shorter lived, typically 10^{-22} seconds. This is the time scale needed by the strong interactions to generate transitions. The resonances were originally viewed as unstable particles themselves, and there were many of them. There were so many that the question arose whether they could fit in something akin to the periodic table of the chemical elements.

The answer came from Gell-Mann and Zweig [1]: the baryons and mesons indeed display repetitive patterns, and these can be ascribed to underlying constituents, the "quarks", particles with spin 1/2. At that time three different quarks species were enough to explain all the known resonances. The three quarks are seen as the members of a triplet, mathematically the fundamental representation of the group $SU(3)$. Two of them, the "up" and "down" (u and d) are members of a doublet in an SU(2) subgroup. The third one is called "strange" (s). It refers to a new property it displays, called *strangeness*, which is conserved in strong interaction prcocesses.

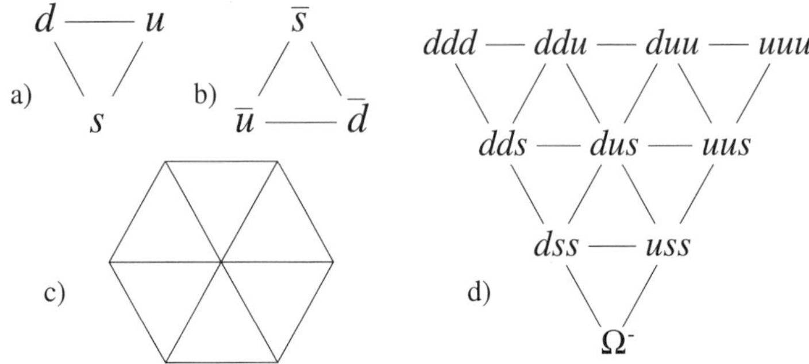

FIG. 1: Quark triplets and baryon multiplets: isospin acts in the horizontal direction, strangeness in the vertical direction. *a)* The fundamental quark triplet. *b)* The antiquark triplet. *c)* This pattern, characteristic for $SU(3)$, is called the octet, to which protons and neutrons with spin 1/2 belong (note that there are two particles at the same point in the center). *d)* The decuplet of baryons with spin 3/2 arises when the three flavors take a fully symmetric position; apparently, the three quarks then all spin in the same direction, so that the spins add up. The particle Ω^- is the (sss) combination.

The mathematical details followed from $SU(3)$ theory. Two quark species, u and d, can be arranged to form the proton (uud) and the neutron (udd). If one or two of these are replaced by the third quark, s, six more cousins of the proton and the neutron can be formed. Together, these are eight baryons with spin 1/2, called the octet representation of $SU(3)$. At the time, all of these octet particles were already known. But when putting together three quarks, $SU(3)$ predicts also a group of ten objects, called a decuplet, that would be each other's look alike. It so happend that nine particles had been discovered that all have spin 3/2. Gell-Mann predicted that there should be a tenth. Soon after his quark proposal, the tenth spin 3/2 baryon was discovered. This particle consists of three s quarks, and is called the Ω^-. The superscript refers to its electric charge which is equal to that of the electron.

One of the striking attributes of quarks is their fractional electric charge, Q. From the Ω^- we deduce that the s quarks must have charge $Q = -1/3$. From the experimentally observed baryons we see that the down quark must have the same fractional charge as the s quark. But the up quark has charge one unit higher: $Q = 2/3$. Mathematically, we can define a quantity that distinguishes the u from the d quark; it is called (the 3rd component of) *isospin*, I_3. In fig. (1), the quantity called I_3 of isospin is on the horizontal axis.

Baryon number $\mathcal{B}$ of the quarks is $1/3$ of that of the proton. Strangeness S is 0 for up and down, and -1 for the strange quark. The so-called Gell-Mann Nishijima formula relates isospin, strangeness and baryon number not only for the baryons and mesons, but also for the quarks:

$$Q = I_3 + \frac{\mathcal{B} + S}{2}. \tag{1}$$

In the seventies and eighties more quark species, called *flavors* were discovered, so that we now not only have isospin I_3 and strangeness S, but also C, B and T, standing respectively for "charm", "bottom" and "top". These new quantum numbers are preserved in the strong interactions, and the group multiplet structure is no longer detemined by the group $SU(3)$, but now by $SU(6)$ (the mathematics of six flavors). The large discrepancies in masses of the top, bottom charm and the older three quarks make that multiplets of hadronic particles with all these different quarks in them are not even approximately degenerate in mass.

Mesons with integer spin are formed by combining quarks and anti-quarks. In the old $SU(3)$ of the three flavor case, this gives octets and singlets, with the same successful agreement with the experimentally observed spectrum.

There remained unsolved questions, though.

One was related to the statistics of quarks [2]. Take the *sss* state Ω^- with spin $3/2$. All the spins of the s quarks apparently are aligned, otherwise they would not add up to get spin $3/2$. But this means that all quarks are in the same state, so the state is symmetric, unless we admit non-zero angular momentum. This is at odds with a rule that so-far was always found to hold in quantum mechanics: *The wave function of multiples of particles must be antisymmetric when we interchange objects with half-odd spin.* This rule, describing the statistics of many-particle systems, called Fermi statistics, is very important for the internal consistency of particle theory, but here, it seemed to be violated.

The solution that was found is to give every s quark an extra "color" charge; there exist exactly three different primary colors. Things must be arranged such that the total color should be "white", or "colorless". This means that in all baryonic states, all three colors occur exactly once. And this implies that the quark wave function is fully antisymmetric under interchange of the colors. Fermi's statistics for the quarks can be fully restored this way! All quark flavors must have this new color degree of freedom.

The next thing that was discovered is that color charge can generate a force just the same way as electric charge generates electromagnetism, by emitting or absorbing photons. Also color charge couples to a particle, called gluon. Just as the photon, the gluon is massless and has spin one. It is however in the octet representation of color $SU(3)$. It does change the color of the quark when it is emitted or absorbed as a "colored photon". This is why this interaction was called *Quantum Chromodynamics* (QCD) [3]. Technically, it is described by a Lagrange function $\mathcal{L}$, built from the octet field strengths of the gluons, and the $SU(3)$ Gell-Mann generators; these are the eight independent traceless hermitean 3×3 matrices λ_a (the octet color index a runs from 1 to 8). Their normalization is usually taken to be $\mathrm{Tr}\lambda_a\lambda_b = \frac{1}{2}\delta_{ab}$.

One then employs the notation:

$$F_{\mu\nu} = F^a_{\mu\nu}\lambda_a \tag{2}$$

and similarly for the vector potential A_μ.

The field strength reads in terms of the potentials:

$$F_{\mu\nu} = \partial_\mu A_\nu - \partial_\nu A_\mu + i[A_\mu, A_\nu]. \tag{3}$$

The last term in the field strength is the commutator term and it implies interactions between the gluons.

The action for QCD now takes the very simple form:

$$S = \int_x \mathcal{L}(x)\mathrm{d}^4x \; ; \qquad \mathcal{L}(x) = \frac{1}{2g^2}\mathrm{Tr}F_{\mu\nu}^2 + \bar{q}_{i,f}\{\gamma_\mu D_\mu(A)_{ij} + \delta_{ij}m_f\}q_{j,f} \; . \tag{4}$$

The quark fields q carry spinor indices, which couple to the Dirac γ matrices. To avoid clutter in the formula we suppressed those. The fundamental color indices i,j are written explicitly and run from 1 to 3. So are the flavor indices f. The covariant derivative equals in matrix form:

$$D_\mu(A) = \partial_\mu + iA_\mu^a\lambda_a \; . \tag{5}$$

Often we use the shorthand D_μ, dropping the reference to the potential. Experiments with deep inelastic scattering had shown the absence of interactions between the quarks at short distances, on the order of 1 femtometer or less. This was reflected in the strength of the coupling g of the gluons at short distance: this coupling strength is not truly constant, but decreases logarithmically when the distance decreases. This important property is called *asymptotic freedom* of QCD, as was first described in detail by Politzer, Gross and Wilczek [4], but observed a year earlier by G. 't Hooft at a conference in Marseille, 1972.

We probe hadronic matter at a high energy scale E. Then the coupling for QCD with N colours and N_f quark flavours in the fundamental colour representation is given by:

$$8\pi^2/g^2(E) = \left(\frac{11}{3}N - \frac{N_f}{2}\right)\log\left(\frac{E}{\Lambda}\right) + \mathcal{O}\left(\log\log(\frac{\mathcal{E}}{*})\right) \; . \tag{6}$$

The scale Λ is a typical hadron scale, a few hundred MeV.

And where these strong couplings become weak, this strong force starts to resemble ordinary electromagnetism. Thus, at very short distances, which means distances tinier than the size of baryons and mesons, the strong force becomes a Coulomb force.

A paramount question is still: if there are quarks, why do we not see them in our particle detectors? That has never been the case. Quarks seem to be confined inside the hadrons, the baryons and the mesons. Why is this so?

The fundamental tenet of QCD is that quarks attract one another with super strong forces until they are arranged into color neutral combinations such as baryons and mesons. Forces between objects that are not color neutral are not only very strong, they do not decrease with distance. This means that colored objects cannot move around freely at all!

How can we understand this phenomenon, called *permanent quark confinement*? The quantitative understanding of confinement is the central subject of this book.

B.   QCD as a dual superconductor

Since at short distance the gluon force is a Coulomb force with logarithmic corrections, such a force can make bound states that look very much like those that we learn about, and later teach about, in a first course of quantum mechanics: the levels of the hydrogen atom. This is exactly what is observed to happen in the bound states of very heavy quarks with their own antiparticles. The analogous situation in electrodynamics is the bound state between an electron and its antiparticle, the positron, which was called "positronium". Therefore, in QCD, we call these objects "quarkonium"

The spectrum of quarkonia could be computed from our theories, and the results have been confirmed by experiment.

For light quarks however, this won't quite work: the distance over which exchanged gluons have to travel can become so large that the force becomes strong. But then the analogy with electromagnetism stops. We have to take recourse to radically different methods.

The method that seems most promising is to start from the idea that the ground state of QCD is physically radically different from what would be obtained if one would apply standard perturbation theory. That is, it looks very different over distances that are large compared to the size of a hadron.

As free quarks have never been observed in any scattering experiment, the binding energy of a quark-anti quark pair must be large. The simplest, educated, guess is that it grows indefinitely, linearly with the distance between the quarks. This would correspond to a physical picture that seems to be right intuitively; it means that there will be color-electric flux lines between the quarks that do not follow at all the pattern that is familiar from the electric field between two Coulombic charges, but instead generate a force that becomes independent of distance.

The field lines generate a color-electric field confined to a narrow tube between the quarks. This flux tube acts as a string connecting the quarks; the string can be stretched to arbitrary lengths, but, due to flux conservation, it cannot break. Only very near a quark, the field lines come together and behave much more like electric field lines surrounding an electric charge. this means that when quarks come very close together, the strong force behaves much more like the conventional Coulomb force.

A situation that is very similar to the one described above is known to occur in a very different branch of physics: inside superconducting materials. When material is super conducting, *magnetic* fields are often screened off completely. Moving magnetic field lines induce electric fields, but these are usually neutralized completely by the supercurrents. These supercurrents then also neutralize any magnetic field, so that magnetic field lines tend to bend around a superconductor.

One might suspect that this needs to be true only for time-dependent magnetic fields, not for static ones, but Maxwell's laws inside a superconductor are a bit more subtle than that. If one would try to 'freeze' a magnetic field in a material while it is cooled below the superconductivity transition temperature, this field is forced to form narrow flux tubes: the Meissner effect. Magnetic flux is conserved, so these flux tubes can be stretched to any length, but they cannot break. If one would try to freeze a *magnetic monopole* inside a superconductor, it would form the end point of a magnetic flux line, which would then drag the monopole out of the material.

Clearly, magnetic monopoles inside a superconductor behave very much like quarks in the QCD vacuum. We only have to interchange *electric* charges with *magnetic* ones and vice versa. This idea was advocated in the mid-seventies by 't Hooft and Mandelstam [5]. The transposition of the idea was that the color electric charge of the quarks is capping the end points of a color electric flux tube. So the ground state of QCD must be such that it supports the formation and stability of color electric flux tubes and that the flux inside the tube precisely matches the flux coming out of a quark. That precise matching means that seen from a distance a meson looks like a quark-anti-quark pair with the flux tube in between them, and no color flux coming out of the meson. The meson is color neutral. A similar picture applies to a baryon; however there, one has to assume that three vortices, all oriented towards one point, can be joined at that single point. This is not possible in a superconductor, not even a magnetic one. The explanation of the three vortex connection point lies entirely in the non-Abelian nature of the underlying theory.

We can also think of the flux tube *without* quarks on its endpoints: a flux tube that ends on its beginning, that is, a closed flux tube. Such a state is called a glueball and such glueballs form a whole new set of states without quarks. Whether they are completely stable is not a priori clear. Probably glueballs are so heavy that they can decay into two or three pions, typically. In our theoretical models we can suppress the quark states. Then, numerical simulations on a lattice version of the pure QCD action [6], the first term in Eq.(4), show convincingly that glueball states must exist. Glueballs are subject of the first half of the reprints.

The physical picture sketched above clarifies why quarks are never seen in scattering experiments. One can imagine pulling the two quarks in a meson apart in an experiment. The energy will increase linearly with the distance because the flux tube has a fixed tension, or energy per unit length. This potential energy increases until there is enough energy for the creation of a new quark anti-quark pair, halfway in the vortex. Then the meson we started off with splits into two mesons.

We just have been expounding the mantra that we live with since 35 years.

What are the facts that underpin the idea of flux tubes? The second part of the research reported in this book gives a very valuable hint: Just as a superconductor owes its behavior to the presence of electrically charged objects that transport the supercurrent (the well-known Cooper pairs of electrons with opposite spins), also the QCD vacuum harbours charged objects, but these are classical configurations with *color-magnetically* charged constituents. To see how this works let us return to the ground state of a superconductor.

The Cooper pairs inside a superconductor are states consisting of electrons bound in an S-wave with little or no momentum. There is something special about this condensate of pairs: it is a state with complete quantum mechanical coherence. In practice this means that we can describe the condensate by a single complex wave function:

$$\psi(\vec{r}) = |\psi(\vec{r})| \exp i\alpha(\vec{r}) \ . \tag{7}$$

The absolute square of this wave function is the density of pairs at the location $\vec{r}$. The phase α is a very interesting quantity and vital for the understanding of the occurrence of supercurrents and the stability of magnetic flux tubes. To see this take a superconductor of type II. Typically such a material can be pierced by thin tubes of magnetic flux.

Let us now examine such a flux tube. There will be a current $\vec{j}$ due to the doubly charged carriers in the condensate. It follows from minimal electromagnetic coupling:

$$\vec{j} = \frac{e}{2m} \left[\psi \left(\frac{\hbar}{i}\vec{\nabla} - (\frac{2e}{c})\vec{A} \right) \psi^* + \psi^* \left(\frac{\hbar}{i}\vec{\nabla} + (\frac{2e}{c})\vec{A} \right) \right] \ . \tag{8}$$

Here, $2e$ is the electric charge of the Cooper pair of electrons. Eq. (8) is the sum of two terms, one proportional to the vector potential, another due to the gradient of the quantum mechanical phase of the condensate:

$$\vec{j} = \frac{2e\psi^*\psi}{mc} \left(2e\vec{A} - c\hbar\vec{\nabla}\alpha \right) \ . \tag{9}$$

Now, consider a large closed loop around the vortex that we are looking at. The current vanishes deep in the superconductor, and the Meissner effect forbids the presence of the magnetic field inside. So there, the line integral of the current in (15) vanishes and gives rise to the relation:

$$0 = \oint d\vec{l}\vec{A} - \frac{\hbar c}{2e} \oint d\vec{l}.\vec{\nabla}\alpha \ . \tag{10}$$

The first term is the total flux $\Phi = \int d\vec{S}.\vec{B}$ through the tube, the second is a multiple of 2π because the condensate has a single valued phase.

So the magnetic flux is quantized:

$$\Phi = n\frac{2\pi\hbar c}{2e}. \tag{11}$$

The *dual superconductor hypothesis* for QCD holds that the gluon fluxes that keep quarks confined are the *color electric analogue* of the magnetic vortex lines inside a superconductor. This hypothesis assumes that one may exchange electric with magnetic forces in the comparison between superconductors and QCD. If the analogy makes sense, we should find color-magnetic monopole charges in the QCD vacuum that play the same role as the Cooper pairs inside a superconductor. Or in other words, we suspect that the ground state of the seemingly innocuous QCD action, Eq.(4), is a quantum mechanical coherent condensate of color magnetic charges.

II.   THE SEARCH FOR MAGNETIC COOPER PAIRS: BEFORE THE CALORON ERA

From the mid seventies on there was a feverish activity in trying to find field configurations that might contain a clue as to what these Cooper pairs might be. The first firm indication that the ground state of QCD was a much richer substratum than simple minded perturbation theory would suggest came with the discovery of instantons.

A.   Instantons

The field equations for theories such as QCD can have very special solutions, very unlike the ones in linear theories such as electrodynamics. A solution that is localized in four Euclidean dimensions was found by Polyakov and coworkers [7]. These solutions are localized in time, so that they do not describe particles in the usual sense, but rather *events* of a limited duration; therefore, they were later called "instantons". When fermions are included in the theory, these field configurations turned out to play a significant role in a problem that had been around for some time, which is the apparent explicit break down of a symmetry called axial $U(1)$ symmetry as discovered by 't Hooft [8]. Below their salient features are briefly outlined. For further reading, the original papers [9] and excellent textbooks [10] are recommended. For simplicity's sake we limit the discussion to the case that the local gauge group is $SU(2)$.

Instantons are gauge field configurations with boundary conditions at infinity telling us that, at infinity, they are pure gauge:

$$A_\mu = i\Omega\partial_\mu\Omega^{-1} \quad \text{at infinity.} \tag{12}$$

We define the dual field strength

$$\widetilde{F}_{\mu\nu} = \frac{1}{2}\epsilon_{\mu\nu\varrho\sigma}F_{\varrho\sigma} \; . \tag{13}$$

Any configuration with this boundary condition has integer topological charge $Q = k$:

$$Q \equiv \frac{1}{16\pi^2}\int d^4x \mathrm{Tr}F_{\mu\nu}\widetilde{F}_{\mu\nu} = k \; . \tag{14}$$

The reason why k is integer can be understood as follows. The density in this integral (14) is known to be the divergence of a current K_μ:

$$\frac{1}{16\pi^2}\mathrm{Tr}F_{\mu\nu}\widetilde{F}_{\mu\nu} = \partial_\mu K_\mu. \tag{15}$$

An explicit expression for this current will be given a little later.

So if the current is non-singular then the only contribution to the topological charge k can come from the three dimensional boundary. Now K_μ is not gauge-invariant, and therefore it is affected by the gauge transformation function $\Omega(x)$ at infinity. In contrast, the integrand in Eq.(14) *is* gauge invariant, and this is why k can only depend on *topological* properties of $\Omega(x)$ at infinity.

If we now take this boundary as a three sphere, then the relevant topological feature of $\Omega(x)$ is the way how Eq.(12) maps the $SU(2)$ gauge group onto this three sphere at infinity; the number k turns out to be the third homotopy number $\Pi_3(SU(2))$.

There is an elegant trick to see how this charge connects to the instanton. Let us take the pure Yang Mills action S; for simplicity, set $g = 1$ [11], and write out the identity

$$\begin{aligned} S &= \frac{1}{2}\int_x \mathrm{Tr}\,F_{\mu\nu}^2 = \frac{1}{4}\int_x \mathrm{Tr}\left((F_{\mu\nu} \pm \widetilde{F}_{\mu\nu})^2 - (\pm)2F_{\mu\nu}\widetilde{F}_{\mu\nu}\right) \\ &= \frac{1}{4}\int_x (F_{\mu\nu} \pm \widetilde{F}_{\mu\nu})^2 + 8\pi^2|k| \; . \end{aligned} \tag{16}$$

So the action is always larger then the topological charge. Only in the case that:

$$F_{\mu\nu} \pm \widetilde{F}_{\mu\nu} = 0 \ , \tag{17}$$

the bound on the action is saturated. That is, to satisfy the bound which is a local minimum of the action, the fields must be dual, $F_{\mu\nu} - \widetilde{F}_{\mu\nu} = 0$ or anti-dual, $F_{\mu\nu} + \widetilde{F}_{\mu\nu} = 0$.

Realizing a local minimum, they satisfy automatically the equations of motion. This also follows from the Bianchi identity, true for any configuration:

$$D_\mu \widetilde{F}_{\mu\nu} = 0 \ . \tag{18}$$

These solutions are called (anti)-instantons with integer topological charge $(-)|k|$. Note that an instanton has a topological charge which is even under charge conjugation, but odd under parity.

Instantons did radically change the notion of the ground state of *any* non-Abelian gauge theory.

Below it is shown that apart from the perturbative vacuum there is an infinite set of vacua $|n\rangle$, n running through the integers. It will be the existence of the instanton solution that allows tunneling between then.

The setting is now one where we put $A_0 = 0$.

All the vacua have in common that the vector potential vanishes. So the vector potential should be a pure gauge at $x_0 = \pm\infty$:

$$A_k = i\Omega\partial_k\Omega^{-1} \ \text{at} \ x_0 = \pm\infty \ . \tag{19}$$

What differentiates them is the winding number n of the gauge transformation $\Omega(\vec{x})$ defined as:

$$w \equiv \frac{1}{24\pi^2} \int_{\vec{x}} \mathrm{Tr}\epsilon_{ijk}\Omega\partial_i\Omega^{-1}\Omega\partial_j\Omega^{-1}\Omega\partial_k\Omega^{-1} = n \ . \tag{20}$$

The integer n results if $\Omega(\vec{x}) \to \mathbf{1}$, when $\vec{x} \to \infty$ and characterizes the vacuum state $|n\rangle$. One would like the vacuum state to be invariant under any of those "big" gauge transforms with non-trivial winding. This is the case for a set of periodic vacua of the type:

$$|\theta\rangle = \sum_n \exp in\theta|n\rangle \ . \tag{21}$$

The new parameter θ is a direct consequence of the existence of "big" gauge transformations, and the need for a groundstate invariant under the latter.

Consider the situation where a gauge field configuration $\vec{A}$ evolves from the vacuum with winding ℓ to the vacuum with winding n. It can be shown that the topological charge of such a configuration is precisely the difference between the two windings. As already mentioned in Eq.(15), the topological charge is the space time integral over the divergence of a current K_μ:

$$K_\mu = \epsilon_{\mu\nu\kappa\lambda}\frac{1}{8\pi^2}\mathrm{Tr}\left(A_\nu\partial_\kappa A_\lambda - \frac{2i}{3}A_\nu A_\kappa A_\lambda\right). \tag{22}$$

As before the topological charge reduces to a surface integral, but now one where only the three dimensional integrals of the density K_0 at large times do contribute.

Now K_0 becomes for the pure gauge, Eq.(19), precisely the winding density in Eq.(20). Hence:

$$Q = n - \ell \ . \tag{23}$$

The interpolating magnetic field $\vec{B}$ *does not vanish* and forms an energy barrier

$$V = \frac{1}{2}\int_{\vec{x}} \mathrm{Tr}\vec{B}^2$$

between the two vacua. At this point the instanton solution of the Euclidean equations of motion with topological charge $Q = \pm 1$ serves as a tunneling configuration. As one usually finds in tunneling events, the amplitude for the tunneling is dominated by the exponent of the action of a solution of the equations in Euclidean space. Here, one finds from Eq.(16) that the tunneling amplitude is proportional to:

$$\exp{-S_0},\tag{24}$$

with $S_0 = \frac{8\pi^2}{g^2}$, or multiples thereof.

It is this tunneling amplitude that determines the order of magnitude of instanton effects.

B. 't Hooft-Polyakov monopoles

A *particle-like* solution of the field equations of a non-Abelian theory also exists, but it involves an adjoint Higgs field. Originally formulated [12] for the $SU(2)$ case, a *magnetic monopole* can be found provided the adjoint Higgs field H ($H = \frac{1}{2}H^a\tau_a$, $a = 1, 2, 3$) spontaneously breaks the local gauge symmetry.

We will now construct this particle-like solution, which is one where the fields are static, and controlled by a Hamiltonian:

$$\mathcal{H} = \mathrm{Tr}(\vec{D}(A)H)^2 + \frac{1}{2}\mathrm{Tr}\,\vec{B}^2 + \lambda(\mathrm{Tr}H^2 - v^2)^2.\tag{25}$$

The potential term minimizes the potential energy of the Higgs field H if it has a vacuum expectation value (VEV), which means that, in the vacuum, $|H| = v \neq 0$.

A subgroup of the local $SU(2)$ gauge symmetry leaves the VEV unaffected, which is the group of rotations around the axis formed by the H field. This is an Abelian subgroup, $U(1)$, just as what we have in electromagnetism. Therefore, this model has an unbroken, long range electromagnetic force, carried by a massless photon. The original $SU(2)$ group was three dimensional, therefore there are two other $SU(2)$ rotations that do not leave the VEV invariant. The photons associated to these rotations, referred to as $W^\pm$, are heavy and electrically charged, and induce only short range forces.

Returning to our particle-like solution, this forms yet another particle besides the ones that form the back bones of the model: a long range photon, the carrier of the unbroken $U(1)$ symmetry, and the heavy $W^\pm$. Our new object is referred to as a heavy soliton. The fields surrounding it are described in the "hedgehog" gauge. This gauge is specified by the behaviour of H and $\vec{A}$ at spatial infinity:

$$H = H(r)\hat{r}_a\frac{\tau_a}{2}v \; ;\tag{26}$$

$$A_i^a = K(r)\epsilon^{aij}\frac{\hat{r}^j}{r} \; .\tag{27}$$

The unit radial vector is $\hat{r}^j$, the functions H and K are regular at the origin and asymptote at large r respectively to 1 and -1.

The adjoint Higgs defines what is the $U(1)$ magnetic field. In particular at spatial infinity it reads $\frac{1}{v}H^a\vec{B}^a$. The total flux follows from the surface integral:

$$m = \frac{1}{4\pi v}\int d\vec{S}.H^a\vec{B}^a = 1 \; .\tag{28}$$

The mass of the monopole is by a standard scaling argument on the order of $\alpha^{-1}m_W$.

An adjoint Higgs field is conspicuously absent from QCD. Nevertheless a periodic version of such a field appears when we heat up QCD. This leads to the genesis of calorons and is the subject of the next sections.

III.   HEATING UP QCD

Meanwhile in the late seventies another aspect of QCD was investigated. Precisely the similarity of QCD with a superconductor suggested that, at sufficiently high temperatures, QCD might undergo the same fate as that of a superconductor: it ceases to be superconducting.

The Cooper pairs loose their quantum mechanical coherence due to temperature fluctuations. Above some temperature T_c they break up into ordinary electrons. The magnetic flux tubes will become unstable.

Likewise, the putative color magnetic charges that are stable at low temperature might loose their coherence above some temperature T_c^{QCD}, that is, they become charges without the coherence needed for forming color electric flux tubes. If this happens, presumably above some critical temperature, the system looses its power to force *all* quarks to form tight bound states with other quarks, allowing them to *deconfine*. That means that, above this temperature, the quarks can move more or less freely with a mean free path much longer than the hadron size. The surrounding vacuum has then turned into what is now known as the *quark-gluon plasma*.

Both temperatures are of the order of the binding energies involved. For the superconductor, these energies are in the order of milli-electron volts; for QCD, they are in the order of a few hundred MeV, so a factor 10^{12} apart. It is remarkable that the same physical ideas apply at such vastly different scales.

So what we dubbed "permanent confinement" in section I is not really permanent, certainly not on timescales like the age of the universe: until shortly after the Big Bang hadrons were *deconfined* into a quark gluon plasma.

In this section we discuss in a nutshell the thermodynamics of a plasma without quarks. It will prove useful for our understanding the precise role of calorons in thermal QCD. The temperature scale T has absorbed the Boltzmann constant k_B.

A.   Asymptotic freedom, the gluon plasma and Debye screening

A straightforward argument in favour of a quark gluon plasma involving only asymptotically free, hence small gauge coupling at very high temperature (or baryon density) was given by Collins and Perry [13].

At high temperature the average kinetic energy of the gluons is of order of the temperature T. According to asymptotic freedom, Eq.(6), the gauge coupling becomes so small,

$$\frac{8\pi^2}{g^2(T)} = \left(\frac{11}{3}N\right) \log\left(\frac{T}{\Lambda_T}\right) + \cdots , \tag{29}$$

that we can neglect any interaction. Again, Λ_T is a hadronic scale much smaller than the ambient temperature T.

Then, at high temperature the system can be described by a gas of non-interacting gluons: a Stefan-Boltzmann gas with the pressure multiplied by the degrees of freedom of gluons:

$$p_{SB} = \left(2(N^2 - 1)\right)\frac{\pi^2}{90}T^4 . \tag{30}$$

This picture of a Stefan-Boltzmann gas is only accurate at temperatures far above hadron scales. Perturbative corrections to this equation of state have been computed, up to and including part of g^6. All of these contributions produce a logarithmic fall off in the pressure.

Later in this section we will see that lattice data [14] are showing us a quite different behaviour when the temperature is below $10T_c$.

At high temperature the long range Coulomb force is modified. Thermal fluctuations screen a test charge and cause the range of the Coulomb force to become that of a Yukawa force:

$$F_{Coulomb} \to F_{Coulomb} \times \exp(-m_D r). \tag{31}$$

A crude estimate of the Debye mass m_D is obtained by scrutiny of the Poisson equation for the scalar potential A_0. Let us for simplicity's sake drop the colour degrees of freedom and take a hot gas of positive and negative charged particles (charge $\pm g$) with a mass much smaller than the temperature and equal densities $n_\pm \equiv n$. Of course the densities depend on T. In the presence of the test charge g_t the $n_\pm$ depend on their distance r from the test particle. The Poisson equation reads:

$$\vec{\nabla}^2 A_0 = -4\pi g_t \delta(\vec{r}) - g(n_+(r) - n_-(r)) . \tag{32}$$

The densities depend on $r = |\vec{r}|$ through the Boltzmann factor on r:

$$n_\pm(r) = n \exp(-(\pm)g A_0(r)/T) . \tag{33}$$

If T is high enough linearization of the Boltzmann factor is allowed. Substitute into Eq.(32) to get:

$$\vec{\nabla}^2 A_0 = -4\pi g_t \delta(\vec{r}) + 2g^2 \frac{n}{T} A_0 . \tag{34}$$

A mass term $2g^2 n/T A_0 \equiv m_D^2 A_0$ in the Poisson equation obtains. For zero mass gluons the density is proportional to T^3, on dimensional grounds. So we expect

$$m_D \sim gT . \tag{35}$$

A perturbative evaluation of the coloured scalar two point function in QCD gives a correct treatment of the colour degrees of freedom. To one loop order one finds:

$$m_D^2 = \frac{g^2 N T^2}{3}. \tag{36}$$

This screening mass is important for the calculation of corrections to the pressure because it tames part of the infra-red divergencies in perturbative contributions. As we will see in section VI the semi-classical contributions to the pressure from the thermal version of instantons are also improved.

B. $\mathbb{Z}(N)$ symmetry in hot QCD and the effective action for the Polyakov loop

Suppose one studies stationary phenomena at a temperature T involving an observable O or a correlator of observables. One is led to examine expectation values for physical states with a Boltzmann factor corresponding to a heat bath at temperature T:

$$\langle O \cdots \rangle_T = \frac{\mathrm{Tr}_{phys}\left\{ O \cdots \exp(-H/T) \right\}}{Z} , \tag{37}$$

with $Z = \mathrm{Tr}_{phys}\left\{ \exp(-H/T) \right\}$ the partition function built from the physical states. The dots indicate eventual other observables.

The physical states are the eigenstates of the Hamiltonian H. They are colourless and so obey Gauss' constraint:

$$\vec{D}(\vec{A}).\vec{E}(\vec{x}) = 0 \ . \tag{38}$$

The reader can verify, using the canonical quantization rules, that Gauss' operator $\vec{D}(A).\vec{E}(\vec{x})$ indeed generates local colour gauge transformations.

In general any matrix element of the operator with Boltzmann weight, $O \cdots \exp(-H/T)$, can be reformulated as a Euclidean path integral with the result for Eq.(37):

$$\langle O \cdots \rangle_T = \int Dq D\bar{q} DA_0 D\vec{A} \ O \cdots \exp(-S)/Z \ . \tag{39}$$

The action S is the QCD action Eq.(4) but now defined with Euclidean time τ.

The real time description uses the statistics of Bose and Fermi fields. This statistics implies that the bosonic fields must be periodic in the Euclidean time direction, with period $1/T$. For fermionic fields , like the quark fields q, there is anti-periodicity.

Due to the heat bath, Lorentz invariance is reduced to three dimensional rotation symmetry, and so A_0 is a scalar. We can say that this scalar is in the adjoint representation of the local gauge group, but only if the gauge transformations do not also depend on the Euclidean time τ. However, a judiciously chosen, non-local combination of this scalar behaves like an adjoint, as we will see presently.

There is an important ingredient in these 'hot gauge theories' that one might overlook. For the sake of discussion, let us drop the quarks in QCD by making them very heavy. Let us consider the general case for N colors, so that the color gauge group is $SU(N)$. What is important then is, that there is the appearance of a discrete symmetry, the *center group* symmetry $\mathbb{Z}(N)$.

Originally discovered by 't Hooft [15], this symmetry has the following origin. A periodic gauge transformation will leave the periodic path integral invariant. However, the periodicity can be relaxed to periodicity modulo an element of the center group. These are the set of group elements that commute with all other group elements. The adjoint gauge potentials will not feel the center group element, so will stay periodic. The quark fields do feel the center group and loose their anti-periodicity, which is why we omit them for the moment. The path integral is then invariant since the gauge fields are, and the action is also invariant.

This center group invariance can be used to distinguish different phases [16] when heating up the system: there is an operator in our gauge theory that is not invariant and gets therefore the meaning of an order parameter like magnetization. This operator is the Polyakov loop $\mathbf{L}$ that runs over a full period $1/T$ of Euclidean time:

$$\mathbf{L}(\tilde{\mathbf{x}}) = \mathcal{P} \exp(i \int_0^{1/T} d\tau A_0(\tau, \vec{x})) \ . \tag{40}$$

Here, the symbol $\mathcal{P}$ stands for 'partial ordering' of the integrand with respect to Euclidean time τ.

Since this is a closed loop, it transfoms under periodic gauge transformations as an adjoint,

$$\mathbf{L}^{\Omega}(\vec{x}) = \Omega(0, \vec{x})\mathbf{L}(\vec{x})\Omega^{-1}(1/T, \vec{x}) \ . \tag{41}$$

and hence its eigenvalues are gauge invariant under infinitesimal local gauge transformations. Its trace is called the Polyakov loop. Now, since $\mathbb{Z}(N)$ only contains gauge transformations that are not infinitesimally close to the identity (except the identity itself), we have, precisely when the periodicity is modulo a center group element:

$$\Omega(1/T, \vec{x}) = \exp\left(ik\frac{2\pi}{N}\right) \Omega(0, \vec{x}) \ , \qquad k = 0, 1, \cdots, N-1 \ , \tag{42}$$

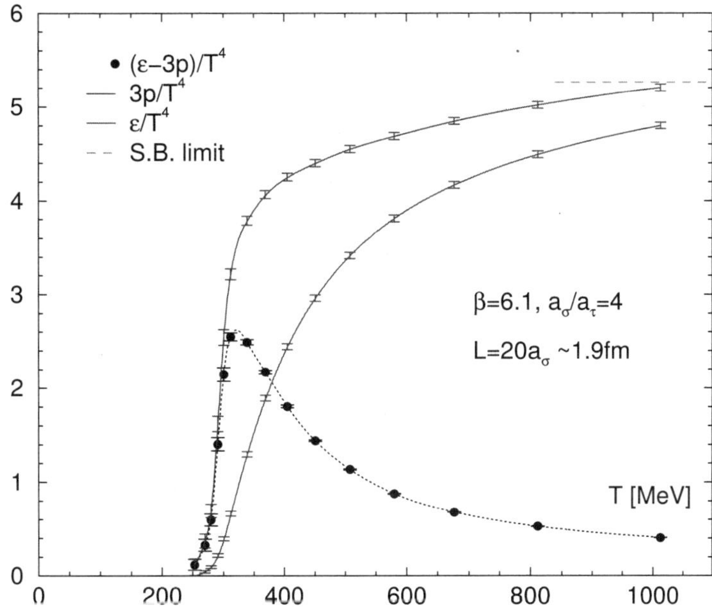

FIG. 2: Pressure, energy density and trace anomaly for three colours $(SU(3))$ as function of temperature above the transition, taken from T. Umeda et al., Phys. Rev. D79, 051501. Note the sharp peak in the trace anomaly just above the critical temperature.

$\mathrm{Tr}\,\mathbf{L}$ will transform with that element. In the confined phase, its thermal expectation value is zero, but beyond T_c it starts to develop a VEV indicating that the $\mathbb{Z}(N)$ symmetry is spontaneously broken.

Phrased differently, $\mathrm{Tr}\,\mathbf{L}$ is an *order parameter*. The hot gauge theory can be described by the *Gibbs free energy* G, which will be a function of this order parameter. In a simplified notation, omitting explicit T, $\vec{x}$ dependence and higher derivative terms :

$$\mathcal{G}(\mathrm{Tr}\,\mathbf{L}) = \int_{\vec{x}} \left((\vec{\nabla}\,\mathrm{Tr}\,\mathbf{L})^2 \mathcal{K}(\mathrm{Tr}\,\mathbf{L}) + \mathcal{V}(\mathrm{Tr}\,\mathbf{L}) \right). \tag{43}$$

If the system is homogeneous the kinetic term is absent and we read the bulk thermodynamics from the minima of the potential term $\mathcal{V}(q,T) \equiv T^4 V(q,T)$ (called the "effective potential" henceforth).

C. Thermodynamics of gauge fields as determined by lattice simulations

In this subsection we discuss the thermodynamic functions like pressure and energy density for $SU(3)$ for temperatures above the critical one, T_c. Their determination at very high temperatures is through the black body radiation: asymptotic freedom tells us we have black body radiation describing our non-interacting gluons.

But lowering the temperature introduces interactions in the glue, and we need lattice simulations. Our starting point is the effective potential $\mathcal{V}(q,T)$ of Eq.(43). Its minima $q_m(T)$ are determined by

$$\frac{\partial \mathcal{V}(q,T)}{\partial q} = 0 \,. \tag{44}$$

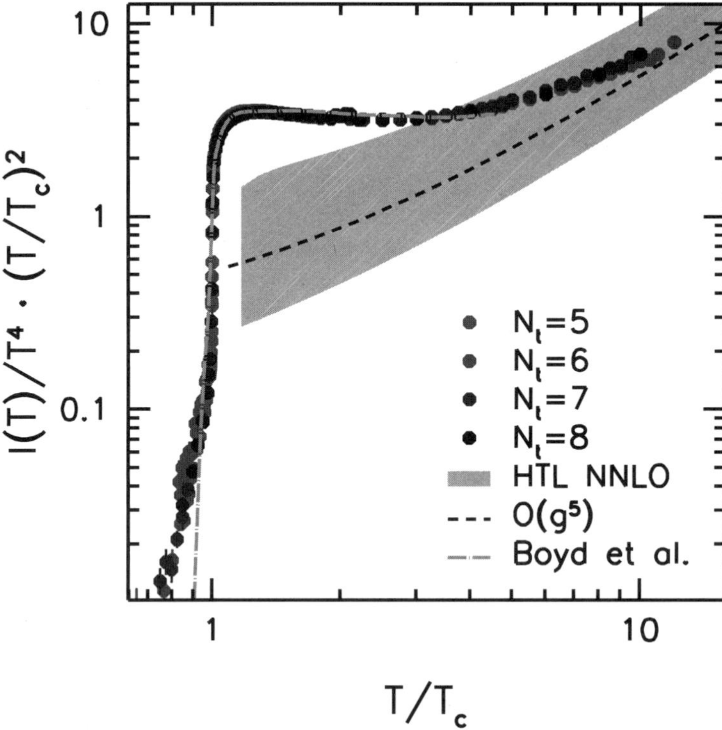

FIG. 3: The rescaled trace anomaly $I(T)$ for $SU(3)$, taken from Sz. Borsanyi et al., arXiv:1204.6184 [hep-lat]. The dashed curve is straightforward perturbation theory. The dark band results from perturbative estimates. The plateau in between $1.15T_c$ and $4T_c$ corresponds to $1/T^2$ fall-off, and is incompatible with the logarithmic perturbative estimates.

From the minima we can find the thermodynamic functions such as pressure

$$p = -\mathcal{V}(q_m(T), T) \, , \tag{45}$$

$$s = \frac{\partial p}{\partial T} \qquad \text{entropy,} \tag{46}$$

$$e = -p + Ts \quad \text{energy density,} \tag{47}$$

$$I(T) = e - 3p \qquad \text{trace anomaly.} \tag{48}$$

They have been computed by lattice simulations and shown in Fig.(2) for the case of three colours.

The trace anomaly $I(T)$ is quite interesting as it measures the amount of gluonic interaction. The Stefan Boltmann contribution drops out in the difference $e - 3p$.

Normalized by the temperature it behaves at high T as the perturbative β-function, so falls off logarithmically.

It has of necessity a peak, but remarkable is that this peak is quite near the transition (at about $T = 1.15T_c$). Equally interesting is the behaviour of the anomaly rescaled by T^2/T_c^2. It shows up to $4T_c$ a plateau, see Fig.(3) [17]. This means it falls off as T^{-2}, much faster than the perturbative β- function. Its behaviour becomes consistent with perturbative estimates above $10T_c$.

What excitations cause this plateau is still not understood.

What about the thermodynamics for a generic number of colours N? Recent lattice simulations [18] show a simple scaling with the number of gluons $N^2 - 1$. So the plateau stays a plateau for all colours [19]. We will see shortly that calorons, the thermal analogues of instantons, cannot cause the plateau because their power law is N-dependent (and that of the plateau is not). But in between the perturbative region and the plateau in Fig.(3) they may play a role. This will be discussed in Section VI.

IV.   A SIMPLE EXAMPLE: $SU(2)$ GLUODYNAMICS

After our first incursion into the thermodynamics of gauge fields we want to take the specific example of $SU(2)$ to get a precise idea of where the calorons may play an important role. The approach will be to find for the effective potential (43) a simple Landau-Ginzburg type of approximation , Eq.(56). The main role of that approximation is

- to reproduce the plateau in the trace anomaly, Fig.(3),

- to provide an understanding of the location of the peak.

- and to have the same Landau-Ginzburg action working for all colours.

The obvious advantage of discussing $SU(2)$ is its mathematical simplicity.

For $SU(2)$ the loop is

$$\mathbf{L} = \exp\left(2\pi i \, \mathbf{q}\right) = \begin{pmatrix} e^{2\pi i q_1} & 0 \\ 0 & e^{2\pi i q_2} \end{pmatrix} . \tag{49}$$

Because of unimodularity $q_1 = -q_2 \equiv q/2$.

The centergroup invariance is $Z(2)$. Changing q into $1 - q$ acts on $\mathrm{Tr}\,\mathbf{L}$ as:

$$\mathrm{Tr}\,\mathbf{L} \to -\mathrm{Tr}\,\mathbf{L} . \tag{50}$$

The fixed point is $q = 1/2$ for which $\mathrm{Tr}\,\mathbf{L} = 0$. This is the confining state where the symmetry is unbroken.

It implies that the potential $\mathcal{V}(\mathrm{Tr}\,\mathbf{L}, T) = \mathcal{V}(q, T)$ in Eq.(43) should be invariant under $q \to 1 - q$.

Let us look in more detail at the behaviour of the effective potential in the high temperature deconfined phase. At very high temperature (small values for the Euclidean period T), the effective potential $V(q, T)$ is well approximated by perturbation theory because of asymptotic freedom.

So let us consider the classical constant background scalar potential q. The classical action is degenerate, it is zero for whatever value of q. So we have to go to one loop precision.

The one loop result [20] is shown in Fig.(4).

$$\mathcal{V}_{pert}(q, T)/T^4 = -\frac{\pi^2}{15} + V_{pert}(q, T) ;$$

$$V_{pert}(q, T) = \frac{4\pi^2}{3} q^2 (1 - q)^2), \ q \bmod 1. \tag{51}$$

The first term is the familiar free energy of the black body radiation due to the gluons, the second is the q dependent potential.

This potential has two degenerate minima, one at $q = 0$ for which $\frac{1}{2}\mathrm{Tr}\,\mathbf{L} = 1$, one at $q = 1$ for which $\frac{1}{2}\mathrm{Tr}\,\mathbf{L} = -1$. This is a sure sign that the Z(2) symmetry is spontaneously broken at this very high temperature where the one loop approximation is valid. The two eigenvalues are coinciding, in either q=0, or q=1. The barrier in between is a quantum effect. Note that the second derivative in the minima gives the Debye screening mass Eq.(36) for the case of two colors.

The maximum is at the $Z(2)$ symmetric point $q = 1/2$.

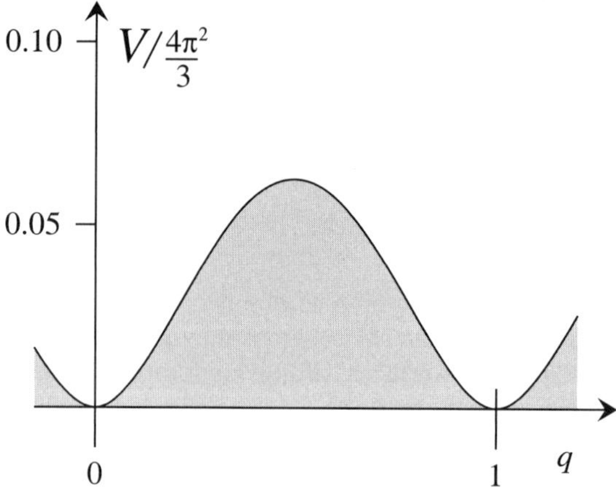

FIG. 4: The effective potential V for $SU(2)$ at infinite temperature.

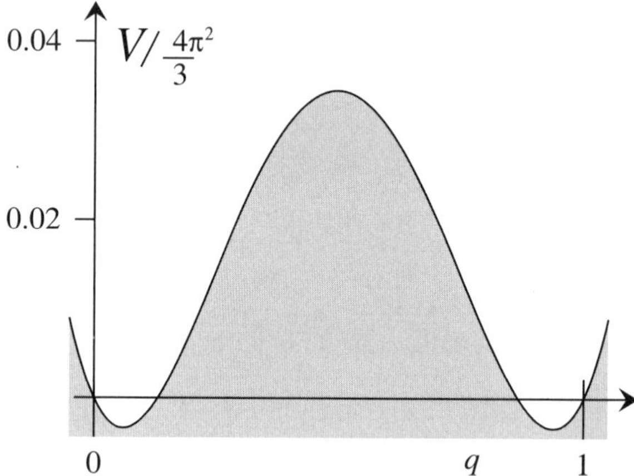

FIG. 5: The effective $SU(2)$ potential V at high T at about $3T_c$. The perturbative Stefan Boltzmann minima from Fig.(4) have been destabilized.

The two loop correction for $SU(2)$ changes Eq.(51) into :

$$\mathcal{V}_{pert}(q,T)\left(1 - 5\frac{g^2 C_2}{(4\pi)^2}\right) . \tag{52}$$

Here C_2 denotes the quadratic Casimir operator in the adjoint representation of $SU(2)$ and takes the value 2.

So this multiplicative correction respects in particular *the location* of the $Z(2)$ (absolute) minima.

It can be argued that to any order in perturbation theory the absolute minima stay at the two $Z(2)$ minima [21].

On the other hand we know that by lowering the temperature the minima have to move towards the $Z(2)$ symmetric point. This is mandatory, because we know from lattice simulations that there

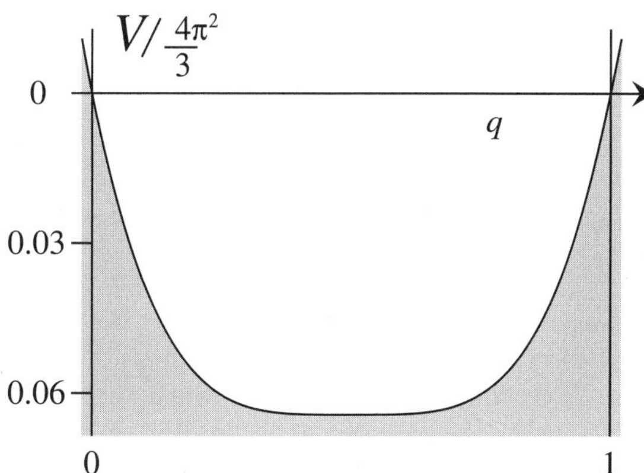

FIG. 6: The effective $SU(2)$ potential V at T_c. The $Z(2)$ conjugate minima have coalesced into the $Z(2)$ invariant minimum at $q = \frac{1}{2}$.

is a transition.

And the question is: how can we get this done?

One simple minded answer is: we add a term *linear* in q to the potential $V(q, T)$ to destabilize the perturbative minimum at $q = 0$. $Z(2)$ invariance requires the term also linear in $1 - q$ for $q \sim 1$.

$$V_{tot} = V_{pert} - \frac{4\pi^2 c_1}{3}\left(\frac{T_c}{T}\right)^p q(1 - q) \,. \tag{53}$$

T_c is a new non-perturbative scale, p can be any positive power, so that the perturbative term dominates at high T. However there are sound reasons to choose $p = 2$.

To see this use the thermodynamic identity:

$$-T\frac{\partial V_{tot}}{\partial T} = \frac{e - 3p}{T^4} \,. \tag{54}$$

It follows from the bulk properties in Eq.(48). Applying this identity to Eq.(53) kills the perturbative contribution. The reason is that the temperature variation through $q_m(T)$ drops out. Only the explicit dependence on T contributes.

This is obvious because the Stefan-Boltzmann term has no scale involved, and the trace anomaly ignores such terms.

So the normalized trace anomaly is, up to a factor 2, the destabilizing non-perturbative term ($q = q_m(T)$):

$$\frac{e - 3p}{T^4} = \frac{8\pi^2 c_1}{3}\left(\frac{T_c}{T}\right)^p q(1 - q) \,. \tag{55}$$

And as the normalized trace anomaly in lattice simulations is found to drop as $1/T^2$, from $1.2T_c$ till $T = 4T_c$, see Fig.(3), we should take $p = 2$.

$Z(2)$ invariance implies the minima are at $q(T)$ and $1/2 - q(T)$, as in Fig.(5). Once we arrive at T_c the minima coalesce at $q(T_c) = 1/2$, see Fig.(6) [22]. The number c_1 is matched such that at $T = T_c$ the confining minimum in Fig.(6) is reached.

Though tailored to destabilize the perturbative vacuum and to reproduce the empirical fall-off of the trace anomaly, it does *not* reproduce the peak at the right location near the critical temperature. This can be cured by adding another term proportional to $(\frac{T_c}{T})^2 q^2 (1-q)^2$. This term introduces the only free parameter c_2 [23]:

$$V_{tot} = V_{pert} - \frac{4\pi^2}{3}\left(\frac{T_c}{T}\right)^2 \left(c_1 q(1-q) + c_2 q^2 (1-q)^2 + c_3 \right). \tag{56}$$

The q independent term c_3 tunes the pressure at T_c to vanish. So only c_2 is a free parameter, and is fixed by the location of the peak in the anomaly.

This simple fourth order $Z(2)$ invariant polynomial constitutes a Landau-Ginzburg approximation to the full effective potential. It reproduces quite well [24] the main features of the non-perturbative region below $10T_c$ in Fig.(2) and Fig.(3), and of the Stefan-Boltzmann regime at very high T.

The number of degrees of freedom at a given energy in the confined phase of an $SU(N)$ gauge theory is the number of glueballs. It is independent of the number of colours, of order $\mathcal{O}(1)$. But in the deconfined phase it is the number of gluons, $\mathcal{O}(N^2)$. It is worth noting that our Ansatz, Eq.(56), produces a pressure that just below the critical temperature is $\mathcal{O}(1)$. However the entropy is negative below T_c, so the model is unphysical in the confined phase.

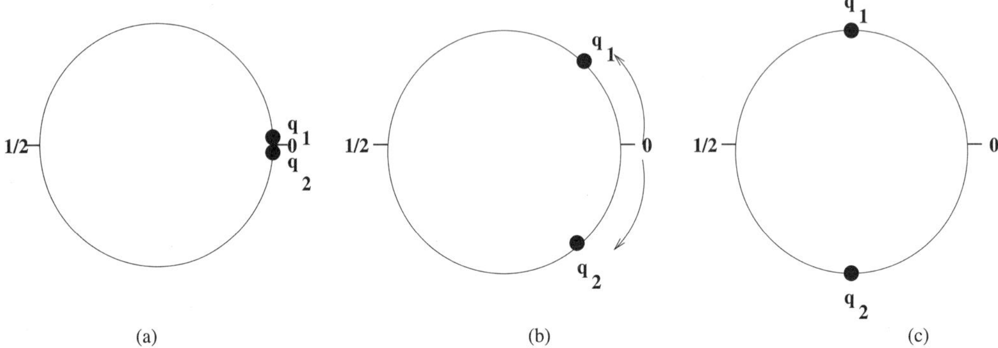

FIG. 7: Eigenvalues q_1 and q_2 of the Polyakov loop start repelling when lowering T. The numbers 0 (1/2) are values of q_1 and q_2 corresponding to $\frac{1}{2}\mathrm{Tr}\,\mathbf{L} = 1$ (-1). In Panel (a) very high T, panel (b) $\infty\langle T \langle T_c$, panel (c) $T = T_c$, and $\frac{1}{2}\mathrm{Tr}\,\mathbf{L} = 0$, maximal repulsion in the confining vacuum.

The eigenvalues $q_1 = -q_2 = q(T)/2$ at the minima move from the perturbative minimum $q = 0$ at very high temperature to $\pm 1/2$, where the $Z(2)$ symmetry is spontaneously broken, to the $Z(2)$ symmetric point at T_c. This is shown in Fig.(7) (a), (b) and (c).

Recapitulating: we have introduced by hand a repulsive term to explain the lattice data. It would be satisfactory to have QCD to provide a repulsive term from first principles.

- So the question is: what QCD mechanism destabilizes the perturbative ground state as in Fig.(5)?

- The main suspect is, not quite unexpected, the semi-classical caloron, as we will argue in the last section VI.

It is now time to introduce the salient features of the caloron.

V. ASSEMBLING THE CALORON

We now embark on a more technical discussion, sticking with the $SU(2)$ gauge group.

The caloron is akin to the instanton already discussed before. But it is adapted to thermal QCD, and its order parameter:

- The caloron has to be periodic in time.

- The caloron configuration obeys a gauge invariant boundary condition: at spatial infinity the Polyakov loop takes the value $\cos(\pi q)$.

- The caloron configuration is a local minimum of the QCD action.

For $q = 0$ (zero holonomy) such a configuration was manufactured [25], based on a multi-instanton formula due to 't Hooft:

$$A_\mu^{mi}(x) \;=\; \frac{1}{2}\eta_{\mu\nu}\partial_\nu \log(\psi(x)) \; ;$$

$$\psi(x) \;=\; 1 + \sum_{j=1}^{M} \frac{\varrho_j^2}{(x - x_j)^2} \; . \tag{57}$$

The $\eta_{\mu\nu}$ symbol equals

$$\eta_{\mu\nu} \;=\; \frac{1}{2i}(\tau_\mu\bar{\tau}_\nu - \tau_\nu\bar{\tau}_\mu) \; ,$$

$$\tau_\mu \;=\; (-i, \vec{\tau}) \; ,$$

$$\bar{\tau}_\mu \;=\; (i, \vec{\tau}) \; . \tag{58}$$

This symbol is self dual if the Levi-Civita tensor obeys

$$\epsilon_{1230} = 1. \tag{59}$$

First the sizes ϱ_j of the instantons are taken the same size ϱ for all, and second their centers x_j are periodically spaced in the time direction $\mathbf{e}_0$, with period $\frac{1}{T}$, creating an infinite periodic array with centers $x_j = a + j\,\frac{1}{T}\mathbf{e}_0$. The summation over all centers gives a simple formula (taking $a = 0$):

$$\psi_{HS}(x) = 1 + \frac{\pi T \varrho^2}{|\vec{x}|}\frac{\sinh(2\pi|\vec{x}|T)}{\cosh(2\pi|\vec{x}|T) - \cos(2\pi x_0 T)} \; . \tag{60}$$

Its topological charge is unity, when integrating over one period, so its action is $\frac{8\pi^2}{g^2}$. This is easily seen by letting the temperature vanish in (60). One recovers the single instanton, M=1, in Eq.(57).

Its contribution at asymptotically large temperature to the pressure has been computed [26]. To do so one has to compute the effect of one loop fluctuations around the periodic instanton, in analogy (but much harder!) with what we did in section IV for the Stefan -Boltzmann pressure. The pressure due to periodic instantons is finite because the contribution of instantons with large radii ϱ is cut off by the Debye screening mass.

In the case of a Polyakov loop with non-trivial values of the phase q, what Pierre van Baal, Kimyeong Lee and coworkers found was a set of degenerate minima of the QCD action, labeled by the phase q (imposed of course only at spatial infinity). For $q = 0$ they retrieved the afore mentioned periodic instanton Eq.(60), with action $8\pi^2/g^2$. Like for the constant background q it warrants the question: does a one loop fluctuation determinant around the caloron lift the degeneracy?

Before attacking these questions we are going to assemble the caloron from its magnetic constituents. They will be self -dual so will carry both magnetic and electric charge. Hence they should be called "dyons". Nevertheless in what follows we will use both the word monopole or dyon for these objects. Below in this section the constituents are constructed.

A.   A static self dual monopole solution

The loop Eq.(49) is for infinite T commuting with all of the gauge group, for finite T only with a $U(1)$ subgroup. Hence the temptation to speak of $\mathbf{L}$ as being an "periodic adjoint Higgs" field, by definition a three dimensional field. But then, in contrast to the 3D adjoint Higgs field of the previous section, one finds not one but *two* monopoles.

With this periodic adjoint Higgs field it is straightforward to construct a static self dual 't Hooft-Polyakov monopole [27], and we will do so in this subsection. From that static solution a second monopole with a time-dependent core is manufactured in the next subsection.

With our definition of the Levi-Civita tensor, Eq.(59), a self dual field strength obeys:

$$
\begin{aligned}
\vec{B} &= \vec{E}, \text{ with} \\
\vec{E} &= \vec{D}(\vec{A})A_0 - \partial_0 \vec{A} \,, \\
\vec{B}_i &= \frac{1}{2}\epsilon_{ijk}F_{jk} \,.
\end{aligned}
\tag{61}
$$

We have for a static potential that the electric field is simply:

$$
\vec{E} = \vec{D}(\vec{A})A_0 \,.
\tag{62}
$$

Substitute this in the action:

$$
\begin{aligned}
S &= \frac{1}{2}\int_0^{1/T} dx_0 \int d^3\vec{x}\, \mathrm{Tr}\,(\vec{E}^2 + \vec{B}^2) \\
&= \frac{1}{2T}\int d^3\vec{x}\, \mathrm{Tr}\left((\vec{D}(\vec{A})A_0 - \vec{B})^2 + 2\vec{D}(\vec{A})A_0.\vec{B}\right).
\end{aligned}
\tag{63}
$$

This looks familiar from Eq.(25). One difference is that the adjoint Higgs is replaced by the scalar A_0. Another that the potential term is absent. This system of equations is called BPS [28], after the discoverers. The absence of the symmetry breaking scale v in the potential is made up for by introducing it in the boundary condition for the solution. The scale v is taken from the Polyakov loop, Eq.(49) so

$$
v = 2\pi T q \,.
\tag{64}
$$

So the equation to solve is:

$$
\vec{D}(\vec{A})A_0 = \vec{B} \,.
\tag{65}
$$

We get as solution the familiar hedgehog form for the scalar potential A_0:

$$
A_0^{BPS}(\vec{x}) = v\hat{r}^a H(r)\frac{\tau_a}{2} \,,
\tag{66}
$$

and for the vector potentials:

$$
A_i^{BPS}(\vec{x}) = \epsilon_{aij}\frac{\hat{r}^j}{r}K(r)\frac{\tau_a}{2} \,,
\tag{67}
$$

H (and K) are known functions. Their precise form [29] is not relevant to the discussion here, except that they are regular at the origin and asymptote to the value 1 (and -1) at spatial infinity.

Then the colour magnetic field becomes at large distance:

$$
B_i^a = \hat{r}^a \frac{\hat{r}^i}{r^2}.
\tag{68}
$$

Let us now calculate the topological charge, using the regularity of the solution at the origin:

$$
\begin{aligned}
Q &= \frac{1}{4\pi^2 T} \int d^3\vec{x} \; \mathrm{Tr} \, \vec{E}.\vec{B} \\
&= \frac{1}{4\pi^2 T} \int d^3\vec{x} \; \mathrm{Tr} \, \vec{D} A_0.\vec{B} \\
&= \frac{1}{4\pi^2 T} \int d\vec{S}.\mathrm{Tr} \, A_0 \vec{B} \\
&= \frac{v}{2\pi T} = q \; .
\end{aligned}
\tag{69}
$$

In the last two equalities we used Eqs.(66), (68) and (64). The self duality implies a positive result and indeed $0 \le q \le 1$. Note the topological charge is taking on any real value in this range.

The magnetic charge m follows from the explicit configuration Eq.(66) and (68):

$$
m^{BPS} = \frac{1}{2\pi v} \int d\vec{S}.\mathrm{Tr} A_0^{BPS} \vec{B}^{BPS} = 1 \; .
\tag{70}
$$

The action of our static (and therefore periodic) monopole is now $S_{BPS} = \frac{8\pi^2 q}{g^2}$ from Eq.(63), restoring the coupling explicitly .

This static monopole is referred to as $M[m, Q]$, m the magnetic charge, Q the topological one.

B.   A time-dependent periodic magnetic charge: the Kaluza-Klein (KK) monopole

There is a big gauge transformation when applied to our monopole that makes it into a new one: the so-called KK monopole. This is precisely possible because A_0 and the Polyakov loop are not a traditional adjoint Higgs field, but periodic.

The big gauge transformation is one which is not strictly periodic but periodic up to a minus sign. This transformation takes a particularly simple form in the hedgehog gauge, Eq.(66) and (67). It reads:

$$
\Omega_1(\tau) = \exp\left(-i2\pi T \tau \hat{\vec{r}}.\frac{\vec{\tau}}{2}\right) \; .
\tag{71}
$$

This transformation is "big" because it is not continuously deformable into the unit transformation: over a full period in the time direction it picks up a minus sign.

To construct a time dependent solution with the Polyakov loop asymptotically equal to $\cos(\pi q)$, we start with its $Z(2)$ transformed BPS counterpart with $q \to 1 - q$.

Applied to our hedgehog and vector potential it gives:

$$
\begin{aligned}
A_0^{KK} &= \Omega_1 A_0^{BPS} \Omega_1^{-1} + \Omega_1 \frac{\partial_\tau}{i} \Omega_1^{-1} \; ,
\end{aligned}
\tag{72}
$$

$$
\vec{A}^{KK} = \Omega_1 \vec{A}^{BPS} \Omega_1^{-1} \; .
\tag{73}
$$

It leaves invariant the radial components of asymptotic $\vec{B}$ fields, as they are proportional to $\hat{\vec{r}}.\vec{\tau}$, but rotates the azimuthal and polar components. The latter are short ranged and produce the core of the monopole. So it is the core of the KK monopole that starts rotating over 2π as time elapses over the period $1/T$. The former are long ranged and stay stationary.

This time dependence of the new configuration is expected to generate electric charge and we will find so. The reason is that the time dependence is a global U(1) transformation.

The monopole hedgehog A_0 stays stationary but suffers a shift from the gradient part of the transformation Eq.(71). At large $\vec{r}$:

$$A_0^{KK} = A_0 - 2\pi T \vec{\hat{r}} \cdot \frac{\vec{\tau}}{2} = -q\, 2\pi T \vec{\hat{r}} \cdot \frac{\vec{\tau}}{2} \ . \tag{74}$$

So the new hedgehog asymptotes with $-q$ instead of q, but with the same value $\cos(\pi q)$ for the Polyakov loop .

This means that the abelian magnetic flux of the KK monopole is opposite in sign to that of the original BPS monopole, because the asymptotic colour magnetic field stays unchanged. We write the two species as $M[m,q]$ and $\widetilde{M}[-m,1-q]$ respectively.

The topological charge density of the KK monopole is unaffected by the big gauge transformation. Hence:

$$Q^{BPS} = q \ , \tag{75}$$
$$Q^{KK} = 1 - q \ . \tag{76}$$

The same is true for the corresponding actions, up to a common factor $\frac{8\pi^2}{g^2}$.

Because of self duality the electric flux changes sign too. Independently this is expected from the time dependence due the big gauge transform Eq.(71). In fact it excites the electric charge zero mode of the dyon.

The translational zero modes are of course three in number, so we have 4 zero modes for each dyon apart.

VI. MONOPOLES AS CONSTITUENTS OF INSTANTONS, AND OF MORE GENERAL BOUND STATES.

Let us recapitulate. We have discussed two types of monopoles , $M[m,q]$ and $\widetilde{M}[-m,1-q]$, and their CP conjugates $\overline{M}[-m,-q]$ and $\overline{\widetilde{M}}[m,-1+q]$. In the table below the possible pairings are shown.

TABLE I: Table of dyon pairings. In each entry the magnetic charge m comes first, then the topological charge Q.

	M	$\widetilde{M}$	$\overline{M}$	$\overline{\widetilde{M}}$
M	2, 2q	0, 1	0, 0	2, -1+2q
$\widetilde{M}$		-2, 2-2q	-2, 1-2q	0, 0
$\overline{M}$			-2 ,-2q	0, -1
$\overline{\widetilde{M}}$				2, -2+2q

One entry in this table stands out because it has a q independent topological charge. It is the entry with M and $\widetilde{M}$ dyons (and an entry with their CP transfoms).

It is magnetically neutral, and has topological charge one (or minus one for its CP transform). It is this particular pairing that turned out to be a local minimum of the action, and is the caloron.

To find the analytic solution for this "bound state" was a major tour de force by van Baal and coworkers [30] and by Kimyeong Lee and collaborators [31].

Note the absence of binding between the constituents: their individual actions add up to the total action $\frac{8\pi^2}{g^2}$ of the composite instanton.

The reader can find useful, pedagogically presented details on the technicalities of the solution in the PhD thesis of Daniel Nogradi [32] and in the lecture notes by Falk Bruckmann [33].

A.   A look at the assemblage of the caloron

Actually the caloron contains the two dyon configurations in an additive way, as is suggested by the additivity of the topological charges. It was pointed out by K. Lee, loc.cit... The very short discussion below is meant to be complementary to that in the reprints [12], [13] and [14], where the basic mathematical tool, the Nahm transform [34], is used.

The caloron potential can conveniently be expressed in terms of the two covariant derivatives $D_\mu(A^{BPS})$, defined as:

$$-i\partial_\mu + A_\mu^{BPS}. \tag{77}$$

The two BPS potentials are those discussed in the previous section, with boundary values q and $1-q$. Then are needed three two by two matrices C_1, C_2 and S. For specifics we refer the reader to Lee's original paper. The vector potential reads in terms of these:

$$A_\mu^{\text{cal}} = C_1^\dagger D_\mu(A^{BPS}(q))C_1 + C_2^\dagger D_\mu(A^{BPS}(1-q))C_2 + S^\dagger \partial_\mu S \ . \tag{78}$$

The first two terms look like gauge transforms of the respective monopoles. But C_1 and C_2 are not unitary! Only if the distance between the monopoles is very large with respect to the monopole sizes one retrieves the monopole configurations modulo gauge transformations. C_1 is a small gauge transformation. C_2 involves a large one like in Eq.(71), because the KK monopole defined in Eq.(73) needs one. The last term is a gradient term that vanishes at large monopole separation. This additivity continues to be true at the action density level, a non-trivial result . And so at large monopole separation the action density is just the sum of the action densities of the monopole constituents, and hence time independent.

On the other hand, when the constituents overlap, for $r << 1/T$, the action density is localized in time, and looks like that of an instanton.

If $q \to 0$ one retrieves as expected the periodic instanton, Eq.(60), for all values of the monopole separation r, with the identification $r = \pi\varrho^2 T$. It was Paolo Rossi, who already in 1979 identified the remaining monopole in the limit of very large radius ϱ [35].

So we have a continuous collection of local minima of the action (calorons), labeled by q and by their eight zero modes, all with the same action $8\pi^2/g^2$. The eight zero modes are precisely those expected for an object of topological charge one. One the other hand they describe precisely the four zero modes of the BPS monopole, and the four zero modes of its KK counterpart.

Note that the label q is *not* labeling a zero mode! The reason is that a variation in q leads to a non-normalizable mode.

The measure associated to the integral over the zero modes was calculated for $SU(2)$ by Kraan and van Baal in their seminal 1998 paper (reprint [12] in the reprint section).

For a discussion of calorons in $SU(N)$ and, more general, in any classical group the reader is referred to the literature [36].

B.   The caloron and its fluctuation determinant

We have analyzed in section IV the fluctuations around a constant Polyakov loop background. They contributed to the effective action a term $\mathcal{V}_{pert}(q)$ with $Z(2)$ conjugate minima at $q = 0, 1$.

Fluctuations around a caloron are of interest because they might induce a potential that destabilizes $\mathcal{V}_{pert}(q)$.

A first observation is on the order of magnitude of the result. It is the typical instanton amplitude, which combines with the asymptotic freedom of the coupling Eq.(6) into:

$$\exp(-8\pi^2/g^2(T)) = (\frac{T^2}{\Lambda_T^2})^{-\frac{11}{6}N}. \tag{79}$$

The scale Λ_T is on the order of T_c.

This is an N-dependent power, which cannot fit the plateau of the anomaly discussed in section IV. Therefore the caloron is a subleading effect in the range of the plateau, typically between T_c and $4T_c$. Clearly the fluctuation determinant (with zero modes taken out):

$$(det(-S''(A_{cal})))^{-1/2} \tag{80}$$

will produce a volume term proportional to $\mathcal{V}_{pert}(q)$, like for the configuration constant in q, discussed in section IV, Eq.(51) . In the case of large separation r of the constituents the caloron looks like a dipole. The electric fields will be Debye screened by the thermal fluctuations and lead to a damping term controlled by $m_D(q)$. Its square is proportional to the double derivative of this potential:

$$m_D^2(q) \sim (\frac{1}{6} - q(1-q)) \tag{81}$$

and is necessarily negative where the potential is concave, around the $Z(2)$ confining point $q = 1/2$. It just means that the fluctuation determinant will blow up and fail to give information in this window of q values. And determination of T_c requires knowledge of the potential at $q = 1/2$!

Magnetic fields are *not* screened by perturbative fluctuations. So the caloron looses its dual properties.

Non-leading terms in the asymptotics of r have been computed [37]. Precisely these non-leading terms constitute part of the correction due to the caloron to the effective potential. The other corrections are provided by the zero mode measure. The authors conclude on the basis of their asymptotics and numerical analysis, that the potential looks qualitatively like the potential in Fig.(5), for small q values and for low enough temperatures, on the order of Λ_T in Eq.(29). So the caloron provides the repulsion of the eigenvalues of the Polyakov loop but only for small enough T (large enough coupling). This is quite interesting but warrants an expansion of the caloron determinant in q and a determination of the coefficient of the leading term. This will answer the question whether the caloron destabilizes $\mathcal{V}_{pert}$ at high temperature.

Let us recall that calorons are present in *all* the classical groups, including those without center-group, like the group $G(2)$ [38]. On the other hand lattice simulations of $G(2)$ [39] show the phases of the Polyakov loop start to repel. In fact, the loop *vanishes* at low enough temperatures, despite the absence of a center symmetry group and consequently of an order parameter. So calorons may supersede the global center symmetry as driving the transition, the sense that they manufacture for small q a destabilizing linear term. Near the transition the q value is about $1/2$, and we have seen that the caloron determinant is not defined in that window.

In the commentary preceding the reprints the reader can find a discussion of very interesting work on the actual realization of caloron configurations on the lattice.

VII.   OTHER DEVELOPMENTS AND CONCLUSION

Apart from the use of calorons for thermal Yang-Mills physics, they have been applied to four dimensional field theories with one compact periodic *space* dimension. A classic is the work [40] on $\mathcal{N} = 1$ supersymmetric field theory. These authors compute the gluino condensate using calorons.

Their motivation is that big calorons with the magnetic charges far apart may mimic a magnetic plasma and therefore lead to semi-classical confinement as in the seminal work of Polyakov in three dimensions [41]. Related to this is the analytic work on periodic gluino zero modes by Garcia-Perez and Gonzalez-Arroyo [42].

The idea of the constituents has generated related developments, by taking various entries in the table, not necessarily the caloron entry .

As an example: there is a combination, $(M[1, q] \widetilde{M}[1, -1 + q]))$, with zero topological charge if $q = 1/2$, and with double magnetic charge. Again there is the conjugate of this combination with opposite double charge. These objects are called "bions" [43]. They have been applied to systems with adjoint Majorana fermions at zero temperature, but with a *periodic* space direction. Also the fermions are taken periodic, so they are non-thermal. As we will see below, a ground state with $q = 1/2$ can be realised.

These periodic fermions are providing the same effective potential as the gauge bosons, Eq.(51), except that the sign is *opposite*. So by adding one Majorana multiplet a vanishing potential results, as supersymmetry tells us. Adding two gives our one loop gauge potential but with an overall sign flip. Hence it has a *minimum* at $q = 1/2$.

In this simple fashion a theory is obtained with broken gauge symmetry but $U(1)$ still unbroken, i.e. with a zero mass photon and heavy vector bosons. Because of the periodicity of the "Higgs" field, i.e. the periodic space component of the vector potential, we have a KK monopole with the same mass as the usual monopole. The author argues that this bionic combination is stable due to the fermionic zero modes.

The bions render the photon massive, in rough analogy with the afore mentioned three dimensional Polyakov model [44]. This is then a four dimensional field theory with one small periodic space dimension, where the Wilson loop shows confining behaviour.

Related is the work [45] where the relation of $\mathcal{N} = 1$ super Yang-Mills theory and thermal Yang Mills theory is exploited through the limit where the mass of the gluino ranges from 0 to ∞.

More recently, a dyon gas, consisting of these (anti-) self dual objects, has been proposed [46] to understand properties of the confining phase. Interestingly, they find a confining potential in the Polyakov loop correlator. This idea was also exploited in the nineties by the work of the Madrid group [47]. However, in the *confining* phase dyons may not be the last word because we know from the old work of 't Hooft (see section III) and lattice simulations by de Forcrand and his group [48] that the behaviour of temporal Polyakov and 't Hooft loops is not related by duality. The latter is screened for *all* temperatures. It would be interesting to determine in this model the behaviour of 't Hooft loop correlators.

Applications to renormalons in $3 + 1$ dimensions (one dimension is again compact) are described in a recent work [49].

Also these new developments make it clear that the road taken by Pierre has led and will continue to lead to fascinating and fruitful developments. The concept of calorons has uncovered hitherto unsuspected views on the origins of the confinement-deconfinement transition.

---

[1] M. Gell-Mann, "The Eightfold Way", Caltech report CTSL-20 (1961), unpub.; G. Zweig, "An SU(3) model for strong interaction symmetry and its breaking II", Published in 'Developments in the Quark Theory of Hadrons'. Volume 1. Edited by D. Lichtenberg and S. Rosen. Nonantum, Mass., Hadronic Press, 1980. pp. 22-101 (1964).

[2] Richard H. Dalitz: Symmetries and the strong interactions , XIII'th International Conference on High Energy Physics, Berkeley CA, 1966, August 31-September 7 (University of California Press, California, 1967).

[3] H. Fritzsch, M. Gell-Mann and H. Leutwyler, "Advantages of the Color Octet Gluon picture", Phys. Lett. 47B (1973) 365.

[4] H.D. Politzer, Phys. Rev. Lett. 30 (1973) 1346; D.J. Gross and F. Wilczek, Phys. Rev. Lett. 30 (1973) 1343.

[5] G. 't Hooft, in High Energy Physics, ed. A. Zichichi (Bologna: Editrice Compositori, 1976); S. Mandelstam, Phys. Lett. B53 (1975) 476.

[6] B. Lucini, M.J. Teper, U. Wenger, JHEP 0406 (2004) 012; hep-lat/0404008.

[7] A.A. Belavin, A. M. Polyakov, A.S. Schwartz, Yu.S. Tyupkin, Phys.Lett. B59 (1975) 85.

[8] G. 't Hooft, Phys.Rev.Lett. 37 (1976) 8-11; Phys.Rev. D14 (1976) 3432-3450, Erratum-ibid. D18 (1978) 2199.

[9] R. Jackiw, C. Rebbi, Phys.Rev.Lett. 37 (1976);C.G. Callan, Jr., R.F. Dashen, D.J. Gross, Phys.Lett. B63 (1976) 334-340.

[10] See e.g.: Quantum Field Theory, Second Edition, by L.H. Ryder, Cambridge University Press,1996; a very pedagogical view on the tunneling of thermal instantons is to be found in: G.V. Dunne, B. Tekin, Phys.Rev. D63 (2001) 085004, hep-th/0011169.

[11] As long as classical solutions are discussed we can set $g = 1$. If its scale dependence becomes relevant we will restore the coupling.

[12] G. 't Hooft, Nucl. Phys. B79 (1974) 276; A.M. Polyakov, JETP Lett. 20 (1974) 194.

[13] J.C. Collins, M.J. Perry, Phys. Rev. Lett. 34 (1975) 1353; for a different line of argument see: N. Cabibbo, G. Parisi, Phys. Lett. B59 (1975) 67-69.

[14] An approach initiated by J. Engels, F. Karsch, H. Satz, I. Montvay, Nucl.Phys. B205 (1982) 545.

[15] G. 't Hooft, Nucl.Phys. B138 (1978) 1.

[16] An early application of this method is in: L.D. McLerran, B. Svetitsky, Phys.Lett. B98 (1981) 195.

[17] Sz. Borsanyi, G. Endrodi, Z. Fodor, S.D. Katz, K.K. Szabo, arXiv:1204.6184 [hep-lat].

[18] M. Panero, Phys. Rev. Lett. 103 (2009) 232001

[19] A narrow region around the transition point shows more structure, in particular at the critical point the order of the transition goes from second order for $N = 2$ to first order for $N = 3$ and larger.

[20] D.J. Gross, R.D. Pisarski and L.G. Yaffe, Rev. Mod. Phys. 53 (1981) 43, N. Weiss, Phys.Rev. D24 (1981) 475.

[21] A. Dumitru, Y. Guo, C.P. Korthals Altes, to appear.

[22] P.N. Meisinger, T.R. Miller, M.C. Ogilvie, Phys. Rev. D65 (2002) 034009; hep-ph/0108009.

[23] A. Dumitru, Y. Guo, Y. Hidaka, C. P.Korthals Altes, R. D. Pisarski, Phys. Rev. D83 (2011) 034022; arXiv:1011.3820 [hep-ph], and], and Phys. Rev D86 (2012) 105017 arXiv:1205.0137 [hep-ph].

[24] A.Dumitru etl al., loc.cit..

[25] B.J. Harrington, H.K. Shephard, Phys. Rev D17 (1978), 2122.

[26] D.J. Gross, R.D. Pisarski and L.G. Yaffe, loc. cit..

[27] K. Lee, P. Yi, Phys. Rev. D 56, (1997), 3711; hep-th970217. D. Diakonov, Prog. Part. Nucl. Phys. 51 (2003) 173; hep-ph/0212026, D. Diakonov , V. Petrov, Phys. Rev. D67 (2003) 105007 ;hep-th/0212018.

[28] E.B. Bogomolny, Sov. J. Nucl. Phys. 24 (1976) 449; M.K. Prasad and C.M. Sommerfield, Phys. Rev. Lett. 35 (1975) 760.

[29] For reference in later sections: $H = coth(vr) - 1/vr$ and $K = -1 + vr/sinh(vr)$.

[30] see this volume: reprints [12], [13 and [14].

[31] K. Lee, C. Lu, Phys. Rev. D.58.025011; hep-th/980210v1.

[32] D. Nogradi, PhD thesis, Leiden University; hep-th/0511125.

[33] F. Bruckmann, PROCEEDINGS 45th Internationale Universitatswochen fuer Theoretische Physik; Edited by C. Gattringer, C.B. Lang. Berlin, Springer, 2007. (Eur. Phys. J., Special Topics 152 (2007) 1-207); arXiv: 0706.2269v1.

[34] W. Nahm, Self-dual monopoles and calorons, in: Lecture Notes in Physics,201 (1984), 189.

[35] P. Rossi, Nucl. Phys B149 (1979), 170.

[36] T. C. Kraan, Commun. Math. Phys. 212, 503 (2000), arXiv: hep-th/9811179; D. Diakonov, N. Gromov, Phys. Rev. D72,025003 (2005), D. Diakonov, V. Petrov, arXiv:hep-th:1011.5636v1.

[37] D. Diakonov , N. Gromov , V. Petrov , S. Slizovskiy, Phys. Rev. D70 (2004) 03600, hep-th/0404042; N. Gromov, hep-th/0701192 and references therein.

[38] K.Y. Lee, Phys. Lett. B 426 (1998) 323: arXiv:hep-th/9802012.

[39] B. H. Wellegehausen, A. Wipf, and C. Wozar, Phys.Rev. D80, 065028 (2009), arXiv:0907.1450 [hep-lat]; Phys.Rev. D83, 016001 (2011), arXiv:1006.2305 [hep-lat]; Phys.Rev. D83, 114502 (2011), arXiv:1102.1900 [hep-lat] and references therein.

[40] N.M. Davies, T.J. Hollowood, V.V. Khoze, M.P. Mattis, Nucl. Phys. B559, 123 (1999), hep-th/9905015; D. Diakonov, V. Petrov, Phys. Rev. D67, 105007 (2003); hep-th/0212018 [hep-th].

[41] A. M. Polyakov, Nucl. Phys.B120, (1977), 429.

[42] M. Garcia-Perez, A. Gonzalez-Arroyo, JHEP 0611 (2006) 091; hep-th/0609058.

[43] M. Unsal, Phys. Rev. D80 (2009) 065001; arXiv:0709.3269; see also: M. Unsal, L.G. Yaffe, JHEP 1008 (2010) 030; arXiv: 1006.2101.

[44] A. M. Polyakov, loc. cit..

[45] E. Poppitz, T. Schaefer, M. Unsal: arXiv:1205.0290.

[46] F. Bruckmann, S. Dinter , E-M. Ilgenfritz, B. Maier, M. Muller-Preussker, M. Wagner, Phys. Rev. D85 (2012) 034502 ; arXiv:1111.3158 [hep-ph].

[47] M. Garcia Perez, A. Gonzalez-Arroyo, P. Martinez, Nucl. Phys. Proc. Suppl. 34, (1994), 228; A. Gonzalez-Arroyo, P. Martinez, Nucl. Phys. B 459, (1996), 337; A. Gonzalez-Arroyo, P. Martinez, A. Montero, Phys. Lett. B 359, (1995), 159; A. Gonzalez-Arroyo, A. Montero, Phys. Lett. B 387, (1996), 823.

[48] P. de Forcrand, M. D'Elia, M. Pepe, Phys. Rev. Lett. 86 (2001) 1438, hep-lat/0007034 and P. de Forcrand, L. von Smekal, Phys. Rev. D66 (2002) 011504, hep-lat/0107018.

[49] P. C. Argyres, M. Unsal, arXiv:1204.1661 [hep-th],P.C. Argyres, M. Unsal. arXiv:1206.1890 [hep-th].

COMMENTARY

There is a natural division of my selection into two categories, roughly coinciding with two periods in time: gauge theory in finite volumes starting with my thesis work in 1982, and going on till 1997. I then got involved in the caloron. That period extends till the present day, while I am writing these notes.

A category apart are the two papers on Topological Field Theory. They were written long before the caloron period, so I included them in the finite volume part.

As the commentaries are meant to give an introduction to the papers I thought it might be useful for the reader to get familiar with recent developments. That is what I did at the end of the caloron commentary section. It clearly carries some prejudice but I hope it may be of use.

The selected papers are numbered chronologically. It respects the two categories mentioned earlier, apart from the review paper on finite volume QCD. This was published during the period that I worked on calorons.

Gauge theory at finite volumes: charting the femto-universe

My interest in gauge theory in finite volumes started with my thesis work in 1982. I was very lucky: my advisor was Gerard 't Hooft.

The title of my thesis was "Twisted boundary conditions: A non-perturbative probe for pure non-Abelian Gauge Theories". At the time 'lattice gauge theory' started becoming an industry and the lattice practitioners needed a benchmark for their simulations.

That benchmark was provided by the validity of asymptotic freedom in small volumes, small with respect to hadronic sizes. Such femto-universes are studied with semi-classical methods. The art of the game was to be able to extrapolate the size of the femto-universe as far as possible without losing control over the approximation. Evidently this extrapolation involved the Gribov horizon, and some of my selected papers deal with this.

Lattice people imposed periodic boundary conditions. Such boundary conditions rendered the finite size effects exponentially small once you are in the confining regime, and was most welcome in numerical work. Important variants like twisted boundary conditions were invented by Gerard 't Hooft[1] shortly before the lattice simulations became popular.

In this arena I worked for quite long, roughly 15 years. Initially I was interested in what the precise topological charge was in case there was a twist in the boundary conditions. Gerard 't Hooft had argued convincingly that it was fractional, but fractional topological charge was not known in the mathematics literature. I discovered with some help of Sander Bais, who was a senior postdoc in Utrecht, that you could also phrase it in adjoint SU(N). Then the topological charge was multiplied with 2N, and becomes integer again. This was known to the mathematicians, but apparently they did not bother writing it down. This and some other results found their realization in my first paper [1], which Raymond Stora eventually accepted in Comm. Math. Physics.

---

[1] A Property of Electric and Magnetic Flux in Nonabelian Gauge Theories. Gerard 't Hooft, Nucl. Phys. B153 (1979) 141.

Another paper which I published in my thesis became important later on for quite different reasons, for example for D-branes and the Born-Infeld action. It was the paper [2], but I worked on the Twisted Eguchi-Kawai model as well, in particular on "hot twists" with Frans Klinkhamer, good for simulating high temperature systems.

My thesis generated interest for the mathematicians in Utrecht as well. Hans Duistermaat was on the committee, and T.A. Springer immediately noticed to Bert van Geemen (who was a friend of mine and was doing a thesis in mathematics) that the twist eating solutions are precisely the representations of the Heisenberg group, and the two of us wrote a paper, "The group theory of twist eating solutions", which was ultimately published [3] in the Proceedings of the Royal Dutch Academy of Sciences, communicated by Gerard 't Hooft. It is included here because of its striking simplicity: the finite Heisenberg groups have two generators Ω_μ and Ω_ν, that commute into a centergroup element:

$$\Omega_\mu \Omega_\nu \Omega_\mu^{-1} \Omega_\nu^{-1} = \exp(2\pi i n_{\mu\nu}/N)I.$$

The reader will recognize in this relation precisely the definition of a single twist eater configuration on a periodic lattice.

In the meantime we published another paper which was called "A Simple Construction of Twist-Eating Solutions"[2], but that one is not included here.

During the last months before the completion of my thesis I worked on "The Energy of Electric Flux on the Hypertorus". This was a preprint[3], of which I was proud. It became the source of the cover for my thesis. But during my first visit to Martin Lüscher in DESY (Hamburg) he convinced me to first try to clean up some issues. I luckily followed his advice. The cleaning up turned out not that easy, so I had the pleasure to make the journey to DESY many times.

In September 1984 I left Utrecht for my first post-doc, at Stonybrook (Peter van Nieuwenhuizen and Chen Ning (Frank) Yang, who visited Leiden as a Lorentz professor, were on the committee for my thesis). I continued working on the energy of electric fluxes with Jeff Koller, who was like me in Stony Brook for three years. We published a version [4] of the preprint, with many things added, called "QCD on a Torus and Electric Flux Energies from Tunneling", but we wrote many more papers, including paper [5]. We made some breakthroughs and could compute the spectrum for volumes as large as $z = m_{0^+}L - 5$, but it is still a finite volume.

As fellow at CERN, from September 1987 till september 1989, I continued the work on finite volume QCD. But three papers are of a very different nature.

The first one is a paper [6] with my friend Peter Braam, a mathematician with a vivid interest in physics but who retired early from his scientific career. It is called "Nahm's Tranformation for Instantons". The paper grew out from a course I taught in Stony Brook, and to which Peter made some great contributions. It took us some time, but we proved (amongst other things) that there were no untwisted charge one instantons on the hypertorus. Peter also played an essential role in a second paper, as he had already thought

---

[2]B. van Geemen and P. van Baal, "A Simple Construction of Twist-Eating Solutions", J. Math. Phys. 27 (1986) 455.

[3]Preprint 84-0571 Utrecht, March 1984.

about many issues concerning the hottest thing at that moment, "Topological Quantum Field Theory" by Ed Witten[4].

Hearing a seminar by Laurant Baulieu about this I could not figure out certain things, and talked to Raymond Stora. I wrote down notes on his blackboard and he asked me if he could copy that (I was a bit startled by this, because it was his own blackboard), but then he came back a few days later and suggested we write a paper about this with Stéphane Ouvry. We called it "On the Algebraic Characterization of Witten's Topological Yang-Mills Theory", and it is included in this volume [7].

At the end of my stay at CERN I was invited once more to Zakopane, tucked away in the beautiful mountains near Kraków in Poland, to give some lectures at the summer school. I decided to talk on "An Introduction to Topological Yang-Mills Theory", because I wanted to learn it even better. It is included here [8] to enjoy, as I did writing it!

When back to Utrecht there was another paper that became important and I separately want to mention it, although it is partly contained in "QCD in a Finite Volume". I called it "More (thoughts on) Gribov copies" [9]. In general the problem of finding the fundamental modular domain is not solved yet (although it is valid for the constant modes), and people keep on struggling with it. Also I published a minor correction, which apparently is hard to find, although it is on the arXiv (my first one). I have included it here so that you can see what is corrected.

In Leiden I became a full professor in December 1992 and worked on a variety of subjects. Margarita (Marga) García Pérez visited me already in Utrecht (she was finishing her thesis in Madrid) and became my first postdoc in Leiden. We published a nice paper with Antonio (Tony) González-Arroyo and my PhD-student Jeroen Snippe, which was called [10] "Instantons from over-improved cooling". It introduced a different, improved action, which had stable instantons. I also wrote a paper at the 29th Ahrenshoop symposium in Buckow, Germany. I would have talked about lattice gauge theory, but I argued to Dieter Lüst, one of the organizers, that it was good to have a program with a broad spectrum, and as proof of that I spoke on new aspects to the Nahm transformation. The paper was called "Instanton moduli for $T^3 \times R$" [11]. It is also included here, but with a lot of mistakes finally corrected.

I summarized the progress on finite volume QCD, which was made in Stony Brook, CERN, Utrecht (where I became KNAW fellow) and finally in Leiden. It was called "QCD in a Finite Volume". This manuscript contains everything I did in finite volumes including lattice applications and fermions. I also made sure that many corrections were included, for example the correction which was found by Claus Vohwinkel that the T_2 state in fig. 3 of paper [5] had in addition two units of electric flux and is now named T_{11}. You can read all about it in [18] "QCD in a Finite Volume".

Calorons with non-trivial holonomy: charting the hot semi-classical universe

The first caloron solution had been found by B.J. Harrington and H.K. Shepard[5], three years after the original work of BPST[6] and 't Hooft[7] This solution supposed trivial boundary

---

[4]E. Witten, Comm. Math. Phys. 117 (1988) 353.

[5]B.J. Harrington and H.K. Shepard, Phys. Rev. D17 (1978) 2122.

[6]A.A. Belavin, A.M. Polyakov, A.S. Schwartz and Yu. S. Tyupkin, Phys. Lett. 59B, 85 (1975).

[7]G. 't Hooft, Phys. Rev. Lett. 37, 8 (1976).

conditions at spatial infinity.

Oddly enough, it took another twenty years before explicit solutions were found with non-trivial values for the thermal Polyakov loop, that is to say with non-trivial holonomy.

These solutions of calorons with non-trivial holonomy are now sometimes also called KvBLL solutions (Kraan-van Baal-Lee-Lu), and I'll try to describe below how it began from my perspective in 1998.

I would like to say beforehand that the work of Kimyeong Lee and his collaborators is providing a beautiful approach to the caloron problem. It has been of great value to the community of practitioners and in particular to me and my collaborators as a complementary attack on the problem.

He invited me to KIAS in Seoul for a workshop on "Quantum Field Theory and Mathematical Physics" where I spoke on May 14, 2003 about "Multi-Caloron Solutions"[8]. Piljin Yi was already a collaborator in Columbia (NY, USA), but also moved to KIAS and both gave a talk at this workshop. Recently I met them again in July 2010 at a Durham symposium. Unfortunately I did not meet Changhai Lu, at the time of the caloron discovery Kimyeong's PhD student, but there were emails back and forth.

So where and when did my part of the caloron venture start?

David Olive, Peter West and I had organised a programme on the "Non-perturbative Aspects of Quantum Field Theory" at the Isaac Newton Institute for Mathematical Sciences (January-June 1997) in Cambridge, England. I gave a talk on "The Nahm transformation for pedestrians" (January 28, 1997). Jerome Gauntlett asked a question which motivated the title for our first manuscript [12]. In that first paper we explicitly (on the first page!) refer to that question posed by Jerome: "concerning the moduli spaces of calorons with non-trivial asymptotic behaviour of the Polyakov loop (non-trivial holonomy)."

Below I give a brief account of the genesis of that first paper.

Back in Leiden I involved Thomas Kraan who had started with me as a PhD-student in January 1996. Initially I left him working alone because I had not much time to investigate the moduli space problem. My first trial at the Newton Institute had given me no explicit solutions.

Thomas made more progress (but with no solutions yet) and had a poster at the Dutch DRSTP meeting, 5-6 June, 1997, (ref. [22] in the first manuscript. But while he collected his results later on (now with explicit solutions!) he came into my office on February 4 1998, saying that Kimyeong also published some results[9], submitted on February 3 and on top of that announced another paper by him and Changhai Lu.

I decided to drop everything and to write a letter with Thomas in three days. The entire calculation which I finished within those three days agreed with what Thomas had done so far, so I was confident that indeed we had explicitly constructed solutions and it appeared [12] on the arXiv a few days later (February 9), hep-th/9802049, and was submitted to Phys. Lett. on February 11.

But while we were waiting for a referee decision Kimyeong and Changhai published their sequel which was submitted to the arXiv on February 16 and to Physical Review on

---

[8]My talk is still available through http://conf.kias.re.kr/lecture/Multi.pdf.

[9]K. Lee, "Instantons and magnetic monopoles on $R^3 \times S^1$ with arbitrary simple gauge groups", Phys. Lett. B426 (1998) 323 [hep-th/9802012].

February 17[10]. So we mentioned this in a "Note added in proof". By the way, their paper has a considerable overlap with what we wrote, so it explains why the results became known as KvBLL solutions. I quote from the Acknowledgments in their paper: "While writing up this paper, we become aware of Ref. [17] which has a considerable overlap with our work."

Our first paper emphasized the exact T-duality, because we finally solved the problem posed by Jerome Gauntlett. But I was very much aware that this was also important for Yang-Mills theories, especially in the confined region where the trace of the holonomy is zero, and the other papers referred only to that application, first for a detailed version of SU(2) and subsequently for SU(N).

The sequel with the details of the SU(2) solutions was easy to write, because it was basically checked already in our first paper, but we added a discussion of the lecture by C.H. Taubes, "Morse theory and monopoles: topology in long-range forces", in Ref.[11], which was held in Cargèse, 1-15 September 1983, where I had been a participant! He explained how the monopole field could be used to create a gauge field with topological charge one. We submitted that second paper to the arXiv on May 25, and waited till we had also finished the SU(N) case, because we realized that the density could be written more easily as

$$-\tfrac{1}{2}\mathrm{tr}F_{\mu\nu}^2 = -\tfrac{1}{2}\partial_\mu^2\partial_\nu^2 \log\psi$$

with

$$\psi = \tfrac{1}{2}\mathrm{tr}(\mathcal{A}_n \cdots \mathcal{A}_1) - \cos(2\pi x_0).$$

The $\mathcal{A}_m$'s are 2×2 matrices,

$$\mathcal{A}_m \equiv \frac{1}{r_m} \begin{pmatrix} r_m & |\vec{y}_{m+1} - \vec{y}_m| \\ 0 & r_{m+1} \end{pmatrix} \begin{pmatrix} \cosh(2\pi\nu_m r_m) & \sinh(2\pi\nu_m r_m) \\ \sinh(2\pi\nu_m r_m) & \cosh(2\pi\nu_m r_m) \end{pmatrix},$$

You can read the details in the paper.

We submitted this one to the arXiv on June 4, and they were published as papers [13] and [14] , both submitted to the journal on the same day, June 8. In both papers we referred to K. Lee and P. Yi's work[12], as it motivated us to generalize our results from SU(3) to SU(N).

Changhai submitted another paper[13]. He put it to the arXiv on June 29 and to the journal on July 14. He wrote a "Noted added. While writing this paper, we noticed the appearance of [17] from which the energy density of SU(3) and Sp(4) monopoles can also be obtained."

This was all within half a year, but I was so excited that I continued working on it. The caloron solution gave many insights into configurations already obtained in finite volume by

[10]K. Lee and C. Lu, "SU(2) calorons and magnetic monopoles", Phys. Rev. D58 (1998) 025011 [hep-th/9802108].

[11]C. H. Taubes, Progress in Gauge Field Theory, ed. G. 't Hooft et al. (Plenum Press, New York, 1984, p. 365).

[12]K. Lee, P. Yi, "Dyons in N=4 supersymmetric theories and three-pronged strings", Phys. Rev. D58 (1998) 066005 [hep-th/9804174].

[13]C. Lu, "Two-monopole systems and the formation of non-Abelian clouds", Phys. Rev. D58 (1998) 125010 [hep-th/9806237].

using twisted boundary conditions[14], the subject of my PhD thesis. This spurred a paper with Marga and Tony (and a PhD-student of his) [15] "Calorons on the lattice - a new perspective", by M. García Pérez, A. González-Arroyo, A. Montero and P. van Baal.

Soon after that came the paper on the (continuum) fermion zero-mode for SU(2) [16], called "Weyl-Dirac zero-mode for calorons", and paper [17], called "Exact fermion zero-mode for the new calorons". This last paper came out in a conference proceeding (Lattice'99 in Pisa), but it contains new results.

Thomas wrote a paper on his own[15] and defended his thesis on March 30, 2000[16].

When Falk Bruckmann came to Leiden as my postdoc, October 2001, we worked on finding calorons with higher charge and published some results [19].

But I had also a new PhD-student, Dániel Nógrádi, who started in December 2000, and the three of us made some essential progress, which culminated in paper [20]. I gave some lectures at a school in Zakopane, May 30 to June 8, 2003, which apparently was well-written, so I included also this paper [21].

Dániel Nógrádi published his thesis and defended it successfully on June 29, 2005, and his thesis is available as hep-th/0511125.

We also collaborated with Ernst-Michael Ilgenfritz and Boris Martemyanov, who were in Leiden for extended visits, but it was getting difficult to make further progress, and I was secretly looking for something else, but on July 31, 2005 I got a major stroke. I have survived, and I was able to write this story. But I am much slower than before.

Recent developments

Probably the first time the caloron solution was referred to as "KvBLL" was in D. Diakonov, "Instantons at work"[17]. He was critical in this paper. But he became even more convinced than I was, when they showed that the free energy has a phase transition at (about) T_c (which I always assumed, but could not prove)[18]. They wrote many papers, even showing that a hyperKähler ansatz for multi-calorons was apparently confining[19]. Mitya Diakonov invited me to a symposium on "Theoretical and Mathematical Physics" at the Euler Institute in St. Petersburg, June 3-8, 2009. Although it was for me a definitely slowed down process, I discussed physics with many people and in particular with Mitya and Victor Petrov.

In Lattice '98 John Negele wrote a nice review[20], with towards the end a chapter on calorons with non-trivial holonomy. There was also lattice work, in part by Ilgenfritz and

---

[14]M. Garcia-Perez, A. Gonzalez-Arroyo, B. Soderberg, Phys. Lett.B235, (1990), 117; A. Gonzalez-Arroyo and A. Montero, Phys. Lett. B **442** (1998) 273 , hep-th/9809037.

[15]T.C. Kraan, "Instantons, Monopoles and Toric HyperKähler Manifolds", Comm. Math. Phys. 212 (2000) 503 [hep-th/9811179].

[16]Available as "http://www.lorentz.leidenuniv.nl/vanbaal/HOME/PUBL/kraan.ps".

[17]D. Diakonov, Prog. Part. Nucl. Phys, 51 (2003) 173 [hep-ph/0212026].

[18]D. Diakonov, N. Gromov, V. Petrov and S. Slizovskiy, "Quantum weights of dyons and of instantons with nontrivial holonomy", Phys. Rev. D70 (2004) 036003, hep-th/0404042.

[19]D. Diakonov and V. Petrov, "Confining ensemble of dyons", Phys. Rev. D76 (2007) 056001, arXiv:0704.3181 [hep-th].

[20]J. Negele, "Intantons, the QCD vacuum, and hadronic physics", Nucl. Phys. B (Proc. Suppl.) 73 (1999) 557, hep-lat/0007026.

Martemyanov, in collaboration with Michael Müller-Preussker. They wrote the first paper with Alexander Veselov[21]. Many papers followed and it was initially called KvB (Kraan-van Baal) solutions as early as June 2002 in Ref.[22], but they called it later also KvBLL calorons. In particular the paper with Philipp Gerhold is interesting, written together with E.-M. Ilgenfritz and M. Müller-Preussker[23].

Christof Gattringer and Stefan Schaefer wrote also an important paper on "New findings for topological excitations in $SU(3)$ lattice gauge theory"[24], which demonstrates clearly how the zero-mode behaves as a function of $\exp(2\pi i\zeta)$ (this is usually -1 for finite temperature), thereby showing that the fermion zero-mode structure of calorons with non-trivial holonomy we constructed was basically right.

We should also mention the work of Michael Davies, Timothy Hollowood, Valetin Khoze and Michael Mattis, "Gluino condensates and magnetic monopoles in supersymmetric gluodynamins"[25]. They only needed the pair of zero-modes of adjoint fermions for every constituent monopole (so for $SU(N)$ it has in total 2N zero-modes for N constituent monopoles). It is periodic in time, to preserve supersymmetry. But the exact result of the zero-modes of adjoint fermions in the case of $SU(2)$ calorons with non-trivial holonomy (made from two constituent monopoles) had to wait until M. García Pérez and A. González-Arroyo "Gluino zero-modes for non-trivial holonomy calorons"[26].

The adjoint fermion zero-modes with anti-periodic boundary conditions in time were also studied by M. García Pérez, A. González-Arroyo and A. Sastre, and are now known in analytic form, "Gluino zero-modes for calorons af finite temperature"[27] and "Adjoint fermion zero-modes for $SU(N)$ calorons"[28] (which also generalizes solutions periodic in time to $SU(N)$). I still remember that nice explanation Marga gave during a visit I made to Madrid, May 24-28, 2009. I thank both her and Tony for the nice hospitality.

Also "multi-calorons" were "revisited" by A. Nakamula and J. Sakaguchi[29]. The authors claim to have the full $SU(2)$ moduli space for $k = 2$. There are more recent reviews by F. Brueckmann[30] and by D. Diakonov[31], but I want to finish with something else.

Mithat Ünsal, whom I met in Florence end of May 2008, has published a paper concerning the mechanism of confinement in QCD-like theories, for example $SU(2)$ with $1 \leq n_f \leq 4$ adjoint Majorana fermions[32]. He argues that there are BPS and KK monopoles (precisely the constituents of the caloron), which have zero-modes under the adjoint fermions. They

---

[21] E.-M. Ilgenfritz, B.V. Martemyanov, A.I. Veselov, M. Müller-Preussker, Lattice 2000, Nucl. Phys. B (Proc. Suppl.) 94 (2001) 407, hep-lat/0011051.

[22] E.-M. Ilgenfritz, B.V. Martemyanov, A.I. Veselov, M. Müller-Preussker and S. Shcheredin, Phys. Rev. D66 (2002) 074503, hep-lat/0206004.

[23] P. Gerbold, E.-M. Ilgenfritz, M. Müller-Preussker, "An $SU(2)$ KvBLL caloron gas model and confinement", Nucl. Phys. B760 (2007) 1, hep-ph/0607315 .

[24] C. Gattringer and S. Schaefer, Nucl. Phys. B654 (2003) 30, hep-lat/0212029.

[25] M. Davies, T. Hollowood, V. Khoze, M. Mattis, Nucl. Phys. B559 (1999) 123, hep-th/9905015.

[26] M. García Pérez, A. González-Arroyo, JHEP 0611 (2006) 091,hep-th/0609058.

[27] M. García Pérez, A. González-Arroyo, A. Sastre, Phys. Lett. B668 (2008) 340, arXiv:0807.2285 [hep-th].

[28] M. García Pérez, A. González-Arroyo, A. Sastre, JHEP 0906 (2009) 065, arXiv:0905.0645 [hep-th].

[29] J. Math. Phys. 51 (2010) 043503, arXiv:0909.1601 [hep-th].

[30] F. Brueckmann, Eur. Phys. J. Special Topics 152 (2007) 61, arXiv:0706.2269 [hep-th].

[31] D.Diakonov, Acta Phys. Pol. B39 (2008) 3365, arXiv:0807.0902 [hep-th], and Nucl. Phys. B (Proc. Suppl.) 195 (2009) 5, arXiv:0906.2456 [hep-ph].

[32] M. Unsal, Phys. Rev. D80 (2009) 065001; arXiv:0709.3269.

then make BPS-$\overline{\text{KK}}$ bound states (instead of BPS-KK). Apparently zero-modes can be quite influential.

There have been many more papers, but I must stop somewhere.

Acknowledgments

First of all I would like to thank Gerard 't Hooft, because without him this reprint volume would never have appeared. I am terribly flattered by it, but of course I know that I have to be modest. At a very early age (I was 50 when I had the stroke) my scientific career was stopped (but I keep on hoping). I also want to thank Chris Korthals Altes very much for luring me into writing something, especially on the finite volume stuff. For calorons I had a lot of correspondence, but I worried to reconstruct things earlier on. (I also know Chris from before the time I became a PhD-student.) Any mistakes are of course entirely my own, but I appreciated the gentle encouragement of Chris, and without that this would not have been written.

I would also like to thank Martin Lüscher, because he had a large impact on my work. Also Chen Ning Yang had a big influence on me, although it was indirect. Finally I want to thank Werner Nahm, especially for his transformation. And of course I am grateful for the many students, postdocs and colleagues who worked with me.

References

[1] P. van Baal, "Some results for SU(N) Gauge Fields on the Hypertorus", *Comm. Math. Phys.* **85** (1982) 529.

[2] P. van Baal, "SU(N) Yang-Mills Solutions with Constant Field Strength on T^4", *Comm. Math. Phys.* **94** (1984) 397.

[3] B. van Geemen and P. van Baal, "The group theory of twist eating solutions", *Proc. K. Ned. Akad. Wet.* **B89** (1986) 39.

[4] P. van Baal and J. Koller, "QCD on a Torus and Electric Flux Energies from Tunneling", *Ann. Phys. (N.Y.)* **174** (1987) 299.

[5] P. van Baal and J. Koller, "SU(2) Spectroscopy in Intermediate Volumes", *Phys. Rev. Lett.* **58** (1987) 2511.

[6] P. J. Braam and P. van Baal, "Nahm's Tranformation for Instantons", *Comm. Math. Phys.* **122** (1989) 262.

[7] S. Ouvry, R. Stora and P. van Baal, "On the Algebraic Characterization of Witten's Topological Yang-Mills Theory", *Phys. Lett.* **B220** (1989) 159.

[8] P. van Baal, "An Introduction to Topological Yang-Mills Theory", *Acta Phys. Polon.* **B21** (1990) 73.

[9] P. van Baal, "More (thoughts on) Gribov copies", *Nucl. Phys.* **B369** (1992) 259. Addendum: P. van Baal, "Topology of the Yang-Mills Configuration Space", hep-lat/9207029.

[10] M. Garcia Perez, A. Gonzalez-Arroyo, J. Snippe and P. van Baal,"Instantons from over-improved cooling", *Nucl. Phys.* **B413** (1994) 535 [hep-lat/9309009].

[11] P. van Baal, "Instanton moduli for $T^3 \times R$", *Nucl. Phys. B (Proc. Suppl.)* **49** (1996) 238 [hep-th/9512223].

[12] P. van Baal and T. C. Kraan, "Exact T-duality between calorons and Taub-NUT spaces", *Phys. Lett.* **B428** (1998) 268), [hep-th/9802049].

[13] P. van Baal and T. C. Kraan, "Periodic Instantons with Non-trivial Holonomy", *Nucl. Phys.* **B533** (1998) 627, [hep-th/9805168].

[14] P. van Baal and T. C. Kraan, "Monopole Constituents inside SU(n) Calorons", *Phys. Lett.* **B435** (1998) 389, [hep-th/9806034].

[15] M. García Pérez, A. González-Arroyo, A. Montero and P. van Baal, "Calorons on the lattice - a new perspective", *JHEP* **06** (1999) 001 [hep-lat/9903022].

[16]]M. García Pérez, A. González-Arroyo, C. Pena and P. van Baal, "Weyl-Dirac zero-mode for calorons", *Phys. Rev.* **D60** (1999) 031901 (Rapid Comm.) [hep-th/9905016].

[17] M. N. Chernodub, T. C. Kraan and P. van Baal, "Exact fermion zero-mode for the new calorons", *Nucl. Phys. B (Proc. Suppl.)* **83-84** (2000), 556, [hep-lat/9907001].

[18] P. van Baal, "QCD in a Finite Volume", in *At the Frontiers of Particle Physics: Handbook of QCD*, Boris Ioffe Festschrift, ed. Mikhail Shifman (World Scientific, Singapore, 2001), Vol. 2, pp. 683-760 [hep-ph/0008206].

[19] F. Bruckmann and P. van Baal, "Multi-Caloron solutions", *Nucl. Phys.* **B645** (2002) 105 [hep-th/0209010].

[20] F. Bruckmann, D. Nógrádi and P. van Baal, "Higher charge calorons with non-trivial holonomy", *Nucl. Phys.* **B698** (2004) 233 [hep-th/0404210].

[21] F. Bruckmann, D. Nógrádi and P. van Baal, "Instantons and constituent monopoles", *Acta Phys. Polon.* **B34** (2003) 5717 [hep-th/0309008].

Commun. Math. Phys. 85, 529–547 (1982)

Communications in
Mathematical
Physics
© Springer-Verlag 1982

Some Results for SU(N) Gauge-Fields on the Hypertorus

Pierre van Baal

Institute for Theoretical Physics, Princetonplein 5, P.O. Box 80.006, NL-3508 TA Utrecht, The Netherlands

Abstract. We show how to prove and to understand the formula for the "Pontryagin" index P for SU(N) gauge fields on the Hypertorus T^4, seen as a four-dimensional euclidean box with twisted boundary conditions. These twists are defined as gauge invariant integers modulo N and labelled by $n_{\mu\nu}$ $(= -n_{\nu\mu})$. In terms of these we can write ($\nu \in \mathbb{Z}$)

$$P = \frac{1}{16\pi^2} \int \mathrm{Tr}(G_{\mu\nu} \tilde{G}_{\mu\nu}) d_4 x = \nu + \left(\frac{N-1}{N}\right) \cdot \frac{n_{\mu\nu} \tilde{n}_{\mu\nu}}{4}.$$

Furthermore we settle the last link in the proof of the existence of zero action solutions with all possible twists satisfying $\dfrac{n_{\mu\nu} \tilde{n}_{\mu\nu}}{4} = \kappa(n) = 0 (\mathrm{mod}\, N)$ for arbitrary N.

1. Introduction

A long standing problem is proving quark confinement in QCD [1]. To simplify the picture a first step in this direction would be to show confinement of static quarks. In this way the problem reduces to an understanding of the behaviour of electric flux strings in quarkless QCD, thus working in pure SU(N) gauge theories, where up to present energies $N = 3$. Usually this boils down to studying the behaviour of the vacuum expectation value of the Wilson loop operator [1, 2].

But some time ago 't Hooft [3] introduced another elegant method for studying flux strings. By putting the gauge fields on a four dimensional euclidean box, one can imitate a quark source on one side and an antiquark source on the other side of the box by introducing socalled twisted boundary conditions. These boundary conditions force electric flux into the box, just as gauge invariance forces an electric flux string in-between a quark and antiquark source. Similarly one can introduce magnetic flux in the box, and the great strength of the method is its electric-magnetic duality properties.

Essential is that all fields transform trivially under the centre Z_N of SU(N). Effectively the gauge group is thus SU(N)/Z_N and the twists are labelled by six

0010-3616/82/0085/0529/$03.80

integers $n_{\mu\nu} = -n_{\nu\mu}$ ($\mu = 1$, 2, 3 or 4) defined modulo N. Here $n_{\mu\nu}$ describes a winding number in the (μ, ν)-plane, as an element of the first homotopy group $\pi_1(\mathrm{SU}(N)/Z_N) \cong Z_N$. The gauge fields are in fact connections for a $\mathrm{SU}(N)/Z_N$ – principal fiber bundle on the hypertorus T^4. This will be discussed in Sect. 2.

The topology of the gauge fields is not completely specified by the twist, but there is also the "2nd Chern class' related to instanton type configurations and given by (the coupling constant g is put equal to 1):

$$P = \frac{1}{16\pi^2} \int d_4 x \, \mathrm{Tr}(G_{\mu\nu}\tilde{G}_{\mu\nu}), \tag{1.1}$$

$$\tilde{G}_{\mu\nu} = \tfrac{1}{2}\varepsilon_{\mu\nu\alpha\beta}G_{\alpha\beta} \quad \text{(the dual of } G_{\mu\nu}\text{)}. \tag{1.2}$$

In Sect. 3 we shall explain why P (or actually $2NP$) should really be called the 1st Pontryagin index. 't Hooft [4] used invariance arguments to show the following identity:

$$P = \nu + \left(\frac{N-1}{N}\right)\kappa(n), \tag{1.3}$$

$$\kappa(n) = \tfrac{1}{4}\tilde{n}_{\mu\nu}n_{\mu\nu} = \tfrac{1}{8}\varepsilon_{\mu\nu\alpha\beta}n_{\alpha\beta}n_{\mu\nu}. \tag{1.4}$$

This will be proved in Sect. 3, where we will resolve the interpretation of this formula, especially what happens if we add a multiple of N to $n_{\mu\nu}$.

For finding solutions to the euclidean equations of motion the importance of P comes from the Schwarz-inequality for the action $S(A)$:

$$S(A) = \tfrac{1}{2}\int \mathrm{Tr}(G_{\mu\nu}G_{\mu\nu})d_4 x \geq 8\pi^2|P|. \tag{1.5}$$

Recently 't Hooft [5] constructed a very special type of solutions with minimal nontrivial action.

In Sect. 4 we complete his proof for the existence of orthogonal twist ($\kappa(n) = 0(\mathrm{mod}\,N)$) zero action configurations for *arbitrary N*.

In an appendix we prove a formula for the 1st Pontryagin (2nd Chern) index for a general four dimensional compact(ifiable) manifold in terms of the transition functions only.

Our primary aim in this paper is a precise understanding of the twist-dependence for the Pontryagin number. For this we need an explicit realization of the topological structure of the fiber bundles over T^4, yielding at the same time a classification of $\mathrm{SU}(N)/Z_N$ bundles over T^4. Classification of fiber bundles is well known to mathematicians, however even for this problem it is a non trivial exercise, using K-theory, as demonstrated by C. Nash in his preprint entitled: "Gauge potentials and bundles over the 4-torus" (St. Patrick's college, January 1982). We thank him for correspondence on this subject.

2. The Structure of the Gauge Fields on the Hypertorus

From the requirement of periodicity for gauge invariant quantities we have [5]: (A_λ is the gauge potential in the fundamental representation)

$$A_\lambda(x_\mu = a_\mu) = [\Omega_\mu]A_\lambda(x_\mu = 0)$$
$$= \Omega_\mu A_\lambda(x_\mu = 0)\Omega_\mu^{-1} - i\Omega_\mu\partial_\lambda\Omega_\mu^{-1}. \tag{2.1}$$

The euclidean box is defined by $0 \leq x_\mu \leq a_\mu$, $\mu = 1, 2, 3, 4$. When the argument of a function on the box is put equal to x_μ, we mean that x_μ is fixed and the other coordinates are arbitrary. $[\Omega_\mu]$ is the action of a SU(N) gauge transformation Ω_μ, independent of x_μ. All other fields ψ satisfy the same formal periodic boundary condition $\psi(x_\mu = a_\mu) = [\Omega_\mu]\psi(x_\mu = 0)$.

As long as all these fields transform trivially under the centre Z_N of SU(N) we have from the consistency of writing $\psi(x_\mu = a_\mu, x_\nu = a_\nu)$ in two ways in terms of $\psi(x_\mu = 0, x_\nu = 0)$ [through $\psi(x_\mu = a_\mu, x_\nu = 0)$ and $\psi(x_\mu = 0, x_\nu = a_\nu)$]:

$$Z_{\mu\nu} = \Omega_\mu(x_\nu = a_\nu)\Omega_\nu(x_\mu = 0)\Omega_\mu^{-1}(x_\nu = 0)\Omega_\nu^{-1}(x_\mu = a_\mu), \qquad (2.2)$$

where $Z_{\mu\nu}$ is an element of the centre Z_N of SU(N). We write $Z_{\mu\nu} \in Z_N$ in terms of the twist integers $n_{\mu\nu} \in \mathbb{Z} \,(\mathrm{mod}\, N)$:

$$Z_{\mu\nu} = \exp(2\pi i n_{\mu\nu}/N). \qquad (2.3)$$

Clearly $n_{\mu\nu} = -n_{\nu\mu}$, they are independent of the coordinates, and are gauge invariant, since under an arbitrary gauge transformation $\Omega(x)$ ($\psi' = [\Omega]\psi$) we have:

$$\Omega_\mu' = \Omega(x_\mu = a_\mu)\Omega_\mu\Omega^{-1}(x_\mu = 0). \qquad (2.4)$$

The gauge functions are really defined modulo the centre of the gauge group, so the gauge group is actually SU(N)/Z_N. The multiple transition functions Ω_μ then take their values in SU(N)/Z_N. In this representation we have (2.2) with $Z_{\mu\nu} = 1$ [the identity in SU(N)/Z_N], which becomes a multiple cocycle condition. We borrowed the terminology from the theory of fiber bundles [6] and we will explain how the above structure defines a SU(N)/Z_N principal fiber bundle on the hypertorus T^4 parametrized by: $0 \leq x_\mu \leq a_\mu$, $\mu = 1, 2, 3, 4$; with $x_\mu = 0$ identified with $x_\mu = a_\mu$.

We will first discuss the case of the two dimensional torus T^2. We need a covering of T^2 with open sets U_i, however it is more advantageous to reduce the overlapping regions to a minimum area. This is done by taking the closure of U_i and reducing their size to a minimum, such that they still cover the manifold completely. We will denote such a covering by $\{U_i^c\}$. For T^n the minimal number of such sets is 2^n and in Fig. 1 we specify the situation for T^2. We can take $\delta = \varepsilon$ (δ and ε are defined in Fig. 1) but it should always be understood in the limit $\delta \downarrow \varepsilon$.

A fiber bundle is specified [6, 7] by the transition functions $\Omega_{ij} = \Omega_{ji}^{-1}$ on $U_i \cap U_j$ (in our case $U_i^c \cap U_j^c$), such that:

$$A_\mu^{(i)}(x) = [\Omega_{ij}(x)]A_\mu^{(j)}(x), \qquad x \in U_i^c \cap U_j^c. \qquad (2.5)$$

And consistency requires these transition functions to satisfy the cocycle condition:

$$\Omega_{ij}(x)\Omega_{jk}(x) = \Omega_{ik}(x), \qquad x \in U_i^c \cap U_j^c \cap U_k^c. \qquad (2.6)$$

In Fig. 1 the relevant transition functions for T^2 are indicated. Gauge transformations are specified by:

$$A_\mu^{(i)'} = [g_i]A_\mu^{(i)}. \qquad (2.7)$$

So

$$\Omega_{ij}' = g_i\Omega_{ij}g_j^{-1}.$$

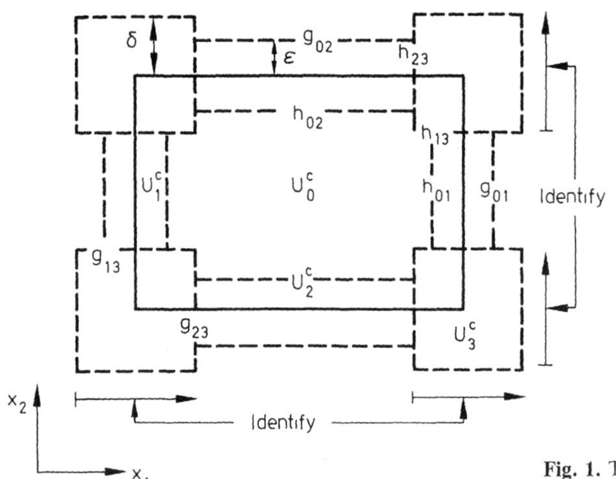

Fig. 1. The fiber bundle structure of T^2

It is now easy to see that when we take the limit $\varepsilon \to 0$, $\delta \to 0$:

$$A_\lambda^{(0)}(x_\mu = a_\mu) = [h_{0\mu}]A_\lambda^{(\mu)} = [h_{0\mu}g_{0\mu}^{-1}][g_{0\mu}]A_\lambda^{(\mu)}$$
$$= [h_{0\mu}g_{0\mu}^{-1}]A_\lambda^{(0)}(x_\mu = 0). \tag{2.8}$$

So we have the following expression for the multiple transition functions Ω_μ in terms of the ordinary transition functions:

$$\Omega_\mu = h_{0\mu}g_{0\mu}^{-1}. \tag{2.9}$$

Using the cocycle conditions (2.6) many times and the fact that $h_{13}, h_{23}, g_{13}, g_{23}$ are independent of the coordinates in the limit $\varepsilon \to 0$, $\delta \to 0$, one derives the consistency condition (2.2) with $Z_{\mu\nu} = 1$. On the other hand it is not difficult to show that given Ω_μ satisfying the consistency condition one can construct a fiber bundle structure in the above sense.

The gauge freedom in (2.7) now reduces to $g_0 = \Omega$, g_μ independent of x_μ, such that:

$$g'_{0\mu} = \Omega(x_\mu = 0)g_{0\mu}g_\mu^{-1},$$
$$h'_{0\mu} = \Omega(x_\mu = a_\mu)h_{0\mu}g_\mu^{-1}. \tag{2.10}$$

This gives with (2.9) the transformation property (2.4). Generalization of the above to T^4 is obvious (also the validity for general T^n and gauge group G with centre Z_G is obvious). We thus proved that 't Hooft's method of introducing $SU(N)$ gauge fields on the hypertorus transforming trivially under the centre Z_N defines a $SU(N)/Z_N$ principal fiber bundle structure on the hypertorus with connection A_μ (in local coordinates). This puts the theory in the right mathematical framework. We would however like to stress that this framework is discussed for the sake of completeness. The essential ingredient is the ansatz (2.9) and the accompanying gauge freedom (2.10), which can be made for $SU(N)/Z_N$ as well as $SU(N)$ multiple transition functions and without the assumption of an underlying bundle structure.

We can use the freedom in choosing either $g_{0\mu}$ or $h_{0\mu}$ to construct a closed loop in $SU(N)/Z_N$. With the aid of the consistency condition one easily checks that the following choice does the job:

$$g_{0\mu}(x_\nu = 0) = 1, \qquad g_{0\mu}(x_\nu = a_\nu) = \Omega_\nu(x_\mu = 0), \qquad (\mu \leftrightarrow \nu). \tag{2.11}$$

The loop along the boundary of T^2 [the solid line in Fig. 1, where $(1, 2)$ is identified with (μ, ν)] is mapped into a closed loop in $SU(N)/Z_N$. Its homotopy type as an element of $\pi_1(SU(N)/Z_N)$ is precisely $n_{\mu\nu}$, since we can pull up the loop to $SU(N)$ but then it jumps by an element of the centre Z_N. Having the same definition for $g_{0\mu}$ and $h_{0\mu}$ as $SU(N)$ functions this jump occurs at (a_μ, a_ν) and is exactly $Z_{\mu\nu} = \exp(2\pi i n_{\mu\nu}/N)$. (The correspondence of $n_{\mu\nu}$ with the homotopy type follows most trivially from the construction of $SU(N)$ as the universal covering group of $SU(N)/Z_N$ [8].)

3. The Pontryagin Number on T^4

The formula (1.1) is the usual form for (minus) the 2nd Chern number for a $SU(N)$ gauge theory[1]. However we saw that in general the gauge theory on T^4 does not define a $SU(N)$-principal fiber bundle structure and thus P need not be an integer. It is nevertheless integer, when there is no twist, since in Sect. 2 we showed that in that case we had a suitable fiber bundle structure. Note that this is consistent with (1.3).

By putting N cubes next to each other in the μ-direction each twist in the (μ, ν)-plane vanishes. From this one can derive that PN^3 is integer. Together with invariance arguments concerning the dependence of P on $n_{\mu\nu}$, the explicit calculation of P for a sample of configurations led 't Hooft [9] to a formula like (1.3).

Let us first put the terminology right. When $n_{\mu\nu} \neq 0$ we have transition functions in $SU(N)/Z_N$ but these are not elements of $GL(k, \mathbb{C})$ for any k, necessary [10] for using Chern classes. However the adjoint representation of $SU(N)$ is a faithful representation of $SU(N)/Z_N$. So provided we transform the gauge potentials to the adjoint representation, we can chose transition functions in $SO(N^2 - 1) \subset GL(N^2 - 1, \mathbb{R})$ and P_{ad} [equals (1.1) but with the gauge fields in the adjoint reprexentation] is just the integer 1st Pontryagin number.

If V_i $i = 1, \ldots, N^2 - 1$ are the generators of $SU(N)$, $\mathrm{Tr}(V_i V_j) = 2\delta_{ij}$, and L_i the generators of the adjoint representation, we can write $\mathrm{ad}(A_\mu) = -iA_\mu^a L_a$ if $A_\mu = A_\mu^a \dfrac{V_a}{2}$; from this one easily finds [11]:

$$P_{ad} = \frac{\mathrm{Tr}(L_a^2)}{\mathrm{Tr}((iV_a/2)^2)} P = 2NP \in \mathbb{Z}. \tag{3.1}$$

With (1.3) this would imply that P_{ad} misses the odd integers. This missing is obvious since the adjoint representation does not cover the whole of $GL(N^2 - 1, \mathbb{R})$.

---

1   See the appendix for a proper definition

We will work alternatively in SU(N) or its adjoint representation, whichever is the most suitable. Translation from one to the other is straightforward.

The Pontryagin index is determined by the topology of the fiber bundle only, which is specified by the transition functions [6]. Therefore it should be a function of these transition functions only, which is generally proved by construction in the appendix. For T^4 we have:

Lemma (3.1).

$$P = \frac{1}{24\pi^2} \sum_\mu \int d_3\sigma_\mu \varepsilon_{\mu\nu\alpha\beta} \mathrm{Tr}((\Omega_\mu \partial_\nu \Omega_\mu^{-1})(\Omega_\mu \partial_\alpha \Omega_\mu^{-1})(\Omega_\mu \partial_\beta \Omega_\mu^{-1}))$$

$$+ \frac{1}{8\pi^2} \sum_{\mu,\nu} \int d_2 S_{\mu\nu} \varepsilon_{\mu\nu\alpha\beta} \mathrm{Tr}((\Omega_\nu^{-1}\partial_\alpha \Omega_\nu)_{x_\mu = a_\mu}(\Omega_\mu \partial_\beta \Omega_\mu^{-1})_{x_\nu = 0})$$

with

$$\int d_3\sigma_1 = \int_0^{a_2} dx_2 \int_0^{a_3} dx_3 \int_0^{a_4} dx_4, \qquad \int d_2 S_{12} = \int_0^{a_3} dx_3 \int_0^{a_4} dx_4$$

etc.

Proof. In the appendix it is shown how to extract this formula from the general case. Here we show it by a direct calculation.

$$P = \frac{1}{16\pi^2} \int d_4 x \, \mathrm{Tr}(G_{\mu\nu}\tilde{G}_{\mu\nu})$$

$$= \frac{1}{16\pi^2} \sum_\mu \int d_3\sigma_\mu [K_\mu(x_\mu = a_\mu) - K_\mu(x_\mu = 0)]$$

with

$$K_\mu = 2\varepsilon_{\mu\nu\alpha\beta} \mathrm{Tr}\left(A_\nu \partial_\alpha A_\beta - \frac{2}{3i} A_\nu A_\alpha A_\beta\right). \tag{3.2}$$

Inserting the boundary condition (2.1) we find:

$$P = \frac{1}{24\pi^2} \sum_\mu \int d_3\sigma_\mu \varepsilon_{\mu\nu\alpha\beta} \left\{ \frac{1}{3} \mathrm{Tr}((\Omega_\mu \partial_\nu \Omega_\mu^{-1})(\Omega_\mu \partial_\alpha \Omega_\mu^{-1})(\Omega_\mu \partial_\beta \Omega_\mu^{-1})) \right.$$

$$\left. + \partial_\nu \mathrm{Tr}\left(\frac{1}{i}(\Omega_\mu^{-1}\partial_\alpha \Omega_\mu) A_\beta(x_\mu = 0)\right) \right\}. \tag{3.3}$$

We use the consistency condition (2.2) to calculate

$$\Omega_\nu^{-1}(x_\mu = 0)(\Omega_\mu^{-1}\partial_\alpha \Omega_\mu)_{x_\nu = a_\nu}\Omega_\nu(x_\mu = 0),$$

which can be used together with the boundary condition (2.1) to show that:

$$\left[\mathrm{Tr}\left(\frac{1}{i}(\Omega_\mu^{-1}\partial_\alpha \Omega_\mu)_{x_\nu = a_\nu} A_\beta(x_\mu = 0, x_\nu = a_\nu)\right) \right.$$

$$\left. - \mathrm{Tr}\left(\frac{1}{i}(\Omega_\mu^{-1}\partial_\alpha \Omega_\mu)_{x_\nu = 0} A_\beta(x_\mu = x_\nu = 0)\right) \right]$$

$$- [\mu \leftrightarrow \nu] = \mathrm{Tr}((\Omega_\nu^{-1}\partial_\alpha \Omega_\nu)_{x_\mu = a_\mu}(\Omega_\mu \partial_\beta \Omega_\mu^{-1})_{x_\nu = 0}) - (\mu \leftrightarrow \nu).$$

Inserting this in (3.3) completes the proof. $\square$

From this lemma one can easily compute P when the multiple transition functions are mutually commuting. By a suitable global gauge transformation they can be simultaneously diagonalized. Here P is gauge invariant so it is certainly invariant under a global transformation. The above configuration, which will be called abelian, can be chosen in $H = U(1)^{N-1}$ the maximal abelian (Cartan) subalgebra of SU(N) generated by:

$$T_a = \operatorname{diag}(1, 1, \ldots, 1, -N + a, 0, \ldots, 0),$$

(The first $N - a$ entries are 1)

$$\operatorname{Tr}(T_a) = 0, \qquad \operatorname{Tr}(T_a T_b) = (N - a + 1)(N - a)\delta_{ab}. \tag{3.4}$$

Write

$$\Omega_\mu = \exp\left(\frac{2\pi i}{N} f_\mu^a T_a\right), \tag{3.5}$$

then the topology is specified by $n_{\mu\nu}^{(a)}$ $a = 1, 2 \ldots N-1$, $n_{\mu\nu}^{(1)} \in \mathbb{Z}$, $n_{\mu\nu}^{(a)} \in N\mathbb{Z}$, $a \neq 1$, and $n_{\mu\nu} = n_{\mu\nu}^{(1)}(\operatorname{mod} N)$. Note that now $n_{\mu\nu}^{(a)}$ are genuine integers and one cannot transform or deform *within H* configurations with different $n_{\mu\nu}^{(a)}$ but equal $n_{\mu\nu}$ into each other. The winding numbers are given by:

$$[f_\mu^a(x_\nu = a_\nu) - f_\mu^a(x_\nu = 0)] - [f_\nu^a(x_\mu = a_\mu) - f_\nu^a(x_\mu = 0)] = n_{\mu\nu}^{(a)}. \tag{3.6}$$

Lemma (3.2). *For an abelian configuration as above with winding numbers* $n^{(a)}$, $a = 1, 2 \ldots N - 1$ *we find*:

$$P = \sum_{a=1}^{N-1} \frac{(N - a + 1)(N - a)}{N^2} \kappa(n^{(a)}) = \left(\frac{N-1}{N}\right)\kappa(n) + \mathbb{Z}.$$

Proof. Insert (3.5) into the formula of Lemma (3.1) to find:

$$P = \sum_{\mu, \nu} \int d_2 S_{\mu\nu} \varepsilon_{\mu\nu\alpha\beta} (\partial_\alpha f_\nu^a)_{x_\mu = a_\mu} (\partial_\beta f_\mu^b)_{x_\nu = 0} \frac{\operatorname{Tr}(T^a T^b)}{2N^2}$$

$$= \varepsilon_{\mu\nu\alpha\beta} \frac{n_{\nu\alpha}^{(a)}}{2} \frac{n_{\mu\beta}^{(b)}}{2} \frac{(N - a + 1)(N - a)}{2N^2} \delta_{ab} = \sum_a \frac{(N - a + 1)(N - a)}{N^2} \kappa(n^{(a)}).$$

Together with $n_{\mu\nu}^{(a)} = \delta_{a1} n_{\mu\nu} + N l_{\mu\nu}^{(a)}$, $l_{\mu\nu}^{(a)} \in \mathbb{Z}$ we find

$$P = \left(\frac{N-1}{N}\right)\kappa(n) + \sum_{a=1}^{N-1} (N - a + 1)(N - a)\kappa(l^{(a)})$$

$$+ \left(\frac{N-1}{4}\right)\varepsilon_{\mu\nu\alpha\beta} n_{\mu\nu} l_{\alpha\beta}^{(1)} = \left(\frac{N-1}{N}\right)\kappa(n) + \mathbb{Z}. \qquad \square$$

We would like to stress that one cannot reach all possible values of P, with an abelian set of multiple transition functions. For example, for $n_{\mu\nu} = 0(\operatorname{mod} N)$ we have $P \in 2\mathbb{Z}$.

We will now make use of the gauge invariance of P by making suitable choices of $g_{0\mu}$ and $h_{0\mu}$, as defined in Sect. 2. This enables us to comprehend the topological structure of the fiber bundle and in the general case to compute P in terms of topological invariants. (An application of this principle was the interpretation of $n_{\mu\nu}$ as winding numbers in Sect. 2.)

Theorem (3.1). *Given arbitrary multiple transition functions* Ω_μ, *we can choose* $h_{0\mu}$ *and* $g_{0\mu}$ *as follows:*

$$h_{0\mu} = U(x_\mu = a_\mu)\omega_\mu, \qquad g_{0\mu} = U(x_\mu = 0), \tag{3.7}$$

$$\omega_\mu = \exp\left(\frac{\pi i}{N}\sum_\nu \frac{n_{\mu\nu}x_\nu}{a_\nu}\, T_1\right), \tag{3.8}$$

where U *is a gauge function on the boundary of the four-dimensional box.*

Proof. First take any Ω_μ, $n_{\mu\nu}$ and try to specify $g_{0\mu}$ and $h_{0\mu}$ on the skeleton of the boundary of the four-dimensional box (the skeleton is defined to be the edges of the 8 cubes specified by $x_\mu = 0$, and $x_\mu = a_\mu$, $\mu = 1, 2, 3,$ and 4). Where ever more than one gauge function is specified on this skeleton, we demand them to be equal.

Let us first use the gauge freedom (2.4) to bring Ω_μ in a suitable form. It is not difficult to see [using that SU(N) is (simply) connected and π_2 (SU(N))$=0$] that one can choose a gauge transformation $\Omega(x)$, such that:

$$\Omega(x=0) = 1,$$
$$\Omega(x_\mu = a_\mu, x_\nu = 0\,; \forall \nu \neq \mu) = \Omega_\mu(0). \tag{3.9}$$

This proves that we can restrict ourselves to:

$$\Omega_\mu(0) = 1, \quad \mu = 1, 2, 3, 4. \tag{3.10}$$

Note that ω_μ is already in this form.

With (2.9) the functions $g_{0\mu}$ and $h_{0\mu}$ on the skeleton are specified by giving $g_{0\mu}$ on the edges of the cube $x_\mu = 0$. Here we work for a moment in the adjoint representation avoiding jumps with elements of Z_N, now hidden in the homotopy type of each square on the skeleton (compare Sect. 2).

If we want to specify U in terms of $g_{0\mu}$ and $h_{0\mu}$ we should have the conditions:

$$h_{0\mu}(x_\nu = a_\nu) = h_{0\nu}(x_\mu = a_\mu),$$
$$h_{0\mu}(x_\nu = 0) = g_{0\nu}(x_\mu = a_\mu), \tag{3.11}$$
$$g_{0\mu}(x_\nu = 0) = g_{0\nu}(x_\mu = 0).$$

If we demand the following condition to be met:

$$g_{0\mu}(x_\nu = a_\nu) = \Omega_\nu(x_\mu = 0)g_{0\mu}(x_\nu = 0), \tag{3.12}$$

the first two conditions in (3.11) are superficial and can be derived from (2.9), (2.2), (3.12) and the last condition in (3.11). Choosing $g_{0\mu}(x_\nu = x_\lambda = 0) = 1$ now completely fixes $g_{0\mu}$ on the skeleton, which is indicated in Fig. 2. It is an easy but tedious exercise to show that at the corners of the cube all definitions of $g_{0\mu}$ represent the same value. For this one uses the consistency condition and the gauge choice (3.10).

Gauge Fields on T^4 537

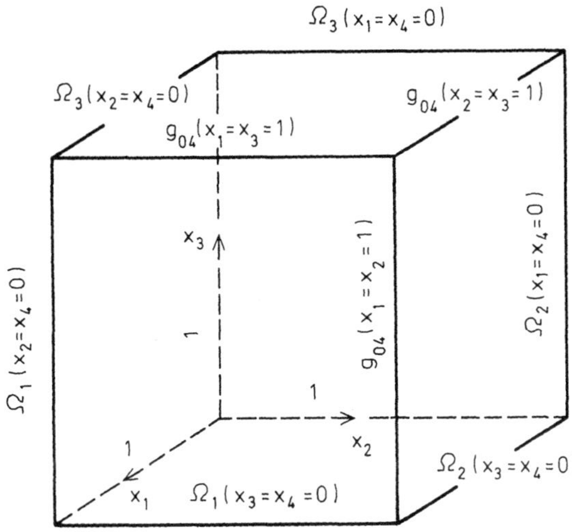

Fig. 2. The choice of $g_{0\mu}$ on the skeleton of the boundary of the four-dimensional box. It is understood that $g_{0\mu}(x_\nu=a_\nu)=\Omega_\nu(x_\mu=0)g_{0\mu}(x_\nu=0)$. We only display g_{04} and take for convenience $a_\mu=1$

We now choose $\tilde{h}_{0\mu}$ and $\tilde{g}_{0\mu}$ in terms of ω_μ as above and introduce $\hat{h}_{0\mu}, \hat{g}_{0\mu}, \hat{\Omega}_\mu$ by "dividing out" the abelian twist carrying configuration

$$\omega_\mu = \tilde{h}_{0\mu}\tilde{g}_{0\mu}^{-1},$$
$$\hat{h}_{0\mu} = h_{0\mu}\tilde{h}_{0\mu}^{-1}, \tag{3.13}$$
$$\hat{g}_{0\mu} = g_{0\mu}\tilde{g}_{0\mu}^{-1}, \qquad \hat{\Omega}_\mu = \hat{h}_{0\mu}\hat{g}_{0\mu}^{-1}.$$

Here $\hat{h}_{0\mu}$ and $\hat{g}_{0\mu}$ are defined in terms of $\hat{\Omega}_\mu$ as above, but since the homotopy type of each square in the (μ, ν) plane for the Ω_μ and ω_μ configuration is equal to $n_{\mu\nu}$, the same homotopy type for the $\hat{\Omega}_\mu$ configuration is cancelled exactly.

So $\hat{h}_{0\mu}, \hat{g}_{0\mu}$ are genuine $SU(N)$ functions specifying U:

$$U(x_\mu=0) = \hat{g}_{0\mu},$$
$$U(x_\mu=a_\mu) = \hat{h}_{0\mu}, \tag{3.14}$$

as a continuous function on the skeleton.

We can extend $U(x_\mu=x_\nu=0)$ restricted to the edges of the cube $x_\mu=0$ continuously to the square $x_\mu=x_\nu=0$, for all six possible combinations fixing U on all sides of the cubes $x_\mu=0$ [and $x_\mu=a_\mu$ by using (3.12) and (2.9)]. Finally, since $\pi_2(SU(N))=0$, we can continuously extend U inside the cubes $x_\mu=0$; (2.9) then fixes U inside the cubes $x_\mu=a_\mu$.

Putting things together we have, restricted to the skeleton:

$$g_{0\mu} = U(x_\mu=0)\tilde{g}_{0\mu}, \qquad h_{0\mu} = U(x_\mu=a_\mu)\tilde{h}_{0\mu}. \tag{3.15}$$

These cannot be extended to the whole of the cubes $x_\mu=0$, respectively $x_\mu=a_\mu$, because of the nontrivial homotopies. However

$$\Omega_\mu = h_{0\mu}g_{0\mu}^{-1} = U(x_\mu=a_\mu)\omega_\mu U(x_\mu=0)^{-1}$$

makes the choice (3.7) possible.

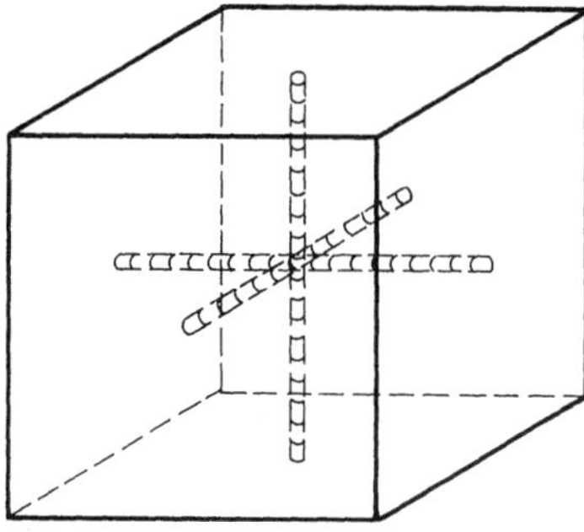

Fig. 3. The singularity structure of $\Lambda(x_\mu = 0)$

We can however make a maximal continuous extension $\tilde{\Lambda}$ of $\tilde{g}_{0\mu}$ and $\tilde{h}_{0\mu}$ to the cubes such that

$$\tilde{\Lambda}(x_\mu = 0) = \tilde{g}_{0\mu} \quad \text{and} \quad \tilde{\Lambda}(x_\mu = a_\mu) = \tilde{h}_{0\mu}. \tag{3.16}$$

We necessarily have line singularities restricted in their topology by (2.9), (3.11), and (3.12). A possible choice is depicted in Fig. 3. We can do the same for $g_{0\mu}$ and $h_{0\mu}$ defining Λ. If we choose the position of the line singularities the same as for $\tilde{\Lambda}$, (3.15) tells us that $\Lambda\tilde{\Lambda}^{-1}$ has a removable singular structure. $\square$

U is a gauge function on the boundary of the four-dimensional hypercube (which is homotopic to S^3) so the homotopy type of U is an element of $\pi_3(\mathrm{SU}(N)) = \mathbb{Z}$ specified by $v(U)$. Given Ω_μ we can have different choices of U belonging to different continuous extensions of U from the skeleton to the boundary of the four-dimensional box. But by construction they can be continuously deformed into each other. So v is a unique function of Ω_μ, which topology is thus completely specified by $n_{\mu\nu}$ and v. The gauge invariance of v is obvious from the transformation property of U [using (2.4)]

$$U' = \Omega U. \tag{3.17}$$

And it is easy to see that any two configurations of Ω_μ with the same $n_{\mu\nu}$ and v are gauge equivalent. Thus P should be a unique function of v and $n_{\mu\nu}$.

Since P is invariant under continuous deformations of the transition functions we have:

$$P(\Omega_\mu) = P(\omega_\mu) + P(\hat{\Omega}_\mu)$$
$$= \frac{(N-1)}{N} \kappa(n) + v, \tag{3.18}$$

where we used Lemma (3.2) and the fact that:

$$v(U) = \frac{1}{24\pi^2} \sum_\mu \int d_3\sigma_\mu \varepsilon_{\mu\nu\alpha\beta} [\mathrm{Tr}((U\partial_\nu U^{-1})(U\partial_\alpha U^{-1})(U\partial_\beta U^{-1}))]_{x_\mu=0}^{x_\mu=a_\mu}. \tag{3.19}$$

In the appendix we will prove that $P(\hat{\Omega}_\mu) = v(U)$ as a simple application of the general formula.

Now we also understand what happens if we add a multiple of N to $n_{\mu\nu}$ (yielding $n'_{\mu\nu}$). We will have different ω'_μ in Theorem (3.1) in constructing U'.

However by applying the theorem to ω'_μ itself we have $\tilde{g}'_{0\mu} = V(x_\mu = 0)\tilde{g}_{0\mu}$, $\tilde{h}'_{0\mu} = V(x_\mu = a_\mu)\tilde{h}_{0\mu}$ [compare (3.15)] and consequently we can continuously deform UV into U', so $v(U') = v(U) + v(V)$. On the other hand applying (3.18) to ω'_μ we have $\left(\dfrac{N-1}{N}\right)\kappa(n') = \left(\dfrac{N-1}{N}\right)\kappa(n) + v(V)$.

So both expressions for $P(\Omega_\mu)$ are equal. However the abelian contribution differs by an integer which is precisely the homotopy type of the gauge transformation (defined on the boundary of the box only) transforming ω_μ into ω'_μ. This is necessarily a nonabelian configuration.

4. Orthogonal Twist, Zero Action Configurations

Let us assume the existence of selfdual solutions for all possible topologies. This then implies that there are solutions with different electric and magnetic fluxes but equal action. The behaviour of $\kappa(n)$ is responsible for this. It boils down to having $n'_{\mu\nu} \neq n_{\mu\nu}(\mathrm{mod}\,N)$ and $\kappa(n) = \kappa(n')(\mathrm{mod}\,N)$.

A striking consequence is then the existence of zero action solutions with nontrivial topology, discovered on the lattice by Groeneveld [16] a.o., and called twist eating configurations by them.

Zero action implies $P = 0$ so these zero action configurations necessarily have $\kappa(n) = 0\,(\mathrm{mod}\,N)$. Since zero action implies $G_{\mu\nu} = 0$ there is a gauge such that $A_\mu = 0$. In this gauge the multiple transition functions Ω_μ are constant. We call the twist $n_{\mu\nu}$ satisfying $\kappa(n) = 0(\mathrm{mod}\,N)$ orthogonal, so we proved [5]:

Theorem (4.1). *There exist zero action, orthogonal twist solutions iff there are* $\Omega_\mu \in \mathrm{SU}(N)$ *such that* [2] $[\Omega_\mu, \Omega_\nu] = \exp(2\pi i n_{\mu\nu}/N)$. $\quad\square$

With the remarks following Theorem 3.1 it is not hard to show uniqueness up to constant gauge transformations! In the box we thus have leading perturbative contributions to the ground state energies in a certain $(\mathbf{e}, \mathbf{m})$ sector.

Ambjørn and Flyvbjerg [14] first showed the existence of zero classical energy solutions in the continuum with arbitrary magnetic flux $(\mathbf{m})$ by constructing $[\Omega_k, \Omega_j] = \exp(2\pi i \varepsilon_{kjl} m_l/N)$, which is in fact in the form of the above theorem for time independent configurations. From the mathematical point of view it is the interplay between the multiconnectedness of the hypertorus and the topology of $\mathrm{SU}(N)/Z_N$ which causes this behaviour [14]. From this formal point of view zero energy solutions on general three spaces and for general gauge groups are studied in Ref. 15. This boils down to the search for zero action solutions in three dimensional euclidean gauge theories.

Recently 't Hooft [5] proved the existence of Ω_μ as in Theorem (5.1) for N not divisible by a prime squared. We will combine his ideas and those of Ref. 14 to complete the proof for general N. The essential part is proving the result for $N = p^e$, $e \in \mathbb{N}$ (p will denote a prime). Let us first review the case where $e = 1$.

Let $P, Q \in \mathrm{SU}(N)$ be such that [5, 14]:

$$[P, Q] = e^{2\pi i/N}. \tag{4.1}$$

---

2   $[\Omega_\mu, \Omega_\nu] = \Omega_\mu \Omega_\nu \Omega_\mu^{-1} \Omega_\nu^{-1}$, the group commutator

Then defining

$$\Omega_\mu = P^{s_\mu} Q^{t_\mu}, \quad s_\mu, t_\mu \text{ integers},$$
$$[\Omega_\mu, \Omega_\nu] = \exp(2\pi i(s_\mu t_\nu - s_\nu t_\mu)/N), \tag{4.2}$$

the problem is reduced to finding s_μ and t_μ such that

$$\hat{n}_{\mu\nu} = s_\mu t_\nu - s_\nu t_\mu, \quad n_{\mu\nu} = \hat{n}_{\mu\nu}(\mathrm{mod}\, N). \tag{4.3}$$

This automatically demands orthogonal twist, because $\kappa(\hat{n}) = 0$. Now one can transform by an $SL(4, \mathbb{Z})$ transformation [*here* we do not need to restrict to $SL(4, Z_N)$]$n_{\mu\nu}$ to the standard form:

$$n'_{12} = n'_{13} = n'_{34} = 0. \tag{4.4}$$

Let $\{X_{\mu\nu}\}$ be an element of $SL(4, \mathbb{Z})$ [or $SL(4, Z_N)$], then:

$$n_{\mu\nu} \to X_{\mu\mu'} X_{\nu\nu'} n_{\mu'\nu'},$$
$$s_\mu \to X_{\mu\mu'} s_{\mu'}, \quad t_\mu \to X_{\mu\mu'} t_{\mu'}, \tag{4.5}$$
$$\kappa(n) \to (\det X)\kappa(n) = \kappa(n).$$

So $\kappa(n) = n'_{14} n'_{23} = 0(\mathrm{mod}\, N)$, if N is prime this implies either $n'_{14} = 0(\mathrm{mod}\, N)$ or $n'_{23} = 0(\mathrm{mod}\, N)$ for which one can solve (4.3) easily [5].

It is not hard to see that in general a necessary condition for the solvability of (4.3) is the condition (for convenience we uniquely label $0 \leq n_{\mu\nu} < N$ if $\mu < \nu$) that the greatest common divisor (g.c.d.) of $n_{\mu\nu}$ and N [notation: g.c.d. $(n_{\mu\nu}, N)$] equals 1. This is because (4.3) would imply the existence of $l_{\mu\nu} \in \mathbb{Z}$ such that $\kappa(n + Nl) = 0$ or solvability of $\dfrac{\kappa(n)}{N} + \dfrac{1}{2} n_{\mu\nu} \tilde{l}_{\mu\nu} + N\kappa(l) = 0$. It is also sufficient because taking $\hat{n} = n + lN$ we can solve for s_μ and t_μ by bringing $\hat{n}$ in the form (4.4) and using $\kappa(\hat{n}) = 0$ as above (now as a genuine equation over $\mathbb{Z}$, and not Z_N).

It remains to prove solvability of $\kappa(n + Nl) = 0$ for $\kappa(n) = 0(\mathrm{mod}\, N)$ if g.c.d. $(n_{\mu\nu}, N) = 1$ or more general:

Lemma (4.1). *If g.c.d.* $(n_{\mu\nu}, N) = 1$ *we can find* $l_{\mu\nu} \in \mathbb{Z}$, *and* $q \in \mathbb{Z}$ *such that* $\kappa(n) = \kappa(n + Nl) + Nq$, $0 \leq \kappa(n + Nl) < N$, *and* $0 \leq n_{\mu\nu} < N$, $\forall \mu < \nu$.

Proof.

$$\kappa(n + Nl) + Nq - \kappa(n) = \tfrac{1}{2} N n_{\mu\nu} \tilde{l}_{\mu\nu} + \tfrac{1}{4} N^2 l_{\mu\nu} \tilde{l}_{\mu\nu} + Nq = 0.$$

Put $m_i = \varepsilon_{ijk} n_{jk}$, $a_i = \varepsilon_{ijk} l_{jk}$, $k_i = n_{i4}$, $b_i = l_{i4}$; so to solve:

$$(\mathbf{m} \cdot \mathbf{b}) + (\mathbf{k} \cdot \mathbf{a}) + N(\mathbf{a} \cdot \mathbf{b}) + q = 0.$$

Transforming to the standard form (4.4) we have g.c.d. $(n_{\mu\nu}, N) = 1$ iff g.c.d. $(n'_{13}, n'_{14}, n'_{24}, N) = 1$ [since we use $SL(4, \mathbb{Z})$ transformations only]. So to solve:

$$m'_2 b'_2 + k'_1 a'_1 + k'_2 a'_2 + N(a'_1 b'_1 + a'_2 b'_2 + a'_3 b'_3) + q = 0,$$

choosing $a'_3 = 1$, $b'_1 = 0$, $b'_3 = b - a'_2 b'_2$ this boils down to solving:

$$b'_2 m'_2 + a'_1 k'_1 + a'_2 k'_2 + bN + q = 0$$

for fixed m'_2, k'_1, k'_2, N. The solvability follows from the elementary algebraic relation (a consequence of Euler's remainder theorem):

$$\sum (q_i \mathbb{Z}) = \text{g.c.d.}(q_i)\mathbb{Z}. \quad \square \tag{4.6}$$

We thus have the following simple corollary:

Corollary. *When g.c.d.* $(n_{\mu\nu}, N) = 1(0 \leq n_{\mu\nu} < N, \mu < \nu)$ *and* $\kappa(n) = 0(\text{mod } N)$ *we can find* Ω_μ *as in (4.2), satisfying the condition in Theorem (4.1).* $\square$

Before proving the existence of Ω_μ for the case $N = p^e$ we want to show that this is sufficient to deal with the case of general N. We follow the method of Ref. 14, splitting general N in its prime factors: $N = \prod_i p_i^{e_i}$ we have ($\oplus$ is the direct sum, $\otimes$ is the tensor product):

$$Z_N \cong \bigoplus_i Z_{N_i}, \qquad N_i = p_i^{e_i},$$
$$\text{SU}(N) \supset \bigotimes_i \text{SU}(N_i). \tag{4.7}$$

We thus decompose $n_{\mu\nu}$ and correspondingly Ω_μ as follows:

$$n_{\mu\nu} = \sum (N p_i^{-e_i}) n_{\mu\nu}^{(i)}, \qquad n_{\mu\nu}^{(i)} \in Z_{N_i},$$
$$\Omega_\mu = \bigotimes_i \Omega_\mu^{(i)}, \qquad \Omega_\mu^{(i)} \in \text{SU}(N_i). \tag{4.8}$$

From the decomposition of Ω_μ and the assumed existence of $\Omega_\mu^{(i)}$ belonging to $n_{\mu\nu}^{(i)}$ we have:

$$[\Omega_\mu, \Omega_\nu] = \left[\bigotimes_i \Omega_\mu^{(i)}, \bigotimes_i \Omega_\nu^{(i)}\right] = \bigotimes_i [\Omega_\mu^{(i)}, \Omega_\nu^{(i)}] = \prod_i \exp(2\pi i n_{\mu\nu}^{(i)}/N_i)$$

$$= \exp(2\pi i (\sum N p_i^{-e_i} n_{\mu\nu}^{(i)})/N) = \exp(2\pi i n_{\mu\nu}/N).$$

There is however one vital thing we forgot to prove. For the existence of $\Omega_\mu^{(i)} \in \text{SU}(N_i)$ we need $\kappa(n_{\mu\nu}^{(i)}) = 0(\text{mod } N_i)$. Let us write $N = N_1 N_2$, g.c.d. $(N_1, N_2) = 1$ then we can uniquely decompose $n_{\mu\nu} \in Z_N$ into $n_{\mu\nu} = N_2 n_{\mu\nu}^{(1)} + N_1 n_{\mu\nu}^{(2)}$, $n_{\mu\nu}^{(i)} \in Z_{N_i}$, from this we have

$$\kappa(n) = N_2^2 \kappa(n^{(1)}) + N_1^2 \kappa(n^{(2)}) (\text{mod } N) = 0(\text{mod } N).$$

This necessarily implies $\kappa(n^{(i)}) = 0(\text{mod } N_i)$, as one can easily deduce from the general solution of $\kappa(n^{(i)})$. Splitting the N_i's up further we can deduce the general case.

We now finish the proof of the existence of zero action solutions for arbitrary orthogonal twist and N by the following theorem:

Theorem (4.2). *If N equals a prime to some positive power $e(N = p^e)$, we can find $\Omega_\mu \in \text{SU}(N)$ such that*

$$[\Omega_\mu, \Omega_\nu] = \exp(2\pi i n_{\mu\nu}/N)$$

for all $n_{\mu\nu} \in Z_N$ satisfying $\kappa(n) = 0(\text{mod } N)$.

Proof. We can write g.c.d. $(n_{\mu\nu}) = r \cdot p^f$, g.c.d. $(r, p) = 1$, $f \in \mathbb{N}$. Let us first reduce the case $2f > e$ to $2f \leq e$. We have both $\kappa(n) = k \cdot p^e$ ($\kappa(n) = 0(\text{mod } N)$) and $\kappa(n) = l \cdot p^{2f}$ (g.c.d. $(n_{\mu\nu}) = r \cdot p^f$), so k contains p^{2f-e} as a factor. Furthermore $f < e$ so we can define $n'_{\mu\nu} = n_{\mu\nu} p^{-(2f-e)}$ and $N' = N p^{-(2f-e)} = p^{2(e-f)}$, and it is easy to deduce

$\kappa(n') = 0 (\mathrm{mod}\, N')$ and g.c.d. $(n'_{\mu\nu}) = r \cdot p^{e-f}$. So $f' = e - f$ and $e' = 2(e - f)$. When we can solve for $\Omega'_\mu \in \mathrm{SU}(N')$ we choose $\Omega_\mu = \Omega'_\mu(\otimes 1_{N'})^{(N-N')/N'}$. Thus we can assume $2f \leq e$ from now on.

There is a pair $(\mu_0, \nu_0)\, \mu_0 < \nu_0$ such that $\tilde{n}_{\mu_0\nu_0} p^{-f}$ is relative prime to p. We split $n_{\mu\nu} \in Z_N$ in $n^{(i)}_{\mu\nu} \in Z_{N_i}$ $(i = 1, 2)$ (with $N_1 = p^{e-f}$, $N_2 = p^f$) according to $n_{\mu\nu} = N_2 n^{(1)}_{\mu\nu} + N_1 n^{(2)}_{\mu\nu}$, with:

$$n^{(1)}_{\mu_0\nu_0} = n_{\mu_0\nu_0} p^{-f} - k p^{e-2f} = - n^{(1)}_{\nu_0\mu_0}, \qquad n^{(2)}_{\mu_0\nu_0} = - n^{(2)}_{\nu_0\mu_0} = k \in Z_{N_2}$$

and

$$n^{(1)}_{\mu\nu} = n_{\mu\nu} p^{-f}, \qquad n^{(2)}_{\mu\nu} = 0$$

for the other pairs (μ, ν). Now

$$\kappa(n^{(1)}) = \left[\frac{\kappa(n)}{N} - k(\tilde{n}_{\mu_0\nu_0} p^{-f}) \right] p^{e-2f}$$

and we can choose k such that $\kappa(n^{(1)}) = 0 (\mathrm{mod}\, N_1)$. We can apply the corollary to find $\Omega^{(1)}_\mu \in \mathrm{SU}(N_1)$. Furthermore $s^{(2)}_\mu = k\delta_{\mu\mu_0}$, $t^{(2)}_\mu = \delta_{\mu\nu_0}$ gives $\Omega^{(2)}_\mu \in \mathrm{SU}(N_2)$ as in (4.2). Putting things together $\Omega_\mu = \Omega^{(1)}_\mu \otimes \Omega^{(2)}_\mu \in \mathrm{SU}(N)$ yields the desired commutation relations. $\square$

The existence of Ω_μ as in Theorem (4.1) for each N are not only important for zero action solutions, 't Hooft [5] also used them extensively in constructing solutions with general twist, and in particular self dual solutions with minimal nontrivial action.

5. Conclusions

We showed that 't Hooft's method for introducing gauge fields on the four-dimensional box is equivalent to a principal fiber bundle on $T^4 = S^1 \times S^1 \times S^1 \times S^1$, with structure group $\mathrm{SU}(N)/Z_N$. It is the invariance of the fields in the box under Z_N which gave us the rich topological structure we found. Equivalently we can take as structure group the adjoint of $\mathrm{SU}(N)$. Then the topological structure associated with Z_N is hidden in winding numbers.

One has a Pontryagin number expressible in the topological invariants of the bundle. A new feature is that one can have configurations with equal Pontryagin number but different topology. Intimately related with this is the existence of twist eating configurations for arbitrary N and orthogonal twist as we proved in detail. The explicit dependence of the Pontryagin number on the magnetic flux **m** is such that one recovers the equivalent of Witten's result [4, 13]; switching on the θ-parameter [12] gives a θ-dependent electric flux. When we can do dynamics in the box it is possible to study things like oblique confinement, introduced by 't Hooft [13] in the continuum theory, in the box.

It is hoped that our complete understanding of the topology of gauge fields on the hypertorus will be helpful in constructing explicit solutions of the euclidean equations of motion. We would like to be able to restrict ourselves to (anti-)self dual solutions. The solutions 't Hooft [5] constructed are only self dual if the ratio of the a_μ satisfy certain conditions. It is thus conjectured that if this is not the case the solutions are unstable.

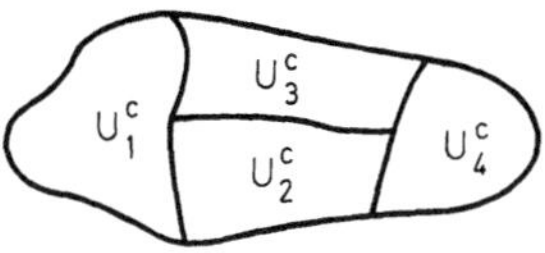

Fig. 4. A generic choice of sets U_i^c satisfying the conditions in the appendix. We suppressed two dimensions

Once we have the classical solutions one can do semiclassical computations of e.g. the relevant free energies, and hope that the weak coupling results can be pushed far enough to be relevant for QCD.

Appendix

Here we prove a general theorem for the 1st Pontryagin or 2nd Chern class of a 4-dimensional compact manifold without boundary (generalization to other manifolds is obvious). We apply the result to a proof of Lemma (3.1) and $P(\hat{\Omega}_\mu) = v(U)$ [compare (3.18) and (3.19)].

Notation. Let M be a compact manifold of real dimension 4 without a boundary and $\{U_i\}$ a covering of M with open sets.

We choose $\{U_i^c\}$ a set of closed sets with $U_i^c \subset U_i$, which still cover M, but satisfy the following properties: $U_i^c \cap U_j^c = \partial U_i^c \cap \partial U_j^c$ and

$$U_i^c \cap U_j^c \cap U_k^c = \partial(U_i^c \cap U_j^c) \cap \partial U_k^c,$$

consequently these are 3 respectively 2-dimensional submanifolds. (See Fig. 4 for a generic situation.) Furthermore our convention is such that $[U_i^c \cap U_j^c]_i$ denotes $U_i^c \cap U_j^c$ with the orientation of ∂U_i^c and $[U_i^c \cap U_j^c \cap U_k^c]_i$ denotes $U_i^c \cap U_j^c \cap U_k^c$ with the orientation of $\partial[U_i^c \cap U_j^c]_i$. A local section of the connection 1-form on U_i is denoted by ω_i $(= iA_\mu^{(i)} dx^{(i)})$, and $\Omega_i = d\omega_i + \omega_i \wedge \omega_i$ is the curvature 2-form on U_i. Here $\wedge$ is the wedge product of p-forms. The wedge product for n equal p-forms is written as $(p\text{-form})^n$. Finally ω_i transforms under a change of coordinate patch, with transition functions g_{ij} defined on $U_i \cap U_j$, according to

$$\omega_j = g_{ij}^{-1} \omega_i g_{ij} + g_{ij}^{-1} dg_{ij}.$$

Lemma (A.1).

$$-\int_M \mathrm{Tr}(\Omega \wedge \Omega) = \tfrac{1}{6} \sum_{i,j} \int_{[U_i^c \cap U_j^c]_i} \mathrm{Tr}((g_{ij} dg_{ij}^{-1})^3)$$

$$+ \tfrac{1}{6} \sum_{i,j,k} \int_{[U_i^c \cap U_j^c \cap U_k^c]_i} \mathrm{Tr}((g_{ij}^{-1} dg_{ij}) \wedge ((dg_{jk}) g_{jk}^{-1}))$$

$$= \tfrac{1}{3} \sum_{i<j} \int_{[U_i^c \cap U_j^c]_i} \mathrm{Tr}((g_{ij} dg_{ij}^{-1})^3)$$

$$+ \sum_{i<j<k} \int_{[U_i^c \cap U_j^c \cap U_k^c]_i} \mathrm{Tr}((g_{ij}^{-1} dg_{ij}) \wedge ((dg_{jk}) g_{jk}^{-1})) \equiv 8\pi^2 P(g_{ij}).$$

Proof.

$$\int_M \mathrm{Tr}(\Omega \wedge \Omega) = \sum_i \int_{U_i^c} \mathrm{Tr}(\Omega_i \wedge \Omega_i) = \sum_i \int_{U_i^c} dK_i$$

$$= \tfrac{1}{2} \sum_{i,j} \int_{[U_i^c \cap U_j^c]_i} (K_i - K_j),$$

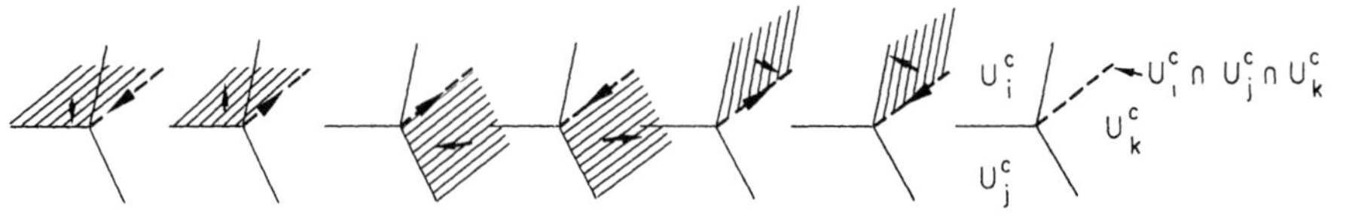

                                                                 P. van Baal

$$\int (ijk) = -\int (jik) = -\int (kji) = \int (jki) = -\int (ikj) = \int (kij)$$

$$\int [U_i^c \cap U_j^c \cap U_k^c]_i = \int (ijk)$$

Fig. 5. Constructing the right sign in the expression for L_{ijk}. We suppressed one dimension

where

$$\mathrm{Tr}(\Omega_i \wedge \Omega_i) = dK_i \quad \text{and} \quad K_i = \mathrm{Tr}(\omega_i \wedge d\omega_i + \tfrac{2}{3}\omega_i \wedge \omega_i \wedge \omega_i).$$

A straightforward calculation gives, using $\omega_j = g_{ij}^{-1}\omega_i g_{ij} + g_{ij}^{-1}dg_{ij}$:

$$K_j - K_i = \tfrac{1}{3}\mathrm{Tr}((g_{ij}dg_{ij}^{-1})^3) + dL_{ij}, \qquad L_{ij} = \mathrm{Tr}(\omega_i \wedge (dg_{ij})g_{ij}^{-1}).$$

Combining these results we find:

$$-\int_M \mathrm{Tr}(\Omega \wedge \Omega) = \tfrac{1}{6}\sum_{i,j} \int_{[U_i^c \cap U_j^c]_i} \mathrm{Tr}((g_{ij}dg_{ij}^{-1})^3) + \tfrac{1}{2}\sum_{i,j} \int_{\partial[U_i^c \cap U_j^c]_i} L_{ij}.$$

$$\sum_{i,j} \int_{\partial[U_i^c \cap U_j^c]_i} L_{ij} = \sum_{i,j,k} \int_{[U_i^c \cap U_j^c \cap U_k^c]_i} L_{ij}$$

$$= \sum_{i,j,k} \tfrac{1}{6}\left\{ \int_{(ijk)} L_{ij} + \int_{(jik)} L_{ji} + \int_{(ikj)} L_{ik} + \int_{(kij)} L_{ki} + \int_{(kji)} L_{kj} + \int_{(jki)} L_{jk} \right\}$$

$$= \sum_{i,j,k} \tfrac{1}{6} \int_{(ijk)} (L_{ij} - L_{ji} - L_{ik} + L_{ki} - L_{kj} + L_{jk}) \equiv \tfrac{1}{6}\sum_{i,j,k} \int_{(ijk)} L_{ijk}.$$

(See Fig. 5, for convenience we defined $(ijk) \equiv [U_i^c \cap U_j^c \cap U_k^c]_i$.)

Expressing ω_j and ω_k in ω_i and using the cocycle condition $g_{ij}g_{jk} = g_{ik}$ to show that

$$2(dg_{ij})g_{ij}^{-1} - 2(dg_{ik})g_{ik}^{-1} = g_{ik}((dg_{kj})g_{kj}^{-1})g_{ik}^{-1} - g_{ij}((dg_{jk})g_{jk}^{-1})g_{ij}^{-1}$$

one finds that the ω_i dependence drops out and we are left with:

$$L_{ijk} = \mathrm{Tr}(g_{ij}^{-1}(dg_{ij}) \wedge (dg_{jk})g_{jk}^{-1}) + \mathrm{Tr}(g_{jk}^{-1}(dg_{jk}) \wedge (dg_{ji})g_{ji}^{-1})$$

$$\equiv \hat{L}_{ijk} + \hat{L}_{jki}, \qquad \hat{L}_{ijk} \equiv \mathrm{Tr}(g_{ij}^{-1}(dg_{ij}) \wedge (dg_{jk})g_{jk}^{-1}).$$

So

$$\tfrac{1}{6}\sum_{i,j,k} \int_{(ijk)} L_{ijk} = \tfrac{1}{6}\sum_{i,j,k} \left\{ \int_{(ijk)} \hat{L}_{ijk} + \int_{(jki)} \hat{L}_{jki} \right\}$$

$$= \tfrac{1}{3}\sum_{i,j,k} \int_{(ijk)} \mathrm{Tr}(g_{ij}^{-1}(dg_{ij}) \wedge (dg_{jk})g_{jk}^{-1}). \qquad \square$$

The 1st Pontryagin Number is defined for a real vector bundle with transition functions $g_{ij} \in \mathrm{GL}(N,\mathbb{R})$ by [10]:

$$P_1(\Omega) = \int_M \frac{\mathrm{Tr}(\Omega \wedge \Omega^\dagger)}{8\pi^2}. \tag{A.1}$$

We have furthermore $\Omega = \frac{i}{2} G_{\mu\nu} dx_\mu \wedge dx_\nu$ (in local coordinates) which puts (A.1) in the form (1.3). Since we will always have an $O(N)$ vector bundle [10], $\Omega^\dagger = -\Omega$ and Lemma (A.1) gives $P_1(\Omega)$ in terms of transition functions only.

The 2nd Chern number is defined for a complex vector bundle with transition functions $g_{ij} \in GL(N, \mathbb{C})$ by [10]:

$$C_2(\Omega) = \frac{1}{8\pi^2} \int_M (\mathrm{Tr}(\Omega \wedge \Omega) - \mathrm{Tr}(\Omega) \wedge \mathrm{Tr}(\Omega)). \tag{A.2}$$

For $SU(N)$ vector bundles $\mathrm{Tr}(\Omega) = 0$ and we have $C_2(\Omega) = -P$, as defined in (1.3) in terms of $G_{\mu\nu}$. Since we claimed a general formula for $C_2(\Omega)$ in terms of the transition functions only we have to deal with the last term in (A.2):

Lemma (A.2).

$$-\int_M \mathrm{Tr}(\Omega) \wedge \mathrm{Tr}(\Omega) = \sum_{i<j<k} \int_{[U_j^c \cap U_j^c \cap U_k^c]_i} \mathrm{Tr}(g_{ij}^{-1} dg_{ij}) \wedge \mathrm{Tr}((dg_{jk})g_{jk}^{-1}) \equiv 8\pi^2 Q(g_{ij}).$$

Proof We can use the same technique as in Lemma (A.1), we now have $\mathrm{Tr}(\Omega_i) = d\,\mathrm{Tr}(\omega_i)$ so

$$\mathrm{Tr}(\Omega_i) \wedge \mathrm{Tr}(\Omega_i) = dK_i, \qquad K_i = \mathrm{Tr}(\omega_i) \wedge d\,\mathrm{Tr}(\omega_i),$$

$$K_j - K_i = dL_{ij}, \qquad L_{ij} = \mathrm{Tr}(\omega_i) \wedge \mathrm{Tr}(g_{ij}^{-1} dg_{ij}),$$

where we used that $d\,\mathrm{Tr}(g_{ij}^{-1} dg_{ij}) = 0$. Again define

$$L_{ijk} = L_{ij} - L_{ji} - L_{ik} + L_{ki} - L_{kj} + L_{jk}.$$

The cocycle condition this time implies

$$\mathrm{Tr}(g_{ij}^{-1} dg_{ij}) + \mathrm{Tr}(g_{jk}^{-1} dg_{jk}) - \mathrm{Tr}(g_{ik}^{-1} dg_{ik}) = 0$$

and also guarantees here absence of ω_i dependence. We find:

$$L_{ijk} = \mathrm{Tr}(g_{ij}^{-1} dg_{ij}) \wedge \mathrm{Tr}(g_{jk}^{-1} dg_{jk}) - \mathrm{Tr}(g_{ik}^{-1} dg_{ik}) \wedge \mathrm{Tr}(g_{kj}^{-1} dg_{kj})$$

$$= \hat{L}_{ijk} - \hat{L}_{ikj}, \qquad \hat{L}_{ijk} \equiv \mathrm{Tr}(g_{ij}^{-1} dg_{ij}) \wedge \mathrm{Tr}(g_{jk}^{-1} dg_{jk}).$$

So as in Lemma (A.1) we have:

$$-\int_M \mathrm{Tr}(\Omega) \wedge \mathrm{Tr}(\Omega) = \tfrac{1}{2} \sum_{i,j} \int_{[U_i^c \cap U_j^c]_i} K_j - K_i$$

$$= \tfrac{1}{12} \sum_{i,j,k} \left(\int_{(ijk)} \hat{L}_{ijk} + \int_{(ikj)} \hat{L}_{ikj} \right). \quad \square$$

Theorem (A.1). (i) *The first Pontryagin number for an* $O(N)$ *vector bundle with transition function* $g_{ij} \in O(N)$ *on a four-dimensional manifold is given by:* $P_1(\Omega) = P(g_{ij})$, *defined in Lemma (A.1).*

(ii) *The second Chern number for a complex vector bundle with transition functions* $g_{ij} \in GL(N, \mathbb{C})$ *on a four-dimensional manifold is given by:* $C_2(\Omega) = Q(g_{ij}) - P(g_{ij})$, Q *defined as in Lemma (A.2).* $\square$

Let us now use this theorem for proving Lemma (3.1). For T^4 we take:

$$U_0^c = \{\varepsilon \leq x_\mu \leq a_\mu - \varepsilon; \mu = 1, 2, 3, 4\},$$

and for $\mu = 1, 2, 3, 4$

$$U_\mu^c = \{|x_\mu| \le \varepsilon, \varepsilon \le x_\nu \le a_\nu - \varepsilon; \ \forall \nu \ne \mu\} \,,$$

taking $\varepsilon \to 0$, with transition functions $h_{0\mu}$ and $g_{0\mu}$ as defined in Sect. 2. We do not need the explicit form of the other sets U_i^c since the relevant intersections can be described by the above alone.

It is obvious that we can read off from Fig. 1 the different contributions to $P(g_{ij})$ (in the limit $\varepsilon \to 0$, $\delta \to 0$. In this figure U_3^c should be labelled different, but there will be no confusion in the following). It is easy to see that we have for T^4:

$$\frac{1}{24\pi^2} \sum_{i<j} \int_{[U_i^c \cap U_j^c]_i} \mathrm{Tr}((g_{ij} dg_{ij}^{-1})^3) = \frac{1}{24\pi^2} \sum_\mu \int_{x_\mu = 0} \mathrm{Tr}((h_{0\mu} dh_{0\mu}^{-1})^3 - (g_{0\mu} dg_{0\mu}^{-1})^3). \quad (A.3)$$

The second term in $P(g_{ij})$ needs somewhat more discussion. We will treat the case where $x \in U_i^c \cap U_j^c$ has fixed x_1 and x_2 and we can read off the contributions from Fig. 1:

$$\mathrm{Tr}(h_{01}^{-1} dh_{01} \wedge (dh_{13}) h_{13}^{-1})_{x_2 = a_2} + \mathrm{Tr}(h_{02}^{-1} dh_{02} \wedge (dg_{23}) g_{23}^{-1})_{x_1 = 0}$$
$$+ \mathrm{Tr}(g_{01}^{-1} dg_{01} \wedge (dg_{13}) g_{13}^{-1})_{x_2 = 0}$$
$$+ \mathrm{Tr}(g_{02}^{-1} dg_{02} \wedge (dh_{23}) h_{23}^{-1})_{x_1 = a_1} - (1 \leftrightarrow 2). \quad (A.4)$$

It is conjectured that in general one can eliminate h_{13}, g_{13}, h_{23}, g_{23} by using $(\mathrm{mod}\, Z_N)$:

$$
\begin{aligned}
h_{13} &= h_{01}^{-1}(x_2 = a_2) h_{02}(x_1 = a_1) h_{23}, \\
h_{23} &= g_{02}^{-1}(x_1 = a_1) h_{01}(x_2 = 0) g_{13}, \\
g_{13} &= g_{01}^{-1}(x_2 = 0) g_{02}(x_1 = 0) g_{23}, \\
g_{23} &= h_{02}^{-1}(x_1 = 0) g_{01}(x_2 = a_2) h_{13}.
\end{aligned}
\qquad (A.5)
$$

We will only specialize to the situations mentioned in the beginning. First we compute $P(\hat\Omega_\mu)$ as defined in Theorem 3.1, see (3.18). By a gauge transformation we can choose $g_{0\mu} = U(x_\mu = 0)$, $h_{0\mu} = U(x_\mu = a_\mu)$. Equation (A.5) implies thus $h_{13} = h_{23} = g_{13} = g_{23}$ and (A.4) is identical zero. So $P(\hat\Omega_\mu)$ equals the contribution in (A.3). Comparing with (3.19) we find $P(\hat\Omega_\mu) = \nu(U)$.

In proving Lemma (3.1) we choose $g_{0\mu} = 1$, $h_{0\mu} = \Omega_\mu$. Using (A.5) we can eliminate h_{13}, g_{13}, h_{23}, g_{23} in the following steps. Put into (A.4) $h_{13} = h_{02}(x_1 = 0) g_{23}$ and $h_{23} = h_{01}(x_2 = 0) g_{13}$ to find for the second term in $P(g_{ij})$:

$$\mathrm{Tr}((\Omega_2^{-1} d\Omega_2)_{x_1 = a_1} \wedge (\Omega_1 d\Omega_1^{-1})_{x_2 = 0})$$
$$+ \mathrm{Tr}((\Omega_2(x_1 = a_1) \Omega_1(x_2 = 0))^{-1} d(\Omega_2(x_1 = a_1) \Omega_1(x_2 = 0)) \wedge g_{13} dg_{13}^{-1}) - (1 \leftrightarrow 2).$$

Furthermore we have $g_{13} = g_{23}$ and using the consistency condition (2.2) the second term cancels the same term with 1 and 2 interchanged. Combining (A.3) with the above result generalized from the $(1, 2)$ to the (μ, ν) contribution, correctly taking care of the orientations, we find the expression of Lemma (3.1).

While completing this work we found that M. Lüscher [17] derived a formula of the type in Lemma (3.1) in a completely different context.

Acknowledgements. We would like to thank Sander Bais for suggesting this problem to us and for many stimulating discussions. Furthermore we thank Gerard 't Hooft for explaining some of his ideas and a critical reading of the manuscript, Richard Ward and Bert van Geemen for discussions on mathematical problems and B. Berg for bringing Ref. 17 to our attention. This work is part of the research programme of the "Stichting voor Fundamenteel Onderzoek der Materie (FOM)", which is financially supported by the "Nederlandse Organisatie voor Zuiver Wetenschappelijk Onderzoek (ZWO".

References

1. Bander, M.: Theories of quark confinement. Phys. Rep. **75**, 205 (1981)
2. Wilson, K.G.: Confinement of quarks. Phys. Rev. **D10**, 2445 (1974)
3. 't Hooft, G.: A property of electric and magnetic flux in non-abelian gauge theories. Nucl. Phys. **B153**, 141 (1979)
4. 't Hooft, G.: Confinement and topology in non-abelian gauge theories. Acta Phys. Austriaca, Suppl. **XXII**, 531 (1980)
   't Hooft, G.: Aspects of quark confinement. Lecture delivered at The Symposium on Perspectives in Modern Field Theories, Stockholm, September 1980
5. 't Hooft, G.: Some twisted self-dual solutions for the Yang-Mills equations on a hypertorus. Commun. Math. Phys. **81**, 267 (1981)
6. Steenrod, N.: The topology of fiber bundles. Princeton Math. Series 14. Princeton: Princeton University Press 1951
   Kobayashy, S., Nomizu, K.: Foundations of differential geometry. New York: Interscience, 1963, Vol. I; 1969, Vol. II
7. Daniel, M., Viallet, C.M.: The geometrical setting of gauge theories of the Yang-Mills type. Rev. Mod. Phys. **52**, 175 (1980)
8. Freudenthal, H., de Vries, H.: Linear Lie groups. London-New York: Academic Press 1969, §28, 29
9. 't Hooft, G.: Private communication
10. Eguchi, T., Gilkey, P.B., Hanson, A.J.: Gravitation, gauge theories and differential geometry. Phys. Rep. **66**, 213 (1980)
11. Macfarlane, A.J., Sudbery, A., Weisz, P.H.: On Gell-Mann's λ-matrices, d- and f-tensors, octets and parametrizations of SU(3). Commun. Math. Phys. **11**, 77 (1968)
12. Callen, C.G., Dashen, R.F., Gross, D.J.: Toward a theory of the strong interactions. Phys. Rev. **D17**, 2717 (1978)
    Jackiw, R., Rebbi, C.: Vacuum periodicity in a Yang-Mills quantum theory. Phys. Rev. Lett. **37**, 172 (1976)
13. Witten, E.: Dyons of charge $e\theta/2\pi$. Phys. Lett. **B86**, 283 (1979)
    't Hooft, G.: Topology of the gauge condition and new confinement phases in non-abelian gauge theories. Nucl. Phys. **B190**, [FS3] 455 (1981)
14. Ambjørn, J., Flyvbjerg, H.: 't Hooft's non-abelian magnetic flux has zero classical energy. Phys. Lett. **B97**, 241 (1980)
15. Isham, C.J., Kunstatter, G.: Yang-Mills canonical vacuum structure in a general three-space. Phys. Lett. **B102**, 417 (1981)
    Kunstatter, G.: Yang-Mills vacua in a general three-space. Imperial College preprint, ICTP/80/81-42
16. Groeneveld, J., Jurkiewicz, J., Korthals Altes, C.P.: Twist as a probe for phase structure. Physica Scripta **23**, 1022 (1981)
17. Lüscher, M.: Topology of lattice gauge theories. Commun. Math. Phys. **85**, 39–48 (1982)

Communicated by R. Stora

Received January 25, 1982

Commun. Math. Phys. 94, 397–419 (1984)

Communications in
**Mathematical
Physics**
© Springer-Verlag 1984

SU(*N*) Yang-Mills Solutions with Constant Field Strength on T^4

Pierre van Baal[*]

Institute for Theoretical Physics, Princetonplein 5, P.O. Box 80006, NL-3508 TA Utrecht, The Netherlands

Abstract. We study for T^4 the class of solutions to the SU(N) Yang-Mills equations with constant field strength. The fluctuation spectrum is explicitly calculated in terms of generalized Riemann theta functions. We show that if these solutions are stable, they are necessarily (anti)-selfdual, in which case we verify the index theorem.

1. Introduction

Euclidean solutions to the classical equations of motion (instantons), play an important role in the nonperturbative analysis of gauge theories. For S^4 the most general solution is known now (ADHM construction [1]) using advanced mathematical results. It was hoped that they could be used to understand confinement, however Coleman's argument shows that instantons have no effect on the Wilson loop, which is used to measure the static quark-antiquark potential [3].

For the torus this argument is no longer valid, because twisted boundary conditions [2] force electric and magnetic flux through the box. Nevertheless one will encounter as always severe infrared problems. Physical quantities are expressed in terms of the running coupling constant, which for small box size L is proportional to $(-\ln L)^{-1/2}$. So if L increases, the running coupling constant increases and perturbation theory breaks down, not only in the perturbative but also in the instanton sectors. Including instantons however can give an earlier signal for the crossover. Moreover the analysis of Lüscher [4] shows that the energy of the ground state is independent of the electric flux **e** (the central sectors) to all orders in perturbation theory. Confinement would be signalled by an energy difference, between the ground state levels in each central sector, proportional to L. Since "twisted" instantons lift the degeneracy they might be crucial to detect this behaviour. This work is intended as a first step in that direction.

We will concentrate on gauge fields with constant curvature. Such soltuions were already considered some time ago by 't Hooft [5] for SU(N) on T^4 with twisted boundary conditions. Only if the sides of the box representing T^4 satisfy certain relations, these solutions are (anti)-selfdual. For small groups and arbitrary compact manifolds [including SU(2) (SO(3)) bundles over T^4] stable extrema of the action are (anti)-selfdual [6]. It is therefore no surprise that in SU(2) an explicit calculation shows that constant solutions which are stable are (anti)-selfdual (the reverse is obvious).

* Address after Sept. 1: Institute for Theoretical Physics, State University of New York at Stony Brook, Stony Brook, NY 11794, USA

                                                    P. van Baal

For constant gauge fields on $\mathbb{R}^4$ this was already established by Leutwyler [7]. The boundary condition is given by requiring the action to differ by a finite amount from that of the constant background field. There is an infinite degeneracy in the fluctuation spectrum, which can be understood by the lack of control over the topology of the field configurations. On the hypercube (T^4) this problem is resolved, but complicated boundary conditions is the price we have to pay. For the boundary conditions that we choose (which are of the abelian type), the Riemann θ-functions provide a *natural* (from the mathematical point of view) solution to this problem. In the case of SU(2) each set of boundary conditions admitting constant solutions is gauge equivalent to an abelian set (Appendix B), but for SU(N), $N > 2$, this is no longer true.

Constant field strength configurations also played an important role in the Copenhagen vacuum picture [8]. One starts from a constant chromomagnetic field in the z-direction. This is certainly not selfdual and thus unstable (unlike in the abelian case). Including the unstable mode yields an effective Higgs type potential. In analogy with the abelian Higgs model one minimizes the potential by assuming the formation of a lattice structure orthogonal to the z-direction, with size of the order of the inverse of the effective Higgs mass ($\simeq \sqrt{2g|B|}$, B the background chromomagnetic field strength). The reason for considering this type of vacuum is that, when including ("static") quantum fluctuations and minimizing with respect to the background magnetic field, the energy is smaller than that of the perturbative vacuum. Furthermore phenomenology indicates $\langle 0|B^2|0\rangle > 0$, and the bag model gives similar results.

It has been shown by Ambjørn and Olesen [8] that their lattice structure is equivalent to imposing twisted boundary conditions on the unit cell in the x, y direction. In essence one considers gauge fields over the 2-torus T^2, the unstable mode is then given by the θ-function on T^2 (up to an exponential factor). It is therefore not really surprising that our analysis on T^4 is also based on θ-functions. Whereas the classical θ-function on the 2-torus (due to Jacobi) is well known, its generalization to a $2n$-torus (due to Riemann) is not.

This article will mainly concentrate on giving the necessary technical details.

As for the 2-dimensional case, there is a specific complex structure on T^4 in which the θ-functions are most easily expressed. This suggests looking for the most general (anti) selfdual solutions on T^4. It is amusing to note that for $\mathbb{C}P_1$ on T^2, the exact instanton solutions are expressed in terms of the Weierstrass σ-function [9a], which has a simple relation to the θ-function [9b]. Also the solution of Gürsey and Tze [10] is based on elliptic functions (the generalization of the Weierstrass ζ-function to quaternions). It seems to be a candidate for an instanton on T^4 with unit Chern number. However, one simply checks that the gauge invariant quantity $\mathrm{Tr}(F_{\mu\nu}^2)$ is not periodic and it can therefore not yield a solution on T^4. Explicitly their solution is of the form: (σ^a the Pauli matrices, η the 't Hooft η symbol [11]):

$$A_\mu = -\tfrac{1}{2}\sigma^a \bar{\eta}_{a\mu\nu}\partial^\nu \ln\varrho, \tag{1.1}$$

with

$$\varrho = \frac{1}{x^2} + \sum_{q\in L}\left\{\frac{1}{(x-q)^2} - \frac{1}{q^2} - \frac{2(x\cdot q)}{q^4} + \frac{x^2}{q^4} - \frac{4(x\cdot q)^2}{q^6}\right\}, \tag{1.2}$$

where L is the lattice (minus the origin) spanned by $q_\mu = \omega_{\mu\nu} n^\nu$, with $n \in \mathbb{Z}^4$ and $\det \omega \neq 0$. ϱ is pseudo periodic: $\varrho(x+q) = \varrho(x) + x \cdot a_q + b_q$, but

$$\mathrm{Tr}(F_{\mu\nu}^2) = - \square\square \ln\varrho, \tag{1.3}$$

with $\square$ the four-dimensional Laplacian, is not periodic.

The organization of the rest of the paper is as follows. In Sect. 2 we will set up the notations and the fluctuation equations. We will use SU(N) gauge fields with twisted boundary conditions, which is equivalent to SO($N^2 - 1$) fiber bundles over T^4 [12]. In Sect. 3 we will give the fluctuation spectrum. Self duality and stability of the solutions will be treated in Sect. 4, together with the index theorem. Section 5 gives a discussion of the results and an outlook for further research. Technicalities on the θ-functions are collected in an appendix.

2. The Solutions and Their Fluctuations

We will first review how to put gauge fields on the hypertorus. These gauge fields will be in the fundamental representation of SU(N) and are hermitian. The curvature or field strength is given by

$$G_{\mu\nu} = \partial_\mu A_\nu - \partial_\nu A_\mu + i[A_\mu, A_\nu]. \tag{2.1}$$

The 4-torus will be labelled by a 4-dimensional hypercube $\{x \in \mathbb{R}^4 | 0 < x_\mu \leq a_\mu\}$, with the standard metric of $\mathbb{R}^4$. It is not really essential that the hypercube is rectangular, but for physical applications [2] it is more convenient. From a mathematical point of view it is more appropriate to view T^4 as $\mathbb{R}^4$ modulo some 4-dimensional lattice L. So with

$$L = \{x \in \mathbb{R}^4 | x = n_\mu a^{(\mu)}; n \in \mathbb{Z}^4\}, \tag{2.2}$$

we have $T^4 = \mathbb{R}^4/L$. Here $a^{(\mu)}$ is a vector in the μ-direction with length a_μ, they form the $\mathbb{Z}$-basis of the lattice L.

To define gauge fields on T^4, we take gauge potentials on $\mathbb{R}^4$ and demand that all gauge invariant quantities which can be formed out of them are periodic over L. This implies that the vector potential satisfies:

$$A_\lambda(x + a^{(\mu)}) = [\Omega_\mu(x)] A_\lambda(x) \equiv \Omega_\mu(x) A_\lambda(x) \Omega_\mu(x)^{-1} - i\Omega_\mu(x) \partial_\lambda \Omega_\mu^{-1}(x). \tag{2.3}$$

In the terminology of fiber bundles over the torus, $\Omega_\mu(x)$ is the elementary cocycle $E_{e(\mu)}(x)$ and in general we have

$$A_\lambda(x + n_\mu a^{(\mu)}) = E_n(x) A_\lambda(x), \tag{2.4}$$

where the cocycles have to satisfy the cocycle condition:

$$E_{n+m}(x) = E_n(x + m_\mu a^{(\mu)}) E_m(x). \tag{2.5}$$

This clearly implies $E_n(x+m) E_m(x) = E_m(x+n) E_n(x)$, and one easily shows that given the elementary cocycles $E_{e(\mu)}$ this identity is satisfies if $E_n(x)$ is defined inductively by (2.5) and if:

$$E_{e(\mu)}(x + a^{(\nu)}) E_{e(\nu)}(x) = E_{e(\nu)}(x + a^{(\mu)}) E_{e(\mu)}(x). \tag{2.6}$$

Since $[Z\Omega_\mu] = [\Omega_\mu]$, for Z an element of the centre of SU(N), identity (2.6) implies:

$$\Omega_\mu(x + a^{(\nu)})\Omega_\nu(x) = Z_{\mu\nu}\Omega_\nu(x + a^{(\mu)})\Omega_\mu(x) \, . \tag{2.7}$$

$Z_{\mu\nu}$ is an element of the centre of SU(N) defining the twist of the bundle [2, 12]. It is a topological invariant, which is conveniently labelled by the twist tensor $n_{\mu\nu}$:

$$Z_{\mu\nu} = \exp(-2\pi i n_{\mu\nu}/N) \, , \tag{2.8}$$

n is an antisymmetric 4×4 matrix with entries in $\mathbb{Z}$ (mod N). Together with the Pontryagin index P_1, this specifies the topology of the fiber bundle:

$$P_1 = -2NC_2 \, , \qquad C_2 = \frac{-1}{16\pi^2} \int_{T_4} \mathrm{Tr}(G_{\mu\nu}\tilde{G}_{\mu\nu})d_4x = \nu + \frac{N-1}{N}\,\mathrm{Pf}(n) \, , \tag{2.9}$$

where $\tilde{G}$ is the dual of G: $\tilde{G}_{\mu\nu} = \frac{1}{2}\varepsilon_{\mu\nu\alpha\beta}G_{\alpha\beta}$. Pf($n$) is the Pfaffian of the twist tensor:

$$\mathrm{Pf}(n) = \tfrac{1}{4}n_{\mu\nu}\tilde{n}_{\mu\nu} = \tfrac{1}{8}n_{\mu\nu}n_{\alpha\beta}\varepsilon_{\mu\nu\alpha\beta} \, , \tag{2.10}$$

and ν is an integer, the "instanton number," see [12] for details. P_1 is always an even integer; C_2 is also integer if there is no twist ($Z_{\mu\nu} = 1$ or equivalently $n_{\mu\nu} = 0 \bmod N$).

We will restrict ourselves here to pure abelian boundary conditions; to be specific [12]:

$$\Omega_\mu(x) = \exp\left(-\frac{\pi i}{N}\sum_\nu \frac{n_{\mu\nu}x_\nu}{a_\nu}T\right) , \tag{2.11}$$

with T the generator in SU(N) which "contains" the centre of SU(N):

$$T = \mathrm{diag}(1, \ldots, 1, 1-N) \, . \tag{2.12}$$

For these configurations $\nu = 0$, or $C_2 = \dfrac{N-1}{N}\,\mathrm{Pf}(n)$. An obvious solution to the Yang-Mills equations of motion:

$$D_\mu G^0_{\mu\nu} = \partial_\mu G^0_{\mu\nu} + i[A^0_\mu, G^0_{\mu\nu}] = 0 \, , \tag{2.13}$$

satisfying the boundary conditions (2.3), (2.11) is:

$$A^0_\mu = -\frac{\pi}{N}\sum_\nu \frac{n_{\mu\nu}x_\nu}{a_\mu a_\nu}T \, , \qquad G^0_{\mu\nu} = \frac{2\pi}{N}\frac{n_{\mu\nu}}{a_\mu a_\nu}T \, . \tag{2.14}$$

One of the reasons which enables us to explicitly compute the fluctuations around these solutions, is that they are sections of certain well-defined U(1) line bundles over T^4.

For this purpose we expand A_μ as follows:

$$A_\mu = A^0_\mu + \sum_{a=1}^{(N-1)^2} b^a_\mu T_a + \sum_{a=1}^{N-1}\sqrt{2}\,\mathrm{Re}(c^a_\mu \Sigma_a) \, , \tag{2.15}$$

where b is real, c is complex, T_a for $a = 1$ up to $(N-1)^2 - 1$ are the generators of SU($N-1$) embedded in SU(N) in the upper left corner, $T_{(N-1)^2} = (2N(N-1))^{-1/2}T$ and $(\Sigma_a)_{kl} = \delta_{ka}\delta_{lN}$. The generators T_a supplemented with

Yang-Mills Solutions with Constant Field Strength 401

$T_{(N-1)^2+2a+1}=\tfrac{1}{2}(\Sigma_a+\Sigma_a^+)$ and $T_{(N-1)^2+2a}=\dfrac{i}{2}(\Sigma_a-\Sigma_a^+)$, form the algebra of SU(N) with normalization $\mathrm{Tr}(T_a T_b)=\tfrac{1}{2}\delta_{ab}$. Using the commutation relations:

$$[T,\Sigma_a]=N\Sigma_a,\quad [T,\Sigma_a^+]=-N\Sigma_a^+,\quad [T,T_a]=0,\qquad (2.16)$$

one finds that the boundary conditions on A_μ are satisfied if and only if:

$$b_\mu^a(x+a^{(\lambda)})=b_\mu^a(x),\quad c_\mu^a(x+a^{(\lambda)})=\exp\left(-\pi i\sum_\nu \frac{n_{\lambda\nu}x_\nu}{a_\nu}\right)c_\mu^a(x).\qquad (2.17)$$

So obviously b is a section of a trivial U(1) line-bundle and c a section of a nontrivial U(1) bundle with its first Chern class in $1-1$ relation with n. Again we can define for this bundle a cocycle $e_n(x)$:

$$c(x+n_\mu a^{(\mu)})=e_n(x)c(x),$$
$$e_{e(\mu)}(x)=\exp\left(-\pi i\sum_\nu \frac{n_{\mu\nu}x_\nu}{a_\nu}\right),\qquad (2.18)$$
$$e_{n+m}(x)=e_n(x+m_\mu a^{(\mu)})e_m(x).$$

Explicitly we have:

$$e_k(x)=\exp\left(-\pi i\sum_{\mu,\nu}\frac{k_\mu n_{\mu\nu}x_\nu}{a_\nu}\right)\alpha(k),$$
$$\alpha(k)=\exp\left(\pi i\sum_{\mu<\nu}k_\mu n_{\mu\nu}k_\nu\right).\qquad (2.19)$$

We are now in a position to formulate the fluctuation equation, by expanding the action around the solution (2.14):

$$S=\tfrac{1}{2}\int d_4 x\,\mathrm{Tr}(G_{\mu\nu}^2)=\tfrac{1}{2}\int d_4 x\{\mathrm{Tr}(G_{\mu\nu}^{02})$$
$$-\mathrm{Tr}(\delta A_\mu D_\nu^2\delta A_\mu-\delta A_\mu D_\mu D_\nu\delta A_\nu+2i\delta A_\mu[G_{\mu\nu}^0,\delta A_\nu])$$
$$+2i\,\mathrm{Tr}(\delta A_\mu D_\nu[\delta A_\mu,\delta A_\nu])+\tfrac{1}{2}\mathrm{Tr}((i[\delta A_\mu,\delta A_\nu])^2)\},\qquad (2.20)$$

where $A_\mu=A_\mu^0+\delta A_\mu$ and $D_\mu=\partial_\mu+i[A_\mu^0,\cdot]$. Furthermore we introduce as usual background gauge fixing, $D_\mu\delta A_\mu=0$, so that in general the action including the Faddeev-Popov fields is given by [11]:

$$S=\int d_4 x\{\tfrac{1}{2}\mathrm{Tr}(G_{\mu\nu}^0)^2+\tfrac{1}{2}\mathrm{Tr}(\delta A_\mu M_A^{\mu\nu}\delta A_\nu)+\mathrm{Tr}(\psi^+ M_{gh}\psi)+\mathcal{O}(\delta A^3)\},\quad (2.21)$$

with

$$M_A^{\mu\nu}=-\delta_{\mu\nu}D_\lambda^2-2i[G_{\mu\nu}^0,\cdot],\qquad M_{gh}=-D_\lambda^2.\qquad (2.22)$$

If we substitute for δA_μ the expression in (2.15) and expand the ghost fields according to:

$$\psi=\sum_{a=1}^{(N-1)^2}\psi^a T_a+\sum_{a=1}^{N-1}\sqrt{2}(\phi^a\Sigma_a+\{\phi^{a+N-1}\Sigma_a\}^\dagger)\qquad (2.23)$$

                                                          P. van Baal

with the appropriate boundary conditions:

$$\psi^a(x+n_\mu a^{(\mu)})=\psi^a(x), \qquad \phi^a(x+n_\mu a^{(\mu)})=e_n(x)\phi^a(x), \tag{2.24}$$

then the action takes the form:

$$S=S_0+\int d_4x\{\tfrac{1}{2}b_\mu^a(M_0\delta^{\mu\nu}\delta_{ab})b_\nu^b+c_\mu^{a*}(M_n\delta^{\mu\nu}-4\pi iF^{\mu\nu})\delta_{ab}c_\nu^b+\psi^{a*}(M_0\delta_{ab})\psi^b$$
$$+\phi^{a*}(M_n\delta_{ab})\phi^b+\mathcal{O}((\delta A_\mu)^3)\}, \tag{2.25}$$

with

$$S_0=\tfrac{1}{2}\int d_4x\,\mathrm{Tr}(G_{\mu\nu}^{02})=\frac{2\pi^2(N-1)}{N}F_{\mu\nu}{}^2V,$$

$$F_{\mu\nu}=\frac{n_{\mu\nu}}{a_\mu a_\nu}, \qquad V=\prod_{\mu=1}^{4}a_\mu,$$

$$M_0=\left(\frac{1}{i}\partial_\mu\right)^2, \tag{2.26}$$

$$M_n=\left(\frac{1}{i}\partial_\mu-\pi F_{\mu\nu}x_\nu\right)^2.$$

Before we will construct the spectrum of these operators, let us briefly discuss the generalization for an arbitrary gauge group H. For convenience we will restrict ourselves to a simple and simply connected group [like SU(N)]. Generalization to semi-simple, multiple connected groups and details can be found with the help of Goddard et al. [13] and Humphreys [14] (see also [20]). One can always choose the following basis of the Lie-algebra of H:

$$T_1,...,T_r,E_{\pm\alpha}{}^{(1)},...,E_{\pm\alpha}{}^{(s)}, \tag{2.27}$$

where r is the rank of H (for SU(N), $r=N-1$, $s=\tfrac{1}{2}N(N-1)$, $E_{-\alpha}=E_\alpha^\dagger$, $T^\dagger=T$) and

$$[T_i,E_\alpha]=\alpha_iE_\alpha, \qquad [T_i,T_j]=0. \tag{2.28}$$

The $\alpha^{(i)}$ are called the roots and they span an r-dimensional space, whose metric is given by the Killing form $\kappa(X,Y)=\mathrm{Tr}_{ad}(XY)$ restricted to the maximal torus (or Cartan subalgebra, spanned by $T_1,...,T_r$)

One can choose normalizations such that this metric is δ_{ij}:

$$\mathrm{Tr}_{ad}(T_iT_j)=2N\,\mathrm{Tr}(T_iT_j)=\delta_{ij}, \qquad \sum_\alpha\alpha_i\alpha_j=\delta_{ij}. \tag{2.29}$$

The abelian boundary conditions are now

$$\Omega_\mu(x)=\exp\left(-\pi i\sum_\nu\frac{n_{\mu\nu}^k x_\nu}{a_\nu}T_k\right), \tag{2.30}$$

and the cocycle conditions imply

$$\exp(2\pi in_{\mu\nu}^kT_k)\in Z(H). \tag{2.31}$$

Using (2.28) one easily finds that this is true if and only if:

$$\exp(2\pi i n^{k}_{\mu\nu}\alpha_{k}) = 1 \quad \text{or} \quad n_{\mu\nu} \cdot \alpha \in \mathbb{Z} \tag{2.32}$$

for all roots α. So for each $\mu, \nu \dfrac{n_{\mu\nu}}{2N}$ is a weight for the "dual" of H [13], which is nothing but the Dirac quantization condition [Eq. (2.31) is the condition for the "single valuedness"]. The solutions of the Yang-Mills equations of motion are:

$$A^{0}_{\mu} = -\sum_{\nu} \frac{n^{k}_{\mu\nu}x_{\nu}}{a_{\mu}a_{\nu}} T_{k}, \qquad G^{0}_{\mu\nu} = 2\frac{n^{k}_{\mu\nu}}{a_{\mu}a_{\nu}} T_{k}. \tag{2.33}$$

Expanding around these solutions to implement the boundary conditions on the fluctuations we find:

$$A_{\mu} = A^{0}_{\mu} + b^{k}_{\mu}T_{k} + \frac{1}{\sqrt{2}}c^{\alpha}_{\mu}E_{\alpha}, \qquad b(x + a^{(\mu)}) = b(x),$$

$$c^{\alpha}_{\mu}(x + a^{(\lambda)}) = \exp\left(-\pi i \sum_{\nu} \frac{\alpha \cdot n\lambda_{\nu}x_{\nu}}{a_{\nu}}\right)c^{\alpha}_{\mu}(x). \tag{2.34}$$

And (2.32) again implies that we have sections of line bundles with a first Chern-index $\alpha \cdot n_{\mu\nu}$. Furthermore we have

$$P_{1} = 2\sum_{k} \mathrm{Pf}(n^{k}_{\mu\nu}) = 2\sum_{\alpha} \mathrm{Pf}(\alpha \cdot n_{\mu\nu}), \tag{2.35}$$

which is again twice an integer.

3. The Fluctuation Spectrum

We will introduce suitable complex coordinates, in which the fluctuation equations obtain an especially simple form. This is also suggested by Leutwyler's [7] analysis, but the amazing thing is that the boundary conditions are compatible with this choice and the whole analysis becomes canonical if one realizes that F [see (2.26)] introduces a positive definite hermitian form H. To be specific we first introduce coordinates:

$$\hat{x} = Sx, \qquad S \in \mathcal{O}(4), \tag{3.1}$$

which brings F in the standard form:

$$SFS^{t} = \hat{F} = \begin{pmatrix} \emptyset & \begin{matrix} f_{1} & 0 \\ 0 & f_{2} \end{matrix} \\ \begin{matrix} -f_{1} & 0 \\ 0 & -f_{2} \end{matrix} & \emptyset \end{pmatrix}, \qquad f_{1} \geqq f_{2} \geqq 0. \tag{3.2}$$

The complex coordinates will be chosen according to

$$z(x) = (z_{1}, z_{2}) = \frac{1}{\sqrt{2}}(\hat{x}_{1} - i\hat{x}_{3}, \hat{x}_{2} - i\hat{x}_{4}), \tag{3.3}$$

and the positive definite hermitian form will be given by

$$H(z, w) = 2(z_{1}f_{1}\bar{w}_{1} + z_{2}f_{2}\bar{w}_{2}) = w^{\dagger}hz, \tag{3.4}$$

404   P. van Baal

with $h = 2\,\text{diag}(f_1, f_2)$[1]. If we furthermore define:

$$(A_{z_1}, A_{z_2}) = \frac{1}{\sqrt{2}}(\hat{A}_1 + i\hat{A}_3, \hat{A}_2 + i\hat{A}_4),\tag{3.5}$$

$$(\partial_{z_1}, \partial_{z_2}) = \frac{1}{\sqrt{2}}(\partial_{\hat{x}_1} + i\partial_{\hat{x}_3}, \partial_{\hat{x}_2} + i\partial_{\hat{x}_4}),\tag{3.6}$$

$$a = (a_1, a_2); \quad a_k = \frac{1}{i}\frac{\partial}{\partial\bar{z}_k} - \frac{i\pi}{2}h_{kl}z_l,\tag{3.7}$$

we find [see (2.26)]

$$M_n = \{a, a^\dagger\},\tag{3.8}$$

$$c_\mu^{a*}(M_n\delta_{\mu\nu} - 4\pi iF_{\mu\nu})c_\nu^a = (c_{z_k}^a)^*(M_n\delta_{kl} - 2\pi h_{kl})c_{z_l}^a + (c_{\bar{z}_k}^a)^*(M_n\delta_{kl} + 2\pi h_{kl})c_{\bar{z}_l}^a.\tag{3.9}$$

Clearly the operators a_k and $a_k^\dagger$ are annihilation and creation operators, since:

$$[a_k, a_l^\dagger] = \pi h_{kl} = 2\pi f_k\delta_{kl}.\tag{3.10}$$

Therefore the spectrum of M_n is given by[2]

$$\lambda_m = 2\pi\sum_{k=1}^{2}(2m_k + 1)f_k, \quad \chi_m = \frac{(a_1^\dagger)^{m_1}(a_2^\dagger)^{m_2}}{\sqrt{m_1!m_2!}}\chi_0,\tag{3.11}$$

where χ_0 is the ground state, uniquely determined by

$$a\chi_0 = \frac{1}{i}\left(\frac{\partial}{\partial\bar{z}_k} + \frac{\pi}{2}h_{kl}z_l\right)\chi_0 = 0,\tag{3.12}$$

and the boundary conditions. Explicitly

$$\chi_0 = e^{-\frac{\pi}{2}H(z,z)}f(z),\tag{3.13}$$

where $f(z)$ is any holomorphic function, such that χ_0 satisfies the boundary conditions.

Let us define unit vectors $e_\mu^{(z_k)}$ and $e_\mu^{(\bar{z}_k)}$, with the properties $e^{(\bar{z}_k)} = (e^{(z_k)})^*$ and $e_{z_l}^{(z_k)} = \delta_{kl}$, $e_{z_l}^{(z_k)} = 0$. Explicitly with respect to the $\hat{x}$ basis in (3.1) we have

$$\hat{e}^{(z_1)} = \frac{1}{\sqrt{2}}(1, 0, -i, 0) = [\hat{e}^{(\bar{z}_1)}]^*, \quad \hat{e}^{(z_2)} = \frac{1}{\sqrt{2}}(0, 1, 0, -i) = [\hat{e}^{(\bar{z}_2)}]^*.\tag{3.14}$$

Then the spectrum of $M_n\delta_{\mu\nu} - 4\pi iF_{\mu\nu}$ is also easily determined:

$$\begin{aligned}\lambda_{m,k}^- &= \lambda_m - 4\pi f_k, \quad &\chi_{m,k}^- &= \chi_m e^{(z_k)}\\\lambda_{m,k}^+ &= \lambda_m + 4\pi f_k, \quad &\chi_{m,k}^+ &= \chi_m e^{(\bar{z}_k)}.\end{aligned}\tag{3.15}$$

---

1   If one prefers to work with $H(z, w) = z^\dagger hw$ one should replace (3.3) by $z = (\hat{x}_1 + i\hat{x}_3, \hat{x}_2 + i\hat{x}_4)/\sqrt{2}$

2   For the time being we assume $f_k \neq 0$, so H non-degenerate

For Leutwyler's analysis [7] the boundary conditions are that χ is square integrabel, but this means that $f(z)$ in (3.13) can be any holomorphic function, whence the infinite degeneracy. In our case f has to satisfy the boundary conditions $f(z+q) = u_q(z)f(z)$, with $u_q(z)$ the appropriate cocycle. In order for (3.13) to admit a non-trivial solution the coycle u_q must necessarily be a holomorphic function. That this turns out to be the case is not really a surprise since the spectrum of a bounded hermitian operator can never be empty. Thus we look for holomorphic sections on a complex torus: $T^4 = \mathbb{C}^2/L$. The canonical objects for these are the (Riemann) θ-functions. For the theory of θ-functions we refer to Igusa [15], whose notations we will roughly follow.

The crucial thing for the complex structure on $\mathbb{C}^2/L$ with hermitian form H to admit θ-functions as holomorphic sections is that

$$E(z, w) = \operatorname{Im} H(z, w) \tag{3.16}$$

restricted to the lattice $L = \{k_\mu \zeta^{(\mu)} | k \in Z^4, \ \zeta^{(\mu)} = z(a^{(\mu)})\}$ takes values in the integers. With respect to the real basis one can simply express the hermitian form explicitly:

$$H(z(x), z(y)) = x_\mu |F|_{\mu\nu} y_\nu + i x_\mu F_{\mu\nu} y_\nu, \tag{3.17}$$

where $|F| = (-F^2)^{1/2}$. To prove this, first express things in the $\hat{x}$ basis of Eq. (3.1), where $(-\hat{F}^2)^{1/2} = \operatorname{diag}(f_1, f_2, f_1, f_2)$. So E is nothing but F and $E/L = n$:

$$E(\zeta^{(\mu)}, \zeta^{(\nu)}) = n_{\mu\nu}, \qquad \zeta^{(\mu)} \cong z(a^{(\mu)}). \tag{3.18}$$

The elementary cocycle of (2.18) can also be expressed in terms of E and z:

$$e_{e(\mu)}(x) = \exp(-\pi i E(\zeta^{(\mu)}, z(x))). \tag{3.19}$$

The boundary condition for χ_0 is therefore

$$\chi_0(z + n_\mu \zeta^{(\mu)}) = \exp(-\pi i E(n_\mu \zeta^{(\mu)}, z)) \alpha(n) \chi_0(z). \tag{3.20}$$

Note that, as it should be, χ_m as defined in (3.11) satisfies the same boundary conditions. For this one uses the property

$$a^\dagger(z+q)e_q(z) = e_q(z)a^\dagger(z), q \in L. \tag{3.21}$$

To make the holomorphic structure visible we finally write down the boundary conditions on $f(z)$:

$$f(z+q) = \alpha(q) \exp\left(\frac{\pi}{2} H(q, q) + \pi H(z, q)\right) f(z)$$

$$\equiv u_q(z) f(z), \qquad q \in L, \tag{3.22}$$

where $\alpha(q)$ is defined in (2.19). It satisfies:

$$\alpha(q+r) = \alpha(q)\alpha(r) \exp(\pi i E(q, r)). \tag{3.23}$$

The holomorphic cocycle u_q is also called an automorphy factor. Any α defined on L satisfying (3.23) and having values in U(1) is called a second degree character strongly associated with E. Then all functions defined by (3.22) are called θ-functions of type (H, α); they form a complex linear space $L(H, \alpha)$ of dimension $|\mathrm{Pf}(n)|$, see [15, Chap. II] for details, some of which are collected in the appendix.

We assumed that H is non-degenerate $(f_1, f_2 \neq 0)$; this is not really necessary. If $f_2 = 0, f_1 \neq 0$, (3.11) still generates part of the spectrum $(m_2 \equiv 0)$, especially the $m = 0$ modes are still given by θ-functions. The dimension of $L(H, \alpha)$ is in this case the greatest common divisor of the entries of n [g.c.d. $(n_{\mu\nu})$], see the appendix for details. There is one θ-function in $L(H, \alpha)$ which has an especially simple form (for H non-degenerate):

$$f_0(z) = \sum_{q \in L} \alpha(q) \exp\left(H(z, q) - \frac{\pi}{2} H(q, q) \right). \tag{3.24}$$

We will call it the intrinsic θ-function [16]. That the analysis of the spectrum of M_n is indeed canonical can be found in [17] and references therein, where the spectrum of the Laplace-Beltrami operator (M_n up to a constant) is constructed.

Let us finally collect our results to give explicitly the eigenfunctions and eigenvalues for the operators in (2.22):

	$M_A \delta A = \lambda \delta A; \; \delta A = \sqrt{2}\,\mathrm{Re}\,\phi$		$M_{gh}\psi = \lambda\psi$	
	ϕ	λ	ψ	λ
$f_i \neq 0$	$e^{(\mu)}\phi^{(p,a)}$	$\sum_\mu \left(\dfrac{2\pi p_\mu}{a_\mu}\right)^2$	$\phi^{(p,a)}$	$\sum_\mu \left(\dfrac{2\pi p_\mu}{a_\mu}\right)^2$
	$e^{(z_k)}\phi^{(m,r,b)}$ $ie^{(z_k)}\phi^{(m,r,b)}$	$2\pi\left(\sum\limits_{i=1}^{2} (2m_i+1)f_i - 2f_k\right)$	$\phi^{(m,r,b)}$	$2\pi \sum\limits_{i=1}^{2} (2m_i+1)f_i$
	$e^{(\bar z_k)}\phi^{(m,r,b)}$ $ie^{(\bar z_k)}\phi^{(m,r,b)}$	$2\pi\left(\sum\limits_{i=1}^{2} (2m_i+1)f_i + 2f_k\right)$	$\phi^{(m,r,b+N-1)}$	$2\pi \sum\limits_{i=1}^{2} (2m_i+1)f_i$
$f_i = 0$	$e^{(\mu)}\phi^{(p,c)}$	$\sum_u \left(\dfrac{2\pi p_\mu}{a_\mu}\right)^2$	$\phi^{(p,c)}$	$\sum_\mu \left(\dfrac{2\pi p_\mu}{a_\mu}\right)^2$
$f_2=0, f_1 \neq 0$ $\lambda < 0 \; only$	$e^{(z_1)}\phi^{(0,r,a)}$ $ie^{(z_1)}\phi^{(0,r,a)}$	$\lambda = -2\pi f_1$		

$$\tag{3.25}$$

where a, b, c, k, m, p, and r have the range:

$$a = 1,\ldots,(N-1)^2; \qquad b = 1,\ldots,(N-1); \qquad c = 1,\ldots,(N^2-1);$$
$$k = 1,2; \qquad m \in \mathbb{N}^2; \qquad p \in \mathbb{Z}^4;$$
$$r_1 = 0\ldots(e_1-1); \qquad r_2 = 0\ldots(e_2-1).$$

We used the following definitions for the normalized eigenfunctions $(\Sigma_a \equiv \Sigma_{a+1-N}$ for $a > N-1)$

$$\phi^{(p,a)} = \frac{1}{\sqrt{V}}\, e^{2\pi i p_\mu x_\mu/a_\mu} \cdot T_a,$$

$$\phi^{(m,r,a)} = \frac{(a_1^+)^{m_1}(a_2^+)^{m_2}}{\sqrt{m_1!m_2!}}\, \chi_r \Sigma_a,$$

$$\tag{3.26}$$

$$e_\nu^{(\mu)} = \delta_{\mu\nu}; \qquad e_\mu^{(z_1)} = \frac{1}{\sqrt{2}}\,{}'(S_{1\mu} - iS_{3\mu}),$$

$$e_\mu^{(z_2)} = \frac{1}{\sqrt{2}}(S_{2\mu} - iS_{4\mu}).$$

The χ_r form an orthogonal basis for $a\chi=0$ (3.12) and are defined in the appendix (A.23).

The generalization to the solutions in (2.33) is straightforward and will therefore not be worked out in detail.

4. Stability and Selfduality

Using the results of the previous section we find $2(N-1)|\mathrm{Pf}(n)|$ negative eigenvalues $(\lambda=2\pi(f_2-f_1))$ if and only if $f_1 \neq f_2$ (by construction we choose $f_1 \geq f_2 \geq 0$). The duality equation is invariant under $\mathcal{O}(4)$ transformations S. If $\det S = -1$ self and antiselfdual configurations are interchanged. Thus the constant field strength solution is (anti) selfdual iff $f_1 = f_2$. (In the $\hat{x}$ coordinates the solution is always antiselfdual). We can conclude from this that the constant solutions are stable if and only if they are (anti)-selfdual. If $f_1 \neq f_2$, the number of negative (and zero) modes is consistent with the lower bound derived by Taubes, Theorem 3.8 [23]. Selfduality imposes for a given twist tensor $n_{\mu\nu}$ constraints on the sides a_μ of the euclidean box which parametrizes the torus (or equivalently the a_μ determine the scale of the coordinates). But there is also a constraint on $n_{\mu\nu}$ itself to admit an (anti)-selfdual solution. It is necessary and sufficient that $\mathrm{sign}(n_{\mu\nu})$ is self or anti-selfdual. If $n_{\mu\nu} \neq 0$ for all $\mu \neq \nu$, a_μ is fixed up to an overall scale (no summation):

$$a_\mu = \left| \frac{n_{\alpha\mu} n_{\mu\beta}}{n_{\alpha\nu} n_{\nu\beta}} \right|^{1/2} a_\nu, \tag{4.1}$$

where μ, ν, α, β are all different. In general the number of undetermined scales is one more than the number of zeros in (n_{14}, n_{24}, n_{34}). These are similar conditions as found by 't Hooft [5]. His boundary conditions are however more complicated for $N \neq 2$. But we are confident that a similar analysis can be performed. One needs a generalization of the θ-function, where the bicharacters $\alpha(k)$ take their values in $\mathrm{SU}(N)$.

We will now use the symmetries of the solution to describe the degeneracies. With (2.14), (3.2), and (3.5) the solution in complex coordinates is given by:

$$A^0_{z_i} = \frac{i\pi}{N} f_i \bar{z}_i T. \tag{4.2}$$

The connection 1-form $A = A_\mu dx_\mu$ is given by $A = A_{z_i} dz_i + A_{\bar{z}_i} d\bar{z}_i$. The solution is therefore obviously invariant under 1) the unitary coordinate transformation $z \to Uz$ which leaves h invariant, and 2) the global gauge transformations which commute with the holonomy group of $(4.2)^3$. These gauge transformations form the group generated by T_a for $a=1\ldots(N-1)^2$ with covering $\mathrm{U}(N-1)$. The unitary transformations leaving h invariant form the group $\mathrm{U}(1) \times \mathrm{U}(1)$ for $f_1 \neq f_2$ and $\mathrm{U}(2)$ for $f_1 = f_2$. In the real $\hat{x}$ coordinates the group $\mathrm{U}(2)$ is $\mathrm{SO}(4) \cap \mathrm{SP}(2)$. Explicitly it is generated by

$$\begin{aligned} \tau_0 &= \sigma_2 \otimes 1_2, & \tau_1 &= \sigma_2 \otimes \sigma_1, \\ \tau_2 &= -1_2 \otimes \sigma_2, & \tau_3 &= -\sigma_2 \otimes \sigma_3, \end{aligned} \tag{4.3}$$

---

3   The holonomy group of A_μ is given by $P\exp\left(i\int_c A_\mu dx_\mu\right)$, where c runs over all closed loops, with a fixed base point, see [18, 20]

where the tensor product $\otimes$ is such that $\hat{F} = -if\sigma_2 \otimes \sigma_3$. τ_0 and τ_3 generate the subgroup $U(1) \times U(1)$. The fluctuations have to be representations of these groups, which explains most of the degeneracies in the eigenvalues.

Assuming from now on selfduality $(f_1 = f_2 \equiv f)$ we find the following eigenvalues (λ) and degeneracies (μ):

	M_A		M_{gh}		
	λ	μ	λ	μ	
$f \neq 0$	$4\pi f|k|$	$4(N-1)|\hat{k} \cdot \mathrm{Pf}(n)|$	$4\pi f|k|; k \neq 0$	$(N-1)|\hat{k} \cdot \mathrm{Pf}(n)|$	(4.4)
	$\sum_\mu \left(\dfrac{2\pi p_\mu}{a_\mu}\right)^2$	$4(N-1)^2$	$\sum_\mu \left(\dfrac{2\pi p_\mu}{a_\mu}\right)^2$	$(N-1)^2$	
$f = 0$	$\sum_\mu \left(\dfrac{2\pi p_\mu}{a_\mu}\right)^2$	$4(N^2-1)$	$\sum_\mu \left(\dfrac{2\pi p_\mu}{a_\mu}\right)^2$	(N^2-1)	

where $k \in \mathbb{Z}$ and $p \in \mathbb{Z}^4$ and $\hat{k} = k + \delta_{k,0}$.

From now on we will only concentrate on the zero-modes. Explicitly they are given by [see (3.25), (3.26)]

	M_A	M_{gh}	
$f \neq 0$	$\sqrt{2}\,\mathrm{Re}(e^{(zk)}\chi_r \Sigma_b)$		
	$\sqrt{2}\,\mathrm{Im}(e^{(zk)}\chi_r \Sigma_b)$	$V^{-1/2}T_a$	(4.5)
	$V^{-1/2}e^{(\mu)}T_a$		
$f = 0$	$V^{-1/2}e^{(\mu)}T_c$	$V^{-1/2}T_c$	

where $k = 1, 2$; $a = 1 \ldots (N-1)^2$; $b = 1 \ldots (N-1)$; $c = 1 \ldots (N^2-1)$; $r_1 = 0 \ldots (e_1 - 1)$; $r_2 = 0 \ldots (e_2 - 1)$ and $\mu = 1, 2, 3$ or 4.

Before we discuss which modes are physical we work out the index of the appropriate operator to check the index theorem, which has been calculated in the literature for arbitrary base manifold and gauge group [18, 19]. If T is the following operator (in the selfdual sector):

$$T: \delta A_\mu \to (-\tfrac{1}{2}\varepsilon_{\mu\nu\alpha\beta}D_{[\alpha}\delta A_{\beta]} + D_{[\mu}\delta A_{\nu]}, D_\mu \delta A_\mu), \tag{4.6}$$

and if $f_{\mu\nu}$ is an anti-selfdual 2-form with values in the Lie algebra and g also takes values in this algebra, we can define the inner products:

$$(\delta A, \delta A') = \mathrm{Tr} \int_{T_4} (\delta A_\mu^\dagger \delta A'^\mu),$$

$$((f, g), (f'g')) = \mathrm{Tr} \int_{T_4} (\tfrac{1}{4}f_{\mu\nu}^\dagger f'^{\mu\nu} + g^\dagger g'). \tag{4.7}$$

With respect to these inner products we can find the adjoint of T:

$$T^\dagger: (f, g) \to -(D_\mu f_{\mu\nu} + D_\nu g). \tag{4.8}$$

After some calculations one finds

$$T^\dagger T = M_A, \tag{4.9}$$

and

$$TT^\dagger(f,g) = (-D^2 f, -D^2 g) = M_{gh}(f,g). \tag{4.10}$$

On T^4, g and f satisfy the same boundary conditions as δA. So we find:

$$\text{index}\, T = \dim \ker T - \dim \ker T^\dagger = \dim \ker T^\dagger T - \dim \ker TT^\dagger$$
$$= \dim \ker M_A - 4 \dim \ker M_{gh}, \tag{4.11}$$

and with the results from Eq. (4.5) this implies:

$$\text{index}\, T = 4(N-1)\,|\text{Pf}(n)|, \tag{4.12}$$

which is exactly the number of *non-constant* zero modes. This corresponds with the general result for T^4 [18], which easily extends to the twisted case too:

$$\text{index}\, T = 4N|C_2| = 2|P_1|. \tag{4.13}$$

To see which modes are physical we first consider the topologically trivial case with zero twist $(n_{\mu\nu}=0)$ and periodic boundary conditions. There we can easily write down the most general solution [4]:

$$A_\mu = \frac{1}{a_\mu}\text{diag}(\varphi_\mu^{(1)}, \ldots, \varphi_\mu^{(N)}), \qquad \sum_{i=1}^N \varphi_\mu^{(i)} = 0, \tag{4.14}$$

where two solutions A and A' of this form are gauge equivalent (allowing *periodic* gauge transformations only) if and only if for all μ and i

$$\varphi_\mu'^{(i)} = \varphi_\mu^{(\sigma(i))}(\text{mod}\, 2\pi), \tag{4.15}$$

with σ some N-permutation. All solutions are gauge equivalent to the above ones. Wilson loops which wind around the torus should be left invariant under gauge transformations; we therefore allow for periodic gauge transformations only[4]. For almost all solutions (4.14), i.e. $\varphi_\mu^{(i)} \neq \varphi_\mu^{(j)}(\text{mod}\, 2\pi)$ for all $i \neq j$, M_A has $4(N-1)$ and M_{gh} has $(N-1)$ zero modes. Those of M_A are to be treated as collective coordinates. Those of M_{gh} are a consequence of the constant gauge transformations which leave A invariant; they are deleted from the spectrum of M_{gh} (see [18] for details, a factor (volume $H_A)^{-1}$ enters the path integral, where H_A is the group which leaves A invariant).

In the topologically nontrivial case with non-zero twist we do not know the most general solution, but we can at least exhibit the most general *abelian* solution:

$$A_\mu = -\frac{\pi}{N} F_{\mu\nu} x_\nu T + \frac{1}{a_\mu}\text{diag}(\varphi_\mu^{(1)}, \ldots, \varphi_\mu^{(N)}), \qquad \sum_{i=1}^N \varphi_\mu^{(i)} = 0, \tag{4.16}$$

where again two solutions are gauge equivalent if and only if for all μ and i:

$$\varphi_\mu'^{(i)} = \varphi_\mu^{(\sigma(i))}(\text{mod}\, 2\pi). \tag{4.17}$$

---

4   Equivalently, allowed gauge transformations (Ω) have to satisfy $\Omega_\mu(x) = \Omega(x+a^{(\mu)})\Omega_\mu(x)\Omega(x)^{-1} \,\text{mod}\, Z_N$, therefore leaving the boundary conditions invariant

                                                   P. van Baal

However in this case the permutation σ should leave N fixed[5]. This constraint on σ comes from the fact that the allowed gauge transformations have to be periodic *and* leave T fixed. We leave it to the reader to show that also for the solutions in Eq. (4.16) the fluctuation spectrum can be found explicitly. In particular for almost all solutions (4.16), [i.e. $\varphi_\mu^{(i)} \neq \varphi_\mu^{(j)} (\mathrm{mod}\,2\pi)$ for all $i \neq j$], M_A has $4(N-1)\,(|\mathrm{Pf}(n)|+1)$ and M_{gh} has $(N-1)$ zero-modes. The non-constant zero-modes of M_A are again given by θ-functions, this time with shifted arguments and modified bicharacters α.

Let us digress to the case of SU(2). We want to determine the number of parameters for an (anti)-selfdual solution if our abelian solution is not (anti)-selfdual. The number of zero-modes in M_A cannot be smaller than the index of T. If it is larger, $\dim \ker M_{gh}$ is necessarily non-zero. So let us investigate solutions ψ [with values in the algebra of SU(2)] of $D_\mu \psi = 0$ (if there are nontrivial solutions the kernel of M_{gh} is not zero). This obviously implies $[D_\mu, D_\nu]\psi = 0$ or equivalently:

$$[G_{\mu\nu}, \psi] = 0. \tag{4.18}$$

We find therefore in SU(2),

$$G_{\mu\nu}(x) = f_{\mu\nu}(x)\psi(x), \tag{4.19}$$

where $f_{\mu\nu}(x)$ is an (anti)-selfdual 2-form with real values. The condition that $G_{\mu\nu}$ satisfies the Bianchi identity and the Yang-Mills equations of motion imply

$$\partial_\mu f_{\mu\nu} = 0, \qquad \partial_{[\mu} f_{\mu\nu]} = 0. \tag{4.20}$$

So $f_{\mu\nu}$ is the field strength of an abelian gauge field b_μ,

$$f_{\mu\nu} = \partial_\mu b_\nu - \partial_\nu b_\mu. \tag{4.21}$$

Obviously the nonabelian gauge potential

$$A_\mu = b_\mu \psi \tag{4.22}$$

gives rise to (4.19). The general form of A_μ is therefore a gauge transformation Ω of (4.22) which leaves (4.19) fixed, so $\Omega\psi\Omega^{-1} = \psi$. This implies that $\Omega\partial_\mu\Omega^{-1}$ is proportional to ψ and thus it only changes b_μ by an abelian gauge transformation and A_μ is still of the form (4.22). Finally we have to implement the boundary conditions with [see (2.11)]

$$\Omega_\mu = \exp\left(\frac{-\pi i}{2} \sum_\nu \frac{n_{\mu\nu} x_\nu}{a_\nu} \sigma_3\right). \tag{4.23}$$

We find:

$$(b_\mu(x + a^{(\lambda)}) - b_\mu(x))\psi(x) = \frac{\pi}{2}\frac{n_{\mu\lambda}}{a_\lambda}\sigma_3. \tag{4.24}$$

---

5   Wilson loops winding around T^4 have to be invariant. See also footnote 4. The appropriate Wilson loop winding once around the torus in the νth direction is now $\mathrm{Tr}\left(P\exp\left(i\oint_{c_x} A_\mu dx_\mu\right)\Omega_\nu(x)\right)$

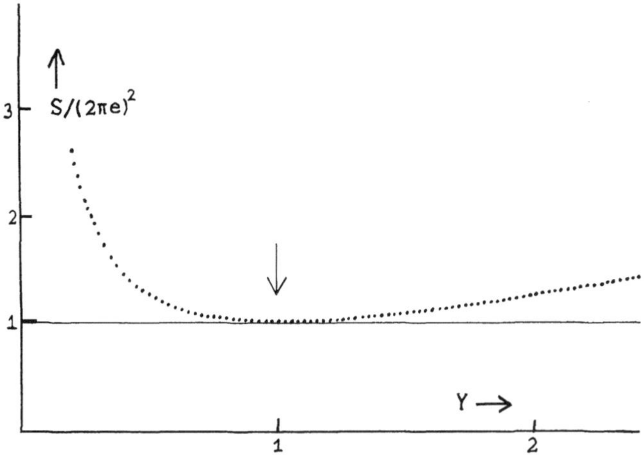

Fig. 1. The action S as a function of y. The dotted line is the abelian and the full line the antiselfdual solution

This is only possible if $\psi(x)$ is a multiple of σ_3, but then A_μ is necessarily of the abelian form [see (4.16)]

$$A_\mu = \left(-\frac{\pi}{2} F_{\mu\nu} x_\nu + \frac{\varphi_\mu^{(1)}}{a_\mu} \right) \sigma_3 . \tag{4.25}$$

Since we considered the case that this solution is not (anti)-selfdual, we conclude that the number of zero modes for M_A, with A (anti)-selfdual is $4(N-1)\,|\mathrm{Pf}(n)|$ and M_{gh} has no zero-modes.

There is a nice intuitive picture which explains the enlargement of the (anti)-selfdual solution manifold for (anti)-selfdual $F_{\mu\nu} = \dfrac{n_{\mu\nu}}{a_\mu a_\nu}$. For convenience fix $n_{13} = n_{24} = e, n_{12} = n_{14} = n_{23} = n_{24} = 0$ and put $y = a_1 a_3/a_2 a_4$ which is variable. For an anti-selfdual solution we must have $y = 1$. The action of the abelian solution in (4.25) is:

$$S_a = 2\pi^2 e^2 (y + y^{-1}) . \tag{4.26}$$

This is depicted in Fig. 1. The unstable solution for $y \neq 1$, joins together with the anti-selfdual solution solution at $y = 1$. At this point the solution manifold is obviously enlarged. As we computed there are $4(N-1)$ extra zero-modes. In [18] one can find that the contribution to the path-integral, in so far as it depends on the coupling constant is

$$g^{-\sigma(F)} \exp\left(\frac{-4\pi^2 |\mathrm{Pf}(n)|}{g^2} \right) , \tag{4.27}$$

where $\sigma(F) = h_1(F) - h_0(F) - (h_1(0) - h_0(0))$, with $h_1 = \dim \ker M_A$ and $h_0 = \dim \ker M_{gh}$. So $\sigma(F) = 4|\mathrm{Pf}(n)|$ if F is (anti)-selfdual and $\sigma(F) = 4|\mathrm{Pf}(n)| - 3$ if F is not. Thus tunneling through (anti)-selfdual solutions in the case that conditions (4.1) are not fulfilled is suppressed by a factor g^3.

412   P. van Baal

5. Discussion

We considered on T^4 abelian solutions of the Yang-Mills equations of motion, satisfying abelian boundary conditions. In Appendix B we prove that this saturates all possibilities for SU(2) (up to a gauge), but for SU(N), $N > 2$, t' Hooft's analysis [5] shows that abelian solutions exist which do not satisfy abelian boundary conditions.

Our computations suggest that the (anti)-selfdual solutions which contain an abelian sector dominate the path integral. Perhaps it is sufficient only to consider these situations. Furthermore in this case the valley in the action functional, describing the local minimum, is widest near the abelian solutions of the form $A_\mu = -\frac{\pi}{N} F_{\mu\nu} x_\nu T$. It is then feasible that one only needs to expand around these solutions. For this situation one can already check that the correct renormalization group behaviour is obtained, by explicitly using the results of Eq. (4.4). Also the quasi-classical expansions for M_A and M_{gh} given in [21] can be used.

However, since the "instantons"[6] on T^4 (after transforming to the $A_0 = 0$ gauge) do not in general represent tunneling between vacuum states, it is not guaranteed that the "instantons" we suspect to dominate the path integral interpolate between states with nonzero overlap with the vacuum states. In that case one is computing energy splits between excited states.

Alternatively one can try to find all euclidean solutions explicitly interpolating between Lüscher's vacuum configurations. This seems as difficult as constructing the general "instantons" on T^4. The subject is presently under investigation.

Finally we hope that our work is also of mathematical interest. It is suggested that θ-functions are the natural objects to construct the "instantons." Since θ-functions are algebraic objects on a torus, an algebraic construction, somewhat similar to S^4 (see [22] for a review), might be possible. Unlike S^4, we have on T^4 a "preferred" complex structure. One can even view our abelian solution in the holomorphic gauge $A_{z_i} = i\pi f_i \bar{z}_i$ as canonically associated with the metric on the U(1)-line bundle, admitting the θ-functions f as sections. The norm n is given by:

$$n(f(z)) = |f(z)| e^{-\frac{\pi}{2} H(z, z)}, \tag{5.1}$$

and the connection in the holomorphic gauge is then found by

$$i A_{z_i} = e^{\frac{\pi}{2} H(z, z)} \frac{\partial}{\partial z_i} e^{-\frac{\pi}{2} H(z, z)}. \tag{5.2}$$

Appendix A

We will discuss the necessary details to prove the following formula [remember that $L(H, \alpha)$ is the linear space of θ-functions of type (H, α), and not the line bundle determined by (H, α)],

$$\dim_{\mathbb{C}} L(H, \alpha) = |\mathrm{Pf}(n)|, \tag{A.1}$$

---

6   Instantons between quotation marks, since we include twist

and to construct explicitly an orthonormal basis for $L(H, \alpha)$. For the time being we assume that H is nondegenerate. An obvious generalization of the intrinsic theta function f_0 (3.24) is:

$$f_t(z) = \sum_{q \in L} \alpha(q) \exp\left(\pi H(z + t, q - t) - \frac{\pi}{2} H(q, q) + \frac{\pi}{2} H(t, t) \right),$$
$$E(t, q) \in \mathbb{Z} \quad \forall q \in L. \tag{A.2}$$

Explicitly the condition for t is:

$$n_{\mu\nu} t_\nu \in \mathbb{Z}. \tag{A.3}$$

Equivalently one can define f_t through:

$$f_t(z) = e^{-\frac{\pi}{2} H(t, t) - \pi H(z, t)} f_0(z + t). \tag{A.4}$$

Therefore t is only defined modulo L. Clearly, t satisfying (A.3) form a lattice L_n which contains L. If (A.1) is correct, then not all these θ-functions can be independent, since L_n/L has $\det(n) = \mathrm{Pf}(n)^2$ points. In the following we will show how to single out the appropriate set.

But first we will discuss the classical θ-functions which will yield a canonical orthonormal base for $L(H, \alpha)$. We remind the reader of the fact that n can be brought in the Frobenius standard form by an $SL(4, \mathbb{Z})$ transformation T [15, p. 71]:

$$n = T^t n^0 T, \qquad n^0 = \begin{pmatrix} & \emptyset & e_1 & 0 \\ & & 0 & e_2 \\ -e_1 & 0 & & \\ 0 & -e_2 & & \emptyset \end{pmatrix} \tag{A.5}$$

$$e_1 = \text{g.c.d.}\,(n_{\mu\nu}), \qquad e_2 = \frac{-\mathrm{Pf}(n)}{e_1} \in e_1 \mathbb{Z}.$$

Let the $\mathbb{Z}$-basis of the lattice L with respects to which n is of this form be $\xi^{(\mu)}$, i.e. $E(\xi^{(\mu)}, \xi^{(\nu)}) = n^0_{\mu\nu}$, or equivalently $\xi^{(\mu)} = T_{\mu\nu} \zeta^{(\nu)}$, and choose a $\mathbb{C}$-basis according to

$$z = \tilde{z}_1 \frac{\xi^{(3)}}{e_1} + \tilde{z}_2 \frac{\xi^{(4)}}{e_2} = U^t \tilde{z}. \tag{A.6}$$

Using $\mathrm{Im}\, H(z, w) = E(z, w)$ and $E(\xi^{(3)}, \xi^{(4)}) = 0$, we find that $\tilde{h} = U^+ h U$ is real and symmetric, so we can define the symmetric $\mathbb{C}$-bilinear form

$$S(z, w) = -\tilde{z} \tilde{h} \tilde{w}. \tag{A.7}$$

Next we introduce the mixed quadratic form,

$$Q(z, w) = H(z, w) + S(z, w), \tag{A.8}$$

                                                            P. van Baal

and transform the θ-function f of type (H, α) to the θ-function θ of type (Q, α):

$$\theta(z) = \exp\left(\frac{\pi}{2} S(z, z)\right) f(z), \tag{A.9}$$

$$\theta(z+q) = \tilde{u}_q(z)\theta(z); \quad q \in L,$$
$$\tilde{u}_q(z) = \exp\left(\frac{\pi}{2} Q(q, q) + \pi Q(z, q)\right)\alpha(q). \tag{A.10}$$

It is obvious that the transformation between (H, α) and (Q, α) is $1-1$. It is introduced because of the property:

$$Q(z, \xi^{(3)}) = Q(z, \xi^{(4)}) = 0, \tag{A.11}$$

or that $Q(z, w)$ vanishes for $\tilde{w}_i$ real. So for $q \in L$ of the form $p_1\xi^{(3)} + p_2\xi^{(4)}$, $\tilde{u}_q(z) = \alpha(q)$, and θ is almost periodic in the real direction. To work this out in more detail, we introduce the so-called characters m and l of α:

$$\alpha(q) = \exp(\pi i B(q, q))\beta(q), \tag{A.12}$$

where $\exp(\pi i B(q, q))$ is a bicharacter as in (2.19) but with respect to the Frobenius bases $(\xi^{(\mu)})$ of L; if we denote the components by $\tilde{q}_\mu(q = \tilde{q}_\mu\xi^{(\mu)})$,

$$B(q, q) = \sum_{i=1}^{2} e_i\tilde{q}_i\tilde{q}_{i+2}. \tag{A.13}$$

$\beta(q)$ is now obviously linear in q and this defines the characters of α:

$$\beta(q) = \exp\left(2\pi i \sum_{k=1}^{2} (m_k\tilde{q}_{k+2} - l_k\tilde{q}_k)\right). \tag{A.14}$$

For α given by (2.19) m and l can always be chosen 0 or $\frac{1}{2}$. With these definitions one easily verifies that the holomorphic function[7]:

$$\theta(z)\exp(-2\pi i m \cdot e^{-1}\tilde{z}) \tag{A.15}$$

is periodic for $\tilde{z}$ real, with periods (e_1, e_2). Therefore θ has the following unique Fourier expansion:

$$\theta(z) = \sum_{r \in \mathbb{Z}^2} c(r)\exp(2\pi i(r+m) \cdot e^{-1}\tilde{z}). \tag{A.16}$$

We still have to satisfy (A.10) for $q = p_1\xi^{(1)} + p_2\xi^{(2)}$,

$$\theta(z+p) = \theta(z)\exp(-2\pi i(p \cdot (z+l) + \tfrac{1}{2}p \cdot \tau p)), \tag{A.17}$$

where we used the following identity:

$$Q(\xi^{(\mu)}, \xi^{(\nu)}) = Q(\xi^{(\nu)}, \xi^{(\mu)}) + 2iE(\xi^{(\mu)}, \xi^{(\nu)}). \tag{A.18}$$

---

7    $e = \begin{pmatrix} e_1 & 0 \\ 0 & e_2 \end{pmatrix}$, $m \cdot \tilde{z} = m_1\tilde{z}_1 + m_2\tilde{z}_2$

Using (A.11) this implies $Q(z, q) = -2ip \cdot z$. On the other hand $E(\xi^{(i)}, \xi^{(j)}) = 0$ $(i, j = 1, 2)$ implies that

$$\tau_{ij} \equiv 2iQ(\xi^{(i)}, \xi^{(j)}) \tag{A.19}$$

is symmetric. τ as defined above is called the period matrix, and one easily verifies that

$$\xi^{(i)} = \sum_{j=1}^{2} \tau_{ij} \frac{\zeta^{(j+2)}}{e_j}, \tag{A.20}$$

and that $\mathrm{Im}\,\tau$ is positive definite,

$$\mathrm{Im}\,\tau = \tilde{h}^{-1}. \tag{A.21}$$

Substituting (A.16) in (A.17) gives a relation between the Fourier coefficients,

$$c(r + ep) = \exp(2\pi i(\tfrac{1}{2}p \cdot \tau p + (r + m) \cdot e^{-1}\tau p + p \cdot l))c(r), \tag{A.22}$$

$$p \in \mathbb{Z}^2$$

and so we find:

$$\theta(z) = \sum_{0 \leq r_i < |e_i|} d(r)\theta_r(z), \tag{A.23}$$

$$\theta_r(z) = \sum_{p \in \mathbb{Z}^2} \exp(\pi i[p + e^{-1}(m + r)] \cdot \tau[p + e^{-1}(m + r)] + 2\pi i(p + e^{-1}(m + r)) \cdot (\tilde{z} + l)).$$

By construction we verified (A.1) since $|\mathrm{Pf}(n)| = |e_1 e_2|$, but furthermore the θ_r form an orthogonal set of θ-functions each with the same length, where the inner product is defined through

$$\langle \theta_1, \theta_2 \rangle = \int_{T^4} d_4 x\, \theta_1(z)\overline{\theta_2(z)} \exp(-\pi \mathrm{Re}(Q(z, z))), \tag{A.24}$$

or if we write things in terms of the original U(1) sections,

$$\chi = \exp\left(-\frac{\pi}{2} Q(z, z)\right)\theta(z), \tag{A.25}$$

it is the standard inner product $\langle \chi_1, \chi_2 \rangle = \int_{T^4} d_4 x\, \chi_1(z)\overline{\chi_2(z)}$. For the proof of

$$\langle \theta_r, \theta_s \rangle = \delta_{rs}\langle \theta_0, \theta_0 \rangle = \delta_{rs}\|\theta_0\|^2 \tag{A.26}$$

we refer the reader to the literature [15, p. 80]. The canonical basis for the zero-modes of M_n is therefore given by the orthonormal set:

$$\chi_r = \exp\left(-\frac{\pi}{2} Q(z, z)\right)\theta_r(z)/\|\theta_0\|, \tag{A.27}$$

$$r = (r_1, r_2), \qquad r_i \in \mathbb{Z}, \qquad 0 \leq r_i < |e_i|,$$

which is still true if H is degenerate $(e_2 = 0)$, e_1 is called the reduced Pfaffian [15][8].

---

8   In (A.6) replace e_2 by 1, then following the analysis one finds (A.23) with all 2 comp. vectors replaced by their first component. So $\theta_r(z)$ is independent of $\tilde{z}_2$

                                                                P. van Baal

There is one disadvantage concerning the above construction and that is its complexity. The intrinsic θ-functions (A.2) are on the other hand in general not orthogonal[9]:

$$\langle f_t, f_s \rangle = \int d_4 x\, f_t(z)\overline{f_s(z)}\, e^{-\pi H(z,z)} = \exp(-\pi i E(t,s)) f_{t-s}(0)/f_1 f_2 , \qquad (A.28)$$

so they all have the same norm, but cannot all be orthogonal. Let us at least constitute a basis for $L(H,\alpha)$, using these intrinsic θ-functions. First we have

$$f_t(z) = e^{-\frac{\pi}{2} S(z,z)} \sum c_{tr} \theta_r(z) . \qquad (A.29)$$

Using (A.4) we find

$$f_t(t) = e^{-\frac{\pi}{2} Q(t,t) - \pi Q(z,t) - \frac{\pi}{2} S(z,z)} \sum c_{or} \theta_r(z+t) . \qquad (A.30)$$

A set of $|\mathrm{Pf}(n)|$ distinct $t \in L_n/L$, such that $Q(\cdot, t) = 0$ is given by

$$t = t_1 \frac{\zeta^{(3)}}{e_2} + t_2 \frac{\zeta^{(4)}}{e_2} ; \qquad 0 \leq t_i < |e_i| , \quad t_i \in \mathbb{Z} . \qquad (A.31)$$

With (A.23) this implies:

$$f_t(z) = e^{-\frac{\pi}{2} S(z,z)} \sum c_{or} e^{2\pi i (m+r)\cdot e^{-1} \tilde{t}} \theta_r(z) . \qquad (A.32)$$

And so

$$c_{tr} = \exp(2\pi i (m+r) \cdot e^{-1}\tilde{t}) c_{or} , \qquad (A.33)$$

$f_t(z)$ form a basis for $L(H,\alpha)$ iff the square metric c_{tr} in (A.29) is nonsingular. Or

$$\det c = \exp\left(\pi i \sum_k m_k(e_k - 1)\right) \prod_r c_{or} \prod_{e_1 > k > l \geq 0} (e^{2\pi i k/e_1} - e^{2\pi i l/e_1})$$

$$\prod_{|e_2| > k > l \geq 0} (e^{2\pi i k/e_2} - e^{2\pi i l/e_2}) \neq 0 . \qquad (A.34)$$

So $c_{or} \neq 0$ $\forall r$ is necessary and sufficient. We leave this to the reader to verify [16, Sect. 4].

Note that $\langle f_t, f_t \rangle = \sum |c_{or}|^2 \|\theta_0\|^2$ independent of t, consistent with (A.28). Using (A.4) only, a somewhat simpler version of (A.28) can be established:

$$\langle f_t, f_s \rangle = e^{-\pi i E(t,s)} \langle f_{t-s}, f_0 \rangle . \qquad (A.35)$$

Also with (A.4) we have $f_{t-s}(0) = e^{-\frac{\pi}{2} H(t-s, t-s)} f_0(t-s)$. So knowledge of $f_0(t)$ enables us to find with the Gramm-Schmidt procedure an orthonormal basis, using (A.28).

We will end this appendix by mentioning a simple consequence of (A.28): All f orthogonal to f_0 have to vanish in $z=0$, and therefore on the whole of L.

---

9   (A.28) is found by explicitly working out the double sums, and using periodicity of the integrand

Appendix B

In this appendix we will answer two questions about $SU(N)/Z_N$ fiber bundles over T^4.

(i) Given a solution of the Yang-Mills equation ($D_\mu G_{\mu\nu}=0$) with constant curvature, is there a gauge in which the boundary conditions are of the abelian type?

(ii) For which values of the twist tensor $n_{\mu\nu}$ and Pontryagin-index P is there a gauge in which the boundary conditions are of the abelian type?

Let us first consider SU(2) and neglect the boundary conditions. In that case it was shown by Leutwyler [7] that all constant curvature solutions are of the abelian type up to a gauge

$$A_\mu = -\tfrac{1}{2}G_{\mu\nu}x_\nu; \qquad G_{\mu\nu} = \pi F_{\mu\nu}\sigma_3 . \tag{B.1}$$

Now we impose the boundary conditions and write

$$\Omega_\nu(x) = \exp\left(\frac{\pi i}{2}\sum_\mu x_\mu F_{\mu\nu}a_\nu\sigma_3\right)\omega_\nu(x) . \tag{B.2}$$

From Eq. (2.3) one deduces

$$D_\mu\omega_\nu = 0 . \tag{B.3}$$

With a similar observation as in (4.18) we find

$$(D_\mu D_\lambda - D_\lambda D_\mu)\omega_\nu = i[G_{\mu\lambda}, \omega_\nu] = 0 . \tag{B.4}$$

So ω_ν commutes with σ_3. Combining with (B.3) this implies that ω_ν is a *constant* gauge function of the form:

$$\omega_\nu(x) = \exp(i\varphi_\nu\sigma_3); \qquad \varphi_\nu \text{ constant.} \tag{B.5}$$

Finally we transform ω_ν to the identity by the gauge transformation

$$\Omega(x) = \exp\left(-i\sum_\mu \frac{x_\mu\varphi_\mu}{a_\mu}\sigma_3\right), \tag{B.6}$$

under which A_μ changes into

$$A_\mu = -\frac{\pi}{2}F_{\mu\nu}x_\nu\sigma_3 + \frac{\varphi_\mu}{a_\mu}\sigma_3 , \tag{B.7}$$

which is exactly the general abelian solution for SU(2) [see (4.14) and (4.16)]. Note that the cocycle condition forces $F_{\mu\nu}$ to be of the form $n_{\mu\nu}/a_\mu a_\nu$.

For SU(N) ($N>2$) we assume the constant curvature solution to be abelian [for SU(2) this is automatically satisfied. Whether this is also true for SU(N), $N>2$, is not relevant for the point we want to make],

$$[G_{\mu\nu}, G_{\lambda\sigma}] = 0 . \tag{B.8}$$

In a suitable gauge we have

$$A_\mu = -\tfrac{1}{2}G_{\mu\nu}x_\nu , \tag{B.9}$$

P. van Baal

with boundary conditions

$$\Omega_\nu(x) = \exp\left(\frac{i}{2}\sum_\mu x_\mu G_{\mu\nu} a_\nu\right)\omega_\nu.\tag{B.10}$$

Again ω_ν is a constant gauge function satisfying

$$[\omega_\nu, G_{\mu\lambda}] = 0.\tag{B.11}$$

Imposing the cocycle condition gives

$$\exp(ia_\mu G_{\mu\nu} a_\nu)\,[\omega_\nu, \omega_\mu] = \exp(2\pi i n_{\mu\nu}/N).\tag{B.12}$$

There is a gauge transformation which brings ω_ν to the identity for all ν, if and only if $[\omega_\mu, \omega_\nu] = 1$. The possibility to have $[\omega_\mu, \omega_\nu] \neq 1$ enabled 't Hooft to construct a constant curvature solution with Chern index $1/N$ [5]. So our first question can only be answered by yes for SU(2). Note that for $[\omega_\mu, \omega_\nu] = 1$ the generalization following Eq. (2.27) is applicable with $A_\mu = A_\mu^0 + \dfrac{1}{a_\mu}\mathrm{diag}(\varphi_\mu^{(1)}...\varphi_\mu^{(N)}); \; \sum \varphi_\mu^{(i)} = 0$ [compare (2.33)] as a general solution.

The second question will be considered for SU(2) only (see [12, Lemma 3.2], also for $N > 2$). The abelian boundary conditions (4.23) uniquely fix C_2 to be $\frac{1}{2}\mathrm{Pf}(n)$. Therefore if $n = 0 \bmod 2$ (no twist) C_2 is always even. There is no constant curvature solution with odd Chern index. For unit Chern index there seems even to be an obstruction for the existence of any solution satisfying the duality equations on T_4 [24]. For C_2 even, there certainly are (anti)-selfdual solutions however. Finally if the twist is non-zero $(n \neq 0 \bmod 2)$ it is not hard to see that each value of C_2 compatible with the given twist can be reached. So to answer the second question: Only boundary conditions yielding $n_{\mu\nu} = 2m_{\mu\nu}$ and $P_1 = 4(2k+1)$, with arbitrary m and k, are not gauge equivalent to abelian boundary conditions.

Acknowledgements. I thank Gerard 't Hooft for a critical reading of the manuscript and for discussions. Furthermore a discussion with C. Taubes is acknowledged. I am deeply indepted to Bert van Geemen who introduced me to the secrets of θ-functions. I also thank him for his continued interest, discussions and suggestions. This work is part of the research programme of the "Stichting voor Fundamenteel Onderzoek der Materie (FOM)," which is financially supported by the "Nederlandse Organisatie voor Zuiver Wetenschappelijk Onderzoek (ZWO)."

References

1. Atiyah, M.F., Drinfield, V.G., Hitchin, N.J., Manin, Yu.I.: Construction of instantons. Phys. Lett. **65** A, 185 (1978)
2. 't Hooft, G.: A property of electric and magnetic flux in non-abelian gauge theories. Nucl. Phys. B **153**, 141 (1979)
3. Coleman, S.: The uses of instantons. In: Zichini, A.: Proceedings of the 1977 International School of Subnuclear Physics, Erice. New York: Plenum Press 1979
4. Lüscher, M.: Some analytical results concerning the mass spectrum of Yang-Mills gauge theories on a torus. Nucl. Phys. B **219**, 233 (1983);
   Lüscher, M., Münster, G.: Weak coupling expansion of low-lying energy values in the SU(2) gauge theory on a torus. Nucl. Phys. B **232**, 445 (1984)
5. 't Hooft, G.: Some twisted self-dual solutions for the Yang-Mills equations on the hypertorus. Commun. Math. Phys. **81**, 267 (1981)

6. Bourguignon, J.P., Lawson, H.B.: Stability and isolation phenomena for Yang-Mills fields. Commun. Math. Phys. **79**, 189 (1981)

7. Leutwyler, H.: Vacuum fluctuations surrounding soft gluon fields. Phys. Lett. **96**B, 154 (1980);
Leutwyler, H.: Constant gauge fields and their quantum fluctuations. Nucl. Phys. B **179**, 129 (1981)

8. Ambjørn, J., Olesen, P.: A color magnetic condensate in QCD. Nucl. Phys. B **170** [FS1], 265 (1980)
A review with further references:
Olesen, P.: On the QCD Vacuum. Physica Scripta **23**, 1000 (1981)

9a. Richard, J.L., Rouet, A.: The CP_1 model on the torus: contribution of instantons. Nucl. Phys. B **211**, 447 (1983)

9b. Erdélyi, A., et al.: Higher trancendental functions, Vol. 2, Chap. 13. New York: McGraw-Hill 1953

10. Gürsey, F., Tze, H.: Complex and quaternionic analyticity in chiral and gauge theories. Ann. Phys. (N.Y.) **128**, 29 (1980)

11. 't Hooft, G.: Computation of quantum effects due to a four-dimensional pseudoparticle. Phys. Rev. D **14**, 3432 (1976)

12. Van Baal, P.: Some results for SU(N) gauge-fields on the hypertorus. Commun. Math. Phys. **85**, 529 (1982)

13. Goddard, P., Nuyts, J., Olive, D.: Gauge theories and magnetic charge. Nucl. Phys. B **125**, 1 (1977)

14. Humphreys, J.E.: Introduction to Lie algebras and representation theory. Berlin, Heidelberg, New York: Springer 1972

15. Igusa, J.: Theta functions. Berlin, Heidelberg, New York: Springer 1972

16. Shimura, G.: Theta functions with complex multiplication. Duke Math. J. **43**, 673 (1976)

17. Hano, J., et al.: Manifolds and Lie groups. In: Progress in Mathematics, Vol. 14, p. 109. Boston, Basel, Stuttgart: Birkhäuser 1981

18. Schwarz, A.S.: Instantons and fermions in the field of instanton. Commun. Math. Phys. **64**, 233 (1979)

19. Schwarz, A.S.: On regular solutions of euclidean Yang-Mills equations. Phys. Lett. **67**B, 172 (1977)

20. Bernard, C.W., Christ, N.H., Guth, A.M., Weinberg, E.J.: Pseudo particle parameters for arbitrary gauge groups. Phys. Rev. D **16**, 2967 (1977)

21. Dyakonov, D.I., Petrov, V.Yu., Yung, A.V.: Quasiclassical expansion of Yang-Mills heat kernels and approximate calculation of functional determinants. Phys. Lett. **130**B, 385 (1983)

22. Atiyah, M.F.: Geometry of Yang-Mills fields, Fermi Lectures. Pisa: Scuola Normale Superiore 1979

23. Taubes, C.H.: Stability in Yang-Mills theories. Commun. Math. Phys. **91**, 235 (1983)

24. Taubes, C.H.: Self-dual connections on 4-manifolds with indefinite intersection matrix. J. Diff. Geom. (to appear) Berkeley Preprint 1984

Communicated by A. Jaffe

Received March 2, 1984

PHYSICS   Proceedings B 89 (1), March 24, 1986

The group theory of twist eating solutions

by Bert van Geemen[1] and Pierre van Baal[2]

[1] *Department of Mathematics, Budapestlaan 6, P.O. Box 80.010, 3508 TA Utrecht, the Netherlands*
[2] *Institute for Theoretical Physics, State University of New York at Stony Brook, Long Island, New York 11794, USA*

Communicated by Prof. G. 't Hooft at the meeting of September 30, 1985

ABSTRACT

In this note we show the relation between solutions to the equation $[\Omega_\mu, \Omega_\nu] = \exp(2\pi i n_{\mu\nu}/N)I$ and the representations of the Heisenberg group, where $\mu = 1,2,\ldots,2g$ and $\Omega_\nu \in SU(N)$. We construct all its irreducible representations and whence all solutions to the above equation for arbitrary g. We give a criterium for existence and uniqueness of solutions.

1. THE SETTING OF TWIST EATING SOLUTIONS

Twisted gauge fields on the hypertorus, both in the continuum and on the lattice posed an interesting mathematical problem, namely finding $SU(N)$ matrices Ω_μ (called twist eating solutions), such that:

$$(1) \qquad \begin{cases} [\Omega_\mu, \Omega_\nu] = \Omega_\mu \Omega_\nu \Omega_\mu^{-1} \Omega_\nu^{-1} \\[2mm] \qquad = \exp(2\pi i n_{\mu\nu}/N) \cdot I \end{cases}$$

n is called the twist tensor, it is skew symmetric with integer entries mod N. The index μ runs from 1 to $2g$ (the dimension of space-time; odd dimensions need not be considered separately). For details and further references see [1, 2] where the full solution of this problem for $g \leq 2$ was found.

By means of a $Sl(2g, \mathbb{Z})$ transformation X, we can always transform n to its standard form:

$$(2) \qquad n^{(0)} = \begin{pmatrix} 0 & & & & e_1 & \\ & & & & & \ddots \\ & & & & & & \ddots \\ -e_1 & & & & & & & e_g \\ & \ddots & & & & & \\ & & \ddots & & & & \\ & & & -e_g & & & & 0 \end{pmatrix}$$

where[1] $e_1|e_2|\ldots|e_g$ and $n = X^t n^{(0)} X$. If $[U_\mu, U_\nu] = \exp(2\pi i n^{(0)}_{\mu\nu}/N)I$, then equation (1) is solved by

$$(3) \qquad \Omega_\mu = \prod_\nu U_\nu^{X_{\nu\mu}}.$$

From now on we assume n to be in the form (2).

This form is not unique since we can add a multiple of N to $n_{\mu\nu}$. However since eq. (3) can be inverted [2], the specific choice of $n^{(0)}$ is irrelevant.

Previously we established that solutions exist iff[2] $Pf(n/N)N$ is an integer, for $g = 1, 2$ (where Pf is the Pfaffian which is a squareroot of the determinant of an even dimensional skew symmetric matrix; $Pf(n^{(0)}) = -\prod_{j=1}^{g}(-e_j)$). In this note we will show that for arbitrary g, $Pf(n/N)N \in \mathbb{Z}$ is a necessary but *not a sufficient* condition for existence of solutions to eq. (1), consistent with the remark in [1] that $Pf(n) = 0 \bmod N$ is not a sufficient condition.

It was also shown for $g \leq 2$ that the solution to equation (1) is unique up to a similarity transformation and multiplication with an element of the centre of $SU(N)$ iff[2]:

$$(4) \qquad \text{g.c.d.}\left(n_{\mu\nu}, Pf\left(\frac{n}{N}\right)N, N\right) = 1.$$

For $g > 2$ this is a sufficient, but *not a necessary* condition. We will show that uniqueness up to a similarity transformation means that the matrices U generate an irreducible representation of the Heisenberg group, in which case there are $N^{2(g-1)}$ $SU(N)$-inequivalent solutions.

It is clear that the group G generated by U has the property that its commutator is in the centre of the group G (G is therefore nilpotent). This is typical of the so-called Heisenberg group. (Indeed also in physics the Heisenberg commutation relation between coordinates and their canonical momenta satisfy the same property).

In the next section we will describe the relevant Heisenberg group and its Schrödinger representation. In section 3 all representations of the Heisenberg group are classified. In section 4 these results are used to construct all solutions to (1) and give the approporiate criteria for existence and uniqueness, based on n.

---

[1] For integer p and q the symbol $p|q$ means that p divides q.
[2] iff = if and only if, g.c.d. = greatest common divisor.

2. THE HEISENBERG GROUP AND ITS CANONICAL REPRESENTATION

In general one can describe the Heisenberg group by the following properties. Let K be an (additive abelian) group, $\mathbb{C}^* = \mathbb{C} - \{0\}$ the multiplicative group of complex numbers and K^* the dual of K, i.e. K^* is the (additive) group of homomorphisms $f : K \to \mathbb{C}^*$. We will denote its elements by x^*. The Heisenberg group is given by $H = \mathbb{C}^* \times K \times K^*$, with a product defined by:

$$(5) \quad \begin{cases} H \times H \to H : (t, x, \cdot x^*) \cdot (s, y, y^*) \\ \qquad = (tsy^*(x), x + y, x^* + y^*). \end{cases}$$

For completeness let us mention the following properties

$$(6) \quad \begin{cases} x^*(a + b) & = x^*(a) \cdot x^*(b) \\ (x^* + y^*)(a) & = x^*(a) \cdot y^*(a) \\ (t, x, y^*)^{-1} & = (t^{-1}y^*(x), -x, -y^*) \\ [(t, x, y^*), (s, u, v^*)] = (v^*(x) \cdot y^*(u)^{-1}, 0, 0). \end{cases}$$

In principle we can restrict ourselves to the subgroup

$$\{t \in \mathbb{C}^* | \exists x \in K, \, y^* \in K^* : t = y^*(x)\} \text{ of } \mathbb{C}^*.$$

When K is finite this makes H a finite group, which is convenient for finding all its representations. In fact when K is finite we must have that $y^*(x)$ is a root of unity for all x and y^*.

We will now introduce the Heisenberg group $H(\delta)$, of type δ, where:

$$(7) \quad \begin{cases} \delta = (d_1, d_2, \ldots, d_g) \\ d_i \in \mathbb{N}, \, d_1 | d_2 \ldots | d_g. \end{cases}$$

The length or norm of δ is given by:

$$|\delta| = d_g.$$

To δ we associate the group $K(\delta)$:

$$(8) \qquad K(\delta) = Z_{d_1} \times Z_{d_2} \times \ldots Z_{d_g}$$

where Z_n stands for $\mathbb{Z}/n\mathbb{Z}$, being the additive group of integers modulo n. To distinguish this from the unimodular group of the n-th roots of unity the latter will be denoted by μ_n, which is a multiplicative group:

$$(9) \quad \begin{cases} \mu_n = \{1, e^{2\pi i/n}, e^{4\pi i/n}, \ldots, e^{2\pi i(n-1)/n}\} \\ \qquad = \{t \in \mathbb{C}^* | t^n = 1\}. \end{cases}$$

The centre of $SU(N)$ is also denoted by μ_N. We define $H(\delta)$ by:

$$H(\delta) = \mu_{|\delta|} \times K(\delta) \times K(\delta)^*.$$

Let $q : \mathbb{Z}^g \to K(\delta)$ be the canonical projection.

Define an isomorphism $K(\delta) \to K(\delta)^*, x \to *(x)$ by:

$$(10) \qquad *(q(x))(q(y)) = \prod_{j=1}^{g} \exp\left(\frac{2\pi i x_j y_j}{d_j}\right)$$

$$\left(\text{so } [(t, q(x), *(q(y))), (s, q(u), *(q(v)))] = \left(\exp\left(2\pi i \sum_k \frac{v_k x_k - y_k u_k}{d_k} \right), 0, 0 \right) \right).$$

We will next construct the canonical representation of $H(\delta)$, called the Schrödinger representation, which is denoted by σ_δ. The representation space will be the $\mathbb{C}$-vector space of functions $f : K(\delta) \to \mathbb{C}$. Define an action of $H(\delta)$ on this vector space of dimension $N = \prod_{j=1}^{g} d_j$ by:

$$(11) \qquad \{[\sigma_\delta(t, x, y^*)](f)\}(z) = t y^*(z) f(x + z).$$

Let us construct explicitely the N-dimensional unitary matrix representation of σ_δ. A $\mathbb{C}$-basis of Func $(K(\delta) \to \mathbb{C})$ is given by

$$(12) \qquad \begin{cases} f_a(x) = 1 & x = a, \quad x, a \in K(\delta) \\ \quad\ \ = 0 & x \neq a \end{cases}$$

or equivalently

$$(13) \qquad \begin{cases} f_{q(a)}(q(x)) = 1 & x_j = a_j \bmod d_j \\ \qquad\quad\ \ = 0 & x_j \neq a_j \bmod d_j \end{cases} \quad \text{(all } j).$$

In an obvious notation: $f_a(x) = \delta_{a, x}$.

It is easy to check that the matrix $\sigma_\delta(t, x, y^*)_{ab}$ is defined by:

$$(14) \qquad [\sigma_\delta(t, x, y^*)](f_b) = \sum_{a \in K(\delta)} \sigma_\delta(t, x, y^*)_{ab} f_a$$

i.e.

$$(15) \qquad \begin{cases} [\sigma_\delta((t, x, y^*) \cdot (s, u, v^*))](f_b) = \\ \quad \sum_{a, c \in K(\delta)} [\sigma_\delta(t, x, y^*)]_{c, a} [\sigma_\delta(s, u, v^*)]_{a, b} f_c. \end{cases}$$

To write down the matrices we use:

$$(16) \qquad (t, q(x), *(q(y))) = (t, 0, 0) \cdot \prod_{j=1}^{g} (1, 0, u_j^*)^{y_j} \cdot \prod_{j=1}^{g} (1, u_j, 0)^{x_j}$$

with

$$(17) \qquad (u_j)_k = \delta_{j, k}.$$

42

If we introduce:

$$(18) \qquad U_{g-j+1} = \hat{U}_j = \sigma_\delta(1, 0, u_j^*), \quad U_{2g-j+1} = \hat{U}_{g+j} = \sigma_\delta(1, u_j, 0)$$

we have (from now on we ignore the difference between x and $q(x)$):

$$(19) \qquad \sigma_\delta(t, x, y) = t \prod_{j=1}^{g} \hat{U}_j^{y_j} \prod_{j=1}^{g} \hat{U}_{g+j}^{x_j}.$$

Explicitly we have:

$$(20) \qquad \begin{cases} (U_{g-k+1})_{a,b} = \exp\left(\dfrac{2\pi i a_k}{d_k}\right)\delta_{a,b} \\[2ex] (U_{2g-k+1})_{a,b} = \delta_{a+u_k,b}. \end{cases}$$

Or as tensor products we can write:

$$(21) \qquad \begin{cases} U_{g-k+1} = 1_{d_1} \otimes \ldots \otimes Q_{d_k} \otimes \ldots \otimes 1_{d_g} \\[2ex] U_{2g-k+1} = 1_{d_1} \otimes \ldots \otimes P_{d_k} \otimes \ldots \otimes 1_{d_g} \end{cases}$$

where 1_n is the n-dimensional identity and Q_n, P_n are the twist matrices satisfying

$$(22) \qquad [P_n, Q_n] = e^{2\pi i/n} \cdot 1_n$$

with

$$Q_n = \text{diag}\,(1, e^{2\pi i/n}, \ldots, e^{2\pi i(n-1)/n})$$

$$(23) \qquad P_n = \begin{pmatrix} 0 & 1 & & & 0 \\ \vdots & 0 & \cdot & & \\ \vdots & & \cdot & \cdot & \\ \vdots & & & \cdot & 1 \\ 1 & \ldots & & & 0 \end{pmatrix}.$$

It is an easy exercise to show that this representation is irreducible. Furthermore we have that U_μ satisfy eq. (2) with:

$$(24) \qquad e_j = -\frac{N}{d_{g-j+1}}, \quad N = \prod_{K=1}^{g} d_k.$$

Note that $Pf(n/N) = -N^{-1}$; therewith we verified for this particular representation: $Pf(n/N)N$ is integer and g.c.d. $(n_{\mu\nu}, N, Pf(n/N)N) = 1$.

3. ALL REPRESENTATIONS OF THE HEISENBERG GROUP $H(\delta)$

In order to find all representations, we can invoke the well known result [3] due to Frobenius for a finite group H:

$$(25) \qquad \sum_\varrho (\dim \varrho)^2 = 0(H)$$

where the summation runs over the irreducible representations ϱ and $0(H)$ is the order (number of elements) of H. We have:

$$(26) \qquad 0(H(\delta)) = |\delta|(\prod_{j=1}^{g} d_j)^2.$$

Denote by $C(\delta)$ the centre of $H(\delta)$

$$(27) \qquad C(\delta) = \{(t,0,0)|t \in \mu_{|\delta|}\}$$

and let ϱ be some irreducible representation of $H(\delta)$ on a vector space V (dim ϱ: = dim V). Then by Schur's lemma:

$$(28) \qquad \varrho(c) = \lambda_{\varrho}(c)I, \qquad \forall c \in C(\delta)$$

with I the identity on V and $\lambda_{\varrho}(c) \in \mathbb{C}$.

Obviously $\lambda_{\varrho}: C(\delta) \to \mathbb{C}$ is a homomorphism which we will call the central character (of ϱ). Note that if two irreducible representations ϱ and ϱ' satisfy $\lambda_{\varrho} \neq \lambda_{\varrho'}$, they can not be equivalent. Since $C(\delta)$ is cyclic of order $|\delta|$ we must have:

$$(29) \qquad \lambda_{\varrho}((t,0,0)) = t^m, \qquad 0 \leq m \leq |\delta| - 1.$$

The following, so-called twisted Schrödinger representation has this central character:

$$(30) \qquad \{[\sigma_{\delta}(m)(t,x,y^*)](f)\}(z) = [ty^*(z)]^m f(x+z).$$

Following the same steps as in the previous section we find the matrix representation of $\sigma_{\delta}(m)$ by:

$$(31) \qquad \begin{cases} U_{g-k+1} = \sigma_{\delta}(m)(1,0,u_k^*) = 1_{d_1} \otimes \ldots \otimes Q_{d_k}^m \otimes \ldots \otimes 1_{d_g} \\ U_{2g-k+1} = \sigma_{\delta}(m)(1,u_k,0) = 1_{d_1} \otimes \ldots \otimes P_{d_k} \otimes \ldots \otimes 1_{d_g}. \end{cases}$$

An invariant subspace W is found by considering each tensor component separately. For the k-th component we have $(z = (1,1,1\ldots1) \in \mathbb{C}^{d_k})$

$$(32) \qquad W_k = \langle z, Q_{d_k}^m z, Q_{d_k}^{2m} z, \ldots, Q_{d_k}^{m(d_k-1)} z \rangle.$$

Using the explicit diagonal form of Q_{d_k} in eq. (23) we find[3]:

$$(33) \qquad \dim W_k = d_k / \text{g.c.d.}(m, d_k).$$

On the other hand, when g.c.d. $(m, d_k) = 1$ the eigenvalues of $Q_{d_k}^m$ are a permutation of those of Q_{d_k}, which implies that there is no non-trivial invariant subspace.

Let us introduce the notation:

$$(34) \qquad \begin{cases} \pi = (p_1, p_2, \ldots, p_g), \qquad p_k = \text{g.c.d.}(m, d_k) \\ \gamma = (c_1, c_2, \ldots, c_g), \qquad c_k = d_k / p_k. \end{cases}$$

---

[3] For $m = 0$ we define g.c.d. $(m, d_k) = d_k$.

From eq. (7) we have that $p_g = 1$ implies that all other p_k equal 1. We conclude: $\sigma_\delta(m)$ is irreducible iff m and $|\delta|$ are relatively prime.

Assume now that ϱ is an irreducible representation of $H(\delta)$ with central character $\lambda_\varrho(t) = t^m$ and with g.c.d. $(m, |\delta| = d_g) \neq 1$. Define π, γ as above and

$$(35) \qquad C(\delta, m) = (\mu_{|\delta|}, \gamma K(\delta), (\gamma K(\delta))^*),$$

where we used the shorthand notation:

$$(36) \qquad \gamma K(\delta) = c_1 Z_{d_1} \times c_2 Z_{d_2} \times \ldots \times c_g Z_{d_g}.$$

Note that elements of $\varrho(C(\delta, m))$ commute with $\varrho(H(\delta))$, hence by Schur's lemma $\varrho(C(\delta, m))$ consists of scalar multiples of the identity. Therefore ϱ restricted to $C(\delta, m)$ is given by ($a, b \in K(\pi)$ and $\gamma \cdot x := (c_1 x_1, \ldots, c_g x_g) \in K(\delta)$):

$$(37) \qquad \varrho(t, \gamma \cdot x, \gamma \cdot y^*) = t^m \chi_{a,b}(x, y) I$$

with

$$(38) \qquad \chi_{a,b}(x, y) = \exp\left(2\pi i \sum_k \frac{x_k a_k + y_k b_k}{p_k}\right).$$

Clearly two irreducible representations with different m, a, b would be inequivalent.

Let us now consider the case where $a = b = 0$. Then ϱ is trivial on the subgroup $C_0(\delta, m)$ of $H(\delta)$:

$$(39) \qquad C_0(\delta, m) = ((\mu_{|\delta|})^{|\gamma|}, \gamma K(\delta), (\gamma K(\delta))^*).$$

Therefore ϱ factors over $C_0(\delta, m)$:

$$(40)$$

$$
\begin{array}{ccc}
H(\delta) & \xrightarrow{\ \varrho\ } & V \\
\downarrow & \nearrow_{\bar\varrho} & \\
H(\delta)/C_0(\delta, m) & &
\end{array}
$$

$\bar\varrho$ is a faithfull representation of $H(\delta)/C_0(\delta, m)$.

On the other hand $H(\delta)/C_0(\delta, m)$ is isomorphic to $H(\gamma)$, which is most easily seen by defining a homomorphism $\phi_\gamma^\delta : H(\delta) \to H(\gamma)$ whose kernel coincides with $C_0(\delta, m)$:

$$(41) \qquad \phi_\gamma^\delta : (t, x, y^*) \to (t^{|\pi|}, \psi(x), \psi^*(y^*))$$

with

$$(42) \qquad
\begin{cases}
\bar x_k &= x_k \bmod c_k \\[1mm]
\psi_k(x) &= (p_k^{-1}|\pi|)\bar x_k \\[1mm]
\psi^*(y^*)(\bar x) &= \exp\left(2\pi i \sum_k \frac{y_k x_k}{p_k^{-1} d_k}\right).
\end{cases}
$$

(Note that: i) $\psi^*(y^*) = \bar{y}^*$ in an obvious way, however one should be careful with this notation when one identifies $q(x)$ with x, ii) $\psi : K(\delta) \to K(\gamma)$ is surjective because $p_i^{-1}|\pi|$ and c_i are relatively prime iii) $\psi(y \cdot x) = 0$ and $\psi^*(y \cdot y^*) = 0$.)

It is now obvious that an irreducible representation with central character t^m is given by: $\varrho_\gamma^\delta = \sigma_\gamma(m/|\pi|) \circ \phi_\gamma^\delta$

$$(43) \qquad \begin{array}{ccc} H(\delta) & \xrightarrow{\;\varrho_\gamma^\delta\;} & V \\[1em] {\scriptstyle\phi_\gamma^\delta}\downarrow & \nearrow {\scriptstyle\sigma_\gamma\left(\frac{m}{|\pi|}\right)} & \\[1em] H(\gamma) & & \end{array}$$

Explicitly (with $f \in \mathrm{Func}(K(\gamma), \mathbb{C})$, $\bar{z} \in K(\gamma)$):

$$(44) \qquad (\varrho(t, x, y^*)\bar{f})(\bar{z})) = (t^{|\pi|}\psi^*(y^*)(\bar{z}))^{m/|\pi|}\bar{f}(\psi(x) + \bar{z}).$$

Finally for $a, b \neq 0$ we can extend $\chi_{a,b}$ uniquely to a 1-dimensional representation of $H(\delta)$, with (necessarily) $m = 0$:

$$(45) \qquad \tilde{\chi}_{a,b}(t, x, y) = \exp\left(2\pi i \sum_k \frac{x_k a_k + y_k b_k}{d_k}\right).$$

The product representation:

$$(46) \qquad \varrho = \tilde{\chi}_{a,b} \cdot \varrho_\gamma^\delta$$

is therefore an irreducible representation for each m, a and b.

We have:

$$(47) \qquad \left\{ \begin{array}{l} \displaystyle\sum_{m=0}^{|\delta|-1} \sum_{a,b \in K(\pi)} \dim(\tilde{\chi}_{a,b} \cdot \varrho_\gamma^\delta)^2 \\[1.5em] = \displaystyle\sum_{m=0}^{|\delta|-1} (\prod_{k=1}^g p_k)^2 (\prod_{k=1}^g c_k)^2 = |\delta| \prod_{j=1}^g d_j^2. \end{array} \right.$$

Using (25) and (26) we see that we have found all irreducible representations. They are in essence all twisted Schrödinger representations.

4. BACK TO THE TWIST

Let us first write down the matrices U_μ for ϱ_γ^δ:

$$(48) \qquad \left\{ \begin{array}{l} U_{g-k+1} = \varrho_\gamma^\delta(1, 0, u_k^*) = 1_{c_1} \otimes \ldots \otimes Q_{c_k}^{m/|\pi|} \otimes \ldots \otimes 1_{c_g} \\[1em] U_{2g-k+1} = \varrho_\gamma^\delta(1, u_k, 0) = 1_{c_1} \otimes \ldots \otimes P_{c_k}^{|\pi|/p_k} \otimes \ldots \otimes 1_{c_g} \end{array} \right.$$

which satisfy eq. (2) with:

$$(49) \qquad e_{g-k+1} = -N\frac{m/p_k}{c_k} = -\frac{Nm}{d_k}, \qquad N = \prod_{k=1}^g c_k : = N_{irr}.$$

Furthermore we have:

$$(50) \qquad Pf\left(\frac{n}{N}\right)N = - \prod_{k=1}^{g} \left(\frac{m}{p_k}\right) \in \mathbb{Z}.$$

Clearly $m_k \equiv m/p_k$ is relatively prime to c_k. Since $c_i | c_k$ for $i < k$, m_k is also relative prime to c_i.

For $g \le 2$ this is easily seen to imply g.c.d. $(n_{\mu\nu}, NPf(n/N), N) = 1$. However, the following example, due to Coste, shows that it is not a necessary condition for irreducibility. Take:

$$(51) \qquad \begin{cases} g = 3, \; N = 2^2 \cdot 7^3 \\ e_3 = 4e_2 = 4e_1 = 2^3 \cdot 3 \cdot 7^2. \end{cases}$$

For this, one explicitly verifies that:

$$(52) \qquad \text{g.c.d.} \left(n_{\mu\nu}, N \cdot Pf\left(\frac{n}{N}\right), N\right) = 2 \ne 1,$$

but that it nevertheless admits a solution which is based on the irreducible representation $\sigma_\delta(m)$ with:

$$(53) \qquad \delta = (7, 28, 28), \quad m = 6.$$

To show that $N \cdot Pf(n/N) \in \mathbb{Z}$ is not a sufficient condition[4] we give the next example, also provided by Coste:

$$(54) \qquad \begin{cases} g = 3, N = 2^2 \cdot 3^6 \\ e_3 = 2^4 e_2 = 2^4 e_1 = 2^4 3^4. \end{cases}$$

This yields $Pf(n/N)N = 1$, but it cannot allow for a twist eating solution. One way to see this is to use the well known result, that

$$(55) \qquad U(k) = \prod_{\mu=1}^{6} U_\mu^{k_\mu}$$

are independent $N \times N$ matrices for $0 \le k_\mu < N_\mu$ if a solution does exist. Where:

$$(56) \qquad N_i = N_{g+i} = c_{g-i+1}.$$

This is an easy generalization of the well known result for $g = 1$ and 2, see e.g. [2].

Therefore we have at least $\prod_\mu N_\mu = N_{irr}^2$ independent $N \times N$ matrices. There can be no more than N^2, so that necessarily

$$(57) \qquad N_{irr} \le N.$$

---

[4] Integrality of $Pf(n/N)N$ in general depends on the choice of n; i.e. it depends on its mod N freedom.

For the case of eq. (54) this bound is easily seen to be violated, hence $Pf(n/N)N \in \mathbb{Z}$ is not sufficient for the existence of solutions to eq. (1).

Since for given e_j and N, c_j and m_j are fixed, one can always find d_j and m such that eq. (7) is satisfied. The simplest choice for m is the smallest common multiple of all m_i, which equals m_1 (see below eq. (50)).

Therefore if N is a multiple of N_{irr} we can write down the following solution:

$$(58) \qquad U_\mu = \Lambda_\mu \times U_\mu^{irr},$$

where: $\Lambda_\mu = \mathrm{diag}\,(\lambda_\mu^{(1)}, \ldots, \lambda_\mu^{(N/N_{irr})}) \in U(1)^{N/N_{irr}}$ and $U_\mu^{irr} \in U(N_{irr})$ coming from $\sigma_\delta(m)$ (see eq. (48)).

We claim that up to a similarity transformation this is the only possible type of solution. Suppose that U_μ is a solution and define:

$$(59) \qquad \omega_\mu = U_\mu^{N_\mu},$$

with N_μ defined in (56) satisfying $Z_{\mu\nu}^{N_\mu} = 1$. This implies that the ω_μ can be simultaneously diagonalized, with U_μ block-diagonal:

$$(60) \qquad [\omega_\mu, \omega_\nu] = [\omega_\mu, U_\nu] = 1.$$

Working in this diagonal gauge one easily verifies that:

$$(61) \qquad \hat{U}_\mu = \Lambda_\mu^{-1} U_\mu, \qquad \Lambda_\mu^g = \omega_\mu$$

satisfy the same equation as U_μ, but such that $\hat{\omega}_\mu = \hat{U}_\mu^{N_\mu} = 1$. Consequently, the $\hat{U}_\mu$ give a representation of the Heisenberg group. With the help of the previous two sections there is no other possibility for $\hat{U}_\mu$ then to be the direct sum of irreducible representation of dimension N_{irr}. And hence N has to be a multiple of N_{irr} and U_μ is of the form of eq. (58). The dimension of the solution manifold to equation (1) up to a similarity transformation is $(N/N_{irr} - 1)$. Given a solution we therefore find[5]:

$$(62) \qquad Pf\left(\frac{n}{N}\right)N = -(\prod_{k=1}^{g} m_k) \cdot \frac{N}{N_{irr}} \in \mathbb{Z}$$

and

$$(63) \qquad e_{g-i+1} = \frac{mN}{d_i} = \frac{N}{N_{irr}} m_i \prod_{k \neq i} c_k,$$

whence:

$$(64) \qquad (N/N_{irr}) | \mathrm{g.c.d.}\left(n_{\mu\nu}, Pf\left(\frac{n}{N}\right)N, N\right).$$

One easily verifies that for $g = 1$ and 2, g.c.d. $(n_{\mu\nu}, Pf(n/N)N, N) = N/N_{irr}$ and the dimension of the solution manifold confirms with previous results [1].

---

[5] For $g = 1$ and 2, g.c.d. $(N_{irr}, \prod_{k=1}^{g} m_k) = 1$, so that $Pf(n/N)N \in \mathbb{Z}$ implies that N is a multiple of N_{irr}.

We also see that g.c.d. $(n_{\mu\nu}, Pf(n/N)N, N) = 1$ implies $N = N_{irr}$, which implies uniqueness of the solution. Since N_{irr} fixes the dimensionality of the solution manifold, we indirectly see that in the case that N is a multiple of N_{irr} for a specific choice of n, N_{irr} does not depend on this choice (because eq. (3) is invertable [2].)

In conclusion, twist eating solutions exist iff N is a multiple of N_{irr} (for which $Pf(n/N)N \in \mathbb{Z}$ is only a necessary condition). A solution is unique up to a gauge and Z_N-factors iff $N = N_{irr}$ iff it corresponds to an irreducible representation (for which g.c.d. $(n_{\mu\nu}, Pf(n/N)N, N) = 1$ is only a sufficient condition), in that case (48) gives an explicit solution for eq. (2), which through a simple rescaling by a phasefactor can be chosen in $SU(N)$

$$\left(\text{replace } Q_n \text{ by } \hat{Q}_n \equiv \exp\left(-\frac{\pi i(n-1)}{n}\right) \cdot Q_n \text{ and similar for } P_n\right).$$

Multiplication with a phasefactor is indeed the only freedom we have, which for $SU(N)$ reduces to multiplication with elements of the centre of $SU(N)$ (isomorphic to μ_N). Using the fact that

$$\text{g.c.d. } (m|\pi|^{-1}, c_k) = \text{g.c.d. } (|\pi|p_k^{\,1}, c_k) = 1$$

we have that $\lambda \cdot \hat{Q}_{c_k}^{m/|\pi|}$ and $\lambda \cdot \hat{P}_{c_k}^{|\pi|/p_k}$ are equivalent to $\hat{Q}_{c_k}^{m/|\pi|}$ and $\hat{P}_{c_k}^{|\pi|/p_k}$ if $\lambda \in \mu_{c_k}$. Therefore all inequivalent solutions to equation (2) are given by (see (48)) $\lambda_k U_{g-k}$ and $\nu_k U_{2g-k}$ with $\lambda_k, \nu_k \in \mu_N/\mu_{c_k}$. Hence there are

$$\prod_{k=1}^{g} 0(\mu_N/\mu_{c_k})^2 = N^{2(g-1)}$$

inequivalent solutions.

5. CONCLUSIONS

In this note new results concern twist eating solutions for more than four dimensions; one might think of applications for TEK-models in the $d \to \infty$ limit, where d is the dimension of space-time. However our main motivation was to show the underlying structure of the Heisenberg group.

ACKNOWLEDGEMENTS

We would like to thank prof. T.A. Springer for a remark which made us see the light. B.v.G. is financially supported by the ZWO project on moduli and P.v.B. by the Dutch National Science Foundation FOM (ZWO). This work was partially supported by NSF grant PHY 81-09110 A-03. One of us (P.v.B.) is also very grateful to Antoine Coste for the counter examples to the $g = 2$ criteria when extended to $g \geq 2$.

REFERENCES

1. Baal, P. van – Communications in Mathematical Physics, **92**, 1 (1983).
2. Baal, P. van – Twisted boundary conditions: A non-perturbative probe for pure non-Abelian gauge theories, thesis, Utrecht, juli 1984.
3. Serre, J.-P. – Représentation lineaires des groupes finis, Hermann, Paris, 1971 (2e éd.).

Reprinted from ANNALS OF PHYSICS
All Rights Reserved by Academic Press, New York and London

Vol. 174, No. 2, March 1987
Printed in Belgium

QCD on a Torus, and
Electric Flux Energies from Tunneling

PIERRE VAN BAAL AND JEFFREY KOLLER

*Institute for Theoretical Physics, State University of New York at Stony Brook,
Stony Brook, New York 11794-3840*

Received May 8, 1986

The energy of 't Hooft-type electric flux in pure QCD on the hypertorus is a non-perturbative effect, caused by tunneling through a quantum-induced potential barrier. We calculate the electric flux energy using the semiclassical approximation. The problem of infrared divergences is attacked directly, and we show how to treat them consistently, using Lüscher's effective Hamiltonian approach. A toy model is first solved to demonstrate the consistency of our approach. Lüscher's effective Hamiltonian for the spatially constant modes in QCD is rederived using a Lagrangian approach with a non-local gauge-fixing term, and we show how this rigorously defines an effective action to all loop orders. The removal of the gauge degrees of freedom from the effective Hamiltonian while avoiding Gribov ambiguities is discussed in detail, as are the numerical methods required to treat this non-integrable system. Some consequences of our final result are noted.   © 1987 Academic Press, Inc.

1. INTRODUCTION

Unbroken gauge theories are plagued by infrared divergences in perturbation theory. However, the infrared region is believed to be crucial for confinement, which is a long distance feature of the theory. Our difficulties are caused by the strong increase of the effective coupling constant at large distances, the counterpart of asymptotic freedom. To be able to do perturbation theory it is therefore useful to use an infrared cutoff, and the most elegant procedure is to consider the theory in a three-dimensional box and impose periodic boundary conditions. These boundary conditions must not break the gauge invariance, so we require only gauge invariant quantities to be periodic; the gauge potential is only periodic up to a gauge transformation. In this way, 't Hooft [1] introduced gauge theories on the hypertorus. If there are no matter fields in the fundamental representation of the gauge group, electric and magnetic flux are defined by a topological property called twist. If we take the hypertorus to be a cube with sides of length L, and assume that when L gets large electric flux tubes form, then the energy of electric flux will be of the form $\sigma \cdot L$, and we recognize σ as the string tension.

However, the energy of electric flux can only be calculated analytically for small L. Lüscher [2] showed how to do perturbation theory on the hypertorus in the zero magnetic flux case. He found that to all orders in perturbation theory there is a

VAN BAAL AND KOLLER

degeneracy in electric flux, i.e., electric flux has no energy in perturbation theory. In a previous paper by one of us [3] it was explained how this degeneracy is lifted by tunneling through a quantum-induced potential, and we called the associated "classical solution" a "pinchon," because it is analogous to an instanton. We expressed the dimensionless ratio of the energy of electric flux ($\Delta E(L)$) and the glueball mas $M_L(0^+)$ in terms of Lüscher's universal expansion parameter $z = M_L(0^+) \cdot L$ [4], which yields a universal (i.e., renormalization group independent) function. The large L (i.e., large z) behaviour is also known if we assume string formation and a finite glueball mass: $\Delta E(L)/M_L(0^+) \sim (\sigma/M(0^+)^2) z$.

We expect the formation of flux tubes to be due to non-perturbative effects. One could thus hope that in a situation where the perturbative contributions are completely absent (unlike in the case of nonzero magnetic flux), there is a smooth behaviour when going from small to large values of z, except of course at the point where the non-perturbative effect sets in.

In a paper [5] where we presented the final result for the short distance expansion of $\mathscr{E}(L) = \Delta E(L)/M_L(0^+)$ for $SU(2)$, this behaviour was indeed observed. Although Monte Carlo calculations [6] have not yet confirmed the onset of tunneling (which we predicted to occur at $z \simeq 1.2$), data is available in the range $z \cong 1.5$ to $z \simeq 8$. Despite the fact that these Monte Carlo calculations were performed on periodic lattices, we were able to identify the energy obtained from time-time correlation functions for spatial Polyakov loops in the *fundamental* representation [6] as the energy of electric flux, which was first defined by 't Hooft by using twisted boundary conditions [1]. Originally these correlation functions were used as an alternative way to calculate the string tension, so one was interested in their large z behaviour; our interpretation provides a well-defined meaning of this quantity for all z values. It becomes an important tool in probing the physics of the QCD vacuum [5].

To understand how one can relate two apparently separate fields, 't Hooft's QCD on a torus with *twisted* boundary conditions, and conventional lattice QCD with *periodic* boundary conditions, we observe that in the Hamiltonian formalism with zero magnetic flux (no twist in the spatial deirection) one also works with periodic boundary conditions. Electric flux, alternatively defined by 't Hooft as a twist in the time direction, is introduced in this case by classifying the physical states as representations of the homotopy group of allowed gauge transformations, as explained in Section 2. Similarly, on the lattice one can define 't Hooft-type electric flux using the transfer matrix approach, as explained in [32]. On the other hand, a spatial Polyakov loop in the fundamental representation creates one unit of electric flux [1]. (Details on this can be found in [8].) Time–time correlation functions of such loops thus select the energy difference between the ground state with one unit of electric flux and the ground state with zero electric flux, which is the way the energy of electric flux was defined by 't Hooft.

This paper is technical in nature. It provides a thorough discussion of the tunneling calculation and gives results contained in [11], hitherto unpublished. It furthermore gives results left out in [3, 5]. The paper is organized as follows: In Section 2 we discuss the classical vacuum. Section 3 presents a toy model closely related to the problem at hand; here we also review the effective Hamiltonian calculation based on Bloch perturbation theory. Section 4 discusses this effective Hamiltonian for $SU(2)$ Yang–Mills on T^3, but also gives a Lagrangian derivation. We work out in great detail the non-local gauge fixing introduced in [3] and show what the consequences are for the ghost interactions. It is important to note that our approach yields a *consistent* split into the spatially constant and spatially varying modes, which allows a proper and gauge-invariant treatment of infrared modes, *without* giving those modes a "mass" term (as would happen for non-zero magnetic flux or for Yang–Mills on a sphere).

Section 5 is devoted to eliminating the gauge degrees of freedom in the effective Hamiltonian. We show that a conventional approach based on gauge fixing leads to Gribov ambiguities. We also show how to decompose this Hamiltonian, which has six degrees of freedom, into a piece along the three-dimensional vacuum valley and a piece orthogonal to it.

Section 6 explains how the tunneling calculation (to the order in which we are interested) reduces to a calculation for a three-dimensional cubic lattice with a well-defined bounded potential. In Section 7 we calculate the appropriate energies using this reduction and summarize our analytic results.

These analytic results involve the perturbative energies and wavefunctions, and a contribution from transverse fluctuations along the pinchon (the instanton path). It has been shown that the effective Hamiltonian (Yang–Mills for spatially constant vectors potentials) is non-integrable [12], so to obtain these quantities, one has to resort to numerical calculations, which are described in detail in Section 8.

The results of these calculations have been presented earlier [5]. Here we describe the methods, closely connected to the route followed by Lüscher and Münster [13] for $SU(2)$.

Finally, there are four appendices, discussing respectively the properties of the one-loop effective potential, the polar decomposition, the multiloop infrared behaviour, and a discussion on generalizing our results to arbitrary gauge group.

2. THE CLASSICAL VACUUM

In this section we will work in the Hamiltonian formalism of Lüscher [2]. Our base space is then the flat torus T^3, or its equivalent $\mathbf{R}^3/\Lambda$, where Λ is the lattice spanned by $\mathbf{a}^{(i)}$, $i = 1, 2, 3$. We will choose the torus symmetric, of length L in each direction ($\mathbf{a}^{(i)} = L\mathbf{e}^{(i)}$), but this is only for convenience, and the calculation could be done on tori of other shapes. If the magnetic flux is zero, as will be assumed throughout, one can always choose a gauge such that the potentials themselves are periodic:

$$\mathbf{A}(\mathbf{x} + \mathbf{a}^{(i)}) = \mathbf{A}(\mathbf{x}). \tag{2.1}$$

VAN BAAL AND KOLLER

Gauss's law then imposes the constraint that physical state vectors are invariant under any periodic, homotopically trivial gauge transformation. However, the full set of all "allowed gauge transformations" (i.e., all those that do not alter the periodicity of any periodic gauge potential), is a larger class also containing gauge transformations only periodic up to an element of the center of the gauge group. For $SU(2)$ this is just $\mathbf{Z}_2 = \{ \pm 1 \}$:

$$\Omega(\mathbf{x} + \mathbf{a}^{(i)}) = (-1)^{k_i} \Omega(\mathbf{x}). \tag{2.2}$$

The homotopy type is thus specified by the three integers $k_j \bmod 2$, and the winding number

$$P = \frac{1}{24\pi^3} \int_{T^3} d^3x \, \varepsilon_{ijk} \, \mathrm{Tr}((\Omega \partial_i \Omega^{-1})(\Omega \partial_j \Omega^{-1})(\Omega \partial_k \Omega^{-1})), \tag{2.3}$$

which (for zero magnetic flux) is an integer.

The physical state vectors are now labeled by the quantum numbers θ (mod 2π) and $\mathbf{e}$ (mod 2), which are conjugate to P and $\mathbf{k}$, in the sense that if we denote the action of a gauge transformation on state vectors by $[\Omega]$, then

$$[\Omega] \, |\mathbf{e}, \theta\rangle = e^{\pi i \mathbf{k} \cdot \mathbf{e} + i\theta P} \, |\mathbf{e}, \theta\rangle. \tag{2.4}$$

The quantum number $\mathbf{e}$ is the electric flux of the state, as defined by 't Hooft [1]. In future we thus distinguish between "periodic gauge transformations," under which physical state vectors are invariant, and the "allowed gauge transformations."

Our first step is to identify the classical vacuum, which will suggest appropriate coordinates for specifying the wavefunctions. The classical potential is the positive definite expression

$$V(\mathbf{A}) = \frac{1}{2g_0^2} \int_{T^3} d^3x \, \mathrm{Tr}(F_{ij}^2), \tag{2.5}$$

$$F_{ij} = \partial_i A_j - \partial_j A_i + i[A_i, A_j], \tag{2.6}$$

so the classical vacuum is the set of the curvature-free ($F_{ij} = 0$) vector potentials. Because T^3 is multiconnected, these are not all gauge equivalent. To see this, consider the Wilson loop

$$W(\mathscr{C}) = \mathrm{Tr} \, \mathrm{P} \exp\left(i \oint_{\mathscr{C}} \mathbf{A} \cdot d\mathbf{x} \right), \tag{2.7}$$

which is invariant under infinitesimal deformations of the loop $\mathscr{C}$ if and only if the colour magnetic field is zero. (Recall that F_{ij} is given by the area derivative [9].) Thus for a vacuum configuration $W(\mathscr{C})$ is completely determined by the homotopy type of the curve $\mathscr{C}$. In a simply connected space $\mathscr{C}$ is always contractible and $W(\mathscr{C}) = 2$. However, for T^3 the homotopy type of $\mathscr{C}$ is specified by three integers n_i, which give the winding numbers around each of the three generating circles of T^3. Hence

$$W(\mathscr{C}) = \mathrm{Tr}(U_1^{n_1} U_2^{n_2} U_3^{n_3}), \qquad (2.8)$$

where

$$U_i = \mathrm{P} \exp\left(i \int_0^L A_i(s\mathbf{e}^{(i)})\, ds \right). \qquad (2.9)$$

Furthermore, $W(\mathscr{C})$ does not depend on the order of U_i's, so these must commute. Let their eigenvalues be $\exp(\pm i\varphi_i)$: then, because $\mathrm{Tr}(U_i) = 2\cos\varphi_i$ is invariant under periodic gauge transformations, vacua with different values of $\cos\varphi_i$ must be inequivalent. Indeed, the vacua can be classified by $\cos\varphi_i$, since we can explicitly construct a curvature free configuration giving any desired value of $\cos\varphi_i$, viz., the constant field

$$A_i = \frac{\varphi_i}{L}\sigma_3 \equiv \frac{C_i}{L}\frac{\sigma_3}{2}, \qquad (2.10)$$

with σ_i the Pauli matrices. Any other curvature-free configuration can be reduced to this form by a periodic gauge transformation. Also, one can identify

$$\varphi_i \cong \varphi_i + 2\pi,$$
$$\varphi \cong -\varphi. \qquad (2.11)$$

This corresponds to Lüscher's gauge-invariant labeling of the torons (i.e., classical vacua). Finally, note that an allowed gauge transformation of the type in Eq. (2.2) transforms $W(\mathscr{C})$ into $(-1)^{\mathbf{k}\cdot\mathbf{n}} W(\mathscr{C})$. Since these transformations are genuine invariances of the Hamiltonian, we make a note of how they transform the vacua:

$$\varphi_i \to \varphi_i + \pi n_i, \qquad C_i \to C_i + 2\pi n_i. \qquad (2.12)$$

These transformations are generated by

$$\Omega^{(j)}(\mathbf{x}) = \exp(-\pi i x_j \sigma_3/L). \qquad (2.13)$$

In summary: the classical "vacuum valley" is parameterized by a vector $\mathbf{C}$, mod 4π, and a symmetry relates vacua $\mathbf{C}$, $-\mathbf{C}$, and $\mathbf{C} + 2\pi\mathbf{n}$ ($\mathbf{n} \in \mathbf{Z}_2^3$).

The next step is to expand the potential around the classical vacuum, so we follow Lüscher and write

$$A_k(\mathbf{x}) = \Omega(\mathbf{x})\left(\frac{C_k\sigma^3}{2L} + g_0 q_k(\mathbf{x}) \right)\Omega^{-1}(\mathbf{x}) - i\Omega(\mathbf{x})\,\partial_i\,\Omega^{-1}(\mathbf{x}), \qquad (2.14)$$

with the gauge condition

$$\int_{T^3} d^3x\, q_k^3(\mathbf{x}) = 0, \qquad D_i q_i(\mathbf{x}) = 0. \tag{2.15}$$

D_i is the covariant derivative in the background vacuum configuration

$$D_k = \partial_k + i\, \mathrm{ad}\left(\frac{C_k \sigma_3}{2L}\right). \tag{2.16}$$

With this parameterization one finds

$$V(\mathbf{A}) = \int_{T^3} \mathrm{Tr}\{(D_i q_k(\mathbf{x}))^2 + 2ig_0(D_i q_j(\mathbf{x}))[q_i(\mathbf{x}), q_j(\mathbf{x})] - \tfrac{1}{2} g_0^2 [q_i(\mathbf{x}), q_j(\mathbf{x})]^2\}. \tag{2.17}$$

The quadratic part of V can be diagonalized by putting

$$q_j(\mathbf{x}) = q_j^3(\mathbf{k})\, e^{2\pi i \mathbf{k}\cdot\mathbf{x}/L}\sigma_3, \qquad\qquad \lambda^3(\mathbf{k}) = \left(\frac{2\pi\mathbf{k}}{L}\right)^2 \quad \mathbf{k}\neq\mathbf{0}$$

$$q_j(\mathbf{x}) = q_j^+(\mathbf{k})\, e^{2\pi i\mathbf{k}\cdot\mathbf{x}/L}\frac{(\sigma_1 + i\sigma_2)}{2}, \qquad \lambda_C^+(\mathbf{k}) = \left(\frac{2\pi\mathbf{k}+C}{L}\right)^2 \tag{2.18}$$

where $q_j^3(\mathbf{k})$ $(\mathbf{k}\neq\mathbf{0})$ and $q_j^+(\mathbf{k})$ satisfy the conditions

$$\mathbf{k}\cdot\mathbf{q}^3(\mathbf{k}) = (2\pi\mathbf{k}+\mathbf{C})\cdot\mathbf{q}^+(\mathbf{k}) = 0. \tag{2.19}$$

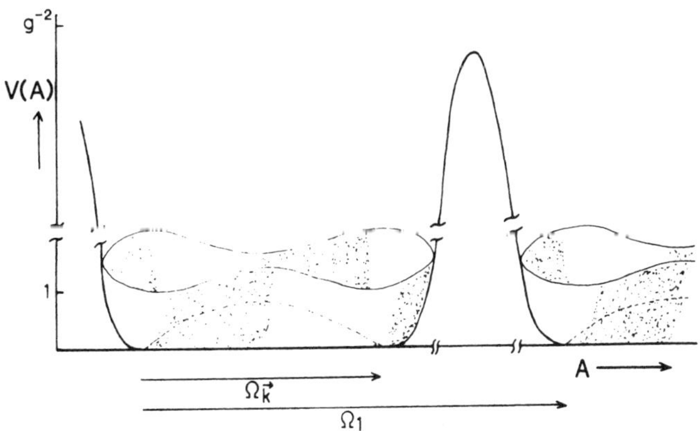

FIG. 1. An impressionistic graph of the potential $V(\mathbf{A})$, with the infinite-dimensional space of field configurations represented as two-dimensional. A classical vacuum valley lies at the minimum of the potential, but the squeezing of the potential in the transverse direction induces a quantum barrier (the dashed curve), causing the perturbative quantum vacua to occur where the potential is widest. Configurations with different winding number are separated by a classical barrier. Allowed gauge transformations Ω map one perturbative vacuum to another.

Note that for $\mathbf{C} = -2\pi\mathbf{n}$ there are modes with zero eigenvalues, which implies that the quadratic approximation fails for these modes. This is the well-known infrared problem, and it means that a naive perturbation expansion will fail. Much of this paper is devoted to treating these "quartic modes" correctly. It is a non-trivial problem to do this, and the same problem reappears in various guises at different places in the calculation. In the next section we will show how to deal with this.

Figure 1 is an attempt to show the shape of the potential $V(\mathbf{A})$. We have reduced the three-dimensional vacuum valley to one dimension, and shown only one of the infinitely many "transverse" directions. The transverse direction we have shown is one with quartic behaviour at $\mathbf{C} = 0$. Thus V is wide at this point, but "pinches" narrower at non-zero $\mathbf{C}$.

One might wonder if it is possible to remove these quartic modes by changing the metric of the torus. However, we see from Eq. (2.17) that the quartic modes are due to a zero eigenvalue for the Laplacian on T^3, and are thus related to the non-triviality of the De Rham cohomology. This depends only on the topology of the space, and not on the metric. For *zero* magnetic flux, there is no way out of the quartic-mode problem. What happens for non-zero magnetic flux is shown in [29].

3. A Toy Model

Let us first remind the reader of a toy model which was analysed in [14]:

$$H = -\frac{g^2}{2}\frac{\partial^2}{\partial x^2} - \frac{g^2}{2}\frac{\partial^2}{\partial y^2} + \frac{1}{2g^2}(x^2-1)^2 y^2. \tag{3.1}$$

This toy model shares with our full problem the feature of a vacuum valley and the breakdown of the quadratic expansion at $x = \pm 1$. However, there is a feature here that is different from the full problem. In [14] it was explained that one can use the one-loop effective potential $V_1(x)$ in the adiabatic approximation. The non-adiabatic corrections could be ignored: where they become important, near the points where the quadratic approximation breaks down, the information is contained in the perturbative wavefunction, which is matched to the non-perturbative part in a higher-dimensional WKB-type approximation, called the *path decomposition expansion*. In the full problem, we have an infinite number of transverse modes, each giving a non-adiabatic correction in the Lagrangian of order $\dot{x}^2$ to the leading order term $(1/g_0^2)\,\dot{x}^2$. Instead of ignoring this correction, it is imperative to note that it should actually give the proper renormalization of the coupling constant. We will come back to this later in a Lagrangian approach (Sect. 4).

Another difference between (3.1) and the full problem is that in (3.1) the breakdown of the quadratic approximation is caused by the same transverse variable y at both $x = \pm 1$. In the QCD problem, a mode which is quartic at $\mathbf{C} = 0$ is perfectly well behaved at $\mathbf{C} = 2\pi\mathbf{n}$. To understand what happens if the breakdown of the quadratic approximation occurs due to different transverse modes, we have constructed another simple toy model,

$$H = -\frac{g^2}{2}\left(\frac{\partial^2}{\partial w^2} + \frac{\partial^2}{\partial x^2} + \frac{\partial^2}{\partial y^2} + \frac{\partial^2}{\partial z^2}\right) + \frac{2}{g^2}\left((z-1)^2 y^2 + (z+1)^2 x^2 + \left(\frac{z^2+3}{z^2+1}\right)^2 w^2\right) - 4.$$

(3.2)

This exhibits the following symmetry:

$$\Pi(w, x, y, z) = (w, y, x, -z).$$

(3.3)

Note that Π^2 is the identity and hence wavefunctions are even or odd under Π. In particular, the twofold-degenerate ground state will be split into an even and an odd state, whose energy difference is determined by tunneling through the one-loop effective potential $V_1(z)$:

$$V_1(z) = \left(|z-1| + |z+1| + \frac{z^2+3}{z^2+1}\right) - 4.$$

(3.4)

Note that the part of the Hamiltonian quadratic in w was added to ensure that $V_1(z)$ has isolated conical minima at $z = \pm 1$. Thus perturbation theory is obtained by expanding around $z = 1$, $x = y = w = 0$ (or $z = -1$, $x = y = w = 0$).

Much of this section is concerned with understanding the relationship between Bloch degenerate perturbation theory [15], and tunneling through an induced potential barrier as discussed in [14]. Given a set of states degenerate in the unperturbed theory, the Bloch approach provides one with an effective Hamiltonian describing the perturbative energies of these states. We will demonstrate that tunneling in the effective theory and in the full theory gives the same result for the energy splitting, in problems of the type we are concerned with.

We thus want to use this example to prove commutativity of the following diagram (BP = Bloch perturbation theory [15], PDX = path decomposition expansion [17]):

$$\begin{array}{ccc} H(x, y, z, w) & \xrightarrow{\ \text{PDX}\ } & \Delta E \\ \ \ {\scriptstyle\text{BP}}\searrow & \nearrow{\scriptstyle\text{PDX}} & \\ & H'_+(y, z) & \\ & H'_-(x, z) & \end{array}$$

(3.5)

Here H' is the effective Hamiltonian in the non-quadratic variables (H'_+ is obtained by expanding around $z = 1$, where x and w are quadratic, whereas H'_- is obtained by expanding around $z = -1$ making y and w behave quadratically). At the end we shall discuss how to combine H'_+ and H'_- into one effective Hamiltonian. For the gauge theory problem, this then implies that we can study the tunneling in a finite dimensional setting, because there is only a finite number of non-quadratic modes.

We will first review Bloch's method for degenerate perturbation theory [15], in a spirit similar to the way Lüscher used it for gauge theories [2]. First we rescale the coordinates via

$$w = g\hat{w}, \qquad x = g\hat{x}, \qquad y = g^{2/3}\hat{y}, \qquad z = 1 + g^{2/3}\hat{z}, \qquad (3.6)$$

so that the hatted coordinates are $O(1)$ for low-lying perturbative states in the well region. We thus find the perturbative expansion (we suppress the suffix $+$)

$$H = H_0 + g^{2/3}H_1,$$

$$H_1 = \sum_{v=0}^{\infty} g^{2v/3}H_1^{(v)}, \qquad (3.7)$$

with the following explicit expressions (a_v constants):

$$H_0 = -\frac{1}{2}\left(\frac{\partial^2}{\partial\hat{w}^2} + \frac{\partial^2}{\partial\hat{x}^2}\right) + 8\hat{w}^2 + 8\hat{x}^2 - 4,$$

$$H_1^{(0)} = -\frac{1}{2}\left(\frac{\partial^2}{\partial\hat{y}^2} + \frac{\partial^2}{\partial\hat{z}^2}\right) + 2\hat{y}^2\hat{z}^2 + 8\hat{z}(\hat{x}^2 - \hat{w}^2), \qquad (3.8)$$

$$H_1^{(1)} = \hat{z}^2(2\hat{x}^2 + \hat{w}^2), \qquad H_1^{(v+2)} = a_v\hat{z}^{v+3}\hat{w}^2.$$

Hence the Hamiltonian separates naturally into an unperturbed piece involving only the quadratic variables, and a perturbation involving the non-quadratic variables. The lowest energy unperturbed wavefunctions are given by

$$\Psi(w, x, y, z) = \phi(y, z)\,\Psi_0(w, x), \qquad (3.9)$$

with $\Psi_0(w, x)$ the groundstate wavefunction for H_0 with zero energy:

$$\Psi_0(w, x) = \frac{2}{g\sqrt{\pi}}\exp\left(\frac{-2(w^2 + x^2)}{g^2}\right), \qquad (3.10)$$

and $\phi(y, z)$ any square-integrable function. The idea of Bloch perturbation theory is that when we turn on the perturbation, $\phi(y, z)$ becomes an eigenfunction of an effective Hamiltonian $H' = \sum_{v=0}^{\infty} g^{2(v+1)/3}H'_v$. In particular, to lowest order $\phi(y, z)$ is an eigenfunction of

$$H'_0 = \langle\Psi_0|\,H_1^{(0)}\,|\Psi_0\rangle = -\frac{1}{2}\left(\frac{\partial^2}{\partial\hat{y}^2} + \frac{\partial^2}{\partial\hat{z}^2}\right) + 2\hat{y}^2\hat{z}^2. \qquad (3.11)$$

(By choosing the z-dependent w-frequency as in Eq. (3.2), we get an H_0' similar to that for Yang–Mills and to the toy model in Eq. (3.1).) Beyond lowest order, the energy eigenstates are no longer direct products as in (3.9), but from the discussion below we see that they are related to direct products by the operator U.

Later we will need to find the higher order terms in the expansion of the effective potential, so we will follow the discussions in [2, 15, 16]. Let P_0 be the projector onto the state (3.10):

$$P_0 \Psi(w, x, y, z) = \Psi_0(w, x) \int dw' \, dx' \, \Psi(w', x', y, z) \, \Psi_0(w', x'). \qquad (3.12)$$

The eigenvalues E_α of H, with the property that $E_\alpha = O(g^{2/3})$ (one of which is the groundstate energy) are determined by constructing an operator R whose eigenvalues are E_α,

$$R \, |\alpha_0\rangle = E_\alpha \, |\alpha_0\rangle, \qquad (3.13)$$

where we *define* R by

$$R = P_0 H U, \qquad (3.14)$$

with U the operator that maps $|\alpha_0\rangle$ to the eigenstate $|\alpha\rangle$ of H with the same energy E_α (and $U|\Psi\rangle = 0$ for all $|\Psi\rangle$ with $P_0|\Psi\rangle = 0$).

$$U \, |\alpha_0\rangle = |\alpha\rangle, \qquad H \, |\alpha\rangle = E_\alpha \, |\alpha\rangle. \qquad (3.15)$$

Therefore combining (3.13) and (3.14), we must have

$$P_0 \, |\alpha\rangle = |\alpha_0\rangle, \qquad (3.16)$$

and one derives from this [15, 16] the Schwinger–Dyson equation for U:

$$U = P_0 + g^{2/3} \frac{(1 - P_0)}{(E_0 - H_0)} (H_1 U - U H_1 U). \qquad (3.17)$$

Note that R is not hermitian. To derive an effective Hamiltonian H' we observe that with P the projector onto the eigenstates $|\alpha>$ of H, one has the following hermiticity property (use $UP_0 = U$ and $PU = U$ [16]):

$$P_0 H U P_0 P P_0 = P_0 H P P_0 = P_0 P H P_0 = P_0 P P_0 U^\dagger H P_0. \qquad (3.18)$$

Thus, define the operator $B: \Omega_0 \to \Omega_0$ (Ω_0 the set of eigenstates of H_0 with eigenvalue 0, i.e., square integrable functions of y and z) by

$$B = P_0 P P_0. \qquad (3.19)$$

Since $B = P_0 + O(g^{2/3})$ it is positive definite on Ω_0 for small enough g, so that we can define H' by

$$H' = B^{-1/2} R B^{1/2}. \tag{3.20}$$

Note that this perturbation theory breaks down for the smallest value of g where B is not invertible.

The operator H' acts on Ω_0 and can hence be expressed as a Hamilton operator in the variables y and z acting on square integrable functions. To lowest order we found the result of Eq. (3.11).

Next we will describe the features of the PDX for the situation where a vacuum valley is present (the line $x = y = w = 0$). In [14], a recipe was given for applying the PDX to a tunneling problem with a vacuum valley. We solve for fixed-z eigenstates of H ($p_z = (1/i)(\partial/\partial z)$, and $n = 1$ labels the ground state),

$$H(p_z = 0)\, \chi^{(n)}_{[z]}(w, x, y) = V_n(z)\, \chi^{(n)}_{[z]}(w, x, y), \tag{3.21}$$

and expand a wavefunction $\Psi(w, x, y, z)$ as

$$\Psi(w, x, y, z) = \sum_n \varphi^{(n)}(z)\, \chi^{(n)}_{[z]}(w, x, y). \tag{3.22}$$

(In this simple example we can obtain the $\chi^{(n)}_{[z]}$ in closed form.) The Hamiltonian in this representation becomes

$$H_{nm} = -\frac{g^2}{2}\left(\delta_{nm}\frac{\partial}{\partial z} - A_{nm}(z)\right)^2 + \delta_{nm} V_n(z), \tag{3.23}$$

with the following definition of A_{nm}:

$$A_{nm}(z) = \langle \chi^{(m)}_{[z]} | \frac{\partial}{\partial z} | \chi^{(n)}_{[z]} \rangle. \tag{3.24}$$

The first ingredient in the recipe for the tunneling energy split is the potential $V_1(z)$, defined in Eq. (3.21) and given in Eq. (3.4). The second ingredient for the PDX as described in [14] is what was called the ground state $\Psi_0^{[0]}$ for the single-well potential $V_{[0]}$ or single-well Hamiltonian $H_{[0]}$. $H_{[0]}$ is H to the lowest order in perturbation theory where degeneracies are lifted. We obviously have

$$\Psi_0^{[0]}(w, x, y, z) = \phi_0^{[0]}(y, z)\, \Psi_0(w, x),$$
$$H_0'\phi_0^{[0]}(y, z) = E_0^{[0]}\phi_0^{[0]}(y, z). \tag{3.25}$$

The final ingredient is the single-well energy $E_0^{[0]}$. Observe, incidentally, that H_0' is identical to the single-well Hamiltonian for Eq. (3.1), which was studied in detail in

[14]. The recipe now instructs one to expand $\Psi_0^{[0]}$ as in (3.22), and thus find the single-well approximation $\varphi_0^{[0]}$ for $\varphi_0^{(1)}(z)$, for z deep within the barrier. As shown in [14], the $n = 1$ term dominates (3.22) in this region, so we find for $|z - 1| \gg g^{2/3}$:

$$\phi_0^{[0]}(y, z) \simeq \varphi_0^{[0]}(z) \frac{1}{\sqrt{g}} \left(\frac{4|z - 1|}{\pi} \right)^{1/4} \exp \left(- \frac{2\,|z - 1|\,y^2}{g^2} \right). \tag{3.26}$$

The energy splitting between the even and odd ground states is now given in terms of what we will call the *tunneling data* $(V_1(z), E_0^{[0]}, \varphi_0^{[0]}(z))$ by [14]:

$$\Delta E(g) = 2g\, |\varphi_0^{[0]}(1 - d)|^2 \sqrt{2(V_1(1 - d) - E_0^{[0]})}$$
$$\times \exp \left(- \frac{1}{g} \int_{d-1}^{1-d} \sqrt{2(V_1(z) - E_0^{[0]})}\, dz \right). \tag{3.27}$$

For $d \simeq g^{1/7}$ this determines $\Delta E(g)$ up to a relative error that vanishes as a positive power of g.

It is immediately clear from Eq. (3.25) that the tunneling data $E_0^{[0]}$ and $\varphi_0^{[0]}$ can equally well be obtained from the single-well ground state $\Psi_0^{[0]}$ of $H_{[0]}$ or from the single-well ground state $\phi_0^{[0]}$ of H_0'. We are now going to show an even stronger result: we can also derive $V_1(z)$ from the effective Hamiltonian H'. This is similar in spirit to the discussion in Appendix A of [5], where it was verified to the fourth order in the coordinates.

Since $V_1(z)$ comes from an infinite resummation of Bloch perturbation theory, we will rearrange things slightly. Instead of rescaling, we will stick to the original variables and write

$$H = H_0 + H_0' + W,$$
$$H_0 = - \frac{g^2}{2} \left(\frac{\partial^2}{\partial w^2} + \frac{\partial^2}{\partial x^2} \right) + \frac{8}{g^2} (w^2 + x^2) - 4,$$
$$H_0' = - \frac{g^2}{2} \left(\frac{\partial^2}{\partial y^2} + \frac{\partial^2}{\partial z^2} \right) + \frac{2}{g^2} y^2 (z - 1)^2, \tag{3.28}$$
$$W = \frac{2}{g^2} [(z + 1)^2 - 4]\, x^2 + \frac{2}{g^2} \left[\left(\frac{z^2 + 3}{z^2 + 1} \right) - 4 \right] w^2.$$

Equation (3.17) can now be rewritten in terms of

$$U(\mathbf{n}) = \langle \mathbf{n} |\, U\, |0 \rangle, \tag{3.29}$$

where $|\mathbf{n}\rangle$ is the set of eigenfunctions of H_0:

$$H_0\, |\mathbf{n}\rangle = E_\mathbf{n}\, |\mathbf{n}\rangle; \qquad E_\mathbf{n} = 4(n_1 + n_2), \tag{3.30}$$

and $\langle w, x \mid 0 \rangle = \Psi_0(w, x)$ as given in Eq. (3.10). The reason that only the matrix element $U(\mathbf{n})$ enters is that $U = UP_0$ (see [16]). We find

$$U(0) = 1,$$
$$U(\mathbf{n}) = \{ U(\mathbf{n}) \langle 0 | W | \mathbf{m} \rangle U(\mathbf{m}) - \langle \mathbf{n} | W | \mathbf{m} \rangle U(\mathbf{m}) + [U(\mathbf{n}), H_0'] \}/E_{\mathbf{n}}, \tag{3.31}$$

whereas the operator R is given by

$$R = H_0' + \langle 0 | W | \mathbf{m} \rangle U(\mathbf{m}). \tag{3.32}$$

The matrix elements of W are easily computed explicitly,

$$\langle \mathbf{n} | W | \mathbf{m} \rangle = \frac{1}{4} \left[\left(\frac{z^2+3}{z^2+1} \right)^2 - 4 \right] X_{n_1 m_1} + \frac{1}{4} \left[(z+1)^2 - 4 \right] X_{n_2 m_2},$$
$$X_{nm} = \sqrt{n(n-1)}\, \delta_{n, m+2} + \sqrt{m(m-1)}\, \delta_{m, n+2} - (2n+1)\, \delta_{nm}, \tag{3.33}$$

but the important point is that they are g-independent. If we next rescale y as

$$\tilde{y} = y/g, \tag{3.34}$$

dependence on g only enters in the term containing H_0' in Eq. (3.31), so we can expand $U(\mathbf{n})$ in powers of g,

$$U(\mathbf{n}) = U_0(\mathbf{n}) + g^2 U_1(\mathbf{n}), \tag{3.35}$$

where $U_0(\mathbf{n})$ satisfies the equation

$$U_0(\mathbf{n}) = \left\{ U_0(\mathbf{n}) \langle 0 | W | \mathbf{m} \rangle U_0(\mathbf{m}) - \langle \mathbf{n} | W | \mathbf{m} \rangle U_0(\mathbf{m}) \right.$$
$$\left. + \left[U_0(\mathbf{n}), \frac{1}{2} \frac{\partial^2}{\partial \tilde{y}^2} + 2\tilde{y}(z^2-1) \right] \right\} \Big/ E_{\mathbf{n}}. \tag{3.36}$$

Since Eq. (3.36) is obtainable from Eq. (3.31) by ignoring $[\partial^2/\partial z^2, U(\mathbf{n})]$, it means we can compute things by deriving the effective Hamiltonian for $H(p_z = 0)$ and considering z as a fixed parameter.

Next, because $H(p_z = 0)$ separates we easily find

$$R - H_0' = |z+1| + \left(\frac{z^2+3}{z^2+1} \right) - 4 + g^2 R_1. \tag{3.37}$$

To obtain H' one observes that B in Eqs. (3.19) and (3.20) is only introduced to obtain a hermitian operator, and since R is hermitian to lowest order, we also obtain

$$H' = H_0' + |z+1| + \left(\frac{z^2+3}{z^2+1} \right) - 4 + g^2 \tilde{H}_1'. \tag{3.38}$$

We can now decompose H' in a similar way along the vacuum valley; we solve for the fixed-z eigenstates of $H'(p_z = 0)$,

$$H'(p_z = 0)\, \chi_{[z]}^{'(n)}(y) = V_n'(z)\, \chi_{[z]}^{'(n)}(y), \tag{3.39}$$

and expand the effective wavefunction $\phi(y, z)$ as

$$\phi(y, z) = \sum_n \varphi^{(n)}(z)\, \chi_{[z]}^{'(n)}(y). \tag{3.40}$$

One therefore finds

$$H_{nm}' = -\frac{g^2}{2}\left(\delta_{nm} \frac{\partial^2}{\partial z^2} - A_{nm}'(z) \right)^2 + \delta_{nm} V_n'(z) + g^2 T_{nm}(z), \tag{3.41}$$

with A_{nm}' defined as in Eq. (3.24) and

$$T_{nm} = \langle \chi_{[z]}^{'(m)} |\, \tilde{H}_1' - \tilde{H}_1'(p_z = 0)\, | \chi_{[z]}^{'(n)} \rangle.$$

One easily deduces (see Eq. (3.4))

$$V_1'(z) = V_1(z) + O(g^2), \tag{3.42}$$

which is what we set out to prove. The $O(g^2)$ correction in (3.42) and the one coming from T_{nm} can be shown to give negligible contributions to $\Delta E(g)$, as should be true for consistency.

In conclusion, one can obtain the tunneling data $\varphi_0^{[0]}(z)$, $E_0^{[0]}$ from H_0' (which is the single-well approximation of H'), and $V_1(z)$ from the one-loop approximation for the effective potential along the vacuum valley, which can be computed in a variety of ways.

The above derivation of the one-loop effective potential was one example showing that H' makes sense beyond perturbation theory. We have two different expressions, $H_\pm'$, related to two different quartic modes. The symmetry Π (Eq. (3.3)) guarantees an intimate connection between these two expressions,

$$H_+'(u, z) = H_-'(u, -z), \tag{3.43}$$

and one easily verifies that the same holds for the operators R and B (as maps from Ω_0 into itself). We can then naturally combine H_+' and H_-' into one effective Hamiltonian:

$$\begin{aligned} H'(u, z) &= H_+'(u, |z|) \\ &= H_-'(u, -|z|). \end{aligned} \tag{3.44}$$

Note that $H_+'(u, z)$ is ill-defined for $z \searrow -1$, but that we do not use it in that region. Now suppose that $\psi(x, y, z, w)$ is a simultaneous eigenvector of H and Π

with eigenvalues E and σ $(=\pm 1)$. The reduced wavefunctions $\psi_+(y, z)$ and $\psi_-(x, z)$ are then related by (use Eqs. (3.12) and (3.16))

$$\psi_+(u, z) = \sigma\psi_-(u, -z), \tag{3.45}$$

and one can show that $\psi_{\pm}(u, z)$ is an eigenfunction of $H'_{\pm}(u, z)$ with eigenvalue E. Let us define the wavefunction $\hat{\psi}(u, z)$ by

$$\hat{\psi}(u, z) = \begin{cases} \psi_+(u, z), & z \geqslant 0 \\ \psi_-(u, z), & z \leqslant 0. \end{cases} \tag{3.46}$$

It is clearly an eigenfunction of H' with eigenvalue E, and by Eq. (3.45) has the property

$$\hat{\psi}(u, z) = \sigma\hat{\psi}(u, -z). \tag{3.47}$$

However, we have to worry about continuity. For example, $\sigma = -1$ forces $\hat{\psi}(u, 0) = 0$ and thus $\psi_+(u, 0) = 0$, which is not automatically guaranteed by the general symmetry properties of ψ under Π. In general, we have

$$\psi(x, y, 0, w) = \sigma\psi(y, x, 0, w), \tag{3.48}$$

but one easily verifies that if $\psi(x, y, 0, w)$ is not even under interchanging x and y, ψ will correspond to an excited state, not degenerate with the ground state for vanishing coupling, and is hence outside Ω_0. To see this, consider the decomposition of the wavefunction in Eq. (3.22). We find in particular that if $\psi(x, y, 0, w)$ is odd under interchanging x and y, *and non-zero*, the wavefunction cannot couple to those modes for which $\chi^{(n)}_{[\dot{\cdot}]}(x, y, w)$ is even under interchange of x and y. But an analysis along the lines of [14] shows that the wavefunctions belonging to Ω_0 do couple to those modes. Thus if $\psi(x, y, 0, w)$ is non-zero and odd under interchanging x and y, then ψ is orthogonal to Ω_0. Consequently, $\hat{\psi}$ is properly defined as an $\mathscr{L}^2$ wavefunction on $\mathbf{R}^2$ with Hamiltonian H' as defined in Eq. (3.44).

We have therefore obtained an effective model in a smaller number of dimensions, allowing one (in principle) to compute the energy split between the even and odd ground states, beyond the semiclassical approximation, using H' as Hamiltonian.

This result carries over in a suitable generalization to our full problem of QCD on a torus, but we will leave the more extensive discussion that this issue deserves to the future, since this paper is restricted to the semiclassical analysis. (For a few additional remarks, see [3].)

4. THE EFFECTIVE LAGRANGIAN

With the example of the previous section to guide us, we can now return to discussing $SU(2)$ Yang–Mills on T^3. The Bloch effective Hamiltonian H' has already been calculated by Lüscher [2], and if we expand about $\mathbf{C}=\mathbf{0}$, then H' is a function of the spatially constant vector potentials ($\mathbf{k}=\mathbf{0}$ in Eq. (2.18)). To lowest order, it is just the Yang–Mills Hamiltonian in the gauge $\mathbf{A}_0=0$, restricted to these constant vector potentials, *with the coupling constant properly renormalized,*

$$H_0' = -\frac{g^2}{2L}\frac{\partial^2}{\partial c_i^{a2}} - \frac{1}{2g^2L}\operatorname{Tr}[c_i, c_j]^2, \tag{4.1}$$

where

$$A_i = \frac{c_i}{L} = \frac{c_i^a \sigma_a}{2L}, \qquad \partial_j c_i = 0 \quad \forall i, j. \tag{4.2}$$

A straightforward generalization of the arguments in the previous chapter tells us that the following Hamiltonian $\hat{H}$ can be used to calculate the lowest order energy splitting between the different electric flux sectors:

$$\hat{H} = H_0'(c_i^a) + V_1'(\mathbf{C}), \qquad |C_i| \leqslant \pi. \tag{4.3}$$

We extend $\hat{H}$ by periodicity to C values outside this range. Here $V_1'(\mathbf{C})$ is the one-loop effective potential along the vacuum valley, where the prime denotes that we have excluded the contribution from the constant modes:

$$V_1'(\mathbf{C}) = V_1(\mathbf{C}) - \frac{2\,|\mathbf{C}|}{L},$$

$$V_1(\mathbf{C}) = \frac{4}{L\pi^2} \sum_{\mathbf{n}\neq 0} \frac{\sin^2(\mathbf{n}\cdot\mathbf{C}/2)}{(\mathbf{n}^2)^2}. \tag{4.4}$$

We have used the fact that up to a constant, the sum of the ($\mathbf{C}$-dependent) eigenvalues of the quadratic fluctuations (see Eq. (2.18)), gives the effective potential

$$V_1(\mathbf{C}) = 2 \sum_{\mathbf{k}\in Z^3} \lambda_{\mathbf{C}}^{+}(\mathbf{k})^{1/2}, \tag{4.5}$$

$$V_1'(\mathbf{C}) = V_1(\mathbf{C}) - 2\lambda_{\mathbf{C}}^{+}(\mathbf{0})^{1/2}. \tag{4.6}$$

In Appendix A we will show how to resum these expressions to obtain the more rapidly converging result

$$V_1(C\mathbf{e}^{(1)}) = \frac{4}{\pi L}\left[\frac{C}{2}\left(\pi - \frac{C}{2}\right) + 2\sum_{n=1}^{\infty} a_n \sin^2(nC/2)\right]. \tag{4.7}$$

The a_n decrease rapidly with n,

$$a_n < \frac{5\pi^2}{n}e^{-2\pi n}, \tag{4.8}$$

and the first few are

$$a_1 = 2.7052746... \cdot 10^{-2}$$
$$a_2 = 1.6048745... \cdot 10^{-5}$$
$$a_3 = 1.6065690... \cdot 10^{-8}$$
$$a_4 = 1.9385627... \cdot 10^{-11}. \tag{4.9}$$

In [5] we explained how one could calculate κ_1, κ_3, and κ_4 in Lüscher's one-loop effective Hamiltonian:

$$LH' = (1 + g^2\kappa_2)\, LH'_0 + \kappa_1 c_k^a c_k^a + \kappa_3 (c_k^a c_k^a c_l^b c_l^b + 2c_k^a c_l^a c_k^b c_l^b)$$
$$+ \kappa_4 (5c_k^a c_k^a c_k^b c_k^b - c_k^a c_k^a c_l^b c_l^b - 2c_k^a c_l^a c_k^b c_l^b). \tag{4.10}$$

We found

$$\kappa_1 = \frac{1}{\pi}\left(-1 + 2\sum_{n=1}^{\infty} n^2 a_n\right),$$
$$\kappa_4 + \frac{3}{2}\kappa_3 = -\frac{1}{12\pi}\sum_{n=1}^{\infty} n^4 a_n,$$
$$\kappa_3 = \frac{4}{45}(4\pi)^{-2}. \tag{4.11}$$

This was part of a consistency check, analogous to the one we performed for the toy model in the previous section.

Before we continue to analyze $\hat{H}$ (Eq. (4.3)) in more detail, it is useful to show how these formulae can be derived using the Lagrangian approach. In this language, we are seeking an effective Lagrangian in the spatially constant vector potentials, obtained by integrating out all other degrees of freedom in the path integral. We start by introducing the projector P onto these states:

$$PA_\mu = \frac{1}{L^3}\int_{T^3} A_\mu. \tag{4.12}$$

A convenient gauge-fixing function for this procedure is then given by [3],

$$\chi = (1 - P)(\partial_\mu A_\mu + i[PA_\mu, A_\mu]) + L^{-1}PA_0, \tag{4.13}$$

so that if we define

$$B_\mu = PA_\mu, \qquad Q_\mu = (1 - P)A_\mu, \tag{4.14}$$

then $\chi = 0$ is equivalent to

$$B_0 = 0, \qquad \partial_\mu Q_\mu + i[B_\mu, Q_\mu] = 0. \tag{4.15}$$

This looks just like the background field gauge-fixing condition. However, there is an important difference from the standard background field approach, in that here B and Q are functions of A. This has consequences for the Faddeev–Popov ghost, as we will see. Also, because B is a dynamical variable, and is not invariant under gauge transformations, we need the first condition in Eq. (4.15) to fix the spatially constant but time varying gauge transformations. Clearly, $B_0 = 0$ is the most convenient choice of gauge for converting the effective Lagrangian to a Hamiltonian.

One can calculate the Faddeev–Popov determinant in the standard way. The variation of χ under an infinitesimal gauge transformation $\Omega = e^{i\epsilon\Lambda}$ is

$$\delta_\Lambda \chi = (1 - P)\{D_\mu(PA)\, D_\mu(\mathbf{A})\, \Lambda + i[P(D_\mu(\mathbf{A})\,\Lambda), A_\mu]\}$$
$$+ L^{-1}\partial_0 P\Lambda + iL^{-1}P[A_0, \Lambda], \tag{4.16}$$

where $D_\mu(A)$ is the covariant derivative

$$D_\mu(\mathbf{A})\,\Lambda = \partial_\mu \Lambda + i[A_\mu, \Lambda]. \tag{4.17}$$

Splitting Λ into $P\Lambda$ and $(1 - P)\,\Lambda = \Lambda'$, we find

$$\delta_\Lambda \chi = (1 - P)\{D_\mu(PA)\, D_\mu(\mathbf{A})\, \Lambda' - [P[A_\mu, \Lambda'], A_\mu]\}$$
$$+ \frac{1}{L}\partial_0 P\Lambda + \frac{i}{L}P[A_0, \Lambda'] + i[\chi, P\Lambda]. \tag{4.18}$$

and introducing the operator $\mathcal{M}$ via

$$\mathcal{M}\Lambda = D_\mu(PA)\, D_\mu(\mathbf{A})\, \Lambda + [A_\mu, P[A_\mu, \Lambda]], \tag{4.19}$$

gives the Fadeev–Popov determinant

$$\Delta(\mathbf{A}) = \left(\int \mathcal{D}\Omega\,\delta(\chi^\Omega) \right)^{-1}$$
$$= \int \mathcal{D}'\Psi\mathcal{D}'\bar\Psi\, d\eta\, d\bar\eta\, \exp\left(\frac{-i}{g_0^2} \int 2\mathrm{Tr}(\bar\Psi\mathcal{M}\Psi) + 2\,\mathrm{Tr}\left(\bar\eta\partial_0\eta + \frac{i}{L}\bar\eta P[A_0, \Psi] \right) \right). \tag{4.20}$$

The prime indicates that $P\Psi = P\bar\Psi = 0$. Here, η and $\bar\eta$ are spatially constant, and can be integrated out explicitly, giving an irrelevant constant. Continuing the standard arguments, we have $1 = \Delta(A^{\Omega_0}) \int \mathcal{D}\Omega\,\delta(\chi^\Omega - E)$, where Ω_0 is defined by $\chi^{\Omega_0} = E$ (see e.g., [18, p. 581]). Inserting this into the generating functional, we get

$$\mathcal{Z} = \int \mathcal{D}A_\mu\, \mathcal{D}'\Psi\, \mathcal{D}'\bar\Psi\, \exp\left(\frac{-i}{g_0^2} \int \frac{1}{2}\mathrm{Tr}(F_{\mu\nu}^2) - 2\,\mathrm{Tr}(\bar\Psi\mathcal{M}\Psi) \right) \delta(\chi - E), \tag{4.21}$$

which is independent of E. Integrating over E with $\int \mathscr{D}'E \exp((i/g_0^2)\int \mathrm{Tr}(E^2))$, where again the prime denotes $PE = 0$), we find

$$\mathscr{Z} = \int \mathscr{D}B_k \int \mathscr{D}'Q_\mu \, \mathscr{D}'\Psi \, \mathscr{D}'\bar{\Psi} \exp\left[\frac{-i}{g_0^2}\int \frac{1}{2}\mathrm{Tr}(F_{\mu\nu}(B+Q)^2)\right.$$

$$\left. - \mathrm{Tr}((D_\mu(B)\,Q_\mu)^2) + 2\,\mathrm{Tr}(\bar{\Psi}D_\mu(B)\,D_\mu(B+Q)\,\Psi) + 2\,\mathrm{Tr}([Q_\mu, \Psi]\,P[Q_\mu, \bar{\Psi}])\right].$$

$$(4.22)$$

One now easily extracts the effective Lagrangian for B, by writing this as

$$\mathscr{Z} \equiv \int \mathscr{D}B_k \exp\left(i\int dt\, \mathscr{L}_{\mathrm{eff}}(B)\right). \tag{4.23}$$

A peculiarity in our expression is the non-local interaction for the ghost, which is of course due to the non-local gauge fixing. In addition, we have a non-linear gauge, which normally needs a counterterm quartic in the ghosts [30]. However, because our non-linearity is only due to the constant modes, simple power counting shows that no quartic ghost divergences arise.

We wish to observe here that although Eq. (4.22) is renormalizable, in principle one needs to verify that gauge invariance is maintained at the quantum level, by using Ward identities. In this way, one could establish perturbative renormalizability *without ignoring infrared difficulties*. Here, we just assume that only one renormalized coupling constant has to be introduced. This assumption, incidentally, is made in the Hamiltonian approach too.

The one-loop calculation of $\mathscr{L}_{\mathrm{eff}}(B)$ is particularly simple, and confirms to that order the above assumption:

$$\exp\left(i\int dt\, \mathscr{L}_{\mathrm{eff}}^{\text{1-loop}}(B)\right) = \int \mathscr{D}'Q_\mu\, \mathscr{D}'\Psi\, \mathscr{D}'\bar{\Psi} \exp\left[\frac{-i}{g_0^2}\int \frac{1}{2}\mathrm{Tr}(F_{\mu\nu}(B)^2)\right.$$

$$\left. + \mathrm{Tr}(Q_\mu \mathscr{W}_{\mu\nu}(B)\,Q_\nu) - 2\,\mathrm{Tr}(\bar{\Psi}D_\mu(B)^2\,\Psi)\right], \tag{4.24}$$

with

$$\mathscr{W}_{\mu\nu}Q_\nu = -D_\sigma(B)^2\,Q_\mu - 2i[F_{\mu\nu}(B), Q_\nu]. \tag{4.25}$$

This is precisely the expression one obtains in a standard background-field calculation, except that the integration over the gauge and ghost fields excludes the constant modes. We can thus follow 't Hooft's approach [19] and write (f_{abc} are the structure constants for the gauge group):

$$\mathrm{Tr}(Q_\mu \mathscr{W}_{\mu\nu}(B)\,Q_\nu) = -\tfrac{1}{2}(\partial_\mu Q_\nu^a)^2 + Q_\alpha^a (N^\mu)_{\alpha\beta}^{ab}\, \partial_\mu Q_\beta^b + \tfrac{1}{2}Q_\mu^a M_{\mu\nu}^{ab} Q_\nu^b,$$

$$-2\,\mathrm{Tr}(\bar{\Psi}D_\mu(B)^2\,\Psi) = -(\partial_\mu \bar{\Psi}^a)(\partial_\mu \Psi^a) + 2\bar{\Psi}^a(\mathscr{N}_\mu)^{ab}\, \partial_\mu \Psi^b + \bar{\Psi}^a \mathscr{M}_0^{ab} \Psi^b,$$

$$(4.26)$$

with

$$(N^\mu)^{ab}_{\alpha\beta} = f_{abc} B^c_\mu \delta_{\alpha\beta},$$

$$M^{ab}_{\mu\nu} = -2f_{abc} F^c_{\mu\nu}(B) + (N_\sigma N^\sigma)^{ab}_{\mu\nu},$$

$$(\mathcal{N}_\mu)^{ab} = f_{abc} B^c_\mu,$$

$$\mathcal{M}^{ab}_0 = (\mathcal{N}_\sigma \mathcal{N}^\sigma)^{ab}.$$

$$(4.27)$$

Note that $N_0 = \mathcal{N}_0 = 0$, and that $f_{abc} \Lambda^c$ is just $i(\mathrm{ad}\,\Lambda)^{ab}$, the matrix for Λ in the adjoint representation. We now want to write down the most general possible one-loop effective action for B, compatible with the remaining symmetries, i.e., the cubic group $O(3, Z)$ and a global background gauge invariance. Clearly $\mathcal{L}^{\text{1-loop}}_{\text{eff}}(B)$ must be a function of $\mathrm{ad}\,\dot{B}_i$, $\mathrm{ad}\,B_i$, and $\mathrm{ad}\,F_{ij}(B)$. If we restrict ourselves to terms no more than quadratic in $\dot{B}$ and quartic in B, which is the order to which Lüscher works [2], we find

$$\mathcal{L}_{\text{eff}}(B) = \frac{1}{2g_0^2} \mathrm{Tr}(F_{\mu\nu}(B)^2) - b_1(\dot{B}^a_i)^2 + b_2(B^a_i)^2 + b_3\,\mathrm{Tr}(F_{ij}(B)^2)$$

$$+ b_4\,\mathrm{Tr}((\mathrm{ad}\,B_i)^4) + b_5 \sum_{i \neq j}^{3} \mathrm{Tr}(\mathrm{ad}\,B_i)^2(\mathrm{ad}\,B_j)^2. \qquad (4.28)$$

This can be put in a more convenient form (see [2] for some tools), which for $SU(2)$ is

$$\mathcal{L}_{\text{eff}}(B) = \frac{1}{2g_0^2} \mathrm{Tr}(F_{\mu\nu}(B)^2) - b_1(\dot{B}^a_i)^2 + b_3'\,\mathrm{Tr}(F_{ij}(B)^2)$$

$$+ b_2(B^a_i)^2 + b_4' B^a_i B^a_i B^b_i B^b_i + b_5'(B^a_i B^a_i B^b_j B^b_j + 2B^a_i B^a_j B^b_i B^b_j). \quad (4.29)$$

Next we observe that when we calculate the *divergent* part of $\mathcal{L}_{\text{eff}}$, the sum over the (non-zero) loop momenta can be replaced by an integral. Thus the divergent parts of the coefficients in Eq. (4.29) are exactly the same as in the infinite volume field theory calculation, i.e., the infinite parts of b_1 and b_3' are equal, and renormalize $(1/2g_0^2)\,\mathrm{Tr}(F^2_{\mu\nu})$ to $(1/2g^2)\,\mathrm{Tr}(F^2_{\mu\nu})$ in the usual way, whereas the other parameters (b_2, b_4', and b_5') are finite.

An alternative approach leading to the same conclusion is to use the quasi-classical expansion, as described for an arbitrary compact manifold in [20]. There one *always* gets the proper renormalization of the coupling constant, essentially because the small-scale behaviour is insensitive to the large-scale details of the smooth manifold on which we formulate our gauge theory.

We can actually get the finite coefficients in Eq. (4.29) without further work, by making some simple observations. First, if the finite parts of b_1 and b_3' are not equal, we can just rescale $B \to (1 + \alpha g^2) B$ to absorb those terms, without changing anything else at this order. Second, note that for $\dot{B} = 0$, $-\mathcal{L}_{\text{eff}}$ is just the effective potential V', which we have already calculated to one loop for abelian B in

Eq. (4.4). A Taylor expansion of V_1' thus fixes the remaining constants, giving $b_2 = L\kappa_1$, $b_4' = 5L^3\kappa_4$, and $b_5' = L^3(\kappa_3 - \kappa_4)$. This argument is clearly the consistency requirement we discussed earlier.

Converting Eq. (4.29) to a Hamiltonian indeed reproduces Lüscher's Hamiltonian, Eq. (4.10), and using the effective potential for arbitrary $SU(N)$ derived in Appendix D, we can apply the same reasoning and reproduce his more general formula. In principle, the results of Appendix D will in this way give the effective Hamiltonian for any simple gauge group, but we have left this to the reader.

We also leave to the reader the task of explicitly doing the background field calculation in Eq. (4.24) [19]. We have found that after integrating over the contiuous energy, one is left with exactly those momentum sums encountered by Lüscher in his Hamiltonian approach. We believe that there is a general one-to-one correspondence between Bloch perturbation theory and this Lagrangian method. In particular, rescaling $B_i^a = g_0^{2/3} c_i^a / L$ and $t = L g_0^{-2/3}\tau$ leaves an expansion in powers of $g_0^{2/3}$, and the energy integral of the propagator is analogous to the fundamental operator in Bloch perturbation theory [15]:

$$\frac{i}{\pi}\int dk_0 \frac{(1 - \delta_{\mathbf{k},\mathbf{0}})}{k_0^2 - \mathbf{k}^2 - i\varepsilon} \Leftrightarrow \frac{1 - P_0}{H_0 - E_0}. \tag{4.30}$$

In all this, the non-local ghost interaction plays a subtle and intriguing role. A nice example of the necessity of this non-local interaction was recently observed in a two-loop calculation [29].

Although the above gauge fixing term allows us to verify Lüscher's results and test our understanding, it is not well suited to a non-perturbative analysis, since it breaks the symmetry which maps $\mathbf{C}$ into $\mathbf{C} + 2\pi\mathbf{n}$. As was noted before in [3], a better choice is to replace everywhere the projector P by the projector P_3:

$$P_3 A_\mu = \frac{1}{L^3}\int_{T^3} \mathrm{Tr}(A_\mu \sigma_3)\frac{\sigma_3}{2}. \tag{4.31}$$

This is invariant under the gauge transformation $\Omega^{(j)}(\mathbf{x})$ in Eq. (2.13). The explicit expression for the gauge-fixing function is then

$$\chi_3 = (1 - P_3)(\partial_\mu A_\mu + i[P_3 A_\mu, A_\mu]) + L^{-1}P_3 A_0, \tag{4.32}$$

and one easily verifies that all steps from Eq. (4.13) to before the one-loop evaluation can be repeated. Thus one can write

$$\mathscr{Z} = \int \mathscr{D}\tilde{B}_k \int \mathscr{D}'\tilde{Q}_\mu \mathscr{D}'\Psi \mathscr{D}'\tilde{\Psi}\exp\left[\frac{-i}{g_0^2}\int \frac{1}{2}\mathrm{Tr}(F_{\mu\nu}(\tilde{B} + \tilde{Q})^2)\right.$$

$$-\mathrm{Tr}((D_\mu(\tilde{B})\tilde{Q}_\mu)^2) - 2\mathrm{Tr}(\tilde{\Psi}D_\mu(\tilde{B})D_\mu(\tilde{B} + \tilde{Q})\Psi)$$

$$\left.-2\mathrm{Tr}([\tilde{Q}_\mu, \Psi]P_3[\tilde{Q}_\mu, \tilde{\Psi}])\right], \tag{4.33}$$

320   VAN BAAL AND KOLLER

with

$$\tilde{B}_k = P_3 A_\mu, \qquad \tilde{Q}_\mu = (1 - P_3) A_\mu, \tag{4.34}$$

and the conditions

$$P_3 A_0 = P_3 \tilde{\Psi} = P_3 \tilde{\tilde{\Psi}} = 0. \tag{4.35}$$

This rigorously defines an effective Lagrangian via the analogue of Eq. (4.23):

$$\mathscr{Z} = \int \mathscr{D}\tilde{B}_k \exp\left(i \int dt \; \mathscr{L}_{\text{eff}}(\tilde{B}) \right). \tag{4.36}$$

Setting as in Eq. (2.11)

$$\tilde{B}_k = \frac{C_k \sigma_3}{2L}, \tag{4.37}$$

we obtain an effective Lagrangian for $\mathbf{C}$ with the symmetries

$$\tilde{\mathscr{L}}_{\text{eff}}(\mathbf{C}) = \tilde{\mathscr{L}}_{\text{eff}}(-\mathbf{C}) = \tilde{\mathscr{L}}_{\text{eff}}(\mathbf{C} + 2\pi\mathbf{n}). \tag{4.38}$$

However, a Feynman graph expansion around $\mathbf{C} = \mathbf{0}$ (or any point gauge-equivalent to $\mathbf{0}$) is not well defined, because the quadratic approximation breaks down there. But one can certainly expand around $\mathbf{C} = \pi\mathbf{e}^{(i)}$, and from Eq. (4.38) one for example easily derives that $L(\mathbf{q}) = \tilde{\mathscr{L}}_{\text{eff}}(\pi\mathbf{e}^{(i)} + \mathbf{q})$ satisfies $L(\mathbf{q}) = L(-\mathbf{q})$ and $L(\mathbf{q} + 2\pi\mathbf{n}) = L(\mathbf{q})$. By arguments similar to those used in deriving $\mathscr{L}_{\text{eff}}$, we find

$$L(\mathbf{q}) = \frac{L\dot{\mathbf{q}}^2}{2g^2}\left(1 + g^2 a_1\right) - a_2 \mathbf{q}^2 - a_3(\mathbf{q}^2)^2 - a_4 \sum_i q_i^4, \tag{4.39}$$

where we work to the same order as in Eq. (4.29). One can then calculate a_2, a_3, and a_4 indirectly as before by noting $L(\mathbf{q}, \dot{\mathbf{q}} = \mathbf{0}) = -V_1(\pi\mathbf{e}^{(i)} + \mathbf{q})$. In general in the classically forbidden region one has

$$\tilde{\mathscr{L}}_{\text{eff}}(\mathbf{C}) = \frac{L\dot{\mathbf{C}}^2}{2g^2} - V_1(\mathbf{C}), \tag{4.40}$$

but this breaks down near the points $\mathbf{C} = 2\pi\mathbf{n}$. As is clear from the toy models, which mimic the quartic mode problem and the V-shape of $V_1(\mathbf{C})$, this approximation is good for $\min_\mathbf{n} |\mathbf{C} - 2\pi\mathbf{n}| \gg g^{2/3}$. The path decomposition expansion tells us that we can deal separately with the clasically forbidden regions and the classically allowed regions. For $\mathbf{C} \to \mathbf{0}$ the dynamics is described by H'_0, and, after proper rescaling, enters in the combination $g^{-2/3}\mathbf{C}$. In Appendix C we will show how this combination can be understood heuristically in a loop expansion for $\tilde{\mathscr{L}}_{\text{eff}}(\mathbf{C})$ by examining the leading infrared ($\mathbf{C} \to \mathbf{0}$) behaviour.

5. The Polar Decomposition of the Effective Hamiltonian

We now address the problem of eliminating the gauge degrees of freedom from Lüscher's effective Hamiltonian, since they complicate the calculations. The gauge-invariant parameter $\mathbf{C}$ is a natural coordinate for the vacuum valley, so we will aim for an expression involving $\mathbf{C}$ and some "transverse" coordinates. Let us first describe an attempt based on the standard Faddeev–Popov gauge-fixing procedure, which turns out to be a nice illustration of the notion of Gribov ambiguities [26].

In the Hamiltonian approach we fix the gauge by $\chi_3 = 0$, restricted to the constant vector potentials, i.e.,

$$\chi = -i[P_3 c_i, c_i] = 0. \tag{5.1}$$

One can easily verify the following expression for the variation of χ under an infinitesimal gauge transformation $\Omega = \exp(i \in \Lambda) = \exp(i \in \Lambda^a \sigma^a / 2)$,

$$\delta_\Lambda \chi = \sum_{a,b=1}^{2} M_{ab} \Lambda^b \frac{\sigma^a}{2} - i \left[\frac{\sigma_3}{2}, \chi \right] \Lambda^3, \tag{5.2}$$

where M is the ghost propagator:

$$M_{ab} = \begin{pmatrix} -(\mathbf{c}^2)^2 + (\mathbf{c}^3)^2 & \mathbf{c}^1 \cdot \mathbf{c}^2 \\ \mathbf{c}^1 \cdot \mathbf{c}^2 & -(\mathbf{c}^1)^2 + (\mathbf{c}^3)^2 \end{pmatrix}, \tag{5.3}$$

with eigenvalues

$$\lambda_\pm(\mathbf{c}) = \tfrac{1}{2}\{3(\mathbf{c}^3)^2 - (c_i^a)^2 \pm \sqrt{((\mathbf{c}^1)^2 - (\mathbf{c}^2)^2)^2 + 4(\mathbf{c}^1 \cdot \mathbf{c}^2)^2}\}. \tag{5.4}$$

Hence Gribov copies are found at $\lambda_\pm(\mathbf{c}) = 0$ or

$$(\mathbf{c}^3)^2 = \tfrac{1}{2}\{(\mathbf{c}^1)^2 + (\mathbf{c}^2)^2 \pm \sqrt{((\mathbf{c}^1)^2 - (\mathbf{c}^2)^2)^2 + 4(\mathbf{c}^1 \cdot \mathbf{c}^2)^2}\} \geqslant 0. \tag{5.5}$$

A naive perturbation expansion around $\mathbf{c} = 0$ will thus encounter severe problems, because Gribov copies occur in any neighbourhood of that point. For $|\mathbf{c}^3|$ sufficiently large they can be avoided in perturbating around $(0, 0, \mathbf{c}^3)$, but their presence is a cause for suspicion, so we will not use this approach.

Instead of using this gauge-fixing method, we will transform to gauge-invariant coordinates using the polar decomposition [12], and thus explicitly eliminate the gauge degrees of freedom. In a sense, Gribov ambiguities are still present, as zeros in a Jacobian. These zeros are related to fixed points under a subgroup of the gauge group. (An analog is spherical coordinates on $\mathbf{R}^3$, with symmetry group $SO(3)$: The origin is left fixed, and the Jacobian vanishes at that point.) However, such zeros do not lead to any singular behaviour, and there is no difficulty in dealing with them, as we will show.

The polar decomposition is given by [27]

$$c_i^a = \sum_{b=1}^{3} \xi^{ab} x_b \eta_{bi}, \tag{5.6}$$

with $\xi, \eta \in SO(3)$. The uniqueness of this decomposition is discussed in Appendix B. Clearly, ξ carries the gauge degrees of freedom and η the rotational degrees. The x_b label the gauge- and rotation-invariant degrees of freedom. Let us first consider gauge-invariant, rotationally-symmetric wavefunctions. Since H_0' is both gauge and rotation invariant, the equation for H_0' acting on such a wavefunction can be written in terms of the x_b only. The potential part is easy:

$$V_0'(c_i^a) = -\tfrac{1}{2}\mathrm{Tr}[c_i, c_j]^2 = \tfrac{1}{4}[(c_i^a c_i^a)^2 - c_i^a c_j^a c_i^b c_j^b]$$
$$= \tfrac{1}{2}(x_1^2 x_2^2 + x_1^2 x_3^2 + x_2^2 x_3^2). \tag{5.7}$$

The kinetic part is a little harder. If we write ξ and η in terms of the $SO(3)$ generators $(L_i)_{jk} = \varepsilon_{ijk}$ as $\zeta = \exp(i\mathbf{y}\cdot\mathbf{L})$ and $\eta - \exp(i\mathbf{z}\cdot\mathbf{L})$, and denote the nonet $(\mathbf{x}, \mathbf{y}, \mathbf{z})$ by u^μ, $\mu = 1,...,9$ we can write

$$\frac{\partial^2}{\partial c_i^{a2}} = \frac{1}{\mathcal{J}}\frac{\partial}{\partial u^\mu} g^{\mu\nu} \mathcal{J} \frac{\partial}{\partial u^\nu}, \tag{5.8}$$

and the metric $g_{\mu\nu}$ is determined by $(\dot{c}_i^a)^2 = g_{\mu\nu}\dot{u}_\mu \dot{u}_\nu$, with $\mathcal{J} = \sqrt{\det(g_{\mu\nu})}$. Using the orthogonality properties $\sum_a \xi^{ab}\dot{\xi}^{ab} = \sum_a \eta_{ab}\dot{\eta}_{ab} = 0$ one easily verifies that $g_{\mu\nu}$ has the form

$$g_{\mu\nu} = \begin{pmatrix} \mathbf{1}_3 & 0 \\ 0 & \hat{g}_{\hat{\mu}\hat{\nu}} \end{pmatrix}, \tag{5.9}$$

where $\hat{g}_{\hat{\mu}\hat{\nu}}$ is a 6×6 symmetric matrix. Clearly the inverse of the metric, $g^{\mu\nu}$, is of the same form. Hence (5.8) acting on gauge-invariant rotation-invariant wavefunctions is of the form $(1/\hat{\mathcal{J}})(\partial/\partial x_i)\,\hat{\mathcal{J}}(\partial/\partial x_i)$. Since $\hat{\mathcal{J}}$ is invariant it can be evaluated at $\xi = \eta = \mathbf{1}$; a simple calculation using the explicit form of the $SO(3)$ generators yields

$$\hat{g}_{\hat{\mu}\hat{\nu}} = \begin{pmatrix} A & B \\ B & A \end{pmatrix},$$

$$A = \begin{pmatrix} x_2^2 + x_3^2 & 0 & 0 \\ 0 & x_1^2 + x_3^2 & 0 \\ 0 & 0 & x_1^2 + x_2^2 \end{pmatrix}, \qquad B = \begin{pmatrix} x_2 x_3 & 0 & 0 \\ 0 & x_1 x_3 & 0 \\ 0 & 0 & x_1 x_2 \end{pmatrix}, \tag{5.10}$$

and we find

$$\mathcal{J}^2 = \det(A^2 - B^2) = \prod_{i>j}(x_i^2 - x_j^2)^2. \tag{5.11}$$

Hence restricted to the gauge and rotation invariant sector we have:

$$H_0' = -\frac{g^2}{2}\left(\prod_{i>j}(x_i^2 - x_j^2)\right)^{-1}\frac{\partial}{\partial x_k}\prod_{i>j}(x_i^2 - x_j^2)\frac{\partial}{\partial x_k} + \frac{1}{2g^2}(x_1^2 x_2^2 + x_1^2 x_3^2 + x_3^2 x_2^3),$$

$$(5.12)$$

and the inner product of two such invariant wavefunctions is (up to a constant)

$$\langle \Psi \mid \Phi \rangle = \int d^3x \left| \prod_{i>j}(x_i^2 - x_j^2) \right| \bar{\Psi}(x)\, \Phi(x). \qquad (5.13)$$

As remarked earlier, like when converting cartesian to polar coordinates we can understand the zeros of the Jacobian as being due to points which are fixed under combined rotations and gauge transformations. A point where some of the x_i are equal is clearly invariant under a nontrivial subgroup of $SO(3) \times SO(3)$.

From the invariance of det $c_i^a = \prod_i x_i$ one sees that any permutation of the x_i or any simultaneous change of the sign of two of the x_i can be achieved by a combined rotation and gauge transformation. This implies we have a 6×4-fold covering of the gauge and rotation invariant configurations by $\mathbf{R}^3$, and it explains why vanishing of the Jacobian at $x_i = x_j$ implies its vanishing at $x_i = -x_j$. It also shows that the absolute value in (5.13) does not lead to singularities, because the wavefunction has a permutation symmetry. Equation (5.12) is hermitian with respect to (5.13) on the Hilbert space of wavefunctions satisfying

$$\psi(x_1, x_2, x_3) = \psi(\pm x_{\pi(1)}, \pm x_{\pi(2)}, \pm x_{\pi(3)}), \qquad (5.14)$$

where π is any 3-permutation. Equivalently, we can restrict x_i as follows:

$$0 \leqslant x_2 \leqslant x_3, \qquad x_1^2 \leqslant x_2^2. \qquad (5.15)$$

The potential V_0' (Eq. (5.7)) of H_0' is zero if and only if two of the three x_i are zero. With the restriction (5.15) this means $x_1 = x_2 = 0$. Hence the vacuum valley is specified by

$$C_i = x_3 \eta_{3i}. \qquad (5.16)$$

Thus $x_3 = |\mathbf{C}|$ and the following parameterization of η is forced upon us:

$$\eta_{3i} = C_i/|\mathbf{C}|,$$

$$\eta_{1i} = a_i \cos \xi - b_i \sin \xi, \qquad (5.17)$$

$$\eta_{2i} = a_i \sin \xi + b_i \cos \xi,$$

324 VAN BAAL AND KOLLER

with $\mathbf{a}$, $\mathbf{b}$, and $\mathbf{C}/|\mathbf{C}|$ mutually orthogonal unit vectors with positive orientation. Explicitly we can write

$$\mathbf{a} = (\cos\varphi\cos\theta,\ \sin\varphi\cos\theta,\ -\sin\theta),$$

$$\mathbf{b} = (-\sin\varphi,\ \cos\varphi,\ 0), \tag{5.18}$$

$$\mathbf{C}/|\mathbf{C}| = (\cos\varphi\sin\theta,\ \sin\varphi\sin\theta,\ \cos\theta),$$

with

$$\theta \in [0,\pi], \qquad \varphi \in [0, 2\pi], \qquad \xi \in [0, 2\pi]. \tag{5.19}$$

Thus (x_3, θ, φ) is a spherical coordinate system for the vacuum valley, and $(x_1\, x_2, \xi)$ are the "transverse" coordinates.

The kinetic term in the general case of a gauge-invariant, but not necessarily rotationally invariant, wavefunction is derived in the appendix,

$$\frac{\partial^2}{\partial c_i^{a2}} = -\frac{1}{2}\left(\prod_{i>j}(x_i^2 - x_j^2)\right)^{-1}\frac{\partial}{\partial x_k}\prod_{i>j}(x_i^2 - x_j^2)\frac{\partial}{\partial x_k} - \frac{1}{4}\sum_{i\neq j\neq k}\frac{x_i^2 + x_j^2}{(x_i^2 - x_j^2)^2}\hat{L}_k^2, \tag{5.20}$$

where $\hat{L}_k$ are the generators for the representation of $SO(3)$ acting on the wavefunction $\psi(x_i, \theta, \varphi, \xi)$ with the inner product

$$\langle \Psi \mid \Phi \rangle = \int_{0 \leqslant x_1, x_2 \leqslant x_3} d^3x\, d\theta\, d\varphi\, d\xi\, \sin\theta\left|\prod_{i>j}(x_i^2 - x_j^2)\right| \bar{\Psi}(c)\, \Phi(c). \tag{5.21}$$

Explicitly these generators are

$$\hat{L}_1 = -i\left[\cot\theta\cos\xi\partial_\xi + \frac{\cos\xi}{\sin\theta}\partial_\varphi + \sin\xi\partial_\theta\right],$$

$$\hat{L}_2 = -i\left[\cot\theta\sin\xi\partial_\xi + \frac{\sin\xi}{\sin\theta}\partial_\varphi - \cos\xi\partial_\theta\right], \tag{5.22}$$

$$\hat{L}_3 = -i\partial_\xi,$$

and a simple calculation shows that indeed $[\hat{L}_i, \hat{L}_j] = i\varepsilon_{ijk}\hat{L}_k$.

Let us finally transform the Jacobian away, by scaling the wavefunction according to

$$\tilde{\Psi} = \sqrt{|\mathcal{J}|}\,\Psi, \qquad \mathcal{J} = (x_2^2 - x_1^2)(x_3^2 - x_1^2)(x_3^2 - x_2^2). \tag{5.23}$$

Noting that $(\partial^2/\partial x_k^2)\,\mathcal{J} = 0$ and $((\partial/\partial x_k)\ln\mathcal{J})^2 = 2\sum_{i\neq j}((x_i^2 + x_j^2)/(x_i^2 - x_j^2)^2)$ we find

$$\tilde{H}_0' = -\frac{g^2}{2L}\frac{\partial^2}{\partial x_k^2} + \frac{g^2}{4L}\sum_{i\neq j\neq k}\frac{x_i^2 + x_j^2}{(x_i^2 - x_j^2)^2}(\hat{L}_k^2 - 1) + \frac{1}{2g^2L}\sum_{i>j}x_i^2 x_j^2. \tag{5.24}$$

Our next task is to define $\tilde{\tilde{H}}$, which is obtained by applying the above transformation to the Hamiltonian $\hat{H}$ of Eq. (4.3). Thus we write

$$\tilde{\tilde{H}} = \tilde{H}'_0 + V'_1(\mathbf{C}) = H(\mathbf{C}) + \tilde{H}_{[\mathbf{C}]}(x_1, x_2, \xi) + T, \qquad (5.25)$$

with (it will become clear below why we make this split)

$$H(\mathbf{C}) = -\frac{g^2}{2L}\frac{\partial^2}{\partial |\mathbf{C}|^2} + \frac{4}{L\pi^2}\sum_{\mathbf{n}\neq 0}\frac{\sin^2(\mathbf{n}\cdot\mathbf{C}/2)}{(\mathbf{n}^2)^2} - \frac{g^2}{2LC^2}(\sin^2\theta\,\partial^2_\varphi + \partial^2_\theta), \quad (5.26)$$

$$\begin{aligned}
\tilde{H}_{[\mathbf{C}]}(x_1, x_2, \xi) = {}&-\frac{g^2}{2L}\left(\frac{\partial^2}{\partial x_1^2} + \frac{\partial^2}{\partial x_2^2}\right) - \frac{g^2}{2L}\left(\frac{x_1^2 + x_2^2}{(x_1^2 - x_2^2)^2}\right)(\partial^2_\xi + 1) \\
&- \frac{g^2}{2L}\cot^2\theta\left[\cos^2\xi\,\frac{(\mathbf{C}^2 + x_1^2)}{(\mathbf{C}^2 - x_1^2)^2} + \sin^2\xi\,\frac{(\mathbf{C}^2 + x_2^2)}{(\mathbf{C}^2 - x_2^2)^2}\right]\partial^2_\xi - \frac{g^2}{L\mathbf{C}^2} \\
&+ \frac{1}{2g^2L}(x_1^2 + x_2^2)\mathbf{C}^2 + \frac{1}{2g^2L}x_1^2 x_2^2 - \frac{2|\mathbf{C}|}{L}.
\end{aligned} \qquad (5.27)$$

One easily verifies that T, implicitly defined by the above three equations, vanishes when ∂_θ and ∂_φ vanish. The fixed-C eigenfunctions $\tilde{\chi}^{(n)}_{[\mathbf{C}]}(x_1, x_2, \xi)$ for $\tilde{H}_{[\mathbf{C}]}(x_1, x_2, \xi)$ satisfy vanishing boundary conditions for $x_1^2 = x_2^2$, $x_1^2 = \mathbf{C}^2$ and $x_2^2 = \mathbf{C}^2$.

The split in Eq. (5.25) of course anticipates that we want to derive an equation along the vacuum valley for the original Yang–Mills Hamiltonian, similar to (3.41) for the toy model; this will be done in the next section. Although in this case, we cannot solve exactly for $\chi^{(n)}_{[\mathbf{C}]}$, we can use perturbation theory [5]. To see what the expansion parameter is we scale x_1 and x_2 as

$$x_i = \frac{gy_i}{|\mathbf{C}|^{1/2}}. \qquad (5.28)$$

The rescaled Hamiltonian becomes

$$\tilde{H}_{[\mathbf{C}]} = \frac{|\mathbf{C}|}{2L}\left\{-\frac{\partial^2}{\partial y_i^2} - \frac{y_1^2 + y_2^2}{(y_1^2 - y_2^2)^2} + y_1^2 + y_2^2 - 2 + K\hat{L}_3^2 + \frac{g^2}{|\mathbf{C}|^3}(y_1^2 y_2^2 - 2)\right\}, \qquad (5.29)$$

with vanishing boundary conditions at $y_1^2 = y_2^2$, $y_1^2 = |\mathbf{C}|^3/g^2$, and $y_2^2 = |\mathbf{C}|^3/g^2$. K is positive definite and can be read off from Eq. (5.27). Comparing with (5.23) it is clear that we can also write

$$\tilde{\chi}^{(n)}_{[\mathbf{C}]} = \sqrt{|x_1^2 - x_2^2|}\,\chi^{(n)}_{[\mathbf{C}]}, \qquad (5.30)$$

VAN BAAL AND KOLLER

with $\tilde{H}_{[C]}$ transformed to $H_{[C]}$,

$$H_{[C]} = \frac{|C|}{2L} \left\{ -\frac{1}{y_1^2 - y_2^2} \frac{\partial}{\partial y_i} (y_1^2 - y_2^2) \frac{\partial}{\partial y_i} + y_1^2 + y_2^2 - 2 + K \left(\frac{1}{i} \partial_\xi \right)^2 \right.$$

$$\left. + \frac{g^2}{|C|^3} (y_1^2 y_2^2 - 2) \right\}, \qquad (5.31)$$

where $\chi_{[C]}$ only has vanishing boundary conditions at $y_1^2 = |C|^3/g^2$ and $y_2^2 = |C|^3/g^2$. The ground state is especially easy to determine; it is clearly independent of ξ, so that $g^2/|C|^3$ will be the expansion parameter;

$$H_{[C]} \chi_{[C]}^{(n)} = W_n(C) \chi_{[C]}^{(n)} \qquad (5.32)$$

therefore has the property that the ground state $(\chi_{[C]}^{(1)}, W_1(C))$ only depends on $|C|$, so $T_{11} \equiv 0$, where

$$T_{nm} = \frac{1}{g^2} \langle \chi_{[C]}^{(n)} | T | \chi_{[C]}^{(m)} \rangle. \qquad (5.33)$$

Furthermore, ignoring the boundary conditions for the ground state would only lead to $O(\exp(-\text{const.}/g))$ corrections, so that these boundary conditions can be replaced by square integrability of $\chi_{[C]}^{(1)}$, and we find ($\chi_{[C]}^{(1)}$ does not depend on ξ, but "ξ-dependence" is present because of the normalization)

$$\chi_{[C]}^{(1)} = \frac{|C|}{\sqrt{4\pi} \, g^2} \exp \left(\frac{-|C| (x_1^2 + x_2^2)}{2g^2} \right) (1 + O(g^2/|C|^3)),$$

$$\qquad (5.34)$$

$$W_1(C) = -\frac{3}{4\pi L} \frac{g^2}{C^2} + L^{-1} O(g^4/|C|^5).$$

$\tilde{\chi}$ is normalized with respect to the standard inner product (cf. (5.21) and Appendix B).

The next section shows how the tunneling problem can be reduced to the tunneling problem for a simple potential (essentially $V_1(C)$) on a cubic lattice (lattice constant 2π) with the lattice momenta restricted to multiples of π.

6. The Reduction Formula

Combining the results of the previous sections we can write (we introduce the factor $|C|$ to convert the measure to $d^3C \, dx_1 \, dx_2 \, d\xi$)

$$\tilde{\Psi}_0(C, x_1, x_2, \xi) = \sum_n |C| \, \varphi_0^{(n)}(C) \, \tilde{\chi}_{|C|}^{(n)}(x_1, x_2, \xi), \qquad (6.1)$$

$$\int_{\substack{|x_1| \leq |C| \\ |x_2| \leq |C|}} d^3C \, dx_1 \, dx_2 \, d\xi \, |\tilde{\Psi}_0(C, x_1, x_2, \xi)/|C||^2 = 1, \qquad (6.2)$$

$$\hat{H}_{nm}(\mathbf{C})\,\varphi_0^{(m)}(\mathbf{C}) = \left[-\frac{g^2}{2L}\left(\delta_{nm}\frac{\partial}{\partial\mathbf{C}} - \mathbf{A}_{nm}(\mathbf{C})\right)^2 + g^2 T_{nm}\left(\mathbf{C},\frac{\partial}{\partial\mathbf{C}}\right) \right.$$

$$\left. + (V_1(\mathbf{C}) + W_n(\mathbf{C}))\,\delta_{nm} \right]\varphi_0^{(m)}(\mathbf{C}) = E_0\,\varphi_0^{(n)}(\mathbf{C}), \qquad (6.3)$$

$$\mathbf{A}_{nm}(\mathbf{C}) = \left\langle \tilde{\chi}_{[\mathbf{C}]}^{(m)} \left| \frac{\partial}{\partial\mathbf{C}} \right| \tilde{\chi}_{[\mathbf{C}]}^{(n)} \right\rangle. \qquad (6.4)$$

The single-well lowest order wavefunction $\Psi_0^{[0]}(x_i)$ is determined by

$$H_0'\,\Psi_0^{[0]}(x_i) = \left(-\frac{g^2}{2L}\frac{1}{\mathscr{J}}\frac{\partial}{\partial x_i}\,\mathscr{J}\,\frac{\partial}{\partial x_i} + \frac{1}{2g^2 L}\sum_{i>j} x_i^2 x_j^2 \right)\Psi_0^{[0]}(x_i)$$

$$= E_0^{[0]}\,\Psi_0^{[0]}(x_i), \qquad (6.5)$$

$$8\pi^2 \int_{\substack{0\leqslant|x_1|\leqslant x_3 \\ 0\leqslant|x_2|\leqslant x_3}} dx_1\,dx_2\,dx_3\,|\mathscr{J}|\,|\Psi_0^{[0]}(x_i)|^2 = 1, \qquad (6.6)$$

$$E_0^{[0]} = \frac{g^{2/3}}{L}\,\varepsilon, \qquad \varepsilon = 4.11672.... \qquad (6.7)$$

The value of ε was determined previously by Lüscher and Münster [13] using Rayleigh–Ritz perturbation theory. The associated expression for $\varphi_0^{(1)}(\mathbf{C})$ is in the single-well approximation given by

$$\varphi_0^{[0]}(\mathbf{C}) = \frac{1}{|\mathbf{C}|}\int dx_1\,dx_2\,d\xi\;\Psi_0^{[0]}(x_i)\,|\mathscr{J}|^{1/2}\,\tilde{\chi}_{[\mathbf{C}]}^{(1)}(x_1, x_2, \xi). \qquad (6.8)$$

Using the approximation (Eq. (5.34)) of $\chi_{[\mathbf{C}]}^{(1)}(x_1, x_2)$ for $|\mathbf{C}| \gg g^{2/3}$ we find

$$\varphi_0^{[0]}(\mathbf{C}) \cong \frac{\sqrt{\pi}\,|\mathbf{C}|^2}{g^2}\int dx_1\,dx_2\,|x_1^2 - x_2^2|\,\Psi_0^{[0]}(x_i)\exp\left(\frac{-|\mathbf{C}|\,(x_1^2 + x_2^2)}{2g^2}\right). \qquad (6.9)$$

This completes the construction of the tunneling data as described in Section 3, with the difference that z is replaced by a three-component vector $\mathbf{C}$.

We next apply the path decomposition expansion, as was described in detail for the toy-models [14]. For this we need to analyze the asymptotic behaviour of $\varphi_0^{[0]}(\mathbf{C})$. For $g^{2/3} \ll |\mathbf{C}| \ll 1$, $\varphi_0^{[0]}(\mathbf{C})$ satisfies the following effective Schrödinger equation:

$$\left[-\frac{g^2}{2L}\frac{\partial}{\partial\mathbf{C}^2} + V_1^{[0]}(\mathbf{C}) \right]\varphi_0^{[0]}(\mathbf{C}) = E_0^{[0]}\,\varphi_0^{[0]}(\mathbf{C}), \qquad (6.10)$$

$$V_1^{[0]}(\mathbf{C}) = \frac{2\,|\mathbf{C}|}{L} = \lim_{\lambda\to 0}\lambda^{-1}V_1(\lambda\mathbf{C}). \qquad (6.11)$$

Using the spherical symmetry, the WKB approximation in $|\mathbf{C}|$ gives the asymptotics

$$\varphi_0^{[0]}(\mathbf{C}) \cong \frac{g^{-1/6}}{|\mathbf{C}|} B \frac{\exp(-(4\,|\mathbf{C}| - 2LE_0^{[0]})^{3/2}/6g)}{(4\,|\mathbf{C}| - 2LE_0^{[0]})^{1/4}}. \tag{6.12}$$

The following observations should make this result clear: Rescaling $x_i = g^{2/3} y_i$ and $\mathbf{C} = g^{2/3}\mathbf{W}$ gives $\Psi_0^{[0]}(x_i) = g^{-1}\hat{\Psi}_0^{[0]}(y_i)$ and $\varphi_0^{[0]}(\mathbf{C}) = g^{-1}\hat{\varphi}_0^{[0]}(\mathbf{W})$, where $\hat{\Psi}$ satisfies (6.6) with x_i replaced by y_i, and $\hat{\Psi}(y_i)$ is independent of g. Consequently $\hat{\varphi}_0^{[0]}(\mathbf{W})$ is also independent of g and it satisfies

$$\left(-\frac{1}{2}\frac{\partial^2}{\partial\mathbf{W}^2} + 2\,|\mathbf{W}|\right)\hat{\varphi}_0^{[0]}(\mathbf{W}) = \varepsilon\hat{\varphi}_0^{[0]}(\mathbf{W}), \tag{6.13}$$

so that the WKB approximation gives

$$\hat{\varphi}_0^{[0]}(\mathbf{W}) \cong \frac{B}{|\mathbf{W}|}\frac{\exp(-\int_{\varepsilon/2}^{|\mathbf{W}|}\sqrt{2(2s-\varepsilon)}\,ds)}{(4\,|\mathbf{W}| - 2\varepsilon)^{1/4}}. \tag{6.14}$$

Substituting $\mathbf{W} = g^{-2/3}\mathbf{C}$ gives Eq. (6.12).

We would now like to construct a three-dimensional quantum mechanics problem completely equivalent (to this order) to the present problem. This can be done as follows: We replace $V_1(\mathbf{C})$ by $\tilde{V}_1(\mathbf{C})$, where

$$\begin{aligned}
\tilde{V}_1(\mathbf{C}) &= LV_1'(\mathbf{C}) + 2g^{2/3}\delta, & |\mathbf{C} - 2\pi\mathbf{n}| \leqslant g^{2/3}\delta \\
\tilde{V}_1(\mathbf{C}) &= LV_1(\mathbf{C}), & \text{elsewhere.}
\end{aligned} \tag{6.15}$$

The g-independent constant δ is chosen so that the lowest eigenvalue for the three-dimensional Hamiltonian with the potential

$$\begin{aligned}
V_\delta(x) &= 2\delta, & |x| \leqslant \delta \\
&= 2\,|x|, & |x| > \delta
\end{aligned} \tag{6.16}$$

is exactly equal to ε:

$$\left(-\frac{1}{2}\frac{\partial^2}{\partial x^2} + V_\delta(|\mathbf{x}|)\right)\hat{\varphi}_\delta(x) = \varepsilon\hat{\varphi}_\delta(x). \tag{6.17}$$

$\hat{\varphi}_\delta(x)$ satisfies the asymptotic equation ($|\mathbf{x}| \gg 1$)

$$\hat{\varphi}_\sigma(\mathbf{x}) = \frac{B_\delta}{|\mathbf{x}|}\frac{\exp(-\int_{\varepsilon/2}^{|\mathbf{x}|}\sqrt{2(2s-\varepsilon)}\,ds)}{(4\,|\mathbf{x}| - 2\varepsilon)^{1/4}}, \tag{6.18}$$

which is purposely of the same form as Eq. (6.14).

Finally, consider the Hamiltonian

$$-\frac{g^2}{2}\frac{\partial^2}{\partial \mathbf{C}^2} + \tilde{V}_1(\mathbf{C}), \tag{6.19}$$

and restrict the wavefunctions to

$$\varphi_\mathbf{e}(\mathbf{C} + 2\pi\mathbf{n}) = (-1)^{\mathbf{e}\cdot\mathbf{n}}\,\varphi_\mathbf{e}(\mathbf{C}), \tag{6.20}$$

and let $E_\delta(\mathbf{e}, g)$ be the associated groundstate energies for each given $\mathbf{e}$. Then (up to relative errors vanishing as a positive power of $g(L)$) the energy of electric flux for $SU(2)$ on the hypertorus T^3 of size L^3 is given by

$$E(\mathbf{e}, L) = \frac{1}{L}\frac{|B|^2}{|B_\delta|^2} E_\delta(\mathbf{e}, g(L)), \tag{6.21}$$

with $g(L)$ the renormalized coupling constant at the scale L, i.e., to leading order

$$g(L) = \left[-\frac{11}{12\pi^2}\ln(\Lambda L)\right]^{-1/2}. \tag{6.22}$$

Hence we have reduced the problem to that of the calculation of the energy for a particle in a cubic lattice with potential $\tilde{V}_1(\mathbf{C})$. Furthermore, we are only interested in the lattice momenta $\mathbf{p} = \mathbf{e}\pi$ (because of the gauge invariance).

It will turn out that the actual value of δ is irrelevant, as it should be, since we introduced it only for a convenient formulation of the problem. We will therefore not specify the value of δ. The next section deals with the calculation of $E_\delta(\mathbf{e}, g)$, and it will be shown (as is obvious from the double-cone toy model in [14]) that $E_\delta(\mathbf{e}, g)$ is of the form $|B_\delta|^2$ times something independent of δ, so that indeed the δ-dependence drops out of Eq. (6.21).

7. The Tunneling Calculation

In the calculation of the energy as a function of the lattice momenta $\mathbf{p}$, for the Hamiltonian of Eq. (6.19) one can apply the tight binding limit of solid state quantum mechanics or the dilute gas picture for instantons. In both cases one finds

$$E_\delta(\mathbf{p}, g) = \left(\sum_{i=1}^{3}\sin^2\frac{p_i}{2}\right)\Delta(g), \tag{7.1}$$

where we normalize such that $E(\mathbf{0}, g) \equiv 0$. $\Delta(g)$ is the energy split between the even and odd wavefunctions, taking only tunneling between two nearest neighbour minima into account. In other words, $\Delta(g)$ is the energy split due to the single pinchon contribution, where we called the instanton for the Hamiltonian of

                              VAN BAAL AND KOLLER

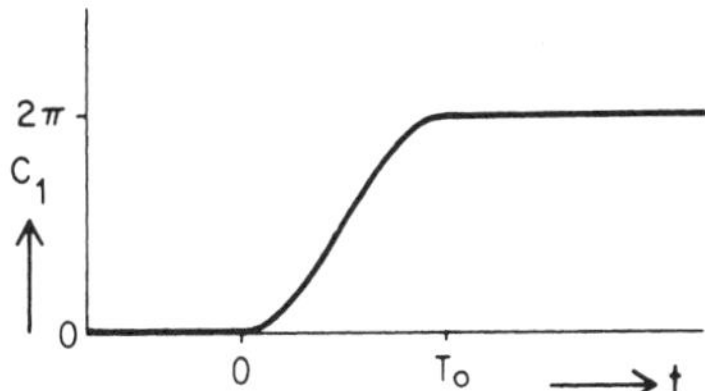

FIG. 2. The pinchon solution $C(t)$ in the approximation $V_1(C) = C(2\pi - C)/\pi$.

Eq. (6.19) a pinchon, because the effective potential $V_1(\mathbf{C})$ is quantum induced and provides a potential barrier due to the pinching of the full potential $V(\mathbf{A})$, between the two minima (cf. Fig. 1). In Fig. 2 we present this pinchon solution in the approximation $a_n = 0$ [3] and in Fig. 3 we sketch the electric field it generates.

Having reduced the problem to that of a double well, we can directly apply techniques developed in [14]. We focus on the domain $-\pi \leqslant C_1 \leqslant 3\pi$, $-\pi \leqslant C_2$, $C_3 \leqslant \pi$, containing two nearest neighbour minima at $\mathbf{C} = \mathbf{0}$ and $\mathbf{C} = 2\pi\mathbf{e}^{(1)}$. It is clear (shifting the origin to $\mathbf{C} = \pi\mathbf{e}^{(1)}$ and rescaling by π) that the conditions of [14] are satisfied for $\tilde{V}_1(\mathbf{C})$. Indeed, in the appendix we will show that $V_1(\mathbf{C}) \geqslant V_1(C_1\mathbf{e}^{(1)})$ for all C_2, C_3, so that the tunneling path is along the C_1 axis. Furthermore, the lowest order single-well potential $\tilde{V}_1^{[0]}(\mathbf{C}) = V_\delta(|\mathbf{C}|)$ is spherically symmetric and thus of the double-cone shape studied in [14]. Accordingly, we can copy the expression for $\Delta(g)$ directly from these toy models ([14], Eq. (4.2)) $(d \gg g^{2/3})$,

$$\Delta(g) = 2g \sqrt{2(\tilde{V}_1(d\mathbf{e}^{(1)}) - g^{2/3}\varepsilon)} \, |\varphi_\delta(d\mathbf{e}^{(1)})|^2$$
$$\times \exp\left(-\frac{1}{g}\int_d^{2\pi - d} \sqrt{2(\tilde{V}_1(s\mathbf{e}^{(1)}) - g^{2/3}\varepsilon)} \, ds\right) \Lambda(d), \qquad (7.2)$$

where $\varphi_\delta(\mathbf{C}) = g^{-1}\hat{\phi}_\delta(g^{-2/3}|\mathbf{C}|)$, (see Eq. (6.18)) and $\Lambda(d)$ is the contribution due to the transverse fluctuations. Hence,

$$\Delta(g) = 2g^{2/3}d^{-2}\Lambda(d)\,|B_\delta|^2 \exp\left(-\frac{1}{g}\int_{g^{2/3}\varepsilon}^{2\pi - g^{2/3}\varepsilon} \sqrt{2(\tilde{V}_1(s\mathbf{e}^{(1)}) - g^{2/3}\varepsilon)} \, ds\right). \quad (7.3)$$

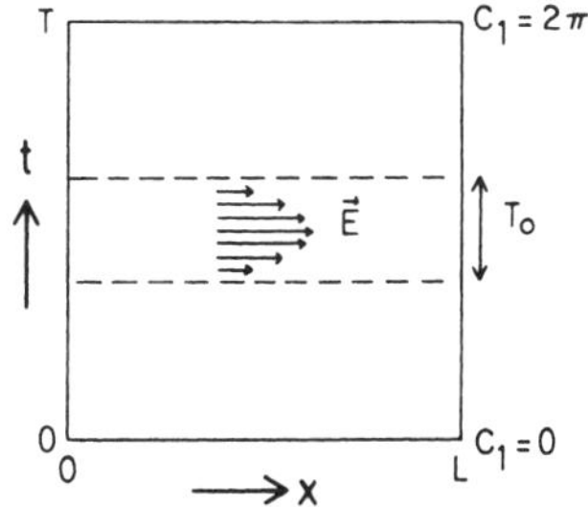

FIG. 3. The electric field $E = \dot{C}(t)$ from the pinchon in Fig. 2.

Thus we are left with the computation of $\Lambda(d)$ which is given by [14] ($\dot{} = d/dt$)

$$\Lambda(d) = 2\pi g [\dot{q}(T) + \phi''_\perp(d)\, q(T)]^{-1},$$

$$\phi''_\perp(d) = -g^2 \frac{\partial^2}{\partial C_2^2} \ln(|\varphi_\delta(\mathbf{C})|)\,|_{\mathbf{C} = d\mathbf{e}^{(1)}}. \tag{7.4}$$

The fluctuation $q(t)$ is a solution of

$$\ddot{q}(t) = \Omega^2(t)\, q(t),$$

$$q(0) = 1, \qquad \dot{q}(0) = \phi''_\perp(d), \tag{7.5}$$

where $\Omega^2(t)$ is the second-order variation of $\tilde{V}_1$ along the pinchon path $\mathbf{C}(t) = C(t)\,\mathbf{e}^{(1)}$,

$$\Omega^2(t) = \frac{\partial^2 \tilde{V}_1}{\partial C_2^2}\,(\mathbf{C}(t)), \tag{7.6}$$

and the pinchon is given by the equation

$$\dot{C}(t) = \sqrt{2(\tilde{V}_1(\mathbf{C}(t)) - g^{2/3}\varepsilon)}, \qquad C(0) = d, \qquad C(T) = 2\pi - d. \tag{7.7}$$

This equation also defines implicitly the tunneling time $T(d)$:

$$T(d) = \int_d^{2\pi - d} \frac{ds}{\sqrt{2(\tilde{V}_1(s\mathbf{e}^{(1)}) - g^{2/3}\varepsilon)}}. \tag{7.8}$$

As noted in [14] and proved further on, we have $\Lambda(d) = \lambda g d^2$, because of the spherical symmetry of $\tilde{V}_1^{[0]}(\mathbf{C})$. As for the double cone we have to calculate $\Lambda(d)$ explicitly to find the constant λ. Furthermore, $\tilde{V}_1$ can be replaced by LV_1 in Eqs. (7.3), (7.6), (7.7), and (7.8), since $d > \delta$ or $(0 < s < 2\pi)$:

$$\tilde{V}_1(s) \equiv \tilde{V}_1(s\mathbf{e}^{(1)}) = \frac{1}{\pi}\left[s(2\pi - s) + 8 \sum_{n=1}^{\infty} a_n \sin^2(ns/2) \right], \tag{7.9}$$

(cf. Eqs. (A.18), (4.7)). Taken together we have

$$\Delta(g) = 2g^{5/3}\lambda\, |B_\delta|^2 \exp\left(-\frac{S}{g} + \frac{\varepsilon T}{g^{1/3}} \right), \tag{7.10}$$

where

$$S = \int_0^{2\pi} \sqrt{2\tilde{V}_1(s)}\; ds, \tag{7.11}$$

$$T = \int_0^{2\pi} \frac{1}{\sqrt{2\tilde{V}_1(s)}}\; ds. \tag{7.12}$$

We are thus left with the calculation of S, T, and λ. The first two can be calculated by explicitly substituting (7.9) in (7.11) and (7.12) and expanding the results in the coefficients a_n. We substitute

$$s = \pi(1 - \sin \theta), \qquad \theta \in \left[-\frac{\pi}{2}, \frac{\pi}{2} \right], \tag{7.13}$$

and find

$$\tilde{V}_1(\theta) = \frac{1}{\pi} \left(\pi^2 \cos^2 \theta + 8 \sum_{n=1}^{\infty} a_n \sin^2 \left(\frac{n\pi}{2} (1 - \sin \theta) \right) \right)$$

$$= \frac{1}{\pi} \left(\pi^2 \cos^2 \theta + 8a_1 \cos^2 \left(\frac{\pi}{2} \sin \theta \right) + 8a_2 \sin^2(\pi \sin \theta) + \cdots \right). \tag{7.14}$$

A straigthforward but lengthy calculation yields

$$S = \frac{1}{2} \pi^2 \sqrt{2\pi} \left[1 + \frac{4a_1}{\pi^2} (1 + J_0(\pi)) + \frac{4a_2}{\pi^2} (1 - J_0(2\pi)) - \frac{4a_1^2}{\pi^3} (2J_1(\pi) + J_1(2\pi)) + \cdots \right], \tag{7.15}$$

$$T = \frac{1}{2} \pi \sqrt{2\pi} \left[1 - \frac{2a_1}{\pi} J_1(\pi) + \frac{4a_2}{\pi^2} J_1(2\pi)) \right.$$

$$\left. + \frac{2a^2}{\pi^3} (4J_1(\pi) + 2J_1(2\pi) - 3\pi J_0(\pi) - \pi J_0(2\pi)) + \cdots \right]. \tag{7.16}$$

J_n is the Bessel function of order n. The numerical values are

$$S = 12.4637..., \tag{7.17}$$
$$T = 3.9186....$$

We now attack the calculation of λ. As in the double cone problem of [14], we can ignore the $g^{2/3}\varepsilon$ terms in Eq. (7.7), because they yield corrections to λ of order $g^{2/3}$. Using the spherical symmetry for $d \ll 1$, we have

$$\phi''_\perp(d) = \frac{1}{d} \sqrt{2\tilde{V}_1(d)} = 2d^{-1/2}, \tag{7.18}$$

with a relative error of $O(gd^{-3/2} + d^{1/2})$, which can be absorbed in the overall error (for $d = g^{1/2}$, $gd^{-3/2} = g^{1/4}$). Because of the isotropy of $\tilde{V}_1(\mathbf{C})$ orthogonal to the pinchon $\mathbf{C}(t)$ we have (Δ the three-dimensional Laplacian)

$$\Omega^2(t) = \frac{1}{2} \left(\Delta - \frac{\partial^2}{\partial C_1^2} \right) \tilde{V}_1 \bigg|_{\mathbf{C}(t)}, \tag{7.19}$$

and $\mathbf{C}(t) = C(t)\,\mathbf{e}^{(1)}$ now satisfies

$$\dot{C}(t) = \sqrt{2\tilde{V}_1(C(t))}, \qquad C(0) = d, \qquad C(T(d)) = 2\pi - d.$$

This allows us to eliminate the pinchon time t in favour of the pinchon coordinate C (from now on we write V for $\tilde{V}_1$, and the arguments apply to any potential with spherically symmetric $V^{[0]}$):

$$\dot{q}(t) = \sqrt{2V(C)}\,\frac{d}{dC}\,q(C) = \sqrt{2V(C)}\,q'(C). \tag{7.20}$$

Hence from Eq. (7.4),

$$\lambda = d^{-2}g^{-1}\Lambda(d) = 2\pi d^{-2}[\sqrt{2V(2\pi - d)}\,q'_d(2\pi - d) + \sqrt{2V(d)}\,d^{-1}q_q(2\pi - d)]^{-1}, \tag{7.21}$$

where $q_d(C)$ satisfies the equations

$$\sqrt{2V(C)}\,\frac{d}{dC}\,\sqrt{2V(C)}\,\frac{d}{dC}\,q_d(C) = \frac{1}{2}\left[\left(\Delta - \frac{\partial^2}{\partial C_1^2}\right)V\right](C)\,q_d(C)$$
$$q_d(d) = 1, \qquad q'_d(d) = d^{-1}. \tag{7.22}$$

Note that (7.21) can be simplified due to the symmetry $V(2\pi - d) = V(d)$, and that $q_d(C)$ depends on d through its initial values. Furthermore, for $d \ll 1$, the spherical symmetry in that region gives $\Delta V = V'' + (2/d)\,V'$, so that $q''_d(d) = 0$. Next, differentiating (7.22) with respect to d shows that $(\partial/\partial d)\,q_d(s)$ satisfies the same differential equation as $q_d(s)$. The boundary conditions are obtained in the same way:

$$\frac{\partial}{\partial d}q_d(s)\bigg|_{s=d} = -q'_d(d) = -\frac{1}{d}q_d(d),$$
$$\frac{\partial}{\partial d}q'_d(s)\bigg|_{s=d} = -q''_d(d) - \frac{1}{d^2} = -\frac{1}{d}q'_d(d). \tag{7.23}$$

Thus the boundary conditions for $(\partial/\partial d)\,q_d(s)$ are $-d^{-1}$ times those for $q_d(s)$, so we have for all s:

$$\frac{\partial}{\partial d}q_d(s) = -\frac{1}{d}q_d(s). \tag{7.24}$$

Hence,

$$\frac{\partial}{\partial d}\lambda^{-1} = \frac{V'(d)}{2V(d)}\lambda^{-1} - q''_d(2\pi - d)\frac{d^2\sqrt{2V(d)}}{2\pi}. \tag{7.25}$$

334   VAN BAAL AND KOLLER

Using the equations of motion (7.22) and the spherical symmetry at $2\pi = d$ we obtain

$$q_d''(2\pi - d) = \frac{V'(d)}{2V(d)}\,(q_d'(2\pi - d) + d^{-1}q_d(2\pi - d)). \tag{7.26}$$

As promised, we find $\partial\lambda/\partial d \equiv 0$. (The relative error due to the breakdown of spherical symmetry is smaller than $O(d^2) = O(g)$ for $d = g^{1/2}$.)

We are finally ready to collect all our analytic results. Inserting the expression for the periodic potential $E_\delta(\mathbf{p}, g)$ (Eq. (7.1)) into the formula for energy of electric flux (Eq. (6.21)), we obtain the energy of electric flux in terms of the double-well energy splitting $\Delta(g)$ (Eq. (7.3)):

$$E(\mathbf{e}, L) = \sum_{i=1}^{3} \sin^2\left(\frac{\pi e_i}{2}\right) 2\lambda\,|B|^2 \frac{g^{5/3}(L)}{L}\exp\left(-\frac{S}{g(L)} + \frac{\varepsilon T}{g(L)^{1/3}}\right). \tag{7.27}$$

The tunneling action S and tunneling time T are given in (7.17), and the constant λ, which accounts for transverse fluctuations, is obtained from Eq. (7.21) by taking the limit $d \to 0$. This involves calculating the transverse fluctuations q_d using Eq. (7.22). Explicitly,

$$\lambda^{-1} = \lim_{d \to 0} \frac{d^{3/2}}{\pi}\,[dq_d'(2\pi - d) + q_d(2\pi - d)],$$

$$\sqrt{2V_1(C)}\,(\sqrt{2V_1(C)}\,q_d(C))' = \tfrac{1}{2}q_d(C)(\Delta V_1(C) - V_1''(C)), \tag{7.28}$$

$$q_d(d) = 1, \qquad q_d'(d) = d^{-1}.$$

The constant B is determined by the asymptotics of the groundstate wave function $\varphi_0^{[0]}(\mathbf{C})$ in the direction of the vacuum valley, Eq. (6.12). This component of the groundstate wavefunction can be obtained from the single-well perturbative groundstate wavefunction $\Psi_0^{[0]}$ by projecting with the "transverse" wavefunction $\tilde{\chi}_{[C]}^{(1)}$ using Eq. (6.8) or (6.9). Thus B is obtained by taking the following limit:

$$B = \lim_{x_3 \to \infty} \sqrt{8\pi}\,x_3^{13/4}\exp\left(\frac{(4x_3 - 2\varepsilon)^{3/2}}{6}\right)\int dx_1\,dx_2\,|x_1^2 - x_2^2|\,\Psi_0^{[0]}(\mathbf{x})$$

$$\times \exp\left(-\frac{1}{2}x_3(x_1^2 + x_2^2)\right), \tag{7.29}$$

with $\Psi_0^{[0]}(\mathbf{x})$ the single-well eigenfunction

$$\left[-\frac{1}{2\mathscr{g}}\frac{\partial}{\partial x_i}\,\mathscr{g}\,\frac{\partial}{\partial x_i} + \frac{1}{2}\,(x_1^2 x_2^2 + x_1^2 x_3^2 + x_2^2 x_3^2)\right]\Psi_0^{[0]}(\mathbf{x}) = \varepsilon\Psi_0^{[0]}(\mathbf{x}), \tag{7.30}$$

normalized according to

$$8\pi^2 \int_{\substack{0 \le |x_1| \le x_3 \\ 0 \le |x_2| \le x_3}} d^3x\, |\mathscr{J}|\, |\Psi_0^{[0]}(\mathbf{x})|^2 = 1. \tag{7.31}$$

Using the fact that (see Appendix B) Eq. (7.31) equals $(2/\pi^2)\int d^9c\, |\Psi_0^{[0]}(\mathbf{x})|^2$, one can alternatively determine B by fitting the wavefunction $\Psi_0(\mathbf{x}) \equiv (2/\pi^2)^{1/2}\, \Psi_0^{[0]}(\mathbf{x})$, normalized as $\int d^9c\, |\Psi_0(\mathbf{x})|^2 = 1$, to the asymptotic form

$$\Psi_0(\mathbf{x}) \underset{\substack{x_3 \gg 1 \\ x_1, x_2 \ll x_3}}{=} \frac{\sqrt{2}\, B \exp(-(4x_3 - 2\varepsilon)^{3/2})/6)\, e^{-(1/2)x_3(x_1^2 + x_2^2)}}{2\pi^{3/2}x_3(4x_3 - 2\varepsilon)^{1/4}}. \tag{7.32}$$

In the next section we will describe the numerical evaluation of the constants B and λ.

8. Calculating the Wave Function

8.1. *Introduction*

In previous sections we showed how the energy split due to tunneling through a potential barrier can be calculated. The g-dependence of the result was completely determined, and we must now find the overall coefficient. It depends on two contributions: a factor λ, accunting for the effect of transverse fluctuations along the tunneling path, and a factor $|B|^2$, representing the amount of penetration of the wavefunction into the barrier. More precisely, we found

$$\Delta E = 2\lambda\, |B|^2\, g^{5/3} \exp(-Sg^{-1} + T\varepsilon g^{-1/3}), \tag{8.1}$$

where λ is determined from the stability equation by (see Sect. 7)

$$\lambda^{-1} = \lim_{d \to 0} \frac{d^{3/2}}{\pi}\, [dq'_d(2\pi - d) + q_d(2\pi - d)], \tag{8.2}$$

and B is the coefficient in the asymptotic form of the single-well wave function deep within the barrier.

The evaluation of λ presents no real problem, amounting to the integration of an ordinary differential equation. We discuss this in Section 8.5, and turn first to calculating B.

At first sight, this is quite an easy calculation, because the single-well Hamiltonian is so simple, being just the kinetic term plus the lowest power term in the potential. Indeed, in many cases the lowest power in the potential is the quadratic one, and we find the familiar harmonic oscillator wavefunction, whose asymptotic form is known exactly. This is the class of tunneling problems that instanton methods can handle. However, if the lowest power in the potential is not quadratic, the problem can be much more complicated. A non-quadratic Hamiltonian cannot in general be separated to give non-interacting normal modes, and we are faced with a true higher dimensional quantum mechnics problem. This is the case in our present problem: we cannot obtain the wavefunction in closed form, and a Raleigh–Ritz approach is necessary.

In Section (8.2) we discuss the single-well Hamiltonian, its transformation to convenient coordinates, and its integrability properties. Section (8.3) considers the choice of basis vectors for our Rayleigh–Ritz calculation, which we will see requires some care. The following section discusses the numerical calculation, and how to extract the coefficient B. Then we return to calculating λ, and put everything together to get the final formula for ΔE.

8.2. *The Single-Well Hamiltonian*

Our single-well Hamiltonian has been examined in other contexts by various authors, because it is interesting in its own right. The Lagrangian for the single well was shown in Section 4 to be the usual Yang–Mills Lagrangian, in the $A_0 = 0$ gauge, with a renormalized coupling constant and spatially constant fields. We can make the following discussion self-contained by simply imposing these conditions. Begin with the standard Yang–Mills action per unit time for $SU(2)$:

$$\mathcal{L} = -\frac{1}{4\pi g^2} \int_{T^3} d^3x \, F_{\mu\nu}^a F^{\mu\nu a}. \tag{8.3}$$

If we now work in the gauge $A_0 = 0$, and make the simplifying assumption that the remaining fields have no space dependence, then we find

$$\frac{g^2}{L^3} \mathcal{L} = \frac{1}{2}\left(\frac{d}{dt} c_i^a\right)^2 + \frac{1}{4}(c_i^a c_j^a)(c_i^b c_j^b) - \frac{1}{4}(c_i^a c_i^a)^2, \tag{8.4}$$

where L^3 is the volume of the box, and

$$A_i^a(\mathbf{x}, t) = c_i^a(t), \qquad A_0^a(\mathbf{x}, t) = 0. \tag{8.5}$$

$\mathcal{L}$ is invariant under spatial rotations and parity, $c_i^a \to c_j^a R_i^j$ ($R \in O(3)$), and also exhibits a remnant of the global gauge invariance, being invariant under $c_i^a \to S_b^a c_i^b$ ($S \in SO(3)$). It is useful to thus think of c_i^a as a square matrix, transforming by $c \to ScR$. In Section 5 we have used the polar decomposition to diagonalize c:

$$
\begin{aligned}
c_i^a &= \eta_b^a \Lambda_j^b \xi_i^j, &\qquad \xi, \eta &\in SO(3), \\
\Lambda_j^b &= \mathrm{diag}(x_1, x_2, x_3), &\qquad x_3 &\geq x_2 \geq |x_1| \geq 0.
\end{aligned}
\tag{8.6}
$$

Under parity, $x_1 \to -x_1$. Here we find it convenient to use a different set of invariant coordinates, (s, x, y), related to (x_1, x_2, x_3) by

$$s \equiv x_1^2 + x_2^2 + x_3^2$$

$$s^2 x = x_1^2 x_2^2 + x_1^2 x_3^2 + x_2^2 x_3^2 \tag{8.7}$$

$$s^3 y = x_1^2 x_2^2 x_3^2.$$

A simple motivation for this choice is to note that the function $f(\lambda) = \det(c^{\mathrm{T}} c + \lambda \mathrm{I})$ is invariant under $c \to ScR$, so the coefficients of the power of λ are invariants of c. Expanding the determinant gives

$$f(\lambda) = \det(c^{\mathrm{T}} c + \lambda \mathrm{I}) = \lambda^3 + s\lambda^2 + s^2 x \lambda + s^3 y. \tag{8.8}$$

We have scaled out appropriate powers of s in the definitions of x and y, so that if we consider the c_i^a as coordinates in a nine-dimensional Euclidean space, then s is a "radial" coordinate,

$$s = \mathrm{Tr}(c^{\mathrm{T}} c) = c_i^a c_i^a = \sum_{i,a} (c_i^a)^2, \tag{8.9}$$

and x and y are "angles." The utility of our new coordinates is that the potential in Eq. (8.4) takes a very simple form:

$$-V(c_i^a) = \tfrac{1}{4}(c_i^a c_j^a)(c_i^b c_j^b) - \tfrac{1}{4}(c_i^a c_i^a)^2 = -\tfrac{1}{2} s^2 x. \tag{8.10}$$

Before converting to Hamiltonian form, we can scale out the g^3/L^2 dependence: just choose the length scale such that $L = g^{2/3}$, and remember that c has units of mass. Then the quantum Hamiltonian is

$$H = -\frac{1}{2} \frac{\partial^2}{(\partial c_i^a)^2} + V(c_i^a) = -\frac{1}{2} \nabla^2 + V, \tag{8.11}$$

where ∇^2 is the nine-dimensional Laplacian. Savvidy [12] has recently shown that this Hamiltonian is non-integrable, which means that there is no operator that commutes with H, other than the known symmetry generators. It is thus impossible to obtain the energy eigenfunctions in closed form, and we *have* to use a numerical method, such as Rayleigh–Ritz perturbation theory [13]. However, let us simplify the problem as much as possible before resorting to such an approach.

We are interested in solutions of

$$H\Psi = E\Psi \tag{8.12}$$

invariant under $c_i^a \to S_b^a c_i^b$, because physical wave functions must be gauge invariant. Also, we will restrict ourselves to the zero angular momentum sector, with positive parity, so that wave functions are also invariant under $c_i^a \to c_j^a R_i^j$ ($R \in O(3)$). It follows then that Ψ is a function of s, x, and y only. The Laplacian ∇^2 can now be written in terms of (s, x, y) as

$$\nabla^2 \Psi(s, x, y) = \left(4s \frac{\partial^2}{\partial s^2} + 18 \frac{\partial}{\partial s} + \frac{1}{s} \tilde{\nabla}^2 \right) \Psi(s, x, y), \tag{8.13}$$

where $\tilde{\nabla}^2$ is the Laplacian on the two-dimensional surface parameterized by the angular variables (x, y):

$$\tilde{\nabla}^2 = (12y + 4x - 16x^2) \frac{\partial^2}{\partial x^2} + 2(8y - 24xy) \frac{\partial^2}{\partial x\, \partial y} + (4xy - 36y^2) \frac{\partial^2}{\partial y^2}$$

$$+ (8 - 44x) \frac{\partial}{\partial x} + (2x - 78y) \frac{\partial}{\partial y}. \tag{8.14}$$

It is clear from the form of V in Eq. (8.10) that H is not separable in our coordinate system, and the statement that H is non-integrable implies that it is not separable in *any* coordinate system. On the other hand, if V were purely a function of s, then the eigenfunction problem would indeed separate into a radial part and an angular part. We are going to use this fact in the next section, when we pick a basis for the Rayleigh–Ritz procedure, so we conclude this section by examining the operator $\tilde{\nabla}^2$ of Eq. (8.14) in detail.

Although $\tilde{\nabla}^2$ is not separable (and possibly non-integrable), we can still solve for the eigenfunctions because they must be polynomials in x and y. To see this, note first that $\tilde{\nabla}^2$ transforms polynomials into polynomials. Now suppose we have a polynomial P containing all powers $x^i y^j$ such that $ai + jb \leq N$ for some positive integers a, b, N. All powers in $\tilde{\nabla}^2 P$ will also have $ai + jb \leq N$, except possibly for terms coming from $y(\partial^2/\partial x^2) P$ and $x(\partial^2/\partial y^2) P$, which change $x^i y^j$ into $x^{i-2} y^{j+1}$ and $x^{i+1} y^{j-1}$, respectively. But both these also satisfy the constraint provided $2a \geq b \geq a$. Thus for fixed N, $\tilde{\nabla}^2$ is a Hermitian operator on a finite set of polynomials and can be diagonalized. It turns out that the correct choice is $a = 2$, $b = 3$. Let

$$P_{\alpha\beta}(x, y) = x^\alpha y^\beta + \sum_{2\gamma + 3\delta < 2\alpha + 3\beta} C_{\gamma\delta} x^\gamma y^\delta, \tag{8.15}$$

for integer α, β, and some coefficients $C_{\gamma\delta}$. Then

$$\tilde{\nabla}^2 P_{\alpha\beta}(x, y) = -(4\alpha + 6\beta)(4\alpha + 6\beta + 7) x^\alpha y^\beta + \sum_{2\gamma + 3\delta < 2\alpha + 3\beta} C'_{\gamma\delta} x^\gamma y^\delta.$$

The $C'_{\gamma\delta}$ are linear combinations of the $C_{\gamma\delta}$, plus some pieces from $\tilde{\nabla}^2 x^\alpha y^\beta$, so we can solve the equations

$$C'_{\gamma\delta} = -(4\alpha + 6\beta)(4\alpha + 6\beta + 7) C_{\gamma\delta}$$

to obtain an eigenfunction $P_{\alpha\beta}(x, y)$ of $\tilde{\nabla}^2$:

$$\tilde{\nabla}^2 P_{\alpha\beta}(x, y) = -(4\alpha + 6\beta)(4\alpha + 6\beta + 7)\, P_{\alpha\beta}(x, y). \tag{8.16}$$

In this way, we can construct a complete set of polynomial eigenfunctions of $\tilde{\nabla}^2$. Clearly, all $P_{\alpha\beta}$ with the same value of $q \equiv 2\alpha + 3\beta$ are degenerate. The combination $2\alpha + 3\beta$ occurs because $s^{2\alpha + 3\beta} P_{\alpha\beta}$ is a polynomial in (x_1, x_2, x_3), by Eqs. (8.7), (8.8), and (8.15).

Finally, we must decide how to normalize $P_{\alpha\beta}$. Introduce an integration measure $d\Omega$ such that integrals in the nine-dimensional Euclidean space are given by $\int d^9 c = \int r^8\, dr \int d\Omega$, where we can identify $r = (c_i^a c_i^a)^{1/2} = \sqrt{s}$. It is easy to show that $\int d\Omega\, \tilde{\nabla}^2 x^i y^j = 0$. (For instance, by using $\int d^9 c \nabla^2 (e^{-s} x^i y^j) = 0$). If we use Eq. (8.14) for $\tilde{\nabla}^2$, this gives a recursion relation for

$$\mathscr{X}_{ij} = \int d\Omega\, x^i y^j, \tag{8.17}$$

which relates $\mathscr{X}_{ij}$ with $2i + 3j \equiv q$ to $\mathscr{X}_{i'j'}$ with $2i' + 3j' < q$:

$$\mathscr{X}_{ij} = \frac{1}{2q(2q+7)} \left[4(1 + i + 4j)\, \mathscr{X}_{i-1,j} + 12i(i-1)\, \mathscr{X}_{i-2,j+1} + 2j(2j-1)\, \mathscr{X}_{i+1,j-1} \right]. \tag{8.18}$$

We can finally evaluate $\mathscr{X}_{00}$ explicitly, since

$$\int d^9 c\, e^{-r^2} = \pi^{9/2} = \int r^8\, dr\, e^{-r^2} \int d\Omega = \tfrac{105}{32} \sqrt{\pi} \int d\Omega,$$

i.e.,

$$\mathscr{X}_{00} = \tfrac{32}{105} \pi^4. \tag{8.19}$$

Thus the $\mathscr{X}$ can be calculated recursively, and this allows one to calculate the inner products $\int d\Omega\, P_{\alpha\beta}^* P_{\gamma\delta}$.

8.3. Preparing for a Rayleigh–Ritz Calculation

We have seen that the single-well Hamiltonian is non-integrable, and must be attacked with numerical methods. This problem has been examined by Lüscher and Münster [13], who have found the lowest energy eigenstates and eigenvalues. They used the Rayleigh–Ritz algorithm, which is:

Given a Hermitian operator H on Hilbert space, and a complete orthonormal set of basis vectors $|n\rangle$, consider the matrix

$$H_{nm}^{(N)} = \langle n|\, H\, |m\rangle \quad n, m \leqslant N.$$

Find the eigenvalues $\lambda_\alpha^{(N)}$ and eigenvectors $v_{\alpha i}^{(N)}$ of this matrix. Then $\lambda_\alpha^{(N)}$ is an approximaton for an eigenvalue λ_α of H, and $|\psi_\alpha^{(N)}\rangle = v_{\alpha i}^{(N)}|i\rangle$ is an approximation for the associated eigenvector $|\psi_\alpha\rangle$ of H. As $N \to \infty$, $\lambda_\alpha^{(N)}$ converges monotonically to λ_α, and $|\psi_\alpha^{(N)}\rangle \to |\psi_\alpha\rangle$ in the Hilbert space norm.

In practice, this method works well, but it is very difficult to put any theoretical bounds on the error in one's answers, because a bad choice of basis vectors could cause very slow convergence. The usual approach is to decide on a particular accuracy, and then methodically increase N until the answers are constant to this accuracy.

For us the problem of errors is quite crucial, because we are interested in the magnitude of the wavefunction in a region where it is decaying exponentially. Because the wavefunction only converges in the $\mathscr{L}^2$ sense as $N \to \infty$, relative errors can be large in regions where the wavefunction is small. We need to take N to very large values to achieve reasonable accuracy, and the method of Lüscher and Münster begins to suffer from rounding inaccuracy before we can reach these values. Lüscher and Münster were well aware of this problem, and to reach $N = 100$ they did sensitive parts of their calculation using extended precision. Let us review the problem.

The authors of [13] start out with an obviously complete but non-orthonormal basis, in our notation

$$\hat{\Psi}_{abc}^\omega = e^{-\omega s/2}s^a x^b y^c. \tag{8.20}$$

The number ω is adjusted to optimize the Rayleigh–Ritz convergence, and is in the range $1 \sim 2$. There is a canonical way to order such a basis: by smallest a, then by smallest $2b + 3c$, then by smallest b. A recursion relation for $\int e^{-\omega s}s^a x^b y^c$ can be found, and the inner product calculated. An orthonormal basis is then constructed by the Gram–Schmidt procedure. The reason for the instability is that for large N, the basis Eq. (8.20) becomes almost degenerate, with

$$\frac{\langle \hat{\Psi}_{abc} \mid \hat{\Psi}_{a'b'c'} \rangle}{|\hat{\Psi}_{abc}| |\hat{\Psi}_{a'b'c'}|} \sim 1. \tag{8.21}$$

The Gram–Schmidt procedure then involves subtracting almost-equal numbers, which produces rounding error in a computer. In our case, this would propagate through the "diagonalization" and "wavefunction reconstruction" steps, to produce large errors in the sensitive part of the wavefunction that we are interested in.

The solution is to orthonormalize to a much greater (preferably infinite) precision. For values of $N \sim 200$, however, orthonormalizing the basis (8.20) to sufficient accuracy as Lüscher and Münster did for $N \sim 100$, would take an enormous amount of computer time and space, so we have taken a slightly different route, which we now describe.

We noted in the previous subsection that the angular Laplacian $\tilde{\nabla}^2$ could be block diagonalized to give sets of degenerate eigenfuctions $P_{\alpha\beta}$ with eigenvalues $-(4\alpha + 6\beta)(4\alpha + 6\beta + 7)$. The size of the blocks is quite small: for $q = 2\alpha + 3\beta \leqslant 16$ we have at most a triple degeneracy. Moreover, the $P_{\alpha\beta}$ are only two-variable polynomials, so an algebraic manipulation program can perform the orthonormalization quite readily.

We now need to take combinations of the $P_{\alpha\beta}$ with orthonormal functions of s to produce a complete basis. A conceptually simple way of doing this is to use the eigenfunctions of the nine-dimensional harmonic oscillator. Thus, replace the potential $V(c)$ in Eqs. (8.4) and (8.10) by

$$V_\omega(c_i^a) = \tfrac{1}{2}\omega^2 c_i^a c_i^a = \tfrac{1}{2}\omega^2 s. \tag{8.22}$$

The eigenvalues equation in the sector we are interested in becomes a separable one:

$$\left[-\frac{1}{2}\left(4s\frac{\partial^2}{\partial s^2} + 18\frac{\partial}{\partial s} + \frac{1}{s}\tilde{\nabla}^2 \right) + \frac{1}{2}\omega^2 s - E_\omega \right] \Psi^\omega(s, x, y) = 0. \tag{8.23}$$

Putting

$$\Psi^\omega(s, x, y) = X_q^\omega(s)\, P_{\alpha\beta}(x, y) \qquad (q = 2\alpha + 3\beta) \tag{8.24}$$

therefore gives us a radial equation

$$\left(-2s\frac{\partial^2}{\partial s^2} - 9\frac{\partial}{\partial s} + \frac{q(2q + 7)}{s} + \frac{\omega^2}{2}s \right) X_q^\omega = E_\omega X_q^\omega, \tag{8.25}$$

the solution to which involves half-integer-order Laguerre polynomials [22]:

$$X_{n,q}^{\omega = 1}(s) = \hat{N}(n, q)\, e^{-(1/2)s} s^q L_n^{2q + 7/2}(s). \tag{8.26}$$

The new quantum number n is related to the eigenvalue $E_{\omega = 1}$ by

$$E_{\omega = 1} = 2n + 2q + \tfrac{9}{2}, \tag{8.27}$$

and $N(n, q)$ is chosen so that

$$\int s^4 d(\sqrt{s})\, (X_{n, q}^{\omega = 1}(s))^2 = 1, \tag{8.28}$$

$$\hat{N}(n, q) = \left[\frac{2n!}{\Gamma(n + 2q + 9/2)} \right]^{1/2}. \tag{8.29}$$

For general ω,

$$E_\omega = \omega E_{\omega = 1}, \qquad X_{n,q}^\omega(s) = \omega^{9/4} X_{n, q}^{\omega = 1}(\omega s). \tag{8.30}$$

342   VAN BAAL AND KOLLER

TABLE I

Orthogonal Eigenfunctions of $\tilde{\nabla}^2$

$$\hat{P}_{0,0} = 1$$

$$\hat{P}_{1,0} = -\tfrac{2}{11} + x$$

$$\hat{P}_{0,1} = \tfrac{4}{663} - \tfrac{1}{17}x + y$$

$$\hat{P}_{2,0} = \tfrac{8}{399} - \tfrac{40}{133}x - \tfrac{4}{7}y + x^2$$

$$\hat{P}_{1,1} = -\tfrac{16}{29325} + \tfrac{4}{345}x - \tfrac{144}{575}y - \tfrac{1}{25}x^2 + xy$$

$$\hat{P}_{0,2} = \tfrac{256}{8554275} - \tfrac{128}{150075}x + \tfrac{212}{6525}y + \tfrac{1}{261}x^2 - \tfrac{6}{29}xy + y^2$$

$$\hat{P}_{3,0} = -\tfrac{80}{44511} + \tfrac{760}{14837}x + \tfrac{4432}{14837}y - \tfrac{6196}{14837}x^2 - \tfrac{21504}{14837}xy + \tfrac{14914}{14837}y^2 + x^3$$

$$\hat{P}_{2,1} = \tfrac{32}{741675} - \tfrac{392}{247225}x + \tfrac{424}{9889}y + \tfrac{128}{9889}x^2 - \tfrac{136}{341}xy - \tfrac{4}{11}y^2 - \tfrac{1}{33}x^3 + x^2y$$

$$\hat{P}_{1,2} = -\tfrac{128}{58908773} + \tfrac{256}{2561251}x - \tfrac{63288}{12806255}y - \tfrac{414}{441595}x^2 + \tfrac{864}{14245}xy - \tfrac{334}{1295}y^2 + \tfrac{3}{1295}x^3 - \tfrac{6}{37}x^2y + xy^2$$

$$\hat{P}_{4,0} = \tfrac{128}{875195} - \tfrac{5888}{875195}x - \tfrac{482688}{6126365}y + \tfrac{119328}{1225273}x^2 + \tfrac{1047168}{1225273}xy - \tfrac{251952}{1225273}y^2 - \tfrac{660160}{1225273}x^3 - \tfrac{399960}{175039}x^2y + \tfrac{365952}{175039}xy^2$$
$$+ x^4$$

$$\hat{P}_{0,3} = \tfrac{256}{1861563795} - \tfrac{960}{124104253}x + \tfrac{2256}{4279457}y + \tfrac{12}{138047}x^2 - \tfrac{1140}{138047}xy + \tfrac{1240}{19721}y^2 - \tfrac{5}{19721}x^3 + \tfrac{15}{533}x^2y - \tfrac{15}{41}xy^2$$
$$+ y^3$$

$$\hat{P}_{3,1} = -\tfrac{1792}{588740115} + \tfrac{6720}{39249341}x - \tfrac{1047936}{196246705}y - \tfrac{481072}{196246705}x^2 + \tfrac{3344816}{39249341}xy + \tfrac{10576720}{39249341}y^2 + \tfrac{1566320}{117748023}x^3 - \tfrac{531400}{1060793}x^2y$$
$$- \tfrac{1267668}{1060793}xy^2 - \tfrac{22416}{25873}y^3 - \tfrac{1}{41}x^4 + x^3y$$

$$\hat{P}_{2,2} = \tfrac{2048}{13801248825} - \tfrac{9216}{920083255}x + \tfrac{92224}{148400525}y + \tfrac{9872}{63600225}x^2 - \tfrac{154144}{12720045}xy - \tfrac{3232}{68757}y^2 - \tfrac{176}{206271}x^3 + \tfrac{376}{5289}x^2y - \tfrac{88}{215}xy^2$$
$$- \tfrac{4}{15}y^3 + \tfrac{1}{645}x^4 - \tfrac{2}{15}x^3y + x^2y^2$$

$$\hat{P}_{5,0} = -\tfrac{256}{22736637} + \tfrac{1920}{2526293}x + \tfrac{10240}{688989}y - \tfrac{1280}{75537}x^2 - \tfrac{874240}{3248091}xy - \tfrac{338080}{3248091}y^2 + \tfrac{141280}{885843}x^3 - \tfrac{5182640}{3248091}x^2y - \tfrac{14320}{98427}xy^2$$
$$- \tfrac{62096}{75537}y^3 - \tfrac{2149870}{3248091}x^4 - \tfrac{232480}{75537}x^3y + \tfrac{77620}{25179}x^2y^2 + x^5$$

$N^2(0,0) = \tfrac{105}{32}\pi^4$	$N^2(1,0) = \tfrac{945}{64}\pi^4$	$N^2(0,1) = \tfrac{1395}{128}\pi^4$
$N^2(2,0) = \tfrac{1395}{256}\pi^4$	$N^2(1,1) = \tfrac{135135}{256}\pi^4$	$N^2(0,2) = \tfrac{2027025}{512}\pi^4$
$N^2(3,0) = \tfrac{45045}{512}\pi^4$	$N^2(2,1) = \tfrac{2027025}{1024}\pi^4$	$N^2(1,2) = \tfrac{34459425}{1024}\pi^4$
$N^2(4,0) = \tfrac{675675}{4096}\pi^4$	$N^2(0,3) = \tfrac{654729075}{2048}\pi^4$	$N^2(3,1) = \tfrac{11486475}{2048}\pi^4$
$N^2(2,2) = \tfrac{654729075}{4096}\pi^4$	$N^2(5,0) = \tfrac{2297295}{8192}\pi^4$	

If we call the orthonormalized $P_{\alpha\beta}$, "$\hat{P}_{\alpha\beta}$" then the complete set of harmonic oscillator eigenfunctions in this sector is

$$\Psi_{n\alpha\beta}(s, x, y) = X_{n,q}(s)\,\hat{P}_{\alpha\beta}(x, y). \tag{8.31}$$

The unnormalized $\hat{P}_{\alpha\beta}$ and their norms $N(\alpha, \beta)$ for $2\alpha + 3\beta \leqslant 10$ are given in Table I.

We are now in a position to evaluate the matrix elements of the Hamiltonian (8.11) exactly in this basis. We can write

$$\begin{aligned} H &= (-\tfrac{1}{2}\nabla^2 + \tfrac{1}{2}\omega^2 s) - \tfrac{1}{2}\omega^2 s + \tfrac{1}{2}s^2 x \\ &= H_\omega - \tfrac{1}{2}\omega^2 s + \tfrac{1}{2}s^2 x, \end{aligned} \tag{8.32}$$

and it merely remains to evaluate the two kinds of matrix elements $\langle \Psi_{n\alpha\beta}|\, s\, |\Psi_{n'\alpha'\beta'}\rangle$ and $\langle \Psi_{n\alpha\beta}|\, s^2 x\, |\Psi_{n'\alpha'\beta'}\rangle$, since H_ω is diagonal in this basis, with

$$\langle \Psi_{n\alpha\beta}|\, H_\omega\, |\Psi_{n'\alpha'\beta'}\rangle = (2n + 4\alpha + 6\beta + \tfrac{9}{2})\,\omega\delta_{nn'}\delta_{\alpha\alpha'}\delta_{\beta\beta'}. \tag{8.33}$$

The quantities $\langle \alpha\beta|\, x\, |\alpha'\beta'\rangle \equiv \int d\Omega\, \hat{P}_{\alpha\beta} x \hat{P}_{\alpha'\beta'}$ can be calculated exactly because the $\hat{P}_{\alpha\beta}$ are polynomials in x and y, and we know $\int d\Omega\, x^i y^j = \mathscr{X}_{ij}$. Moreover, by using the generating function for Laguerre polynomials, one can calculate

$$\langle n, q|\, s\, |n', q'\rangle \equiv \int s^4 d\sqrt{s}\, X_{n,q}(s)\, s X_{n',q'}(s)$$

$$= (2n + 2q + \tfrac{9}{2})\,\delta_{nn'} - \sqrt{n+1}\,\sqrt{n + 2q + \tfrac{9}{2}}\,\delta_{n,n'-1}$$

$$- \sqrt{n}\,\sqrt{n + 2q + \tfrac{7}{2}}\,\delta_{n,n'+1}, \tag{8.34}$$

and

$$\langle n, q|\, s^2\, |n', q'\rangle \equiv \int s^4 d(\sqrt{s})\, X_{n,q}(s)\, s^2 X_{n',q'}(s)$$

$$= \frac{\sqrt{\pi}}{2^{q+q'+6}}\,(-1)^{n+n'}$$

$$\times \sum_{t=t_{\min}}^{t=t_{\max}} \binom{q'-q+2}{n-t}\binom{m-m'+2}{n'-t}\frac{(2(t+q+q')+11)!!}{2^t t!}, \tag{8.35}$$

where $t_{\max} = \min(n, n')$ and $t_{\min} = \max(n - q' + q - 2,\, n' - m + m' - 2, 0)$. The obvious requirement $t_{\min} \leqslant t_{\max}$ then implies the matrix elements are zero unless

$$-2 - q + q' \leqslant n' - n \leqslant 2 - q + q', \tag{8.36}$$

and the sum in Eq. (8.35) runs over at most 5 values of t. Equations $(8.33)-(8.35)$ are sufficient to evaluate all matrix elements $\langle n\alpha\beta|\, H\, |n'\alpha'\beta'\rangle$ exactly. Reinstating the ω dependence gives

$$\langle n\alpha\beta|\, H\, |n'\alpha'\beta'\rangle = \delta_{nn'}\delta_{\alpha\alpha'}\delta_{\beta\beta'}E_\omega - \tfrac{1}{2}\omega\delta_{\alpha\alpha'}\delta_{\beta\beta'}\langle nq|\, s\, |n'q'\rangle$$

$$+ \frac{1}{2\omega^2}\langle nq|\, s^2\, |n'q'\rangle\langle \alpha\beta|\, x\, |\alpha'\beta'\rangle. \tag{8.37}$$

One needs to decide on an order for the $|n\alpha\beta\rangle$. We order first by lowest $n + 2\alpha + 3\beta$, then by lowest n, then by lowest α. This choice can be understood as an ordering in increasing powers of the variables s, x, y: $|n\alpha\beta\rangle \sim e^{-(1/2)s}s^{n+2\alpha+3\beta}x^\alpha y^\beta$. In our calculation we have chosen the bound $n + 2\alpha + 3\beta \leqslant 16$, which allows 204 orthonormal basis vectors. Increasing the bound to say 17 would only increase this to 237, and not change the result significantly.

344

8.4. *The Wavefunction in the Barrier Region*

Once a suitable basis is chosen, the Rayleigh–Ritz algorithm can be applied, using readily available computer programs for matrix diagonalization. We list here the various steps in our calculation. The algebraic manipulation was done using the program SMP [28], and floating point calculations used routines in the IMSL library.

(1) The angular eigenfunctions $P_{\alpha\beta}$ for $q = 2\alpha + 3\beta \leqslant 16$ were found by explicitly solving Eq. (8.16) with the ansatz Eq. (8.15). The result was a set of polynomials in (x, y) with rational coefficients.

(2) The integrals $\int d\Omega\, x^i y^j = \mathcal{X}_{ij}$ were calculated exactly from the recursion relation Eq. (8.18), and the initial condition Eq. (8.19). We need all $\mathcal{X}_{ij}$ for $2i + 3j \leqslant 34$.

(3) In each subspace of fixed $q = 2\alpha + 3\beta$, the Gram–Schmidt orthonormalization procedure was applied to the functions $P_{\alpha\beta}$. Because the degeneracy was at most triple ($q - 12, 14, 15, 16$), it was feasible to do this exactly using algebraic manipulation. The lowest order orthogonal $\hat{P}_{\alpha\beta}$ and the squares of their norms are given in Table I. Note that after orthogonalization, there is no significance to how we label the degenerate $\hat{P}_{\alpha\beta}$ for a given $2\alpha + 3\beta$.

FIG. 4. Non-zero matrix elements $\langle n\alpha\beta |\, xs^2\, | n'\alpha'\beta' \rangle$ are shown as black squares. The 204 vectors $|n\alpha\beta\rangle$ are canonically ordered as described in the text. The large scale structure can be understood from the derivation in the text, but the many zeros within the black "islands" are a mystery.

(4)　The Laguerre polynomials $L_n^{2q+7/2}(s)$ were constructed for $n+q \leqslant 16$, using their explicit definition [22].

(5)　The reduced matrix elements $\langle \alpha\beta| \, x \, |\alpha'\beta'\rangle$, $\langle nq| \, s \, |n'q'\rangle$, and $\langle nq| \, s^2 \, |n'q'\rangle$ were evaluated. Many of these are zero, so only those expected to be non-zero were calculated.

(6)　The full matrix elements $\langle n\alpha\beta| \, s^2 x \, |n'\alpha'\beta'\rangle$ were formed. In Fig. 4 we have plotted the positions of all non-zero elements as black squares, where the vectors are ordered canonically.

(7)　The matrix elements were converted to 64-bit floating-point numbers and stored.

(8)　The wave functions $\Psi_{n\alpha\beta}$ were calculated, and the coefficients in the polynomial part (i.e., in $e^{s/2}\Psi_{n\alpha\beta}$) were converted to floating point and stored.

(9)　A program using IMSL routines (and written in the language C) read in the matrix elements and wavefunction coefficients.

(10)　For any given ω, and number of basis vectors N, the Hamiltonian was constructed and the lowest eigenvalues and eigenvectors found. (Figure 5 is a plot of the groundstate wavefunction.)

(11)　Using the basis wavefunction coefficients from step (8), the eigenvectors were converted to polynomial functions of (s, x, y) with floating-point coefficients.

(12)　The value of B was extracted from the wavefunction in various ways, as will be described later in this section.

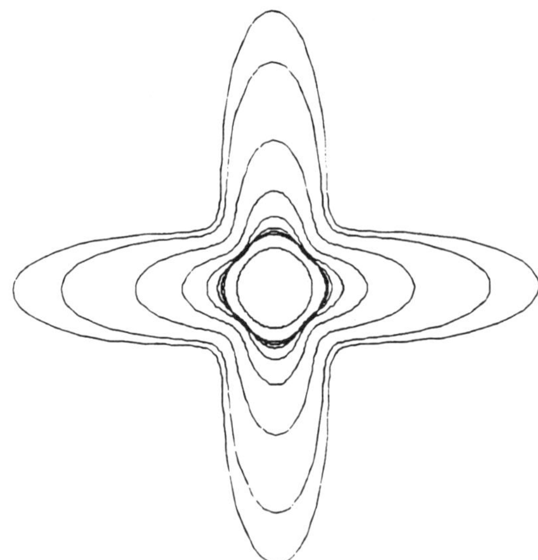

FIG. 5.　Plot showing the single-well wavefuction $\Psi_0^{[0]}(x_1, x_2, x_3)$. We have taken a spherical coordinate system for (x_1, x_2, x_3) and plotted the probability per unit solid angle as a function of φ, for θ in steps of $10°$. The outermost contour is $\theta = 90°$. It is instructive to compare with the three-dimensional plots of the energy surface in [12].

The total amount of computer time exceeded 50 h on a Ridge 32 (comparable in speed to a VAX 11/780). To insure reliability of our results, we made the following consistency checks at each stage:

(1)   Each $P_{\alpha\beta}$ satisfies Eq. (8.16) identically.

(2) $\int d\Omega \, P_{\alpha\beta} = 0$ if $\alpha \neq 0$ or $\beta \neq 0$ checks the $\mathcal{X}_{ij}$, as does $\int P_{\alpha\beta} P_{\alpha'\beta'} = 0$ if $2\alpha + 3\beta \neq 2\alpha' + 3\beta'$.

(3)   The 30 $\hat{P}_{\alpha\beta}$ are orthogonal after application of the Gram–Schmidt procedure.

(4)   All Laguerre polynomials satisfy the defining equations identically.

(5) $xs^2 \, |n\alpha\beta\rangle = \sum_{n'\alpha'\beta'} |n'\alpha'\beta'\rangle\langle n'\alpha'\beta'| \, xs^2 \, |n\alpha\beta\rangle$ checks the accuracy of the matrix elements. This test is limited to the first ~ 140 vectors, because beyond that point matrix elements involving vectors outside the set of 204 that we work with are required for the completeness relation (see Fig. 4).

(6)   Converting to floating point at earlier points in the calculation produces consistent results.

(7)   Results scale correctly with ω, and accuracy improves with increasing N.

We therefore believe our result is trustworthy.

One finds that the results are fairly insensitive to the value of ω if $\omega \sim 1.5$–1.6. Any ω in this range gives results consistent with the error bars we quote.

We must now compare the asymptotic form of the Rayleigh–Ritz eigenfunction with the predicted asymptotic form. We are interested in the properties of the eigenfunction along the vacuum valley, which is the direction $x = y = 0$. However, we will find it convenient to work for a while in the polar decomposition variables (x_1, x_2, x_3) of Eqs. (8.6) and (8.7), where the vacuum valley is along $x_1 = x_2 = 0$.

Recall that in Section 6 we derived the following "adiabatic decomposition" of a spherically symmetric wavefunction

$$\Psi(x_1, x_2, x_3) = [(x_3^2 - x_1^2)(x_3^2 - x_2^2)]^{-1/2} \sum_{n=1}^{\infty} \Phi^{(n)}(x_3) \, \chi_{[x_3]}^{(n)}(x_1, x_2), \quad (8.38)$$

where here $\Phi^{(n)}(|\mathbf{C}|) - |\mathbf{C}| \, \varphi^{(n)}(\mathbf{C})$ of Section 6. This expression is useful because the $\Phi^{(n)}$ are known to be exponentially decaying functions of x_3. As was shown in some detail for the toy models in [14], the asymptotic form of the wavefunction for large x_3 is given by the $n = 1$ term. The $\chi_{[x_3]}^{(n)}$ are otrhonormal eigenfunctions of a Hamiltonian in which x_3 is regarded as an adiabatic parameter.

Now the forms of $\Phi^{(1)}$ and $\chi_{[x_3]}^{(1)}$ are both known for large x_3. $\chi_{[x_3]}^{(1)}$ is known to be normalized for any value of the "parameter" x_3, but what is not known, and what this whole section is trying to find, is the normalization of $\Phi^{(1)}$. From Eq. (5.34), we have to first order in $1/x_3^3$,

$$\chi_{[x_3]}^{(1)} = \frac{1}{\sqrt{2}} x_3 \left(1 - \frac{[2x_1^2 x_2^2 x_3^2 + x_3(x_1^2 + x_2^2) - 3]}{16 x_3^3} \right) e^{-x_3(x_1^2 + x_2^2)/2}. \quad (8.39)$$

Here we adopt the simpler normalization conventions of [5], which differ by a factor of $\sqrt{2\pi}$ in $\chi^{(1)}_{[x_3]}$ and Ψ (because we drop ξ in the measure for the spherical sector). Thus $\chi^{(1)}_{[x_3]}(x_1, x_2) = \sqrt{2\pi}\, \chi^{(1)}_{[x_3]}(x_1, x_2, \xi)$ and $\Psi(x_1, x_2, x_3) = \sqrt{2\pi}\, \Psi^{[0]}_0(x_1, x_2, x_3)$. Let ε be the true energy eigenvalue of $\Psi(x_1, x_2, x_3)$. Then, as shown in Section 6 and [5], $\Phi^{(1)}$ is known to satisfy

$$\left(-\frac{1}{2}\frac{\partial^2}{\partial x_3^2} + 2x_3 - \frac{1}{2x_3^2} + O(x_3^{-5}) \right)\Phi^{(1)}(x_3) = \varepsilon\Phi^{(1)}(x_3). \tag{8.40}$$

This equation determines the asymptotic behaviour of $\Phi^{(1)}(x_3)$. If for convenience we relable $x_3 = r$, then we find from Eq. (6.12)

$$\Phi^{(1)}(r) \to B(4r - 2\varepsilon)^{-1/4} \exp\left(-\frac{(4r - 2\varepsilon)^{3/2}}{6} \right) \qquad \text{as} \quad r \to \infty \tag{8.41}$$

(where only the lowest order term $(2x_3)$ in Eq. (8.40) has been used so far). This is the *definition* of B. The simplest way, therefore, to extract B from the wavefunction $\Psi(x_1, x_2, x_3)$ is to take the limit of (cf. Eq. (7.32))

$$B(r) = (4r - 2\varepsilon)^{1/4} \exp\left(\frac{(4r - 2\varepsilon)^{3/2}}{6} \right) \frac{\Psi(r, 0, 0)}{r^2\chi^{(1)}_{[r]}(0, 0)} \tag{8.42}$$

as $r \to \infty$.

It turns out, though, that this approach is too simplistic. In Fig. 6, we plot the function on the right of Eq. (8.42), for the lowest energy wave function Ψ, calculated using $\omega = 1.5$ and $N = 204$ in the Rayleigh–Ritz program. It obviously does not have a non-zero limit as $r \to \infty$, but we cannot expect it to have a meaningful limit: all our basis functions are polynomials times $e^{-\omega r^2/2}$, and a finite sum of such terms can never accurately reproduce the different exponential behaviour of Eq. (8.41). Thus there is a limitation on the Rayleigh–Ritz method's ability to reproduce asymptotic behaviour correctly. The problem therefore is to extract B without taking the limit in Eq. (8.42).

Suppose, therefore, that we look at $B(r)$ for some finite value of r, where the Rayleigh–Ritz wave function is accurate (say $r \leqslant r_{\max}$). $B(r)$ may not be constant for $r < r_{\max}$, for two reasons. First, $\Phi^{(1)}(r)$ may not yet have attained its asymptotic form, Eq. (8.41). Second, for finite r, terms involving $\Phi^{(2)}$, $\Phi^{(3)}$, etc. are still contributing to the wavefunction Ψ, by Eq. (8.38). Nevertheless, we do expect $B(r)$ to be approximately constant for r large enough, where these effects are small, say in a region $r_{\min} \leqslant r \leqslant r_{\max}$. Clearly, $r_{\min}$ is intrinsic to our problem, but $r_{\max}$ depends on how many basis vectors we use in the Rayleigh–Ritz calculation. Looking at Fig. 6 shows that $r_{\max} \sim 6$, and since the plateau in $B(r)$ is not very flat, we deduce that the asymptotic form of Ψ has not been attained very accurately yet. This uncertainty in the precise height of the plateau is the dominant contribution to our error in the determination of B.

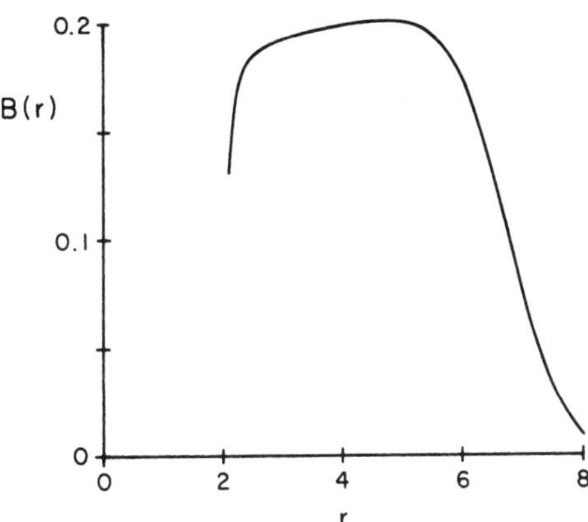

FIG. 6. $B(r)$ extracted from the wavefunction using the naive definition Eq. (8.42). There is no flat plateau.

The fact that the plateau in Fig. 6 has a definite slope suggests that a systematic correction could be made to improve our accuracy. We have found two ways to do this. First, one can find the leading correction to the asymptotic form of $\Phi^{(1)}(r)$, and divide that into $B(r)$. Second, one can use the orthogonality of the $\chi^{(n)}_{[r]}(x_1, x_2)$ for different n to project out the $\Phi^{(1)}$ contribution to Ψ in Eq. (8.38).

For the correction to the asymptotic form of $\Phi^{(1)}(r)$, we just use Eq. (8.40), but now with the $-1/2x_3^2$ term included. (Note that this is an $O(r^{-3})$ correction relative to the $2x_3$ term.) In practice, we made the ansatz

$$\Phi(r) = Bf(r)(4r - 2\varepsilon)^{-1/4} \exp\left(-\frac{(4r - 2\varepsilon)^{3/2}}{6}\right),$$

$$f(\infty) = 1,$$

$$(8.43)$$

and numerically integrated Eq. (8.40) back from $r \sim 100$ to get $f(r)$. The integration is stable all the way down to $r \sim 2.1$.

Projection with the $\chi^{(1)}_{[r]}$ term also has to be carried out to order r^{-3}, and for this we use Eq. (8.39). Taking into account the normalization of the $\chi^{(n)}_{[x_3]}$, we can thus project out $\Phi^{(1)}$ from Eq. (8.38) (cf. Eq. (6.8) and [5, Eq. (23)] with $N = \pi^{-3/2}$):

$$\Phi^{(1)}(x_3) = \int_{-x_3}^{x_3} dx_1 \int_{-x_3}^{x_3} dx_2 \, |x_1^2 - x_2^2| \, [(x_3^2 - x_1^2)(x_3^2 - x_2^2)]^{1/2}$$

$$\times \, \Psi(x_1, x_2, x_3) \, \chi^{(1)}_{[x_3]}(x_1, x_2).$$

$$(8.44)$$

In Fig. 7 we have plotted two curves. The curve labeled (I) is $B(r)$ improved by both methods. That labeled (II) is just the $B(r)$ of Fig. 6 divided by $f(r)$ of Eq. (8.43), i.e., $B(r)$ improved only by the first method.

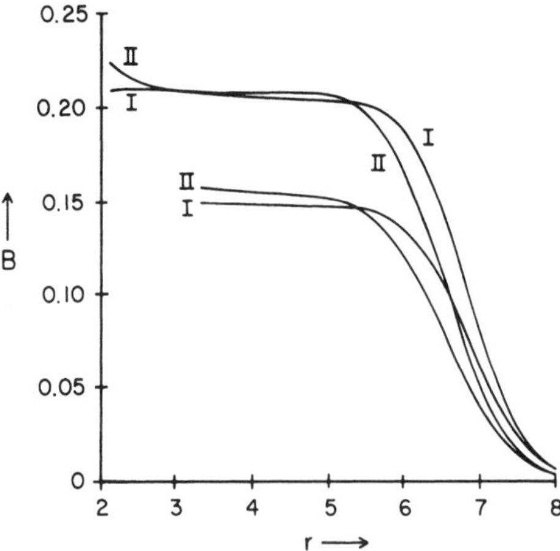

FIG. 7. Improved $B(r)$ for the ground state (upper two curves) and 0^+ state (lower two curves). Curves labelled I are improved by both methods discussed in the text. Those labeled II have been improved only by the first method.

We have included in the computer program code to use either or both of these methods, and have examined results for various numbers of basis vectors and values of ω. Some of these curves $B(r)$ are shown in Fig. 8. We calculated the integral in Eq. (8.44) by expanding the square root to the appropriate order in x_3^{-3}. The type (I) curves were obtained by integrating this correction explicitly. The curves merging into the middle curve of Fig. 8 were instead obtained by approximating the square root by its weighted average (i.e., the right hand side of (8.44) with Ψ replaced by $\chi_{[x_3]}^{(1)}$). In general, the curves coincide between $r \sim 2.9$ and $r \sim 3.5$, and

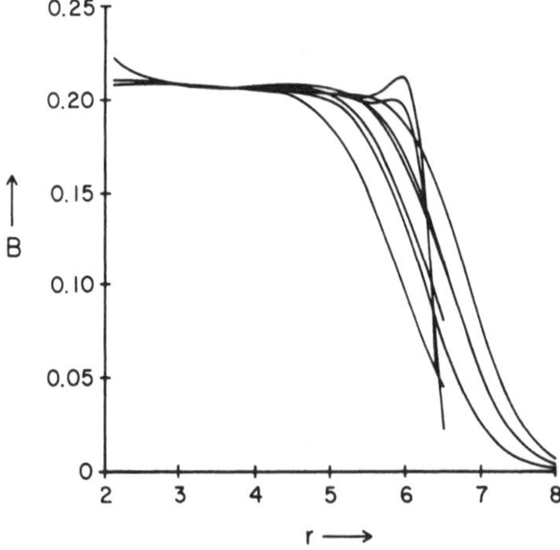

FIG. 8. $B(r)$ for the ground state, for a variety of values of N ($100 \leqslant N \leqslant 204$) and ω ($1.5 \leqslant \omega \leqslant 1.6$), showing that curves overlap between $r = 2.9$ and $r = 3.5$.

differ slightly for larger r. It is clear that the Rayleigh–Ritz wavefunction is inaccurate beyond $r \sim 5$, whereas for $r \leqslant 2.9$ curves only coincide within each class of corrections, showing that for small x_3, $\Phi^{(n)}(x_3)$ $(n > 1)$ gives a significant contribution to Ψ. We thus measure B between $r = 2.9$ and $r = 3.5$.

From the curves in Fig. 7, we can thus extract the following values for the lowest energy state and the first excited state:

$$(1) \quad \varepsilon = 4.116719735 \pm 1 \times 10^{-9}$$
$$B = 0.2063 \pm 8 \cdot 10^{-4} \tag{8.45}$$

$$(2) \quad \varepsilon' = 6.3863588 \pm 1 \times 10^{-7}$$
$$B' = 0.1480 \pm 8 \cdot 10^{-4}. \tag{8.46}$$

The energies were obtained directly from the Rayleigh–Ritz calculation. The errors in ε are estimated from the variation in ε as the number of basis vectors is changed, and the error in B is the uncertainty in the height of the plateau between $r = 2.9$ and $r = 3.5$. For the first excited state, correcting by just the first method is inadequate for any value of r, because a much larger fraction of the wavefunction is contained in $\Phi^{(n)}$ $(n > 1)$; therefore we *must* project out the $\Phi^{(1)}$ term first, via Eq. (8.44).

In summary, we saw that the Rayleigh–Ritz method became inaccurate before we reached the asymptotic region, but by including corrections we could understand the wavefunction in the non-asymptotic region, and still determine the constant B. It was nevertheless imperative that we used at least 200 basis vectors.

8.5. Calculation of λ

In Section 7, we defined the constant λ through the stability equation

$$\lambda^{-1} = \lim_{d \to 0} \frac{d^{3/2}}{\pi} \left[dq'_d(2\pi - d) + q_d(2\pi - d) \right], \tag{8.47}$$

where $q_d(s)$ satisfies the differential equation

$$W(s)(W(s) q'_d(s))' = q_d(s) U(s), \tag{8.48}$$

$$q_d(d) = 1, \qquad q'_d(d) = \frac{1}{d}. \tag{8.49}$$

U and W are known functions of s, given by Eqs. (7.28), (A.18), and (A.30).

The only difficulty in evaluating λ is that $q_d(2\pi - d)$ diverges as $d \to 0$. We have numerically integrated Eq. (8.48) from d to $2\pi - d$ for various values of d, using a fourth-order Runge–Kutta algorithm. We plot the results for $\lambda(d)$ in Fig. 9. The limit $d \to 0$ is now easily obtained by extrapolation or, equivalently, by linear regression. Thus we find

$$\lambda = 0.69970 \pm 0.00001. \tag{8.50}$$

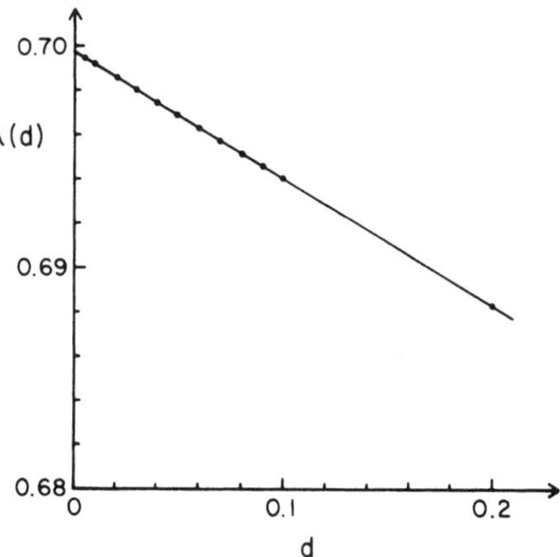

FIG. 9. Evaluation of the constant $\lambda = \lim_{d \to 0} \lambda(d) = 0.69970 \pm 0.00001$. Because it is defined as the difference of two divergent functions, $\lambda(0)$ cannot be evaluated directly. Points marked were evaluated using numerical integration.

Inserting the values for λ, ε, and B into Eq. (8.1) thus gives the final answer for ΔE for the two states we have considered [5]:

$$(1) \qquad \varepsilon = 4.116719735 \pm 1 \times 10^{-9}$$

$$\Delta E = 0.05956 g^{5/3} L^{-1} \exp(-12.4637 g^{-1} + 16.132 g^{-1/3}) \qquad (8.51)$$

$$(2) \qquad \varepsilon = 6.3863588 \pm 1 \times 10^{-7}$$

$$\Delta E = 0.03065 g^{5/3} L^{-1} \exp(-12.4637 g^{-1} + 25.026 g^{-1/3}). \qquad (8.52)$$

These are the semi-classical expressions for the tunneling energy splits. In the next section we will discuss these results, but for a more complete discussion of the physics involved, we refer the reader to [5], where we used them without a detailed derivation.

9. DISCUSSION AND CONCLUSION

The expression for the energy of electric flux in terms of L and $g(L)$ in general depends on the renormalization scheme used, although to the order displayed in Eq. (8.51) it is actually scheme independent. To compare analytic and numerical calculations it is convenient to derive an expression in terms of scheme-independent parameters. Pure QCD in a box has only two free parameters: the box size L, and a mass scale Λ. Thus the theory is completely fixed by specifying some energy (say the mass of the 0^+ glueball, $M_L(0^+)$), plus the box size in those units:

352   VAN BAAL AND KOLLER

$z = L \cdot M_L(0^+)$. There are of course other choices of parameterization, but this is the one used by Lüscher [4], and we adopt it here. Note that our definitions in no way require $M_L(0^+)$ to be the mass gap (i.e., the *lowest* mass state). In perturbation theory, z has been calculated by Lüscher and Münster [13]:

$$z = 2.2696390 \cdot g^{2/3} - 0.7975278 \cdot g^{4/3} + O(g^2) \qquad (9.1)$$

If we now eliminate g from our expressions in favour of z, and express all energies in units of $M_L(0^+)$, we obtain physical, renormalization-scheme-independent functions. Not only are these functions physically meaningful, but, as emphasized in previous work [5], we expect these functions of z to have a smoother behaviour than functions of g as we go from small to large box size, for the following reason. As L gets large, $g(L)$ gets large, and higher order terms in the β-function start to affect the behaviour of g. Therefore g can vary rapidly, and its behaviour is highly renormalization-scheme-dependent. Thus any physical quantity, expressed as a function of g or L, could acquire a complicated behaviour just to compensate for this irregular g behaviour.

Rewriting Eq. (8.51) in terms of the renormalization-scheme-independent parameters gives

$$\mathscr{E}(z) \equiv \frac{\Delta E}{M_L(0^+)} = 0.00767 \cdot z^{3/2} \cdot \exp(-42.6169 \cdot z^{-3/2} + 34.2001 \cdot z^{-1/2}). \qquad (9.2)$$

We have discussed the implications of this result in detail in another paper [5]. Note that $\mathscr{E}(z)$ increases suddenly by several orders of magnitude at $z \sim 1.2$. Thus for $z < 1.2$, the energy of electric flux is exponentially small, because the box is so small that all interactions are weak. However, as we increase the size of the box, stronger interactions can take place, and tunneling between the degenerate vacua occurs. The effect of tunneling becomes sizeable at around $z \simeq 1.2$. For boxes with $z \geqslant 1.2$, the degeneracy between states with different electric flux is lifted completely. We wish to emphasize, however, that this transition is smooth, and is thus *not* a phase transition. In particular, it should not be confused with the deconfining phase transition in finite-temperature QCD. It is a smooth but rapid transition associated with the restoration of the Z_2 symmetry of the vacuum.

Because $\mathscr{E}(z)$ is a physical function, it is possible to relate our results to those obtained in other ways, such as from Monte Carlo calculations on a lattice. In particular, by taking recent Monte Carlo results [6, 7], and assuming that they are in a region where lattice artifacts are small, we can obtain an estimate of $\mathscr{E}(z)$ for $z \geqslant 1.5$ [5]. On the other hand, from the discussion in the previous paragraph, it is clear that our analytic result is valid for z values up to about 1.2. At that point, where the tunneling becomes important, the assumptions inherent in our semiclassical approximation break down. Taking our analytic result and the Monte Carlo data together thus provides a description of $\mathscr{E}(z)$ over an almost continuous range, from $z = 0$ to $z \sim 8$, and the remarkably simple picture which emerges is described in [5]. The dominant feature in $\mathscr{E}(z)$ appears to be the onset of the tunneling that we have just calculated, and *no other rapid transitions seem to occur.* We also noted in [5] that string formation appears to set in around $z = 5$, so that only beyond $z = 5$ will $\mathscr{E}(z)$ approach its asymptotic value. There could be variables which are more sensitive to an expected transition around $z = 5$, such as the expectation value of the absolute value of spatial Polyakov loop in the fundamental representation. For a further discussion of this important issue, see [29].

It is our conjecture that we will be able to calculate $\mathscr{E}(z)$ for z values up to ~ 2 by going beyond the semiclassical approximation. This would completely bridge the gap between our present analytic calculation and the Monte Carlo data, and even provide some overlap, which would be a very significant advance.

Instead of repeating discussions contained in [5], we will conclude by describing in this article a classic textbook tunneling example (see, e.g., [31]), the double harmonic oscillator. The Hamiltonian is given by

$$H = -\frac{1}{2}\frac{\partial^2}{\partial x^2} + \frac{1}{2}|x^2 - a^2|. \tag{9.3}$$

Here, the quantity analogous to the electric flux is the energy difference between the even and odd ground states Δv; the first excited state in the even sector, with energy v_1 above the ground state, is analogous to the 0^+ glueball in QCD. The parameter a is adjustable: when $a \gg 1$ the wells are far apart and tunneling is suppressed, but as a decreases, the tunneling sets in. Thus $1/a$ is analogous to L. In Fig. 10 we plot $\mathscr{E}(z) = \Delta v/v_1$ vs. $z = v_1(a)/(\sqrt{2}\,a)$ for this simple model, using data from the exact solution of [31]. The reader is invited to compare Fig. 10 with Fig. 2 of [5], the energy of electric flux for $SU(2)$ QCD on the hypertorus. If we could show that Fig. 2 of [5] is asymptotically linear while Fig. 10 is asymptotically constant, we would be explaining confinement.

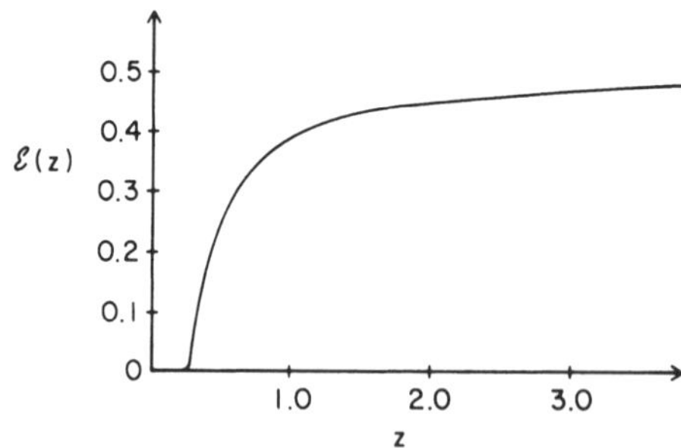

FIG. 10. The analog of $\mathscr{E}(z)$ for the double harmonic oscillator. This should be compared with Fig. 2a of [5].

VAN BAAL AND KOLLER

Appendix A

In this appendix we will derive some of the properties of the one-loop effective potential $V_1(\mathbf{C})$. Here we will put $L \equiv 1$ and hence study

$$V_1(\mathbf{C}) = \frac{4}{\pi^2} \sum_{\mathbf{n} \neq 0} \frac{\sin^2(\mathbf{n} \cdot \mathbf{C}/2)}{(\mathbf{n}^2)^2}. \tag{A.1}$$

We will first discuss the use of lattice sums [21] to express $V_1(\mathbf{C})$ in rapidly converging sums, valid for all $\mathbf{C}$. After that we construct more efficient expressions when restricting (A.1) to one of the coordinate axes. Finally the extremal properties of $V_1(\mathbf{C})$ are discussed.

The lattice sum technique is based on splitting $\sum f(\mathbf{n})$ into $\sum f(\mathbf{n}) g(\mathbf{n})$ and $\sum f(\mathbf{n})(1 - g(\mathbf{n}))$. Here $g(\mathbf{n})$ is a rapidly -decreasing auxiliary function such that $f(\mathbf{x})(1 - g(\mathbf{x}))$ behaves very smoothly for $|\mathbf{x}| \to 0$. So if $f(\mathbf{n})$ is some more slowly decreasing series, $f(\mathbf{n}) g(\mathbf{n})$ becomes rapidly decreasing, whereas the second part behaves similarly after Poisson resummation. One obtains the result [21]

$$\sum_{\mathbf{n} \neq 0} \frac{e^{2\pi i \mathbf{X} \cdot \mathbf{r}(\mathbf{n})}}{|\mathbf{r}(\mathbf{n}) - \mathbf{R}|^{2p}}$$

$$= \frac{1}{\Gamma(p)} \left\{ \sum_{\mathbf{n} \neq 0} \frac{\Gamma(p, \pi |\mathbf{r}(\mathbf{n}) - \mathbf{R}|^2)}{|\mathbf{r}(\mathbf{n}) - \mathbf{R}|^{2p}} e^{2\pi i \mathbf{X} \cdot \mathbf{r}(\mathbf{n})} - \frac{\gamma(p, \pi \mathbf{R}^2)}{\mathbf{R}^{2p}} \right.$$

$$\left. + \frac{\pi^{2p - 3/2}}{V_a} \sum_{\mathbf{n}} |\mathbf{x}(\mathbf{n}) - \mathbf{X}|^{2p - 3} \Gamma\left(\frac{3}{2} - p, \pi |\mathbf{x}(\mathbf{n}) - \mathbf{X}|^2\right) e^{2\pi i (\mathbf{x}(\mathbf{n}) - \mathbf{X}) \cdot \mathbf{R}} \right\}, \tag{A.2}$$

with $\mathbf{r}(\mathbf{n})$ running over the lattice spanned by $\mathbf{a}_i$, $\mathbf{x}(\mathbf{n})$ running over the dual lattice, $\gamma(p, x)$ and $\Gamma(p, x)$ the incomplete gamma function [22, p. 260]:

$$\mathbf{r}(\mathbf{n}) = \sum n_i \mathbf{a}_i, \qquad \mathbf{x}(\mathbf{n}) = \sum n_i \mathbf{b}_i,$$

$$\tag{A.4}$$

$$\mathbf{a}_i \cdot \mathbf{b}_j = \delta_{ij}, \qquad V_a = \mathbf{a}_1 \cdot (\mathbf{a}_2 \wedge \mathbf{a}_3),$$

$$\Gamma(p, x) = \int_x^\infty e^{-t} t^{p-1} \, dt,$$

$$\tag{A.5}$$

$$\gamma(p, x) = \Gamma(p) - \Gamma(p, x).$$

To compute $V_1(\mathbf{C})$ we take the limit $\mathbf{R} \to 0$, $\mathbf{X} = \mathbf{C}/2\pi$, $\mathbf{a}_i = \mathbf{b}_i = \mathbf{e}^{(i)}$ (hence $V_a = 1$) and substitute $p = 2$:

$$V_1(\mathbf{C}) = 4 \sum_{\mathbf{n} \neq 0} \left[\frac{1 + \pi \mathbf{n}^2}{(\pi \mathbf{n}^2)^2} + 2 \right] e^{-\pi \mathbf{n}^2} \sin^2(\mathbf{n} \cdot \mathbf{C}/2)$$

$$- 4\pi \sum_{\mathbf{n}} \left[|\mathbf{n}| \, \mathrm{erfc}(\sqrt{\pi} \, |\mathbf{n}|) - |\mathbf{n} - \mathbf{C}/2\pi| \, \mathrm{erfc}(\sqrt{\pi} \, (|\mathbf{n} - \mathbf{C}/2\pi|)) \right], \tag{A.6}$$

with

$$\text{erfc}(x) = \frac{1}{\sqrt{\pi}}\, \Gamma(\tfrac{1}{2}, x^2) = \frac{2}{\sqrt{\pi}} \int_x^\infty e^{-t^2}\, dt. \tag{A.7}$$

In going from (A.2) to (A.6) we applied the Poisson resummation formula:

$$\sum_{\mathbf{n}} \left[e^{-\pi \mathbf{n}^2} - e^{-\pi(\mathbf{n} - \mathbf{C}/2\pi)^2} \right] = 2 \sum_{\mathbf{n} \neq 0} \sin^2\left(\frac{\mathbf{n} \cdot \mathbf{C}}{2}\right) e^{-\pi \mathbf{n}^2}. \tag{A.8}$$

Similarly we compute the Laplacian of the one-loop effective potential

$$\Delta V_1(\mathbf{C}) = \frac{2}{\pi^2} \sum_{\mathbf{n} \neq 0} \frac{\cos(\mathbf{n} \cdot \mathbf{C})}{\mathbf{n}^2}, \tag{A.9}$$

in which case we substitute $p = 1$ in Eq. (A.2),

$$\Delta V_1(\mathbf{C}) = \frac{2}{\pi} \left[-1 + \sum_{\mathbf{n} \neq 0} \frac{e^{-\pi \mathbf{n}^2}}{\pi \mathbf{n}^2} \cos(\mathbf{n} \cdot \mathbf{C}) + \sum_{\mathbf{n}} \frac{\text{erfc}(\sqrt{\pi}\,|\mathbf{n} - \mathbf{C}/2\pi|)}{|\mathbf{n} - \mathbf{C}/2\pi|} \right], \tag{A.10}$$

or we can apply the Laplacian directly to Eq. (A.6).

A special role is played by the restriction of $V_1(\mathbf{C})$ and $\Delta V_1(\mathbf{C})$ to $C_2 = C_3 = 0$, for which (A.6) and (A.10) are a bit clumsy. We can write ($\mathbf{C} = C\mathbf{e}^{(1)}$):

$$V_1(\mathbf{C}) = \frac{4}{\pi^2} \sum_{n=1}^\infty b_n \sin^2(nC/2), \tag{A.11}$$

$$b_n = \sum_{\mathbf{k} \in Z^2} \frac{2}{(\mathbf{k}^2 + n^2)^2}. \tag{A.12}$$

To compute b_n we note that

$$g_n(\mathbf{y}) = \sum_{\mathbf{k} \in Z^2} \frac{e^{i\mathbf{y} \cdot \mathbf{k}}}{n^2 + \mathbf{k}^2} \tag{A.13}$$

satisfies the equation

$$\left[-\frac{\partial^2}{\partial y^2} + n^2 \right] g_n(\mathbf{y}) = (2\pi)^2 \sum_{\mathbf{k}} \delta(\mathbf{y} - 2\pi\mathbf{k}), \tag{A.14}$$

which is solved by the modified Bessel function [22, p. 374],

$$g_n(\mathbf{y}) = 2\pi \sum_{\mathbf{k}} K_0(n\,|\mathbf{y} - 2\pi\mathbf{k}|), \tag{A.15}$$

and we find

$$b_n = \lim_{y \to 0} -\frac{1}{n}\frac{d}{dn} g_n(\mathbf{y}) = \frac{2\pi}{n^2} + \frac{(2\pi)^2}{n} \sum_{\mathbf{k} \neq 0} |\mathbf{k}| \, K_1(2\pi n \, |\mathbf{k}|). \qquad (A.16)$$

We can exactly calculate the slowly converging part

$$\sum_{n=1}^{\infty} \frac{2\pi}{n^2} \sin^2(nC/2) = \pi C(2\pi - C)/4, \qquad C \in [0, 2\pi], \qquad (A.17)$$

periodically extended outside the interval $[0, 2\pi]$ ($\partial^2/\partial C^2$ of the left- and right-hand sides of (A.17) are identical). Hence

$$V_1(C\mathbf{e}^{(1)}) = \frac{C(2\pi - C)}{\pi} + \frac{8}{\pi} \sum_{n=1}^{\infty} a_n \sin^2(nC/2), \qquad (A.18)$$

$$a_n = \frac{2\pi}{n} \sum_{\mathbf{k} \neq 0} |\mathbf{k}| \, K_1(2\pi n \, |\mathbf{k}|). \qquad (A.19)$$

The a_n form a very rapidly decreasing series, and one easily shows that $a_n < 5\pi^2 n^{-1} e^{-2\pi n}$. The values of the first four a_n are

$$\begin{aligned}
a_1 &= 2.7052746... \times 10^{-2} \\
a_2 &= 1.6048745... \times 10^{-5} \\
a_3 &= 1.6065690... \times 10^{-8} \\
a_4 &= 1.9385627... \times 10^{-11}.
\end{aligned} \qquad (A.20)$$

Similar arguments show that we can write ($\mathbf{C} = (C, y_1, y_2)$):

$$\Delta V_1(\mathbf{C}) = \frac{8}{\pi} \sum_{n=1}^{\infty} c_n \cos(nC) \mid \frac{8}{\pi} \sum_{n=1}^{\infty} \cos(nC) \, K_0(2\pi n \, |\mathbf{y}|) + \frac{2}{\pi} \sum_{\mathbf{k} \neq 0} \frac{e^{i\mathbf{k} \cdot \mathbf{y}}}{\mathbf{k}^2}, \quad (A.21)$$

$$c_n = \sum_{\mathbf{k} \neq 0} K_0(2\pi n \, |\mathbf{k}|). \qquad (A.22)$$

Now the limit $\mathbf{y} \to 0$ is not easily taken because of logarithmic singularities. However, for $|\mathbf{y}| < \pi$, applying the Laplacian to the last two terms in (A.21) is easily seen to give a sum of δ-functions at $\mathbf{C} = 2\pi n \mathbf{e}^{(1)}$, and this is sufficient to show that

$$(\Delta V_1)(C\mathbf{e}^{(1)}) = -\frac{16}{\pi} \sum_{n=1}^{\infty} c_n \sin^2(nC/2) + \sum_{n=-\infty}^{\infty} \frac{4}{|C - 2\pi n|} + Q, \qquad (A.23)$$

where Q is an infinite constant. Introducing the logarithmic derivative of the Gamma function

$$\psi(C) = \frac{d}{dC}(\ln \Gamma(C)) \tag{A.24}$$

which satisfies [22]

$$-\psi(x+1) = \gamma + \sum_{n=1}^{\infty}\left[\frac{1}{n+x} - \frac{1}{n}\right], \tag{A.25}$$

we can write (A.23) for $C \in [0, 2\pi]$ as

$$(\Delta V_1)(Ce^{(1)}) = -\frac{16}{\pi}\sum_{n=1}^{\infty} c_n \sin^2(nC/2) + \frac{4}{C}$$
$$-\frac{2}{\pi}\left[\psi\left(1+\frac{C}{2\pi}\right) + \psi\left(1-\frac{C}{2\pi}\right)\right] + Q'. \tag{A.26}$$

Q' is now a finite constant and it can be fixed by comparing (A.26) with (A.10) or with the expansion of $V_1(C)$ to fourth order in C in terms of a_n (Eqs. (4.10) and (4.11)) which gives

$$(\Delta V_1)(C) = \frac{4}{|C|} + 6\kappa_1 + 60\kappa_3 C^2 + O(C^4). \tag{A.27}$$

Combining this with the expansion of $\psi(1+x)$,

$$-\psi(1+x) = \gamma + \sum_{n=1}^{\infty} \zeta(n+1)(-x)^n, \tag{A.28}$$

we find ($\gamma = 0.57721566...$, Euler's constant):

$$Q' = 6\kappa_1 - \frac{4\gamma}{\pi} = -\frac{6}{\pi} - \frac{4\gamma}{\pi} + \frac{12}{\pi}\sum_{n=1}^{\infty} n^2 a_n. \tag{A.29}$$

Exhibiting explicitly the pole terms we can write

$$(\Delta V_1)(Ce^{(1)}) = \frac{4}{C} + \frac{4}{2\pi - C} - \frac{8}{\pi} + \frac{12}{\pi}\sum_{n=1}^{\infty} n^2 a_n - \frac{16}{\pi}\sum_{n=1}^{\infty} c_n \sin^2(nC/2)$$
$$-\frac{2}{\pi}\left[\psi\left(1+\frac{C}{2\pi}\right) - \psi(1) + \psi\left(2-\frac{C}{2\pi}\right) - \psi(2)\right] \quad C \in [0, 2\pi], \tag{A.30}$$

358   VAN BAAL AND KOLLER

and periodically extended outside the interval $[0, 2\pi]$. The first four coefficients for c_n (Eq. (A.22)) are given by

$$c_1 = 3.9029359... \times 10^{-3}$$

$$c_2 = 4.9074404... \times 10^{-6}$$

$$c_3 = 7.4739898... \times 10^{-9}$$ $$\text{(A.31)}$$

$$c_4 = 1.2102679... \times 10^{-11}.$$

They satisfy the same upperbound as the coefficients a_n.

Note that the function in square brackets in Eq. (A.30) is symmetric about $C = \pi$ and vanishes at $C = 0$ and $C = 2\pi$. It can therefore be written as $\sum_{n=1}^{\infty} \frac{1}{8} d_n \sin^2(nC/2)$, with

$$d_n = -\frac{32}{\pi} \int_0^{2\pi} \cos(nC)\, \psi\left(1 + \frac{C}{2\pi}\right) dC = -64 Ci(2\pi n), \qquad \text{(A.32)}$$

$(Ci(2\pi n) = -g(2\pi n) \sim 1/(2\pi n)^2$, see [22]) and thus

$$(\Delta V_1)(C e^{(1)}) = \frac{8\pi}{C(2\pi - C)} + \frac{12}{\pi} \sum_{n=1}^{\infty} n^2 a_n - \frac{8}{\pi} - \frac{16}{\pi} \sum_{n=1}^{\infty} (c_n + d_n) \sin^2(nC/2), \qquad \text{(A.33)}$$

or (one can sum the slowly converging part of d_n using Eq. (A.17))

$$\Omega^2 = \frac{1}{2}\left(\Delta - \frac{\partial^2}{\partial C^2}\right) V_1(C e^{(1)})$$

$$= \frac{4\pi}{C(2\pi - C)} - \frac{3}{\pi} + \frac{2}{\pi} \sum_{n=1}^{\infty} n^2 a_n + 4 \sum_{n=1}^{\infty} (n^2 a_n - 2c_n - 2d_n) \sin^2(nC/2). \qquad \text{(A.34)}$$

We finally address the properties of $V_1(\mathbf{C}) - V_1(C_1 e^{(1)})$ which can be written as $(\mathbf{C} = (C, y_1, y_2),\ \mathbf{n} - (n, k_1, k_2))$:

$$V_1(\mathbf{C}) - V_1(C e^{(1)}) = \frac{4}{\pi} \sum_{\mathbf{n} \neq 0} \frac{\cos(nC)\sin^2(\mathbf{k} \cdot \mathbf{y}/2)}{(\mathbf{n}^2)^2}. \qquad \text{(A.35)}$$

In this case we find the sum over n to be

$$\frac{4}{\pi^2} \sum_n \frac{\cos(nC)}{(n^2 + \mathbf{k}^2)^2} = -\frac{2}{\pi^2 |\mathbf{k}|} \frac{d}{dx}\left[\sum_n \frac{\cos(nC)}{n^2 + x^2}\right]_{x = |\mathbf{k}|}$$

$$= -\frac{2}{\pi^2 |\mathbf{k}|} \frac{d}{dx}\left[\frac{\pi}{x} \frac{\cosh(x(\pi - C))}{\sinh(\pi x)}\right]_{x = |\mathbf{k}|}, \qquad C \in [0, 2\pi]. \qquad \text{(A.36)}$$

The result is

$$V_1(\mathbf{C}) = V_1(C\mathbf{e}^{(1)}) + \sum_{\mathbf{k} \neq 0} f_{|\mathbf{k}|}(C) \sin^2(\mathbf{k} \cdot \mathbf{y}/2), \tag{A.37}$$

with $f_{|\mathbf{k}|}(2\pi - C) = f_{|\mathbf{k}|}(C)$ and periodically extended outside the interval $[0, 2\pi]$. Explicitly $f_x(C)$ can be written as

$$f_x(C) = \frac{2}{\pi x^3} \frac{\cosh(x(\pi - C)) + xC \sinh(x(\pi - C))}{\sinh(x\pi)} + \frac{2\cosh(xC)}{x^2 \sinh^2(x\pi)}, \tag{A.38}$$

which is positive definite. Hence $V_1(\mathbf{C}) \geqslant V_1(C_1\mathbf{e}^{(1)})$ for all $\mathbf{C}$, and equal iff $C_2 = C_3 = 0 \bmod 2\pi$.

Consequently for fixed C_1, $C_2 = C_3 = 0 \bmod 2\pi$ are minima. This is in accordance with

$$\left. \frac{\partial V_1(\mathbf{C})}{\partial C_i} \right|_{C_i = 0 \bmod \pi} = 0, \qquad \mathbf{C} \neq \mathbf{0} \bmod 2\pi, \tag{A.39}$$

which can be derived from the symmetries of $V_1(\mathbf{C})$: $V_1(\mathbf{C} + 2\pi\mathbf{n}) = V_1(-\mathbf{C}) = V_1(\mathbf{C})$. The extrema of V_1 are hence given by $\mathbf{C} = \mathbf{0} \bmod \pi$, where $\mathbf{C} = \mathbf{0} \bmod 2\pi$ are absolute minima ($V_1(\mathbf{C}) = 2\,|\mathbf{C}| + O(\mathbf{C}^2)$), $\mathbf{C} = \pi(1, 1, 1) \bmod 2\pi$ are maxima and the other extrema are saddle points. Except for the cone-shaped minima, $V_1(\mathbf{C})$ is C^∞. In Fig. 11 we sketch the behaviour of $V_1(\mathbf{C})$ in one eighth of the unit cell (i.e., $[0, \pi]^3$).

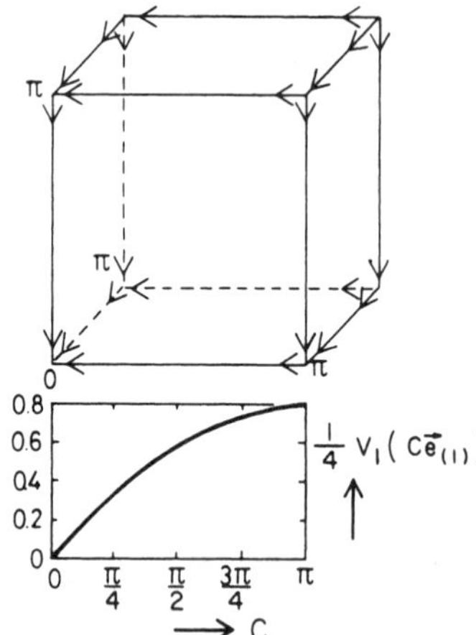

FIG. 11. Picture showing the behaviour of $V_1(\mathbf{C})$. The corners are extrema and arrows give the direction of decrease of the potential. The behaviour of V_1 along the C_1 axis is plotted.

APPENDIX B

Here we will derive the polar decomposition retaining gauge and rotational degrees of freedom. (There is some overlap with [27], and we choose our definitions so as to reproduce the standard commutation relations in Eq. (B.26).) To keep the algebra simple, we parameterize ζ and η by their S_3 coordinates of $SU(2)$ after steriographic projection (giving a double covering for both $SO(3)$ factors)

$$\zeta^{ab} = \tfrac{1}{2}\,\mathrm{Tr}(R\sigma^a R^\dagger \sigma^b) \tag{B.1}$$

$$\eta_{ij} = \tfrac{1}{2}\,\mathrm{Tr}(S^\dagger \sigma_i S\sigma_j) \tag{B.2}$$

$$R = ((1-\mathbf{r}^2) + 2i\mathbf{r}\cdot\sigma)/(1+\mathbf{r}^2), \qquad \mathbf{r} \in R^3 \tag{B.3}$$

$$S = ((1-\mathbf{s}^2) + 2i\mathbf{s}\cdot\sigma)/(1+\mathbf{s}^2), \qquad \mathbf{s} \in R^3. \tag{B.4}$$

The S_3 coordinates are given by

$$\hat{r}_0 = (1-\mathbf{r}^2)/(1+\mathbf{r}^2), \qquad \hat{\mathbf{r}} = 2\mathbf{r}/(1+\mathbf{r}^2), \tag{B.5}$$

$$\hat{s}_0 = (1-\mathbf{s}^2)/(1+\mathbf{s}^2), \qquad \hat{\mathbf{s}} = 2\mathbf{s}/(1+\mathbf{s}^2), \tag{B.6}$$

in terms of which we can write

$$\zeta^{ab} = (\hat{r}_0^2 - \hat{\mathbf{r}}^2)\,\delta_{ab} + 2\hat{r}_a \hat{r}_b - 2\hat{r}_0 \hat{r}_k \varepsilon_{kab}, \tag{B.7}$$

$$\eta_{ij} = (\hat{s}_0^2 - \hat{\mathbf{s}}^2)\,\delta_{ij} + 2\hat{s}_i \hat{s}_j - 2\hat{s}_0 \hat{s}_k \varepsilon_{kij}. \tag{B.8}$$

The next step is to calculate $\dot{c}_i^a \dot{c}_i^a$ for $c_i^a = \sum_{b=1}^{3} \zeta^{ab} x_b \eta_{bi}$ to deduce the metric $g_{\mu\nu}$, where we label $x^\mu = (\mathbf{x}, \mathbf{r}, \mathbf{s})$, and a dot denotes differentiation w.r.t. the geodesic parameter. So

$$\dot{x}^\mu g_{\mu\nu} \dot{x}^\nu = \sum_{b=1}^{3} \dot{x}_b \dot{x}_b + 2 \sum_{b,c=1}^{3} \dot{\zeta}^{ab} x_b \eta_{bi} \zeta^{ac} x_c \dot{\eta}_{ci} + \sum_{b=1}^{3} [(\dot{\zeta}^{ab})^2 + (\dot{\eta}_{ib})^2]\, x_b^2. \tag{B.9}$$

We have

$$\begin{aligned}
\dot{\zeta}^{ab}\zeta^{ac} &= \tfrac{1}{4}\,\mathrm{Tr}(\sigma^a (R^\dagger \sigma^b \dot{R} + \dot{R}^\dagger \sigma^b R))\,\mathrm{Tr}(\sigma^a R^\dagger \sigma^c R) \\
&= \tfrac{1}{2}\,\mathrm{Tr}((R^\dagger \sigma^b \dot{R} + \dot{R}^\dagger \sigma^b R)\, R^\dagger \sigma^c R) \\
&= -i\,\mathrm{Tr}(\dot{R} R^\dagger \sigma^a)\,\varepsilon_{abc},
\end{aligned} \tag{B.10}$$

and, similarly,

$$\eta_{ik}\dot{\eta}_{jk} = i\,\mathrm{Tr}(\dot{S} S^\dagger \sigma_k)\,\varepsilon_{kij}. \tag{B.11}$$

Introducing the $SO(3)$ matrices U and V by

$$U_{ia} = -\frac{i}{4}(1+\mathbf{r}^2)\,\mathrm{Tr}\left(\frac{\partial R}{\partial r_i}R^\dagger\sigma_a\right),\qquad(\mathrm{B}.12)$$

$$V_{ia} = \frac{i}{4}(1+\mathbf{s}^2)\,\mathrm{Tr}\left(\frac{\partial S}{\partial s_i}S^\dagger\sigma_a\right),\qquad(\mathrm{B}.13)$$

we can write

$$g_{\mu\nu} = \begin{pmatrix} 1_3 & 0 & 0 \\ 0 & A & B \\ 0 & B^\dagger & C \end{pmatrix},\qquad(\mathrm{B}.14)$$

with

$$A_{ij} = \frac{16}{(1+\mathbf{r}^2)^2}\sum_{b=1}^{3}U_{ib}U_{jb}(\mathbf{x}^2-x_b^2),\qquad(\mathrm{B}.15)$$

$$B_{ij} = \frac{16}{(1+\mathbf{r}^2)(1+\mathbf{s}^2)}\sum_{b=1}^{3}\frac{U_{ib}V_{jb}}{x_b}\cdot 2x_1x_2x_3,\qquad(\mathrm{B}.16)$$

$$C_{ij} = \frac{16}{(1+\mathbf{s}^2)^2}\sum_{k=1}^{3}V_{ik}V_{jk}(\mathbf{x}^2-x_k^2).\qquad(\mathrm{B}.17)$$

Hence $g = \det g_{\mu\nu} = \det A\,\det(C-B^\dagger A^{-1}B)$ and because $U,V\in SO(3)$ they drop out of the expression for g:

$$g = \left(\frac{16}{(1+\mathbf{r}^2)(1+\mathbf{s}^2)}\right)^6\prod_{i>j}(x_i^2-x_j^2)^2.\qquad(\mathrm{B}.18)$$

Also $g^{\mu\nu}$ is easily determined:

$$g^{\mu\nu} = \begin{pmatrix} 1_3 & 0 & 0 \\ 0 & E & F \\ 0 & F^\dagger & G \end{pmatrix}.\qquad(\mathrm{B}.19)$$

If we introduce the shorthand notation,

$$W_b = \tfrac{1}{2}\sum_{i,j=1}^{3}\varepsilon_{bij}^2(x_i^2-x_j^2)^2,\qquad(\mathrm{B}.20)$$

we can write

VAN BAAL AND KOLLER

$$E_{ij} = \frac{(1+\mathbf{r}^2)^2}{16} \sum_{b=1}^{3} U_{ib} U_{jb} \frac{(\mathbf{x}^2 - x_b^2)}{W_b}, \tag{B.21}$$

$$F_{ij} = -\frac{(1+\mathbf{r}^2)(1+\mathbf{s}^2)}{16} \sum_{b=1}^{3} \frac{U_{ib} V_{jb}}{x_b W_b} \cdot 2x_1 x_2 x_3, \tag{B.22}$$

$$G_{ij} = \frac{(1+\mathbf{s}^2)^2}{16} \sum_{k=1}^{3} V_{ik} V_{jk} \frac{(\mathbf{x}^2 - x_k^2)}{W_b}. \tag{B.23}$$

Using $(\partial/\partial r_i)((1+\mathbf{r}^2)^{-2} U_{ib}) = (\partial/\partial s_i)((1+\mathbf{s}^2)^{-2} V_{ib}) = 0$ we finally find

$$\frac{\partial^2}{\partial c_i^a \partial c_i^a} = \frac{1}{\sqrt{g}} \frac{\partial}{\partial x_\mu} \sqrt{g}\, g^{\mu\nu} \frac{\partial}{\partial x_\nu}$$

$$= \frac{1}{\mathcal{J}} \frac{\partial}{\partial x_i} \mathcal{J} \frac{\partial}{\partial x_i} - \sum_{b=1}^{3} \frac{(\mathbf{x}^2 - x_b^2)}{W_b} (\hat{T}_b^2 + \hat{L}_b^2) - \sum_{b=1}^{3} \frac{4x_1 x_2 x_3}{x_b W_b} \hat{T}_b \hat{L}_b, \tag{B.24}$$

with $\mathcal{J} = \prod_{i>j}(x_i^2 - x_j^2)$ and

$$\hat{T}_b = i\frac{(1+\mathbf{r}^2)}{4} U_{jb} \frac{\partial}{\partial r_j}, \qquad \hat{L}_b = -i\frac{(1+\mathbf{s}^2)}{4} V_{jb} \frac{\partial}{\partial s_j}, \tag{B.25}$$

which are easily seen to be representations of $SO(3)$:

$$[\hat{L}_i, \hat{L}_j] = i\varepsilon_{ijk} \hat{L}_k, \qquad [\hat{T}_i, \hat{T}_j] = i\varepsilon_{ijk} \hat{T}_k, \qquad [\hat{L}_i, \hat{T}_j] = 0. \tag{B.26}$$

We can rewrite (B.24) in the elegant form

$$\frac{\partial^2}{\partial c_i^a \partial c_i^a} = \frac{1}{\mathcal{J}} \frac{\partial}{\partial x_i} \mathcal{J} \frac{\partial}{\partial x_i} - \frac{1}{4} \sum_{i \neq j \neq k} \left[\left(\frac{\hat{L}_k + \hat{T}_k}{x_i - x_j} \right)^2 + \left(\frac{\hat{L}_k - \hat{T}_k}{x_i + x_j} \right)^2 \right], \tag{B.27}$$

and the normalization of the wave function is related by

$$\int d^9c\, |\psi(c)|^2 = \frac{1}{v} \int d^3x\, d^3r\, d^3s\, \sqrt{g}\, |\psi(\mathbf{x}, \mathbf{r}, \mathbf{s})|^2$$

$$= \frac{16^3}{384} \int \frac{d^3r}{(1+\mathbf{r}^2)^3} \int \frac{d^3s}{(1+\mathbf{s}^2)^3} \int d^3x\, |\mathcal{J}|\, |\psi(\mathbf{x}, \mathbf{r}, \mathbf{s})|^2. \tag{B.28}$$

$v = 384$ is the degree of the map $(\mathbf{x}, \mathbf{r}, \mathbf{s}) \mapsto c_i^a = \zeta_{ab} x_b \eta_{bi}$ decomposed as follows: a factor 2 for each $SU(2)$ (covering $SO(3)$ twice), a factor 4 for the set (ξ, η) leaving $\mathbf{x}$ fixed (hence ξ and η are equal and diagonal with ± 1 on the diagonal), a factor 6 for the pairs (ξ, η) which permute the x_b, and a factor 4 for the pairs (ξ, η) which change the signs of x_b (det $c = \prod_{i=1}^{3} x_i$; thus sign changes occur in pairs). In total $v = 2 \cdot 2 \cdot 4 \cdot 6 \cdot 4 = 384$ (one easily verifies, e.g., that Eq. (B.28) is satisfied for $\psi(c) = 1$ (0) for $c^2 < 1$ (> 1)). These symmetries will be needed to show the gauge and rotational invariance of (B.27) (i.e., to show explicitly that $[\partial^2/\partial c_i^a \partial c_i^a, \hat{L}_k] =$

$[\partial^2/\partial c_i^a \, \partial c_i^a, \hat{T}_k] = 0)$. We will not discuss this, or the interesting consequences for the spectrum at non-zero angular momentum. (For example, in [2] it was shown that gauge invariant wave functions with $J = 1$ do not exist.)

Finally, we discuss the explicit parameterization of η_{ij} (Eqs, (5.17) and (5.18)). This parameterization is equivalent to the following one for S^3:

$$\hat{s}_\mu = \left(\cos\left(\frac{\theta}{2}\right) \cos\left(\frac{1}{2}(\varphi - \xi)\right), \ -\sin\left(\frac{\theta}{2}\right) \sin\left(\frac{1}{2}(\xi + \varphi)\right), \right.$$
$$\left. \sin\left(\frac{\theta}{2}\right) \cos\left(\frac{1}{2}(\xi + \varphi)\right), \ \cos\left(\frac{\theta}{2}\right) \cos\left(\frac{1}{2}(\varphi - \xi)\right) \right). \tag{B.29}$$

Using (B.4), (B.6), (B.13), and (B.25) we find

$$\hat{L}_b = \frac{i}{2}\left[\hat{s}_0 \frac{\partial}{\partial \hat{s}_0} + \hat{s}_k \varepsilon_{kbi} \frac{\partial}{\partial \hat{s}_i} \right], \tag{B.30}$$

which after some calculation yields (5.22).

Appendix C

In this appendix we will give a heuristic derivation of the contribution to the effective potential along the vacuum valley due to the leading infrared behaviour in a multiloop expansion for the effective Lagrangian introduced in Eq. (4.36). Hence we take **C** to be constant in time. One can easily specify the Feynman rules in terms of **C** and there is no need to expand in powers of the background field. If we call σ_3 the neutral component and $(\sigma_1 \pm \sigma_2)/\sqrt{2}$ the charge components, we have two types of propagators:

$$= \frac{1}{k_0^2 - ((2\pi \mathbf{k})/L)^2}, \qquad \mathbf{k} \neq \mathbf{0};$$
$$\longrightarrow = \frac{1}{k_0^2 - ((2\pi \mathbf{k} + \mathbf{C})/L)^2}, \tag{C.1}$$

where we suppress factors of i, 2π and spacetime indices. We also make no distinction between vector and ghost particles. The vertices are of the following type:

$$g\left(k_0, \left(\frac{2\pi \mathbf{k} + \mathbf{C}}{L}\right)\right), \qquad g^2,$$
$$g\left(k_0, \left(\frac{2\pi \mathbf{k}}{L}\right)\right), \qquad g^2. \tag{C.2}$$

We are now interested in the leading infrared behaviour and will only consider the part of the diagram with $\mathbf{k} = \mathbf{0}$, hence only the last (four-point) vertex in (C.2) will occur. Let us point out that this is precisely what one would obtain if an effective Lagrangian in $\mathbf{C}$ were derived from the Yang–Mills Lagrangian in the sector of spatially constant vector potentials, and indeed one easily verifies that in this sector no three-point vertices are present.

If we next rescale k_0 by $|\mathbf{C}|/L$, each propagator has a factor $L^2/\mathbf{C}^2$ and each loop integration a factor $|\mathbf{C}|/L^4$. A vacuum graph with V_4 vertices, P propagators and l loops therefore contributes

$$\mathscr{G}(V_4, P, l) \propto L^3 g^{2V_4} \left(\frac{|\mathbf{C}|}{L} \right)^{l - 2P} L^{-3l} \tag{C.3}$$

to the effective potential. Using the well-known relations

$$P = 2V_4, \qquad l = 1 + P - V_4, \tag{C.4}$$

one finds

$$\mathscr{G}(V_4, P, l) = L^{-1} g^{2/3} (g^{-2/3} |\mathbf{C}|)^{1 - 3V_4}. \tag{C.5}$$

One indeed recognizes the one-loop result $L^{-1} |\mathbf{C}|$ ($V_1(\mathbf{C}) = 2L^{-1} |\mathbf{C}|$ for $|\mathbf{C}| \ll 1$) and a two-loop contribution behaves as $g^2 L^{-1} |\mathbf{C}|^{-2}$. In conclusion, the leading infrared behaviour of the effective potential to all loop orders yields a contribution of the form

$$V_{\text{eff}} = g^{2/3} L^{-1} f(|\mathbf{C}| \, g^{-2/3}), \tag{C.6}$$

and it is no accident that W_1 in Eq. (5.34) is exactly of this form.

Appendix D

In this appendix we discuss the generalization to $SU(N)$ for zero magnetic flux $\mathbf{m}$ in such a way that the generalization to any simple, simply connected group is obvious.

The vector potential is again chosen periodic, and the remaining gauge invariance is specified by a gauge function, periodic up to an element of the center:

$$\Omega(\mathbf{x} + L\mathbf{e}^{(j)}) = \exp(2\pi i k_j / N) \, \Omega(\mathbf{x}), \tag{D.1}$$

where $\mathbf{k} \in \mathbf{Z}_N^3$ and $P \in \mathbf{Z}$ (Eq. (2.3)) specify the homotopy type, with conjugate variables $\mathbf{e}$ (integer mod N) and θ (mod 2π). $\mathbf{e}$ and θ label the physical state vectors

$$[\Omega] \, |\mathbf{e}, \theta\rangle = e^{2\pi i \mathbf{k} \cdot \mathbf{e}/N + i\theta P} \, |\mathbf{e}, \theta\rangle. \tag{D.2}$$

The classical vacuum is again given by the curvature-free connections (Ω *not* periodic)

$$A_k(\mathbf{x}) = -i\Omega(\mathbf{x})\,\partial_k\Omega^{-1}(\mathbf{x}). \tag{D.3}$$

Imposing periodicity on $\mathbf{A}$ is easily seen to imply that the

$$\Omega_j(\mathbf{x}) \equiv \Omega^{-1}(\mathbf{x}+L\mathbf{e}^{(j)})\,\Omega(\mathbf{x}) \tag{D.4}$$

are independent of $\mathbf{x}$ and mutually commute

$$\Omega_j\Omega_k = \Omega^{-1}(\mathbf{x}+L\mathbf{e}^{(j)}+L\mathbf{e}^{(k)})\,\Omega(\mathbf{x}) = \Omega_k\Omega_j. \tag{D.5}$$

Therefore, we can simultaneously diagonalize Ω_k, $k=1,2,3$ by a constant gauge transformation U,

$$\Omega_k = U\exp(i\Phi_k)\,U^{-1}, \tag{D.6}$$

where Φ_k is real, traceless, and diagonal. Using Eq. (D.4) this is seen to imply that

$$\tilde{\Omega}(\mathbf{x}) = \Omega(\mathbf{x})\,U\exp\left(i\sum_k x_k\Phi_k/L\right) \tag{D.7}$$

is a periodic gauge function, such that A_k is gauge equivalent to Φ_k/L:

$$A_k(\mathbf{x}) = \tilde{\Omega}(\mathbf{x})(\Phi_k/L)\,\tilde{\Omega}(\mathbf{x})^{-1} - i\tilde{\Omega}(\mathbf{x})\,\partial_k\tilde{\Omega}(\mathbf{x})^{-1}. \tag{D.8}$$

Again as in Section 2, the gauge invariance of Φ can be established by considering the Wilson loop

$$W(\mathscr{C}) = \mathrm{Tr}\left(\prod_k \exp(i\Phi_k)^{n_k}\right), \tag{D.9}$$

where n_l are the winding numbers of the curve $\mathscr{C}$, but apart from these, $\mathscr{C}$ is arbitrary. $W(\mathscr{C})$ is invariant under: (i) $\Phi_k \to \Phi_k + 2\pi$ and (ii) permutations of the eigenvalues of Φ_k (for all k simultaneously). The associated gauge transformations are easily written down. Including invariance of $W(\mathscr{C})$ up to an element of the center of the gauge group, we also have invariance under the gauge transformation

$$\Omega_\mathbf{k}(\mathbf{x}) = \mathrm{diag}(e^{2\pi i\mathbf{k}\cdot\mathbf{x}N/L},...,e^{2\pi i\mathbf{k}\cdot\mathbf{x}N/L},e^{2\pi i(1-N)\mathbf{k}\cdot\mathbf{x}N/L}), \tag{D.10}$$

which leaves Φ_j diagonal. Hence Φ_j is to be identified with $\Phi_j + 2\pi\Theta_j$ whenever

$$\exp(2\pi i\Theta_j) \in \mathbf{Z}_N \tag{D.11}$$

($\mathbf{Z}_n$ the center of $SU(N)$). This condition on Θ is well known from monopoles [23], and is solved by elements of the dual weight lattice of $SU(N)$.

                     VAN BAAL AND KOLLER

To be more precise, we consider the canonical (complex) basis for the Lie algebra (l is the rank and m the dimension of the group (resp. $N-1$ and N^2-1 for $SU(N)$), $r = \frac{1}{2}(m-l)$):

$$T_1, T_2,..., T_l, E_{\pm \alpha^{(1)}},..., E_{\pm \alpha^{(r)}}, \tag{D.12}$$

$$[T_i, E_\alpha] = \alpha_i E_\alpha. \tag{D.13}$$

The T_i span the Cartan sub-algebra and can be chosen diagonal, hermitian, and traceless; they generate the maximal torus which contains the center of the gauge group and we can consequently expand

$$\boldsymbol{\Phi} = \mathbf{c}^a T_a; \qquad \boldsymbol{\Theta} = \mathbf{t}^a T_a. \tag{D.14}$$

Note $\exp(2\pi i \Theta_k) \in \mathbf{Z}_N$ if and only if it commutes with all group elements or, iff for all α

$$\exp(2\pi i \Theta_k) E_\alpha \exp(-2\pi i \Theta_k) = E_\alpha. \tag{D.15}$$

The α form an l-dimensional root system. The metric on the Euclidean space spanned by these roots is ($(\alpha, \beta) \equiv \alpha_i g^{ij} \beta_j$, $g^{ij} = g_{ij}^{-1}$):

$$g_{ij} = \mathrm{Tr}_{\mathrm{ad}}(T_i T_j). \tag{D.16}$$

Let $\Delta = \{\alpha^{(1)}, \alpha^{(2)},..., \alpha^{(l)}\}$ be a basis [24] for the root lattice. The associated *dual* weight lattice (i.e., the weight lattice for the dual root lattice spanned by $\alpha^v = 2\alpha/(\alpha, \alpha)$) is defined by its basis $\{\bar{\lambda}^{(1)}, \bar{\lambda}^{(2)},..., \bar{\lambda}^{(l)}\}$ such that

$$(\bar{\lambda}^{(a)}, \alpha^{(b)}) = \bar{\lambda}_i^{(a)} g^{ij} \alpha_j^{(b)} = \delta_{ab}. \tag{D.17}$$

Since Eq. (D.15) is easily seen to imply that $\exp(2\pi i t_k^a \alpha_a) = 1$ for each root α we find that $\mathbf{t}_a$ is an element of the dual weight lattice:

$$\mathbf{t}^b = g^{ba} \mathbf{t}_a; \qquad \mathbf{t}_a = \mathbf{n}^{(i)} \bar{\lambda}_a^{(i)}, \quad n_k^{(i)} \in \mathbf{Z}. \tag{D.18}$$

If we express $\mathbf{c}^a$ in terms of the basis for the dual weight lattice,

$$\boldsymbol{\Phi} = \mathbf{c}^a T_a = \mathbf{z}^{(i)} g^{ab} \bar{\lambda}_b^{(i)} T_a = \mathbf{z}^{(i)} (\bar{\lambda}^{(i)}, T), \tag{D.19}$$

then the gauge transformation (D.10) corresponds to shifting $\mathbf{z}^{(i)}$ over multiples of 2π, whereas the gauge transformations which permute the eigenvalues of $\boldsymbol{\Phi}$ correspond to the action of the Weyl group (see also [10]). Let us remind the reader that the Weyl group is generated by the Weyl reflections (α a root)

$$\sigma_\alpha(\beta) = \beta - \frac{2(\beta, \alpha)}{(\beta, \beta)} \alpha, \tag{D.20}$$

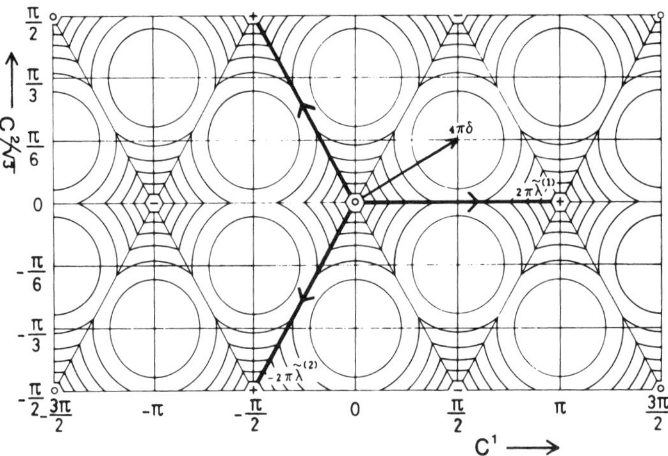

FIG. 12. Equipotential lines for the $SU(3)$ one-loop potential V_1 restricted to one dimension ($\mathbf{c}^a = c^a\mathbf{e}^{(1)}$). The homotopically nontrivial gauge transformations $\Omega_\mathbf{k}$ with $\mathbf{k} = \pm\mathbf{e}^{(1)} \bmod 3$ connecting minima of type 0 and type $\pm$ correspond to translations over $2\pi\tilde\lambda^{(1)}$ and $2\pi\tilde\lambda^{(2)}$, respectively. The broad lines correspond to the pinchons connecting type **0** and type $+$ vacua. The point $4\pi\delta$ corresponds to the maximum M of V_1, and equipotentials are plotted in steps of $M/7$.

and that an element of the Weyl group is an automorphism of the root system. The above-mentioned gauge transformations form the normalizer $N_\mathscr{G}(H)$ of the Cartan subalgebra H (the span $\{T_i\}_{i=1\cdots l}$) in $\mathscr{G}$:

$$N_\mathscr{G}(H) = \{g \in \mathscr{G} \mid ghg^{-1} \in H, \forall h \in H\}. \tag{D.21}$$

Note that the normalizer contains the centralizer $C_\mathscr{G}(H)$ of H in $\mathscr{G}$ which leaves Φ invariant.

$$C_\mathscr{G}(H) = \{g \in \mathscr{G} \mid ghg^{-1} = h, \forall h \in H\}. \tag{D.22}$$

Hence (see [25, Theorem 4.9.1]) we have an isomorphism between the Weyl group and $N_\mathscr{G}(H)/C_\mathscr{G}(H)$ generated by the map $g \mapsto \mathrm{Ad}(g)|_H$, with $\mathrm{Ad}(g)\,h = ghg^{-1}$.

There exists an elegant description for the action of the Weyl group. Recall that the fundamental Weyl chamber (w.r.t. the basis Δ) is defined by y such that $(y, \alpha^{(i)}) > 0$ for all i, or $y_i > 0$ for all i, where y_i are the components of y w.r.t. the basis for the dual weight lattice ($y = y_i\tilde\lambda^{(i)}$). Any y can be written as an element of the Weyl group acting on the closure of the fundamental Weyl chamber. For $SU(3)$ this is sketched in Fig. 12, to which we will return in a while.

As for $SU(2)$, the one-loop effective potential is given by

$$V_1(\Phi) = \mathrm{Tr}((-D_i^2(\Phi))^{1/2}), \tag{D.23}$$

with

$$D_k(\Phi) = \partial_k + i\,\mathrm{ad}(\Phi_k/L). \tag{D.24}$$

The spectrum of $-D_i^2(\Phi)$ is explicitly given by

$$-D_i^2(\Phi)\, e^{2\pi i\mathbf{k}\cdot\mathbf{x}/L} T_a = \left(\frac{2\pi\mathbf{k}}{L}\right)^2 e^{2\pi i\mathbf{k}\cdot\mathbf{x}/L} T_a$$

$$-D_i^2(\Phi)\, e^{2\pi i\mathbf{k}\cdot\mathbf{x}/L} E_\alpha = \left(\frac{2\pi\mathbf{k}+\mathbf{c}^a\alpha_a}{L}\right)^2 e^{2\pi i\mathbf{k}\cdot\mathbf{x}/L} E_\alpha, \tag{D.25}$$

and we can copy the result for $V_1(\Phi)$ directly from $SU(2)$:

$$\begin{aligned} V_1(\Phi) &= \frac{2}{\pi^2 L}\sum_\alpha\sum_{\mathbf{n}\neq 0}\frac{\sin^2((1/2)\,\mathbf{n}\cdot\mathbf{c}^a\alpha_a)}{(\mathbf{n}^2)^2}\\ &= \frac{2}{\pi^2 L}\sum_\alpha\sum_{\mathbf{n}\neq 0}\frac{\sin^2((1/2)\,\mathbf{n}\cdot\mathbf{z}^{(i)}(\lambda^{(i)},\alpha))}{(\mathbf{n}^2)^2}. \end{aligned} \tag{D.26}$$

We can explicitly verify the invariance of $V_1(\Phi)$ under Weyl transformations W using $\{W\alpha\}-\{\alpha\}$, and under translations of $z_k^{(i)}$ over 2π, because a basis $\varDelta$ for a root lattice is defined such that for any root α we have $\alpha=\sum_{i=1}^l k_i\alpha^{(i)}$ with k_i all positive (or all negative) integers.

As for $SU(2)$ we have that the quadratic approximation breaks down for those $\mathbf{z}^{(i)}$ where $-D_i^2(\Phi)$ has a vanishing eigenvalue. Since this vanishing eigenvalue for $SU(2)$ causes $V_1(\mathbf{z})$ to behave as $2\,|\mathbf{z}|/L$ we see that for general gauge group this breakdown occurs at $\mathbf{z}^{(i)}=0\ (\mathrm{mod}\ 2\pi)$ for one or more values of i. This corresponds precisely to the boundaries of the Weyl chambers. This is obvious since boundaries of Weyl chambers are associated with points left invariant by one or more elements of the Weyl group. This is equivalent to the fact that the associated value of Φ has a smaller holonomy group than those Φ not related to the boundary. As we discussed in detail this leads to the breakdown of the quadratic approximations, and extra zeros in the Faddeev–Popov determinant. For $SU(2)$ these singularities were isolated, but in general they are of codimension 3. Although of zero measure, they strongly dominate the quantum behaviour, as caustics do in optics.

Away from these singularities the effective one-loop Lagrangian is given by

$$\mathcal{L}=\frac{1}{g(L)^2}\,\mathrm{Tr}\,(\Phi^2)-V_1(\Phi). \tag{D.27}$$

It should be plausible, but we will give no proof, that in *lowest* order the energy splitting due to electric flux is determined by the Lagrangian in (D.27). The situation is more complicated than for $SU(2)$; it will turn out that the pinchon solution is contained in the set of singular points. Still, one can show that $V_1(\Phi)$ can be regularized by $O(g^{2/3})$ corrections by taking the quartic interactions into account. This, however, leaves the leading result unaltered.

Before specializing to $SU(N)$ we first derive a few general properties of $V_1(\Phi)$. $V_1(\Phi)\geqslant V_1(\Phi_1 e^{(1)})$ follows because $V_1(\Phi)$ is related to the $SU(2)$ potential. It means that pinchon solutions are again obtained by "dimensional reduction." Restricted to one dimension we find[1]:

$$V_1(\Phi e^{(1)}) = \frac{1}{\pi L} \sum_{\alpha > 0} \left[z_\alpha (2\pi - z_\alpha) + 8 \sum_{n=1}^{\infty} a_n \sin^2(n z_\alpha / 2) \right], \qquad \text{(D.28)}$$

$$z_\alpha = c^a \alpha_a = z^{(i)} (\lambda^{(i)}, \alpha) \in [0, 2\pi], \qquad \text{(D.29)}$$

extended periodically in z_α with periods 2π in the obvious way. If we introduce the so-called maximal weight δ

$$\delta = \tfrac{1}{2} \sum_{\alpha > 0} \alpha = \sum_{i=1}^{l} \lambda^{(i)} \qquad \text{(D.30)}$$

where $\lambda^{(i)}$, $i = 1,...,l$ form the basis of the weight lattice defined by

$$\langle \lambda^{(i)}, \alpha^{(j)} \rangle = \delta_{ij}, \qquad \text{(D.31)}$$

with (for α, β roots always integer)

$$\langle \alpha, \beta \rangle = 2 \frac{(\alpha, \beta)}{(\beta, \beta)}, \qquad \text{(D.32)}$$

and if we use the "completeness" relation

$$\sum_{\alpha} \alpha_i \alpha_j = g_{ij} \qquad \text{(D.33)}$$

(the sum is over all roots), then for $z_\alpha \in [0, 2\pi]$,

$$V_1(\Phi e^{(1)}) = \frac{1}{2\pi L} ((4\pi\delta)^2 - (c - 4\pi\delta)^2) + \frac{8}{\pi L} \sum_{n=1}^{\infty} \sum_{\alpha > 0} a_n \sin^2(n z_\alpha / 2). \quad \text{(D.34)}$$

This, in the approximation $a_n \equiv 0$, is indeed a very simple expression. Also, the kinetic part is simple since

$$\mathrm{Tr}(\dot{\Phi}^2) = \dot{c}^a \dot{c}^b \, \mathrm{Tr}(T_a T_b) = \dot{c}^a \dot{c}^b K g_{ab}. \qquad \text{(D.35)}$$

The constant K only depends on the group and g_{ab} can always be chosen proportional to δ_{ab}.

We will now specialize to $SU(N)$ for which $K = 2N$. The associated algebra is of type A_N [24]. All roots are of equal length $(\alpha, \alpha) = N^{-1}$ and for $\alpha > 0$ are given by $\alpha^{(i)}$ and $\alpha^{(i)} + \alpha^{(j)}$, $i \neq j$. It hence follows that

$$\lambda^{(i)} = 2N \lambda^{(i)}, \qquad \delta = \frac{1}{2N} \sum_{i=1}^{N-1} \lambda^{(i)} \qquad \text{(D.36)}$$

---

[1] Any root can be written as $\sum n_i \alpha^{(i)}$ with all n_i positive ($\alpha > 0$) or negative ($\alpha < 0$).

370   VAN BAAL AND KOLLER

and

$$\mathscr{L} = \frac{L\dot{\mathbf{z}}^{(i)}(A_N^{-1})_{ij}\,\dot{\mathbf{z}}^{(j)}}{g^2(L)} - \sum_{i=1}^{N-1} V_1(\mathbf{z}^{(i)}) - \sum_{i>j=1}^{N-1} V_1(\mathbf{z}^{(i)}+\mathbf{z}^{(j)}), \qquad \text{(D.37)}$$

where $(A_N)_{ij} = \langle \alpha^{(i)}, \alpha^{(j)} \rangle$ is the Cartan matrix with 2 on the diagonal and -1 on both subdiagonals, and $V_1(\mathbf{z})$ is the $SU(2)$ potential. Note that $(\bar{\lambda}^{(i)}, \bar{\lambda}^{(j)}) = (2N)^2(\lambda^{(i)}, \lambda^{(j)})$ and $\alpha^{(i)} = \langle \alpha^{(i)}, \alpha^{(j)} \rangle \lambda^{(j)} = (A_N)_{ij}\lambda^{(j)}$, which leads to $(\bar{\lambda}^{(i)}, \bar{\lambda}^{(j)}) = 2N(A_N^{-1})_{ij}$. This proves the expression for the kinetic part, whereas the potential part follows from the above-mentioned explicit form of the roots α. In Fig. 12 we exhibit the potential for $SU(3)$ by choosing units according to:

$$\Phi = \text{diag}\left(\frac{2c^1}{3} + \frac{2c^2}{\sqrt{3}}, \frac{2c^1}{3} - \frac{2c^2}{\sqrt{3}}, -\frac{4c^1}{3}\right). \qquad \text{(D.38)}$$

From this it is obvious that the pinchon has one and only one pair (i, k) such that $z_k^{(i)} \neq 0$. This is easily seen to generalize to arbitrary N (with a little imagination). The energy splitting due to electric flux can therefore be computed by restricting $\mathscr{L}$ as

$$\mathscr{L} = \frac{L\dot{z}^2(A_N^{-1})_{11}}{g^2(L)} - (N-1)\,V_1(z\mathbf{e}^{(1)}). \qquad \text{(D.39)}$$

With $(A_N^{-1})_{11} = \det(A_{N-1})/\det(A_N) = (N-1)/N$ we find

$$E(\mathbf{e}, L) = \frac{1}{L}\sum_{i=1}^{3}\sin^2\left(\frac{\pi e_i}{N}\right)\exp\left(-\left(\frac{2(N-1)^2}{Ng^2(L)}\right)^{1/2} S + O(g^{-1/3})\right), \qquad \text{(D.40)}$$

with S the $SU(2)$ pinchon action.

Note that we expect $O(g^{2/3})$ corrections to V_1 in Eq. (D.39), and hence the $O(g^{-1/3})$ is expected to have contributions from this modification. Equation (D.40) would also imply that the effect does not survive in the $N \to \infty$ limit.

ACKNOWLEDGMENTS

We would like to thank Martin Lüscher for his interest and providing his notes on the Rayleigh–Ritz analysis. We are also grateful to Fred Goldhaber for moral support. This work was supported in part by NSF grants PHY-85-07627 and DMS-84-05661.

Note added in proof. Recently the $SU(3)$ effective Hamiltonian of Lüscher [2] was numerically analysed by Weisz and Ziemann [33], giving results similar to those found by Lüscher and Münster [13] for $SU(2)$. We also extended our result for arbitrary $SU(N)$ (Eq. (D.40)) beyond the leading order, predicting for $SU(3)$ the onset of tunneling at $z \sim 1.6$ [34]. Reference 34 also contains a further discussion of an expected crossover at $z \sim 5$. It is only beyond this value of z that one can expect formation of flux tubes and the glueball to behave as a particle state.

REFERENCES

1. G. 'T HOOFT, *Nucl. Phys. B* **153** (1979), 141.
2. M. LÜSCHER, *Nucl. Phys. B* **219** (1983), 233.
3. P. VAN BAAL, *Nucl. Phys. B* **264** (1986), 548.
4. M. LÜSCHER, *Phys. Lett. B* **118** (1982), 391.
5. J. KOLLER AND P. VAN BAAL, *Nucl. Phys. B* **273** (1986), 387.
6. B. BERG AND A. BILLOIRE, *Phys. Lett B* **166** (1986), 203.
7. B. BERG, A. BILLOIRE, AND C. VOHWINKEL, Tallahassee, preprint.
8. P. VAN BAAL, "Twisted Boundary Conditions: A Non-perturbative Probe for Pure Non-Abelian Gauge Theories, Thesis, Utrecht, July 1984.
9. Y. NAMBU, *Phys. Lett. B* **80** (1979), 372; A. A. MIGDAL, *Ann. Phys.* **126** (1980), 279.
10. E. WITTEN, *Nucl. Phys. B* **202** (1982), 253.
11. P. VAN BAAL, The energy of electric flux on the hypertorus, Utrecht, internal report, March 1984, unpublished.
12. G. K. SAVVIDY, *Nucl. Phys. B* **246** (1984), 302.
13. M. LÜSCHER AND G. MÜNSTER, *Nucl. Phys. B* **232** (1984), 445.
14. P. VAN BAAL AND A. AUERBACH, *Nucl. Phys. B* **275** [FS17] (1986), 93.
15. C. BLOCH, *Nucl. Phys. B* **6** (1958), 329.
16. E. G. FLORATOS AND D. PETCHER, *Nucl. Phys. B* **252** (1985), 689.
17. A. AUERBACH AND S. KIVELSON, *Nucl. Phys. B* **257** [FS14] (1985), 799.
18. C. ITZYKSON AND J. B. ZUBER, "Quantum Field Theory," McGraw–Hill, New York, 1980.
19. G. 'T HOOFT, *Nucl. Phys. B* **62** (1973), 444.
20. D. I. DYAKONOV, V. YU. PETROV, AND A.V. YUNG, *Phys. Lett. B* **130** (1983), 385.
21. B. R. A. NIJBOER AND F. W. DE WETTE, *Physica* **23** (1957), 309.
22. M. ABRAMOWITZ AND I. STEGUN, "Handbook of Mathematical Functions," Dover, New York, 1978.
23. P. GODDARD, J. NUYTS, AND D. OLIVE, *Nucl. Phys. B* **125** (1977), 1.
24. J. E. HUMPHREYS, "Introduction to Lie Algebras and Representation Theory," Springer, Berlin/Heidelberg/New York, 1972.
25. V. S. VARADARADJAN, "Lie Groups, Lie Algebras and their Representations," Prentice–Hall, Englewood Cliffs, N.J., 1977.
26. V. GRIBOV, *Nucl. Phys. B* **139** (1978), 1; W. NAHM, Gribov copies and instantons, Cern preprint TH.3114, July 1981.
27. G. K. SAVVIDY, *Phys. Lett. B* **159** (1985), 325.
28. S. WOLFRAM *et al.*, SMP Reference Manual, Inference Corp., Los Angeles, 1983.
29. T. H. HANSSON, P. VAN BAAL, AND I. ZAHED, Stony Brook preprint, ITP-SB-86-49, July 1986; *Nucl. Phys. B*, to be published.
30. J. ZINN-JUSTIN, *Nucl. Phys. B* **246** (1984), 246.
31. E. MERZBACHER, "Quantum Mechanics," Wiley, New York, 1961.
32. C. BORGS AND E. SEILER, *Commun. Math. Phys.* **91** (1983), 329.
33. P. WEISZ AND V. ZIEMANN, Weak coupling expansion of low-lying energy values in $SU(3)$ gauge theory on a torus, Hamburg preprint, September 1986.
34. P. VAN BAAL AND J. KOLLER, *Phys. Rev. Lett.* **57** (1986), 2783; P. van Baal, *in* "Proceedings of Lattice Gauge Theory 1986," (H. Satz, Ed.), Plenum, New York, 1987, to appear.

Printed by the St. Catherine Press Ltd., Tempelhof 41, Bruges, Belgium

VOLUME 58, NUMBER 24   PHYSICAL REVIEW LETTERS   15 JUNE 1987

SU(2) Spectroscopy in Intermediate Volumes

Jeffrey Koller and Pierre van Baal

Institute for Theoretical Physics, State University of New York at Stony Brook, Stony Brook, New York 11794
(Received 13 April 1987; revised manuscript received 18 May 1987)

We present results for the low-lying energies of pure SU(2) gauge theory in a cubic volume up to a size of five glueball masses, based on an analytic-variational continuum calculation. We observe as lowest mass a $T_2(2^+)$ state, roughly half the mass of the almost degenerate $A_1(0^+)$ and $E(2^+)$ states. The $A_1(0^-)$ state shows a pronounced volume dependence. There is fair agreement with existing finite-volume Monte Carlo results.

PACS numbers: 11.15.Ha, 11.15.Tk

This Letter will give the results of our calculations for the low-lying energy spectrum of pure SU(2) gauge theory in a finite cubic volume. It goes beyond the earlier perturbative[1] and semiclassical calculations[2] and overlaps fully with existing Monte Carlo data based on (adjoint) Polyakov loop correlations.[3] Our main motivation was and still is to *understand* confinement, expecting this to have a purely gluonic origin. The challenge is to use only the Yang-Mills Lagrangean as input, without any free parameters except for the mass scale. We believe that our present calculations bring us close to the point where confinement (string formation) sets in, and provide the proper tools for the calculation of glueball masses and string tensions and for the unlocking of the "secrets" of confinement.

Nevertheless, of more immediate use will be the comparison of our results with Monte Carlo calculations, which is why we present these results in this form and defer technical details to a future publication. Realizing our responsibility in establishing a reference frame, we have taken great care to demonstrate stability of our results. We base these results on a Rayleigh-Ritz calculation, where we provide both upper and lower bounds on the energy. We also include a two-loop correction. To facilitate the comparison of our results and the lattice Monte Carlo data in a finite volume, we convert to dimensionless ratios (except in Fig. 2, curves c). As discussed elsewhere,[4] this removes the ambiguity due to our lack of precise information on the nonperturbative β function.[5] We consider the parameters z_R and $\mathcal{E}_R^{(i)}$, being respectively the linear dimension L of the cubic volume and the energy of 't Hooft–type electric flux[6] in units of some mass associated to an irreducible representation R of the cubic group. This group has the representations[7] $R = \{A_1, A_2, E, T_1, T_2\}$ with dimensions

$d_R = \{1, 1, 2, 3, 3\}$ occurring with both parities. The lowest angular momentum these representations couple to are $\{0, 3, 2, 1, 2\}$, respectively. The electric flux is classified by the vector $\mathbf{e} \in Z_2^3$ and using the cubic symmetry we get three different energies, determined by the number of components of $\mathbf{e}$ equal to 1 (mod 2) (we call this the number of "units" of electric flux, i.e., $\mathcal{E}^{(i)}$ is the energy of i units of electric flux). Lattice Monte Carlo data for the volumes where we expect our result to be accurate is available in the form of z_{E^+}, $(\mathcal{E}_E^{(1)}/z_{E^+})^{1/2}$ and $z_{A_1^+}/z_{E^+}$.[3] In the Monte Carlo calculations energy differences are measured by use of time-time correlations for spatial Polyakov loops on elongated lattices (adjoint loops for "glueballs" and fundamental loops for "string tensions," the quotation marks are explained in Ref. 4).

We first briefly outline our approach, which is based almost entirely on Lüscher's effective Hamiltonian for the spatially constant gauge fields,[1] with, however, the crucial addition of proper boundary conditions. These boundary conditions reflect the nontrivial topology of configuration space (ignoring this leads to a Gribov ambiguity[8]), and is also related to the fact that a zero-momentum configuration represented by a spatially constant gauge potential is not a gauge-invariant notion. The boundary conditions are the remnant, in an adiabatic approximation, of the patching of coordinate systems in configuration space. This adiabatic approximation can be explicitly tested in our calculational scheme and will be discussed in more detail elsewhere. The coordinates of Lüscher's effective Hamiltonian are the spatially constant vector potentials c^a_i with a the SU(2) color indices and i the space indices. The symmetries are given by $c^a_i \rightarrow \xi^{ab} c^b_j \eta_{ij}$, with $\xi \in \mathrm{SO}(3)$ and $\eta \in \mathrm{O}(3,Z)$. The effective Hamiltonian is given in terms of the minimally subtracted renormalized coupling constant g at the scale L by

$$LH_{\mathrm{eff}} = -\tfrac{1}{2}(1/g^2 + a_1)^{-1}\partial^2/\partial c^a_i{}^2 + V_T(c) + V_1(c) + V_2(c) + \cdots. \tag{1}$$

Here V_T is the transverse potential vanishing along Abelian configurations, called the vacuum valley,[9] while V_1 and V_2

VOLUME 58, NUMBER 24   PHYSICAL REVIEW LETTERS   15 JUNE 1987

are the one- and two-loop contributions to the effective potential along the vacuum valley,

$$V_1(c) = \frac{4}{\pi^2} \sum_{\mathbf{n}\neq 0} \frac{\sin^2(\mathbf{n}\cdot\mathbf{r}/2)}{(\mathbf{n}^2)^2} - 2|\mathbf{r}|, \quad r_i = \left(\sum_{a=1}^{3} c^a{}_i c^a{}_i\right)^{1/2}, \tag{2a}$$

$$V_2(c) = \frac{g^2}{32}\{[\Delta V_1(c)]^2 + 2\Delta V_1(c)\Delta V_1(0)\}, \quad \Delta = \sum_{i=1}^{3} \frac{\partial^2}{\partial r_i^2}, \tag{2b}$$

$$V_T(c) = \frac{1}{4}\left[\frac{1}{g^2} + \alpha_2\right]\sum_{ija} F^a{}_{ij}F^a{}_{ij} + \alpha_3 \sum_{i,j,k,a,b} F^a{}_{ij}F^a{}_{ij}c^b{}_k c^b{}_k + \alpha_4 \sum_{i,j,a,b} F^a{}_{ij}F^a{}_{ij}c^b{}_j c^b{}_j + \alpha_5[\det(c)]^2 + \cdots, \tag{2c}$$

where $\alpha_1 = 2.181\,0429\times10^{-2}$, $\alpha_2 = 7.571\,4590\times10^{-3}$, $\alpha_3 = -1.113\,0266\times10^{-4}$, $\alpha_4 = -2.147\,5176\times10^{-4}$, $\alpha_5 = -1.277\,5652\times10^{-3}$, and $F^a{}_{ij} = \epsilon_{abd}c^b{}_i c^d{}_j$. We have extended Lüscher's effective Hamiltonian to higher orders relevant for our nonperturbative analysis. The need for the one-loop term in Eq. (2a) was discussed previously.[2] The terms proportional to α_3, α_4, and α_5 in Eq. (2c) were obtained from a one-loop calculation and the two-loop term of Eq. (2b) follows from a straightforward extension of previous two-loop calculations for chromomagnetic vacuum energies.[10] We made a polynomial fit to eighth order in r_i for $V_1(c)$ and $V_2(c)$, accurate to one-third of 1% for r_i smaller than π. Terms of sixth order are the appropriate Taylor coefficients, whereas the eighth-order term is considered as a correction. Finally the boundary conditions are given by

$$\left[\frac{\partial}{\partial r_i}\right]^{1-e_i} r_i \Psi_{\mathbf{e}}(c)\big|_{r_i = \pi} = 0. \tag{3}$$

We use spherical coordinates (r_i, θ_i, ϕ_i) for the vectors $(c^1{}_i, c^2{}_i, c^3{}_i)$, and the problem becomes that of three interacting particles enclosed in a sphere. Gauge invariance requires the total "angular momentum" to be zero.

Next we analyze this Hamiltonian using the Rayleigh-Ritz technique and in general we take a basis of the form

$$\langle c | l_1 l_2 l_3 n_1 n_2 n_3\rangle = \sum_m W(l_1 l_2 l_3 m_1 m_2 m_3)\prod_i \chi_{l_i n_i}(r_i) Y_{l_i m_i}(\theta_i, \phi_i). \tag{4}$$

Here W are the Clebsch-Gordan-Wigner coefficients for combining three SU(2) irreducible representations into a (color) singlet, thereby resolving the problem of eliminating the gauge degrees of freedom; each element of our basis is gauge invariant. The $\chi_{l_i n_i}(r_i)$ are one-particle radial eigenfunctions and satisfy the boundary conditions $(d/dr_i)^{1-e_i}r_i\chi_{l_i n_i}(r_i)\big|_{r_i=\pi}=0$. For simplicity we chose $\chi_{l_i n_i}(r_i) = j_{l_i}(w_{n_i}^{(e_i)}r_i)$. These spherical Bessel functions allow us to calculate the Hamiltonian [Eq. (1)] with respect to the basis in Eq. (4) *algebraically* in the "momenta" $w_{n_i}^{(e_i)}$. The latter are easily calculated to a high precision. Since we expect the wave function to become less peaked for large g, this should be a good basis for that region. Furthermore, we restricted the basis by suitable projections to the various representations of the cubic group (or whatever symmetry is left in the presence of nonzero electric flux), and most states considered are ground states in these particular sectors. Typically we took 350–450 basis vectors, chosen such that they contain the most relevant coefficients of the wave function, and diagonalized the Hamiltonian using the IMSL (International Mathematical Subroutine Library) routines. This provides an upper bound on the energies. To make absolutely sure our results were accurate we also calculated a lower bound, using Temple's inequality.[11] (It requires one to calculate in addition the matrix of H^2.)

Figure 1 gives for $g \geq 0.8$ our results for the low-lying energy values (in units of $1/L$, L being the physical size of the cubic volume). States are labeled by their representation [e_i^+ denotes the (ground) state with i units of electric flux]. Energies were calculated for steps of 0.2 in g. The full lines result from the Hamiltonian [Eq. (1)], excluding the two-loop contribution, the terms of order 6 in the transverse part of the potential (V_T) and the terms of order 8 in the one-loop contribution (V_1).

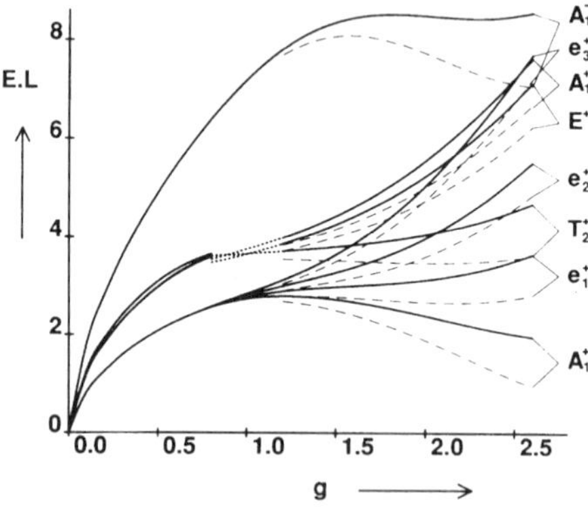

FIG. 1. The energies in units of $1/L$, labeled by their representation, for a few low-lying levels as a function of the renormalized coupling constant. The dotted parts of the curves use only the upper bound. Elsewhere upper and lower bound coincide within the thickness of the curves. Below $g = 0.8$ we show the perturbative result (Ref. 1). The dashed curves give the results including the higher-order corrections.

The upper and lower bounds coincide within the thickness of these lines. Where the lower bound becomes less accurate, but we still expect the upper bound to be reasonable, we dotted the curve, using the value of the upper bound. The full lines for $g \leq 0.8$ gives the perturbative result of Lüscher and Münster.[1] Tunneling for the A_1^+ excited and the E^+ ground states is expected to set in around $g = 0.5$, which explains the reasonably large deviation for these states, but otherwise the agreement is impressive. The dashed curves give the result for the full effective Hamiltonian [Eq. (1)]. The eighth-order one-loop term, the sixth-order transverse part and the two-loop term contribute approximately with the ratios 0.1:2:3.

We observe the remarkable fact that the ground-state energy decreases beyond $g \sim 1.3$, corresponding to $z_{E^+} \sim 1.5$. This might indicate an instability of the vacuum under formation of magnetically neutral domains, which was speculated on in Ref. 10. We believe this to be a significant step towards understanding the long-range behavior of the Yang-Mills vacuum and certainly deserves more attention than we can give it here.

In Fig. 2, curves a and b, we compare our results for the string tension $(\mathscr{E}_E^{(1)}/z_{E^+})^{1/2}$ and the mass ratio $z_{A_1^+}/z_{E^+}$ as functions of z_{E^+} with existing Monte Carlo results.[3] Figure 2 is obtained as follows from Fig. 1: $z_{E^+} = L(E_{E^+}^{(0)} - E_{A_1^+}^{(0)})$, $z_{e_1^+} = L(E_{e_1^+}^{(0)} - E_{A_1^+}^{(0)})$, $z_{A_1^+} = L(E_{A_1^+}^{(1)} - E_{A_1^+}^{(0)})$, and $\mathscr{E}_E^{(1)} = z_{e_1^+}/z_{E^+}$. We see that the results agree reasonably well. (For the string tension, compare also Fig. 2 of the first paper in Ref. 2, but note the E/A_1 mixing.[3,12]) The mass ratio in our calculations is to good accuracy identically 1.1, whereas the Monte Carlo data are somewhat higher for z_{E^+} between 1.5 and 3.5. The systematic errors in our method are the higher-order corrections in g and the nonadiabatic

corrections, whereas in the Monte Carlo method the time scales used might not be large enough compared with the relevant energy differences.[4,12] Furthermore, the lattice approximation (the finite a correction) will also lead to systematic errors (nonuniversality). Initial expectations[2] were that our result would only be reliable up to about $z_{E^+} = 2$, but surprisingly z and $\mathscr{E}$ are stable within (2–4)% under the higher-order corrections discussed previously, all the way to $z_{E^+} = 5$.

From Fig. 1 one deduces that z_{E^+} and $z_{A_1^+}$ are almost constant functions of g around $g = 1$, which is largely due to the substantial lowering of the excited energies due to the tunneling phenomenon. It is therefore not permissible to convert an expansion in g to an expansion in z, as was done both in Refs. 1 and 2. But it does explain the sharp onset of tunneling in Fig. 2, curves b, at $z_{E^+} \sim 0.9$ and the reasonable accuracy of our semiclassical prediction[2] for this onset ($z_{A_1^+} \sim 1.0$).

In Fig. 2, curves c, we convert our values for $z_{E^+}(g)$ to the physical mass by

$$M(E(2^+))$$
$$= z_{E^+}(g)(11g^2/24\pi^2)^{51/121}\exp(12\pi^2/11g^2)\Lambda_{\text{MS}},$$

where MS denotes the minimal-subtraction scheme (the three-loop result of Tarasov, Vladimirov, and Yu[13] suggests an error of 6% or less), giving a plateau[1,14] at $M(E(2^+)) \sim 16\Lambda_{\text{MS}} \sim 119\Lambda_L$, which is below the Monte Carlo result.[12] This is probably due to higher-order corrections in the β function.[5]

Finally Fig. 3 gives some new predictions. First, a $T_2(2^+)$ mass surprisingly lower than the $E(2^+)$ and the $A_1(0^+)$ mass. Second, a pronounced z dependence of the $A_1(0^-)$ mass. Still, we do *not* consider it prudent to predict the lowest glueball state to have $J^P = 2^+$ at large z. Only when we understand how rotational invariance is restored at large z will we be able to tell *which* E^+ doublet combines with this T_2^+ triplet state into a $J^P = 2^+$ quintet,[1] and what its final mass will be. More

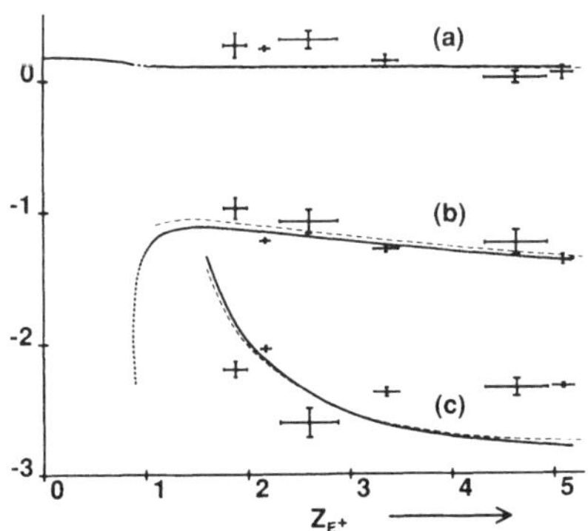

FIG. 2. Our results compared to lattice Monte Carlo results (Refs. 3 and 12) for (curves a) $\ln(z_{A_1^+}/z_{E^+})$, (curves b) $\ln(\mathscr{E}_E^{(1)}/z_{E^+})^{1/2}$, and (curves c) $\ln[M(E(2^+))/(250\Lambda_{\text{MS}})]$. The conventions for the curves are as in Fig. 1.

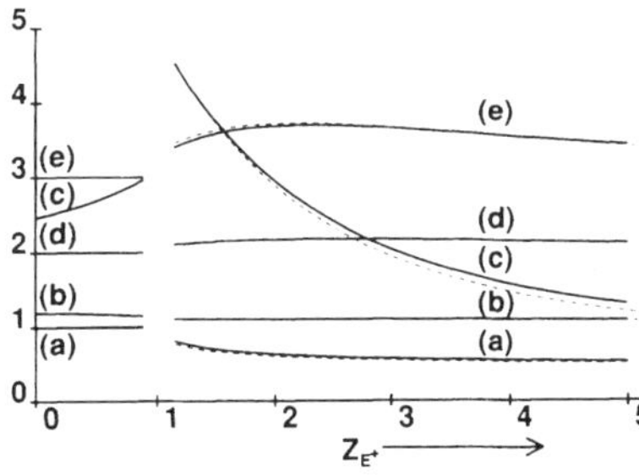

FIG. 3. Results for the following mass ratios: (curves a) $z_{T_2^+}/z_{E^+}$, (curves b) $z_{A_1^+}/z_{E^+}$, (curves c) $z_{A_1^-}/z_{E^+}$, and ratios of electric-flux energies: (curves d) $\mathscr{E}^{(2)}/\mathscr{E}^{(1)}$, (curves e) $\mathscr{E}^{(3)}/\mathscr{E}^{(1)}$. The conventions for the curves are as in Fig. 1.

Volume 58, Number 24   PHYSICAL REVIEW LETTERS   15 June 1987

accurately measurable with lattice Monte Carlo techniques, we believe, is our prediction for the energy ratios of different amounts of electric flux. For two (three) units of electric flux we propose the use of two (three) spatial Polyakov loops, in different directions. The time-time correlation of such an operator should then allow one to estimate the energy of two (three) units of electric flux. These ratios can be an important tool for testing the string picture, because such a picture predicts[6] $\mathcal{E}^{(i)}/\mathcal{E}^{(1)} = z_{e_i+}/z_{e_1+} = \sqrt{i}$, which deviates considerably from the values which we find below $z_{E+} = 5$. For the lattice we suggest plotting this ratio as a function of $\sqrt{z_{e_1}}$, because z_{e_1} can be measured more accurately than z_E. Since there are various indications[2,4] that string formation sets in for $z_{E+} \gtrsim 5$, it would be important to see if and how quickly these ratios of electric-flux energies settle to their expected asymptotic values.

In conclusion, we feel confident that we have demonstrated the nature of the nonperturbative dynamics of $SU(2)$ gauge theory in a finite but reasonably large volume. Of course, the physically interesting domain is still beyond the distances that we have probed, but we are getting close.

We thank Andreas Gocksch, Anna Hasenfratz, Julius Kuti, Greg Kilcup, Peter Lepage, Don Petcher, Robert Pisarski, Pietro Rossi, Yue Shen, George Sterman, and Hank Thacker for discussions. One of us (P.v.B) especially wishes to thank Peter Woit and Yair Arian for investigating volume dependence in their topological susceptibility calculations (unpublished). Furthermore, we are grateful to Gerard 't Hooft and Martin Lüscher for their encouragement. This work was supported in part by National Science Foundation Grants No. PHY-85-07627 and No. DMS-84-05661.

---

[1]M. Lüscher, Nucl. Phys. **B219**, 233 (1983); M. Lüscher and G. Münster, Nucl. Phys. **B232**, 445 (1984).

[2]J. Koller and P. van Baal, Nucl. Phys. **B273**, 387 (1986); P. van Baal and J. Koller, Ann. Phys. (NY) **174**, 299 (1987).

[3]B. Berg, A. Billoire, and C. Vohwinkel, Phys. Rev. Lett. **57**, 400 (1986).

[4]P. van Baal and J. Koller, Phys. Rev. Lett. **57**, 2783 (1986); P. van Baal, in Proceedings of Lattice Gauge Theory Conference, Brookhaven National Laboratory, NY, 1986, edited by H. Satz (Plenum, New York, to be published).

[5]A. Hasenfratz, P. Hasenfratz, U. Heller, and F. Karsch, Phys. Lett. **143B**, 193 (1984).

[6]G. 't Hooft, Nucl. Phys. **B153**, 141 (1979).

[7]H. Bethe, Ann. Phys. (Leipzig) **3**, 133 (1929); B. Berg and A. Billoire, Nucl. Phys. **B221**, 109 (1983).

[8]V. Gribov, Nucl. Phys. **B139**, 1 (1978); W. Nahm, in Proceedings of the Fourth Warsaw Symposium on Elemental Partical Physics, Warsaw, 1981, edited by Z. Ajduk (to be published), p. 275.

[9]P. van Baal, Nucl. Phys. **B264**, 548 (1986).

[10]T. H. Hansson, P. van Baal, and I. Zahed, State University of New York at Stony Brook Report No. ITP-SB-86-49, 1986 (to be published).

[11]M. Reed and B. Simon, *Methods of Modern Mathematical Physics* (Academic, New York, 1972), Vol. 4.

[12]B. Berg, in Proceedings of Lattice Gauge Theory Conference, Brookhaven National Laboratory, NY, 1986, edited by H. Satz (Plenum, New York, to be published).

[13]O. V. Tarasov, A. A. Vladimirov, and A. Yu. Zharkov, Phys. Lett. **93B**, 429 (1980).

[14]M. Lüscher, Phys. Lett. **118B**, 391 (1982).

Commun. Math. Phys. 122, 267–280 (1989)

Communications in
**Mathematical
Physics**
© Springer-Verlag 1989

Nahm's Transformation for Instantons

Peter J. Braam[1,*] and Pierre van Baal[2]

[1] Merton College, Oxford OX1 4JD, United Kingdom
[2] CERN, Theory Division, CH-1211 Geneva 23, Switzerland

Abstract. We describe in mathematical detail the Nahm transformation which maps anti-self dual connections on the four-torus $(S^1)^4$ onto anti-self-dual connections on the dual torus. This transformation induces a map between the relevant instanton moduli spaces and we show that this map is a (hyperKähler) isometry.

Introduction

This paper deals with "magical" properties of $U(n)$ anti-self-dual (asd) connections A on a C^n—bundle F over a four-torus T^4. The "witchcraft" starts by introducing a family of Dirac operators coupled to (F, A) parametrized by the dual torus $\hat{T}^4$. The families index turns out to be a bundle $\hat{F} \to \hat{T}^4$ (under a genericity assumption on A) and comes equipped with a natural connection $\hat{A}$, which is again asd (Theorem 1.5). This is Nahm's transform. Doing it again to $(\hat{F}, \hat{A})$, we obtain $(\hat{\hat{F}}, \hat{\hat{A}})$ and a unitary equivalence $(F, A) \sim (\hat{\hat{F}}, \hat{\hat{A}})$. In other words the square of Nahm's transform is the identity (Theorem 2.8 & 2.9). This was discovered by the authors and independently by Schenk [21], and relies heavily on some ideas of Nahm [19, 20] The transformation now induces a map of moduli spaces of (generic) asd connections $N : \mathcal{M}(F) \to \mathcal{M}(\hat{F})$. The spaces $\mathcal{M}(F)$ and $\mathcal{M}(\hat{F})$ carry a hyperKähler metric and N turns out to be a hyperKähler isometry, as was conjectured by S. K. Donaldson (Theorem 3.4).

This transformation has been around for awhile, and the fact that its square is the identity was announced by Nahm [19, 20] in the early eighties (see also Corrigan and Goddard [6]). However, for mathematicians it is not so easy to understand Nahm's work. In a way, the torus case treated here is the simplest version of Nahm's transform. Nahm originally developed his transformation for instantons invariant under subgroups of R^4, different from the 4-dimensional lattice (such as R or Z). For time-invariant instantons this was used extensively by Hitchin

* Address until September 1, 1989, Department of Mathematics University of Utah, Salt Lake City, Utah 84112, USA

P. J. Braam and P. van Baal

[13] and Hurtubise–Murray [15] to study magnetic monopoles. In these cases the direct analytical attack is still missing.

Two further comments are in place. In algebraic geometry, Mukai [16] discovered the same transformation, but now in the category of coherent sheaves on abelian varieties. Mukai [18] applied this to the study of moduli spaces of stable bundles on abelian varieties and this could be of future use in mathematical physics. Antony Maciocia pointed out that one can use Mukai's work to give an alternative proof of the isometry property of the transform.

Secondly, it should be pointed out that Nahm's work is very reminiscent of certain methods used in the theory of completely integrable systems. Also there the solutions of an associated set of linear equations is a key ingredient, see e.g. Segal–Wilson [22] and Duistermaat–Grünbaum [12].

1. Connections on T^4 and the Nahm Transform

Let $\Lambda \subset R^4$ be a lattice of rank 4. Then $T^4 = R^4/\Lambda$ inherits from R^4 the structure of an oriented Riemannian manifold. Clearly T^4 is a four-dimensional torus, i.e. T^4 is diffeomorphic to $(S^1)^4$. We shall assume the volume to be 1. Let $\hat{R}^4$ denote the dual space of R^4 and define the dual lattice Λ^* as

$$\Lambda^* = \{\mu \in \hat{R}^4; \mu(\lambda) \in Z, \forall \lambda \in \Lambda\}.$$

Cohomology classes on T^4 can be represented by constant, or equivalently, invariant differential forms, therefore

$$H^1(T^4; R) = \hat{R}^4, \quad H^1(T^4; Z) = \Lambda^*.$$

We can also define a dual torus $\hat{T}^4 = \hat{R}^4/\Lambda^*$. Points in $\hat{T}^4$ parametrize the unitary flat connections on the trivial line bundle $T^4 \times C = \mathscr{L} \to T^4$, up to gauge equivalence, because

$$\mathrm{Repr}(\pi_1(T^4), S^1) = H^1(T^4; R)/H^1(T^4; Z) = \hat{T}^4.$$

Denote by $\pi: T^4 \times \hat{R}^4 \to T^4$ the projection. On the trivial line bundle $\pi^* \mathscr{L} \to T^4 \times \hat{R}^4$ we have a universal connection 1-form given by

$$\omega(x, z) = 2\pi i \sum_{\mu=1}^{4} z_\mu dx_\mu,$$

where x_μ and z_μ are dual linear coordinates on R^4 and $\hat{R}^4$. Clearly $\omega(x, z)$ and $\omega(x, z + \lambda)$ are, for $\lambda \in \Lambda^*$, gauge equivalent connections on T^4; the gauge transformation $g(x) = \exp(-2\pi i \lambda(x))$ satisfies $g \cdot \omega_z = \omega_{z+\lambda}$. This gives a quotient bundle $\mathscr{P} \to T^4 \times \hat{T}^4$ with connection ω. The bundle $\mathscr{P}$ is called the Poincaré bundle and the curvature of ω equals

$$\Omega = 2\pi i \sum_\mu dz_\mu \wedge dx_\mu. \tag{1.1}$$

There is a canonical isomorphism $\hat{\hat{T}}^4 = T^4$ and under this isomorphism $\hat{\hat{\mathscr{P}}} = \mathscr{P}^*$, the dual of $\mathscr{P}$. In order to see this we use the gauge transformation $g(x, z) = \exp(-2\pi i x \cdot z)$ on $R^4 \times \hat{R}^4$ to transform $\omega(x, z)$ into $-2\pi i \sum_\mu x_\mu dz_\mu$, which is minus the universal connection on $R^4 \times \hat{T}^4$.

Let Q be a principal $U(n)$-bundle over T^4 equipped with a connection A. Let

$F = Q \times_{U(n)} C^n$ and define a family of connections A_z on F by $A_z = A + 2\pi i \mathrm{Id}_n \sum_\mu z_\mu dx_\mu$. Clearly A_z is the connection $A \otimes 1 + \mathrm{Id}_n \otimes \omega$ on $\pi^*(F \otimes \mathscr{L})$ restricted to $T^4 \times \{z\}$. Because the modification lies in the centre of $U(n)$ the curvature of A_z equals that of A. In the same way we can equip $\mathscr{P} \otimes \pi^* F$ with a connection A_z.

If $S^+, S^- \to T^4$ denote the spin-bundles of T^4 then we have Dirac operators:

$$D_{A_z}^+ : \Gamma(T^4, S^+ \otimes F) \to \Gamma(T^4, S^- \otimes F),$$
$$D_{A_z}^- : \Gamma(T^4, S^- \otimes F) \to \Gamma(T^4, S^+ \otimes F).$$

A connection A on F is said to be 1-irreducible if there are no covariantly constant $s \in \Gamma(T^4, F)$ for any A_z, $z \in \hat{T}^4$; this concept is due to Donaldson–Kronheimer [10]. If the holonomy of A_z is an open set in $U(n)$ (for all z) then A is certainly 1-irreducible.

Lemma 1.1. *If A is 1-irreducible and anti-self dual then* $\ker D_{A_z}^+ \subset \Gamma(T^4, S^+ \otimes F)$ *is zero for all z.*

Proof. Use the Weitzenboch formula:

$$D_{A_z}^- D_{A_z}^+ = \nabla_{A_z}^* \nabla_{A_z}, \tag{1.2}$$

see Atiyah–Hitchin–Singer [3], §6. ∎

For 1-irreducible A the vector spaces $\hat{F}_z = \ker D_{A_z}^-$ build up a smooth vector bundle $\hat{F}$ over $\hat{T}^4$. Let $\hat{H} \to \hat{T}^4$ be the bundle of Hilbert spaces with fiber $\hat{H}_z = L^2(T^4, S^- \otimes F \otimes \mathscr{P}_{|T^4 \times \{z\}})$. Then $\hat{F}$ is a subbundle of $\hat{H}$. Observe further that the "horizontal-component" d/dz of the connection $A \otimes 1 + \mathrm{Id}_n \otimes \omega$ induces a (flat) connection $\hat{d}$ on $\hat{H} \to \hat{T}^4$, and that over $\hat{R}^4$, $\hat{\pi}^* \hat{F}$ is a Hermitian subbundle of the trivial, flat Hilbert space bundle with fiber $L^2(T^4, S^- \otimes F)$. This will be important for computational purposes. Let

$$P : \hat{H} \to \hat{F}$$

denote the L^2—projection, then it is well known (see e.g. Atiyah [1]) that:

$$\hat{\nabla} = P\hat{d} : \Gamma(\hat{T}^4, \hat{F}) \to \Gamma(\hat{T}^4, \Lambda^1 \otimes \hat{F})$$

defines a connection $\hat{A}$ on $\hat{F}$.

Definition 1.2. The Nahm transform $\mathscr{N}$ of (Γ, A) is the pair of vector bundle and connection $(\hat{F}, \hat{A})$.

First we shall study the topology of $\hat{F}$. It is not very hard to prove that, topologically, $U(n)$ bundles $F \to T^4$ are determined by

$$n = \mathrm{rk}(F), \quad c_1(F) \in H^2(T^4, Z), \quad c_2(F) \in H^4(T^4, Z).$$

For $n > 1$ and any $c_1(F)$, $c_2(F)$ a bundle F can be constructed, and if F is a line bundle ($n = 1$) then $c_2(F) = 0$ is the only constraint on these data. If one assumes that F carries an anti-self dual connection then $c_1(F)$ must be an anti-self dual class in $H^2(T^4, R) \cap H^2(T^4, Z)$. For this the three dimensional eigenspace of $*$ in $H^2(T^4, R)$ with eigenvalue -1, must have a non-empty intersection with the lattice $H^2(T^4, Z)$. This depends on the Riemannian structure, and for a generic flat torus the only integral anti-self dual class is 0. Concerning the topological invariants of $\hat{F}$ we have:

Proposition 1.3. *Let A be a 1-irreducible* asd *connection on F. For $\hat{F}$ we have*

$$\mathrm{rk}(\hat{F}) = -\tfrac{1}{2}p_1(F_R) = c_2(F) - \tfrac{1}{2}c_1^2(F),$$

$$c_1(\hat{F}) = \frac{1}{4\pi^2}\,\Omega \cup \Omega \cup c_1(F)/[T^4],$$

$$p_1(\hat{F}_R) = -2\,\mathrm{rk}(F) \quad \text{or} \quad c_2(\hat{F}) = \mathrm{rk}(F) + \tfrac{1}{2}c_1^2(F).$$

Here $/$ denotes slant product, i.e. integration over T^4, $\cup$ denotes cup product and $(F_R, \hat{F}_R)$ are the underlying real bundles of $(F, \hat{F})$.

Proof. The Atiyah–Singer index theorem for families (see Atiyah–Singer [4]) asserts that

$$ch(\hat{F}) = \mathrm{rk}(\hat{F}) + c_1(\hat{F}) + \tfrac{1}{2}p_1(\hat{F}) = -ch(\mathscr{P}) \cup ch(F) \cup \hat{\mathscr{A}}(T^4)/[T^4],$$

with $\hat{\mathscr{A}}(X)$ the $\hat{\mathscr{A}}$—genus of X. This proves the proposition. ∎

This proposition has a nice corollary (see Mukai [16], [18] and also Schenk [21]):

Corollary 1.4. *There exist no* asd *connections A on a bundle F with $c_1(F) = 0$, $c_2(F) = 1$.*

Proof. It is easy to see that such a pair (F, A) would be 1-irreducible. Thus $\hat{F}$ is a vector bundle of rank 1 with $c_2 \neq 0$, which is impossible. ∎

If one allows a twist such 1-instantons do exist. More precisely let $Q \to T^4$ be an $SO(3)$ bundle with non-zero second Stiefel-Witney class ($w_2(Q) \neq 0$, i.e. there is twist). Such a bundle has a unique flat connection A_0, up to $SO(3)$-gauge transformations for which all cohomology groups of the Atiyah–Hitchin–Singer complex [3], §6 vanish. Using the construction of Taubes [23] and Donaldson [11] one can "attach" a localized 1-instanton from S^4 to this flat connection in order to obtain a new asd connection on a bundle $Q' \to T^4$ with $w_2(Q') = w_2(Q)$ and $p_1(Q') = -4$. The dimension of the hyperKähler moduli space $\mathscr{M}$ of these connections is 8. The eight parameters asymptotically equal centre and scale of the attached, localized instanton ($T^4 \times (0, \varepsilon)$) and finally the attaching parameters in $SO(3)/Z_2$, where Z_2 is the stabilizer of A_0. Consequently $\mathscr{M}$ modulo the 4-torus, acting by translations, is a hyperKähler 4-manifold, which is presumably related to a $K3$-surface.

The Nahm transformation has a "magic" property:

Theorem 1.5. *Let A be* asd *and 1-irreducible then $\hat{A}$ is* asd *on $\hat{F}$.*

Proof. It will be expedient to introduce a more detailed notation. Let $\hat{f}^j(z) = \psi_z^j(x) \in L^2(T^4, S^- \otimes \mathscr{L}_z \otimes F)$ with $z \in \hat{R}^4$, $j = 1, \ldots, -\tfrac{1}{2}p_1(F)$ be an orthonormal framing of $\hat{F}$ on $\hat{R}^4$. Note that $\mathscr{L}$ is trivialized on $T^4 \times \hat{R}^4$, using the connection $\omega = 2\pi i z_\mu dx_\mu$, therefore we map into the fixed vector space $L^2(T^4, S^- \otimes \mathscr{L}_0 \otimes F)$, which is isomorphic to $L^2(T^4, S^- \otimes F)$.

Then for a section $\hat{s}(z) = \sum_j \hat{s}_j(z)\hat{f}^j(z)$, $\hat{s}_j \in C^\infty(\hat{R}^4)$ we have (with $P: L^2(T^4, S^- \otimes F) \otimes \hat{\Lambda}^1 \to \hat{F} \otimes \hat{\Lambda}^1$ the projection):

$$\hat{V}\hat{s} = P\hat{d}\hat{s} = (1 - D_{A_z}^+ G_z D_{A_z}^-)[\hat{d}(\hat{s}_j(z)\psi_z^j(x))], \tag{1.3}$$

because, using Hodge theory,

$$P = (1 - D_{A_z}^+ G_z D_{A_z}^-)$$

with $G_z = (D_{A_z}^- D_{A_z}^+)^{-1}$ acting on $L^2(T^4, F \otimes S^-)$. For P we also have a pedestrian expression. If $r \in L^2(T^4, F \otimes S^-)$, then

$$Pr = \sum_j \langle \hat{f}^j, r \rangle \hat{f}^j, \tag{1.4}$$

where $\langle , \rangle$ denotes the $L^2(T^4, F \otimes S^-)$ inner product. So

$$\hat{\nabla}\hat{s} = (\hat{d}\hat{s}_j + \hat{A}_{jk}\hat{s}_k)\hat{f}^j$$

with $\hat{A}_{jk}$ the connection matrix of $\hat{A}$ equal to

$$\hat{A}_{jk} = \langle \hat{f}^j, \hat{d}\hat{f}^k \rangle = \langle \psi_z^j, \hat{d}\psi_z^k \rangle. \tag{1.5}$$

For the curvature one easily finds:

$$\hat{F}_{ij} = \hat{d}\hat{A}_{ij} + \hat{A}_{ik} \wedge \hat{A}_{kj} = \langle \hat{d}\psi_z^i, \wedge \hat{d}\psi_z^j \rangle + \langle \psi_z^i, \hat{d}\psi_z^k \rangle \wedge \langle \psi_z^k, \hat{d}\psi_z^j \rangle.$$

Now $\langle \hat{d}\psi_z^i, \psi_z^k \rangle = - \langle \psi_z^i, \hat{d}\psi_z^k \rangle$, so

$$\begin{aligned}
\hat{F}_{ij} &= \langle \hat{d}\psi_z^i, \wedge \hat{d}\psi_z^j \rangle - \langle P\hat{d}\psi_z^i, \wedge \hat{d}\psi_z^j \rangle \\
&= \langle D_{A_z}^+ G_z D_{A_z}^- \hat{d}\psi_z^i, \wedge \hat{d}\psi_z^j \rangle = \langle G_z D_{A_z}^- \hat{d}\psi_z^i, \wedge D_{A_z}^- \hat{d}\psi_z^j \rangle.
\end{aligned}$$

But

$$D_{A_z}^- \hat{d}\psi_z^i = [D_{A_z}^-, \hat{d}]\psi_z^i = - \Omega \cdot \psi_z^i. \tag{1.6}$$

The dot in this equation stands for Clifford multiplication, using the fact that there is a covariantly constant framing of S^+, S^- and T^*T^4. Thus G_z acts on $L^2(T^4, S^+ \otimes F) \simeq S^+ \otimes_C L^2(T^4, F)$ as $\mathrm{Id}_{S^+} \otimes_C \mathscr{G}_z$ with $\mathscr{G}_z = (\nabla_{A_z}^* \nabla_{A_z})^{-1}$ acting on $L^2(T^4, F)$. This shows that G_z commutes with Clifford multiplication. Thus we obtain

$$\hat{F}_{ij} = - \langle \Omega \wedge \Omega \cdot \psi_z^i, G_z \psi_z^j \rangle \tag{1.7}$$

with $\Omega \wedge \Omega \cdot \psi_z^i = (2\pi)^2 \sum_{\mu,\nu} dz_\mu \wedge dz_\nu dx_\mu \cdot dx_\nu \cdot \psi_z^i$ which is an element of $\hat{\Lambda}_-^2 \otimes S^- \otimes F$. ∎

The formal manipulations in this proof will appear again and will sometimes be left to the reader.

There is a holomorphic version of the Nahm transformation, which has been discovered by Mukai [16, 18]. For this one needs to choose a holomorphic structure on T^4. The Poincaré bundle now appears as a holomorphic line bundle on $T^4 \times \mathrm{Pic}^0(T^4)$; here $\mathrm{Pic}^0(T^4) \simeq \hat{T}^4$ in a canonical way. A bundle F carrying a 1-irreducible asd connection is holomorphic and stable; conversely every stable holomorphic bundle carries a unique asd connection inducing the holomorphic structure, see Donaldson [8]. Mukai's "Fourier functor" is defined as:

$$\mathscr{F}(F) = R^1 \hat{\pi}_*(\mathscr{P} \otimes \pi^* F),$$

where $\hat{\pi}: T^4 \times \hat{T}^4 \to \hat{T}^4$ is the projection. If F is a stable vector bundle then also $\mathscr{F}(F)$ will be stable and we shall prove (see also Donaldson–Kronheimer [10]) that $\mathcal{F}(F)$ is the holomorphic bundle underlying $(\hat{F}, \hat{A})$. In fact the fibre $\mathcal{F}(F)_z$, $(z \in \hat{T}^4)$ equals $H^1(T^4, F \otimes L)$, where $L = \mathscr{P}|_{T^4 \times z}$. By Hodge theory this space is canonically isomorphic to the kernel of $\bar{\partial}_{F \otimes L} \oplus \bar{\partial}^*_{F \otimes L} : \Gamma(T^4, \Lambda^{0,1} \otimes F \otimes L) \rightarrow \Gamma(T^4, (\Lambda^0 + \Lambda^{0,2}) \otimes F \otimes L)$. However, this operator is precisely the Dirac operator $D^-_{A_z}$ discussed earlier (see Atiyah [2]). This establishes an isomorphism $\mathcal{F}(F) \rightarrow \hat{F}$ as bundles. It remains to show that the holomorphic structure on $\mathcal{F}(F)$ coincides with that of $\hat{F}$. This was proved in a general context by Bismut–Gillet–Soulé [7] (Proposition 3.10 and Theorem 3.11).

It order to prove our next proposition we need the quaternionic structure on spinors on 4-manifolds. There is an anti-linear, covariantly constant map

$$\varepsilon : S^\pm \rightarrow S^\pm$$

satisfying $\varepsilon^2 = -1$. Tensoring this with the anti-linear map: $E \rightarrow E^*$ for any vector bundle E gives an anti-linear map $\hat{\varepsilon} : S^\pm \otimes E \rightarrow S^\pm \otimes E^*$ which still commutes with Clifford multiplication.

Proposition 1.6. *Assume that (F, A) is asd and 1-irreducible. Then $\mathcal{N}(F^*, A^*)$ is isomorphic to $((-1)^* \hat{F}^*, (-1)^* \hat{A}^*)$. Here A^* denotes the connection on the dual bundle F^* of F, and -1 is the inversion map on $\hat{T}^4$.*

Proof. Let $\psi_z \in L^2(T^4, F \otimes \mathscr{L}_z \otimes S^-)$ be a harmonic spinor for $D^-_{A_z}$. Then $\hat{\varepsilon}(\psi_z) \in L^2(T^4, F^* \otimes \mathscr{L}_{-z} \otimes S^-)$ is a harmonic spinor for $D^-_{A^*_{-z}}$. Indeed, let e_μ be a covariantly constant orthonormal framing of T^*T^4 then:

$$D^-_{A^*_{-z}}(\hat{\varepsilon}\psi_z) = e_\mu \cdot \left(\frac{\partial}{\partial x_\mu} (\hat{\varepsilon}\psi_z) - A^t_\mu \hat{\varepsilon}\psi_z - 2\pi i z_\mu \hat{\varepsilon}\psi_z \right)$$

$$= e_\mu \cdot \hat{\varepsilon}\left(\frac{\partial}{\partial x_\mu} + A_\mu \psi_z + 2\pi i z_\mu \psi_z \right)$$

$$= \hat{\varepsilon}(e_\mu \cdot \nabla^\mu_{A_z} \psi_z) = \hat{\varepsilon}(D^-_{A_z} \psi_z) = 0.$$

Therefore we have a complex linear bundle map $(-1)^* \hat{F}^* \rightarrow \mathcal{N}(F^*)$. It is trivial to show that this preserves connections. ∎

In Mukai [16] the reader can also find a convolution type relation between $\mathcal{F}(E \otimes F)$ and $\mathcal{F}(E) \otimes \mathcal{F}(F)$, for two vector bundles E and F.

We end this section with a remark of a different nature. The Nahm transformation can be studied more generally for asd connections A on R^4, invariant under a subgroup of translations $\Lambda \subset R^4$. Possible choices of Λ include $\Lambda = \{0\}$, these are the ordinary instantons on S^4. $\Lambda = R$ gives rise to monopoles, whereas $\Lambda = Z$ will correspond to the so-called calorons. In this paper, we are restricting ourselves to instantons on the 4-torus, for which $\Lambda = Z^4$. For the more generalized situations mentioned above, going through a construction similar to ours, we see that the Nahm transformation gives rise to new instantons on $\hat{R}^4$, which are invariant under

$$\Lambda^* = \{\alpha \in \hat{R}^4; \alpha(\lambda) \in Z, \forall \lambda \in \Lambda\}.$$

If $\Lambda = \{0\}$ this is closely related to the celebrated ADHM construction, see Nahm [20] and Donaldson–Kronheimer [10]. In the case of magnetic monopoles ($\Lambda = R$),

Nahm's transformation has been a powerful tool to understand moduli spaces, see Hitchin [13], Donaldson [9], Hurtubise–Murray [15] and Atiyah–Hitchin [5]. In a forthcoming paper we shall consider a Nahm transformation for instantons on $T^3 \times R$, which can be studied as monopoles on T^3 with some singularities. Instantons on $T^3 \times R$ are intimately related to problems of confinement, see 't Hooft [14].

2. The Square of Nahm's Transform

In this section we shall prove that there is a canonical isomorphism $(F, A) \simeq (\hat{\hat{F}}, \hat{\hat{A}})$ of Hermitian bundles with asd connections. Before starting in earnest, we give an outline of the ingredients of the proof.

a. We express harmonic spinors for $\hat{D}^-_{\hat{A}_x}$ in terms of those for $D^-_{A_z}$, using the Green operator G_z.

b. It is shown that $\hat{A}$ is 1-irreducible, by explicitly giving the inverse $\hat{G}_x$ of $\hat{\hat{\Delta}}_{\hat{A}_x}$.

c. To prove that the spinors found in a. form an orthonormal framing of $\hat{F}$ we need $\hat{F} \otimes S^-$-innerproducts of spinors for $\hat{D}_{\hat{A}_x}$. These appear as the Laplacian in z applied to the z-dependent operator G_z. Here we use the first term of the celebrated expansion of $G_z(x_1, x_2)$.

d. Finally we can compare A and $\hat{\hat{A}}$ using the same method and the second term in the expansion.

Recall that $\hat{F}$ was defined as a subbundle of $\hat{H} \to \hat{T}^4$. So we have a canonical element:

$$\Psi \in \Gamma(T^4 \times \hat{T}^4, \hat{\pi}^* \hat{F}^* \otimes \pi^* F \otimes S^- \otimes \mathcal{P}),$$

such that $\Psi(\hat{f})(\hat{f} \in \hat{F}_z)$ is annihilated by $D^-_{A_z}$ on $T^4 \times \{z\}$. If $G_z = (D^-_{A_z} D^+_{A_z})^{-1}$ then tensoring over $C^\infty(\hat{T}^4)$ with $\mathrm{id}_{\hat{F}}$ we obtain a section:

$$G\Psi \in \Gamma(T^4 \times \hat{T}^4, \hat{\pi}^* \hat{F}^* \otimes \pi^* F \otimes S^- \otimes \mathcal{P}). \tag{2.1}$$

We shall use the metric to let $TT^4 \simeq T^*\hat{T}^4$ act on $S^\pm$, thereby identifying $S^\pm$ with $\hat{S}^\pm$.

Proposition 2.1. *For $f \in F^*_x$ the section $G\Psi(f) \in \Gamma(\{x\} \times \hat{T}^4, \hat{\pi}^* \hat{F}^* \otimes S^- \otimes \mathcal{P})$ lies in* $\ker \hat{D}^-_{\hat{A}^*}$.

Proof. First we will state the following

Lemma 2.2. $G_z \nabla^\mu_{A_z} G_z = (1/4\pi i)(\partial G_z/\partial z_\mu)$, *where* x_μ, z_μ *are dual linear coordinates on* $R^4, \hat{R}^4$ *and* $\nabla^\mu_{A_z} s = \nabla_{A_z}(s)(\partial/\partial x_\mu)$.

Proof of the lemma follows from $[\nabla^*_{A_z} \nabla_{A_z}, (\partial/\partial z_\mu)] = 4\pi i \nabla^\mu_{A_z}$. ∎

As before, let a local framing $\hat{f}^j(z)$ of $\hat{F}$ be given, together with the framing of $\mathcal{P}$ on $T^4 \times \hat{R}^4$. Then $\hat{R}^4 \ni z \to \Psi(\hat{f}^j(z)) = \psi^j_z \in \Gamma(T^4, F \otimes \mathcal{L} \otimes S^-)$ and $(G\Psi)(\hat{f}^j) = G_z \psi^j_z$. If $f \in F^*_x$ and $\hat{f}^*_j$ is the framing of $\hat{F}^*$ dual to $\hat{f}^j$ then $(G\Psi)(f) = \sum_j (G_z \psi^j_z)(f)\hat{f}^*_j$, and now we have the connection matrix $\hat{A}^*_{ij} = -\hat{A}^t_{ij}$ for $\hat{F}^*$ at our disposal (observe that in this trivialization the connection on $\mathcal{P}$ has no dz components). Let $e_\mu, \hat{e}_\mu$

be dual orthonormal framings of TT^4, $T\hat{T}^4$ which can be identified, using the metric. Following Atiyah–Hitchin–Singer [3] we choose the relations in the Clifford algebra such that $e_\mu e_\nu + e_\nu e_\mu = -2\delta_{\mu\nu}$, so the inner products on $S^\pm$ satisfy $\langle e_\mu s_+, s_-\rangle_{S^-} = -\langle s_+, e_\mu s_-\rangle_{S^+}$. Then we have

$$(\hat{D}^-_{\hat{A}*}(G\,\Psi)(f))^i = \hat{e}_\mu \cdot \left\{ \delta_{i,j}\frac{\partial}{\partial z_\mu} - \left\langle \psi^j_z, \frac{\partial}{\partial z_\mu}\psi^i_z \right\rangle \right\}(G_z\psi^j_z)(f)$$

$$= \hat{e}_\mu \cdot \left\{ \frac{\partial}{\partial z_\mu}(G_z\psi^i_z)(f) - G_z(1 - D^+_{A_z}G_z D^-_{A_z})\frac{\partial\psi^i_z}{\partial z_\mu}(f) \right\}.$$

Using Lemma 2.2, $D^+_{A_z} = e_\mu \nabla^\mu_{A_z}$ and $D^-_{A_z}(\partial\psi^j_z/\partial z_\mu) = -2\pi i e_\mu \psi^j_z$ this equals:

$$\hat{e}_\mu \cdot (\delta_{\mu,\nu} - \tfrac{1}{2}e_\nu e_\mu)\left(\frac{\partial G_z}{\partial z_\nu}\psi^i_z\right)(f),$$

which is easily seen to vanish, using $\sum_\mu e_\mu e_\nu e_\mu = 2e_\nu$. ∎

This is quite a remarkable formula as it relates the spinors for A with those for $\hat{A}$, connections on bundles over different manifolds! In homological algebra such results can be obtained by using spectral sequences or, in a more sophisticated version, derived categories. Donaldson–Kronheimer [10] and Mukai [16] proceed in this way.

Reasoning as in the proof of Proposition 1.6 one sees that

$$u = 4\pi\hat{e}(G\,\Psi) \tag{2.2}$$

is a map from F_x to the harmonic spinors for $\hat{D}^-_{\hat{A}}$ acting on $\Gamma(\hat{T}^4, \mathscr{P}_x \otimes S^- \otimes \hat{F})$. We shall now show that it is actually an isometry $F \to \hat{F}$. Our reasoning closely follows the exposition given by Schenk [21].

First we shall establish that $(\hat{F}, \hat{A})$ is 1-irreducible, by showing that $\hat{V}^*_{\hat{A}}\hat{V}_{\hat{A}}$ has an inverse. The inverse can actually be written down quite explicitly. Recall that the maps $\hat{R}^4 \to L^2(T^4, S^- \otimes F): z \to \psi^j_z = \Psi(\hat{f}^j(z))$ describe a section of $\hat{\pi}^*\hat{F}^* \otimes \pi^*F \otimes S^- \otimes \mathscr{P}$ over $T^4 \times \hat{T}^4$. Assume that for $\lambda \in \Lambda^*$, $\lambda \to \hat{g}_\lambda(z)_{ij}$ is a set of automorphy factors for $\hat{F}$, then we may therefore assume that on $\hat{R}^4$:

$$\psi^i_{z+\lambda}(x) = \exp(-2\pi i\lambda(x))\hat{g}_\lambda(z)_{ij}\psi^j_z(x).$$

Sections of $\hat{F}$ satisfy similar relations with $\exp(-2\pi i\lambda(x))$ left out. Thus the object:

$$\hat{G}(z_1, z_2)_{ij} = \frac{1}{4\pi^2}\sum_{\mu\in\Lambda}\int_{T^4} dx \exp(2\pi i\mu(x))\frac{\langle \psi^i_{z_1}(x), \psi^j_{z_2}(x)\rangle}{|\mu - z_1 + z_2|^2} \tag{2.3}$$

represents for $z_1 - z_2 \notin \Lambda^*$ an element of $\hat{F}_{z_1} \otimes \hat{F}^*_{z_2}$. Here $\langle,\rangle$ is the Hermitian product on $F \otimes S^-$, which is antilinear in the first argument so as to correctly give the linear map $\hat{G}: \hat{F}_{z_2} \to \hat{F}_{z_1}$. Observe that the summation is over Fourier coefficients of a smooth function. These decrease rapidly and this ensures locally uniform convergence of all derivatives in the summation.

In order to prove that $\hat{G}$ is the inverse of $\hat{\Delta}_{\hat{A}} \equiv \hat{V}^*_{\hat{A}}\hat{V}_{\hat{A}}$, we will need two lemmas. We shall encounter sections of $\hat{F}$ of the form $\sum_j \langle \psi^j_z, s\rangle \hat{f}^j(z)$, where

$s \in L^2(T^4, S^- \otimes F)$. The following two lemmas give information about covariant derivatives of these sections.

Lemma 2.3.

$$(\hat{\nabla}^\mu_{\hat{A}*})_{ij}\psi^j_z = -2\pi i D^+_{A_z} G_z e_\mu \psi^i_z.$$

Equivalently one has for arbitrary $s \in L^2(T^4, F \otimes S^-)$,

$$(\hat{\nabla}^\mu_{\hat{A}})_{ij}\langle \psi^j_z, s \rangle = -2\pi i \langle \psi^i_z, G_z e_\mu D^-_{A_z} s \rangle,$$

where the inner product is for $L^2(T^4, F \otimes S^-)$.

Proof. Substitute $(\hat{\nabla}^\mu_{\hat{A}*})_{ij} = \delta_{ij}(\partial/\partial z_\mu) - \langle \psi^j_z, (\partial \psi^i_z/\partial z_\mu) \rangle$, and proceed as in the proof of Proposition 2.1. ∎

The next lemma will also play an important role in Sect. 3.

Lemma 2.4.

$$(\hat{\Lambda}_{\hat{A}*})_{ij}\psi^j_z = (4\pi)^2 G_z \psi^i_z,$$

or equivalently for any $s \in L^2(T^4, F \otimes S^-)$,

$$(\hat{\Delta}_{\hat{A}})_{ij}\langle \psi^j_z, s \rangle = (4\pi)^2 \langle \psi^i_z, G_z s \rangle.$$

Proof. Employing the previous lemma, we can write

$$(\hat{\Delta}_{\hat{A}*})_{ij}\psi^j_z = (\hat{\nabla}_{\hat{A}*})_{ik}(2\pi i D^+_{A_z} G_z e_\mu \psi^k_z)$$

$$= \frac{\partial}{\partial z_\mu}(2\pi i D^+_{A_z} G_z e_\mu \psi^i_z) - 2\pi i D^+_{A_z} G_z e_\mu P \frac{\partial \psi^i_z}{\partial z_\mu}.$$

Using $[(\partial/\partial z_\mu), D^+_{A_z}] = 2\pi i e_\mu$ and Lemma 2.2, together with the by-now-standard manipulations will finish the proof. ∎

Proposition 2.5. $\hat{G}$ *is the integral kernel of the inverse of* $\hat{\Delta}_{\hat{A}}$. *Therefore* $(\hat{F}, \hat{A})$ *is 1-irreducible.*

Proof. Assume $\hat{\Delta}_{\hat{A}}$ acts on the z_1 variable. First we show $(\hat{\Delta}_{\hat{A}}\hat{G})(z_1, z_2) = 0$ if $z_1 - z_2 \notin \Lambda^*$. This goes by brute force: Applying Lemma 2.3 to the section $s(x) = \exp(2\pi i \mu(x))\psi^j_{z_2}(x)$ gives

$$(\hat{\nabla}^\alpha_{\hat{A}})_{ik}\langle \psi^k_{z_1}, s \rangle = (2\pi)^2(\mu - z_1 + z_2)_\alpha \langle \psi^i_{z_1}, G_{z_1} \psi^j_{z_2} \rangle,$$

and combining with the result of Lemma 2.4,

$$(\hat{\Delta}_{\hat{A}})_{ik} \int_{T^4} dx \exp(2\pi i \mu(x))|\mu - z_1 + z_2|^{-2} \langle \psi^k_{z_1}, \psi^j_{z_2} \rangle$$

gives

$$\int_{T^4} dx \exp(2\pi i \mu(x)) \langle \psi^i_{z_1}, \psi^j_{z_2} \rangle \hat{d}^* \hat{d} |\mu - z_1 + z_2|^{-2},$$

which is zero for $z_1 - z_2 \notin \Lambda^*$. This shows that $\int_{T^4} dz_1 \hat{G}(z_1, z_2)(\hat{\Delta}_{\hat{A}} f)(z_1) = 0$ for f supported in $\hat{T}^4 - \{z_2\}$. To finish the proof one simply has to observe three points:

Near z_2

a) $\hat{\Delta}_{\hat{A}} = \hat{d}^*\hat{d} + $ something bounded.

b) $\hat{G}(z_1, z_2) = (\mathrm{Id}/(4\pi^2|z_1 - z_2|^2) + R_{z_2}(z_1))$ with $\hat{\Delta}_{\hat{A}} R_{z_2}$ integrable.

c) $(\mathrm{Id}/(4\pi^2|z_1 - z_2|^2))$ is the Green's function for $\hat{d}^*\hat{d}$.

These three points are easy to see and quickly lead to a proof of the Green's function formula. The 1-irreducibility follows from the fact that $\hat{A}$ is asd. ∎

We shall now show that $F_x^* \to L^2(\hat{T}^4, \hat{F}^* \otimes \mathscr{P}_{|\{x\} \times \hat{T}^4} \otimes S^-) : f \to 4\pi G\,\Psi(f)$ is an injective isometry. To study Nahm's transform on F we used $\psi_z^j(x)$, which mapped $\hat{F}_z$ into the fixed vector space $L^2(T^4, S^- \otimes F)$, see the proof of Theorem 1.5. We now have to map F_x into the fixed vector space $L^2(\hat{T}^4, S^- \otimes \hat{F})$. This requires a trivialisation of the pullback of $\mathscr{P}$ over $R^4 \times \hat{T}^4$, and this was discussed in Sect. 1. It follows that

$$x \to v_x(z) = \exp(-2\pi i z(x))\hat{\varepsilon}(G_z \psi_z^j(x))\hat{f}^j(z)$$

describes a map $F_x \to L^2(\hat{T}^4, \hat{F} \otimes S^-)$. In this trivialisation the connection of the pullback of $\hat{\mathscr{P}} \otimes \pi^*\hat{F}$ over $R^4 \times \hat{T}^4$ equals: $\hat{A}_x = \hat{A} + 2\pi i x_\mu dz_\mu$.

Just as $\langle \psi_{z_1}(x), \psi_{z_2}(x) \rangle_{F \otimes S^-}$ gives a map $\hat{F}_{z_2} \to \hat{F}_{z_1}$, we can likewise use $\langle v_{x_1}(z), v_{x_2}(z) \rangle_{\hat{F}^* \otimes S^-}$ as a map $F_{x_2} \to F_{x_1}$. We have

Lemma 2.6. $\langle v_{x_1}(z), v_{x_2}(z) \rangle_{\hat{F}^* \otimes S^-} = \exp(2\pi i z(x_1 - x_2))\hat{d}^*\hat{d}G_z(x_1, x_2)$.

Proof. Since

$$v_x(\hat{f}_j^*(z)) = 4\pi \exp(-2\pi i z(x))\hat{\varepsilon}\left\{ \int_{T^4} dy\, G_z(x, y)\psi_z^j(y) \right\}$$

and $\langle \psi_z^j, f \rangle_{L^2(T^4, S^- \otimes F)} \psi_z^j = Pf$ we find that

$$\langle v_{x_1}(z), v_{x_2}(z) \rangle_{\hat{F}^*} = (4\pi)^2 \exp(2\pi i z(x_1 - x_2)) G_z P G_z \tag{2.4}$$

acting on $\Gamma(T^4, S_- \otimes F)$. As explained in the proof of Theorem 1.5 this operator is really a constant matrix acting on S^-, tensored over C with a pseudo-differential operator acting on $L^2(T^4, F)$. Thus it makes sense to take the spinor trace tr_{S^-} of Eq. (2.4) which gives:

$$\langle v_{x_1}(z), v_{x_2}(z) \rangle_{S^- \otimes \hat{F}^*} = 8\pi^2 \exp(2\pi i z(x_1 - x_2))(\mathscr{G}_z^2 + \mathscr{G}_z \nabla_{A_z}^\mu \mathscr{G}_z \nabla_{A_z}^\mu \mathscr{G}_z)$$

acting on $L^2(T^4, F)$. A repeated application of Lemma 2.2 gives the required result. ∎

That the map $u : F \to \hat{\hat{F}}$ is an isometry follows from:

Proposition 2.7. *For all $x \in T^4$,*

$$\int_{\hat{T}^4} dz \langle u_x^i(z), u_x^j(z) \rangle_{\hat{F} \otimes S^-} = \delta_{ij},$$

with $u_x^j(z) = u(f^j(x))$ for $f^j(x)$ an orthonormal framing of F_x.

Proof. We obviously have:

$$\langle u_x^i(z), u_x^j(z) \rangle_{\hat{F} \otimes S^-} = \lim_{x_1 \to x_2} \langle v_{x_1}^i(z), v_{x_2}^j(z) \rangle_{\hat{F} \otimes S^-}.$$

Combining this with the identity obtained in Lemma 2.6 we find:

$$\int_{\hat{T}^4} dz \langle u_x^i(z), u_x^j(z) \rangle = \lim_{x_1 \to x_2} \int_{\hat{T}^4} dz \exp(2\pi i z(x_1 - x_2)) \hat{d}^* \hat{d} G_z^{ij}(x_1, x_2).$$

From the definition of the Green's function one easily finds the short distance expansion:

$$G_z^{ij}(x, y) = \frac{1}{4\pi^2 |x-y|^2} \left\{ \delta_{ij} - \sum_\mu (A_\mu^{ij}(x) + 2\pi i z_\mu \delta_{ij})(x_\mu - y_\mu) + \mathcal{O}(|x-y|^2) \right\}. \quad (2.5)$$

Partially integrating twice will give the desired result. ∎

Collecting the results we have:

Theorem 2.8. $4\pi \hat{e}(G \, \Psi) \in \Gamma(T^4 \times \hat{T}^4, \pi^* F^* \otimes \hat{\pi}^* \hat{F} \otimes \hat{\mathcal{P}} \otimes S^-)$ *gives a Hermitian isometry* $F \to \hat{F}$. ∎

Theorem 2.9. *Under the isomorphism of Theorem 2.8,* $A = \hat{\hat{A}}$.

Proof. By definition (see Eqs. (1.5) and the discussion above Lemma 2.6)

$$\hat{\hat{A}}_{ij}^\mu = \int_{\hat{T}^4} dz \left\langle v_x^i(z), \frac{\partial v_x^j(z)}{\partial x_\mu} \right\rangle = \lim_{x_1 \to x_2} \int_{\hat{T}^4} dz \left\langle v_{x_1}^i(z), \frac{\partial v_{x_2}^i(z)}{\partial x_2^\mu} \right\rangle.$$

Using Lemma 2.6 and Eq. (2.5), together with partially integrating twice will give the desired result. ∎

In some sense it seems that the Nahm transformation is perfect. It preserves self-duality, its square is the identity and as we shall see next, it gives a hyperKähler isometry between moduli spaces.

3. Metric Properties of Nahm's Transform

We have seen that the Nahm transformation gives a diffeomorphism from the space $\mathcal{M}'(F)$ of 1-irreducible asd connections modulo gauge transformations to $\mathcal{M}'(\hat{F})$. These spaces are open subsets of the moduli spaces $\mathcal{M}(F), \mathcal{M}(\hat{F})$ of instantons and it is well known that $\mathcal{M}(F), \mathcal{M}(\hat{F})$ are smooth manifolds away from the reducible connections (see Atiyah–Hitchin–Singer [3]). Moreover $\mathcal{M}(F)$ (and also of course $\mathcal{M}(\hat{F})$) are supplied with a Riemannian metric as follows. A tangent vector $X \in T_A \mathcal{M}'$ can be uniquely represented by an element of $\Gamma(T^4, \Lambda^1 \otimes (Q \times_{\mathrm{Ad}} u(n)))$, also denoted by X, satisfying

$$d_A^* X = 0 \quad \text{(Coulomb gauge condition)},$$
$$P_+ d_A X = 0 \quad \text{(deformation equation)}. \quad (3.1)$$

The L^2-metric on $\Lambda^1 \otimes (Q \times_{\mathrm{Ad}} u(n))$ now induces a Riemannian metric on $\mathcal{M}(F)$. We shall show here that Nahm's transformation is an isometry with respect to this metric on the moduli spaces.

Suppose a is a tangent vector to $A \in \mathcal{M}'(F)$. The first step is to compute the infinitesimal change in the spinors, $\delta \psi_z^j(a)$. It is easily seen that $\delta \psi_z^j$ should satisfy $D_{A_z}^- \delta \psi_z^j = -a \cdot \psi_z^j$, where a acts on ψ_z^j by Clifford multiplication tensored with the fundamental representation of $U(n)$. One can normalize the $\delta\psi^j$ by requiring $\langle \delta\psi^j, \psi^i \rangle = 0$, for all i, j. The normalized $\delta\psi^j$ is

$$\delta\psi_z^j = -D_{A_z}^+ G_z(a \cdot \psi_z^j). \tag{3.2}$$

Combining this with Eq. (1.5) for the Nahm transformation, we obtain the following formula for the infinitesimal change $\hat{a} \in \Gamma(\hat{T}^4, \hat{\Lambda}^1 \otimes (\hat{Q} \times_{\mathrm{Ad}} u(k)))(k = \mathrm{rk}(\hat{F}))$, which is the derivative of the Nahm transformation applied to a:

$$\begin{aligned}
\hat{a}_{ij} &= -\langle D_{A_z}^+ G_z(a \cdot \psi_z^i), \hat{d}\psi_z^j \rangle - \langle \psi_z^i, \hat{d}D_{A_z}^+ G_z(a \cdot \psi_z^j) \rangle \\
&= 2\pi i \langle a \cdot \psi^i, G_z(dz^k e_k \cdot \psi_z^j) \rangle + 2\pi i \langle dz^k e_k \cdot \psi^i, G_z(a \cdot \psi^j) \rangle \\
&= \langle a \cdot \psi_z^i, \Omega \cdot G_z \psi_z^j \rangle - \langle \Omega \cdot G_z \psi_z^i, a \cdot \psi_z^j \rangle. \tag{3.3}
\end{aligned}$$

The inner products are all with respect to $L^2(T^4, F \otimes S^-)$. The deformation of $\hat{A}$ could also be given by describing how the bundle $\hat{F}$ varies as a subbundle of $\hat{H} \to \hat{T}^4$, but this would not be a suitable description to compare metrics.

Proposition 3.1. *If A is asd and if a satisfies the Coulomb gauge condition (3.1) then also $\hat{a}$, given by Eq. (3.3), satisfies the gauge condition.*

Proof. Using the identities $\sum_\mu e_\mu D_{A_z}^{\pm} e_\mu = 2D_{A_z}^{\mp}$, one finds easily that in the framing ψ^i,

$$(\hat{d}_{\hat{A}}^* \hat{a})_{ij} = \sum_\mu 4\pi^2 \left\langle \psi_z^i, G_z \sum_\mu e_\mu(D_{A_z}^- \circ a - a \circ D_{A_z}^+) e_\mu \cdot G_z \psi_z^j \right\rangle \in \Gamma(\hat{T}^4, T^*\hat{T}^4 \otimes \mathrm{End}\,\hat{E}),$$

where $\circ$ denotes composition of operators. Now with $\phi \in \Gamma(T^4, S^- \otimes F)$,

$$\sum_\mu e_\mu \cdot D_{A_z}^-(a \cdot e_\mu \cdot \phi) - e_\mu \cdot a \cdot D_{A_z}^+(e_\mu \cdot \phi) = \sum_\mu e_\mu \cdot (D_{A_z}^- a) \cdot e_\mu \cdot \phi + 2 \sum_{\mu, \alpha \neq \beta} e_\mu e_\alpha e_\beta e_\mu \cdot (a^\beta \nabla_{A_z}^\alpha \phi),$$

where we substituted $D_{A_z}^- = e_\mu \nabla_{A_z}^\mu$ and $a = e_\alpha a^\alpha$. But $D_{A_z}^- a = 0$ because the Dirac operator on 1-forms combines the deformation equation and the Coulomb gauge condition of Eq. (3.1), see Atiyah–Hitchin–Singer [3], §6. Finally it is a relation in the Clifford algebra of R^4 that: $\sum_\mu e_\mu e_\alpha e_\beta e_\mu = 0$ if $\alpha \neq \beta$. ∎

Proposition 3.2. *The Nahm transformation preserves the L^2—metric, provided the $u(n)$ metric is normalized as $\langle X, Y \rangle = -\mathrm{tr}(XY)$.*

Proof. Let $a \in T_A \mathcal{M}(F)$ and $\hat{a} \in T_{\hat{A}} \mathcal{M}(\hat{F})$ be arbitrary elements, satisfying Eqs. (3.1) and its analogue for $\hat{F}$. We denote by dN the derivative of the Nahm transformation $N: \mathcal{M}(F) \to \mathcal{M}(\hat{F})$, i.e. $dN(a)$ is given by Eq. (3.3). We have to show

$$\langle dN(a), \hat{a} \rangle_{\mathcal{M}(\hat{F})} = \langle a, dN(\hat{a}) \rangle_{\mathcal{M}(F)}, \tag{3.4}$$

since the right-hand side equals $\langle a, (dN)^{-1}(\hat{a}) \rangle_{\mathcal{M}(F)}$ as $N^2 = 1$. Now Eq. (3.4) would follow from

$$\int_{\hat{T}^4} dz \langle a \cdot \psi_z^i, 2\pi i e_\mu \cdot G_z \psi_z^j \rangle_{L^2(T^4, F \otimes S^-)} \hat{a}_{ji}^\mu$$

$$= \int_{T^4} dx\, a_{\alpha\beta}^\mu \langle \hat{a} \cdot v_x^\beta, 2\pi i e_\mu \cdot \hat{G}_x v_x^\alpha \rangle_{L^2(\hat{T}^4, \hat{F} \otimes S^-)}, \tag{3.5}$$

where we used Eqs. (3.3) for the derivative of the Nahm transformation. The Green's function $\hat{G}_x$ is related to $\hat{G}$ in Eq. (2.3) by the gauge transformation $g(x, z)$ introduced in Sect. 1: $\hat{G}_x(z_1, z_2) = \exp(-2\pi i z_1(x))\hat{G}(z_1, z_2)\exp(2\pi i z_2(x))$. Using now both the indices $i, j, \ldots$ for $\hat{F}$ and $\alpha, \beta, \ldots$ for F we can rewrite Eqn. (3.5), multiplying both sides with $-2i$, as

$$\int_{T^4 \times \hat{T}^4} dx\, dz \langle a^{\alpha\beta} \cdot \psi^i_\beta, \hat{a}_{ji} \cdot \hat{\varepsilon}^{-1} u^\alpha_j \rangle_{S^-} = \int_{T^4 \times \hat{T}^4} dx\, dz \langle \hat{a}_{ji} \cdot u^\alpha_i, 4\pi a_{\alpha\beta} \cdot (\hat{G} u^\beta)_j \rangle_{S^-}, \qquad (3.6)$$

or erasing the indices

$$\int_{T^4 \times \hat{T}^4} \langle \hat{a}u, a 4\pi \hat{G} u \rangle = \int_{T^4 \times \hat{T}^4} \langle a\psi, \hat{a}\hat{\varepsilon}^{-1} u \rangle.$$

Finally, and crucially, we make use of Lemma 2.4 which can be restated as

Lemma 3.3. $\Psi = \hat{\varepsilon}^{-1}(4\pi \hat{G} u)$.

Proof. This follows directly from Lemma 2.4, Proposition 2.5 and the fact that $\hat{\varepsilon}^{-1}\hat{G}\hat{\varepsilon} = \hat{G}$. ∎

Using this to rewrite the right-hand side of Eq. (3.6) as

$$\int_{T^4 \times \hat{T}^4} \langle \hat{a}_{ji} \cdot u^\alpha_i, a_{\alpha\beta} \cdot \hat{\varepsilon}\psi^j_\beta \rangle_{S^-},$$

the fact that $\hat{\varepsilon}$ commutes with Clifford multiplication and that $\langle v, \hat{\varepsilon}^{-1} w \rangle = \langle w, \hat{\varepsilon}v \rangle$, one easily establishes equality with the left-hand side of Eq. (3.6). ∎

The tangent bundle of T^4 can be equipped with a hyperKähler structure; that is, there are three integrable complex structures I, J, K on T^4 which satisfy quaternionic relations, and are all compatible with the orientation such that the metric is Kähler for each of them. It is a special feature of the anti-selfduality equations that also $\mathcal{M}(F)$ is a hyperKähler manifold where the complex structures on $T_A \mathcal{M}(F) \subset \Gamma(T^4, \Lambda^1 \otimes (Q \times_{\mathrm{Ad}} u(n)))$ are simply given by those on Λ^1. We refer the reader to Atiyah–Hitchin [5], Mukai [17].

Theorem 3.4. $N: \mathcal{M}'(F) \to \mathcal{M}'(\hat{F})$ *is a hyperKähler isometry.*

Proof. It remains to show that the derivative $dN: a \to \hat{a}$ of Nahm's transformation commutes with a complex structure. This follows immediately from:

$$\hat{a}_{ij} = \langle G_z \psi^i_z, \Omega \cdot a \cdot \psi^j_z \rangle - \langle \Omega \cdot a \cdot \psi^i_z, G_z \psi^j_z \rangle,$$

which is equivalent to Eq. (3.3), but makes more explicit how Ω connects the two spaces of one-forms, in terms of which the complex structures are defined. ∎

A. Maciocia pointed out to us that alternatively one can deduce the results of this section in the algebro-geometric setup as follows. Choosing a Kähler structure on $\mathcal{M}$ defines a complex analytic complex symplectic form on $\mathcal{M}$, see Mukai [17]. This form is the natural bilinear form on $H^1(T^4, \mathrm{End}_0 F)$ with values in $H^2(T^4, \mathrm{End}\, F) = C$. It can be shown that this pairing is preserved under the Fourier transform, thereby establishing that Nahm's transform is a hyperKähler isometry.

Acknowledgements. Both authors would like to thank Simon Donaldson, Antony Maciocia, and Werner Nahm for discussions and the ITP at Stony Brook, CERN and Merton College for hospitality. P.J.B. is partially supported by a C&C Huygens fellowship of the Netherlands Organization for Advancement of Science.

References

1. Atiyah, M. F.: Geometry of Yang-Mills fields, Fermi lectures. Scuola Normale Superiore, Pisa 1979
2. Atiyah, M. F.: Classical groups and classical differential operators on manifolds: Differential operators on manifolds, pp. 6–48. CIME, Verenna 1975
3. Atiyah, M. F., Hitchin, N. J., Singer, I. M.: Selfduality in four-dimensional Riemannian geometry. Proc. Roy. Soc. Lond. Ser. A, **362**, 425–461 (1978)
4. Atiyah, M. F., Singer, I. M.: The index of elliptic operators IV. Ann. Math. **93**, 119–138 (1971)
5. Atiyah, M. F., Hitchin, N. J.: The geometry and dynamics of magnetic monopoles. Porter Lectures, Princeton 1988
6. Corrigan, E., Goddard, P.: Construction of instanton and monopole solutions and reciprocity. Ann. Phys. (NY) **154**, 253–279 (1984)
7. Bismut, J.-M., Gillet, H., Soulé, C.: Analytic torsion and holomorphic determinant bundles. III Quillen metrics on holomorphic determinants. Commun. Math. Phys. **115**, 301–351 (1988)
8. Donaldson, S. K.: Anti self-dual Yang-Mills connections over complex algebraic surfaces and stable vector bundles. Proc. Lond. Math. Soc. **50**, 1–26 (1985)
9. Donaldson, S. K.: Nahm's equations and the classification of monopoles. Commun. Math. Phys. **96**, 387–407 (1984)
10. Donaldson, S. K.: Kronheimer, P. B.: A book on Yang-Mills theory and 4-manifolds. Oxford, University Press (to appear)
11. Donaldson, S. K.: Connections, cohomology and the intersection forms of 4-manifolds. J. Diff. Geom. **24**, 275–341 (1986)
12. Duistermaat, J. J., Grünbaum, B.: Differential equations in the spectral parameter. Commun. Math. Phys. **103**, 177 (1986)
13. Hitchin, N. J.: On the construction of monopoles. Commun. Math. Phys. **89**, 145–190 (1983)
14. 't Hooft. G.: A property of electric and magnetic flux in non-Abelian gauge theories. Nucl. Phys. **B153**, 141 (1979)
15. Hurtubise, J., Murray, M : The moduli space of $SU(N)$ monopoles and Nahm's equations. Princeton preprint 1988
16. Mukai, S.: Duality between $D(X)$ and $D(\hat{X})$ with its application to Picard sheaves. Nagoya Math. J. **81**, 153–175 (1981)
17. Mukai, S.: Symplectic structure of the moduli space of sheaves on an Abelian or K3 surface. Invent. Math. **77**, 101–116 (1984)
18. Mukai, S.: Fourier functor and its application to the moduli of bundles on an Abelian variety. Adv. Studies in Pure Math. **10** (1987), Alg. Geometry Sendai, 515–550 (1985)
19. Nahm, W.: Monopoles in quantum theory. In: Proceedings of the Monopole Meeting, Trieste (eds.) Craigie, e.a. Singapore: World Scientific 1982
20. Nahm, W.: Self-dual monopoles and calorons. In: Lecture Notes in Physics. vol. **201**. Denardo, G., et al. (eds.). Berlin, Heidelberg, New York: Springer 1984
21. Schenk, H.: On a generalized fourier transform of instantons over flat tori. Commun. Math. Phys. **116**, 177–183 (1988)
22. Segal, G., Wilson, G.: Loop groups and the equations of KdV type. Publ. Math. IHES 61, 1985
23. Taubes, C. H.: Self-dual connections on 4-manifolds with indefinite intersection matrix. J. Diff. Geom. **19**, 517–560 (1984)

Communicated by A. Jaffe

Received August 22, 1988

Volume 220, number 1,2   PHYSICS LETTERS B   30 March 1989

ON THE ALGEBRAIC CHARACTERIZATION
OF WITTEN'S TOPOLOGICAL YANG–MILLS THEORY

Stéphane OUVRY
Division de Physique Théorique [1], IPN, and LPTPE, F-91406 Orsay Cedex, France

Raymond STORA
LAPP, B.P. 909, F-74019 Annecy-le-Vieux, France

and

Pierre VAN BAAL
CERN, CH-1211 Geneva 23, Switzerland

Received 11 November 1988

We interpret in terms of "basic" cohomology the recently proposed supersymmetric, supergauge invariant formulation of topological Yang–Mills theory. Our interpretation shows that this formulation leads to the correct observables.

1. Introduction

In a recent series of articles [1,2], Witten investigates the expression of various topological invariants in terms of local field theory. The first examples of this sort we know of are due to Schwarz [3] who gives a field theory expression for the Ray–Singer analytic torsion [4], and are related to the quantization of differential forms [5]. The situations considered by Witten are of a more exotic type and lead to essentially non-linear theories, to be treated in the weak coupling regime. In principle, to obtain the sort of results one expects, a rigorous treatment of the renormalized perturbation expansion ought to be sufficient for a rigorous mathematical construction. Here, we shall be concerned with gauge fields and the recently discovered Donaldson invariants [6].

The following construction owes much to seminars by Singer, Baulieu [7] and Braam [8]. However, since local field theory is to be used [9], we find it necessary to characterize the model by a complete set

of Ward identies. We believe that ref. [7], as well as subsequent proposals [10] are incomplete in this respect. The solution is to be found in an article by Horne [11]. The purpose of this note is to explain why, in more geometrical terms.

2. The differential algebra

As suggested in Witten's paper [1] (eq. (2.41)) and emphasized in ref. [7], one wishes to gauge-fix a topological invariant, e.g.

$$S_{\mathrm{inv}}(A) = \int_{M} \mathrm{tr}(F(A) \wedge F(A)) , \qquad (1)$$

where $F((A)$ is the curvature of a connection A on a principal G-bundle P(M, G), over a compact 4-manifold M, without boundary, and tr is an invariant polynomial over Lie G. The group G is assumed to be compact.

The action S_{inv} is, by essence, invariant under arbitrary variations of A:

[1] Laboratoire Paris 6-Paris 11 associé au CNRS.

Volume 220, number 1,2　　　　　PHYSICS LETTERS B　　　　　30 March 1989

$$\delta A = \psi \, . \tag{2}$$

From now on, all fields are differential forms on M, taking values in ad(Lie G). One insists on gauge-fixing S_{inv}, leaving the gauge freedom pending till the end (because of the known Gribov difficulty), localizing the system on the self-dual connections. The corresponding gauge fixing term is

$$S^{(1)} = \int_{M} \mathrm{tr}(b \wedge b - b \wedge F^{-} - \bar{\psi} \wedge (D\psi)^{-}) \, , \tag{3}$$

where F^{-} is the antiself-dual part of $F(A)$ – for some metric g on M –, b and $\bar{\psi}$ are antiself-dual two-forms and $(D\psi)^{-}$ is the antiself-dual part of $D\psi$, the covariant differential of the one-form ψ.

The new action $S_{\mathrm{inv}} + S^{(1)}$ is invariant under the Slavnov symmetry:

$$s_1 A = \psi, \quad s_1 \psi = 0, \quad s_1 \bar{\psi} = b, \quad s_1 b = 0 \tag{4}$$

and satisfies

$$S^{(1)} = s_1 \int_{M} \mathrm{tr}(\bar{\psi} \wedge (b - F^{-})) \, . \tag{5}$$

It is still gauge invariant, i.e. invariant under

$$\delta A = D\omega, \quad \delta \psi = [\psi, \omega] \, , \tag{6}$$

$$\delta \bar{\psi} = -[\omega, \bar{\psi}], \quad \delta b = [b, \omega] \, , \tag{7}$$

where $\omega \in$ Lie $\mathscr{G}$, $\mathscr{G}$ being the gauge group of P(M, G). This yields the nilpotent s_2 operation:

$$s_2 A = \psi - D\omega, \quad s_2 \psi = [\psi, \omega], \quad s_2 \bar{\psi} = -[\omega, \bar{\psi}] + b,$$

$$s_2 b = [b, \omega], \quad s_2 \omega = -\tfrac{1}{2}[\omega, \omega] \, . \tag{8}$$

The action $S_{\mathrm{inv}} + S^{(1)}$ is invariant under s_2 and does not depend on ω (the Faddeev–Popov ghost for Lie $\mathscr{G}$). Except for the $\int b \wedge b$ term, it is invariant under (φ is odd)

$$\delta b = [\bar{\psi}, \varphi], \quad \delta \psi = D\varphi \, , \tag{9}$$

which yields the nilpotent operation s (now with φ even):

$$sA = \psi - D\omega, \quad s\psi = [\psi, \omega] - D\varphi \, ,$$

$$s\bar{\psi} = -[\omega, \bar{\psi}] + b, \quad sb = [b, \omega] - [\bar{\psi}, \varphi] \, ,$$

$$s\varphi = [\varphi, \omega], \quad s\omega = -\tfrac{1}{2}[\omega, \omega] + \varphi \, . \tag{10}$$

It is easy to modify $S^{(1)}$ in such a way that it is invariant under (10). Following eq. (5) we find

$$\hat{S}^{(1)} = s \int_{M} \mathrm{tr}(\bar{\psi} \wedge (b - F^{-}))$$

$$= \int_{M} \mathrm{tr}(b \wedge b - b \wedge F^{-} - \bar{\psi} \wedge (D\psi)^{-} + \bar{\psi} \wedge [\bar{\psi}, \varphi]) \, . \tag{11}$$

Notice that in eq. (10), sb needs a φ dependent term in order for s to be nilpotent. The φ invariance can be gauge fixed in a gauge invariant way using the gauge function D*ψ:

$$\hat{S}^{(2)} = \int_{M} \mathrm{tr}(^{*}\beta \wedge D^{*}\psi + ^{*}\bar{\varphi} \wedge (D^{*}D\varphi + [^{*}\psi, \psi])) \, . \tag{12}$$

Including $\hat{S}^{(2)}$ in the action, one gets the Slavnov symmetry defined by (10), together with

$$s\bar{\varphi} = \beta + [\bar{\varphi}, \omega], \quad s\beta = -[\omega, \beta] + [\varphi, \bar{\varphi}] \, , \tag{13}$$

such that we have the following expression:

$$\hat{S}^{(1)} + \hat{S}^{(2)} = s \int_{M} \mathrm{tr}(^{*}\bar{\varphi} \wedge D^{*}\psi + \bar{\psi} \wedge (b - F^{-})) \, . \tag{14}$$

A few remarks are in order:

(i) $S_{\mathrm{inv}} + \hat{S}^{(1)} + \hat{S}^{(2)}$ is not quite the most general ω independent, gauge invariant, renormalizable action invariant under s – actually of the form $S_{\mathrm{inv}} + s S_{\mathrm{gf}}$: one may add an extra term compatible with ghost number neutrality and renormalizability as in refs. [1,2,10,11] and of the form $s \int \mathrm{tr}(\beta[\varphi, \bar{\varphi}])$. Both ω independence and gauge invariance are essential.

(ii) Changing generators according to

$$\psi' = \psi - D\omega, \quad b' = b - [\omega, \bar{\psi}] \, ,$$

$$\varphi' = \varphi - \tfrac{1}{2}[\omega, \omega], \quad \beta' = \beta + [\bar{\varphi}, \omega] \, , \tag{15}$$

the s-operation assumes the form

$$sA = \psi', \quad s\psi' = 0, \quad s\bar{\psi} = b', \quad sb' = 0 \, ,$$

$$s\omega = \varphi', \quad s\varphi' = 0, \quad s\bar{\varphi} = \beta', \quad s\beta' = 0 \, . \tag{16}$$

It therefore has vanishing cohomology as well as vanishing cohomology mod d. The desired cohomology [1,2] is, however, not the local cohomology of s mod d, but its restriction to ω independent, gauge invariant objects. What is involved is equivariant cohomology [8] [*1], or rather its original form, namely "basic" cohomology [13], which is exactly adapted to the present local field theory context, as we shall see.

(iii) It is interesting to observe that s can be split into a sum of two anticommuting differentials and that the algebra can be cast in a supersymmetric form, which is *distinct* from that of ref. [11], if we insist that, as we will demonstrate, s generates the supersymmetry. However, the superfield content will be that of ref. [11] (without imposing the gauge condition $\omega = 0$). Details are given in section 4.

3. The "basic" cohomology of s

The differential algebra defined by the structure eqs. (10) and (13) has the following property, which makes it a differential algebra with an action of the gauge Lie algebra; for $\lambda \in$ Lie $\mathscr{G}$ define

$$\delta_\lambda \psi = [\psi, \lambda], \quad \delta_\lambda A = D\lambda,$$

$$\delta_\lambda b = [b, \lambda], \quad \delta_\lambda \bar{\psi} = -[\lambda, \bar{\psi}],$$

$$\delta_\lambda \varphi = [\varphi, \lambda], \quad \delta_\lambda \omega = -[\lambda, \omega],$$

$$\delta_\lambda \bar{\varphi} = [\bar{\varphi}, \lambda], \quad \delta_\lambda \beta = -[\lambda, \beta].$$
(17)

Define also for $\lambda \in$ Lie $\mathscr{G}$, ι_λ by

$$\iota_\lambda A = \iota_\lambda \psi = \iota_\lambda \bar{\psi} = \iota_\lambda b = \iota_\lambda \varphi = \iota_\lambda \bar{\varphi} = \iota_\lambda \beta = 0,$$

$$\iota_\lambda \omega = \lambda.$$
(18)

One can easily check that

$$\delta_\lambda = \iota_\lambda s + s \iota_\lambda$$
(19)

and one has the classical [13] commutation rules

$$[\iota_\lambda, \iota_\mu]_+ = 0, \quad [\delta_\lambda, \delta_\mu]_- = \delta_{[\lambda,\mu]}, \quad [\delta_\lambda, \iota_\mu]_- = \iota_{[\lambda,\mu]},$$

$$[s, \iota_\lambda]_+ = \delta_\lambda, \quad [s, \delta_\lambda]_- = 0.$$
(20)

This makes $\{\delta_\lambda, \iota_\mu | \lambda, \mu \in$ Lie $\mathscr{G}\}$ into a graded Lie algebra. Recall that $S_{\text{tot}} = S_{\text{inv}} + \hat{S}^{(1)} + \hat{S}^{(2)}$ fulfils

$$sS_{\text{tot}} = \delta_\lambda S_{\text{tot}} = \iota_\mu S_{\text{tot}} = 0, \quad \lambda, \mu \in \text{Lie } \mathscr{G}.$$
(21)

In technical terms S_{tot} is a "basic" [13] local functional for the differential structure (10), (13), with the Lie $\mathscr{G}$ action defined by (17), (18).

[*1] This is an amplification of a remark by Braam (see ref. [8]). See also refs. [7,12].

Now let us turn λ and μ into ghosts, in the usual fashion (λ odd and μ even) and define

$$W = \delta + \iota,$$
(22)

where δ and ι are obtained by (17), (18) on all fields except for A, ω, λ and μ for which

$$WA = -D\lambda, \quad W\omega = -[\lambda, \omega] - \mu,$$

$$W\lambda = -\tfrac{1}{2}[\lambda, \lambda], \quad W\mu = [\mu, \lambda].$$
(23)

One easily shows that $W^2 = 0$. Adjoining λ and μ as new generators to our differential algebra, we still have a choice to define $s\lambda$ and $s\mu$. In particular, if we define s on λ and μ by

$$s\lambda = \mu, \quad s\mu = 0,$$
(24)

we obtain

$$[s, W]_+ = 0.$$
(25)

The comparison with the supersymmetric formalism of ref. [11] is now straightforward. In terms of the primed variables defined in eq. (15), one may introduce the superfields

$$\mathscr{A}_x = A + \theta\psi', \quad \mathscr{A}_\theta = \omega + \theta\varphi',$$

$$\Psi = \bar{\psi} + \theta b', \quad \Phi = \bar{\varphi} + \theta\beta'.$$
(26)

Then one has

$$s = \partial/\partial\theta.$$
(27)

The supergauge transformation ghost

$$\Lambda = \lambda + \theta\mu,$$
(28)

fulfils $W\Lambda = -\tfrac{1}{2}[\Lambda, \Lambda]$ and s still acts on Λ by $\partial/\partial\theta$. W acts on all fields by supergauge transformations, with $\mathscr{A}_x$, $\mathscr{A}_\theta$ a superconnection and Ψ, Φ transforming under the adjoint representation.

In terms of the unprimed variables the action and local cohomology mod d are characterized by ω independence and gauge invariance, as we have already remarked. In terms of the primed variables and the supersymmetric formulation of ref. [11], this is equivalent to supersymmetry (invariance under $\partial/\partial\theta$) and supergauge invariance. So, this equivalence proves in particular that the supersymmetric supergauge invariant cohomology is identified with the "basic" cohomology, which is known to be correct [1,2,7,8]. We refer to ref. [11] for the s-invariant gauge fixing of W.

Volume 220, number 1,2　　　　PHYSICS LETTERS B　　　　30 March 1989

4. An alternative supersymmetry

In this section we discuss the alternative supersymmetry mentioned at the end of section 2. One may split s in eqs. (10) and (13) as

$$s = \sigma + w, \tag{29}$$

with

$$\sigma A = \psi, \quad \sigma\psi = 0, \quad \sigma\bar\psi = b, \quad \sigma b = 0,$$
$$\sigma\omega = \varphi, \quad \sigma\varphi = 0, \quad \sigma\bar\varphi = \beta, \quad \sigma\beta = 0 \tag{30}$$

and

$$w\psi = [\psi, \omega] - \mathrm{D}\varphi, \quad wA = -\mathrm{D}\omega,$$
$$wb = [b, \omega] - [\bar\psi, \varphi], \quad w\bar\psi = -[\omega, \bar\psi],$$
$$w\varphi = [\varphi, \omega], \quad w\omega = -\tfrac{1}{2}[\omega, \omega],$$
$$w\bar\varphi = [\bar\varphi, \omega], \quad w\beta = -[\omega, \beta] + [\varphi, \bar\varphi]. \tag{31}$$

One can easily check that $\sigma^2 = w^2 = [\sigma, w]_+ = 0$. This structure suggests the use of a supersymmetric formalism. Let

$$A = A + \theta\psi, \quad \Omega = \omega + \theta\varphi,$$
$$\bar\Psi = \bar\psi + \theta b, \quad \bar\Phi = \bar\varphi + \theta\beta. \tag{32}$$

Then, in terms of the superfields:

$$\sigma = \partial/\partial\theta, \tag{33}$$

and w is a supergauge transformation:

$$wA = -\mathrm{D}(A)\Omega, \quad w\Omega = -\tfrac{1}{2}[\Omega, \Omega],$$
$$w\bar\Psi = -[\Omega, \bar\Psi], \quad w\bar\Phi = [\bar\Phi, \Omega], \tag{34}$$

where the covariant differential $\mathrm{D}(A)$ is given by

$$\mathrm{D}(A)\Omega = \mathrm{d}\Omega + [A, \Omega]. \tag{35}$$

Eq. (34) defines a differential superalgebra with a super Lie algebra action in terms of

$$\Lambda = \lambda + \theta\mu, \quad \lambda, \mu \in \text{Lie } \mathcal{G}. \tag{36}$$

We define δ_Λ according to

$$\delta_\Lambda A = \mathrm{D}(A)\Lambda, \quad \delta_\Lambda\Omega = [\Omega, \Lambda],$$
$$\delta_\Lambda = [\bar\Psi, \Lambda], \quad \delta_\Lambda = [\bar\Phi, \Lambda] \tag{37}$$

and ι_Λ according to

$$\iota_\Lambda A = \iota_\Lambda \bar\Psi = \iota_\Lambda \bar\Phi = 0, \quad \iota_\Lambda \Omega = \Lambda. \tag{38}$$

Then we have

$$\delta_\Lambda = [w, \iota_\Lambda]. \tag{39}$$

5. Concluding remarks

The algebraic set-up proposed in ref. [11] has been shown to describe the "basic" cohomology adapted to the characterization of a perturbative treatment [9] of the situation described by Witten [1,2] in terms of equivariant cohomology [8] (see also footnote 1). There are two heavy technical problems to be dealt with:

(i) Perturbative renormalization theory for a field theory associated with an arbitrary compact manifold without boundary in a particular topological sector.

(ii) The proper treatment of different vacua and the inclusion in the $s-W$ operation of global zero modes, that ought to make the theory not completely empty.

Acknowledgement

It is a pleasure to acknowledge the warm hospitality of the following institutions: the LAPP theory group, Merton College, the mathematics department of Salt Lake City and the CERN theory division, where most of this work was performed. We wish to thank J.H. Horne for sending his preprint communicating additional comments, which led to the interpretation given here. We are also grateful to M.F. Atiyah, L. Baulieu, P. Braam, S.K. Donaldson, J.M.F. Labastida, D. Montano and I.M. Singer for illuminating discussions, as well as for communicating their work prior to publication.

References

[1] E. Witten, Commun. Math. Phys. 117 (1988) 353.

[2] E. Witten, Commun. Math. Phys. 118 (1988) 601; Phys. Lett. B 206 (1988) 601.

[3] A.S. Schwarz, Lett. Math. Phys. 2 (1978) 247; Commun. Math. Phys. 67 (1978) 1.

[4] D.B. Ray and I.M. Singer, Adv. Math. 7 (1971) 145.

[5] J. Thierry-Mieg, Harvard preprint HUMTP 79/B86 (1979), unpublished;

Volume 220, number 1,2   PHYSICS LETTERS B   30 March 1989

H. Hata, T. Kugo and N. Ohta, Nucl. Phys. B 178 (1981) 527;
J. Thierry-Mieg and L. Baulieu, Nucl. Phys. B 228 (1983) 259;
L. Baulieu and J. Thierry-Mieg, Phys. Lett. B 144 (1984) 221.

[6] S.K. Donaldson, J. Diff. Geom. 18 (1983) 269; 26 (1987) 397; Polynomial invariants for smooth four-manifolds, Oxford preprint.

[7] L. Baulieu and I.M. Singer, communications at the LAPP Meeting on Conformal field theories and related topics (Annecy-le-Vieux, March 1988); to be published.

[8] P. Braam, Seminar at the CERN Theory Division (April 1988); at the RCP25 Meeting (Strasbourg, June 1988); Floer homology groups for homology three-spheres, Utrecht preprint, Nr. 484 (November 1987).

[9] R. Stora, CIME lectures (July 1988).

[10] J.M.F. Labastida and M. Pernici, Phys. Lett. B 212 (1988) 56;
R. Brooks, D. Montano and J. Sonnenschein, preprint SLAC-Pub-4630 (May 1988).

[11] J.H. Horne, Superspace versions of topological theories, Princeton preprint PUPT-1096 (June 1988).

[12] V. Mathai and D. Quillen, Topology 1 (1986) 85.

[13] H. Cartan, in: Colloque de Topologie (Espaces fibrés) (C.B.R.M. Bruxelles) pp. 57–71;
W. Greub, S. Halperin and R. Vanstone, Connections, curvature and cohomology, Vol. III (Academic Press, New York, 1976).

Vol. **B21** (1990) *ACTA PHYSICA POLONICA* No 2

AN INTRODUCTION TO TOPOLOGICAL YANG-MILLS THEORY*

By P. van Baal**

CERN — Geneva

(Received July 17, 1989)

In these lecture notes we give a "historical" introduction to topological gauge theories. Our main aim is to clearly explain the origin of the Hamiltonian which forms the basis of Witten's construction of topological gauge theory. We show how this Hamiltonian arises from Witten's formulation of Morse theory as applied by Floer to the infinite dimensional space of gauge connections, with the Chern-Simons functional as the appropriate Morse function(al). We therefore discuss the De Rham cohomology, Hodge theory, Morse theory, Floer homology, Witten's construction of the Lagrangian for topological gauge theory, the subsequent BRST formulation of topological quantum field theory and finally Witten's construction of the Donaldson polynomials.

PACS numbers: 11.15.Tk

1. Introduction

Topological quantum field theories are field theories that have at most a finite number of degrees of freedom (in particular there are no propagating physical states), so a legitimate question is then, why one should be interested in them. Probably a more than adequate justification comes from the way these theories were first discovered by Witten [1] in the context of Yang-Mills gauge theories. This justification is, however, mathematical in nature and is a prime example of the extremely fruitful interactions between physics and mathematics. Crudely stated, because there is no dynamics, the quantum field theory can only be sensitive to invariants of the basis manifold on which the theory is defined. As we will see, a topological quantum field theory can be defined as a field theory on a smooth manifold which is independent of additional structures, such as a metric, on the basis manifold. Thus appropriate observables will have expectation values independent of the metric, and will hence give invariants. As will be discussed in the last Section, for Yang-Mills

---

* Presented at the XXIX Cracow School of Theoretical Physics, Zakopane, Poland, June 2–12, 1989.

** Address: Institute for Theoretical Physics, Princetonplein 5, P. O. Box 80.006, NL-3508 TA Utrecht, The Netherlands.

74

gauge theory over a four manifold, these will lead to the invariants considered by Donaldson [2], who used them to distinguish different differential structures. Hence these are called differential invariants, although one also encounters the name topological invariants (which are of course, strictly speaking, weaker).

By now, many types of topological field theories have been constructed [3, 4]. There are two reasons why some of them might be interesting from the physical point of view. One argument by Witten [1] is that these theories have general covariance built in without having to integrate over the space of all metrics. It is this integration over all metrics which forms the obstacle to finding a quantum theory of gravity. The hope [1] is now that one might be able to find a suitable topological quantum field theory in which this general covariance is spontaneously broken, and a dynamical theory which includes gravity might arise. However, one should realize that since topological field theories have more or less by definition no dynamics, it will be very hard to find a mechanism for this spontaneous breaking of the general covariance, unless one "tinkers by hand" with the theory. The other reason for its physical significance lies in the beautiful connection of the pure Chern–Simons theories in three dimensions with the conformal field theories in two dimensions [4]. Hence, one might envisage this as a means to classify conformal field theories [5]. Whether this will give a complete classification is not clear yet. From the mathematical side, this connection between three and two dimensions is relevant for the knot invariants (Jones polynomials) [6]. The description of knots in three dimensions is more natural than its traditional two-dimensional formulation. This higher ventage point might resolve also many riddles related to the connection between Yang-Baxter equations, braid groups and conformal field theories [7]. It is therefore not surprising that this corner of topological quantum field theories is attracting most attention. However, we will concentrate ourselves in these lecture notes on topological Yang-Mills theory, because this is where the development started.

It is maybe instructive to sketch the history of topological Yang-Mills theory. The development of these ideas grew out of the study of harmonic forms in increasingly complex situations. Traditionally, harmonic forms have played a very important role both in mathematics and in physics. For example in three dimensions, they form the solutions to the Laplace equations, which are essential in the study of electrostatics and as we will show, the solutions to the Maxwell equations can be seen as harmonic 2-forms. From the mathematics point of view, harmonic forms are important for the study of topological invariants. Hodge theory relates the number of independent harmonic p-forms to the cohomology $H^p(M)$ of the compact and smooth manifold M, which in turn is dual to the homology $H_p(M)$. The homology can be defined for any orientable manifold by a topological construction. We will review homology and De Rham cohomology in Section 2, to make these lectures more or less self contained. The reason is that (co-)homology is the central theme in all these developments. In five different settings we will introduce operators whose square is zero (they are called (co-)boundary operators). This is all one needs to define a (co-)homology.

Morse theory gives another way of studying the topology of a manifold, by studying the critical points for a generic function (called a Morse function) on the manifold M. For example, one can easily determine the Euler number of a manifold from Morse theory, but in general it gives only inequalities for the Betti numbers B_p, where B_p is the dimension of $H^p(M)$ (or by Hodge theory the number of harmonic p-forms on M). This is where physics re-entered mathematics, due to Witten's [8] analysis of supersymmetry breaking. Especially his study of a supersymmetric non-linear sigma model led to a formulation of Morse theory, based on studying harmonic forms constructed from an exterior algebra intertwined with a Morse function. This exterior algebra was basically equivalent to the supersymmetry algebra (in the zero-momentum sector) and the harmonic forms are the zero-energy states for the Hamiltonian. The Witten index, which is a measure for supersymmetry breaking (to break supersymmetry the Witten index needs to be zero), can thus be shown to be the Euler characteristic. Topology was therefore in the way for supersymmetry breaking. However, Witten [9] realized that his formulation of Morse theory, which was based on tunnelling in supersymmetric quantum mechanics, was a powerful mathematical tool. Instead of bounds (weak Morse inequalities), it allowed one to obtain the Betti numbers directly from Morse theory (related to the strong Morse inequalities). This was based on constructing a homology based on tunnelling, as we will review in some detail in Section 3. Another instance where physical questions have strongly stimulated mathematical development has been the study of instantons [10]. These are solutions to the (anti-)self-duality equations for non-Abelian gauge theories on four manifolds. Many powerful mathematical ideas were bundled in algebraically constructing the set of all solutions on S^4 for a given topological charge k (Pontryagin or Chern class), called the moduli space $\mathcal{M}_k$. This is the Atiyah–Drinfeld–Hitchin–Manin [11] construction. The instanton moduli spaces were used by Donaldson [2] to construct powerful differential invariants for four-manifolds. Four-dimensional manifolds are particularly notorious for their difficulty in classifying differential structures. For example, in five or more dimensions, fixing the topology will fix the differential structure up to finitely many possibilities (see for a review [12]). Note that the self-duality equations, in a sense, generalize the study of harmonic two-forms and it is therefore natural (maybe with some hindsight) that the added non-Abelian group structure will lead to more refined invariants.

Since in the Hamiltonian formulation of gauge theories, one has a three-manifold as a basis manifold, Floer (after Taubes) [13] asked himself whether one could similarly construct invariants for three manifolds. His answer was affirmative, beautifully combining the Yang–Mills gauge theories with Witten's analysis of Morse theory. We will outline this development in Section 4. Basically, it amounts to considering the exterior algebra on the infinite dimensional manifold of gauge equivalence classes of connections (gauge potentials) on the three-manifold. Then he intertwines this, as in Witten's finite dimensional analysis, with a Morse function(al), for which he chooses the Chern–Simons functional. The resulting homology amounts to studying the zero-energy solutions of a "supersymmetric" Hamiltonian, whose bosonic part is nothing but the pure Yang–Mills Hamiltonian. Not surprisingly, the tunnelling analysis in this infinite dimensional formulation of Morse theory is precisely described by the (anti-)self-duality equations. For a review see also [14].

Atiyah and Donaldson now realized that there was a connection between this Floer homology and the Donaldson invariants, for particular four-manifolds. Since the Floer

76

homology is naturally connected to field theory and since instantons also play a natural role, Atiyah [15] asked the question whether it would be possible to find the Lorentz invariant formulation of the Floer Hamiltonian, that is to find a Lagrangian, whose Hamiltonian will be the one which comes from the Floer theory. The euclidean formulation on closed four manifolds is then likely to lead to Donaldson invariants. This question was answered by Witten [1] in the affirmative (as well as [4] Atiyah's question whether there was a three-dimensional field theory, which would lead to the formulation of the Jones polynomials). Thus were born topological field theories. One thing is for sure, despite the fact that many details will still need to withstand the test of mathematical rigour, it will have a large impact on various areas of mathematics. The excitement is based on the fact that Witten's work gives explicit formulas for Donaldson and Jones polynomials, and especially in the latter case it provides many clues for generalizations.

In Section 5 we will discuss Witten's construction of the topological Yang-Mills action. It was later realized [16], that there was an underlying BRST symmetry and that one could view Witten's Lagrangian as coming from the gauge fixing of an action which is given purely by the topological charge $\left(\dfrac{1}{8\pi^2} \displaystyle\int_M \mathrm{Tr}(F \wedge F) \right)$ which is obviously independent of the metric. We also mention the underlying equivariant or basic cohomology [17] which can be formulated in terms of this BRST or Slavnov symmetry [18]. Finally in Section 6 we will consider Witten's construction [1] of the Donaldson polynomials. We will not attempt to describe Donaldson's [2] original formulation.

These notes are mainly intended for a readership of physicists. We attempt to use as much as possible physical intuition to outline the various developments, in order to hide the author's inadequacy in achieving mathematical rigour. Nevertheless, he hopes that these lecture notes will contribute, not only to his own, but also to the reader's understanding of this new field at the borderline between physics and mathematics.

2. De Rham cohomology and Hodge theory

As the best known example of cohomology we discuss De Rham cohomology, denoted by $H^p(M, R)$, where M is an n dimensional manifold and p runs from 0 to n. (The R stand for the real numbers, but other types like integer cohomology will not concern us here). The manifold actually needs a differential structure to define De Rham cohomology, but the invariants will turn out not to depend on this structure. We can now define the space of smooth p-forms Λ_p, where we recall that a p-form ω can locally be written as:

$$\omega = \omega_{i_1,i_2,\ldots,i_p} dx^{i_1} \wedge dx^{i_2} \wedge \ldots \wedge dx^{i_p}. \tag{1}$$

The indices i_j run from 1 to n, $\omega_{i_1,i_2,\ldots,i_p}$ is antisymmetric in its indices and we use Einstein's summation convention throughout this paper. The space of 0-forms are simply the set of all (differential) functions on M and the volume form for M is an example of a n-form. We can now introduce the exterior derivative $d: \Lambda_n \to \Lambda_{n+1}$ through the following local definition:

$$d\omega = \frac{\partial \omega_{i_1, i_2, \cdots, i_p}}{\partial x^j} dx^j \wedge dx^{i_1} \wedge \ldots \wedge dx^{i_p}. \tag{2}$$

For later purposes we rewrite this in an operator form as:

$$d\omega = a^{*j} \frac{\partial \omega}{\partial x^j}, \tag{3}$$

where the operator a^* is defined locally as:

$$a^{*j}\omega = dx^j \wedge \omega. \tag{4}$$

The exterior derivative d is called a coboundary operator (for reasons which will become clear shortly) and it satisfies the important property that it squares to zero: $d^2 = 0$. This implies the following crucial relation

$$\operatorname{im} d|_{\Lambda_{p-1}} \subset \ker d|_{\Lambda_p}, \tag{5}$$

where $\operatorname{im} d$ is the image of the operator d, i.e. $\operatorname{im} d|_{\Lambda_{p-1}} = \{\omega \in \Lambda_p | \exists \lambda \in \Lambda_{p-1}, \; \omega = d\lambda\}$ and $\ker d$ is the kernel of the operator d, i.e. $\ker d|_{\Lambda_p} = \{\omega \in \Lambda_p | d\omega = 0\}$. The following definition of De Rham cohomology will therefore make sense:

$$H^p(M, R) = \ker d / \operatorname{im} d \cap \Lambda_p. \tag{6}$$

In the future, as in Eq. (6), we will assume the space on which d acts implicitly defined. Thus $H^p(M, R)$ is the set of closed ($d\omega = 0$), non-exact p-forms (ω cannot be written as $d\lambda$). The Betti numbers $B_p(M)$ are equal to the dimension of $H^p(M, R)$, that is the number of independent closed, non-exact p-forms. An example for a well-known topological i nvariant is the Euler characteristic:

$$\chi(M) = \sum_{q=0}^{n} (-1)^q B_q(M). \tag{7}$$

The best way to see that the Betti numbers are topological invariants is to note that the De Rham cohomology $H^p(M, R)$ is dual to the real homology $H_p(M, R)$, which can be defined purely topologically (provided M has an orientation). We will be a bit sloppy in its description and refer the reader to standard mathematics textbooks for more details on singular or simplicial homology. One considers the so-called cell complex C_p of p-dimensional oriented subspaces (cells) embedded in the manifold M. They can be formally added and multiplied with real numbers. Multiplying with -1 will change the orientation. Two cells can join into one if their orientations match along the boundaries. We can now define the boundary operator $\partial : C_p \rightarrow C_{p-1}$ (taking the boundary of a set reduces its dimension) which squares again to zero ($\partial^2 = 0$), since the boundary of a boundary is empty. Completely analogous to the De Rham cohomology one can now define the homology by:

$$H_p(M, R) = \ker \partial / \operatorname{im} \partial \cap C_p, \tag{8}$$

78

which is called the set of p-cycles, that is the set of closed p-dimensional cells, that are not the boundary of some $(p+1)$-dimensional cell. Note that in homology ∂ decreases the grading by one whereas in cohomology d increases the grading by one. This is of course a more or less artificial distinction between homology and cohomology. However, there is a more precise relation between cohomology and homology, which shows that they are actually dual to each other and that the exterior derivative is the dual of the boundary operator. This is why the exterior derivative is also called a coboundary operator. The duality is established with the following bilinear form:

$$H_p(M, R) \times H^p(M, R): (\gamma, \omega) \to \oint_\gamma \omega \in R. \tag{9}$$

To show that this is indeed a proper map we have to show that $\oint_\gamma \omega$ does not change if we add the boundary of a $(p+1)$-cell α to γ or if we add the exterior derivative of a $(p-1)$-form λ to ω. This follows by Stokes' law from the fact that ω is a closed p-form ($d\omega = 0$) and γ is a closed p-cell ($\partial\gamma = 0$). To be more precise, Stokes' theorem for integration yields:

$$\oint_{\partial\alpha} \omega = \oint_\alpha d\omega = 0 \quad \text{and} \quad \oint_\gamma d\lambda = \oint_{\partial\gamma} \lambda = 0, \tag{10}$$

which is easily seen to prove the above statements. One can also show that the bilinear form is non-degenerate, which establishes the duality between homology and cohomology and shows that the Betti numbers are topological invariants. Stokes' theorem also illustrates that the boundary operator is dual to the exterior derivative with respect to the bilinear form defined in Eq. (9), i.e. $\oint_\gamma d\omega = \oint_{\partial\gamma} \omega$.

Before we discuss the Hodge theory we introduce a terminology which will be of use later. Two p-cycles $\gamma_{1,2}$ are called homologous if their difference $\gamma_1 - \gamma_2$ is the boundary of a $(p+1)$-dimensional cell, which we illustrate in Fig. 1. As we have just seen, the integral of a closed p-form over γ_1 equals its integral over γ_2. That is, the integral only depends on the homology class of the p-cycle.

To discuss the Hodge theory, which basically states that B_p is the number of harmonic

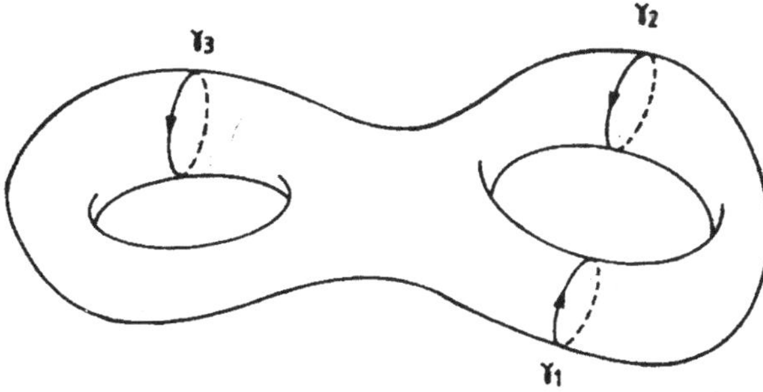

Fig. 1. We illustrate here that the 1-cycles γ_1 and γ_2 are homologous. This is because cutting the figure along both of these curves we get two 2-cells. For each of these it is true that the boundary of the 2-cell is $\pm(\gamma_1 - \gamma_2)$ (the simplest 2-cell with this property is the tube). However, γ_1 and γ_3 are clearly not homologous

p-forms, we first have to define what is meant by a harmonic p-form. For this we need a metric g_{ij} on the manifold M, which is needed to define the adjoint $d*$ of d. If we talk about an adjoint operator, we need an inner product on the space of p-forms. This inner product is defined with the help of the Hodge $*$ operator $* : \Lambda_p \to \Lambda_{n-p}$, which in local coordinates reads as follows:

$$*\omega = \frac{1}{(n-p)!} \, \omega^{i_1,i_2,\cdots,i_p} \varepsilon_{i_1,\cdots,i_p,j_1,\cdots,j_{n-p}} dx^{j_1} \wedge \ldots \wedge dx^{j_{n-p}}. \tag{11}$$

Note that we used the metric to raise the indices for ω. The inner product is now given by:

$$\Lambda_p \times \Lambda_p : (\alpha, \beta) \to \langle \alpha, \beta \rangle = \int_M \alpha \wedge *\beta. \tag{12}$$

Since $\alpha \wedge * \beta$ is a n-form the integral is well defined and the adjoint of d follows from:

$$\langle \alpha, d*\beta \rangle = \langle d\alpha, \beta \rangle, \quad \alpha \in \Lambda_p, \quad \beta \in \Lambda_{p+1}. \tag{13}$$

We have the following useful properties, which involve the Hodge $*$ operator on p-forms:

$$d* = -(-1)^{n(p-1)} * d * \quad \text{and} \quad *^2 = (-1)^{p(n-p)}. \tag{14}$$

It is now easily verified that in local coordinates one finds the following expression for $d*\omega$, when ω is a p-form:

$$d*\omega = - \sum_{j=1}^{p} (-1)^j g^{l i_j} \frac{\partial \omega_{i_1,\cdots,i_p}}{\partial x^l} dx^{i_1} \wedge \ldots \wedge \widehat{dx^{i_j}} \wedge \ldots \wedge dx^{i_p}, \tag{15}$$

where the hat over x^{i_j} means that this differential should be eliminated from the wedge product. Again for later purposes we can rewrite this in an operator formulation:

$$d*\omega = -a^i \frac{\partial \omega}{\partial x^i}. \tag{16}$$

The a^i act on a basis element $dx^{i_1} \wedge \ldots \wedge dx^{i_p} \in \Lambda_p$ by simply leaving out dx^i from this wedge product, after anti-commuting dx^i to the left (giving zero if dx^i does not occur). The $a^i(x)$ form a basis for the tangent space $T_x M$ and act by exterior multiplication on the cotangent space $T_x^* M$, for which $a^{*i}(x)$ forms the basis dual to $a^i(x)$. We leave it to the reader to verify that these operators satisfy the Dirac algebra:

$$\{a^i, a^j\} = \{a^{*i}, a^{*j}\} = 0 \quad \text{and} \quad \{a^i, a_j^*\} = \delta_j^i. \tag{17}$$

In the next Section we will indeed see that it is natural to call a^i creation and a^{*j} annihilation operators. After all this preparation we are ready to define harmonic p-forms as those p-forms ω which satisfy the Laplace equation:

$$(dd* + d*d)\omega = 0. \tag{18}$$

80

One easily verifies that if ω is a 0-form, Eq. (18) indeed reduces to the normal (covariant) Laplace equation, we are all so familiar with. In particular on a flat space, where the metric is constant, using Eqs. (3), (16), (17) one finds $dd^* + d^*d = \partial^2/\partial x_i^2$.

To prove the Hodge theorem, we first note that

$$\Lambda_p = \ker d \oplus \operatorname{im} d^*, \tag{19}$$

since $d\omega = 0$ iff for all $\alpha \in \Lambda_{p+1}$ we have $\langle d\omega, \alpha \rangle = 0$, which is equivalent to $\langle \omega, d^*\alpha \rangle = 0$, or $\omega \in (\operatorname{im} d^*)^\perp$. Using now the definition of $H^p(M, R)$ in Eq. (6) we find the following direct sum decomposition for Λ_p:

$$\Lambda_p = \operatorname{im} d \oplus H^p(M, R) \oplus \operatorname{im} d^*. \tag{20}$$

This then implies that $H^p(M, R)$ lies in the intersection of the orthogonal complements of $\operatorname{im} d$ and $\operatorname{im} d^*$, i.e. $H^p(M, R) = \ker d \cap \ker d^*$. Hence it remains to prove that ω is harmonic iff $d\omega = d^*\omega = 0$, but this follows from the fact that if ω is harmonic, one has $0 = \langle \omega, (dd^* + d^*d)\omega \rangle = \langle d^*\omega, d^*\omega \rangle + \langle d\omega, d\omega \rangle$, and from the fact that $\langle \alpha, \alpha \rangle$ is always strictly positive, except for $\alpha = 0$.

Thus we have the interesting result that although one had to define a metric and a differential structure on the manifold M in order to define harmonic forms, the number of harmonic forms is actually independent of the choice of metric and differential structure, they are topological invariants. Harmonic forms occur naturally in physics, as we mentioned in the introduction. There we promised to show that the solutions to the Maxwell equations are equivalent to harmonic 2-forms. The Maxwell equations are defined in terms of the curvature $F_{\mu\nu}$ on four-dimensional space-time, where $E_i = F_{0i}$ is the electric and $B_i = \varepsilon_{ijk}F_{jk}/2$ the magnetic field. These Maxwell equations are equivalent to $d^*F = 0$, where the 2-form F is given in local coordinates by $F = F_{\mu\nu}dx^\mu \wedge dx^\nu/2$. On the other hand, we know that we can write the curvature in terms of a connection 1-form $A: F = dA$, where $A = A_\mu dx^\mu$ and A_0 is called the scalar potential, whereas A_i is called the vector potential in electrodynamics. This fact is easily seen to imply that the curvature, or field strength, F satisfies the constraint of the Bianchi identities $dF = 0$. We have just seen that $dF = d^*F = 0$ indeed implies that F is harmonic.

3. Morse theory and supersymmetric quantum mechanics

In this Section we will discuss Morse theory as an alternative way to study topology of a manifold. In particular we will consider Witten's [9] formulation of Morse theory to prepare us for an attempt to understand Floer homology [13]. For a more detailed discussion of the mathematical aspects of Morse theory we refer to the literature [19].

If $h: M \to R$ is a generic function with isolated critical points $P_1, P_2, ..., P_q$, where a critical point is defined as a zero of the gradient vector field (i.e. $\partial h(P_i)/\partial x_k = 0$ for $k = 1, 2, ..., n = \dim M$), then the Morse index $\mu(P_i)$ of a critical point P_i is the number of negative eigenvalues for the Hessian of h evaluated at the point P_i (the Hessian for a function h is given by the matrix $\partial^2 h(x)/\partial x_i \partial x_j$). We define M_p as the number of critical points with Morse index p. The weak Morse inequality states the following result [19]: $M_p \geqslant B_p$.

One can prove stronger versions of the Morse inequalities [19],

$$\sum_{p=0}^{n} (M_p - B_p)t^p = (1+t) \sum_{p=0}^{n} Q_p t^p, \qquad Q_p \geqslant 0. \tag{21}$$

They will not concern us here, but as an illustrative example we give a useful corollary of these inequalities. Namely, by substituting $t = -1$ one gets an equality for the Euler characteristic (compare with Eq. (7))

$$\chi(M) = \sum_{q=0}^{n} (-1)^q M_q. \tag{22}$$

It is instructive to illustrate the topological nature of this formula in an example. In Fig. 2 we have sketched a two-dimensional manifold M with two holes, such that its Euler characteristic is $\chi(M) = 2 - 2H = -2$, where H is the number of holes (or handles). Let us choose for h the height function (the gravitational potential) with respect to the horizontal plane. Then there are 6 critical points, of which the lowest one is obviously stable and has Morse index 0, the highest one is unstable with Morse index 2 and the remaining four critical points are saddle points, with one stable and one unstable direction, hence these have Morse index 1 (see Fig. 2). Therefore one has the result: $\sum_{p} (-1)^p M_p = M_0 - M_1 + M_2 = 1 - 4 + 1 = -2$, which is indeed the Euler characteristic. We can now give a rule of thumb which shows that it is indeed a topological invariant by pressing with our thumb

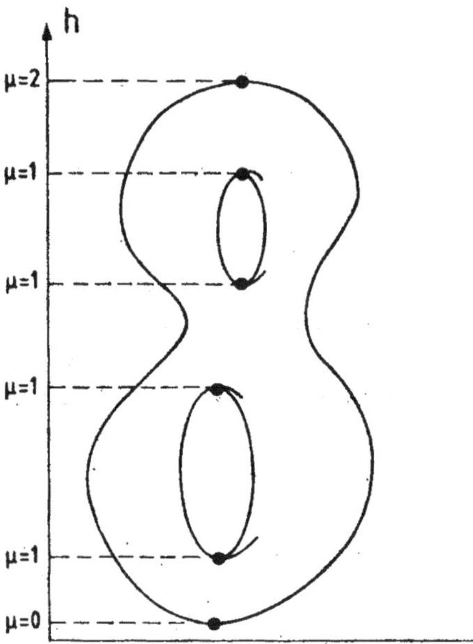

Fig. 2. Here we illustrate Morse theory, where h is the height function. We indicate the critical points and their Morse indices μ

82

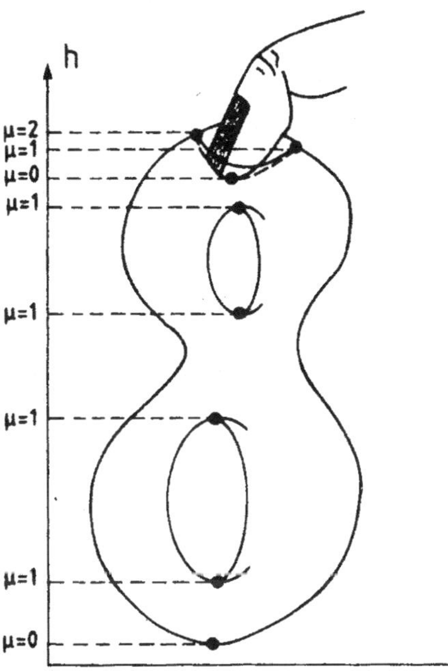

Fig. 3. The same as figure 2, but now deformed by pressing with our thumb at the maximum. This illustrates that the Euler characteristic, calculated by using Morse theory will not change under this deformation

at the maximum. Generically, this will create one additional local minimum and one additional saddle point (see Fig. 3). Therefore $\delta M_0 = 1$, $\delta M_1 = 1$, $\delta M_2 = 0$, but $\delta(\sum_p (-1)^p M_p) = 0$. It could happen that we press our thumb in such a way that the maximum will become degenerate with the saddle point, i.e. the maximum is obtained along a closed curve (however, try it and you will see that this is very hard to arrange, generically this will therefore not occur). There is so-called degenerate Morse theory which is able to deal with these situations. However, typically, this degeneracy is unstable against small perturbations, and for things to make sense topologically, definitions should not depend on these perturbations. Thus the only situation where one really has to worry about degenerate critical points is when there is a symmetry (for example axial symmetry along the vertical in Fig. 2) and perturbations are required to respect the symmetry. In that case one talks about equivariant Morse theory, this is however all we will say about it.

As we have already hinted at when we introduced the operators a^i and a^{*i}, there is an underlying supersymmetry in the exterior algebra. After all, the exterior derivative squares to zero, as does a supercharge Q, since it is an anticommuting object. Q changes the fermion number, whereas d changes the grading (defined by the p-forms). This then implies that one can identify p with the fermion number F. To make things more precise we have the following identifications:

$$(-1)^p \leftrightarrow (-1)^F,$$

$$d \leftrightarrow Q_1,$$

$$d^* \leftrightarrow Q_2 = Q_1^*,$$

$$dd^* + d^*d \leftrightarrow 2H = \{Q_1, Q_2\}. \tag{23}$$

Furthermore, since B_p is precisely the number of zero-energy p-forms, and since due to the supersymmetry, non-zero energy eigenstates will always occur in pairs of fermions and bosons, the Euler characteristic is exactly equal to the Witten index:

$$\sum_p (-1)^p B_p = \mathrm{Tr}\,((-1)^F). \tag{24}$$

This connection is more or less the reason Witten [9] discovered his formulation of Morse theory in terms of the exterior algebra, which will occupy the rest of this Section.

Motivated by removing degeneracies in a study of the Witten index for supersymmetric sigma models [8], Witten introduced the following modification of the exterior algebra:

$$d \rightarrow d_t = e^{-th}de^{th}, \quad d^* \rightarrow d_t^* = e^{th}d^*e^{-th}. \tag{25}$$

One can then define $B_p(t)$ as the number of harmonic forms for the exterior algebra d_t, i.e.

$$B_p(t) = \dim \{\ker (d_t d_t^* + d_t^* d_t) \cap \Lambda_p\}. \tag{26}$$

Clearly, $B_p(t)$ will depend continuously on t, however, $B_p(t)$ is a discrete function, hence it is independent of t and one can therefore find $B_p = B_p(0)$ by studying the vacua (zero-energy states) of the Hamiltonian

$$H_t = \tfrac{1}{2}(d_t d_t^* + d_t^* d_t), \tag{27}$$

in the limit of $t \rightarrow \infty$. This has a tremendous advantage, since we will find the wave functions in this limit to be highly peaked around the critical points of the Morse function h, such that in the lowest non-trivial order the wave function is a p-form with a harmonic oscillator type coordinate dependence, centered at a critical point. This immediately implies that in this approximation (which will neglect tunnelling) the number of vacua, which are in Λ_p, equals precisely M_p. The higher-order analysis can only have the effect of lifting the energy of some (or all) of the states thus constructed. This gives Witten's very simple proof of the weak Morse inequality, $B_p (=$ the number of exact zero energy states$) \leqslant M_p (=$ the number of approximate zero energy states).

Let us now make these considerations more precise by calculating in some detail H_t from the exterior algebra and by performing the large t asymptotic expansion for the eigenstates of this Hamiltonian. In the next Section it is this part of the analysis which allows generalization to the infinite dimensional context of the Yang-Mills configuration space, connected to the Floer homology. Let us first remind ourselves of the operator expressions for the exterior algebra:

$$d\omega = a^{*i}\frac{\partial \omega}{\partial x^i}, \quad d^*\omega = a^i\frac{\partial \bar\psi}{\partial x^i}, \quad \{a^i, a^{*j}\} = g^{ij}. \tag{28}$$

84

With the help of these we have the following results:

$$d_t = e^{-th}de^{th} = d + ta^{*i}\partial h/\partial x^i,$$

$$d_t^* = e^{th}d^*e^{-th} = d^* + ta^i\partial h/\partial x^i, \tag{29}$$

which, when substituted in the expression for H_t, Eq. (27) gives:

$$2H_t\omega = (dd^* + d^*d)\omega + t^2 g^{ij}\frac{\partial h}{\partial x^i}\frac{\partial h}{\partial x^j}\omega + t[a^{*i}, a^j]D_iD_jh\omega. \tag{30}$$

To obtain this result we used:

$$d\left(a^i\frac{\partial h}{\partial x^i}\omega\right) = a^{*j}\frac{\partial}{\partial x^j}\left(a^i\frac{\partial h}{\partial x^i}\right)\omega - a^i\frac{\partial h}{\partial x^i}d\omega,$$

$$d^*\left(a^{*j}\frac{\partial h}{\partial x^j}\omega\right) = a^i\frac{\partial}{\partial x^i}\left(a^{*j}\frac{\partial h}{\partial x^j}\right)\omega - a^{*j}\frac{\partial h}{\partial x^j}d\omega. \tag{31}$$

Finally D_i stands for the covariant derivative. By definition one has $D_ih = \partial h/\partial x^i$ and $\frac{\partial}{\partial x^j}(a^iD_ih) = a^iD_jD_ih$, which completes the derivation of Eq. (30). The Hamiltonian has now acquired a potential term proportional to the square of the gradient of h. Its minima are therefore the critical points of the Morse function and for large t these minima become increasingly localized. We can expand around a critical point, and choose locally flat coordinates, such that $g_{ij}(x) = \delta_{ij} + \mathcal{O}(x^2)$ (which means that the connection Γ^i_{jk} vanishes to $\mathcal{O}(x)$) and $h(x) = h(0) + \lambda_i x_i^2/2 + \mathcal{O}(x^3)$. Note that the number of negative λ_i is precisely the Morse index of the critical point we are expanding about. Thus we get the following expansion for the Hamiltonian around the critical points:

$$2H_t = \sum_i \left\{ -\frac{\partial^2}{\partial x_i^2} + t^2\lambda_i^2 x_i^2 + t\lambda_i[a_i^*, a_i] \right\}, \tag{32}$$

which everybody will recognize as the n-dimensional harmonic oscillator (in diagonal form), plus something that commutes with that, which is easily seen from the following properties (no summation over i):

$$[a_i^*, a_i]dx^{i_1} \wedge dx^{i_2} \wedge ... \wedge dx^{i_p} = \pm dx^{i_1} \wedge dx^{i_2} \wedge ... \wedge dx^{i_p}, \tag{33}$$

with the eigenvalue $+1$ if $i \in \{i_1, i_2, ..., i_p\}$ and -1 otherwise. From this one immediately obtains the spectrum for the Hamiltonian:

$$E_t = \tfrac{1}{2}t\sum_i (|\lambda_i|(1+2N_i) + \lambda_i n_i) + \mathcal{O}(t^0), \qquad n_i = \pm 1, \qquad N_i \in N. \tag{34}$$

This could only give zero energy if $N_i = 0$ for all i and if $n_i = -\text{sign }\lambda_i$. Since the number of negative eigenvalues λ_i is precisely the Morse index p for the critical point we are expanding about, we have p indices i for which $n_i = +1$ and the eigenfunction is therefore a p-form. In the approximation we are currently working with, each critical point gives a suitable groundstate wave function whose energy vanishes to order t, and as asserted before there are M_p such wave functions in Λ_p.

One might think that higher order perturbation theory will remove the degeneracy among the zero-energy states, however, supersymmetry will actually guarantee that the energy will vanish to all orders in perturbation theory, and only tunnelling effects will be able to remove some of the degeneracies. The number of harmonic forms is determined by topology and this is consistent with the fact that perturbation theory is a local expansion, which is blind to the topology of the manifold. Tunnelling will involve paths that do probe large portions of the manifold and should be able to distinguish which critical points are removable. These considerations therefore motivate that one can refine the weak Morse inequalities by studying tunnelling for the Hamiltonian H_t. This is what we will consider next.

Before we discuss the tunnelling analysis, it is instructive to give the supersymmetric non-linear sigma-model. Its action is given by [8]:

$$S = \tfrac{1}{2}\int d^2x\{g_{ij}(\phi)\partial_\mu\phi^i\partial^\mu\phi_j + ig_{ij}(\phi)\bar\psi^i\gamma^\mu D_\mu\psi^j + \tfrac{1}{12}R_{iklj}(\phi)\bar\psi^i\psi^l\bar\psi^k\psi^j$$

$$-g^{ij}(\phi)\frac{\partial h}{\partial\phi^i}\frac{\partial h}{\partial\phi^j} - D_iD_jh(\phi)\bar\psi^i\psi^j\}. \tag{35}$$

where the covariant derivative of the spinor fields is given by:

$$D_\mu\psi^i = \partial_\mu\psi^i + \Gamma^i_{jk}(\phi)\partial_\mu\phi^j\psi^k. \tag{36}$$

For the Witten-index computation all non-zero energy levels have boson-fermion degeneracy, so that one can restrict oneself to the zero-momentum sector, which gives a superesymmetric quantum mechanics equivalent to the exterior algebra which we considered before. To be more precise, since ψ is a Majorana fermion, in the representation of the Dirac-matrices where $\gamma_0 = \text{diag}(1, -1)$, ψ^i has two components, which are each others conjugate. If we call the upper component a^i, then Q and Q^* in this supersymmetric quantum mechanics are exactly equal to d and d^*. Note that ϕ^i are the coordinates on the target manifold M, which are to be identified with x^i in our earlier discussion of the exterior algebra. Originally Witten introduced the Morse function h to get rid of a degeneracy in the classical potential (which is zero for all ϕ^i if $h = 0$). In the superfield formulation of the supersymmetric sigma model h appears simply as a magnetization:

$$S = \tfrac{1}{2}\int d^2x d^2\theta\{g_{ij}(\Phi)\bar D\Phi^iD\Phi^j + h(\Phi)\}, \tag{37}$$

where the superfield and superderivative are given by:

$$\Phi^i = \phi^i + \bar\theta\psi^i + \bar\theta\theta F^i/2, \qquad D_\alpha = \partial/\partial\bar\theta_\alpha - i(\bar\theta\gamma^\mu)_\alpha\partial_\mu. \tag{38}$$

We will now discuss the tunnelling calculation. For $|P_i\rangle$ the perturbative vacua we constructed before, we can compute the matrix elements $\langle P_i|d_t|P_j\rangle$, which will allow us to calculate the Hamiltonian with respect to this basis. This matrix element can be calcu-

86

lated approximately by an instanton calculation, but will in general be zero due to the fermionic zero-modes. $\langle P_i|d_t|P_j\rangle$ will only be non-zero if the instantons relevant for the tunnelling from P_i to P_j have exactly one fermionic zero-mode, which will be absorbed by d_t. Due to the supersymmetry, the number of fermionic zero-modes equals the number of bosonic zero-modes. There is always at least one bosonic zero-mode, which is related to the invariance of the instanton solution under time-translation. We cannot have more, otherwise the matrix element of interest would vanish. Therefore, the only instantons relevant for computing $\langle P_i|d_t|P_j\rangle$ correspond to tunnelling paths that are isolated. They form a discrete moduli space, see Fig. 4 for an example.

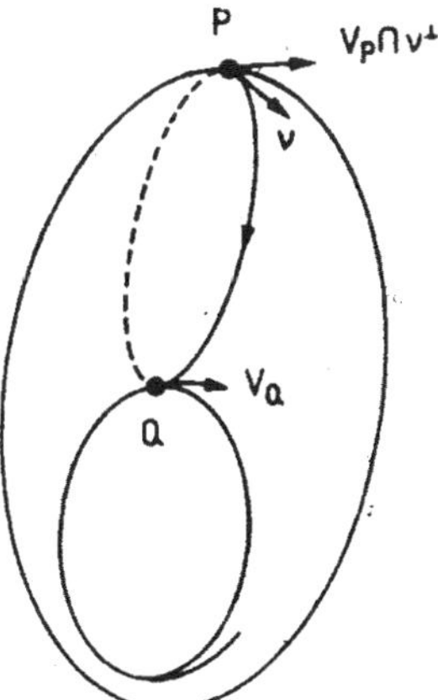

Fig. 4. Here we illustrate the tunnelling paths between two critical points, which differ in Morse index by one. In this example we see that there are two instantons. We also indicate the tangent space $V_P \cap v^\perp$, which is transported along the tunnelling path, so that we can compare its orientation with V_Q. In this example both instantons contribute with the same sign

We can now make use of a version of the index theorem [20] relevant to the present tunnelling analysis. It states that the number of fermionic zero-modes is equal to the spectral flow of the Hessian $D_i D_j h(x)$ along the tunnelling path from P_i to P_j; in more mundane terminology, this spectral flow is the number of eigenvalues that change sign, when following the Hessian along the tunnelling path from one critical point to the other. Since the number of negative eigenvalues at a critical point equals its Morse index, the only tunnelling paths we need to consider are those between critical points for which the respective Morse indices differ by 1. To be even more precise the instanton path has to go from P_i, with Morse index $p+1$, down the gradient lines of h to P_j, with Morse index p. One easily verifies that the instantons are determined by the equations:

$$\frac{dx^i(\tau)}{d\tau} = -g^{ij}(x(\tau))\frac{\partial h(x(\tau))}{\partial x^j}\,, \tag{39}$$

with the boundary conditions $x(-\infty) = P_i$ and $x(\infty) = P_j$. The total action is found to be $S = t(h(P_i)-h(P_j)) > 0$.

87

The steepest descent approximation for the instanton calculation [8, 9, 21] gives us:

$$\langle P_i | d_t | P_j \rangle = \sum_r K(r) \exp \left[-t(h(P_i) - h(P_j)) \right], \tag{40}$$

where r runs over the discrete set of instantons and $K(r)$ is the prefactor, which one obtains by calculating the determinant of the quadratic piece in the action, when expanding around the instantons (taking into account the transverse fluctuations) and a factor coming from the zero-modes. One can now show that the supersymmetry will cancel the following two contributions to $K(r)$ up to a possible overall minus sign, namely those coming from the fermionic and the bosonic transverse fluctuations and the contribution coming from the fermionic and bosonic zero-mode. Hence for each tunnelling path, $K(r) = \pm 1$, so that one has the following result:

$$\langle P_i | d_t | P_j \rangle = n(P_j, P_i) \exp \left[-t(h(P_i) - h(P_j)) \right], \tag{41}$$

where $n(P_j, P_i)$ is an integer. To determine the sign that each instanton will contribute requires a more careful analysis of the instanton calculation and involves the parallel transport of a certain tangent frame along the tunnelling path, which allows one to compare the orientation of the frame at P_i, to the frame at P_j. These notions all occur quite naturally in the WKB analysis involved in the tunnelling calculation [22], but we will only state its result (see Fig. 4). Let P_i be the critical point with index $p+1$, and V_i the $(p+1)$-dimensional subspace of the tangent space at P_i, spanned by the eigenvectors of the Hessian with a negative eigenvalue (V_i is the tangent space of the unstable manifold associated with P_i). Let v the tangent vector to the tunnelling path at P_i, then the tangent frame we wish to transport along the tunnelling path is $V_i \cap v^\perp$, which is p-dimensional and can therefore be compared with the p-dimensional space V_j (the tangent space to the unstable manifold, associated to the critical point P_j). The orientations of the space V_i and V_j are the ones induced respectively by the $(p+1)$-form and the p-form, which arise as the eigenfunctions in the perturbative analysis. This prescription gives us a unique way of determining $n(P_j, P_i)$ and Witten used these to define a new cohomology (called twisted cohomology [9]):

$$\delta |Q\rangle = \sum_{P \in W_{p+1}} n(Q, P) |P\rangle, \tag{42}$$

where $Q \in W_p$ and the W_p form the so-called Witten complex:

$$W_p = \{ |P\rangle \, | \, \mu(P) = p \}. \tag{43}$$

Thus $\delta : W_p \to W_{p+1}$ and the matrix elements of δ are precisely those of d_t, with respect to the set of perturbative vacua $= \bigcup_p W_p$. Consequently $\delta^2 = 0$ and we can form a cohomology. The instanton calculation proves that if $(\delta\delta^* + \delta^*\delta)|\chi\rangle = \lambda|\chi\rangle$, with $\lambda \neq 0$, then $|\chi\rangle$ has non-zero energy. It has required some work to rigorously prove that the converse is also true, that is when $\lambda = 0$ the corresponding eigenstate (which is in the twisted cohomology) does indeed have zero energy (Witten [9] only supplies some intuitive arguments). This then shows that $B_p = \dim ((\ker \delta / \operatorname{im} \delta) \cap W_p)$, establishing that the Betti numbers can be determined from Morse theory. For the strong Morse inequalities, see Ref. [23].

88

4. Floer homology

We consider now Y to be a three-manifold over which there is defined a Yang-Mills gauge theory, that is one associates to Y a fibre bundle, with fibres in a compact gauge group, for which we will choose SU(2). The analogue of the manifold M is now played by the configuration space, $\mathscr{C} = \mathscr{A}/\mathscr{G}$ which is the space $\mathscr{A}$ of connections A on the manifold Y, modulo the SU(2) gauge transformations $\mathscr{G}$ (maps g of Y into SU(2) where g acts as follows: $^{g}A = gAg^{\dagger}+gdg^{\dagger}$). The infinite-dimensional configuration space $\mathscr{C}$ is a manifold away from the so-called reducible connections (A is reducible if there is a gauge function not in the centre of the gauge group, which leaves the connection invariant: $\exists g \notin Z_2, {}^{g}A = A$). The reducible connections, of which 0 is an obvious example, give rise to conic (orbifold-type) singularities. However, $\mathscr{C}$ is smooth enough to allow for a formulation of Morse theory. We will sketch here some of the results without going into too much details and we refer to Floer's original paper [13] (of which the introduction is quite readable for a physicist), or to Braam's review [14]. Atiyah's paper [15] provides the grand scheme of how everything is to fit together.

Floer takes as a Morse function the Chern-Simons functional $\left(A = A_{\mu}^{a} \dfrac{\sigma_a}{2_i} dx^{\mu}\right)$:

$$h(A) = \int_{Y} \mathrm{Tr}\,(A \wedge dA + \tfrac{2}{3} A \wedge A \wedge A) = \int_{Y} d_3 x \varepsilon^{ijk}\, \mathrm{Tr}\,(A_i \partial_j A_k + \tfrac{2}{3} A_i A_j A_k). \tag{44}$$

We note that h is not quite an appropriate function on $\mathscr{C}$, since it is not exactly invariant under gauge transformations,

$$h(^{g}A) = h(A) - \tfrac{1}{3} \int_{Y} \mathrm{Tr}\,((g^{\dagger}dg)^3) = h(A) - 8\pi^2 \deg\,(g: Y \to \mathrm{SU}(2)). \tag{45}$$

Therefore one rather considers h as a function on $\mathscr{A}/\mathscr{G}^0$, where $\mathscr{G}^0$ is the connected component of the space of gauge transformations. The critical points of h are easily found to be the curvature free configurations, called flat connections in the mathematical terminology. They are the classical minima of the Yang-Mills potential. Indeed one easily verifies:

$$\frac{\partial h(A)}{\partial A_i^a(x)} = -\varepsilon^{ijk}F_{jk}^a(x)/2 = -B_i^a(x), \tag{46}$$

where $F_{ij}^a(x) = \partial_i A_j^a(x) - \partial_j A_i^a(x) + \varepsilon_{abc}A_i^b(x)A_j^c(x)$ is the curvature (as a 2-form it is given by $F = dA + A \wedge A = F_{\mu}^a \dfrac{\sigma_a}{4_i} dx^{\mu} \wedge dx^{\nu}$). Similarly it is very easy to determine the gradient flow of this Morse function, whose solutions will describe the tunnelling in this infinite-dimensional analogue of the exterior algebra (further on we will explicitly give the relevant supersymmetric Hamiltonian). We find:

$$\frac{\partial A_i^a(x;\tau)}{\partial \tau} = B_i^a(x;\tau). \tag{47}$$

These are exactly the anti-self-duality equations if we extend the connection A over Y, to a connection over $M = Y \times R$, with of course τ being identified with the fourth extra coordinate. One easily verifies that Eq. (47) is equivalent to $F = - *F$. The Hodge $*$ is defined with respect to the metric of M, which is in an obvious way obtained from the metric of Y, by defining $g_{44} = 1, g_{i4} = 0$ for $i = 1, 2, 3$. Thus we see that the Yang-Mills instantons occur in a natural way in Floer's analysis.

Similar to the Witten complex, one can now consider what we will call the Floer complex. It is simply the set of gauge equivalence classes of the flat connections $\mathcal{V}$. A flat connection is classified by its holonomy along a loop, the well-known Wilson loop $P \exp (\oint_\gamma A)$, and one can show that this holonomy only depends on the homotopy of the loop γ. The argument is that under a small deformation of the loop, the Wilson loop changes proportional to the curvature (this is obvious for an abelian gauge group, but can be suitably extended to non-Abelian groups [24]) and is hence zero. Since the homotopy does not change under continuous deformation, the holonomy stays constant under these changes. The Floer complex is therefore in one-to-one relation with the set of SU(2) representations of the fundamental group $\pi_1(Y)$. In general, however, the space of equivalence classes of flat connections is not discrete. My favourite example is of course the case that Y is the three-torus T^3, where it can be shown that $\mathcal{V}$ is the orbifold T^3/Z_2 [25]. Also in general, the Floer complex will contain reducible connections (since 0 is always a flat connection). For these reasons one imposes some constraints on the manifolds Y to be considered. One requires them to be so-called homology three spheres, which means that their (integer) first homology $H_1(Y, Z)$ is zero. This is sufficient to show that $\pi_1(Y)$ is finite and to guarantee that the only reducible connection in the now finite Floer complex is the 0 connection.

There are, however, two (related) technical problems one has to deal with. Firstly, for h to be an appropriate Morse function we were actually working on $\mathcal{A}/\mathcal{G}^0$, whereas for the Floer complex, we considered dividing out all gauge transformations. Thus, we have to convolute the Floer complex with $\mathcal{G}/\mathcal{G}^0 \sim \pi_3(SU(2))$. The second problem is that the Hessian of the Chern-Simons functional is no longer an elliptic operator, it is actually the covariant derivative operator, which is of Dirac-type and is unbounded from below. Therefore, it is impossible to define the Morse index, since there will always be an infinite number of negative eigenvalues. One can, however, define a relative Morse index, by declaring the Morse index related to a preferred flat connection equal to 0. The most obvious choice is the 0 connection, but it is reducible and needs special care [13, 14]. In the physical terminology, together with the fact that the Hessian is a Dirac-type operator this choice corresponds to a choice of vacuum, or Dirac-sea. The relative Morse index follows now directly from the index theorem [20] for the Hessian, similar to what we saw in the previous ection.

From instanton calculations on $M = Y \times S^1$, we know that the spectral flow for tunnelling from a flat connection A to a flat connection ^gA is equal to $8k$, when k is the winding number of the gauge transformation g. It is essential that M is a compact manifold, which is achieved because we tunnel to *gauge equivalent* configurations. This implies that one can divide out the action of $\mathcal{G}/\mathcal{G}^0$, provided the grading due to the relative Morse index is defined modulo 8. This brings us back to a finite Floer complex, with a grading defines modulo 8 and, as for Witten's formulation of Morse theory for a finite dimensional

90

manifold, Floer was able to define a boundary operator ∂, which squares to zero and defines a homology, where the eight homology groups

$$HF^i(Y) = (\ker \partial / \mathrm{im}\, \partial) \cap F^i. \tag{48}$$

are called the Floer groups, with F^i the Floer complex of the flat connections which have a Morse index i (mod 8) with respect to the 0 connection. Floer has proved that these homology groups lead to invariants for the manifold Y (i.e. do not dependent on the choice of metric on Y).

We end this Section with the promised construction of the supersymmetric Hamiltonian, whose zero-energy ground states should correspond to the Floer groups (or to $\ker (\partial^*\partial + \partial\partial^*)$). As in the previous Section this Hamiltonian should be given by the Laplacian on $\mathscr{A}$ in terms of the twisted exterior algebra on $\mathscr{A}$. The standard exterior algebra on $\mathscr{A}$ is given by (compare Eqs. (3.16)):

$$d = \int_Y d^3x\, \psi_i^a(x)\, \frac{\delta}{\delta A_i^a(x)}, \qquad d^* = - \int_Y d^3x\, \bar{\psi}_i^a(x)\, \frac{\delta}{\delta A_i^a(x)}, \tag{49}$$

where $\psi_i^a(x)$ form a basis for the exterior algebra of $\mathscr{A}$ at $A(T_A^*\mathscr{A})$ and $\bar{\psi}_i^a(x)$ form the dual basis (in $T_A\mathscr{A}$). They are anticommuting *spin-one* fields and satisfy a Dirac algebra:

$$\{\psi_i^a(x), \psi_j^b(y)\} = \{\bar{\psi}_i^a(x), \bar{\psi}_j^b(y)\} = 0, \quad \{\psi_i^a(x), \bar{\psi}_j^b(y)\} = g_{ij}(x)\delta^{ab}\delta_3(x-y). \tag{50}$$

The twisted exterior algebra is now given, in terms of the Chern-Simons functional which Floer chose as his Morse function, by:

$$d_{e^{-2}} = \exp(-h(A)/e^2)d \exp(h(A)/e^2), \quad d_{e^{-2}}^* = \exp(h(A)/e^2)d^* \exp(-h(A)/e^2). \tag{51}$$

For reasons which will become obvious, we have chosen e^{-2} instead of t and we are now interested in the limit $e \to 0$. The Hamiltonian to consider is given in terms of the twisted Laplacian on $\mathscr{A}$ by:

$$2e^{-2}H = d_{e^{-2}}d_{e^{-2}}^* + d_{e^{-2}}^*d_{e^{-2}}. \tag{52}$$

A straightforward computation, exactly analogous to Eq. (30), gives the supersymmetric Hamiltonian:

$$H = \int_Y d^3x\, \mathrm{Tr}\left(e^2\vec{\Pi}^2 + \frac{1}{e^2}\vec{B}^2 + \varepsilon^{ijk}\psi_i D_j\bar{\psi}_k\right), \tag{53}$$

where $D_j = \partial_j + \mathrm{ad}\, A_j$ is the covariant derivative in the adjoint representation and $\Pi_i^a(x) = -i\dfrac{\delta}{\delta A_i^a(x)}$ are the canonical momenta. Restriction to the gauge invariant configuration space $\mathscr{C}$ is easily achieved by imposing Gauss' law, in the same way as this is done in the Hamiltonian theory for ordinary Yang-Mills theory in the $A_0 = 0$ gauge. We now see that e plays the role of the coupling constant and that the large t asymptotic analysis of the previous section corresponds to the weak coupling expansion. Finally we note, as mentioned in the introduction, that the bosonic part of the Hamiltonian is exactly the ordinary Yang-Mills Hamiltonian.

5. The Lagrangian for topological Yang-Mills theory

As we mentioned in the introduction, there is a relation of the Floer groups to the Donaldson invariants, which arise when one "cuts" a closed four-manifold M into two pieces along a three-manifold. That is, each half has the same three manifold Y as a boundary. For the interested reader we refer to Atiyah's paper [15] for more details. Here we only remark, that due to this relation one anticipates that there should be a Lorentz-invariant formulation in four dimensions, which reduces to the Floer theory on a manifold of type $M = Y \times R$ in the Hamiltonian formulation, but which on a closed four manifold, would be intimately connected to the Donaldson invariants. In physical terminology this means that one should look for a Lagrangian formulation of the Hamiltonian in Eq. (53). This is the challenge Atiyah [15] put to the physics community and we will discuss how Witten addressed the question.

He introduced a U-quantum number, which corresponds to the Floer groups (and is conserved mod 8), with the following assignment:

$$U(A) = 0, \quad U(\psi) = 1, \quad U(\bar{\psi}) = -1. \tag{54}$$

One now likes to fit $(A, \psi, \bar{\psi})$ into a Lorentz multiplet, which will somehow have to play a role in the instanton calculation on the four manifold M. There is an obvious difficulty with this, which is related to the fact that ψ and $\bar{\psi}$ are anticommuting spin-one fields, which already indicates that the supersymmetry, alluded to in the previous Section is not quite standard (as we will see, it is more natural to see ψ as a ghost field and to talk about a Slavnov or BRST symmetry).

If we consider deformations δA along a given instanton moduli space $\mathcal{M}_k$, then in order for $A + \delta A$ to still be a solution it has to satisfy the deformation equations:

$$D_\alpha \delta A_\beta - D_\beta \delta A_\alpha - \varepsilon_{\alpha\beta\mu\nu} D^\mu \delta A^\nu = 0, \quad D_\alpha \delta A^\alpha = 0. \tag{55}$$

The first equation is simply the deformation of the self-duality equations, the second is a gauge condition. The δA form a tangent vector to the moduli space $\mathcal{M}_k$, which in the physical terminology is called a zero-mode. Usually $\mathcal{M}_k$ is considered the space of all anti-self-dual solutions (or anti-instantons). A simple change of orientation of M together with some field redefinitions will give Witten's [1] results, but here we follow the notations of Ref. [18]. We want every bosonic zero-mode to be cancelled by an anti-commuting zero-mode, which will be established by the following term in the Lagrangian:

$$4\bar{\psi}^{\mu\nu} D_\mu \psi_\nu + \beta D_\mu \psi^\mu. \tag{56}$$

Here $\bar{\psi}_{\mu\nu}$ is an anti-commuting anti-self-dual tensor field, related to the fields in the previous Section by $\bar{\psi}_i = \varepsilon_{ijk}\bar{\psi}^{jk}/2$. We therefore see that we had to introduce β and ψ_0 as new

92

fields. One should compensate those by commuting ghost of ghost fields $\bar{\phi}$ and ϕ to cancel these added degrees of freedom.

Witten [1] conjectured the following Lagrangian that could be related to the Hamiltonian of Eq. (53):

$$\mathcal{L} = \mathrm{Tr}\,\left\{ -\tfrac{1}{2}\,F_{\mu\nu}F^{\mu\nu} + 4\bar{\psi}^{\mu\nu}D_\mu\psi_\nu + \beta D_\mu\psi^\mu + \bar{\phi}D_\mu^2\phi + \bar{\phi}[\psi_\mu,\,\psi^\mu] \right.$$

$$\left. + \phi[\bar{\psi}_{\mu,\nu},\,\bar{\psi}^{\mu\nu}] + (D_\mu^{bg}A^\mu)^2 + \bar{c}D_\mu^{bg}D^\mu c + \ldots \right\}, \tag{57}$$

where the dots stand for terms higher order in the fields, D_μ^{bg} is the background covariant derivative for the background gauge, where the last two terms describe the standard Faddeev–Popov gauge fixing, with c, $\bar{c}$ the normal ghost fields. The main motivation for this expression is largely that in a naive instanton calculation (ignoring for a moment the zero-modes, by assuming that the moduli space is a single point) the partition function is given by

$$Z = \frac{\mathrm{Pf}\,(D)}{\sqrt{\det\,(\Delta)}}. \tag{58}$$

In this equation $\mathrm{Pf}(D)$ is the Pfaffian of the antisymmetric operator D, appearing in Eq. (56). The Pfaffian comes from the Grassmann integration over the anti-commuting fields and up to a factor ± 1 is equal to the square root of the determinant of the operator. The factor in the denominator is of course coming from the Gaussian integration over the commuting fields. The operator Δ is the one occurring in Eq. (55), which is, by construction, related to the operator D through the anti-commuting symmetry $\delta A_\mu = \varepsilon\psi_\mu$, where ε is in this case a *scalar* Grassmann variable. In the BRST language this defines an anti-commuting operation s, such that $sA_\mu = \psi_\mu$. This means that in Eq. (58) the denominator will cancel against the nominator, up to a factor ± 1. Note that the Faddeev–Popov determinant cancels exactly against the determinant coming from the integration over the ghost of ghost fields $\bar{\phi}$, ϕ, which are assumed to be complex fields. Requiring the quadratic part of the action to be invariant fixes the symmetry for the other fields and the U quantum numbers for ϕ and $\bar{\phi}$ to be $U(\phi) = 2$ and $U(\bar{\phi}) = -2$. It also almost fixes the form of the higher order part of the Lagrangian (Witten showed that there is a choice which corresponds exactly to a so-called "twisted" version of $N = 2$ super Yang–Mills theory). The action in Eq. (57) presents potential pitfalls. It is not obviously positive definite and it might be plagued by Gribov ambiguities. As we will see, most of the fields are ghost or ghost of ghost fields, and the action is actually constant. Thus, we could phrase the question as whether the physicist's description of how to deal with the Lagrangian in Eq. (57) leads to a normalizable integration measure on the configuration space (in a fixed topological sector).

Let us now come back to the result of the partition function in Eq. (58), which more or less by construction is equal to ± 1 in the case that $\mathcal{M}_k$ exists of one point. Similarly, when $\mathcal{M}_k$ is discrete, one finds $Z = \sum_{\mathcal{M}_k} (-1)^{n_i}$ and Witten argues [1] that this is precisely one of the Donaldson invariants (by studying how the sign of the Pfaffian is determined). The strength of Witten's analysis [1] is the way one shows that the partition function is an invariant, i.e. does not depend on the metric. This property is at the heart of what topological field theories are about. In Witten's original analysis this invariance seems to come out of the blue, however, furtheron we will analyse the deeper reasons for this. At this point we will only specify the properties, which will guarantee the topological nature of the theory. Firstly the anti-commuting symmetry s is a BRST or Slavnov symmetry, whose square is zero, $s^2 = 0$. Witten [1] did not take ordinary gauge fixing into account, implicitly working on the space $\mathscr{A}/\mathscr{G}$, rather than on $\mathscr{A}$. In that case s^2 is only zero on gauge invariant states, which is all one needs anyhow. The advantage is furthermore that in this way one does not have to address the issue of Gribov ambiguities in the usual gauge fixing. The second property we already alluded to before, is that the action is actually constant. This is expressed by the fact that one can write the action $S = \int_M \sqrt{g}\mathscr{L}$ as ($F = \frac{1}{2}F_{\mu\nu}dx^\mu \wedge dx^\nu$):

$$S = 2 \int_M \mathrm{Tr}\,(F \wedge F) + sS_{\mathrm{gf}}, \tag{59}$$

where S_{gf} is an integral over a local gauge invariant polynomial of the fields. Finally and crucially, one can show [1] that the energy-momentum tensor $T_{\mu\nu}$, which is obtained by varying the action with respect to the metric g, is BRST-trivial: $T_{\mu\nu} = s\lambda_{\mu\nu}$, where again λ is a local gauge invariant polynomial in the fields (this actually follows from Eq. (59)). Following Witten [1], the fact that $s^2 = 0$ and $T = s\lambda$ will lead to the fact that the partition function is independent of the metric:

$$\delta_g Z = -\frac{1}{e^2}\int \mathscr{D}A\,\delta_g S \exp\left(-\frac{1}{e^2}S\right) = -\frac{1}{2e^2}Z\left\langle s\int_M \sqrt{g}\,\delta g_{\mu\nu}\lambda^{\mu\nu}\right\rangle = 0. \tag{60}$$

It is crucial to observe here that one has to assume that there is no anomaly in the energy-momentum tensor, i.e. one has to make sure that quantum corrections do not spoil the BRST-triviality of the energy-momentum tensor. This has successfully been shown to one-loop in Ref. [26], but it is expected that the symmetries are strong enough to allow one to derive Ward identities that will establish the BRST-triviality to all orders. Similarly one can now ask which operators $\mathcal{O}$ will have invariant expectation values, i.e. satisfy the equation:

$$\delta_g\langle\mathcal{O}\rangle = \left\langle\delta_g\mathcal{O} - \frac{1}{2e^2}\mathcal{O}s\int_M \sqrt{g}\,\delta g_{\mu\nu}\lambda^{\mu\nu}\right\rangle = 0. \tag{61}$$

We see that a sufficient condition is that $\delta_g\mathcal{O} = s\varrho$ for some gauge invariant operator ϱ (we will, however, only consider operators that do not depend on the metric) and that $\mathcal{O}$ is BRST-trivial, i.e. $s\mathcal{O} = 0$. When, however, $\mathcal{O}$ itself is the s of something ($\mathcal{O} = s\varrho$, for some *gauge invariant* operator ϱ) then its expectation value is zero and does not lead to anything useful. In conclusion, the interesting observables, which have invariant expectation values are those with a non-trivial equivariant (or basic) cohomology [17]:

$$\mathcal{O} \in \ker \hat{s}/\mathrm{im}\,\hat{s}, \tag{62}$$

94

where $\hat{s}$ restricts s to the *gauge invariant* set of operators (which are polynomials in the fields). However, it is not known whether *all* non-trivial observables are of this type.

Until now, we only considered the case where the moduli space is discrete, but in the examples mostly known to physicists like for S^4, this is not the case. Then in general the partition function will be zero, due to the presence of anti-commuting zero-modes. To have a non-zero result one needs to consider operators which will cancel all these zero-modes and this is what the quantum number U is useful for. One can show [1], that the operator in question should be a polynomial in the fields, with a net value for U (which is additive), equal to the number of zero-modes. This is all quite similar to 't Hooft's construction of the effective action for the breaking of chiral $U_A(1)$ through instantons [27]. Again, by the index theorem the number of zero-modes is equal to the dimension of the moduli space:

$$U(\mathcal{O}) = \dim(\mathcal{M}_k) = 8k - \tfrac{3}{2}(\chi(M) + \sigma(M)), \tag{63}$$

where $\chi(M)$ is the Euler characteristic and $\sigma(M)$ is the so-called signature of M, e.g. $\chi(S^4) = 2$, $\sigma(S^4) = 0$. In the instanton calculation the expectation value of the operator $\mathcal{O}$ is calculated by first integrating out the non-zero modes, including the ϕ and $\bar{\phi}$ field, after which $\mathcal{O}$ takes the form

$$\mathcal{O} = \Phi_{i_1,i_2,\cdots,i_n}(a_k)\theta^{i_1}\theta^{i_2}\ldots\theta^{i_n}, \qquad U(\mathcal{O}) = n, \tag{64}$$

where a_k are parameters describing the instantons (moduli parameters) and θ^k the zero-modes for ψ. The expectation value of $\mathcal{O}$ then reduces to an integration over moduli space, with the canonical measure $d\mu$ defined on $\mathcal{M}_k$ [1]. Actually, since at least for a physicist, it can be easily shown that the expectation values of the operators are independent of the coupling constant (if $\partial\mathcal{O}/\partial e = 0$), one can work in the weak-coupling limit, and the only integration over non-zero modes that survives in the limit $e \to 0$, is the integration over the fields ϕ and $\bar{\phi}$. From Eq. (57) we see that $[\psi_\mu, \psi^\mu]$ acts as a source ϕ. Thus [1], one replaces in $\mathcal{O}$ ϕ by

$$\phi(x) = -\int_M \sqrt{g}\, G(x, y)\, [\psi_\mu(y), \psi^\mu(y)], \tag{65}$$

ψ by their zero-modes and A by the self-dual connections, where $G(x, y)$ is the Green function for $D_\mu D^\mu$. This defines therefore explicit formulas for the invariants in terms of integrals over the moduli space, which we will discuss in the next Section.

We will end this Section by a discussion of the BRST formulation of topological Yang–Mills theory [16, 18, 28, 29, 30]. The strategy is to start with an action which is independent of the metric (to guarantee the topological properties). This will in general have large symmetries. In the case of Yang–Mills theory, one starts [16] with an action that is proportional to the topological charge:

$$S_0 = 2\int_M \mathrm{Tr}\,(F \wedge F). \tag{66}$$

This action is clearly invariant under an arbitrary variation of the gauge field (provided we stay in the same topological sector). This symmetry needs to be gauge fixed: $\mathcal{L} = \mathcal{L}_0$

$+sV_{\mathrm{gf}}$. The normal gauge fixing (dividing out the action of the gauge group $\mathscr{G}$) can and should be done separately [28]

$$\mathscr{L} = \mathscr{L}_0 + sV_{\mathrm{gf}} + s_g V_{\mathrm{FP}}, \tag{67}$$

see also [30]. Both s and s_g generate Slavnov or BRST symmetries and satisfy $s^2 = 0$, $s_g^2 = 0$ and $\{s, s_g\} = 0$.

To be more precise, the invariance is given by $\delta A_\mu = \psi_\mu$, which we fix by the gauge condition $F^- = (F^- * F) = 0$. Note that this does not completely fix the gauge, but this is on purpose, since we wish to be left with the (finite set of) degrees of freedom that describe the instantons, which form exactly the kernel of this gauge condition. Thus, as is familiar in the BRST formalism, ψ_μ becomes a ghost. We also have a Lagrange multiplier field $b_{\mu\nu}$, which will enforce the constraint $F^- = 0$ and is therefore an anti-self-dual tensor field. Finally one completes the set of fields by the anti-ghost $\bar{\psi}_{\mu\nu}$. However, the variations of A in the direction of the gauge orbit for which $\psi_\mu = D_\mu\phi$ with ϕ some function in the adjoint representation of the gauge group, is a redundant symmetry (because it will be described by dividing out the action of the gauge group $\mathscr{G}$) and needs therefore to be removed too. This symmetry is then fixed by $D_\mu\psi^\mu = 0$. The field ϕ will hence be the ghost of a ghost (and is a commuting field). We will call the Lagrange multiplier, used to enforce this gauge condition β and the anti-ghost of the ghost $\bar{\phi}$. This completes the description of the field content and explains the origin of all the fields in Witten's [1] original formulation. It is easy to find the appropriate gauge fixing Lagrangian $\mathscr{L}_{\mathrm{gf}} = sV_{\mathrm{gf}}$ of Eq. (67)

$$\mathscr{L}_{\mathrm{gf}} = s\,\mathrm{Tr}\,(\bar{\phi}D_\mu\psi^\mu + \bar{\psi}^{\mu\nu}(b_{\mu\nu} - F^-_{\mu\nu})) + s\,\mathrm{Tr}\,(\beta[\phi, \bar{\phi}]), \tag{68}$$

where the last term can be freely added; however, with it the Lagrangian can be seen as a "twisted" version of $N = 2$ supersymmetric Yang–Mills [1]. The Slavnov or BRST symmetry s is given by the following formula [18] (for ease of notation we suppressed the indices):

$$\begin{aligned}
sA &= \psi', & s\psi' &= 0, & \omega' &= \psi - D\omega, \\
s\bar{\psi} &= b', & sb' &= 0, & b' &= b - [\omega, \bar{\psi}], \\
s\bar{\phi} &= \beta', & s\beta' &= 0, & \beta' &= \beta - [\omega, \bar{\phi}], \\
s\omega &= \phi', & s\phi' &= 0, & \phi' &= \phi - [\omega, \omega]/2.
\end{aligned} \tag{69}$$

We have written the action of s in terms of shifted fields to demonstrate the fact that the ordinary local s-cohomology is trivial. Any operator which is in the kernel of s, can also be written as the s of some other operator. That would not leave any interesting observables. As we saw, it is not the local, but the equivariant cohomology that determines the interesting observables. Related to this is that in Eq. (69) the field ω appears, which is a ghost-field, with values in the adjoint representation of the gauge group. It plays the role of an infinitesimal gauge transformation on which the action does not depend. This is crucial for defining equivariant cohomology, which goes back to Cartan [17], who called it basic cohomology. For this cohomology one restricts s to the gauge invariant and ω independent

96

operators. It is this cohomology which is non-trivial, leading to the interesting observables that will be discussed in the next section. To complete the discussion, the ordinary gauge fixing does not interfere with the above construction and is as usual formulated as follows:

$$\mathscr{L}_{\text{FP}} = s_g V_{\text{FP}} = s_g \bar{c}(b - D_\mu^{\text{bg}} A^\mu), \tag{70}$$

$$s_g A_\mu = D_\mu c, \quad s_g c = -[c, c]/2, \quad s_g \bar{c} = b, \quad s_g b = 0. \tag{71}$$

Let us emphasize again that Gribov ambiguities will in general be present, but to the point of dividing out the gauge group, this seems to be only a technical handicap. As we said before, and which is the attitude taken by Witten [1], one can entirely work on $\mathscr{A}/\mathscr{G}$, where this issue need not be addressed. It is, however, possible that Gribov will still take revenge through the gauge condition $D_\mu \psi^\mu = 0$, but we have nothing sensible to say on this right now.

6. Donaldson polynomials

In this Section we will finally be able to construct the Donaldson polynomials. As observed before, we need operators with a net U charge equal to the dimension of the moduli space and we will build these operators as a product of operators $\mathcal{O}_i$, which are non-trivial elements of the equivariant s-cohomology with charge U_i. Thus $\mathcal{O} = \Pi \mathcal{O}_i$ with $\Sigma U_i = \dim(\mathscr{M}_k)$. The simplest such non-trivial element of the equivariant s-cohomology is:

$$W_0 = \tfrac{1}{2}\operatorname{Tr}(\phi^2(x)), \quad U(W_0) = 4. \tag{72}$$

Note that $W_0 = sW = \tfrac{1}{2}s\operatorname{Tr}(\phi\omega - \tfrac{1}{3}\omega^3)$, so indeed $sW_0 = 0$, but nevertheless it is a non-trivial element of the equivariant s-cohomology, due to the ω dependence of W. We now need to verify explicitly that W_0 is indeed an invariant, by demonstrating that it does not depend on the coordinate $x \in M$. For this it is sufficient to show that dW_0 is s-trivial, i.e. there exists a 1-form W_1 such that $dW_0 = sW_1$. This is easily checked:

$$dW_0 = \operatorname{Tr}(\phi D\phi) = -s\operatorname{Tr}(\phi\psi) = sW_1, \quad W_1 = -\operatorname{Tr}(\phi\psi). \tag{73}$$

In this equation D is the covariant differential and ψ is the ghost 1-form $\psi_\mu dx^\mu$. From Eq. (73) we see that W_1 itself is not s-trivial, however, since it is a 1-form on M, we can integrate it over a 1-cycle γ_1, which by the use of Stokes' law, will show that $\oint_{\gamma_1} W_1$ is s-trivial:

$$s\oint_{\gamma_1} W_1 = \oint_{\gamma_1} dW_0 = \oint_{\partial\gamma_1} W_0 = 0. \tag{74}$$

However, in order for $\oint_{\gamma_1} W_1$ to have a chance to be an invariant, it should not depend on the particular choice of the (closed) 1-cycle γ_1, but only on its homology class. For this it is sufficient to show that the integral of sW_1 over the boundary of a 2-cell α_2, is zero. Again by Stokes' law, this is equivalent to demanding dW_1 to be s-trivial. One easily finds this to be the case:

$$dW_1 = -\mathrm{Tr}\,(\phi D\psi + D\phi \wedge \psi) = s\mathrm{Tr}\,(\tfrac{1}{2}\psi \wedge \psi - \phi F) = sW_2,$$

$$W_2 = \mathrm{Tr}\,(\tfrac{1}{2}\psi \wedge \psi - \phi F). \tag{75}$$

Now W_2 is a 2-form on M and its integral over a 2-cycle γ_2 (i.e. a closed two-dimensional cell) should be s-trivial, we leave this to the reader to verify by using Stokes' law. Again demanding that $\oint_{\gamma_2} W_2$ will depend only on the homology class of the 2-cycle γ_2, leads to the requirement that dW_2 is s-trivial, and sure enough things work out beautifully:

$$dW_2 = \mathrm{Tr}\,(D\psi \wedge \psi - D\phi \wedge F) = s\,\mathrm{Tr}\,(\psi \wedge F) = sW_3, \quad W_3 = \mathrm{Tr}\,(\psi \wedge F). \tag{76}$$

It starts to get boring, but hang on, we are almost at the end. Again W_3 itself would not lead to an observable but it is a 3-form on M and we can integrate over a 3-cycle γ_3, which in order to lead to an invariant should only depend on the homology of γ_3, leading to the condition that dW_3 is s-trivial. So for the last time:

$$dW_3 = \mathrm{Tr}\,(D\psi \wedge F) = -\tfrac{1}{2}s\,\mathrm{Tr}\,(F \wedge F) = sW_4, \quad W_4 = \mathrm{Tr}\,(F \wedge F). \tag{77}$$

We recognize that the 4-form is precisely proportional to the Pontryagin class and the integral over the only 4-cycle, which is the manifold M, is thus proportional to the topological charge and is clearly an invariant. And if one really wants to go to the bottom of it, the Bianchi identity $DF = 0$, will show that $s\int_M W_4 = 0$.

We have thus constructed a chain of observables $\mathcal{O}_i = \oint_{\gamma_i} W_i$, with $U_i = 4 - i$, where γ_0 is a point x and $\gamma_4 = M$. They form a map, which we will call the Donaldson map, from the homology $H_k(M, R)$ of the base manifold M into the equivariant s-cohomology of the field theory. As we observed in the previous Section around Eq. (65), these observables descent to objects on the moduli. Those objects are polynomials in the zero-modes θ_j (with coefficients that are functions on the moduli space) of a degree which is exactly the U quantum number of the operator in question. The θ_j can be naturally interpreted as differential forms on the moduli space $\mathcal{M}_k$ and therefore $\mathcal{O}_i$ will become a $(4-i)$-form on $\mathcal{M}_k$ after substituting in $\mathcal{O}_i \phi$ by Eq. (65), ψ by their zero-modes and A by the self-dual connections. This also means that s, after this restriction to the moduli space, should play the role of the exterior derivative on $\mathcal{M}_k$. Although this seems obvious, the author believes that this has not yet been established in a sufficiently clear way in the literature up to now, but let us here assume it to be the case. Then the Donaldson map is a map from $H_i(M, R)$ into $H^{4-i}(\mathcal{M}_k, R)$ and a Donaldson polynomial is simply the wedge product of these elements in the cohomology of the moduli space, such that the total degree is the dimension of the moduli space. The invariant is obtained by integrating over moduli space. If Φ^{γ_k} is the image of the Donaldson map for $\mathcal{O}_k = \oint_{\gamma_k} W_k$, then the Donaldson invariant is given by:

$$\int_{\mathcal{M}_k} \Phi^{\gamma_{k1}} \wedge \Phi^{\gamma_{k2}} \wedge \ldots \wedge \Phi^{\gamma_{km}}, \quad \sum_i (4 - k_i) = \dim(\mathcal{M}_k). \tag{78}$$

98

This gives very explicit formulas for the Donaldson invariants, which is the strength of Witten's [1] construction. Not surprisingly, there exists a topological version of quantum mechanics, which will give the Euler characteristic as an invariant [31]. It provides an interesting framework in which many of the above manipulations can be defined more rigorously.

I am grateful to Laurent Baulieu, Iz Singer and above all to Peter Braam for their stimulating seminars and discussions, which originally triggered my interest in this subject. I am especially grateful to Stephane Ouvry and Raymond Stora for generously involving me in a collaboration. I also thank Luis Alvarez-Gaumé, Danny Birmingham, Robbert Dijkgraaf, Ennio Gozzi, José Labastida, Jean-Michel Maillet, Antti Niemi and Silvio Sorella for preprints, comments and discussions and Raymond Stora for commenting on the manuscript. These lectures were prepared while visiting DESY during the month of May. I thank the Theory Group and in particular Martin Lüscher for the warm hospitality. Finally, it is a pleasure to thank the organizers of the 29th Kraków School of Theoretical Physics for inviting me again to Zakopane at such a historic time in Polish history.

REFERENCES

[1] E. Witten, *Comm. Math. Phys.* **117**, 353 (1988).

[2] S. Donaldson, *J. Diff. Geom.* **18**, 269 (1983); **26**, 397 (1987); Polynomial Invariants for smooth four-manifolds, Oxford preprint.

[3] E. Witten, *Comm. Math. Phys.* **118**, 411 (1988); E. Witten, *Phys. Lett.* **B206**, 601 (1988); J. Sonnenschein, Topological Quantum Field Theories, Moduli Spaces and Flat Gauge Connections, SLAC-preprint SLAC-PUB-4970, April 1989, and references therein.

[4] E. Witten, *Comm. Math. Phys.* **121**, 351 (1989); R. Dijkgraaf, E. Witten, Topological Gauge Theories and Group Cohomology, Princeton preprint IASSNS-HEP-89/33, June 1989.

[5] G. Moore, N. Seiberg, Taming the Conformal Zoo, Princeton preprint, IASSNS-HEP-89/6, January 1989.

[6] V. F. R. Jones, *Bull. AMS* **12**, 103 (1986); *Ann. Math.* **126**, 335 (1987).

[7] J. Fröhlich, Statistics of Fields, The Yang-Baxter Equation and the Theory of Knots and Links, Zürich preprint ETH-88-0173, March 1988; E. Witten, Gauge Theories and Integrable Lattice Models, Princeton preprint, IASSNS-HEP-89-11, February 1989.

[8] E. Witten, *Nucl. Phys.* **B202**, 253 (1982).

[9] E. Witten, *J. Diff. Geom.* **17**, 661 (1982).

[10] A. A. Belavin, A. M. Polyakov, A. S. Schwarz, Y. A. Tyupkin, *Phys. Lett.* **59B**, 85 (1975).

[11] M. F. Atiyah, V. Drinfield, N. Hitchin, Y. I. Manin, *Phys. Lett.* **A65**, 185 (1978).

[12] R. Friedman, J. W. Morgan, *J. Diff. Geom.* **27**, 297 (1988).

[13] A. Floer, *Comm. Math. Phys.* **118**, 215 (1988).

[14] P. Braam, Floer Homology Groups for Homology Three-Spheres, Utrecht preprint, Math. 484, November 1987.

[15] M. F. Atiyah, *New Invariants of 3 and 4 Dimensional Manifolds*, in: *The Mathematical Heritage of Hermann Weyl*, ed. R. Wells, A.M.S., Providence 1988.

[16] L. Baulieu, I. Singer, Topological Yang-Mills Symmetry, preprint LPTHE-88/18, and talks presented at the Annecy Meeting on Conformal Field Theory, Annecy, France, March 1988; J. M. F. Labastida, M. Pernici, *Phys. Lett.* **212B**, 56 (1988).

[17] H. Cartan, in: *Colloque de Topologie (Espaces fibrés)*, C.B.R.M. Bruxelles, p. 57.

[18] S. Ouvry, R. Stora, P. van Baal, *Phys. Lett.* **220B**, 159 (1989).

[19] R. Bott, in: *Recent Developments in Gauge Theories*, ed. G. 't Hooft, e.a., Plenum, New York 1980; J. Milnor, *Morse Theory, Ann. of Math. Studies* **51**, Princeton U.P., 1973.

[20] M. F. Atiyah, I. Singer, *Ann. of Math.* **87**, 484 (1968).

[21] S. Coleman, in: *The Whys of Subnuclear Physics*, ed. A. Zichichi, Plenum, New York 1977.

[22] L. Schulman, *Techniques and Applications of Path Integrals*, Wiley, New York 1981.

[23] H. L. Cycon, R. G. Froese, W. Kirsch, B. Simon, *Schrödinger Operators: with Application to Quantum Mechanics and Global Geometry*, Springer, Berlin 1987.

[24] A. A. Migdal, *Ann. Phys.* **126**, 279 (1980).

[25] P. van Baal, *Acta Phys. Pol.* **B20**, 295 (1988).

[26] D. Birmingham, M. Rakowski, G. Thompson, Renormalization of Topological Field Theory, Trieste preprint IC/88/387, November 1988.

[27] G. 't Hooft, *Phys. Rev.* **D14**, 3432 (1976); *Phys. Rep.* **142**, 357 (1986).

[28] J. H. Horne, *Nucl. Phys.* **B318**, 22 (1989).

[29] R. Brooks, D. Montano, J. Sonnenschein, SLAC preprint SLAC-PUB-4630, May 1988; S. P. Sorella, CERN preprint TH-5284/89, February 1989; J. M. Maillet, A. Niemi, CERN preprint TH-5322/89, March 1989.

[30] R. Myers, Gauge Fixing Topological Yang-Mills, Santa Barbara preprint NSF-ITP-89-45.

[31] J. M. F. Labastida, Morse Theory Interpretation of Topological Quantum Field Theories, CERN preprint TH-5240/88, November 1988.

Nuclear Physics B 369 (1992) 259–275
North-Holland

NUCLEAR
PHYSICS B

More (thoughts on) Gribov copies

Pierre van Baal *

Institute for Theoretical Physics, Princetonplein 5, P.O. Box 80006, 3508 TA Utrecht, The Netherlands

Received 3 June 1991
(Revised 15 August 1991)
Accepted for publication 19 September 1991

Inspired by an unambiguous observation of Gribov copies on the lattice an old continuum example due to Henyey is re-analysed in terms of bifurcations at the Gribov horizon. This gives yet another proof that there are in general Gribov copies within the horizon. Using results by Semenov-Tyan-Shanskii, Franke, Dell'Antonio and Zwanziger on a possible fundamental modular domain we argue that its boundary has both Gribov copies and points that coincide with the Gribov horizon.

1. Introduction

One way strong interactions in non-abelian gauge theories manifest themselves is by a spreading of the wave functionals over such a large portion of configuration space, that the issue of Gribov copies [1] can no longer be ignored. To have a chance to understand the non-perturbative dynamics of the low-energy physics a better understanding of the physical configuration space is essential, particularly for describing how physics depends on the quantum numbers associated to the homotopically non-trivial gauge transformations. These issues are intimately connected to the dynamical questions involving tunnelling between different vacua. To make this a little more explicit we review an old argument due to Jackiw et al. [2] in somewhat more generality. (In much of this paper we will be primarily interested in the hamiltonian formulation, in which case the gauge fields are defined over a three-dimensional space, but some of the discussion is relevant for the four-dimensional situation too.) Let g be a homotopically non-trivial gauge transformation, such that $A_i = [g]\Theta_i = -ig\partial_i g^{-1}$ satisfies the Coulomb gauge $\partial_i A_i = 0$ (where throughout this paper Θ will denote the zero connection). Then the Faddeev–Popov operator (our gauge fields will be hermitian Lie algebra elements and $\mathrm{ad}\, X(Y) \equiv [X, Y]$)

$$\mathrm{FP}(A) = -\partial_i D_i(A) = -\partial_i(\partial_i + i\, \mathrm{ad}\, A_i) \tag{1}$$

* KNAW fellow.

has a vanishing eigenvalue. The argument is simple: Constant gauge transformations leave the gauge condition invariant. The obviously related zero modes, that correspond to constant Lie algebra elements, are removed by defining the theory modulo constant gauge transformation and demanding the wave functionals to be trivial representations of the gauge group G. As g is a homotopically non-trivial gauge transformation, gXg^{-1} cannot be constant for all constant Lie algebra elements X, but it is nevertheless a zero mode of FP(A). Thus the Gribov horizon will contain gauge copies of the classical vacuum, making it more urgent to understand the problem of gauge fixing.

It seemed therefore that Gribov copies are always associated to homotopically non-trivial gauge transformations. It was first demonstrated in an example by Henyey [3] that this is not always true. But probably the most simple example is provided by gauge fields on the torus in the abelian zero-momentum sector. For definiteness let us take G = SU(2) (with the Pauli matrices τ_i as generators of the algebra and L the size of the torus) and $A_i = (C_i/2L)\tau_3$, then the gauge transformation $g_{(k)} = \exp(-\pi i x_k \tau_3/L)$ maps C_k to $C_k + 2\pi$. As $g_{(k)}$ is anti-periodic it is homotopically non-trivial (they are 't Hooft's twisted gauge transformations [4]). However, $g_{(k)}^2$ will map C_k to $C_k + 4\pi$. As this gauge transformation is periodic and abelian, it is contractable to the identity and provides an example of gauge copies by homotopically trivial gauge transformations. This example also proves that in general there are gauge copies *within* the Gribov horizon, which occurs [5,6] at $|C_k| = 2\pi$. For example $\mathbf{C} = \mathbf{D} + (-\pi, 0, 0)$ is a gauge copy of $\mathbf{C} = \mathbf{D} + (\pi, 0, 0)$ and both occur inside the horizon if $|D_i| < \pi$. (It would seem the simplest case arises by taking $\mathbf{D} = \mathbf{0}$, however, in that case both configurations are related by a constant gauge transformation, which are to be divided out.)

Still, one might conjecture (as in the above example) that such copies within the horizon are always associated to homotopically non-trivial gauge transformations. Henyey's [3] example will be shown to provide an explicit example of gauge copies within the horizon, that are related by homotopically trivial gauge transformations (and that are related to a copy outside the horizon). In sect. 1 we will first give an argument based on Morse theory that in certain cases shows how, when moving from inside the horizon to outside (where the lowest eigenvalue of FP(A) flips sign) two additional copies will be created for which FP(A) is positive and which are therefore inside the horizon. As these copies coalesce at the horizon, the gauge transformation relating the two can be deformed to the identity. Our interest in this example was stirred by an unambiguous observation of Gribov copies on a lattice [7]. The structure of the relevant gauge transformations was analysed and some of them would have non-zero winding number, whereas others were typically of the form expected from ref. [3].

That there are Gribov copies within the Gribov horizon was first demonstrated by Semenov-Tyan-Shanskii and Franke [8] and analysed in more detail by Dell'Antonio and Zwanziger [9,10]. These authors also provide a recipe that will (almost) uniquely fix an element of the gauge orbit. In sect. 4 we make more precise in what sense this forms a fundamental modular domain. The interior is a convex subset of the transverse connections, that contains no Gribov copies. The boundary will contain Gribov copies, but can at most coincide with the horizon on a subset of the boundary of codimension one. The gauge transformations that relate the copies provides the identifications at the boundary that makes it into a fundamental modular domain for configuration space. These considerations are both relevant for the hamiltonian formulation and for the recent path integral formulation on this fundamental modular domain by Zwanziger [11], Parinello and Jona-Lasinio [12]. In the case of SU(2) gauge theory on the torus the intersection of this fundamental modular domain with the constant abelian gauge fields is given by $|C_k| \leqslant \pi$ and $g_{(k)}$ provides the identification of the (opposite) points on the boundary. In this case the boundary is regular, but we will argue that in general so-called singular boundary points (i.e. those that coincide with the horizon) do occur.

It should be mentioned that for the torus in the presence of fields in the fundamental representation (quarks) only periodic gauge transformations are allowed. In that case it is easily seen that the intersection of the fundamental modular domain with the constant abelian gauge fields is given by the domain $|C_k| \leqslant 2\pi$, whose boundary coincides with the Gribov horizon. Nevertheless, this domain is a covering space of the domain $|C_k| \leqslant \pi$ and fields in the fundamental representation can be defined on the latter if allowed to be multi-valued (an alternative formulation in terms of coordinate patches is described in ref. [6]). Later we will see that even for pure gauge theories, the wave functionals are still multi-valued, but in that case the multi-valuedness is restricted to a phase factor determined by the periodicity condition that arises as a consequence of the identifications at the boundary of the fundamental modular domain. The relevant "Bloch momenta" give the topological quantum numbers.

In sect. 5 we conclude with a discussion on the implications for the hamiltonian formulation. We also re-emphasize [13] that for supersymmetric gauge theories on a torus the Gribov copies are essential for a proper evaluation of the Witten index, despite the fact that one can take the coupling constant (and hence the volume) as small as one likes. The problem remains open (despite the claim made in ref. [14]) and turns out to be related to the non-trivial issue of constructing Dirac vacuum bundles that incorporate the identifications associated to the Gribov copies. It seems to require one to introduce a multi-valued vacuum wave functional that incorporates the non-conservation of chiral particle number, induced by the chiral U(1) anomaly [15]. Note that this multi-valuedness is again of a different nature as discussed in the previous paragraph (the fermions in supersymmetric gauge theories are in the adjoint representation). These issues certainly deserve further study. However, this paper will concentrate on the pure gauge sector.

 P. van Baal / More (thoughts on) Gribov copies

2. Morse theory and bifurcation of copies at the horizon

The Coulomb gauge condition $\partial_i A_i = 0$ can be formulated in terms of an action principle [8,9,16] (M is the manifold over which the gauge theory is defined, which we will mostly take three dimensional and compact. For $\mathbb{R}^3$ we assume the fields to have an asymptotics that allows compactification to the three sphere.)

$$I(g; A) \equiv \|[g]A\|^2 = \int_M \text{Tr}\big(\{[g]A_i\}^2\big) = \int_M \text{Tr}\big(\{A_i + ig^{-1}\partial_i g\}^2\big), \qquad (2)$$

whose critical points satisfy the gauge condition and whose hessian is precisely the Faddeev–Popov operator. This is most easily established by observing that

$$I(hg; A) = I(h; [g]A). \qquad (3)$$

Writing $h = e^X$ and using

$$e^{-X}\partial_i\, e^X = \frac{1 - \exp(-\text{ad } X)}{\text{ad } X}(\partial_i X)$$

$$= \partial_i X + \tfrac{1}{2}[\partial_i X,\, X] + \tfrac{1}{6}[[\partial_i X,\, X],\, X] + \ldots, \qquad (4)$$

one finds

$$I(e^X; A) = \|A\|^2 - 2i\int_M \text{Tr}(X\partial_i A_i) + \int_M \text{Tr}(X^\dagger \text{FP}(A)X)$$

$$+ \tfrac{1}{3}i\int_M \text{Tr}(X[[A_i,\, X],\, \partial_i X]) + \ldots. \qquad (5)$$

We note that for $g \in G$ a constant group element $I(g; A) = I(1; A)$, thus $F_A(g)$ defined by

$$F_A(g) = I(g; A) \qquad (6)$$

is for generic A a Morse function [18] on $\mathscr{G}/G$, where $\mathscr{G}$ is the group of local gauge transformations, i.e. the set of functions $g(x)$ that map M to G. (For Morse theory in relation to supersymmetric quantum mchanics, see ref. [19].) The critical points of the Morse function are precisely the gauge functions g for which $[g]A$ is transverse (i.e. $\partial_i[g]A_i = 0$.) The hessian of the Morse function at the critical point is precisely the Faddeev–Popov operator $\text{FP}([g]A)$. The Morse index μ for the critical points, is defined as the number of negative eigenvalues of the hessian. As F_A depends continuously on A (in the norm implicitly defined in eq. (2)), the alternating sum of the Morse indices over the critical points (which is the Euler characteristic of the manifold on which F_A is defined) is conserved.

This argument needs to be treated with caution, as $\mathscr{G}/G$ is not compact and there will in general be infinitely many critical points. Nevertheless, the property is topological in nature and is invariant under continuous deformations of F_A. If we will be interested in a local neighborhood of a particular critical point, we can if necessary deform F_A such that when changing A, there will only happen something to the critical points in this particular neighborhood. As the type of none of the other critical points will change, the change is finite and the alternating sum over the critical points in the chosen neighborhood will be conserved. However, in general singularities occur at reducible connections (these are connections for which there exists a non-trivial gauge function g, such that $[g]A = A$). Thus $\mathscr{G}$ does not act freely and although dividing out the constant gauge transformations will remove some of the singularities, it can be shown that $\mathscr{G}/G$ does not act freely either. Nevertheless, the presence of the remaining singularities does not affect our arguments. (For example, for SU(2) gauge theory on S^3 it can be shown that up to gauge transformations $A = 0$ is the only reducible connection.)

The Gribov region Ω^0 is defined as the set of transverse gauge fields for which the Faddeev–Popov operator is (strictly) positive. Its boundary $(\partial\Omega,\ \Omega \equiv \Omega^0 \cup \partial\Omega)$ is the Gribov horizon, where the lowest eigenvalue vanishes. It is well known that Ω^0 is convex [8,17]. This is essentially because FP(A) is linear in A. Thus if $A_{(1)}$ and $A_{(2)}$ are in Ω^0,

$$\mathrm{FP}\big(sA_{(1)} + (1 - s)A_{(2)}\big) = s\,\mathrm{FP}\big(A_{(1)}\big) + (1 - s)\mathrm{FP}\big(A_{(2)}\big) > 0 \qquad (7)$$

for all $s \in [0, 1]$. Thus, if A is on the horizon, the Morse index $\mu(sA)$ will be zero for $0 \leqslant s < 1$ and one for $s > 1$, at least in a small neighborhood of $s = 1$. Let us assume that as s approaches one from below, there is a sufficiently small neighborhood of the identity in $\mathscr{G}/G$ such that $F_A(g)$ has no further extrema, except the one at $g = 1$. The only way the alternating sum of Morse indices can be preserved is (when s increases beyond one) if two additional extrema are created, that are both local minima, see fig. 1. Thus for $s > 1$ F_A has extrema at $g_0(s) = 1$, $g_1(s)$ and $g_2(s)$, where $\lim_{s \to 1} g_1(s) = \lim_{s \to 1} g_2(s) = 1$, such that $g_{(1,2)}$ are homotopically trivial gauge transformations. Clearly $[g_{(1,2)}(s)]sA \in \Omega^0$ and are thus points inside the Gribov horizon (inside and outside is of course well defined since Ω is convex) that are related by a homotopically trivial gauge transformation.

Thus under the assumption that $g = 1$ is the only extremum of F_A in a sufficiently small neighborhood in $\mathscr{G}/G$ there will be a bifurcation at the horizon of solutions to the Coulomb gauge condition into two stable and one unstable solution. The assumption just mentioned, however, is easily seen to imply that the third-order term in eq. (5) has to vanish at the horizon for X the zero mode of the hessian (apart from a factor $\frac{1}{3}$ this term coincides with the definition of μ in ref. [9]). Generically this will not be the case, in which instance there will be a saddle point that coalesces with the local minimum at the horizon and the situation is as

 P. van Baal / More (thoughts on) Gribov copies

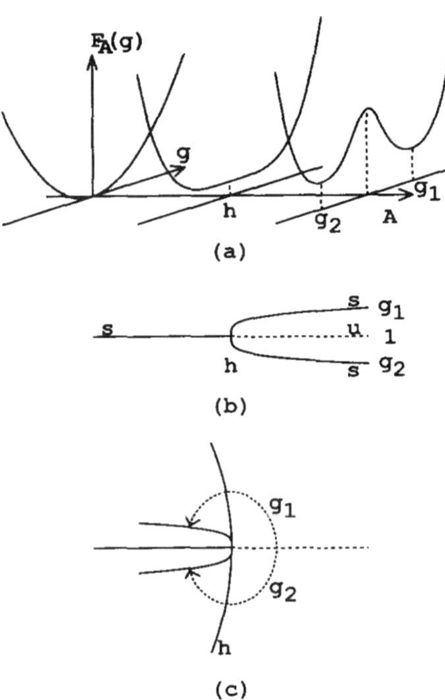

Fig. 1. Here we sketch the bifurcation at the horizon by three different representations. In (a) we show what happens to $F_A(g)$ when we pass the horizon (h), (b) is the well-known bifurcation picture, where s stands for stable (a local minimum) and u stands for unstable (a saddle point). Finally, in (c) we sketch the situation regarding the Gribov region.

sketched in fig. 2, i.e. the saddle point with Morse index 1 will turn into a local minimum, such that the alternating sum of the Morse indices is again conserved. Note that, for $s > 1$ this local minimum is a homotopically trivial copy of the saddle point sA, such that the region just outside the Gribov horizon has copies just inside the Gribov horizon, which is what was already proved by Gribov [1]. In refs. [8,9] it is, however, proven that if the third-order term of eq. (5) at the horizon is non-zero for X the zero mode of the hessian, then sA for $0 \leqslant 1-s < \epsilon$ (with ϵ sufficiently small) cannot be an absolute minimum of F_{sA} and hence also in this case there will be a gauge copy within the horizon (in general by a large, possibly homotopically non-trivial gauge transformation).

3. The explicit example

As the Morse theory arguments of sect. 2 are formal and not easily made rigorous, it is useful to consider an explicit example, where all the features of the bifurcation can be checked in detail. We will consider Henyey's example [3] of an

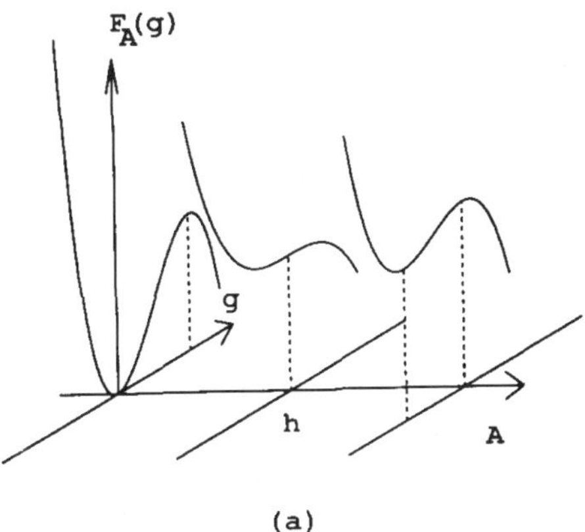

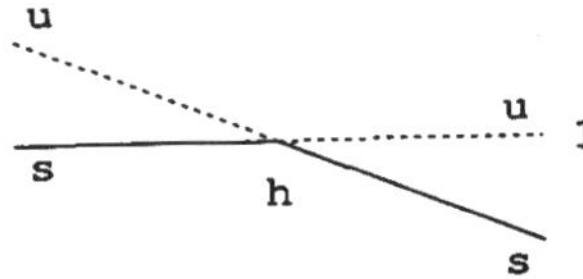

Fig. 2. Here we sketch the generic situation at the horizon (h), when a (local) minimum coalesces with a saddle point.

axial symmetric abelian gauge field for SU(2) gauge theory on $\mathbb{R}^3$, which obviously satisfies $\partial_i A_i = 0$,

$$A = a(r, \theta)\hat{\phi}\tau_3, \qquad \hat{\phi} = (-\sin\phi, \cos\phi, 0),$$

where (r, θ, ϕ) are spherical coordinates. The following gauge transformation:

$$g = \exp(\alpha(\cos\phi\tau_1 + \sin\phi\tau_2)) \tag{8}$$

will leave the Coulomb gauge condition invariant if and only if [3]

$$2r^2 \sin^2\theta\partial_i^2\alpha + \sin(2\alpha)(2a(r, \theta)r\sin\theta - 1) = 0. \tag{9}$$

Rather then solving this equation for α, Henyey's strategy was to choose α and solve for $a(r, \theta)$. He took

$$\alpha(r, \theta) = b(r)r\sin\theta, \tag{10}$$

which through eq. (9) yields

$$a(r, \theta) = \frac{1}{2r\sin\theta} - \frac{b(r) + r^2\sin^2\theta\left(\dfrac{\mathrm{d}^2b(r)}{\mathrm{d}r^2} + \dfrac{4}{r}\dfrac{\mathrm{d}b(r)}{\mathrm{d}r}\right)}{\sin(2rb(r)\sin\theta)}, \tag{11}$$

defining a proper gauge field, provided $b(r)$ satisfies the following three conditions [3]: (i) it is regular at the origin with vanishing first derivative ($\mathrm{d}b(0)/\mathrm{d}r = 0$); (ii) $2rb(r) < \pi$ for all r; (iii) $r^3 b(r)$ is bounded.

We note, as expected, that g comes always into pairs as a is invariant under a change of sign of b. Thus this example $g_1 = g_2^{-1} = g$, cf. fig. 1. To investigate the bifurcation structure of the Gribov copies, we replace b by βb, with β a constant. It satisfies the above-mentioned conditions provided $|\beta| \leqslant 1$. Explicitly we therefore have

$$A(\beta) = a(\beta)\hat{\phi}\tau_3,$$

$$a(\beta) = \frac{1}{2r \sin \theta} - \frac{b + r^2 \sin^2 \theta \left(\dfrac{\mathrm{d}^2 b}{\mathrm{d}r^2} + \dfrac{4}{r}\dfrac{\mathrm{d}b}{\mathrm{d}r} \right)}{\beta^{-1} \sin(2r\beta b \sin \theta)},$$

$$g(\beta) = \exp\left(i\beta rb \sin \theta \left(e^{i\phi}\tau_- + e^{-i\phi}\tau_+ \right) \right). \tag{12}$$

As $g(0)$ is the identity, the two gauge copies $[g(\pm\beta)]A(\beta)$ will coalesce at $\beta = 0$, and it should be such that $FP(A(0))$ has a zero mode. We even know in advance what this zero mode should be

$$X_0 \propto \frac{\partial g}{\partial \beta}(0) = br \sin \theta \left(e^{-i\phi}\tau_+ + e^{i\phi}\tau_- \right). \tag{13}$$

This is easily checked explicitly. Introducing the function $f(r)$ by

$$f(r) = -\frac{a(\beta = 0)}{r \sin \theta} = \frac{1}{2b}\left(\frac{\mathrm{d}^2 b}{\mathrm{d}r^2} + \frac{4}{r}\frac{\mathrm{d}b}{\mathrm{d}r} \right), \tag{14}$$

one finds

$$FP(A(0))X_0 = -\left(\partial_i^2 X_0 - if(r)\left[\tau_3, \partial_\phi X_0 \right] \right) = \left(-\partial_i^2 + 2f(r) \right)X_0 = 0.$$

We now address the issue of the spectrum of $FP(A(\beta))$ for $\beta \neq 0$. We do this in perturbation theory to order β^2, restricting ourselves to the mode that coincides with X_0 at $\beta = 0$. As

$$FP(A(\beta)) = FP(A(0)) - i\beta^2 \frac{\Delta a}{r \sin \theta}\, \mathrm{ad}\,\tau_3\,\partial_\phi + \ldots, \tag{15}$$

with

$$\Delta a = \tfrac{1}{2}\partial_\beta^2 a(\beta = 0) = -\tfrac{1}{6}b^2 r \sin \theta (2 + fr^2 \sin^2 \theta), \tag{16}$$

the relevant eigenvalue is obtained from first-order perturbation theory. With the innerproduct $\langle X \mid Y \rangle = \int_M \text{Tr}(X^\dagger Y)$ this becomes (using some partial integrations in the last step) to second order in β

$$\lambda = \frac{\beta^2}{\langle X_0 \mid X_0 \rangle} \langle X_0 \mid -\frac{i\Delta a}{r \sin \theta} \text{ad } \tau_3 \, \partial_\phi \mid X_0 \rangle$$

$$= \frac{\beta^2}{3} \frac{\int dr \, d\theta \, r^4 b^4 \sin^3 \theta (2 + f(r) r^2 \sin^2 \theta)}{\int dr \, d\theta \, r^4 b^2 \sin^3 \theta}$$

$$= -\frac{2\beta^2}{5} \frac{\int dr \, r^2 \left(r^2 b^4 + \left\{ \frac{dr^2 b^2}{dr} \right\}^2 \right)}{\int dr \, r^4 b^2}. \tag{17}$$

We thus confirm that $A(\beta)$ always corresponds to an unstable solution to the gauge condition.

It remains to show that the corresponding eigenvalues for the Gribov copies, $[g(\beta)]A(\beta)$, stay positive. Explicitly we find

$$[g(\beta)]A(\beta) = A_+(\beta) + A_-(\beta),$$

$$A_+(\beta) = \left\{ a(\beta) + \sin^2(\beta\alpha) \left(\frac{1}{r \sin \theta} - 2a(\beta) \right) \right\} \hat{\phi} \tau_3,$$

$$A_-(\beta) = -\beta \left\{ \frac{\sin(2\beta\alpha)}{2\beta\alpha} D(A(\beta))(X_0) + \left(1 - \frac{\sin(2\beta\alpha)}{2\beta\alpha} \right) X_0 \, \partial \ln(\alpha) \right\}. \tag{18}$$

One easily verifies that $A_-(\beta)$ does not contribute to second order in β, since

$$[A_-(\beta), \partial X_0] = 0. \tag{19}$$

Thus to second order in β the eigenvalue λ' of FP($[g(\beta)]A(\beta)$), that reduces to zero at $\beta = 0$, is given by

$$\lambda' = \frac{\beta^2}{\langle X_0 \mid X_0 \rangle} \langle X_0 \mid \frac{2i\Delta a}{r \sin \theta} \text{ad } \tau_3 \, \partial_\phi \mid X_0 \rangle = -2\lambda, \tag{20}$$

where we used the remarkable coincidence that $(A_+ \equiv a_+ \hat{\phi} \tau_3)$

$$a_+(\beta) = a(\beta) - 3\frac{\sin^2(\beta\alpha)}{\alpha^2}\Delta a = a(\beta = 0) - 2\beta^2 \Delta a + O(\beta^4). \tag{21}$$

We thus confirm the bifurcation picture, but in general $A(\beta = 0)$ need not be on the Gribov horizon as $\mathrm{FP}(A(\beta = 0))$ might have negative eigenvalues. To see this note that

$$\mathrm{FP}(A(0)) = -\partial_i^2 + if(r)\,\mathrm{ad}\,\tau_3\,\partial_\phi \tag{22}$$

commutes with $\mathrm{ad}\,\tau_3$, L_z and L^2. Thus we can decompose the eigenfunctions as

$$X = X_+(r)Y_{lm}(\theta, \phi)\tau_+ + X_+^*(r)Y_{lm}^*(\theta, \phi)\tau_- + X_3(r)Y_{lm}(\theta, \phi)\tau_3. \tag{23}$$

Restricted to X_3 the hessian is positive definite and only the "charged" sector will be relevant, for which the hessian reduces to

$$H_{l,m} = -\frac{1}{r^2}\frac{\mathrm{d}}{\mathrm{d}r}r^2\frac{\mathrm{d}}{\mathrm{d}r} + \left(\frac{l(l+1)}{r^2} - 2mf(r)\right). \tag{24}$$

For given m and l this is a one-dimensional potential problem. It has a zero eigenvalue for $l = -m = 1$, with $X_+ = b$. Thus, if b has nodes there are lower (and hence negative) eigenvalues.

As an example without a node, we take [3]

$$b(r) = K\left(r^2 + r_0^2\right)^{-3/2}, \qquad f(r) = -\frac{15r_0^2}{\left(r_0^2 + r^2\right)^2}. \tag{25}$$

The hamiltonian (24) can only have negative eigenvalues if its potential can become negative. This only leaves $(l, m) = (2, -2)$ and $(1, -1)$. As b has no nodes $H_{1,-1} \geqslant 0$, whereas $H_{2,-2} - H_{1,-1} \geqslant 0$ proves that also $H_{2,-2} \geqslant 0$. In conclusion, eq. (25) provides an example for which $A(\beta = 0)$ is at the Gribov horizon, which concludes this section.

4. The fundamental modular domain

Let us consider the set

$$\Lambda = \left\{A \mid F_A(g) \geqslant F_A(1), \quad \forall g \in \mathscr{G}\right\}. \tag{26}$$

Clearly one has Λ a subset of Ω. Furthermore, it was proven by Semenov-Tyan-Shanskii and Franke [8] and Dell'Antonio and Zwanziger [10] that Λ covers *all* gauge orbits. That is, given a connection A (with finite norm $\|A\|$), there exists a $g \in \mathscr{G}^c$ for which $F_A(g)$ is at its absolute minimum. The space $\mathscr{G}^c$ is the completion of $\mathscr{G}$ with respect to the norm $\|\delta g\|_1^2 = \|\delta g\|^2 + \|d\delta g\|^2$, in which δg is viewed as a complex $N \times N$ matrix (in ref. [8] one works on $\mathbb{R}^3$ and a slightly different norm is used, so as to eliminate the constant gauge transformations). Let $\Lambda^0 \subset \Lambda$ be the set where the minimum is unique, i.e. if $A \in \Lambda^0$ then $F_A(g) > F_A(1)$ for *all* non-constant g. Both Λ^0 and Λ are convex [8,9], for let $A_{(1)}, A_{(2)} \in \Lambda^0(\Lambda)$ then

$$\|[g](sA_{(1)} + (1-s)A_{(2)})\|^2 - \|sA_{(1)} + (1-s)A_{(2)}\|^2$$

$$= s\left(\|[g]A_{(1)}\|^2 - \|A_{(1)}\|^2\right) + (1-s)\left(\|[g]A_{(2)}\|^2 - \|A_{(2)}\|^2\right) \quad (27)$$

shows that $sA_{(1)} + (1-s)A_{(2)} \in \Lambda^0(\Lambda)$ for $s \in [0, 1]$. Also clearly $A = \Theta = 0 \in \Lambda^0$. In ref. [9] it was proved that the boundary of Λ^0 ($\partial\Lambda^0$) is contained in Λ, and that Λ is closed (apply the "Frist step" lemma [9] to Λ^0 and Λ), which is basically a continuity argument. Alternatively, take $A \in \Lambda - \Lambda^0$, and use eq. (27) for $A_{(1)} = 0$ and $A_{(2)} = A$, which implies that $sA \in \Lambda^0$ for all s, $0 \leqslant s < 1$ and hence $A \in \partial\Lambda^0$. Thus the boundary still exists of transverse gauge fields A such that $F_A(g)$ reaches its absolute minimum at $g = 1$, but this minimum need not be unique or might be on the Gribov horizon.

To analyse these two options, take $A \in \Omega - \Lambda$, then the ray sA will cross the boundary of Λ where necessarily an absolute minimum turns into a relative minimum. At the boundary $F_A(g)$ therefore has degenerate absolute minima, that are related by in general large gauge transformations. We will call these points on $\partial\Lambda$ *regular boundary points*. The remaining points of the boundary will necessarily be on the Gribov horizon and are called *singular boundary points*. That the set of regular boundary points is non-empty is easily established by taking the example of $A = [g]0$ for g homotopically non-trivial, such that A is on the Gribov horizon. The path sA will pass $\partial\Lambda$ at (necessarily) a regular point (as the absolute minimum of F_A does not occur at constant gauge functions, but at g^{-1}, where it vanishes). The gauge transformation that relates the two copies at the boundary of Λ is essentially g^{-1} and is thus homotopically non-trivial (as mentioned before, in the torus example the fundamental modular domain restricted to the abelian constant modes is given by $|C_k| \leqslant \pi$, modulo the action of the Weyl group $\mathbf{C} \to -\mathbf{C}$ and all boundary points are easily seen to be regular [5,6]). Strictly speaking therefore, also Λ is *not* a fundamental modular domain. Yet it will be, once we have appropriately identified the boundary points. It is these boundary identifications that will give the fundamental modular domain the topology of the full configuration space which is the basis of Singer's [20] argument why for gauge theories on a compact three- or four-dimensional manifold M, there necessarily have to be gauge copies.

Observe that homotopical non-trivial gauge transformations are in one-to-one correspondence with non-contractable loops in configuration space, which give rise to conserved quantum numbers. The quantum numbers are like the Bloch momenta in a periodic potential and have to be representations of the homotopy group of gauge transformations. On the fundamental modular domain the non-contractable loops arise through identifications of boundary points (as is quite explicit for the torus in the zero-momentum sector). Although slightly more hidden, the fundamental modular domain will therefore contain all the information relevant for the topological quantum numbers (i.e. it does not have to be "put in by hand"). Sufficient accurate knowledge of the boundary identifications will allow, however, for an efficient and natural projection on the various superselection sectors (i.e. by choosing the appropriate "Bloch wave functionals"). All these features were at the heart of the finite-volume analysis on the torus [5] and we see that they can in principle naturally be extended to the full theory, thereby including the desired θ-dependence. In ref. [6] we proposed formulating the hamiltonian theory on coordinate patches with homotopically non-trivial gauge transformations as transition functions. We can shrink these patches almost to Λ (and their associated gauge copies, with the homotopically non-trivial gauge transformations that relate the inequivalent classical vacua). If there would be no singular boundary points we would avoid any points on the Gribov horizon, thereby defining open sets that do cover the whole configuration space. These two formulations are therefore equivalent.

However, it is essential to note that the topology of the configuration space is *not* described entirely by non-contractable loops. One also needs to consider non-contractable spheres of any dimension. It is the non-contractable spheres that in general will be responsible for singular boundary points in the fundamental modular domain. As the interior is convex, non-contractable d-spheres can only arise if the boundary contains a $(d-1)$-sphere on which all points are identified. These correspond to gauge orbits for which F_A is degenerate along this $(d-1)$-sphere embedded in $\mathscr{G}$. Thus, these A necessarily coincide with the horizon. One can introduce a regular coordinate patche in the neighborhood of these singular points to eliminate the singularities, as observed by Singer and Nahm [20,21]. In this case, though, transition functions can not be described in terms of gauge transformations [6]. For example, when $G = SU(2)$ these non-contractable spheres do actually occur [20]. Thus it seems that we cannot avoid singular boundary points (as was suggested by fig. 1 of ref. [11]). We do not claim that our arguments present a proof for the existence of singular boundary points, as we implicitly assumed that the topology of configuration space is unaffected by the completion with respect to the norm $\| A \|$. These issues can be quite intricate, as for example the winding number of a C^1 gauge transformation, when defined through

$$\nu(g) = \frac{1}{24\pi^2} \int_M \mathrm{Tr}\left((g^{-1}\, \mathrm{d}g)^3 \right), \tag{28}$$

is not continuous in the norm $\|\delta g\|_1$. However, continuity is sufficient to define homotopy types. Relevant for this issue is the Sobolev embedding theorem $W_p^m \subset C^k$ for $k < m - n/p$ (see e.g. ref. [22]), where n is the dimension of the manifold, W_p^m is the Sobolev space of functions for which the first m distributional derivatives are in L^p, and C^k is the set of k times continuously differentiable functions. In the one-dimensional case, functions in W_p^m for $p > 1$, $m \geqslant 1$ are guaranteed to be continuous (as can be easily deduced explicitly by using Hölder's inequality). In the present case $n = 3$, $m = 1$ and $p = 2$, which unfortunately does *not* imply continuity. Indeed, for example, one can construct a series of maps $g_n : S^3 \cong \mathrm{SU}(2) \to \mathrm{SU}(2)$ that for all n have winding number zero but converges in the norm $\|\delta g\|_1$ to the identity map with winding number one. Yet, one must remember that the gauge fields in the fundamental modular domain satisfy the Coulomb condition (in the weak sense) from which one might deduce stronger smoothness properties. Thus, it is possible that using more sophisticated results from functional analysis will allow one to make stronger claims than we are willing to commit ourselves to here. One should address these issues primarily in the light of the physically more relevant dynamical questions and in that context we certainly intend to come back to this in the future.

In conclusion, it might be that the "hole" in configuration space, due to a non-contractable loop or sphere, is of zero size in the norm $\|A\|$. We consider this unlikely, but cannot exclude it. Actually, it might be the mechanism through which lattice gauge theories in the continuum limit reproduce the various topological sectors associated with the winding numbers. By this we mean that by demanding the fields to be "smooth", we will "cut" the necessary holes in configuration space. This requires an appropriate understanding of what it means to take the continuum limit, for which the "dislocations" [23] play an important role. Concerning gauge fixing in lattice gauge theory we only wish to remark that one can similarly impose a Coulomb (or Landau) type gauge through an action principle [24]. Our statement that Λ requires identifications at the boundary is equally valid in this case. Also one has to realize that, although there are no homotopically non-trivial gauge transformations associated with the winding number, twisted gauge transformations [4] are still well defined and cannot be deformed to the identity.

5. Discussion

We have reconsidered the proposed [8,10–12] fundamental modular domain Λ consisting of the absolute minima of the norm $\|[g]A\|$ on a gauge orbit and shown it is a fundamental modular domain provided *necessary* gauge identifications at the boundary are taken into account. As these identifications determine uniquely the topology of the configuration space, it can be argued, due to the presence of non-contractable spheres in the configuration space (this is true both for three- and four-dimensional compact manifolds M over which the gauge theory is defined) that in general there will be so-called singular boundary points, on which the Faddeev–Popov determinant will vanish. One can still attempt to formulate the standard hamiltonian [25] on <u>this fundamental</u> modular domain. Usually one rescales the wave functional with $\sqrt{\det'(\mathrm{FP}(A))}$ which will be strictly positive everywhere, expect at a subset of the boundary to the fundamental modular domain of codimension 1, where its vanishing is associated to a coordinate singularity due to a non-contractable sphere in configuration space. As topological quantum numbers are only associated to non-contractable loops, it might be that it is sufficient to simply demand the (rescaled) wave functional to vanish at the singular boundary points. This requires further study as subtle effects can complicate the issue, especially in the presence of fermions. We only need to remind the reader of the global SU(2) anomaly [26], which on a three-dimensional manifold M will be associated with a two-dimensional non-contractable sphere in configuration space. In this context we can recommend ref. [27] for a clear description of the issue of anomalies in the hamiltonian formulation.

The issue of gauge copies has played an important role in the analysis of the spectrum of the low-energy states for SU(2) gauge theory on the torus in a finite volume. As was mentioned before, in the sector of the abelian constant modes $A_k = (C_k/2L)\tau_3$, which forms the "vacuum or toron valley" along which the classical energy vanishes, Λ is described by $|C_k| \leqslant \pi$. The gauge transformation that maps $C_k = -\pi$ to $C_k = \pi$ is given by $g_{(k)} = \exp(-\pi i\tau_3 x_k/L)$, which due to its anti-periodicity is homotopically non-trivial and provides the required boundary identifications. In this subsector all boundary points are easily seen to be regular. The "Bloch momenta" label 't Hooft's electric flux quantum numbers [4] $\Psi(C_k = -\pi) = exp(\pi i e_k)\,\Psi(C_k = \pi)$. Note that the phase factor is not arbitrary, but ± 1. This is because $g_{(k)}^2$ is homotopically trivial. It thus looks like as if we have to put the topological structure in by hand after all, however, one should realize that considering a slice of Λ will obscure some of the topological features. A loop that winds around the slice twice is contractable in Λ as soon as it is allowed to leave the slice. Indeed including the lowest modes transverse to this slice will make the $\mathbb{Z}_2$ nature of the relevant homotopy group evident [5,6]. It shows that not dynamically motivated truncations can obscure things. For example, a recent reduction to spherically symmetric gauge fields [28] is only of limited value if the non-spherical fluctuations (which can in principle lead to Gribov horizons) are not taken into account.

In weak coupling Lüscher [29] showed unambiguously that the wave functionals are localized around $A = 0$, that they are normalizable and that the spectrum is discrete. In this limit the spectrum is insensitive to the boundary identifications (giving rise to a degeneracy in the topological quantum numbers). At stronger coupling the wave functional spreads out over the vacuum valley and the boundary conditions drastically change the spectrum [5]. In supersymmetric gauge theory the situation is even more dramatic. In the bosonic case, what localizes the wave functional in weak coupling is an induced potential barrier due to the zero-point fluctuations of the modes transverse to the vacuum valley. Due to the supersymmetry this induced barrier is expected to be canceled exactly by the fermionic contribution and the wave functional is expected to spread out over the whole vacuum valley. The problem is, however, that transverse fluctuations become singular near $A = 0$, preventing a reduction to the vacuum valley. As in the bosonic sector one can hope that a truncation to the zero-momentum sector will be possible, but as the wave functional is expected to spread out over the vacuum valley and as the gauge copies that thus arise will bring one outside of the zero-momentum sector, this would not be a consistent truncation either. Indeed, it was rigorously proven in ref. [13] that the spectrum of the zero-momentum Yang–Mills hamiltonian is continuous down to zero energy. Apart from the fact that such a continuous spectrum would make the Witten index ill defined, it is not compatible with the fact that the theory was originally defined in a finite volume. We consider this as a strong indication for the spreading of the wave functional beyond the Gribov copies. The appropriate identifications due to these copies should lead to the desired discrete spectrum. In the bosonic sector there is a *dynamical* reduction to a finite number of degrees of freedom [5,6]. But in the fermionic sector it requires one to construct vacuum Dirac bundles that incorporate the identifications at the boundary of the fundamental modular domain. Our problem is, that this does not seem to allow for a dynamical reduction to a finite number of degrees of freedom. As the results of ref. [14] rely on the truncation to the zero-momentum sector, without addressing the Gribov copy problem, we do not understand how the results in that paper can solve the problem of constructing the zero-energy states (unfortunately the construction of the wave function in ref. [14] is rather implicit and incomplete, which makes it hard to pin down exactly what might make it unsuitable as a zero-energy ground-state wave function). Thus it will remain an interesting and unfortunately open problem, whose solution will shed light on the discrepancy between the naive Witten index calculation [30] on the torus for $O(N)$ ($N > 6$) (giving an index equal to the rank of $O(N)$ plus one) and the value deduced from the gluino condensate calculations in an infinite volume [31] (yielding the value $N - 2$). When it persists, this would imply that the Witten index in this case will have discontinuities, which will have interesting consequences for the non-perturbative vacuum in these theories.

I thank Daniel Zwanziger for sending me ref. [10] and Maarten Golterman, Hari Dass, Andreas Kronfeld, Morton Laursen, Peter Lepage, Jeff Mandula, Paul Mackenzie, Michael Ogilvie, Stepen Sharpe, Jan Smit, Arjan van der Sijs and Jac Verbaarschot for discussions on Gribov copies and horizons at various occasions. I also thank Bernard de Wit, Hiroshi Itoyama, Bob Razzaghe-Ashrafi, Arkadi

Vainshtein and in particular Mikhail Shifman for discussions on the Witten index calculation on the torus. I am grateful to Karl Isler for his insights about the Dirac vacuum bundles.

Many useful comments on the manuscript and during seminars have helped me to clarify a number of issues, for which I thank Sidney Coleman, Jan de Boer, Philippe de Forcrand, Jim Hetrick, Key-Fei Liu, Herman Verlinde and Daniel Zwanziger. I am grateful to the Cern Theory Division for their hospitality during my visit in July. This research has been made possible by a fellowship of the Royal Netherlands Academy of Arts and Sciences.

References

[1] V. Gribov, Nucl. Phys. B139 (1978) 1
[2] R. Jackiw, I. Muzinich and C. Rebbi, Phys. Rev. D17 (1978) 1576
[3] F.S. Henyey, Phys. Rev. D20 (1979) 1460
[4] G. 't Hooft, Nucl. Phys. B153 (1979) 141
[5] J. Koller and P. van Baal, Nucl. Phys. B302 (1988) 1
[6] P. van Baal, *in* Probabilistic methods in quantum field theory and quantum gravity ed. P.H. Damgaard et al. (Plenum, New York, 1990) p. 131; *in* Frontiers in non-perturbative field theory, ed. Z. Horvath et al. (World Scientific, Singapore, 1989) p. 204
[7] Ph. de Forcrand et al., Nucl. Phys. B (Proc. Suppl.) 20 (1991) 194
[8] M.A. Semenov-Tyan-Shanskii and V.A. Franke, Zapiski Nauchnykh Seminarov Leningradskogo Otdeleniya Matematicheskogo Instituta im. V.A. Steklov AN SSSR, 120 (1982) 159 [Translation: (Plenum, New York, 1986) p. 1999]
[9] G. Dell'Antonio and D. Zwanziger, *in* Probabilistic methods in quantum field theory and quantum gravity, ed. P.H. Damgaard et al. (Plenum, New York, 1990) p. 107
[10] G. Dell'Antonio and D. Zwanziger, Commun. Math. Phys. 138 (1991) 291
[11] D. Zwanziger, Nucl. Phys. B345 (1990) 461
[12] C. Parinello and G. Jona-Lasinio, Phys. Lett. B251 (1990) 175
[13] B. de Wit, M. Lüscher and H. Nicolai, Nucl. Phys. B320 (1989) 135
[14] H. Itoyama and B. Razzaghe-Ashrafi, Nucl. Phys. B354 (1991) 85
[15] G. 't Hooft, Phys. Rev. Lett. 37 (1976) 8; Phys. Rev. D14 (1976) 3432
[16] K. Wilson, *in* Recent developments in gauge theories, ed. G.'t Hooft et al. (Plenum, New York, 1980) p. 363;
G. 't Hooft, Nucl. Phys. B190 [FS3] (1981) 455
[17] D. Zwanziger, Phys. Lett. B114 (1982) 337; Nucl. Phys. B209 (1982) 336
[18] J. Milnor, Morse theory, Ann. Math. Stud. 51 (Princeton Univ. Press, Princeton NJ, 1973)
[19] E. Witten, J. Diff. Geom. 17 (1982) 661;
P. van Baal, Acta Phys. Pol. B21 (1990) 73
[20] I. Singer, Commun. Math. Phys. 60 (1978) 7
[21] W. Nahm, *in* Proc. IV Warsaw Symp. on Elementary particle physics, ed. Z. Ajduk (Warsaw, 1981) p. 275
[22] Y. Choquet-Bruhat et al., Analysis, manifolds and physics (North-Holland, Amsterdam, 1977) p. 412
[23] D.J.R. Pugh and M. Teper, Phys. Lett. B224 (1989) 159
[24] J. Mandula and M. Ogilvie, Phys. Lett. B185 (1987) 127
[25] N.M. Christ and T.D. Lee, Phys. Rev. D22 (1980) 939
[26] E. Witten, Phys. Lett. B117 (1982) 324
[27] P. Nelson and L. Alvarez-Gaumé, Commun. Math. Phys. 99 (1985) 103

[28] D. Schütte, The problem of gauge fixing for spherically symmetric Yang–Mills fields, Bonn preprint (April 1991)
[29] M. Lüscher, Nucl. Phys. B219 (1983) 233;
     M. Lüscher and G. Münster, Nucl. Phys. B232 (1984) 445
[30] E. Witten, Nucl. Phys. B202 (1982) 253
[31] M. Shifman and A. Vainshtein, Nucl. Phys. B296 (1988) 455

Proceedings of ISATQP-Shanxi
1992, pp.133-136

TOPOLOGY OF THE YANG-MILLS CONFIGURATION SPACE

PIERRE VAN BAAL*
Institute for Theoretical Physics, Princetonplein 5, P.O.Box 80.006, NL-3508 TA Utrecht, The Netherlands

ABSTRACT

It will be described how to uniquely fix the gauge using Coulomb gauge fixing, avoiding the problem of Gribov copies. The fundamental modular domain, which represents a one-to-one representation of the set of gauge invariant degrees of freedom, is a *bounded* convex subset of the transverse gauge fields. Boundary identifications are the only remnants of the Gribov copies, and carry all the information about the topology of the Yang-Mills configuration space. Conversely, the known topology can be shown to imply that (on a set of measure zero on the boundary) some points of the boundary coincide with the Gribov horizon. For the low-lying energies, wavefunctionals can be shown to spread out "across" certain parts of these boundaries. This is how the topology of Yang-Mills configuration space has an essential influence on the low-lying spectrum, in a situation where these non-perturbative effects are *not exponentially suppressed.*

The write-up of my contribution to the International Symposium on Advanced Topics of Quantum Physics will be a short summary, with adequate references, where most of the material I have presented can be found. The observation concerning Henyey's gauge copies have not been published before.

Gauge fixing has remained an essential ingredient for studying non-abelian gauge theories[1], despite the many attempts of finding gauge invariant variables. This is not too surprising, as the Yang-Mills configuration space is topologically non-trivial[2]. There exists no choice of (affine) coordinates suitable for the whole manifold. Gauge fixing can be conveniently seen as just one particular choice of coordinates[3]. They are only locally defined, and one would need different coordinate patches and transition functions to describe the whole manifold[4]. Trying to extend the coordinate patch to the whole manifold has as a consequence that the gauge condition, beyond a certain distance in the configuration space, no longer uniquely fixes the gauge, as was first discovered by Gribov[5]. Analysing this in a finite spatial volume using the Hamiltonian formulation ($A_0 = 0$ and using the Coulomb gauge to fix the time-independent gauge parameters) has demonstrated the usefulness of this approach. In that case the transition functions were described by topologically non-trivial gauge transformations[6]. The reason for its success lies in the fact that asymptotic freedom in four dimensions guarantees that the effective coupling constant in a small volume is small[7], and that non-perturbative effects will only be of importance for a few low-energy modes.

More recently, a different way of describing the Yang-Mills configuration space was developed, by using a functional method to not only satisfy the Coulomb gauge condition $\partial_i A_i = 0$, but also pick from the different Gribov copies a unique configuration. The collection of these configurations should thus form a fundamental modular domain, in other words, it should form a one-to-one mapping with the Yang- Mills configuration space. The relevant functional is the L^2-norm of the gauge field: $\| A_i \|^2 = \int_M Tr(A_i^\dagger A_i)$. For each gauge invariant field configuration this gives a Morse function on the gauge orbit. One easily verifies that stationary points of this Morse function satisfy the Coulomb gauge condition and that the Hessian at the stationary point is precisely given by the Faddeev-Popov

---

^{0}* KNAW fellow

operator, whose determinant measures the volume of the gauge orbit. This procedure to fix the gauge has been known already for quite some time[8]], as well as the obvious conclusion that the absolute minimum provides the natural way of choosing a unique representative[9], whose recent rediscovery[10] has resulted in a recurrent interest in this problem.

Gribov's original conjecture[5] was that one could find a unique representative by demanding the Faddeev-Popov operator to be positive. This region in the space of connections can be shown to be convex and is called the Gribov region. In a finite volume one can show that in each direction of configuration space, the distance of its boundary to the origin is finite[11]. However, this would mean that this choice is only unique if the above discussed Morse function has all its minima to be absolute minima. The existence of a Gribov horizon, where the Faddeev-Popov determinant vanishes (we will be reserving the name horizon for the more stringent condition that the lowest eigenvalue of the Faddeev-Popov operator vanishes), does imply that points just outside the horizon have copies just inside, as Gribov showed. But more importantly, taking the global nature of the Morse function into account, implies two possibilities. Either near the horizon the points inside the Gribov region correspond to local minima, which therefore have to be copies of an absolute minimum, by definition also inside the Gribov region[9,10]. This happens if the third order term of the expansion of the Morse function around the stationary point, in the direction where the Hessian vanishes, is non-vanishing. Or, in the case this third order term vanishes, the (local) minimum bifurcates at the horizon in two local minima and one saddle point[12]. As the two local minima in the above bifurcation coalesce at the horizon, the relevant gauge transformation has to be topologically trivial. By definition, these gauge copies lie inside the Gribov region.

Care is required if the gauge configuration is reducible, as in that case the gauge group does not act freely. However, the gauge subgroup that leaves the reducible connection fixed, generally does not support topologically non-trivial maps. Nevertheless, reducible connections give rise to singularities in the physical configuration space[13], where the Morse theory arguments might not be valid. These singularities turn out to be particularly treacherous in the case of an abelian connection, which are the simplest examples of reducible connections. As the Coulomb gauge does not fix the constant gauge transformations, one has to still divide out these constant gauge transformations to obtain the fundamental modular domain.

With this in the back of our minds let us reconsider an old example due to Henyey[14], for gauge copies (inside the Gribov horizon) that are related by a homotopically trivial gauge transformation. As was explained in ref.[12], his was an example related to a bifurcation of local minima at the horizon. Explicitly Henyey's example was given by [14] an abelian $SU(2)$ connection on compactified IR^3:

$$\vec{A}(\beta) = a(\beta)\hat{\phi}\tau_3,$$

$$a(\beta) = \frac{1}{2r\sin\theta} - \frac{b + r^2\sin^2\theta\left(\dfrac{d^2b}{dr^2} + \dfrac{4}{r}\dfrac{db}{dr}\right)}{\beta^{-1}\sin(2r\beta b\sin\theta)},$$

$$g(\beta) = \exp(i\beta r b\sin\theta(e^{i\phi}\tau_- + e^{-i\phi}\tau_+)).$$

where (r, θ, ϕ) are the spherical coordinates, τ_i the Pauli matrices, b an arbitrary function of r and $g(\pm\beta)$ defines the two gauge copies $[g(\pm\beta)]A(\beta)$, that will coalesce at $\beta = 0$, which is easily verified to coincide with the Gribov horizon for a suitable choice of the radial function $b^{[12]}$. Unfortunately, close inspection shows that, since $g(-\beta) = \tau_3 g(\beta)\tau_3$ and since $A(\beta)$ commutes with τ_3, the two copies $[g(\pm\beta)]A(\beta)$ are related by a constant gauge transformation. As we still have to divide out these constant gauge transformations, Henyey's example does not really provide an example of two gauge copies inside the Gribov horizon, that are related by a topologically trivial gauge transformation.

Fortunately, however, the general Morse theory argument is sufficiently strong not to have to worry, that bifurcations of the above type do not occur. To find an example one simply should look for an irreducible connection on the Gribov horizon, where the third order term in the expansion around the stationary point of the Morse function vanishes in the direction of the zero eigenvector of the Hessian. In a recent paper, that analysed gauge fields on a three-sphere[15], such configurations featured prominently as the so-called sphaleron modes (u, v). Part of the Gribov horizon is given by the equation $u + v = 1.5$ (for u and $v \in [0, 1.5]$) and one easily verifies these configurations are not reducible and have a Morse function that vanishes to third order in the direction of the zero eigenvector of the Faddeev-Popov operator.

We now come back to the issue of the fundamental modular domain. It can be shown that the collection of absolute minima of the Morse function is also convex[10]. One would thus expect it to be contractable, which is in flagrant contradiction with the assumption that it is a in one-to-one correspondence with the physical configuration space. The catch is that the absolute minimum of the Morse function need not be unique. Using the convexity of the set of absolute minima (also called Λ) it is not hard to show[12] that A has a boundary, which if it does not coincide with the Gribov horizon, necessarily corresponds to a degenerate absolute minimum of its Morse function. The other absolute minimum is necessarily a gauge copy and also a point on the boundary. Thus the set of absolute minima will only become a fundamental modular domain if these appropriate boundary identifications are taken into account. In this way it will support the known non-trivial topology of the physical configuration space. On the other hand, the known non-trivial topology implies the existence of non-contractable n-spheres. Those can only arise through boundary identifications if all points of a suitable $n - 1$-dimensional subspace of the boundary are identified (gauge equivalent). These points consequently coincide with the Gribov horizon. These "singular" boundary points, however, form a subset of the boundary of zero measure (which does not mean they are unimportant). Note that as the Gribov region is bounded in each direction, the same holds for the fundamental modular domain.

The dynamical consequence in the context of gauge theories in a finite volume is that at increasing volume (i.e. increasing coupling constant) the wavefunctional starts to spread out over configuration space, and unavoidably will start to overlap with the boundary of the fundamental modular domain. Important is that the size of the volume controls this process. At very small volumes the boundary is irrelevant (for a torus typically below $0.1 - 0.01 fm^3$), whereas at increasing volume first the lowest energy modes start to be sensitive to the boundary identifications. In this way the energy of electric flux and the low-lying glueball spectrum were calculated in volumes up to $1 fm^3$ in the geometry of a torus[12,16], which agreed with the lattice Monte Carlo results obtained in the same volume within the 2% statistical error of the latter[17]. For the torus geometry the zero energy modes are parameterized by the constant abelian modes, $A_i = C_i \tau_3/(2L)$ (where L is the length of the torus). The Gribov horizon and the boundary of the fundamental modular domain in C- space are given by cubes, centered at $C = 0$ and respectively with sides of length 4π and 2π[12]. A recent analysis on the three-sphere is aimed at going to larger volumes[15].

References

[1] C.N. Yang and R.L. Mills, *Phys. Rev.*, **96** (1954), 191.

[2] I. Singer, *Comm. Math. Phys.*, **60** (1978), 7.

[3] O. Babelon and C. Viallet, *Comm. Math. Phys.*, **81** (1981), 515.

[4] W. Nahm, in: IV Warsaw Symp. on Elem. Part. Phys., ed. Z. Ajduk, p.275 (Warsaw, 1981).

[5] V. Gribov, *Nucl. Phys.*, **B139** (1978), 1.

[6] P. van Baal, in: Probalilistic Methods in Quantum Field Theory and Quantum Gravity, ed. P. Damgaard et al (Plenum, New York, 1990) p.131.

[7] M. Lüscher, *Phys. Lett.*, **118B** (1982), 391; *Nucl. Phys.*, **B219** (1983), 233.

[8] K. Wilson, in: Recent developments in Gauge Theories, ed. G. 't Hooft, et al (Plenum Press, New York, 1980) p.363; G. 't Hooft, *Nucl. Phys.*, **B190** (1981), 455; T. Maskawa and H.Nakajima, *Prog. Theor. Phys.*, **60** (1978), 1526.

[9] M.A. Semenov-Tyan-Shanskii and V.A. Franke, Zapiski Nauchnykh Seminarov Leningradskogo Otdeleniya Matematicheskogo Instituta im. V.A. Steklov AN SSSR, 120(1982)159. Translation: (Plenum Press, New York, 1986) p.1999.

[10] G. Dell'Antonio and D Zwanziger, in: Probabilistic Methods in Quantum Field Theory and Quantum Gravity, ed. P.H. Damgaard et al, (Plenum Press, New York, 1990) p.107; *Comm. Math. Phys.*, **138** (1991), 291.

[11] G. Dell'Antonio and D. Zwanziger, *Nucl. Phys.*, **B326** (1989), 333; D. Zwanziger, Critical limit of lattice gauge theory, NYU preprint, December 1991.

[12] P. van Baal, *Nucl. Phys.*, **B369** (1992), 259.

[13] S. Donaldson and P. Kronheimer, The geometry of four-manifolds (Oxford University Press, 1990); D. Freed and K. Uhlenbeck, Instantons and four-manifolds, M.S.R.I. publications, Vol 1 (Springer, New York, 1984).

[14] F.S. Henyey, *Phys. Rev.*, **D20** (1979), 1460.

[15] P. van Baal and N.D. Hari Dass, "The theta dependence beyond steepest descent", Utrecht preprint, THU-92/03, January 1992.

[16] J. Koller and P. van Baal, *Nucl. Phys.*, **B302** (1988), 1.

[17] P. van Baal, *Phys. Lett.*, **B224** (1989), 397.

Nuclear Physics B413 (1994) 535–552
North-Holland

Instantons from over-improved cooling

Margarita García Pérez [a], Antonio González-Arroyo [b], Jeroen
Snippe [a] and Pierre van Baal [a]

[a] *Instituut-Lorentz for Theoretical Physics, University of Leiden, PO Box 9506, NL-2300 RA
Leiden, The Netherlands*
[b] *Departamento de Física Teórica C-XI, Universidad Autónoma de Madrid, 28049 Madrid, Spain*

Received 20 September 1993
Accepted for publication 18 October 1993

Lattice artefacts are used, through modified lattice actions, as a tool to find the largest
instantons in a toroidal geometry $[0, L]^3 \times [0, T]$ for $T \to \infty$. It is conjectured that the
largest instanton is associated with tunnelling through a sphaleron. Existence of instantons
with at least eight parameters can be proven with the help of twisted boundary conditions
in the time direction. Numerical results for $SU(2)$ gauge theory obtained by cooling are
presented to demonstrate the viability of the method.

1. Introduction

Since the time of the discovery of instantons [1] in non-abelian gauge theories,
as vacuum to vacuum quantum mechanical tunnelling events [2], their role in
strongly interacting theories has been controversial, both in the continuum [3]
and in the lattice formulation [4]. For the continuum this has been mainly due to
applying semiclassical techniques, which cannot be justified at strong coupling.
In the lattice formulation the main problems were the instantons localised at the
scale of the lattice cut-off for which topological charge cannot be defined unambiguously [5], and which have actions considerably lower than the continuum
action of $8\pi^2$. Strictly speaking, there are no locally stable solutions on a lattice using the standard Wilson action [6], because this lattice action decreases
when the instanton becomes more localized [7], as we will demonstrate also
from analytic considerations. On a trial and error basis, different (improved)
lattice actions were considered, some of them indeed giving rise to stable lattice
solutions [8]. This paper will provide the proper framework to understand the
stability.

It is not too difficult to understand the reason of the instability. At finite
lattice sizes the lattice action deviates from the continuum and this deviation
is larger for stronger fields. For the Wilson action, as we will show, the lattice
artefacts make the action decrease as compared to the continuum. In the continuum, instantons have a scale (or size) parameter ρ, on which the action does

     M. García Pérez et al. / Instantons from over-improved cooling

not depend. But the smaller ρ becomes, the larger the fields get, which makes the lattice action *decrease*. On dimensional grounds one easily argues that (generically) $S_{\text{latt}}(a, \rho) = 8\pi^2(1 + (a/\rho)^2 d_2 + O(a/\rho)^4)$ for $\rho \gg a$, which will be demonstrated in more detail further on. For the Wilson action [6] $d_2 < 0$, explaining the instability. Hence, one simply modifies the action, such that $d_2 > 0$, in order to get stable solutions for the maximal value of ρ allowed by the volume $[0, L]^3$, which is kept finite. As we are interested in the classical solutions to the equations of motion, the modified action need not be of the type of an improved action [9], for which typically one wants to achieve $d_2 = 0$, as in that case (as we will show) the $(a/\rho)^4$ term might still destabilize the solution.

We deliberately want to keep $d_2 > 0$, which we will hence call over-improvement. The reason is, that our motivation for embarking on this project was to find the instantons with the largest scale ρ. This presumably will correspond to tunnelling over the lowest energy barrier, separating two classical vacua. The configuration that corresponds to the lowest barrier height is then conjectured to be a sphaleron (which exists due to the fact that we keep the volume finite). A sphaleron [10] is by definition a saddle point of the energy functional with precisely one unstable direction, which corresponds to the direction of tunnelling. In this way we use the instantons to map out the part of the energy functional relevant for the dynamical region where a semiclassical analysis of tunnelling amplitudes will break down. We refer to a pilot study [11] on $S^3 \times \mathbb{R}$ for readers interested in this issue, and for an explanation of the relevance of the geometry $T^3 \times \mathbb{R}$ which is studied in this paper. This geometry allows us to find the instantons using the lattice approximation. For simplicity we restrict ourselves to $SU(2)$ pure gauge theories.

2. On the existence of continuum solutions

The geometry $T^3 \times \mathbb{R}$, in particular in a lattice formulation, can be seen as a limiting case of an asymmetric four torus $[0, L]^3 \times [0, T]$. The only known solutions have constant curvature [12] and hence cannot correspond to vacuum to vacuum tunnelling, furthermore their topological charge is at least 2. Actually, it can be proven rigorously [13], that for T finite, no regular charge-1 self-dual solutions can exist on a four-torus (we will illustrate this with our numerical results). As soon as we allow for twisted boundary conditions [14], existence of minimal non-trivial topological charge instanton solutions can be proven. One distinguishes two cases, depending on the properties of the twist tensor $n_{\mu\nu} \in \mathbb{Z}_2$.

When $\frac{1}{8}\epsilon_{\mu\nu\lambda\sigma} n_{\mu\nu} n_{\lambda\sigma} = 1 \bmod 2$, the topological charge is half-integer. The minimal action allowed by the topological bound is therefore $4\pi^2$, corresponding to topological charge $1/2$. As twist is also well defined on the lattice [15], and in the above situation (called non-orthogonal twist) does not allow for zero-action configurations, these instantons cannot "fall through the lattice". Indeed, the index theorem predicts in this case four parameters ($8 \times$ topological charge), which have to correspond to the position parameters. The charge-1/2 instanton hence has fixed size and cannot shrink due to lattice artefacts. Impressively accurate results [16] were obtained for this case using the well known cooling method [7,17] to find a solution of the (lattice) equations of motion, whose smoothness and scaling with the lattice volume leaves no room to doubt it provides an accurate approximation to the continuum solution with action $4\pi^2$. In the continuum, existence of smooth non-trivial (but not necessarily self-dual) solutions was proven by Sedlacek [18], whereas theorem 3.2.1. of ref. [19] states that the moduli space of self-dual solutions is isomorphic with a four-torus.

When $\frac{1}{8}\epsilon_{\mu\nu\lambda\sigma}n_{\mu\nu}n_{\lambda\sigma} = 0 \bmod 2$, also called an orthogonal twist, there are "twist-eating" [15] configurations, i.e. configurations that have zero action *and* are compatible with twisted boundary conditions (see also ref. [21]). For SU(2), it is not too difficult to show that as long as $n_{\mu\nu} \neq 0 \bmod 2$ for some μ and ν, this twist eating configuration is unique [22], up to a global gauge transformation if a twist is introduced as in ref. [15,16] and multiplication with elements of the center of the gauge group. With twisted boundary conditions as originally defined by 't Hooft [14], such a global gauge transformation would even change the boundary conditions, and as an SO(3) bundle the twist-eating configuration is unique. (For SU(N) it can be proven [23] that out of the N^4 center elements that can multiply the twist, only N^2 give rise to gauge inequivalent configurations.) Under this condition it can be shown [19] that there are instanton solutions with 8 parameters (its moduli space, when dividing out the trivial translation parameters, is even related to a K3 surface [19,20]) using Taubes' [24] technique of glueing a localized instanton (with scale, position and global gauge parameters) to the "twist-eating" flat connection (i.e. zero action configuration). As the latter is *not* invariant under global gauge transformations, the global gauge parameters of the localized instantons are genuine parameters of the moduli space (see also ref. [25]).

The reason twisted boundary conditions are useful, is that at finite T there are no exact instantons on T^4 with periodic boundary conditions, but there are exact solutions for any non-trivial twist in the time direction. As $T \to \infty$ these solutions are also solutions on $T^3 \times \mathbb{R}$. This comes about as follows. Since at $T \to \infty$ the action can only stay finite if for $|t| \to \infty$ the energy density goes to zero, we deduce from a vanishing magnetic energy that up to a gauge

$$A_j(\boldsymbol{x}, t \to \pm\infty) = iC_j^{\pm}\sigma_3/2L, \tag{1}$$

where $C_j^{\pm} \in [0, 4\pi]$ ($A_0 = 0$) parametrizes the vacuum or toron valley [26], whose gauge invariant observables are best described by the Polyakov-line ex-

538   *M. García Pérez et al. / Instantons from over-improved cooling*

pectation values

$$P_i \equiv \tfrac{1}{2}\mathrm{Tr}\left(\mathrm{P}\exp\left(\int_0^L A_i(\boldsymbol{x},t)\mathrm{d}x_i\right)\right) = \cos\left(C_i/2\right), \qquad (2)$$

(for the proper definiton in the presence of twist, see ref. [16].) In the vacuum valley, P_i is space independent and the vanishing of the electric energy at $t \rightarrow \pm\infty$ also requires P_i (or $C_i^\pm$) to be asymptotically time independent. Instanton solutions on $T^3 \times \mathbb{R}$ are hence characterized by the boundary conditions $C_i^\pm$ at $t \rightarrow \pm\infty$. It is these general instantons that are physically relevant. It is not clear if solutions exist with arbitrary boundary values. Approaching $T \rightarrow \infty$, by using periodic boundary conditions (which would impose $C_i^+ = C_i^-\bmod 4\pi$ up to a periodic gauge transformation) does not allow us to prove existence. As long as T is finite there are no solutions [13] and the proof of non-existence breaks down as $T \rightarrow \infty$. On the other hand, with twist in the time direction, $n_{0i} = 1$, even at T finite there is in the continuum an eight parameter set of exact instanton solutions, which at $T \rightarrow \infty$ will correspond to $C_i^+ = (2\pi - C_i^-)\bmod 4\pi$ (again up to a periodic gauge transformation). For localized instantons, asymptotically the field has to coincide with the unique flat connection, which fixes the possible values of $C_i^\pm$ to π, but at the other extreme, as the instanton in the spatial direction extends up to the "boundary" of the torus, the regions $t \rightarrow +\infty$ and $t \rightarrow -\infty$ no longer are connected, which will relax the fact that $P_i = 0$ ($C_i^\pm = \pi$). Although we have no proof, it is reasonable to assume that the eight parameters for the instantons close to the maximal size are described by ρ, the four position parameters and the three vacuum valley parameters C_i^+ (or C_i^-). Note that for $P_i \rightarrow 0$ as $t \rightarrow \pm\infty$, the solution is both compatible with twisted *and* periodic boundary conditions *at infinite* T. In any case we have now learned that on $T^3 \times \mathbb{R}$ (i.e. with free boundary conditions at $t \rightarrow \pm\infty$) there are, at least eight and at most eleven continuous parameters that describe the instanton solutions for vacuum to vacuum tunnelling.

3. The lattice actions and cooling

Let us start with discussing the standard Wilson action [6]

$$S = \sum_{x,\mu,\nu}\mathrm{Tr}\left(1 - \nu\,\square\,\right) = \sum_{x,\mu,\nu}\mathrm{Tr}(1 - U_\mu(x)U_\nu(x+\hat\mu)U_\mu^\dagger(x+\hat\nu)U_\nu^\dagger(x)), \qquad (3)$$

where $U_\mu(x)$ are SU(2) group elements on the link that runs from x to $x + \hat\mu$, the latter being the unit vector in the μ direction. To derive the equations of motion, we observe that S depends on $U_\mu(x)$ through the expression

$$S(U_\mu(x)) = \mathrm{Tr}(1 - U_\mu(x)\tilde U_\mu^\dagger(x)) + \mathrm{Tr}(1 - U_\mu^\dagger(x)\tilde U_\mu(x)), \qquad (4)$$

where

$$\tilde{U}_\mu(x) = \sum_{\nu \neq \mu} \left(\nu \boxed{}^\mu + \nu \boxed{}_\mu \right)$$

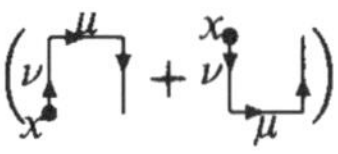

$$= \sum_{\nu \neq \mu} (U_\nu(x) U_\mu(x + \hat{\nu}) U_\nu^\dagger(x + \hat{\mu}) + U_\nu^\dagger(x - \hat{\nu}) U_\mu(x - \hat{\nu})$$

$$\times U_\nu(x + \hat{\mu} - \hat{\nu})), \tag{5}$$

which is independent of $U_\mu(x)$. Hence, $S(e^X U_\mu(x)) - S(U_\mu(x)) = O(X^2)$ for any Lie algebra element X, implies

$$\mathrm{Tr}[\sigma_i(U_\mu(x)\tilde{U}_\mu^\dagger(x) - \tilde{U}_\mu(x)U_\mu^\dagger(x))] = 0, \tag{6}$$

where σ_i are the Pauli matrices. This is easily seen to imply that $U_\mu(x)\tilde{U}_\mu^\dagger(x)$ is a multiple of the identity, and as $\tilde{U}_\mu$ is the sum of $SU(2)$ matrices, it can be written as $\tilde{U}_\mu = a_0 + i\mathbf{a} \cdot \boldsymbol{\sigma}$, with $a_\mu \in \mathbb{R}^4$. If we define $\|\tilde{U}_\mu\| = (a_\mu^2)^{1/2}$, eq. (6) is seen to imply

$$U_\mu(x) = \pm\tilde{U}_\mu(x)/\|\tilde{U}_\mu(x)\|. \tag{7}$$

As we are only interested in stable solutions (i.e. local minima of the action), the plus sign in eq. (7) is the relevant one. The process of iteratively finding the solution to the equations of motion is called cooling [17], as in all cases it is devised such that the action is lowered after each iteration. The easiest is to simply choose $U_\mu'(x) = \tilde{U}_\mu(x)/\|\tilde{U}_\mu(x)\|$ since the fixed point of this iteration is clearly a solution to the equations of motion. An optimal way to sweep through the lattice is to divide for each μ the links $U_\mu(x)$ in two mutually exclusive checkerboard patterns Π_μ^i such that all links on a particular pattern Π_μ^i (i.e. for fixed i and μ) can be changed simultaneously, which is a well known trick to vectorize this procedure. At the cost of roughly a factor two in memory-use, vectorization is also achieved for the modified action we have considered so far for our numerical simulations:

$$S(\varepsilon) = \frac{4-\varepsilon}{3} \sum_{x,\mu,\nu} \mathrm{Tr}\left(1 - \nu\boxed{}_\mu\right) + \frac{\varepsilon-1}{48} \sum_{x,\mu,\nu} \mathrm{Tr}\left(1 - \nu\boxed{}_\mu\right). \tag{8}$$

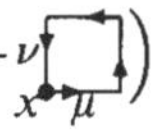

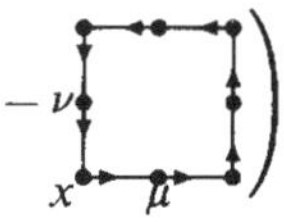

The meaning of the parameter ε will become clear in the next section. For ease of our numerical studies we have not considered modified single plaquette actions (see also the next section for a discussion on the adjoint and Manton actions).

4. Lattice artefacts

To calculate the effect of the discretization on the solutions of the equations of motion we first take a smooth continuum configuration (not necessarily a solution) $A_\mu(x)$. For definiteness we put $L = 1$, and N_s the number of lattice points in the spatial direction such that $a = 1/N_s$. We put this configuration on the lattice by defining:

$$U_\mu(x) = \text{Pexp}\left(\int_0^a A_\mu(x + s\hat{\mu})\,\mathrm{d}s\right). \tag{9}$$

The value of the plaquette thus corresponds to parallel transport around a square and can easily be proven to be given by [27] ($D_\mu = \partial_\mu + A_\mu(x)$ the covariant derivative in the fundamental representation)

$$\text{Tr}\left(\nu\,\square\right) = \text{Tr}\left(e^{aD_\mu(x)}e^{aD_\nu(x)}e^{-aD_\mu(x)}e^{-aD_\nu(x)}\right). \tag{10}$$

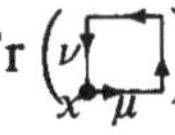

The proof simply amounts to observing that if $A_\mu(x) = A_\mu$, i.e. A_μ is space-time independent, then

$$\text{Tr}\left(\nu\,\square\right) = \text{Tr}\left(e^{aA_\mu}e^{aA_\nu}e^{-aA_\mu}e^{-aA_\nu}\right)$$

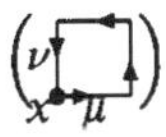

and eq. (10) is the only way to make this formula gauge invariant under arbitrary (i.e. x-dependent) gauge transformations. Using the Campbell–Baker–Hausdorff formula, eq. (10) can be expressed in terms of products of covariant derivatives $\mathcal{D}_\mu$ (in the adjoint representation) acting on the curvature $F_{\mu\nu} \equiv [D_\mu, D_\nu] = \partial_\mu A_\nu - \partial_\nu A_\mu + [A_\mu, A_\nu]$, e.g. $\mathcal{D}_\mu F_{\mu\nu} = [D_\mu, [D_\mu, D_\nu]]$. As the action involves a sum over all x, μ and ν, things can be considerably simplified by computing, what we will call, the clover average

$$\left\langle\text{Tr}\left(\nu\,\square\right)\right\rangle_{\text{clover}} = \frac{1}{4}\text{Tr}\left(\text{\large ⊞}\right)$$

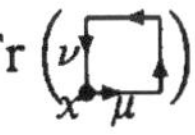

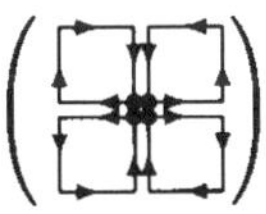

$$= \frac{1}{4}\text{Tr}\left[e^{-aD_\mu}e^{-aD_\nu}e^{aD_\mu}e^{aD_\nu} + e^{-aD_\mu}e^{aD_\nu}e^{aD_\mu}e^{-aD_\nu}\right.$$

$$\left.+ e^{aD_\mu}e^{-aD_\nu}e^{-aD_\mu}e^{aD_\nu} + e^{aD_\mu}e^{aD_\nu}e^{-aD_\mu}e^{-aD_\nu}\right]$$

$$= \text{Tr}\left[1 + \frac{a^4}{2}F_{\mu\nu}^2(x) - \frac{a^6}{24}\left((\mathcal{D}_\mu F_{\mu\nu}(x))^2 + (\mathcal{D}_\nu F_{\mu\nu}(x))^2\right)\right.$$

$$+ \frac{a^8}{24}\left\{F_{\mu\nu}^4(x) + \frac{1}{30}\left((\mathcal{D}_\mu^2 F_{\mu\nu}(x))^2 + (\mathcal{D}_\nu^2 F_{\mu\nu}(x))^2\right)\right.$$

$$\left.\left.+ \frac{1}{3}\mathcal{D}_\mu^2 F_{\mu\nu}(x)\mathcal{D}_\nu^2 F_{\mu\nu}(x) - \frac{1}{4}(\mathcal{D}_\mu \mathcal{D}_\nu F_{\mu\nu}(x))^2\right\}\right]$$

$$+ O(a^{10}) + \text{total derivative terms}, \tag{11}$$

for which the multiple Campbell–Baker–Hausdorff expansion of eq. (10) is required to $O(a^6)$, obtained with the aid of the symbolic manipulation program FORM [28]. The clover average allows one to ignore many terms (all those odd in any of the indices) in evaluating the trace of the exponent.

Eq. (11) was also derived using the non-abelian Stokes formula [29] ($s_0 \equiv 1$)

$$U_{\mu\nu}(x) \equiv U_\mu(x)U_\nu(x+\hat{\mu})U_\mu^\dagger(x+\hat{\nu})U_\nu^\dagger(x)$$

$$= \mathrm{P}\exp\left[a^2\int_0^1 \mathrm{d}s\int_0^1 \mathrm{d}t\, \mathcal{F}_{\mu\nu}(x+as\hat{\mu}+at\hat{\nu})\right]$$

$$\equiv 1 + \sum_{n=1}^{\infty}\prod_{i=1}^{n}\int_0^{s_{i-1}}\mathrm{d}s_i\int_0^1 \mathrm{d}t_i\, a^2\mathcal{F}_{\mu\nu}(x+as_1\hat{\mu}+at_1\hat{\nu})$$

$$\times \dots a^2\mathcal{F}_{\mu\nu}(x+as_n\hat{\mu}+at_n\hat{\nu})\,, \tag{12}$$

where $\mathcal{F}_{\mu\nu}(y)$ equals $F_{\mu\nu}(y)$ up to the backtracking loop that connects y to x, or

$$V(s,t) = \mathrm{P}\exp\left[a\int_0^s A_\mu(x+a\tilde{s}\hat{\mu})\mathrm{d}\tilde{s}\right]\mathrm{P}\exp\left[a\int_0^t A_\mu(x+as\hat{\mu}+a\tilde{t}\hat{\nu})\mathrm{d}\tilde{t}\right]\,,$$

$$\mathcal{F}_{\mu\nu}(x+as\hat{\mu}+at\hat{\nu}) = V(s,t)F_{\mu\nu}(x+as\hat{\mu}+at\hat{\nu})V^\dagger(s,t)\,. \tag{13}$$

To obtain the result of eq. (11) one now expands $\mathcal{F}_{\mu\nu}(x+as\hat{\mu}+at\hat{\nu})$ around the point x, making use of the identity

$$\partial_\mu^n\partial_\nu^m\mathcal{F}_{\mu\nu}(x) = \mathcal{D}_\mu^n\mathcal{D}_\nu^m F_{\mu\nu}(x)\,. \tag{14}$$

Note that the ordering of the covariant derivatives in the r.h.s. of eq. (14) is essential. Also crucial is that the path ordering $U(s,t) \equiv \mathrm{P}\exp(\int_s^t A(u)\mathrm{d}u)$ (where $A(t) = \hat{e}_\mu A_\mu(x+t\hat{e})$ for some unit vector $\hat{e}$) is compatible with the covariant derivative, i.e. $\hat{e}_\mu D_\mu(x+s\hat{e})U(s,t) = 0 = \hat{e}_\mu D_\mu(x+t\hat{e})U^\dagger(s,t)$ (in this respect we have corrected the formula in ref. [29]). Inserting the Taylor expansion of $\mathcal{F}_{\mu\nu}(x+as\hat{\mu}+at\hat{\nu})$ with respect to (s,t) in eq. (12), gives the result of eq. (11). A very useful check is that the symmetry implied by $U_{\mu\nu}(x) = U_{\nu\mu}^\dagger(x)$, not explicit at intermediate steps of the calculation, is respected by the final result.

Using eq. (9,11), one finds to $\mathrm{O}(a^{10})$ for the modified action $S(\varepsilon)$

$$S(\varepsilon) = \sum_{x,\mu,\nu}\mathrm{Tr}\left[-\frac{a^4}{2}F_{\mu\nu}^2 + \frac{\varepsilon a^6}{24}\left((\mathcal{D}_\mu F_{\mu\nu}(x))^2 + (\mathcal{D}_\nu F_{\mu\nu}(x))^2\right)\right.$$

$$-\frac{(15\varepsilon-12)a^8}{72}\left\{F_{\mu\nu}^4(x) + \frac{1}{30}\left((\mathcal{D}_\mu^2 F_{\mu\nu}(x))^2 + (\mathcal{D}_\nu^2 F_{\mu\nu}(x))^2\right)\right.$$

$$\left.\left.+\frac{1}{3}\mathcal{D}_\mu^2 F_{\mu\nu}(x)\mathcal{D}_\nu^2 F_{\mu\nu}(x) - \frac{1}{4}(\mathcal{D}_\mu\mathcal{D}_\nu F_{\mu\nu}(x))^2\right\}\right]\,. \tag{15}$$

Obviously, $S(\varepsilon = 1)$ corresponds to the Wilson action, and the sign of the leading lattice artefacts are simply reversed by changing the sign of ε. Most of the numerical results were obtained for $\varepsilon = -1$, but ε is useful in the initial cooling from a random configuration. By keeping $\varepsilon > 0$ as long as $S > 8\pi^2$, and only switching to $\varepsilon = -1$ when $S \sim 8\pi^2$, we can avoid the solution to get

       M. García Pérez et al. / Instantons from over-improved cooling

stuck at higher topological charges. Once we set $\varepsilon = -1$, we have yet to see an instanton fall through the lattice. We will come back to these issues when discussing the numerical results. Also note that, as $\mathrm{Tr}_{\mathrm{ad}}(U) = |\mathrm{Tr}(U)|^2 - 1$, one finds that the Wilson action in the adjoint representation (S_{ad}) satisfies $S_{\mathrm{ad}} = 4S(\varepsilon = 1) + \mathrm{O}(a^8)$ and does not allow us to change the sign of the a^6 term. The same holds for the Manton action [30] which by definition agrees to $\mathrm{O}(a^8)$ with the Wilson action.

In the past, more complicated improved actions were considered [9,31], for which we will present the result similar to eq. (15), as it allows us to predict whether or not they give rise to stable solutions [8]. It also allows comparison with earlier results by Lüscher and Weisz [31] obtained from a perturbative analysis. In the following, the coefficients in front of the c_i are to match with the definitions of ref. [31]. The averages $\langle \cdots \rangle$ are similar to the clover average above, but include now also averaging over all orientations of the loops. After some algebra one finds

$$
\begin{aligned}
S(\{c_i\}) &\equiv \sum_x \mathrm{Tr}\left\{ c_0 \left\langle 1 - \Box \right\rangle + 2c_1 \left\langle 1 - \boxed{} \right\rangle \right.\\
&\quad \left. + 4c_2 \left\langle 1 - \diagbox \right\rangle + \frac{4}{3}c_3 \left\langle 1 - \hexbox \right\rangle \right\}\\
&= -\frac{a^4}{2}(c_0 + 8c_1 + 16c_2 + 8c_3) \sum_{x,\mu,\nu} \mathrm{Tr}(F_{\mu\nu}^2(x))\\
&\quad + a^6\left(c_2 + \frac{c_3}{3}\right) \sum_{x,\mu,\nu,\lambda} \mathrm{Tr}(\mathcal{D}_\mu F_{\mu\lambda}(x)\mathcal{D}_\nu F_{\nu\lambda}(x))\\
&\quad + \frac{a^6}{12}(c_0 + 20c_1 + 4c_2 - 4c_3) \sum_{x,\mu,\nu} \mathrm{Tr}(\mathcal{D}_\mu F_{\mu\nu}(x))^2\\
&\quad + a^6\frac{c_3}{3} \sum_{x,\mu,\nu,\lambda} \mathrm{Tr}((\mathcal{D}_\mu F_{\nu\lambda})^2) + \mathrm{O}(a^8)\,.
\end{aligned}
\tag{16}
$$

One can therefore achieve tree-level improvement by choosing [31] $c_0 + 8c_1 + 16c_2 + 8c_3 = 1$, $c_0 + 20c_1 + 4c_2 - 4c_3 = 0$ and $c_2 = c_3 = 0$. Note that the condition $c_2 + c_3/3 = 0$ only applies off-shell, since on-shell

$$
\sum_{x,\mu,\nu,\lambda} \mathrm{Tr}(\mathcal{D}_\mu F_{\mu\lambda}(x)\mathcal{D}_\nu F_{\nu\lambda}(x)) = 0\,.
$$

Iwasaki and Yoshié [8] considered cooling for the Symanzik improved action, that is $c_0 = \frac{5}{3}$, $c_1 = -\frac{1}{12}$ and $c_{2,3} = 0$, for which the a^6 term vanishes. The a^8 term will have to be computed to settle stability. From eq. (15) one sees that the a^6 term has a definite sign. This is no longer the case for the a^8 term. The same holds for the Symanzik improved action:

$$S_{\text{Symanzik}} = \sum_{x,\mu,\nu} \text{Tr}\left[-\frac{a^4}{2}F_{\mu\nu}^2(x) + \frac{a^8}{24}\left\{F_{\mu\nu}^4(x) + \frac{1}{3}\mathcal{D}_\mu^2 F_{\mu\nu}(x)\mathcal{D}_\nu^2 F_{\mu\nu}(x)\right.\right.$$

$$\left.\left. -\frac{1}{4}(\mathcal{D}_\mu \mathcal{D}_\nu F_{\mu\nu}(x))^2 + \frac{4}{15}(\mathcal{D}_\mu^2 F_{\mu\nu})^2\right\}\right] + \text{O}(a^{10}). \tag{17}$$

To decide in these cases if the lattice admits a stable solution (i.e. its action increases with decreasing ρ), one can compute the lattice action using explicitly the topological charge-1 instanton solution with scale ρ. Eqs. (15)–(17) are only valid as long as $a \ll \rho$ because for $\rho \sim a$ the expansion in powers of a no longer converges. For $\rho \ll L$ to a good approximation we can substitute the infinite-volume continuum instanton solution:

$$A_\mu(x) = -i\frac{\eta_{\mu\nu}^a x_\nu \sigma_a}{(x^2 + \rho^2)}, \tag{18}$$

with $\eta_{\mu\nu}^a$ the self-dual 't Hooft tensor [33]. When $\rho \sim L$ the solution will of course be modified by the boundary effect. Substituting eq. (18) we find

$$S(\varepsilon) = 8\pi^2\left\{1 - \frac{\varepsilon}{5}(a/\rho)^2 - \frac{15\varepsilon - 12}{210}(a/\rho)^4 + \text{O}(a/\rho)^6\right\},$$

$$S_{\text{Symanzik}} = 8\pi^2\left\{1 - \frac{17}{210}(a/\rho)^4 + \text{O}(a/\rho)^6\right\}, \tag{19}$$

we thus confirm the observation of Iwasaki and Yoshié [8] that the Symanzik tree-level improved action has no stable instanton solutions. Since $S(\varepsilon = 0) = 8\pi^2\{1 + \frac{2}{35}(a/\rho)^4 + \text{O}(a/\rho)^6\}$ we predict even at $\varepsilon = 0$ the lattice to have stable solutions, which we have verified for the case with twisted boundary conditions in the time direction (see below).

Iwasaki and Yoshié [8] also considered cooling for Wilson's choice [32] (W) of $c_0 = 4.376, c_1 = -0.252, c_2 = 0$ and $c_3 = -0.17$ and for (R) $c_0 = 9, c_1 = -1$ and $c_2 = c_3 = 0$. To O(a^8) these actions effectively correspond respectively to $\varepsilon = -2.704$ and $\varepsilon = -11$, which for the case (W) we computed by substituting the continuum instanton solution. Indeed, they see stability up to 250 sweeps in both cases.

5. Non-leading lattice artefact corrections

In presenting eq. (19) we have replaced the sum over the lattice points by an integral and ignored the fact that on the lattice the equations of motion are modified. Both effects turn out to be small, the first exponential in ρ/a, the other gives a correction to the expression for $S(\varepsilon)$ in eq. (19) proportional to

$\varepsilon^2 (a/\rho)^4$ (whereas the correction to S_{Symanzik} is proportional to $(a/\rho)^8$, which also holds for $S(\varepsilon = 0)$).

We wish to compute $\sum_x f(x) = \sum_n f(na)$, for which we can use its Fourier decomposition

$$a^4 \sum_x f(x) = a^4 \sum_{n \in \mathbb{Z}^4} \sum_k e^{ik \cdot na} \tilde{f}(k)$$

$$= a^4 N_s^3 N_t \sum_{p \in \mathbb{Z}^4} \tilde{f}(\frac{2\pi p}{a}) = \sum_{p \in \mathbb{Z}^4} \int e^{-2\pi i p \cdot x/a} f(x) \mathrm{d}^4 x . \tag{20}$$

The terms with $p \neq 0$ give the error one makes, when replacing the lattice sum by an integral. For $a \ll \rho \ll L$ and $f(x) = -\frac{1}{2}\text{Tr}(F_{\mu\nu}^2(x + x_0))$ one finds explicitly (using eq. (18))

$$-\frac{a^4}{2} \sum_{x,\mu,\nu} \text{Tr}(F_{\mu\nu}^2(x + x_0))$$

$$= 8\pi^2 \left[1 + \sum_{p \in \mathbb{Z}^4 \setminus \{0\}} 2\pi^2 p^2 (\rho/a)^2 \cos(2\pi p \cdot x_0/a) K_2(2\pi|p|\rho/a) \right]$$

$$= 8\pi^2 \left[1 - 8\pi^2 (\rho/a)^{3/2} e^{-2\pi\rho/a}(1 + O(a/\rho)) \right] , \tag{21}$$

(with K_2 the modified Bessel function [34]). Here we have taken x_0 to coincide with a point on the *dual* lattice, $2x_0^\mu = a$ for all μ, as this minimizes the action.

To estimate the shift in the equations of motion due to the lattice artefacts, we again consider $\rho \gg a$, such that the action can, in a good approximation, be given by

$$\tilde{S}(\varepsilon, A_\mu) = \sum_{\mu,\nu} \int \mathrm{d}^4 x \left\{ -\frac{1}{2}\text{Tr}(F_{\mu\nu}^2(x)) + \frac{\varepsilon a^2}{12}\text{Tr}((\mathcal{D}_\mu F_{\mu\nu}(x))^2) \right.$$

$$-a^4 \frac{(15\varepsilon - 12)}{72} \text{Tr} \left[F_{\mu\nu}^4(x) + \frac{1}{15}(\mathcal{D}_\mu^2 F_{\mu\nu}(x))^2 \right.$$

$$\left. \left. + \frac{1}{3}\mathcal{D}_\mu^2 F_{\mu\nu}(x)\mathcal{D}_\nu^2 F_{\mu\nu}(x) - \frac{1}{4}(\mathcal{D}_\mu \mathcal{D}_\nu F_{\mu\nu}(x))^2 \right] \right\} + O(a^6), \tag{22}$$

which implies the equations of motion

$$\sum_\nu \mathcal{D}_\nu F_{\nu\mu} = \varepsilon a^2 H_\mu + O(a^4),$$

$$H_\mu \equiv -\sum_\nu (\tfrac{1}{12}\mathcal{D}_\nu^3 F_{\nu\mu} + \tfrac{1}{6}[F_{\mu\nu}, \mathcal{D}_\mu F_{\mu\nu}] + \tfrac{1}{12}\mathcal{D}_\mu^2 \mathcal{D}_\nu F_{\nu\mu}). \tag{23}$$

As eq. (22) breaks the scale invariance, there will in general not be solutions close to eq. (18). Variation with respect to ρ no longer leaves the action invariant. Still, since this variation corresponds to a near zero-mode, it makes sense to expect quasi-stability under cooling. The action changes only slowly in the direction of this near zero-mode but is predominantly lowered in those directions that leave the curvature square integrable and are spanned by the non-zero modes of the quadratic fluctuation operator for the action, which in the background gauge corresponds to

$$\mathcal{M}_{\lambda\sigma} = \delta_{\lambda\sigma}\mathcal{D}_\mu^2 + 2\,\mathrm{ad}\,F_{\lambda\sigma}\,. \tag{24}$$

If $\mathcal{P}$ is the projection operator on the normalizable non-zero modes of $\mathcal{M}$ one has (at $\rho = 1$)

$$A_\mu^{(\varepsilon)} = A_\mu^{(0)} + \varepsilon a^2 \mathcal{P}\mathcal{M}_{\mu\nu}^{-1}\mathcal{P}H_\nu + \mathrm{O}(a^4) \equiv A_\mu^{(0)} + \varepsilon a^2 A_\mu^{(1)} + \mathrm{O}(a^4),$$

$$H_\mu(x) = \frac{16}{(1+x^2)^5}\psi_\mu^{(3/2,5/2)}(x) + \frac{8}{(1+x^2)^5}\psi_\mu^{(1/2,1/2)}(x), \tag{25}$$

where $H_\mu(x)$ is evaluated by substituting for $A_\mu^{(0)}(x)$ the continuum solution $A_\mu(x)$ given in eq. (18). For convenience we introduced the quantities $\psi_\mu^{(l,j)}(x)$:

$$\psi_\mu^{(1/2,1/2)}(x) = i\sum_{\nu,a}\eta_{\mu\nu}^a x_\nu \sigma_a, \quad \psi_\mu^{(3/2,3/2)}(x) = i\sum_{\nu,a}\eta_{\mu\nu}^a x_\nu \sigma_a(x^2 - 6x_\mu^2),$$

$$\psi_\mu^{(3/2,5/2)}(x) = i\sum_{\nu,a}\eta_{\mu\nu}^a x_\nu \sigma_a(3x^2 - 3x_\mu^2 - 5x_\nu^2), \tag{26}$$

which are eigenfunctions of the angular momentum operators L_1^2 and J^2 (as defined in ref. [33] $L_1^a = -\frac{1}{2}i\eta_{\mu\nu}^a x_\mu \partial_\nu$, $J^a = L_1^a + \mathrm{ad}(\sigma_a/2)$). To compute $\mathcal{M}_{\mu\nu}^{-1}\mathcal{P}H_\nu$ one can use for $\mathcal{M}_{\mu\nu}^{-1}$ the explicit expression [35]

$$\mathcal{M}_{\mu\nu}^{-1} \equiv \bar{\eta}_{\mu\lambda}^a \mathcal{D}_\lambda (\mathcal{D}_\alpha^{-2})\bar{\eta}_{\nu\sigma}^a \mathcal{D}_\sigma$$

(in the gauge $\mathcal{D}_\mu(\mathcal{P}H_\mu) = 0$). The result, which can be verified by applying $\mathcal{M}_{\mu\nu}$, is found to be

$$\begin{aligned}
\mathcal{M}_{\mu\nu}^{-1}\mathcal{P}H_\nu = {}& \left(\frac{\log(1+x^2)}{5(1+x^2)^2} - \frac{1}{3(1+x^2)^3} + \frac{1}{10(1+x^2)}\right)\psi_\mu^{(1/2,1/2)}(x) \\
&+ \left(\frac{2\log(1+x^2)}{5x^8(1+x^2)^2} - \frac{6+3x^2-x^4}{15x^6(1+x^2)^3}\right)\psi_\mu^{(3/2,3/2)}(x) \\
&+ \left(\frac{2(3+5x^2)\log(1+x^2)}{5x^8(1+x^2)^2} - \frac{6+13x^2+4x^4}{5x^6(1+x^2)^3}\right)\psi_\mu^{(3/2,5/2)}(x).
\end{aligned} \tag{27}$$

The algebraic manipulation program Mathematica [36] was useful in obtaining and checking these results. Despite its appearance, this result is regular at $x \to 0$. However, it contains a non-normalizable deformation (since $\psi_\mu^{(1/2,1/2)}(x)/(1+x^2) = -A_\mu^{(0)}(x)$), which would make the action diverge and should be removed

 M. García Pérez et al. / Instantons from over-improved cooling

by projecting on normalizable deformations:

$$A_\mu^{(1)} = \mathcal{P}\mathcal{M}_{\mu\nu}^{-1}\mathcal{P}H_\nu = \mathcal{M}_{\mu\nu}^{-1}\mathcal{P}H_\nu - \frac{23 + 17x^2}{60(1 + x^2)^2}\psi_\mu^{(1/2,1/2)}(x),$$

$$\mathcal{M}_{\mu\nu}A_\nu^{(1)} = H_\mu - \frac{8\psi_\mu^{(1/2,1/2)}(x)}{5(1 + x^2)^3}. \tag{28}$$

One easily verifies that $A_\mu^{(1)}$ and $\mathcal{M}_{\mu\nu}A_\nu^{(1)}$ are both square integrable and orthogonal to the zero-mode $\partial A_\mu^{(0)}/\partial\rho|_{\rho=1} = 2\psi_\mu^{(1/2,1/2)}(x)/(1 + x^2)^2$.

We can now substitute $A_\mu^{(\varepsilon)} = A_\mu^{(0)} + \varepsilon a^2 A_\mu^{(1)} + O(a^4)$ in eq. (15), to obtain the shift in the action. It will be useful for verifying eq. (32) if we evaluate

$$\tilde{S}(\tilde{\varepsilon}, A_\mu^{(\varepsilon)}) = \tilde{S}(\tilde{\varepsilon}, A_\mu^{(0)}) - 2\varepsilon\tilde{\varepsilon}a^4 \int d^4x \mathrm{Tr}(A_\mu^{(1)}(x)H_\mu(x))$$

$$+ \varepsilon^2 a^4 \int d^4x \mathrm{Tr}(A_\mu^{(1)}(x)\mathcal{M}_{\mu\nu}A_\nu^{(1)}(x)) + O(a^6). \tag{29}$$

The equations of motion for $A_\mu^{(0)}$ where used to simplify the term linear in the shift $A_\mu^{(\varepsilon)} - A_\mu^{(0)} = \varepsilon a^2 A_\mu^{(1)} + O(a^4)$, which makes it evident that the $O(a^4)$ term in $A_\mu^{(\varepsilon)}$ will only contribute $O(a^6)$ to eq. (29). To evaluate the action used for cooling, one simply equates $\tilde{\varepsilon}$ to ε. Evaluating the integrals, reintroducing the ρ dependence using trivial dimensional arguments, gives

$$\tilde{S}(\tilde{\varepsilon}, A_\mu^{(\varepsilon)}) = 8\pi^2 \left\{ 1 - \frac{\tilde{\varepsilon}}{5}(\frac{a}{\rho})^2 - \left[\frac{15\tilde{\varepsilon} - 12}{210} + \frac{284\varepsilon\tilde{\varepsilon}}{2625} - \frac{179\varepsilon^2}{5250}\right](\frac{a}{\rho})^4 \right.$$

$$\left. + O(\frac{a}{\rho})^6 \right\}. \tag{30}$$

At $\varepsilon = \tilde{\varepsilon} = -1$ the shift due to the modified equations of motion in the $(a/\rho)^4$ term is 58%.

Strictly speaking our expression for the ρ-dependence of the lattice action is only valid for $\rho/a \gtrsim 2$ and $\rho \ll L$, since we are using the continuum infinite volume solution as the zero-order approximation. But even for $\rho \sim L/2$ it is not unreasonable to expect the order of magnitude of the corrections to be given by eqs. (21) and (30).

6. Numerical results and discussion

This section discusses the numerical results obtained as described in section 3, mainly to illustrate the viability of our ideas. A more detailed and careful analysis will be left for a future publication. So far we have worked mainly on lattices of size $N_s^3 \times N_t$, with $N_s = 7$ or 8 and with $N_t = 3N_s$ to $4N_s$. At $\varepsilon = -1$ (see eq. (8)) we settle to an action near $8\pi^2$, and we have seen stability for up to 6000 sweeps. The same is true for $\varepsilon = 0$ in the presence of a twist (but not without twist,

TABLE 1

Numerical results obtained by cooling with $S(\varepsilon)$ and twist $n_{0i} = (1,1,1)$.

$N_s \times N_t$	ε	$S_{1\times 1}$	$S_{1\times 2}$	$S_{2\times 2}$	$S_{1\times 3}$
$7^3 \times 21$	-1	0.982591	0.957050	0.928823	0.918105
$7^3 \times 21$	0	0.982287	0.956437	0.927908	0.917109
$8^3 \times 24$	-1	0.986720	0.967122	0.945887	0.936619
$8^3 \times 24$	0	0.986529	0.966736	0.945310	0.935976

where our configuration ultimately decays to the vacuum at $\varepsilon = 0$, but note that in that case there are no regular instanton solutions). Apart from the total action we compute separately the sum over the $n \times m$ plaquettes, denoted by $S_{n\times m}$ and averaging over the two orientations if $n \neq m$. $S_{n\times m}$ is normalized by dividing by $8\pi^2 n^2 m^2$, such that for an infinite lattice and $\rho/a \to \infty$, $S_{n\times m} \to 1$. When we perform cooling with the action of eq. (8) we should take $A_\mu = A_\mu^{(0)} + \varepsilon a^2 A_\mu^{(1)}$ in calculating $S_{n\times m}$. Eq. (11) for $S_{1\times 1}$ easily leads to the general result for $S_{n\times m}$ by inserting for each index $\mu(\nu)$ a factor $n(m)$. Together with eq. (29) one deduces, that to $O(a^6)$

$$S_{n\times m} = 1 - \frac{(n^2 + m^2)}{2}\alpha a^2$$
$$- \left(m^2 n^2 \beta_1 - \frac{m^4 + n^4}{2}\beta_2 + \frac{n^2 + m^2}{2}\varepsilon\gamma - \varepsilon^2\delta\right)a^4, \qquad (31)$$

up to the discretization error implied by eq. (21), which for the lattices we are considering can be estimated to be not bigger than 10^{-6}. This formula holds for sufficiently smooth configurations, i.e. $\alpha a^2 \ll 1$, even if the configuration has non-vanishing action over the entire spatial volume. It is these configurations that are of interest to us and which deviate considerably from localized instantons (eq. (18)) for which $\rho \ll L$. From eqs. (19, 30) we easily deduce for those localized instantons the results

$$\alpha a^2 = \frac{1}{5}\left(\frac{a}{\rho}\right)^2, \qquad \beta_1 a^4 = \frac{29}{630}\left(\frac{a}{\rho}\right)^4, \qquad \beta_2 a^4 = \frac{2}{63}\left(\frac{a}{\rho}\right)^4,$$
$$\gamma a^4 = \frac{284}{2625}\left(\frac{a}{\rho}\right)^4, \qquad \delta a^4 = \frac{179}{5250}\left(\frac{a}{\rho}\right)^4. \qquad (32)$$

From the numerical results we have obtained $S_{1\times 1}$, $S_{1\times 2}$, $S_{2\times 2}$ and $S_{1\times 3}$ on two lattices of size respectively $7^3 \times 21$ and $8^3 \times 24$, for $\varepsilon = 0$ and $\varepsilon = -1$ with a twist $n_{0i} = (1,1,1)$ (see table 1). From these we extract the coefficients in eq. (31), whose values are summarized in table 2 (the error due to neglecting the $O(a^6)$ term is of the order of $(n^6 + m^6)(\alpha a^2)^3$).

It is interesting to analyse the untwisted case in more detail to illustrate the difficulty in having self-dual solutions at finite T. In fig. 1a we plot the total elec-

548 *M. García Pérez et al. / Instantons from over-improved cooling*

TABLE 2

Coefficients appearing in eq. (31) extracted from the numerical results in table 1, using $S_{1\times1}$, $S_{2\times2}$ and $S_{1\times2}$ (the latter at $\varepsilon = -1$ only).

$N_s \times N_t$	αa^2	β_1/α^2	β_2/α^2	γ/α^2	δ/α^2
$7^3 \times 21$	0.01761	0.96	0.63	0.66	0.32
$8^3 \times 24$	0.01340	1.01	0.64	0.71	0.35
$\rho \ll L$	$0.2(a/\rho)^2$	1.15	0.79	2.70	0.85

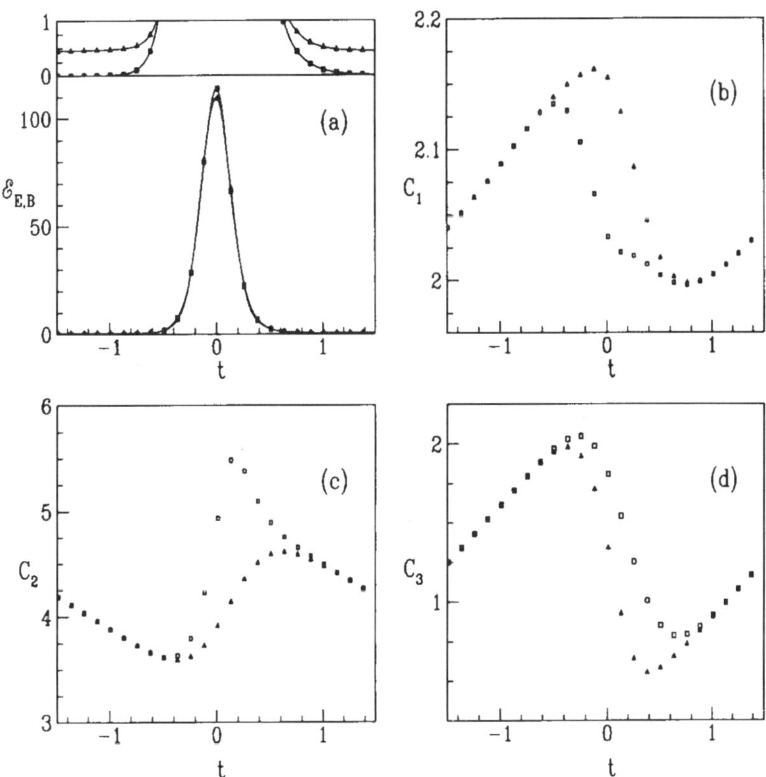

Fig. 1. Numerical results (after scaling appropriately with N_s) for the case of an $8^3 \times 24$ lattice without twist, obtained from over-improved cooling at $\varepsilon = -1$. In (a) the electric ($\mathcal{E}_E(t)$ triangles) and magnetic ($\mathcal{E}_B(t)$ squares) energies are plotted. In the upper part of this figure the tails are plotted at an enlarged scale. In (b-d) are plotted $C_i(t) \equiv 2a\cos(P_i(t))$ through two distinct spatial points on the lattice.

tric and magnetic energies $\mathcal{E}_{E,B}(t)$, in fig. 1b the Polyakov line $P_1(t)$ through two particular points x and similarly for $P_{2,3}(t)$ in figs. 1c,d. We see two features that are intimately related. First, where $\mathcal{E}_B(t) \to 0$, the electric energy $\mathcal{E}_E(t) \to$ const. Second for the same t values where this occurs $C_i(t)$ ($P_i(t) = \cos(C_i(t)/2)$) is

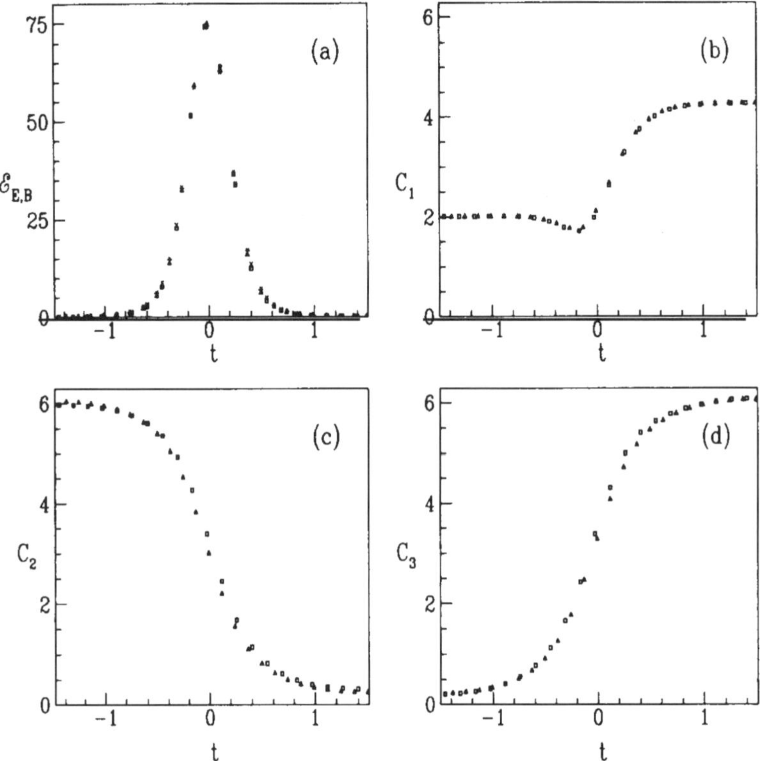

Fig. 2. Numerical results (after scaling appropriately with N_s) for the cases of a $7^3 \times 21$ (squares) and an $8^3 \times 24$ (triangles) lattice with twist $n_{0i} = (1, 1, 1)$, obtained from over-improved cooling at $\varepsilon = -1$. Fig. 2a contains four data sets. Two for $\mathcal{E}_E(t)$ with the above mentioned symbols and two (crosses for $N_s = 7$ and stars for $N_s = 8$) for $\mathcal{E}_B(t)$. Figs. 2b–d exhibit $C_i(t)$, through the spatial lattice point with maximal E_i^2 at $t = 0$.

linear in t and x independent. These are precisely the equations of motion when restricting to the vacuum valley. Classically motion on this valley, which itself has the geometry of a three-torus, is free. On the lattice this motion is described by the action

$$\sum_t 4N_s^3 \left[1 - \cos\left(\frac{C_k(t+1) - C_k(t)}{2N_s} \right) \right] . \tag{33}$$

One easily checks that the values of $C_k(t+1) - C_k(t)$ obtained from figs. 1b–d quite accurately reproduce through eq. (33) the value for $\mathcal{E}_E(t)$. Clearly the electric tail destroys the self-duality. Suppose that at $T \to \infty$ the solution describes tunnelling from C_i^- to C_i^+ and $C_i^+ \neq C_i^-$, then at finite T the periodic boundary conditions force $C_i(t)$ to linearly interpolate between C_i^+ and C_i^- over

                 M. García Pérez et al. / Instantons from over-improved cooling

a time $T - T_0$, if T_0 is the time interval over which $\mathcal{E}_B(t) \neq 0$. Thus the action, even in the continuum, would be bigger than $8\pi^2$ by a number proportional to $1/(T-T_0)$ except when there are solutions with $C_i^- = C_i^+$ for $T \to \infty$. These can certainly not be excluded, in particular as $C_i^+ = C_i^- = \pi$ is compatible with a twist $n_{0i} = 1$, but if these are very localized instantons, the lattice artefacts might make their action so big, that the lattice will prefer the least localized solutions with $C_i^+ \neq C_i^-$. If T is not big enough the lattice will find a compromise between these two cases. There are indications that the largest instanton prefers $C_i^+ \neq C_i^-$ and from the numerical results with twist $n_{0i} = (1,1,1)$ presented in fig. 2, the preferred values seem to be such that two of the C_i go from 0 to 2π and one goes from $2\pi/3$ to $4\pi/3$. We compare (after appropriate scaling with N_s) for $N_s = 7$ and 8, using over-improved cooling at $\varepsilon = -1$, in fig.2a the electric and magnetic energy profiles, and in figs. 2b–d the values of $C_i \equiv 2a\cos(P_i)$, at the spatial lattice point with maximal energy (to be precise, with maximal E_1^2). From this we deduce $\mathcal{E}_E = \mathcal{E}_B$ to a high accuracy, consistent with self-duality, and the excellent scaling with N_s. The results in fig. 2 are obtained after roughly 6600 cooling sweeps, which is necessary since the dependence of the lattice action on $C_i^\pm$ is rather weak (at $\varepsilon = 0$ too weak to observe) and the configuration only slowly reaches the minimum of the lattice action. We have verified that the approach to this minimum is exponential, as is illustrated in fig. 3, where we plot the total action and the maximum of $\mathcal{E}_B(t)$ (i.e. $\mathcal{E}_B(0)$) as a function of the number of cooling sweeps. We see indeed that the maximal energy along the tunnelling path decreases under cooling, which is mainly due to the increasing size, as otherwise the action should depend more strongly on the number of cooling sweeps. (For the Wilson action one sees a dramatic increase of $\mathcal{E}_B(0)$ under cooling, until the action suddenly drops to zero.) With boundary conditions that fix the link variables at $t = 0$ and $t = T$ to the vacuum configurations, the approach to the minimum action is much faster.

Elsewhere we will publish a more detailed analysis of the scaling properties, as well as testing our conjecture to be able to find a sphaleron. Also numerical results with fixed boundary conditions, that allow us to investigate if solutions exist for arbitrary $C_i^\pm$ will be presented elsewhere. This paper mainly served the purpose to describe the formalism, and demonstrate the large amount of control obtained in this way in studying instanton solutions on a torus.

It would be interesting to repeat this analysis for the two-dimensional O(3) model, for which the instantons on a torus are exactly known [37], in the light of the "perfect" lattice action recently considered by Hasenfratz and Niedermayer [38]. But as we have shown, appropriate deviations from a "perfect" action can be quite helpful.

Finally, over-improvement might be an efficient tool to measure the topological susceptibility, as the action to generate a statistical ensemble need not be the same as the one used to measure the topological charge.

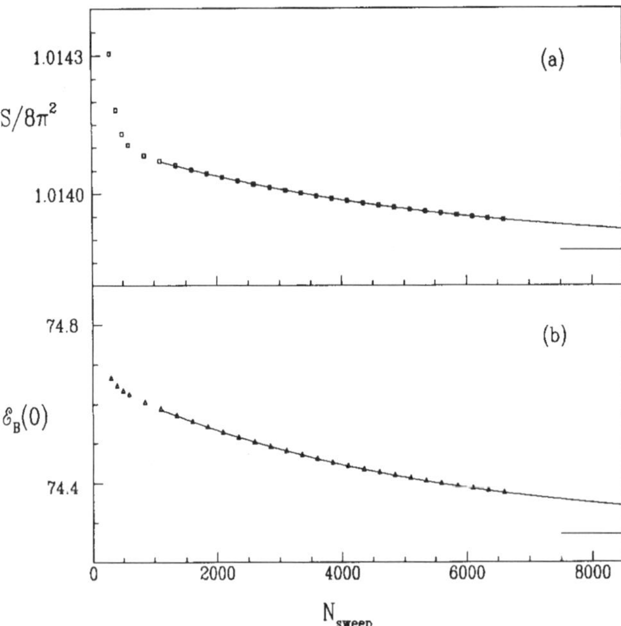

Fig. 3. The history of the action $S(\varepsilon = -1)$ and the maximal magnetic energy $\mathcal{E}_B(t = 0)$ as a function of the number of cooling sweeps for an $8^3 \times 24$ lattice with twist $n_{0i} = (1, 1, 1)$, together with their exponential fits. The short lines on the right indicate the asymptotic values following from these fits.

We have benefited from discussions with Wolfgang Bock, Maarten Golterman, Peter Hasenfratz, Jim Hetrick, Arjan Hulsebos, Ferenc Niedermayer, Erhard Seiler and Peter Weisz. In particular we thank Peter Braam and Jan Smit for discussions on respectively the moduli spaces and improved actions. One of us (P.v.B.) wishes to take this opportunity to thank Erhard Seiler and Peter Weisz for hospitality at the Max Planck Institute in Munich, Dan Freedman and Ken Johnson for hospitality at MIT and Tony González-Arroyo for hospitality at UAM in Madrid.

This work was supported in part by grant number AEN 90-0272, financed by CICYT, and by grants from Stichting voor Fundamenteel Onderzoek der Materie (FOM) and Stichting Nationale Computer Faciliteiten (NCF) for use of the CRAY Y-MP at SARA. M.G.P. was supported by a Human Capital and Mobility EC fellowship.

References

[1] A.A. Belavin et al., Phys. Lett. B59 (1975) 85
[2] G. 't Hooft, Phys. Rev. Lett. 37 (1976) 8 ;

552   *M. García Pérez et al. / Instantons from over-improved cooling*

      R. Jackiw and C. Rebbi, Phys. Rev. Lett. 37 (1976) 172
[3] C.G. Callen et al., Phys.Rev. D17 (1978) 2717
[4] A.S. Kronfeld, Nucl. Phys. B (Proc. Suppl.) 4 (1988) 329;
     M. Teper, Nucl. Phys. B (Proc. Suppl.) 20 (1990) 159
[5] M. Lüscher, Commun. Math. Phys. 85 (1982) 39; Nucl. Phys. B200 [FS4] (1982) 61
[6] K. Wilson, Phys. Rev. D10 (1974) 2445
[7] M. Teper, Phys. Lett. B162 (1985) 357; B171 (1986) 81, 86
[8] Y. Iwasaki and T. Yoshié, Phys. Lett. B131 (1983) 159
[9] K. Symanzik, Nucl. Phys. B226 (1983) 187, 205;
     M. Lüscher and P. Weisz, Phys. Lett. B158 (1985) 250
[10] F.R. Klinkhamer and M. Manton, Phys. Rev. D30 (1984) 2212
[11] P. van Baal and N.D. Hari Dass, Nucl. Phys. B385 (1992) 185
[12] G. 't Hooft, Commun. Math. Phys. 81 (1981) 267;
     P. van Baal, Commun. Math. Phys. 94 (1984) 397
[13] P.J. Braam and P. van Baal, Commun. Math. Phys. 122 (1989) 267
[14] G. 't Hooft, Nucl. Phys. B153 (1979) 141; Acta Physica Aust. Suppl. XXII (1980) 531;
     P. van Baal, Commun. Math. Phys. 85 (1982) 529
[15] J. Groeneveld et al., Physica Scripta 23 (1981) 1022;
     A. González-Arroyo and M. Okawa, Phys. Rev. D27 (1983) 2397
[16] M. Garcia Pérez, A. González-Arroyo and B. Söderberg, Phys. Lett. B235 (1990) 117;
     M. García Pérez and A. González-Arroyo, J. Phys. A26 (1993) 2667;
     M. García Pérez, et al., Phys. Lett. B305 (1993) 366
[17] B. Berg, Phys. Lett. B104 (1981) 475;
     J. Hoek, M. Teper and J. Waterhouse, Nucl. Phys. B288 (1987) 589
[18] S. Sedlacek, Commun. Math. Phys. 86 (1982) 515
[19] P. Braam, A. Maciocia and A. Todorov, Invent. Math. 108 (1992) 419
[20] F. Bogomolov and P. Braam, Commun. Math. Phys. 143 (1992) 641
[21] J. Ambjørn and H. Flyvbjerg, Phys. Lett. B97 (1980) 241
[22] P. van Baal, Commun. Math. Phys. 92 (1983) 1
[23] P. van Baal, Phys. Lett. B140 (1984) 375
[24] C. Taubes, J. Diff. Geom. 19 (1984) 517
[25] S. Donaldson and P. Kronheimer, The geometry of four manifolds (Oxford Univ. Press,
     Oxford, 1990);
     D. Freed and K. Uhlenbeck, Instantons and four-manifolds, M.S.R.I. publ. Vol. I (Springer,
     New York, 1984)
[26] P. van Baal and J. Koller, Ann. Phys. (N.Y.) 174 (1987) 299
[27] P. van Baal, Theorem 4 in the set of theorems appended to the Ph. D. thesis Twisted
     boundary conditions: a non-perturbative probe for pure non-abelian gauge theories, Utrecht,
     July 1984
[28] J.A.M. Vermaseren, Symbolic manipulation with FORM, distributed by CAN, Amsterdam,
     ISBN 90-74116-01-9 (1991)
[29] I. Aref'eva, Teor. Mat. Fiz. 43 (1980) 111 (translation p. 353)
[30] N. Manton, Phys. Lett. B96 (1980) 328
[31] M. Lüscher and P. Weisz, Commun. Math. Phys. 97 (1985) 59 [Erratum: 98 (1985) 443]
[32] K. Wilson, *in* Recent developments in gauge theories, ed. G. 't Hooft et al. (Plenum Press,
     New York, 1980) p.363
[33] G. 't Hooft, Phys. Rev. D14 (1976) 3432
[34] M. Abramowitz and I. Stegun, Handbook of Mathematical Functions (Dover, New York,
     1972) p.374
[35] L.S. Brown, et al., Phys. Lett. B71 (1977) 103
[36] S. Wolfram, et al., Mathematica (Addison-Wesley, New York, 1991)
[37] J.L. Richard and A. Rouet, Nucl. Phys. B211 (1983) 447
[38] P. Hasenfratz and F. Niedermayer, Perfect lattice action for asymptotically free theories,
     Bern preprint, BUTP-93/17, July 1993

ELSEVIER

Nuclear Physics B (Proc. Suppl.) 49 (1996) 238–249

Instanton moduli for $T^3 \times \mathbb{R}$

Pierre van Baal [a]

[a]Instituut-Lorentz for Theoretical Physics,
University of Leiden, PO Box 9506,
NL-2300 RA Leiden, The Netherlands.

We review the specific problems that arise when studying instantons on a torus. We discuss how the Nahm transformation shows that no exact charge one instanton on T^4 can exist. However, taking one of the directions (the time) to infinity, it can be shown that vacuum to vacuum tunnelling solutions exist. A precise description of the moduli space for $T^3 \times \mathbb{R}$, studied numerically using lattice techniques, remains an interesting open problem. New is an explicit application of the Nahm transformation to (anti-)selfdual constant curvature solutions on T^4 and a discussion of its properties relevant to instantons on $T^3 \times \mathbb{R}$.

1. INTRODUCTION

Instantons and monopoles have become important mathematical tools to study invariants of four dimensional manifolds [1,2]. This is due to the fact that their moduli space (the parameter or solution space) depends on the space-time on which the self-duality equations are studied. To some extent we can compare [3] the situation with the study of harmonic forms on a manifold and its relation to the DeRham cohomologies. Like in that case, also here there are certain spaces for which there are no solutions to the equations studied.

The torus T^4 turns out to be an example where the moduli space for instantons of unit charge is empty. For higher charges existence was proved by Taubes [4] quite some time ago. The general proof for existence takes a small localized instanton from $\mathbb{R}^4$ that is matched smoothly to the flat connection of the trivial bundle. In particular if the moduli space of flat connections has non-zero dimension, there can be obstructions against this procedure. This generally occurs when $H_2(M, \mathbb{Z})$ is non-trivial, as the trace of the Wilson loop for a flat connection can be non-trivial when the loop cannot be contracted to a point.

Using the Nahm transformation [5], it can be proven that indeed charge one instantons do not exist for T^4. The Nahm transformation will be reviewed in the next section (see also refs. [6]).

It maps self-dual solutions on T^4 to self-dual solutions on the dual torus [7,8]. This map is an involution, i.e. its square is the identity, and it preserves metric and hyperKähler structures of the moduli spaces [9].

¿From the physical point of view it is somewhat disturbing, in particular when scaling up the volume to sizes much bigger than $1/\Lambda_{QCD}$, that the existence would depend on the geometry of space-time. The point is of course that one can get arbitrarily close to a solution, for sufficiently large volumes, but an exact solution is only achieved in the singular limit. Nevertheless, the obstruction against the existence of a solution is relatively mild. For twisted boundary conditions [10], the existence of charge one instantons can be understood from the fact that any twist removes the continuous degeneracy in the moduli space of (four dimensional) flat connections, removing the obstruction to glueing in a localized instanton [11]. Twisting the boundary conditions in the time direction provides the proper framework for understanding the existence of charge one instanton solutions on $T^3 \times \mathbb{R}$, as was confirmed by numerical lattice studies [12].

The physical picture is most simply explained at infinite time. To keep the action of the imaginary time solution finite, the magnetic (and electric) energy should vanish at either end. This means that the connections at these ends are flat connections on T^3, whose moduli space is

P. van Baal / Nuclear Physics B (Proc. Suppl.) 49 (1996) 238–249

an orbifold (for SU(2) it is T^3/Z_2), most conveniently parametrized by the trace of the Wilson loops along the three generators of the three-torus. When twisting in the time direction, at least one of these Wilson loops is required to have opposite signs at the two ends, somewhat misleadingly this can be compared to anti-periodic boundary conditions. Apparently, instanton solutions are not compatible with periodic boundary conditions. From the Hamiltonian point of view, in the $A_0 = 0$ gauge, there is no difference between the twisted and periodic case, except that in the periodic case the trace of the Wilson loops at $t = \pm\infty$ are identical and for the twisted case opposite in sign. In particular the sphaleron giving the saddle point, or minimal barrier that separates two vacua, is obtained by solving the static Yang-Mills equations on the (untwisted) three-torus [13]. This sphaleron would not exist in an infinite volume, but the finite volume breaks the classical scale invariance, which also implies that the scale parameter of the instanton is no longer associated to a symmetry, but it is still a non-trivial moduli parameter.

An interesting open problem remains if there are solutions on $T^3 \times \mathbb{R}$ that can not be compactified to T^4, with twist in the time.

2. THE NAHM TRANSFORMATION

It is convenient to view T^4 as $\mathbb{R}^4/\Lambda$, where $\Lambda = \oplus_{\mu=1}^4 \mathbb{Z}e_\mu$ is a four dimensional lattice. The connection one-forms $\omega(x) = A_\mu(x)dx_\mu$ are invariant up to a gauge transformation under translation over a lattice vector. These gauge transformations, also called cocycles, satisfy cocycle conditions so as to assure one has an appropriate principal fiber bundle over the torus:

$$\begin{aligned}
\omega(x+\lambda) &= g_\lambda(x)(\omega(x)+d)g_\lambda^{-1}(x), \\
g_{\lambda+\mu}(x) &= g_\lambda(x+\mu)g_\mu(x), \quad \lambda, \mu \in \Lambda.
\end{aligned} \quad (1)$$

We will consider anti-selfdual connections with gauge group $SU(N)$ whose curvature satisfies

$$\Omega = d\omega + \omega \wedge \omega = -{}^*\Omega. \quad (2)$$

In local coordinates the curvature is given in terms of the field strength by $\Omega = \frac{1}{2}F^{\mu\nu}dx_\mu \wedge dx_\nu$, whereas ${}^*\Omega$ is defined similarly in terms of the

dual of the field strength, $\tilde{F}^{\mu\nu} = \frac{1}{2}\varepsilon^{\mu\nu\lambda\sigma}F_{\lambda\sigma}$. The topological charge k is given by the integral over the second Chern class, $k = \int \mathrm{Tr}\left(\frac{\Omega}{2\pi i} \wedge \frac{\Omega}{2\pi i}\right)$.

The Nahm transformation, $\hat{\omega} \equiv \mathcal{N}\omega$, will define a connection on the dual torus $\hat{T}^4 = \mathbb{R}^4/\hat{\Lambda}$, where

$$\hat{\Lambda} = \{\mu \in \mathbb{R}^4 | <\mu, \lambda> \in \mathbb{Z}, \, \forall \lambda \in \Lambda\} \quad (3)$$

is the lattice dual to Λ. We will show that $\hat{\omega}$ is a $SU(k)$ anti-selfdual connection with topological charge N. The rank of the gauge group and the topological charge are therefore interchanged under the Nahm transformation. If we denote by $\mathcal{M}_{N,k}$ the moduli space of $SU(N)$ charge k instantons, the Nahm transformation induces a map between moduli spaces, $\mathcal{N} : \mathcal{M}_{N,k} \to \mathcal{M}_{k,N}$, which is an involution that preserves the natural metric and hyperKähler structure of the moduli space [9]. The dimension of the moduli space, $4Nk$, is indeed symmetric under interchanging k and N.

We will now explain the essential ingredients of the Nahm transformation. The starting point is that a charge k (anti-)instanton has k (negative) positive chirality zero-modes for the massless Dirac equation, which is most conveniently written in the Weyl format. We introduce the four unit quaternions σ_μ, with $\sigma_0 = 1_2$ and $\sigma_i = -i\tau_i$, where τ_i are the usual Pauli-matrices. The Weyl operators are given by $D^- \equiv D$ and $D^+ \equiv -D^\dagger$, with

$$D \equiv \sigma_\mu D_\mu(A) = \sigma_\mu(\partial_\mu + A_\mu). \quad (4)$$

Hence, in the background of a charge k instanton there are k independent solutions to $D\Psi = 0$. For $\Psi(x)$ to be defined as a two-spinor on the torus one requires $\Psi(x + \lambda) = g_\lambda(x)\Psi(x)$. One now adds a spectral parameter $z_\mu \in \mathbb{R}^4$ in the form of a flat abelian connection

$$\omega_z = \omega + 2\pi i z_\mu dx^\mu, \quad (5)$$

which leaves the curvature unchanged, $\Omega_z = \Omega$, and in particular anti-selfdual. Hence there is a smooth family of k fermionic zero-modes

$$D_z \Psi_z^{(i)}(x) = \sigma_\mu(\partial_\mu + A_\mu + 2\pi i z_\mu)\Psi_z^{(i)}(x) = 0. \quad (6)$$

¿From this family one construct a connection $\hat{\omega}$ by

$$\hat{A}_\mu^{ij}(z) = \int_{T^4} d^4x \, \Psi_z^i(x)^\dagger \frac{\partial}{\partial z_\mu}\Psi_z^{(j)}(x). \quad (7)$$

This can be seen to form a connection on the dual torus using the fact that the $\Psi_z^{(i)}(x)$ form a complete orthogonal set of solutions of the Weyl equation and the observation that $e^{-2\pi i x \cdot \hat{\lambda}}\Psi_z^{(i)}(x) \in \ker D_{z+\hat{\lambda}}$, such that

$$\Psi_{z+\hat{\lambda}}^{(i)}(x) = e^{-2\pi i x \cdot \hat{\lambda}}\Psi_z^{(j)}(x)S_{\hat{\lambda}}^{ji}(z), \qquad (8)$$

with $S_{\hat{\lambda}}(z)$ a unitary $k \times k$ matrix, which defines the (inverse) of the cocycle for $\hat{\omega} \equiv \hat{A}_\mu(z)dz_\mu$ as a connection on $\hat{T}^4$

$$\hat{A}_\mu(z + \hat{\lambda}) = S_{\hat{\lambda}}^{-1}(z)(\hat{A}_\mu(z) + \frac{\partial}{\partial z_\mu})S_{\hat{\lambda}}(z). \qquad (9)$$

We can use the Atiyah-Singer family index theorem [14] to relate the Chern character of the bundle E, associated with the connection ω to the Chern character of the bundle $\hat{E}$, associated with the connection $\hat{\omega}$. For this it is necessary to assume that ω is 1-irreducible or WFF (without flat factors [1]), which is equivalent to stating that $\operatorname{coker} D_z = \ker D_z^+ = 0$, implying that $\hat{E}_z \equiv \ker D_z$ is smooth (remember that the index theorem states that $k = \dim \ker D^- - \dim \ker D^+$). One can view ω_z as a connection over $T^4 \times \hat{T}^4$, where the abelian part has a curvature $2\pi i dz_\mu \wedge dx_\mu$ (forming the so-called Poincaré bundle $\mathcal{P}$ [9]). It now follows that

$$\operatorname{ch}(\hat{E}) = \int_{T^4} \operatorname{ch}(E) \wedge \operatorname{ch}(\mathcal{P}). \qquad (10)$$

This is easily seen to interchange the rank and topological charge between the original and Nahm bundles. Also the first Chern classes will be related, but we will assume that E is a SU(N) bundle, for which the first Chern class vanishes. The family index theorem guarantees that the Nahm bundle $\hat{E}$ also has vanishing first Chern class, from which it follows that $\hat{E}$ is a SU(k) bundle with topological charge N. If, however, the first Chern class does not vanish (in which case the topological charge should be determined from the first Pontryagin class), one has [9]

$$c_1(\hat{E}) = -\int_{T^4} (dz_\mu \wedge dx_\mu)^2 \wedge c_1(E). \qquad (11)$$

A direct corollary is now that charge 1 instantons cannot exist on T^4. Suppose they would exist. The Nahm bundle would give rise to a U(1) bundle of charge N, which is impossible as the first Chern class vanishes and U(1) bundles have always vanishing second Chern class.

The family index theorem only provides topological information on the Nahm bundle obtained from an anti-selfdual connection ω. To demonstrate that $\hat{\omega}$ is also an anti-selfdual connection, a little more work is needed. As some of the ingredients will be important for the formulation of the Nahm transformation on $T^3 \times \mathbb{R}$ we provide a few details necessary to understand this beautiful result due to Nahm, who originally introduced his transformation for the study of monopoles, as a generalization [5,6] of the ADHM construction [15]. A crucial ingredient is formed by the Weitzenböck formula

$$D_z^- D_z^+ = D_\mu^2(A_z) + \sigma_{[\mu}\sigma_{\nu]}^\dagger F_{\mu\nu}, \qquad (12)$$

using $\sigma_\mu \sigma_\nu^\dagger = \delta_{\mu\nu} + \sigma_{[\mu}\sigma_{\nu]}^\dagger$. Since $\sigma_{[\mu}\sigma_{\nu]}^\dagger \equiv \sigma_i \eta_{\mu\nu}^i$ is a selfdual tensor, we see that $D_z^- D_z^+ = D_\mu^2(A_z)$. Its kernel is trivial for ω WFF and the Greens function $G_z = (D_z^- D_z^+)^{-1}$ commutes with the quaternions σ_μ.

It is now remarkably simple to show that $\hat{\omega}$ is also anti-selfdual

$$\hat{\Omega}^{ij}(z) = \hat{d}\hat{\omega} + \hat{\omega} \wedge \hat{\omega} = <\hat{d}\Psi_z^{(i)}|1 - P|\hat{d}\Psi_z^{(j)}>, (13)$$

where

$$\hat{d} \equiv dz_\mu \frac{\partial}{\partial z_\mu}, \qquad P \equiv |\Psi_z^{(k)}><\Psi_z^{(k)}|. \qquad (14)$$

One easily shows that

$$\begin{aligned} P &= 1 - D_z^+ G_z D_z, \qquad (15)\\ D_z \hat{d}\Psi_z^{(i)} &= [D_z, \hat{d}]\Psi_z^{(i)} = -2\pi i \sigma_\mu \Psi_z^{(i)} dz_\mu, \end{aligned}$$

such that

$$\hat{F}_{\mu\nu}^{ij}(z) = 8\pi^2 < \Psi_z^{(i)}|\sigma_{[\mu}^\dagger \sigma_{\nu]} G_z|\Psi_z^{(j)}> . \qquad (16)$$

There are two essential ingredients that enter these manipulations. First, on T^4 we can perform partial integrations without picking up boundary terms. Second, $G_z = (D_z^- D_z^+)^{-1}$ commutes with the quaternions. Anti-selfduality immediately follows from the fact that $\sigma_{[\mu}^\dagger \sigma_{\nu]} \equiv \sigma_i \bar{\eta}_{\mu\nu}^i$ is an anti-selfdual tensor. When one or more space-time directions become non-compact, boundary

P. van Baal/Nuclear Physics B (Proc. Suppl.) 49 (1996) 238–249

terms complicate the construction [5,6]. For T^4, applying the Nahm transformation the second time (in case of non-compact directions this requires modification), it can be shown that $\mathcal{N}^2\omega = \omega$, and the explicit form of $\hat{\Psi}_z(z)$ in terms of $\Psi_z(x)$ allows one to show that metric and hyperKähler structures of the moduli spaces are preserved under $\mathcal{N}$ [9].

We note that in the case of twisted boundary conditions [10], it is not possible to construct the Nahm transformation, as the fields need to be invariant under the center of the gauge group, which is not the case for the fermionic zero-modes required for the construction of $\hat{\omega}$. Nevertheless, for gauge theories on $T^3 \times \mathbb{R}$, where we have periodic boundary conditions in the space directions and none in the time directions, we can attempt to construct a variant of the Nahm transformation, as will be discussed in the last section.

3. EXPLICIT EXAMPLE

In general no explicit (anti-)selfdual connections on T^4 are known. However, for some choices of the periods Λ (or equivalently for some choices of metrics), constant curvature solutions exist [16]. A complete classification for SU(2) was given in ref. [17]. Under deformations of Λ these connections remain solutions, but are no longer (anti-)selfdual. In absence of twist, that is as proper SU(2) (rather than $SU(2)/Z_2$=SO(3)) bundles, their topological charge is always even. They have U(1) holonomy, as there exists a gauge in which they are abelian. Consequently they are examples of so-called reducible connections, giving rise to singular points in the moduli space [1]. The term reducible derives from the fact that these bundles decompose in the direct sum of U(1) line bundles $E = L \oplus L^{-1}$, see ref. [1]. As the Nahm transformation preserves the metric structure of the moduli spaces, reducible connections should be mapped to reducible connections. This implies that the Nahm transformation of a (anti-)selfdual constant curvature connection is expected to be also a (anti-)selfdual constant curvature connection, to be illustrated in this section with the help a simple explicit example for SU(2) and topological charge 2.

We consider on $T^4 = \mathbb{R}^4/\mathbb{Z}^4$ the connection

$$A_\mu(x) = -\tfrac{1}{2}\pi i n_{\mu\nu}x_\nu \tau_3, \quad n_{03} = n_{21} = 2, \quad (17)$$

with the cocycles given by

$$g_\lambda(x) = \alpha(\lambda)\exp(\lambda_\mu A_\mu(x)). \quad (18)$$

The so-called bi-characters α are given by

$$\alpha(\lambda) \equiv \exp(-\tfrac{1}{2}\pi i \sum_{\mu<\nu} \lambda_\mu n_{\mu\nu}\lambda_\nu). \quad (19)$$

As in the study of the fluctuations around the constant curvature connections [17], the zero-modes (and for that matter the whole spectrum) of the Weyl operators can be expressed in terms of theta functions. This is most simply seen by introducing complex coordinates

$$y_1 = \frac{x_3 + ix_0}{\sqrt{2}} \quad , \quad y_2 = \frac{x_1 + ix_2}{\sqrt{2}}$$

$$u_1 = \frac{z_3 + iz_0}{\sqrt{2}} \quad , \quad u_2 = \frac{z_1 + iz_2}{\sqrt{2}} \quad (20)$$

and suitable creation and annihilation operators

$$a_i = -i\sqrt{2}\left(\frac{\partial}{\partial \bar{y}_i} + \pi(y_i + 2iu_i)\right),$$

$$b_i = -i\sqrt{2}\left(\frac{\partial}{\partial y_i} + \pi(\bar{y}_i + 2i\bar{u}_i)\right), \quad (21)$$

which satisfy the commutation relations

$$[a_i, a_j^\dagger] = 4\pi\delta_{ij}, \quad [b_i, b_j^\dagger] = 4\pi\delta_{ij}, \quad (22)$$

with all other commutators trivial. Hence all eigenfunctions can be constructed, as for a harmonic oscillator, in terms of the functions $\chi_z(x)$ and $\chi^*_{-z}(x)$, annihilated by respectively a_i and b_i,

$$\chi_z(x) = \exp(-\tfrac{1}{2}\pi(x + nz)^2)\theta_u(y), \quad (23)$$

where $\theta_u(y)$ is holomorphic in the complex coordinates y_i. Introducing isospin projection operators $I_\pm \equiv \tfrac{1}{2}(1 \pm \tau_3)$ one easily finds that

$$D_z = \begin{pmatrix} a_1 & a_2^\dagger \\ a_2 & -a_1^\dagger \end{pmatrix}\otimes I_+ + \begin{pmatrix} b_1^\dagger & b_2 \\ b_2^\dagger & -b_1 \end{pmatrix}\otimes I_-. \quad (24)$$

With $D_z D_z^\dagger = (a_i^\dagger a_i + 4\pi)\otimes I_+ + (b_i^\dagger b_i + 4\pi)\otimes I_-$, one finds that the cokernel of D_z is trivial such that the Nahm transformation is well defined.

The splitting of D_z in isospin-up and down components is a direct consequence of the fact that A is a reducible connection. The associated line bundle L has a section $s(x) = \chi_z(x)$ and the cocycle condition $\Psi_z(x + \lambda) = g_\lambda(x)\Psi_z(x)$ reduces to

$$\chi_z(x + \lambda) = \alpha(\lambda) \exp(-\tfrac{1}{2}\pi i\lambda_\mu n_{\mu\nu}x_\nu)\chi_z(x), \quad (25)$$

which implies that $\theta_u(y+q_c)$ equals $\theta_u(y)$ up to a q and z-dependent holomorphic factor, as required for holomorphic $\theta_u(y)$. With q_c we indicate the complex two-vector constructed form $q \in \mathbb{Z}^4$ as in eq. (19) (cmp. $x_c = y$ and $(nz)_c = 2iu$). From the general theory on θ-functions one finds that $\dim \ker a_i = 1$, such that $\chi_z(x)$ is unique up to a (z-dependent) factor. For other choices of $n_{\mu\nu}$ it can be easily shown that there are $\sqrt{\det(\tfrac{1}{2}n)}$ such functions, as required by the index theorem, see ref. [17] and references therein. Explicitly

$$\chi_0(x) = \sum_{q \in \mathbb{Z}^4} \alpha(q)e^{-\tfrac{1}{2}\pi\{ixnq+(x+q)^2\}}, \quad (26)$$

$$\theta_0(y) = \sum_{q \in \mathbb{Z}^4} \alpha(q)e^{2\pi y\bar{q}_c - \tfrac{1}{2}\pi q^2},$$

from which we can construct a smooth family of zero-modes

$$\chi_z(x) = e^{-\pi izz}\chi_0(x + \tfrac{1}{2}nz). \quad (27)$$

Its norm is independent of z and found to be [17] $\chi_0(0) = (1.66925368)^2$, obtained by working out the sum over $\mathbb{Z}^4 \times \mathbb{Z}^4$ in $\int_{T^4} \chi_z^*(x)\chi_z(x)$, such that one of the sums allows one to extend the integral over the unit cell to an integral over $\mathbb{R}^4$, and evaluate the gaussian integral over x.

We therefore find two normalized zero-modes $\Psi_{ab}^{(i)}(x; z) = <a, b, x|\Psi_z^{(i)}>$ for the Weyl operator D_z. The spinor and isospin indices are denoted by a and b. The only non-zero components are

$$\Psi_{11}^{(1)}(x; z) = h(z)\chi_z(x)/\sqrt{\chi_0(0)},$$

$$\Psi_{22}^{(2)}(x; z) = h^*(z)\chi_{-z}^*(x)/\sqrt{\chi_0(0)}, \quad (28)$$

where $h(z)$ is an arbitrary phase. This phase ambiguity is equivalent to a gauge transformation in the subgroup generated by σ_3 (assigning a different phase factor to $\Psi_z^{(2)}$ will introduce in addition

a U(1) gauge component, generated by $i\sigma_0$). We will choose $h(z) = 1$.

One easily shows that eq. (25) holds[1], and that furthermore (note that $\alpha(\tfrac{1}{2}nq) = \alpha(q)$)

$$\chi_{z+q}(x) = \alpha(q)e^{-2\pi iqx}e^{-\tfrac{1}{2}\pi iqnz}\chi_z(x). \quad (29)$$

¿From this one easily deduces that (see eq. (8))

$$S_\lambda(z) = \alpha(\lambda) \exp(-\tfrac{1}{2}\pi i\lambda_\mu n_{\mu\nu}z_\nu\tau_3). \quad (30)$$

As $S_\lambda(z) = \hat{g}_\lambda^{-1}(z)$, we see that the cocycle for the Nahm bundle is the inverse of the original cocycle (see eq. (18)). The topological charge is not affected by this inversion. This can also be seen from the explicit computation of the connection $\hat{\omega}$ and its curvature $\hat{\Omega}$. Using the fact that

$$\partial_\mu\chi_0(x) = \tfrac{1}{2}\pi in_{\mu\nu}x_\nu\chi_0(x), \quad (31)$$

it is almost trivial to show that

$$\hat{A}_\mu(z) = -A_\mu(z), \quad \hat{F}_{\mu\nu}(z) = \tilde{F}_{\mu\nu}(z). \quad (32)$$

We could have anticipated the fact that the curvature of the Nahm bundle is associated to the dual of (and hence minus) the curvature of the original reducible bundle. The line bundle L of its reducible U(1) component has topological charge 1, determined from the square of the first Chern class, $c_1(L) = \tfrac{1}{4}n_{\mu\nu}dx_\mu \wedge dx_\nu$. The above construction has given us the Nahm transformation for this line bundle. Therefore we can compute $c_1(\hat{L})$ from eq. (11), which is seen to give the desired result, $c_1(\hat{L}) = \tfrac{1}{4}\bar{n}_{\mu\nu}dz_\mu \wedge dz_\nu$.

The generalization to arbitrary constant curvature connections is now straightforward and is left to the reader.

---

[1]It should be noted that there are many representations of the θ-function by a lattice sum. Any holomorphic function $r(y)$, under some mild conditions that guarantee convergence of the lattice sum, gives a proper solution by defining $\tilde{\theta}(y) = \sum_{q \in \mathbb{Z}^4} \alpha(q)e^{2\pi y\bar{q}_c - \tfrac{1}{2}\pi q^2}r(y+q_c)$. For example, this leads to the following alternative choice for a family of solutions, $\tilde{\chi}_z(q) = \sum_{q \in \mathbb{Z}^4} \alpha(q)e^{-\tfrac{1}{2}\pi\{ixnq+(x+q+nz)^2\}}$. The normalized solution $\tilde{\chi}_z(x)/\tilde{\chi}_z(-nz)^{\frac{1}{2}}$ is necessarily related to $\chi_z(x)/\chi_0(0)^{\frac{1}{2}}$ by a phase $h(z)$. As $\tilde{\chi}_z(-nz)$ vanishes whenever $z_0 = z_3 = \tfrac{1}{2}$ or $z_1 = z_2 = \tfrac{1}{2}$, $h(z)$ represents a singular gauge transformation, also signaled by the fact that $\tilde{\chi}_z$ gives rise to a trivial cocycle, $\tilde{S}_\lambda(z) = 1$. Yet, with $\hat{\omega} = \tfrac{1}{4}i\partial_\mu \log \tilde{\chi}_z(-nz)n_{\mu\nu}dz_\nu$ one does retrieve the correct curvature, using that $\partial_{u_i}\partial_{a_i} \log \tilde{\chi}_z(-nz) = -2\pi\delta_{ij}$.

P. van Baal / Nuclear Physics B (Proc. Suppl.) 49 (1996) 238–249

4. NUMERICAL RESULTS

To warm up for our discussion of instantons on $T^3 \times \mathbb{R}$ it is useful to first consider the case of instantons for the O(3) non-linear sigma model in two dimensions. The model is described by a field that lives on S^2, parametrized through stereographic projection by a complex function u(x+iy). Instantons [18] on $\mathbb{R}^2$ are meromorphic functions $u(z) = c \prod_i (z - a_i)/(z - b_i)$, where the topological charge equals the number of poles. A standard result in complex function theory says that a periodic meromorphic function with only one pole cannot exist, ruling out the existence of charge 1 instantons on T^2. Higher charged instantons are constructed from the Weierstrass σ-functions, see ref. [19].

¿From the Hamiltonian point of view, the classical vacuum is degenerate along the constant functions on the circle S^1. Its moduli space (the equivalent of the space of flat connections on T^3) is equal to S^2, parametrized by a complex number. Charge 1 instantons, corresponding to vacuum to vacuum tunnelling events do, however, exist for $S^1 \times \mathbb{R}$. Taking $S^1 = \mathbb{R}/2\pi\mathbb{Z}$ and introducing $z \equiv t + ix$ one finds [20,21]

$$u(z) = -(c + de^z)/(a + be^z), \quad ad - bc \neq 0. \quad (33)$$

It follows that $u(t \to \infty) = -d/b$ and $u(t \to -\infty) = -c/a$, two *different* points in the classical vacuum. As for $\mathbb{R}^2$ the charge 1 instanton has six parameters, e.g. the complex ratios b/a, c/a and d/a, two of which are associated to translations, and three to O(3) rotations. The scale parameter is no longer associated to a symmetry (as S^1 has a fixed size). A suitable parametrization of charge one instantons [21] is found in terms of the polar angle $\varphi \in [0, \pi]$

$$u(z) = \operatorname{ctg}(\tfrac{1}{2}\varphi) - e^z / \sin(\tfrac{1}{2}\varphi), \quad (34)$$

chosen such that $u(t \to \infty) = \infty$ (the north pole) and $u(t \to -\infty) = \operatorname{ctg}(\tfrac{1}{2}\varphi)$, whereas $u(0)$ is real and the maximum of the energy occurs at $t = 0$. This is seen to fix all symmetries. Other solutions can be found by applying symmetry transformations to this basic solution. Apparently the angle φ is related to the size of the instanton. On the other hand, φ is the geodesic distance between the points $u(t \to \pm\infty)$ in the classical vacuum. A suitable definition for the size ρ of the instanton is given by [21] $\rho(\varphi) = \sin(\tfrac{1}{2}\varphi)$. We explicitly see how the instanton is becoming singular if one tries to impose periodic boundary conditions, $\varphi \to 0$. On the other hand for antipodal $u(t \to \pm\infty)$, which are like anti-periodic boundary conditions, the instanton has maximal size. In this limit the energy profile is constant on S^1 and the configuration at $t = 0$ corresponds to a static solution of the equations of motion with one unstable mode in the direction of tunnelling. This configuration is therefore a sphaleron, which gives the minimal energy barrier to be taken along the tunnelling path.

The situation for $T^3 \times \mathbb{R}$ is seen to be quite similar. In the absence of exact results we have used the lattice approximation to study the instanton solutions numerically. Within the theory of finite volume gauge theories [22] these instanton solutions provide an essential ingredient to go to larger volumes. Typically instantons become dynamically relevant for the low-lying glueball spectrum in volumes around one half to one cubic fermi. The finite volume sphaleron [13] sets the energy scale beyond which tunnelling effects are no longer exponentially suppressed. The tunnelling path through the sphaleron [12] gives important information on the degrees of freedom that need to be accounted for non-perturbatively. For a recent review of the dynamical aspects see ref. [23].

The standard Wilson action for the lattice is a discrete approximation to the continuum action. This discretization breaks the scale invariance, such that the action of approximate instanton solutions will weakly depend on the moduli parameters. In particular the action is decreasing as the scale parameter of the instanton is lowered. As a consequence, looking for solutions by lowering the action towards a local minimum (called cooling [25]), moves the configuration to (approximate) instantons of ever smaller size, at some point "falling through the lattice". To reverse the shrinking of instantons under cooling we have introduced over-improved cooling [12]. This is cooling with a modified lattice action, which reverses the size of the scaling violations (rather than removing them to second order in the lattice spacing). This is achieved by adding to the standard Wilson action plaquettes of size 2×2, with a relative coefficient of $-1/40$. As a consequence, the cooling now increases the size of the instanton, until it cannot grow further due to the finite volume. Even on a lattice this provides an exact solution. By scaling-up the number of lattice points a rather good approximation to the continuum solution can be found, because lattice artefacts are small for large instantons.

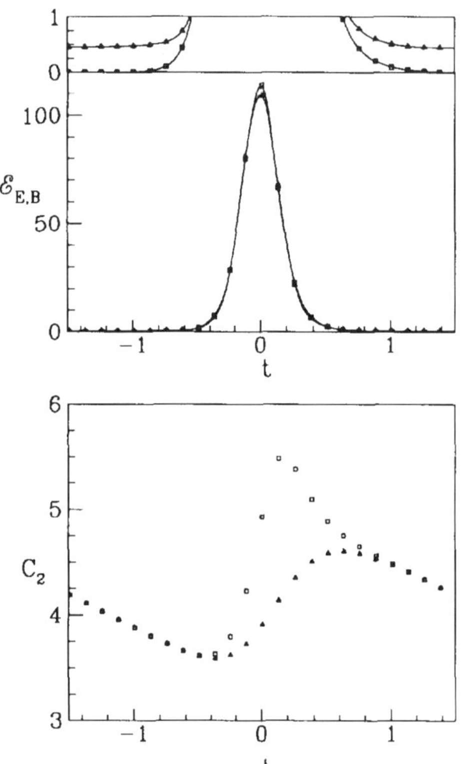

Figure 1. Numerical results [12] (after scaling appropriately with N_s) for the case of an $8^3 \times 24$ lattice with periodic boundary conditions, obtained from over-improved cooling. In the top figure electric ($\mathcal{E}_E(t)$ triangles) and magnetic ($\mathcal{E}_B(t)$ squares) energies are plotted. The tails are plotted at an enlarged scale in the inset. In the lower figure we plot $C_2(t)$ through two distinct spatial points on the lattice.

We have clearly observed that the instanton does not like to have periodic boundary conditions in the time direction. As we discussed before, the obstruction against the existence of charge one instantons on T^4 seems to imply that the configuration becomes peaked in the limit that it tends to a solution. For the O(3) model this persists even when time has an infinite extent. With the over-improved action this effect is countered by the tendency of instantons to grow under cooling. Indeed we found these effects to balance each other at some point. Nevertheless, the resulting lattice solution clearly shows deviations from self-duality, see fig. 1.

As we already described in the introduction, the classical vacua on T^3 are given by flat connections. They can be parametrized by abelian constant vector potentials $A_i^{\text{flat}} = iC_i\tau_3/2L$ for a three-torus of size L. Not all of these are gauge inequivalent, and it is most convenient to parametrize these flat connections by the Wilson loops that close because of the periodic boundary conditions (also called Polyakov loops) for each of the three generating circles of the torus

$$\begin{aligned}
P_i(x) &\equiv \tfrac{1}{2}\operatorname{Tr}\operatorname{Pexp}\{\int_0^L A_i(x + s\hat{e}_i)ds\} \\
&\equiv \cos(C_i(x)/2).
\end{aligned} \tag{35}$$

One easily finds that $P_i^{\text{flat}}(x) = \cos(C_i/2)$. For large extensions T in the time direction, the finite action of an instanton requires the potential energy to go to zero at both ends, such that A_i will approach a flat connection. As cooling brings the configuration to a solution of the equations of motion, one easily finds that restricted to the set of flat connections, the electric field $(\partial_t C_i)$ has to vanish too. Self-duality would require C_i to become time-independent. From the lower part of figure 1 it is clear that this is not the case for our lattice solution. Indeed, it can be verified that the slope of $C_i(t)$ (only shown for $i = 2$) is entirely responsible for the electric energy in the tail regions where the magnetic energy vanishes to a high accuracy. It should be noted that for increasing T (N_t lattice units) this so-called electric tail will go down as $1/T^2$ and from these numerical studies we can not rule out if instantons do exist when $T \to \infty$.

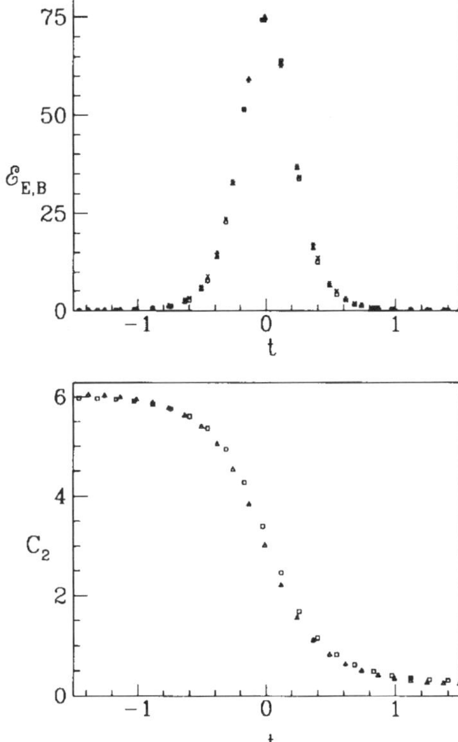

Figure 2. Numerical results [12] for lattices with $N_s = 7$ and 8 (and $N_t = 3N_s$), properly scaled with N_s and using twisted boundary conditions ($n_{0i} = 1$). In the top figure electric ($\mathcal{E}_E(t)$ squares at $N_s = 7$ and triangles at $N_s = 8$) and magnetic ($\mathcal{E}_B(t)$ crosses at $N_s = 7$ and stars at $N_s = 8$) energies are plotted. In the lower part we plot $C_2(t)$ through a given spatial point.

Nevertheless, the situation is dramatically different when twisting the boundary conditions in the time-direction, as demonstrated in fig. 2. Lattice artefacts are seen to be very small by comparing results on lattices with $N_s = 7$ and $N_s = 8$. The electric tail has completely disappeared, as can also be seen from the behaviour of $C_i(t)$, and the solution is perfectly self-dual. Twisting the boundary conditions can be most simply achieved in the $A_0 = 0$ gauge by applying an anti-periodic gauge transformation [10],

$$g(x) = \exp\{\pi i (x_1 + x_2 + x_3)\tau_3/L\} \qquad (36)$$

at $t = T$, properly transposed to the lattice. It has the effect of changing the sign of all Wilson loops, $P_i(t = T) = -P_i(t = 0)$. One can also apply a twist such that, say, only P_1 is (anti-)periodic. In all these cases instantons can be shown to exist. Assuming that solutions persist under taking the limit $T \to \infty$, this shows existence of instanton solutions on $T^3 \times \mathbb{R}$.

It is tempting to expect that solutions will exist for any value of P_i^{flat}. It would give a natural way of counting the moduli parameters, adding a scale parameter and the four translation parameters. It should be said that this is not the way one counts the parameters when glueing in a localized solution from $\mathbb{R}^4$ to the flat four-dimensional connection. In this case there are three so-called attachment parameters [4,11]. Assuming the conjecture to be true, however, we can prove the existence of an instanton solution on $T^3 \times \mathbb{R}$ with $P_i = 0$ at *both* ends. We have shown that this solution cannot be compactified, as it is also compatible with periodic boundary conditions. It might, however, show the same behaviour on $T^3 \times \mathbb{R}$ in approaching $P_i(t \to \pm\infty) = 0$ as the O(3) instanton for $\varphi \to 0$. Over-improved cooling with twisted boundary conditions will lead to a solution with a fixed value of P_i^{flat}, driven there by the well defined (but hard to control) scaling violations.

Fixing P_i^{flat} at the the two ends is easily implemented on the lattice. For the continuum, in the $A_0 = 0$ gauge, one would need to apply a gauge transformation with unit winding number to the constant abelian representation of the flat connection at $t - T$ to admit an instanton. On the lattice a smooth configuration can only be specified by requiring gauge invariant quantities to be smooth. A quantity like the vector potential (on the lattice encoded in link variables) is in general not smooth, unless one imposes in addition a "smoothing" gauge condition, like the Coulomb gauge. It implies that configurations can be in a singular gauge, with the singularity "hidden" between the meshes of the lattice. In this way the periodic boundary conditions of the lattice are no obstacle for having smooth instanton configurations.

Initially, fixed boundary conditions were introduced to search for instantons with minimal energy barrier to locate the sphaleron[2]. We can, however, now also study solutions with $|P_i|$ different at both ends. Our results indicate that, like for the case of periodic boundary conditions, solutions exist that become self-dual when T is chosen sufficiently large. If this is also true in the continuum remains to be seen, as we have insufficient control over interchanging the limits N_t and $N_s \to \infty$.

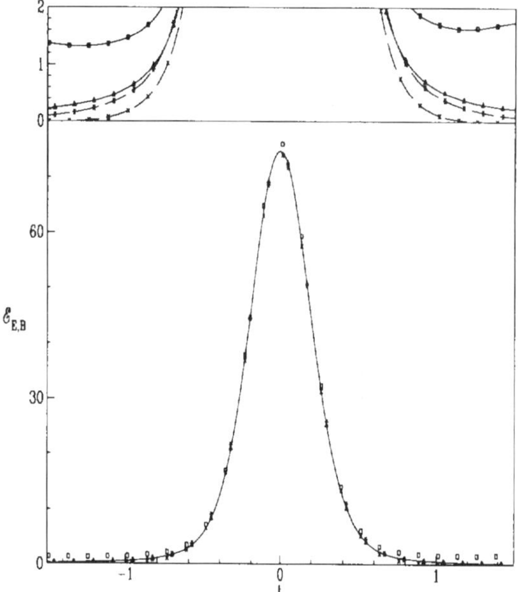

Figure 3. The electric and magnetic energies obtained after over-improved cooling for a lattice of spatial size $N_s = 8$ with boundary conditions fixed in the time direction to $P_i = -1$ at one and $P_i = 1$ at the other end [13]. Results are for $N_t = 24$ (squares for $\mathcal{E}_E$ and crosses for $\mathcal{E}_B$, respectively the upper and lower curves in the inset) and $N_t = 48$ (triangles for $\mathcal{E}_E$ and stars for $\mathcal{E}_B$, respectively the middle upper and lower curves in the inset).

---

[2] The sphaleron was also constructed more directly using saddle-point cooling [13]. It was found to have the remarkable property that the gauge invariant part of the magnetic field, $\mathrm{Tr}\, B_k^2(\vec{x})$, is independent of the direction k for each point $\vec{x}$ separately. Cubic invariance only requires this to be the case for the average. So far we have been unable to find an analytic expression for this sphaleron.

Since counting the number of moduli parameters (based on index theorems) in general requires compact four-dimensional manifolds, it is not ruled out that the moduli space of charge one instantons on $T^3 \times \mathbb{R}$ is of higher dimension than eight. We should point out, however, that despite of the similarities with the non-linear sigma model, there is a marked difference. Different points in the vacuum valley are not related by a symmetry. Indeed, the shape of the static potential $V(A)$ depends strongly on the position within the vacuum valley. Our analysis has, however, established that the instanton that corresponds to tunnelling through the sphaleron is associated to $P_i(t \to \pm \infty) = \pm 1$. These are rather special points of the vacuum valley, with directions in field space where $V(A)$ shows quartic, rather than quadratic behaviour. This also means that the approach of the tunnelling path to these points in the vacuum valley is not guaranteed to be exponential in time, as seems to be confirmed by our numerical results illustrated in fig. 3. Note that by forcing the configuration to be in the vacuum valley at either end (with $P_i^{\mathrm{flat}} = \pm 1$) we also destroy the self-duality, in spite of the fact that the configuration in fig. 3 is compatible with twisted boundary conditions. At finite T, instantons do no longer correspond to vacuum-to-vacuum tunnelling solutions (it takes an infinite time to exactly reach the vacuum).

It might very well be that in the continuum only instantons with $P_i(t \to \pm \infty) = \pm 1$ will be exact solutions. At the other extreme there seems to be the possibility of having three additional moduli parameters. ¿From the physical point of view the most urgent problem would be to get some analytic hold on the instanton solution that tunnels through the sphaleron, so as to be able to take these effects into account beyond a semiclassical approximation, mandatory when the wave functionals of states that appear in the low-lying spectroscopy will no longer be exponentially suppressed near the sphaleron. For $S^3 \times \mathbb{R}$ such a study [26] has been recently shown to be feasible [27]. Also from the mathematical point of view, the properties of the moduli space for $T^3 \times \mathbb{R}$ present an interesting challenge. In the next section we make some general observations about the Nahm transformation in this context.

5. THE NAHM TRANSFORMATION ON $T^3 \times \mathbb{R}$

When time extends to infinity, the partial integration involved, going from eq. (13) to (16) can pick up a boundary term,

$$\hat{F}^{ij}_{\mu\nu}(z) = 8\pi^2 < \Psi^{(i)}_z | \sigma^\dagger_{[\mu} \sigma_{\nu]} G_z | \Psi^{(j)}_z > \qquad (37)$$

$$- 4\pi i \oint \frac{\partial \Psi^{(i)}_z(x)^\dagger}{\partial z_{[\mu}} \sigma^\dagger_\lambda \sigma_{\nu]} \Big(G_z \Psi^{(j)}_z \Big)(x)\, d^3_\lambda x.$$

This boundary term in general destroys the anti-selfduality of $\hat{\Omega}$. It is useful to review the situation for $\mathbb{R}^4$, so as to get some insight in how the Nahm transformation needs to be modified in the presence of these boundary terms [5,6]. If $\Psi(x)$ is a normalized zero-mode of the Weyl operator D, also $e^{-2\pi i z x}\Psi(x)$ is normalized and easily seen to be a zero-mode of D_z. Hence

$$\Psi_z(x) = e^{-2\pi i z x}\Psi(x), \qquad (38)$$
$$G_z(x,y) = e^{2\pi i z(y-x)}G(x,y).$$

Compactifying to S^4 and using conformal invariance allows for a regular description of the gauge field at infinity, after a suitable gauge transformation $g(x)$ (with unit winding number). From this a solution of the Weyl equation is seen to have the following asymptotic behaviour [5,6]

$$\Psi(x) = \sigma^\dagger_\mu x_\mu g(x)\frac{\alpha}{\pi}|x|^{-4}+\mathcal{O}(|x|^{-4}), \qquad (39)$$

$$(G\Psi)(x) = -\tfrac{1}{4}\sigma^\dagger_\mu x_\mu g(x)\frac{\alpha}{\pi}|x|^{-2}+\mathcal{O}(|x|^{-3}),$$

where α, associated to each zero mode, is a constant factor with spinor and isospin (i.e. group) indices. We easily deduce from this that the Nahmtransformed connection is z-independent and that its curvature is no longer anti-selfdual

$$\hat{A}^{ij}_\mu = -2\pi i < \Psi^{(i)}|x_\mu|\Psi^{(j)}>, \qquad (40)$$

$$\hat{F}^{ij}_{\mu\nu} = 8\pi^2 < \Psi^{(i)}|\sigma^\dagger_{[\mu}\sigma_{\nu]}G|\Psi^{(j)}> +\pi^2\alpha^\dagger \sigma_{[\mu}\sigma^\dagger_{\nu]}\alpha.$$

The boundary term in this case has added a self-dual contribution to the anti-selfdual part that was already present. Nevertheless, one can proceed to attempt to perform the Nahm transformation a second time. The Weyl operator now becomes algebraic, $\hat{D}^-_x = \sigma_\mu(\hat{A}_\mu + 2\pi i x_\mu)$, but

$\hat{D}^-_x \hat{D}^+_x$ no longer commutes with the quaternions, since $\hat{F}_{\mu\nu} = [\hat{A}_\mu, \hat{A}_\nu]$ has a self-dual component. In this way, applying the Nahm transformation again will not give an anti-selfdual connection. However, one can modify the Weyl operator as follows [5,6]

$$\hat{\tilde{D}}^-_x \equiv \Big(2\pi\sigma_2\alpha, \hat{D}^-_x\Big) = \Big(2\pi\sigma_2\alpha, \sigma_\mu(\hat{A}_\mu+2\pi i x_\mu)\Big).(41)$$

This leads to the desired commutation with the quaternions (tr is the trace on the spinor indices)

$$\hat{\tilde{D}}^-_x \hat{\tilde{D}}^+_x = 4\pi^2\alpha\otimes\alpha^\dagger +(\hat{A}_\mu+2\pi i x_\mu)^2+\sigma_{[\mu}\sigma^\dagger_{\nu]}\hat{F}_{\mu\nu}$$

$$= (\hat{A}_\mu+2\pi i x_\mu)^2 -2\pi^2\text{tr}\,(\alpha\otimes\alpha^\dagger), \qquad (42)$$

using $\sigma_{[\mu}\sigma^\dagger_{\nu]}\hat{F}_{\mu\nu}/\pi^2 = \sigma_{[\mu}\sigma^\dagger_{\nu]}\alpha^\dagger \sigma_{[\mu}\sigma^\dagger_{\nu]}\alpha = -4\tau_i\alpha^\dagger\tau_i\alpha$ and the completeness relation for the Pauli matrices, $\tau^{ab}_i\tau^{cd}_i = \delta_{ad}\delta_{bc}-\tfrac{1}{2}\delta_{ab}\delta_{cd}$. The Nahm transformation can be shown to be complete and coincides with the algebraic ADHM [15] construction.

It is also worthwhile to discuss the construction for monopoles [5], which are time independent self-dual solutions on $\mathbb{R}^3$. Here A_0 plays the role of the Higgs field, required to approach a fixed length at infinity, which can always be scaled to one. For SU(2) we therefore can choose $A_0 \to -\tfrac{1}{2}ig(x)\tau_3 g^{-1}(x)$, for $r = |\vec{x}| \to \infty$, with $g(x)$ a non-trivial gauge transformation. As for the instantons, one easily sees that $\hat{A}_\mu(z)$ is independent of z_i. The remaining parameter $z = z_0$ would be expected to have infinite range. But for $\Psi \in \ker D_z$, the asymptotic behaviour of the Weyl equation is seen to require $\det(A_0 + 2\pi i z) > 0$, in order for it to admit normalizable zero-modes [5,29], thus $z \in (-1/4\pi, 1/4\pi)$. Amazingly, the BPS [28] monopole solution can be retrieved by putting $\hat{A}_\mu(z) = 0$ as a solution to the one dimensional duality equations on this interval and by performing the Nahm transformation [5].

For $T^3 \times \mathbb{R}$, eq. (37) will only give a non-trivial boundary term from the $\lambda = 0$ contribution. For convenience we will choose $T^3 = \mathbb{R}^3/\mathbb{Z}^3$, i.e. $L = 1$. Like for $\mathbb{R}^4$, the z_0 dependence associated with the non-compact time parameter, becomes trivial, $\Psi_z(x) = \exp(-2\pi i z_0 t)\Psi_{\vec{z}}(x)$. The Weyl equation, with $D_{\vec{z}}$ its spatial part, reduces to

$$D_z\Psi_z(x) = e^{-2\pi i z_0 t}(\partial_0 + D_{\vec{z}})\Psi_{\vec{z}}(x), \qquad (43)$$

in the $A_0 = 0$ gauge, and we can study the zero-modes in terms of the spectral decomposition of the spatial part of the Weyl equation, which for a flat connection reduces, up to a gauge, to

$$D_{\vec{z}}(A^{\text{flat}})e^{2\pi i \vec{p}\cdot\vec{x}}\Psi_{\vec{z}}^s(\vec{p}) = e^{2\pi i \vec{p}\cdot\vec{x}}E_{\vec{z}}^s(\vec{p})\Psi_{\vec{z}}^s(\vec{p}), \quad (44)$$

where $s = \pm\frac{1}{2}$ is the isospin of the solution and $\vec{p} \in \mathbb{Z}^3$ gives the momentum. The momentum and isospin eigenstates satisfy the spinor equation

$$\vec{\tau}\cdot(2\pi\vec{p}+s\vec{C}+2\pi\vec{z})\Psi_{\vec{z}}^s(\vec{p}) = E_{\vec{z}}^s(\vec{p})\Psi_{\vec{z}}^s(\vec{p}), \quad (45)$$

with positive and negative energy eigenvalues given by

$$E_{\vec{z}}^s(\vec{p}) = \pm\|2\pi\vec{p}+s\vec{C}+2\pi\vec{z}\|. \quad (46)$$

When $A = A^{\text{flat}}$, the zero-modes can be decomposed according to

$$\Psi_{\vec{z}}(x) = \sum_{\vec{p},s,v} a_{\vec{p}}^{s,v}(t)e^{2\pi i \vec{x}\cdot\vec{p}}\Psi_{\vec{z}}^{s,v}(\vec{p}), \quad (47)$$

where v labels the positive or negative energy states[3].

At fixed $\vec{C}$, for all but a finite number of values of $\vec{z}$ within one unit cell of $\hat{T}^3 = \mathbb{R}^3/\mathbb{Z}^3$, the spectrum of $D_{\vec{z}}(A^{\text{flat}})$ has a gap, in which case the zero-modes of the Weyl equation are expected to decay exponentially in time. There would in this case be no boundary term contributing to eq. (37). As long as $\vec{C} \neq \vec{0} \bmod 2\pi$, any instanton solution will approach A^{flat} exponentially in time and the conclusion that $\Psi_{\vec{z}}$ decays exponentially in time will hold. If, however, A^{flat} is associated to one of the quartic points in the vacuum valley, $\vec{C} = \vec{0} \bmod 2\pi$, this might require more care. Assuming for the moment that this will cause

---

[3] In the background of an instanton configuration which connects the flat connections at either end, each eigenstate of $D_{\vec{z}}(\vec{A}(t,\vec{x}))$ evolves smoothly as a function of t from one state in the spectrum determined by eq. (46) at $t = -\infty$ to another at $t = \infty$, of course with the appropriate values of $\vec{C}$ inserted at either end. From the relation between the index theorem and spectral flow, we still expect that generically there is exactly one such state Ψ_t (with eigenvalue E_t) that moves from negative energy at $t = -\infty$ to positive energy at $t = \infty$. Eq. (43) would imply that in an adiabatic approximation $\Psi_{\vec{z}} \sim \exp(-E_t t)\Psi_t$ and the spectral flow provides a natural explanation for the existence of a normalizable zero-mode. We will not attempt to make this more precise here.

no problems for the behaviour of the zero-modes of the Weyl equation, also in this case boundary terms will be absent for $\vec{z} \neq \vec{0} \bmod \frac{1}{2}$. As a consequence eq. (16) will remain valid almost everywhere, with a correction that has support at a finite number of points only. From the theory of distributions, this correction term should be expressible in terms of delta functions. This provides the interesting suggestion to study the BPS equations, $B_k = D_k A_0$, in the presence of point-like sources. Performing the Nahm transformation away from these finitely many point-sources might provide us with a method of constructing instanton solutions on $T^3 \times \mathbb{R}$. Future work will tell if we require more detailed information on these singularities in order to successfully construct non-trivial solutions.

6. CONCLUSION

We have reviewed the role of the Nahm transformation in constructing solutions of the self-duality equations for gauge theories, with an emphasis on the applications to T^4 and in particular $T^3 \times \mathbb{R}$. On T^4 we explicitly illustrated this transformation for the class of (reducible) constant curvature solutions, which form special (singular) points in the moduli space. The quest for explicit instanton solutions on a torus is motivated by the dynamical study of glueball spectroscopy in small to intermediate volumes. Numerical lattice studies of these instantons have therefore been developed recently and have been reviewed here too. These numerical studies have led to interesting conjectures on the moduli space of charge one instantons on $T^3 \times \mathbb{R}$. It provides the motivation for a renewed attack on the study of the Nahm transformation for this setting. We showed how singularities arise due to certain boundary terms (like for $\mathbb{R}^4$), related to the asymptotic behaviour of the chiral zero-modes of the Dirac-Weyl equation. The study of the BPS equations on T^3 with finitely many sources is hoped to give us an analytic technique to construct the charge one instanton on $T^3 \times \mathbb{R}$, using the Nahm transformation.

P. van Baal/Nuclear Physics B (Proc. Suppl.) 49 (1996) 238–249

ACKNOWLEDGEMENTS

I am grateful to Gerhard Weigt and Dieter Lüst for their generous invitation to the Buckow meeting. I thank Andreas Wipf for discussions concerning instantons for the O(3) model. I would like to also thank my many collaborators for their interest in these problems. In particular I am grateful to Peter Braam and Margarita García Pérez for many stimulating and helpful discussions over the years. The numerical aspects of this work were supported in part by a grant from "Stichting Nationale Computer Faciliteiten (NCF)" for use of the CRAY C98 at SARA.

REFERENCES

1. S. Donaldson and P. Kronheimer, The geometry of four manifolds (Oxford University Press, 1990).
2. E. Witten, J. Math. Phys. 35 (1994) 5101; Monopoles and four-manifolds, and references therein, hep-th/9411102.
3. P. van Baal, Acta Phys. Pol. B21 (1990) 73.
4. C. Taubes, J. Diff. Geom. 19 (1984) 517.
5. W. Nahm, Phys. Lett. B90 (1980) 413; Construction of all self-dual monopoles by the ADHM method, In:"Monopoles in quantum field theory", eds. N. Craigie, e.a. (World Scientific, Singapore, 1982); Self-dual monopoles and calorons, Lect. notes in Phys., vol 201, eds. G. Denardo, e.a. (Springer, Berlin, 1985).
6. E. Corrigan and P. Goddard, Ann. Phys. (NY) 154 (1984) 253.
7. P. van Baal, "Complex structures in gauge theories", graduate lectures, Stony Brook, spring 1986, unpublished notes.
8. H. Schenk, Comm. Math. Phys. 116 (1988) 177.
9. P.J. Braam and P. van Baal, Comm. Math. Phys. 122 (1989) 267.
10. G. 't Hooft, Nucl. Phys. B153 (1979) 141.
11. P.J. Braam, A. Maciocia and A. Todorov, Inv. Math. 108 (1992) 419
12. M. García Pérez, A. González-Arroyo, J. Snippe and P. van Baal, Nucl. Phys. B413 (1994) 535; Nucl. Phys. B(Proc.Suppl.)34 (1994) 222.
13. M. García Pérez and P. van Baal, Nucl. Phys. B429 (1994) 451.
14. M.F. Atiyah and I.M. Singer, Ann. Math. 93 (1971) 119.
15. M.F. Atiyah, V. Drinfield, N. Hitchin and Y.A. Tyupkin, Phys. Lett. A65 (1978) 185; M.F. Atiyah, Geometry of Yang-Mills fields, Fermi lectures, (Scuola Normale Superiore, Pisa, 1979).
16. G. 't Hooft, Comm. Math. Phys. 81 (1981) 267.
17. P. van Baal, Comm. Math. Phys. 94 (1984) 397.
18. A.A. Belavin and A.M. Polyakov, JETP Lett. 22 (1975) 245.
19. J.-L. Richard and A. Rouet, Nucl. Phys. B211 (1983) 447.
20. E. Mottola and A. Wipf, Phys. Rev. D39 (1989) 588.
21. J. Snippe, Phys. Lett. B335 (1994) 395.
22. J. Koller and P. van Baal, Nucl. Phys. B302 (1988) 1; P. van Baal, Acta Phys. Pol. B20 (1989) 295.
23. P. van Baal, Global issues in gauge fixing, to appear in the proceedings of the ECT* workshop "Non-perturbative approaches to QCD", July 10 - 29, 1995, Trento, Italy, hep-th/9511119.
24. C. Michael, G.A. Tickle and M.J. Teper, Phys. Lett. 207B (1988) 313; P. van Baal, Phys. Lett. 224B (1989) 397.
25. B. Berg, Phys. Lett. 104B (1981) 475; J. Hoek, M. Teper and J. Waterhouse, Nucl. Phys. B288 (1987) 589.
26. P. van Baal and N. D. Hari Dass, Nucl. Phys. B385 (1992) 185; P. van Baal and B. van den Heuvel, Nucl. Phys. B417 (1994) 215.
27. B. van den Heuvel, Glueball spectroscopy on S^3, Leiden preprint INLO-PUB-10/95, hep-lat/9509019, Phys.Lett. B in press; Nucl. Phys. B(Proc.Suppl.)42 (1995) 823.
28. M.K. Prasad and C.M. Sommerfield, Phys. Rev. Lett. 35 (1975) 760; E.B. Bogomolny, Sov. J. Nucl. Phys. 24 (1976) 861.
29. C.Callias, Comm. Math. Phys. 62 (1978) 213; R. Bott and R. Seeley, Comm. Math. Phys. 62 (1978) 235.

4 June 1998

PHYSICS LETTERS B

Physics Letters B 428 (1998) 268–276

Exact T-duality between calorons and Taub-NUT spaces

Thomas C. Kraan [1], Pierre van Baal [2]

Instituut-Lorentz for Theoretical Physics, University of Leiden, P.O. Box 9506, NL-2300 RA Leiden, The Netherlands

Received 11 February 1998
Editor: P.V. Landshoff

Abstract

We determine all $SU(2)$ caloron solutions with topological charge one and arbitrary Polyakov loop at spatial infinity (with trace $2\cos(2\pi\omega)$), using the Nahm duality transformation and ADHM. By explicit computations we show that the moduli space is given by a product of the base manifold $R^3 \times S^1$ and a Taub-NUT space with mass $M = 1/\sqrt{8\omega(1-2\omega)}$, for $\omega \in [0,\frac{1}{2}]$, in units where $S^1 = R/Z$. Implications for finite temperature field theory and string duality between Kaluza-Klein and H-monopoles are briefly discussed. © 1998 Elsevier Science B.V. All rights reserved.

1. Introduction

Properties of self-dual solutions to the Yang-Mills equations of motion have played an important role in understanding both the physical and mathematical properties of gauge theories. The last couple of years these solutions also feature prominently in the description of dualities in supersymmetric theories and in string theories, in particular for extensions to D-branes and M-theory.

We will present the calorons [1], which are instantons at finite temperature defined on $R^3 \times S^1$, in an explicit and simple form for topological charge one. The reader interested in the physical applications, like for finite temperature field theory, should skip the mathematically oriented introduction below and go directly to Section 2. Sections 3 and 4 can be skipped as well. A more detailed description will be published elsewhere.

We were inspired to pursue the case of calorons by a question posed one year ago by J. Gauntlett concerning the moduli space of calorons with non-trivial asymptotic behaviour of the Polyakov loop (non-trivial holonomy) [2]. These calorons appear as the non-trivial component of H-monopoles [3]. Oddly enough explicit solutions with non-trivial holonomy were not known. With trivial holonomy they can be obtained as an infinite periodic array of instantons, all oriented parallel in group space [1]. Recent work on the T-duality between Kaluza-Klein and H-monopoles in string theory [4] made us aware that our construction explicitly provides the classical duality transformation. It can be formulated without their embedding in string theory. We nevertheless hope this result can contribute to resolving some of the puzzles that seem to be involved in the relevant string dualities. There will be many experts better equipped than we are in addressing these stringy issues.

The Nahm transformation [5,6], also known as Mukai transformation [7] when considered as a map-

[1] E-mail: tckraan@lorentz.leidenuniv.nl.

[2] E-mail: vanbaal@lorentz.leidenuniv.nl.

ping between holomorphic vector bundles, maps self-dual fields on R^4/Λ to self-dual fields on R^4/Λ^*. Here Λ is an integer lattice and Λ^* is its dual. For the gauge group $U(N)$ this Nahm transformation interchanges the rank N and the topological charge (also mapping the first Chern class to its Hodge dual), as follows from a family index theorem [8]. The family parameter is defined in terms of the moduli space of flat $U(1)$ connections, $\omega_z = 2\pi i z_\mu dx_\mu$, which when added to the self-dual $U(N)$ connection does not change the curvature. This gives rise to a family of zero-modes for the chiral Dirac operator (the Weyl operator). The vector bundle defined over the (dual) space of flat connections thus obtained, has itself a self-dual connection. Monopoles, calorons, and instantons on R^4 can all be considered to arise from suitably chosen limits of lattices Λ. In particular for R^4 seen as a torus, all of whose sides are sent to infinity, the dual space is a single point (periods L are mapped to $1/L$ and it is in this sense that the Nahm transformation is in fact a T-duality mapping [9]). This explains the algebraic nature of the Atiyah-Drinfeld-Hitchin-Manin (ADHM) construction [10] and can be used as most elegant and straightforward derivation [6,11,12]. It can be shown that the Nahm transformation is an involution; applied twice it gives the identity operation. Furthermore it preserves the metric and hyperKähler structure of the moduli spaces [8].

In the process of sending certain periods to infinity, boundary terms arise that destroy the self-duality of the Nahm bundle, but this can be repaired by suitably extending the Weyl operator on the dual space [6,11,13]. A particularly interesting feature arises on non-compact four dimensional manifolds for which infinity has the topology of T^d, where d can be either 1, 2, or 3. These correspond respectively to instantons on $R^3 \times S^1$, $R^2 \times T^2$ and $T^3 \times R$. Note that in the latter case infinity actually contains two disconnected three dimensional tori [13]. As the solutions have finite action the connection at infinity is flat, parametrised by the Polyakov loops winding around the d circles. Combined with the flat $U(1)$ connection that is added to perform the Nahm transformation, the Weyl operator reduced to the asymptotic T^d generically has a gap. This is guaranteed to be the case as long as the combined flat $U(N)$ connection on T^d is without flat factors [12], i.e.

does not reduce to $U(1) \oplus U(N-1)$ with $U(1)$ trivial. For $SU(2)$ it is easily seen that the Weyl operator reduced to the asymptotic T^d will have a zero eigenvalue at 2^d values of z. If the holonomy in the direction i is in the center of the gauge group, the values for the components z_i of these points will coincide. As soon as (at least one of) the Polyakov loops is non-trivial, the symmetry is spontaneously broken to $U(1)$. The zero-modes of the *reduced* Weyl operator lead to non-exponential decay of the zero-modes for the *full* Weyl operator and a partial integration, required in computing the curvature of the Nahm bundle, will pick up a boundary term. These boundary terms can only occur at the 2^d points mentioned above, and therefore are distributions, indeed for the calorons easily seen to be delta functions [13]. This was already realised long ago by Nahm himself [6], but up to now this has not led to explicit construction of solutions.

While finishing this paper we became aware of Ref. [14] in which some of the same issues are addressed.

2. The solutions

For calorons we compactify the time direction by periodic identification. One requires the gauge fields to be periodic up to a gauge transformation. By a suitable choice of gauge, where A_0 tends to zero at infinity, and the topological charge k is realised by the winding number of the gauge transformation that describes A_i at spatial infinity, one has

$$A_\mu(x, x_0 + 1) = \exp(2\pi i \boldsymbol{\omega} \cdot \boldsymbol{\tau}) A_\mu(x, x_0)$$
$$\times \exp(-2\pi i \boldsymbol{\omega} \cdot \boldsymbol{\tau}), \qquad (1)$$

with τ_i the Pauli matrices. We have chosen units such that the period in the time direction equals one. Its proper value, where relevant, can be reinstated later on dimensional grounds.

The Polyakov loop, $P(x) \equiv P \exp(\int_0^1 dt A_0(x))$, is seen to satisfy

$$P_\infty \equiv \lim_{|x| \to \infty} P(x) = \exp(2\pi i \boldsymbol{\omega} \cdot \boldsymbol{\tau}). \qquad (2)$$

For the periodic case ($\omega \equiv |\boldsymbol{\omega}| = 0$) the caloron solutions are well known [1], but to this date no solutions for the general case were known. Although it was argued that for the case of non-trivial values of P_∞ these solutions are not important in the finite temperature partition function [15], it might be worthwhile to reinvestigate this issue now the solutions are known explicitly. As the finite temperature partition function requires the physical, i.e. gauge invariant, components of the fields to be periodic, we do in principle have to include also the configurations with P_∞ non-trivial.

We will first give the explicit solution before discussing its construction. Using a rotation, we can achieve $\hat{\omega} \cdot \tau = \omega \cdot \tau / \omega = \tau_3$, with $\omega \in [0, \frac{1}{2}]$. The solution is written in terms of one real ($\phi(x)$) and one complex ($\chi(x)$) function, and in terms of the (anti-)self dual 't Hooft tensors [16] $(\overline{\eta}^i_{\mu\nu})\eta^i_{\mu\nu}$ (with our conventions of $t = x_0$, $\varepsilon_{0123} = -1$)

$$\eta^i_{j0} = -\eta^i_{0j} = \overline{\eta}^i_{0j} = -\overline{\eta}^i_{j0} = \delta_{ij}, \quad \eta^i_{jk} = \overline{\eta}^i_{jk} = \varepsilon_{ijk}.$$

$$(3)$$

We find

$$A_\mu(x) = \frac{i}{2}\tau_3\overline{\eta}^3_{\mu\nu}\partial_\nu\log\phi$$

$$+ \frac{i}{2}\text{Re}\left\{(\tau_1 + i\tau_2)\left(\overline{\eta}^1_{\mu\nu} - i\overline{\eta}^2_{\mu\nu}\right)\partial_\nu\chi\right\}\phi,$$

$$(4)$$

where

$$\phi = \psi/\hat{\psi},$$

$$\chi = \pi\rho^2\psi^{-1}e^{4\pi i\omega x_0}$$

$$\times\left(s^{-1}\sinh(4\pi s\omega)e^{-2\pi ix_0} + r^{-1}\sinh(4\pi r\overline{\omega})\right),$$

$$\overline{\omega} \equiv \tfrac{1}{2}(1 - 2\omega),$$

$$\hat{\psi} = \cosh(4\pi s\omega)\cosh(4\pi r\overline{\omega})$$

$$+ \frac{\left(r^2 + s^2 - \pi^2\rho^4\right)}{2rs}\sinh(4\pi s\omega)\sinh(4\pi r\overline{\omega})$$

$$- \cos(2\pi x_0),$$

$$\psi = \hat{\psi}$$

$$+ \pi\rho^2\left(s^{-1}\sinh(4\pi s\omega)\cosh(4\pi r\overline{\omega})\right)$$

$$+ r^{-1}\sinh(4\pi r\overline{\omega})\cosh(4\pi s\omega)$$

$$+ \frac{\pi^2\rho^4}{rs}\sinh(4\pi s\omega)\sinh(4\pi r\overline{\omega}).$$

$$(5)$$

The two radii, r and s, that appear in Eq. (5) are defined by

$$r^2 = \left(x + 2\pi\omega\rho^2a\right)^2, \quad s^2 = \left(x - 2\pi\overline{\omega}\rho^2a\right)^2,$$

$$a = \hat{\omega},$$

$$(6)$$

and in a sense the solution can be seen as being built from a suitable combination of two dyons (BPS monopoles) of opposite charge, best understood in terms of an old construction by Taubes involving non-contractible loops in Yang-Mills configuration space [17]. The parameter ρ is related to the scale of the instanton solution, and the two constituent BPS monopoles are separated by a distance $\pi\rho^2$. Their mass ratio approaches $\omega/\overline{\omega}$ for large ρ, when the solution becomes static (in a suitable gauge). Some of these features are illustrated in Fig. 1. For $\omega = 0\,\text{mod}\,\frac{1}{2}$, $P_\infty = \pm 1$, the gauge symmetry is no longer broken to the $U(1)$ subgroup generated by $\hat{\omega} \cdot \tau$. In this case one of the two radii will drop out of the problem and the solution is spherically symmetric. The relation to the BPS monopole for large ρ at $\omega = 0$ can already be found in Ref. [18]. The constituent monopole description is also the basis for the results in Ref. [14], and seems to be the natural framework for discussing the situation for arbitrary gauge groups, going back to the work of Nahm [6].

For the case $P_\infty = \pm 1$ one finds $\chi = \chi^* = 1 - \phi^{-1}$ and $A_\mu = \frac{1}{2}i\tau_j\overline{\eta}^j_{\mu\nu}\partial_\nu\log\phi$, which is in the form of the celebrated 't Hooft ansatz [19], and for which $\phi^{-1}\partial^2_\mu\phi = 0$. For non-trivial values of ω such a simple characterisation is not readily available. Nevertheless, the expression of $\text{tr}F^2_{\mu\nu}(x) = -\partial^2_\mu\partial^2_\nu\log\phi$, derived for the 't Hooft ansatz [19], has a remarkable generalisation to the case of non-trivial ω,

$$\text{tr}F^2_{\mu\nu}(x) = -\partial^2_\mu\partial^2_\nu\log\psi.$$

$$(7)$$

This equation was used for constructing Fig. 1. We have also computed numerically the curvature directly from Eq. (4), checking the self-duality and verifying Eq. (7).

Finally we note that ω should not be considered as part of the moduli. For each ω one has a different set of solutions. This is particularly clear when we transform to the periodic gauge, for which $A_0 = 2\pi i\omega \cdot \tau$ at $|x| \to \infty$. For each choice of ω we have

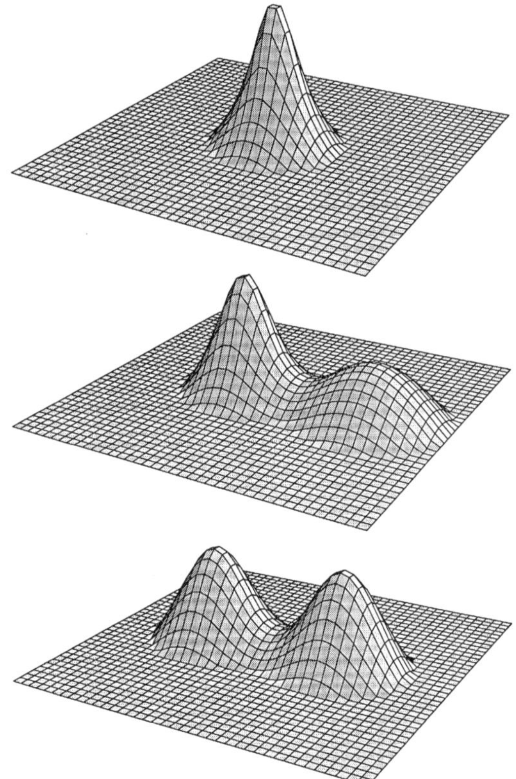

Fig. 1. Profiles for calorons at $\omega = 0$, 0.125, 0.25 (from top to bottom) with $\rho = 1$. The axis connecting the lumps, separated by a distance π (for $\omega \neq 0$), corresponds to the direction of $\hat{\omega}$. The other direction indicates the distance to this axis, making use of the axial symmetry of the solutions. Vertically is plotted the action density, at the time of its maximal value, on equal logarithmic scales for the three profiles. The profiles were cut off at an action density below $1/e$. The mass ratio of the two lumps is approximately $\omega / \overline{\omega}$, i.e. zero (no second lump), a third and one (equal masses), for the respective values of ω.

an eight dimensional moduli space with as parameters the position of the caloron, which can be obtained by translating our solution in space and time, the scale ρ and a combined rotation and gauge transformation (keeping P_∞ fixed).

3. The construction

Rather than presenting the Nahm transformation [5,6], we make a shortcut by describing the ADHM construction [10], and showing how from an infinite periodic array of instantons (no longer oriented parallel in group space [15]) we can obtain the caloron. Input from the Nahm transformation comes at the point where we interpret the quaternionic valued matrices and vectors that appear in the ADHM construction (collectively known as ADHM data) as the Fourier coefficients with respect to z, appearing in the Nahm transformation. The main advantage of this approach is that already much is known about the calculus of multi-instantons in the ADHM formalism [11], in particular also for computing the metric on the moduli space [20]. The ADHM data are obtained after applying the Nahm transformation. Applying this transformation for the second time yields the construction of the self-dual field in terms of the ADHM parameters [13].

Specifically, for charge k $SU(2)$ instantons the ADHM data are given by a quaternionic valued vector $\lambda = (\lambda_1, \lambda_2, \cdots, \lambda_k)$ and a symmetric quaternionic valued $k \times k$ matrix B. We parametrise the quaternions as linear combinations of the unit quaternions, $\sigma_0 = 1_2$ and $\sigma_j = i\tau_j$. The vector λ is directly related to the asymptotic behaviour of the zero-modes for the Weyl operator, which gives rise to the boundary terms mentioned in the introduction (see Ref. [6,11,13] for details). This will be seen to be responsible for the announced delta function singularities in the case of calorons. The matrix B is related directly to the connection for the Nahm bundle. In order for the ADHM data to describe a self-dual connection, they have to satisfy a quadratic relation which states that $B^\dagger B + \lambda^\dagger \lambda$ is a non-singular symmetric $k \times k$ matrix whose entries are real (i.e. proportional to σ_0). Alternatively one may state that $B^\dagger B + \lambda^\dagger \lambda$ has to commute with the quaternions.

We replace B by $B - x$, with $x = x_\mu \sigma_\mu$ (a $k \times k$ unit matrix is implicit in our notation). The quadratic ADHM relation obviously remains valid. We note that x corresponds precisely to adding the flat $U(1)$ connection to the Nahm connection when applying the Nahm transformation for the *second* time. The self-dual gauge field is now given by [10]

$$A_\mu(x) = \frac{u^\dagger(x)\left(\partial_\mu u(x)\right) - \left(\partial_\mu u^\dagger(x)\right)u(x)}{2\left(1 + u^\dagger(x)u(x)\right)},$$

$$u^\dagger(x) = \lambda(B - x)^{-1}. \tag{8}$$

T.C. Kraan, P. van Baal / Physics Letters B 428 (1998) 268–276

There remains a redundancy which can be related to the gauge invariance for the Nahm bundle,

$$\lambda \to q\lambda T, \quad B \to T^{-1}BT, \quad A_\mu(x) \to qA_\mu(x)\bar{q}, \tag{9}$$

where q is a unit quaternion ($q\bar{q} = |q|^2 = 1$), i.e. a constant gauge transformation (we use $\bar{x}$ to denote the conjugate quaternion $x^\dagger$, note $\bar{q} = q^{-1}$), and T is an orthogonal $k \times k$ matrix with real entries. This can be used to count the number of moduli of a charge k instanton, being $8k - 3$. Including the q as moduli gives $8k$ parameters, forming a hyperKähler manifold [12,21].

The boundary condition $A_\mu(x+1) = \exp(2\pi i \boldsymbol{\omega} \cdot \boldsymbol{\tau})A_\mu(x)\exp(-2\pi i \boldsymbol{\omega} \cdot \boldsymbol{\tau})$ is compatible with the algebraic nature of the ADHM construction, and can be implemented by

$$\lambda_n = \exp(2\pi i n\boldsymbol{\omega} \cdot \boldsymbol{\tau})\zeta, \quad B_{m,n} = B_{m-1,n-1} + \delta_{m,n}, \tag{10}$$

with ζ an arbitrary quaternion, such that

$$u_{m+1}(x+1) = u_m(x)\exp(-2\pi i \boldsymbol{\omega} \cdot \boldsymbol{\tau}). \tag{11}$$

We note that this means $k = \infty$. Indeed, $A_\mu(x)$ viewed as a solution on R^4 with unit topological charge per period has an infinite total topological charge. For trivial holonomy ($\omega \equiv |\boldsymbol{\omega}| = 0 \bmod \frac{1}{2}$) it is seen that the quadratic constraint on the ADHM data is solved by choosing $B_{m,n} = (m + \xi)\delta_{m,n}$, with ξ an arbitrary quaternion, which describes the position of the caloron. The caloron size is given by $\rho = |\zeta|$ and ζ/ρ represents a constant gauge transformation.

The major obstacle for non-trivial holonomy was satisfying the non-linear constraint. This can be solved most easily by introducing a Fourier transformation [22]. Let us first give the solution in the matrix representation

$$B_{m,n} = (m + \xi)\delta_{m,n} + \hat{A}_{m,n},$$

$$\hat{A}_{m,n} = i\bar{\zeta}\hat{\boldsymbol{\omega}} \cdot \boldsymbol{\tau}\zeta \frac{\sin(2\pi\omega(m-n))}{m-n}(1 - \delta_{m,n}). \tag{12}$$

It has the right number of parameters, 8 in total, where $q \equiv \zeta/\rho$ is split in a $U(1)$ part commuting with P_∞, describing the residual $U(1)$ gauge invari-

ance, and a part $SU(2)/U(1)$ describing a rotation of the vector $\boldsymbol{\omega}$, compensated by a gauge transformation to ensure that P_∞, or equivalently the periodicity condition, is unaltered.

It is advantageous to first give some general results, valid for arbitrary instantons, giving a more efficient way of representing $A_\mu(x)$. The derivation is straightforward and will be given elsewhere. It is well known that in this problem two Green's functions appear [6,11]. One is associated to the quadratic ADHM relation

$$f_x = \left(\Delta^\dagger(x)\Delta(x)\right)^{-1}, \quad \Delta^\dagger(x) = \left(\lambda^\dagger,(B-x)^\dagger\right),$$

$$\Delta^\dagger(x)\Delta(x) = (B-x)^\dagger(B-x) + \lambda^\dagger\lambda. \tag{13}$$

$\Delta^\dagger(x)$ is a $k \times (k+1)$ dimensional quaternionic matrix. Self-duality implies that f_x commutes with the quaternions. The other Green's function is given by

$$G_x = \left((B-x)^\dagger(B-x)\right)^{-1}. \tag{14}$$

From the definition of $u(x)$ it follows that

$$\phi \equiv 1 + \lambda G_x \lambda^\dagger = 1 + u^\dagger(x)u(x). \tag{15}$$

One finds the following compact result

$$A_\mu(x) = -\tfrac{1}{2}\phi\partial_\nu\left(\phi^{-1}\left(\lambda\bar{\eta}_{\nu\mu}G_x\lambda^\dagger\right)\right), \tag{16}$$

where $\bar{\eta}_{\mu\nu} \equiv \sigma_i\bar{\eta}^i_{\mu\nu}$ (the 't Hooft tensors [16] may be defined through $\bar{\eta}_{\mu\nu} = \tfrac{1}{2}(\bar{\sigma}_\mu\sigma_\nu - \bar{\sigma}_\nu\sigma_\mu)$ and $\eta_{\mu\nu} = \tfrac{1}{2}(\sigma_\mu\bar{\sigma}_\nu - \sigma_\nu\bar{\sigma}_\mu)$).

By choosing B diagonal and the entries of λ real, this result immediately leads to the subclass of solutions that are expressed in terms of the 't Hooft ansatz [19]. The quadratic ADHM condition is obviously satisfied, and $A_\mu(x) = \tfrac{1}{2}\bar{\eta}_{\mu\nu}\partial_\nu\log\phi$.

We note that the Green's functions f_x and G_x are intimately related. In particular

$$G_x\lambda^\dagger = \phi f_x\lambda^\dagger, \tag{17}$$

which implies that (see also Ref. [23]; their conventions relate to ours by $u \to \lambda^\dagger$, $\phi \to \phi^{-1}$)

$$\phi = \left(1 - \lambda f_x\lambda^\dagger\right)^{-1}, \quad A_\mu(x) = -\tfrac{1}{2}\phi\partial_\nu\left(\lambda\bar{\eta}_{\nu\mu}f_x\lambda^\dagger\right). \tag{18}$$

As a consequence we will only need to know f_x. This Green's function is simpler to determine than

G_x, since f_x is proportional to σ_0. A standard computation, that lies at the heart of showing that the curvature obtained after applying the Nahm transformation is self-dual [6,11,8] (for which it is crucial that f_x commutes with the quaternions), yields

$$F_{\mu\nu} = 2\phi^{-1}u^\dagger\eta_{\mu\nu}f_x u. \tag{19}$$

We have now reduced the explicit computation of instanton solutions to the computation of f_x. Incidentally, it can be verified that all infinite sums involved in expressions that appear for the calorons are convergent. Eq. (19) demonstrates that it gives a self-dual solution. We now discuss the Fourier transformation, in terms of which one solves for the quadratic ADHM constraint and for the Green's function f_x. One defines

$$\hat{\lambda}(z) = \sum_m \exp(2\pi imz)\lambda_m$$
$$= (P_+\delta(z-\omega) + P_-\delta(z+\omega))\zeta,$$
$$\delta(z'-z)\hat{D}(z) = \sum_{m,n}\exp(2\pi i(mz'-nz))B_{m,n},$$
$$\tag{20}$$

where $P_\pm = \frac{1}{2}(1 \pm \hat{\omega}\cdot\tau)$. Parametrising $B_{m,n}$ as before in terms of ξ and $\hat{A}_{m,n}$, with $\delta(z'-z)\hat{A}(z) = \sum_{m,n}\exp(2\pi i(mz'-nz))\hat{A}_{m,n}$, we find

$$\hat{D}(z) = \frac{1}{2\pi i}\frac{d}{dz} + \xi + \hat{A}(z). \tag{21}$$

Thus, B has been turned into a differential operator, precisely the Weyl operator appearing in the Nahm transformation, with $\hat{A}(z) \equiv \hat{A}_\mu(z)\sigma_\mu$ the connection for the Nahm bundle [13] (up to factors $2\pi i$, to match with the conventions of the ADHM construction). The Nahm transformation would require $\hat{D}^\dagger\hat{D}$ to commute with the quaternions, which is equivalent to saying that the curvature of the Nahm connection is self-dual. Due to the boundary terms discussed in the introduction, this self-duality is violated at a finite number of points [13] and the presence of $\lambda^\dagger\lambda$ in the quadratic ADHM relation is precisely so as to correct for the violations of self-duality, in accordance with the expectations expressed in the introduction. After Fourier transformation this quadratic relation reads, with a slight abuse of notation,

$$(\Delta^\dagger(x)\Delta(x))(z)$$
$$\equiv (\hat{D}(z) - x)^\dagger(\hat{D}(z) - x) + \hat{\Lambda}(z), \tag{22}$$

where

$$\delta(z'-z)\hat{\Lambda}(z) = \hat{\lambda}^\dagger(z')\hat{\lambda}(z),$$
$$\hat{\Lambda}(z) = \bar{\zeta}(P_+\delta(z-\omega) + P_-\delta(z+\omega))\zeta. \tag{23}$$

The condition that $\Delta^\dagger(x)\Delta(x)$ has to commute with the quaternions is now seen to lead to the equation

$$d\hat{A}(z)/dz = \pi\bar{\zeta}\hat{\omega}\cdot\boldsymbol{\sigma}\zeta(\delta(z+\omega) - \delta(z-\omega)), \tag{24}$$

which is solved by (eliminating an arbitrary additive constant that can be absorbed in ξ by imposing $\int_0^1 dz\hat{A}(z) = 0$),

$$\hat{A}(z) = \bar{\zeta}\hat{\omega}\cdot\boldsymbol{\sigma}\zeta\hat{A}^{(0)}(z),$$
$$\hat{A}^{(0)}(z) = \pi(1 - 2\omega - \chi_\omega(z)), \tag{25}$$

where $\chi_\omega(z) = 1$ for $\omega < z < 1 - \omega$ (requiring $\omega \in [0,\frac{1}{2}]$) and 0 elsewhere. Fourier transformation of $\hat{A}(z)$ yields the result in Eq. (12).

As x and ξ always occur in the combination $x - \xi$, we absorb ξ by a translation in x. The computation of f_x now reduces to a one-dimensional quantum mechanical problem on the circle,

$$\left\{\left(\frac{1}{2\pi i}\frac{d}{dz} - x_0\right)^2 + r^2\chi_\omega(z) + s^2(1 - \chi_\omega(z))\right.$$
$$\left. + \frac{1}{2}\rho^2(\delta(z+\omega) + \delta(z-\omega))\right\}f_x(z,z')$$
$$= \delta(z - z'), \tag{26}$$

where the radii r and s were given in Eq. (6) (note that here $a\cdot\boldsymbol{\sigma} = \bar{\zeta}\hat{\omega}\cdot\boldsymbol{\sigma}\zeta/|\zeta|^2$). We will present the explicit analytic solution for $f_x(z,z')$ elsewhere, but it should be noted that, due to the particular form of $\hat{\lambda}(z)$, only $f_x(\omega,\omega) = f_x(-\omega,-\omega)$ and $f_x(\omega,-\omega) = f_x(-\omega,\omega)^*$, respectively real and complex functions of x_μ, will occur in the evaluation of $A_\mu(x)$ (see Eq. (18)). To obtain Eq. (4) from Eq. (18), one moves $\bar{\eta}_{\mu\nu}$ through $\hat{\lambda}(z)$, which for non-trivial ω do not commute.

We close this section by quoting a useful and remarkable result [11,20],

$$\mathrm{tr}F_{\mu\nu}^2(x) = -\partial_\mu^2\partial_\nu^2\log\det(f_x), \qquad (27)$$

which leads to the result given in Eq. (7). Although for the caloron $\log\det(f_x)$ is divergent, $\partial_\mu\log\det(f_x)$ is well defined.

4. The geometry of moduli space

The moduli space of the self-dual solutions is given by the ADHM data, or equivalently by the Nahm connections. The latter are self-dual connections on the dual space and the Nahm transformation provides precisely a T-duality [9]. It has been well established that this transformation preserves the metric and hyperKähler structure of the moduli spaces [8,12]. By computing this metric on the moduli space we can determine its geometry. We use results due to Osborn [20], which can be readily transposed to the case of the calorons and become particularly elegant after the Fourier transformation.

Since we have a closed expression for $A_\mu(x)$ in terms of the ADHM parameters, we can compute the variations $\delta A_\mu(x)$ with respect to the moduli in terms of variations of the ADHM data, summarised in terms of $\delta\Delta(x)$. The metric is obtained by computing $\|\delta A\|^2 = -\int d_4 x\,\mathrm{tr}(P\delta A_\mu(x))^2$, where P is the projection on the transverse gauge fields, achieved by applying an infinitesimal gauge transformation such that $\delta A'_\mu(x)$ satisfies the background gauge condition $D_\mu\delta A'_\mu(x) = 0$. It can be shown that [20]

$$D_\mu\delta A_\mu(x)$$

$$= \phi^{-1}u^\dagger f_x\sigma_\mu\big((\delta\Delta^\dagger)\Delta - \Delta^\dagger(\delta\Delta)\big)\bar\sigma_\mu f_x u, \qquad (28)$$

which vanishes if and only if $\frac{1}{2}\mathrm{tr}\big((\delta\Delta^\dagger)\Delta - \Delta^\dagger(\delta\Delta)\big) = 0$. This condition is precisely the background gauge condition for the Nahm connection (for T^4 this is an exact statement [8], whereas in general λ provides the corrections due to the asymptotic behaviour of the chiral zero-modes of the Weyl operator). In particular this implies that the background gauge condition is preserved under the Nahm, or T-duality, transformation. Projection to a transverse variation of the connection can therefore be achieved by applying an infinitesimal gauge transfor-

mation to the ADHM data as given in Eq. (9) ($q = 1$). Under such an infinitesimal gauge transformation, $T = \exp(\delta X) = 1 + \delta X + \cdots$,

$$\delta_X\lambda = \lambda\delta X, \quad \delta_X B = [B, \delta X], \quad \delta X^\dagger = -\delta X. \qquad (29)$$

Replacing $\delta\Delta$ by $C_X \equiv \delta\Delta + \delta_X\Delta$, the background gauge condition gives an equation for δX in terms of $\delta\Delta$,

$$\tfrac{1}{2}\mathrm{tr}\big(B^\dagger[B,\delta X] - [B^\dagger,\delta X]B + 2\delta X\Lambda + \Delta^\dagger\delta\Delta$$

$$- \delta\Delta^\dagger\Delta\big) = 0. \qquad (30)$$

For the caloron, with δX preserving the periodicity (11), the transformation of $\delta\Delta$ to a transverse variation can now be reformulated after Fourier transformation as

$$-\frac{1}{4\pi^2}\frac{d^2\delta\hat{X}(z)}{dz^2}$$

$$+ |\zeta|^2\big(\delta(z-\omega) + \delta(z+\omega)\big)\delta\hat{X}(z)$$

$$= \frac{i}{4}\mathrm{tr}\big((\delta\zeta\bar\zeta - \zeta\delta\bar\zeta)\hat\omega\cdot\boldsymbol{\sigma}\big)$$

$$\times\big(\delta(z-\omega) - \delta(z+\omega)\big), \qquad (31)$$

where $\delta(z'-z)\delta\hat{X}(z) = \sum_{m,n}\exp(2\pi i(mz' - nz))\delta X_{m,n}$. Solving for $\delta\hat{X}(z)$ gives

$$\delta\hat{X}(z) = -\pi i\frac{\mathrm{tr}\big((\delta\zeta\bar\zeta - \zeta\delta\bar\zeta)\hat\omega\cdot\boldsymbol{\sigma}\big)}{1 + 4\pi^2\omega(1-2\omega)|\zeta|^2}$$

$$\times\int_0^z dz'\hat{A}^{(0)}(z'), \qquad (32)$$

which is a zig-zag function (periodic and odd in z), with discontinuous derivatives at $z = \pm\omega$.

The following miraculous formula due to Osborn [20] allows us to compute $\|\delta A\|^2$,

$$\mathrm{tr}\big(P\delta A_\mu(x)\big)^2$$

$$= \tfrac{1}{2}\partial_\mu^2\,\mathrm{tr}\,\mathrm{Tr}\big(C_X^\dagger(2 - \Delta(x)f_x\Delta^\dagger(x))C_X f_x\big). \qquad (33)$$

Since the right-hand side of this equation is smooth and a total derivative, the integration over space and time is completely determined by the behaviour for $r = |x| \to \infty$. In this limit we may replace $f_x(z,z')$ by

$\pi r^{-1}\exp\left(-2\pi r|z-z'|+2\pi ix_0(z-z')\right)$ (near $z=z'$, properly extended as a function on $S^1 \times S^1$). From this we find

$$\|\delta A\|^2 = 2\pi^2 \operatorname{tr} \int_0^1 dz \Big\{ d\hat{B}^\dagger(z)\, d\hat{B}(z)$$
$$+ 2d\hat{\lambda}^\dagger(z)\int_0^1 dz'\, d\hat{\lambda}(z') \Big\}, \qquad (34)$$

where

$$d\hat{\lambda}(z) \equiv P_+\,\delta(z-\omega)\big(\delta\zeta + \zeta\delta\hat{X}(\omega)\big)$$
$$+ P_-\,\delta(z+\omega)\big(\delta\zeta - \zeta\delta\hat{X}(\omega)\big),$$

$$d\hat{B}(z) \equiv \delta\xi + \delta\hat{A}(z) + \frac{1}{2\pi i}\frac{d\delta\hat{X}(z)}{dz}. \qquad (35)$$

For the metric one finds the explicit result

$$\|\delta A\|^2 = 4\pi^2|\delta\xi|^2 + 8\pi^2(1+R^2)|\delta\zeta|^2$$
$$- 2\pi^2 R^2 |\zeta|^2 \left(1 + \frac{1}{1+R^2}\right)(\hat{\omega}\cdot\delta\boldsymbol{\sigma})^2, \qquad (36)$$

where $(\zeta \equiv y_\mu\sigma_\mu)$

$$R^2 = \pi^2|\zeta|^2/M^2, \quad M^{-2} = 8\omega(1-2\omega),$$

$$\tfrac{1}{2}\delta\sigma_j = \eta_{\mu\nu}^j |\zeta|^{-2} y_\mu\, dy_\nu. \qquad (37)$$

One readily recognises, putting $\delta\xi = 0$, the Taub-NUT metric [24] with mass M. For $\hat{\omega}$ in the third direction this metric is given by [25]

$$ds^2 = \left(1 + \frac{x^2}{16M^2}\right)\left(dx^2 + \tfrac{1}{4}x^2\left(d\sigma_1^2 + d\sigma_2^2\right)\right)$$
$$+ \tfrac{1}{4}x^2 d\sigma_3^2 \Big/ \left(1 + \frac{x^2}{16M^2}\right), \qquad (38)$$

where we identify $x^2 = 8\pi^2\rho^2$. We note that the Taub-NUT space is a self-dual Einstein manifold [26] and that it has a hyperKähler structure [27], inherited from the hyperKähler structure of $R^3 \times S^1$.

5. Conclusions

We have found the explicit charge one $SU(2)$ caloron solutions with the Polyakov loop at spatial infinity non-trivial. Previously only solutions for which the latter was trivial were known [1]. Those

were argued to dominate in the instanton contribution to the finite temperature partition function [15], a question that can now more directly be addressed and is perhaps of physical significance.

We have shown that the moduli space of these solutions forms a Taub-NUT space, providing an exact classical T-duality between H-monopoles and Kaluza-Klein monopoles [4]. Indeed it is well-known that the Taub-NUT metric describes (the spatial part of) the Kaluza-Klein monopole [28] with compactification radius $4M$. Most importantly we have related the holonomy to the compactification radii involved in the dual descriptions.

6. Note added in proof

For clarity we emphasise the obvious fact that the Taub-NUT space is a double cover of the moduli space of framed instantons. The relevant identification, $\zeta \to -\zeta$, corresponds to the Z_2 gauge invariance and leaves the gauge field unaltered. It has the origin as a fixed point (resulting in an orbifold singularity for the moduli space). Indirect arguments concerning the nature of the moduli space can be found in Ref. [2,29]. See also the results in Ref. [30], which appeared after the completion of our paper.

Acknowledgements

This work was started when one of us (P.v.B.) was visiting the Newton Institute during the first half of 1997. He thanks the staff for their hospitality and gratefully acknowledges discussions with Jerome Gauntlett, Nick Manton, Werner Nahm, Hiraku Nakajima, David Olive and Erik Verlinde. T.C.K. was supported by a grant from the FOM/SWON Association for Mathematical Physics.

References

[1] B.J. Harrington, H.K. Shepard, Phys. Rev. D 17 (1978) 2122; D 18 (1978) 2990.

[2] J.P. Gauntlett, J.A. Harvey, J. Liu, Nucl. Phys. B 409 (1993) 363; J.P. Gauntlett, J.A. Harvey, S-Duality and the Spectrum of Magnetic Monopoles in Heterotic String Theory, hep-th/9407111.

[3] R. Rohm, E. Witten, Ann. Phys. (NY) 170 (1986) 454.

[4] R. Gregory, J.A. Harvey, G. Moore, Unwinding Strings and T-duality of Kaluza-Klein and H-monopoles, hep-th/9708086.

[5] W. Nahm, Phys. Lett. B 90 (1980) 413; The construction of all self-dual multimonopoles by the ADHM method, in: N. Craigie et all. (Eds.), Monopoles in quantum field theory, World Scientific, Singapore, 1982.

[6] W. Nahm, Self-dual monopoles and calorons in: G. Denardo et al. (Eds.), Lect. Notes in Physics. 201, 1984, p. 189.

[7] S. Mukai, Invent. Math. 77 (1984) 101.

[8] P.J. Braam, P. van Baal, Commun. Math. Phys. 122 (1989) 267.

[9] M.R. Douglas, G. Moore, D-branes, quivers, and ALE instantons, hep-th/9603167; D.-E. Diaconescu, Nucl. Phys. B 503 (1997) 220, hep-th/9608163; J.-S. Park, Nucl. Phys. B 493 (1997) 198, hep-th/9612096; F. Hacquebord, H. Verlinde, Nucl. Phys. B 508 (1997) 609, hep-th/9707179.

[10] M.F. Atiyah, N.J. Hitchin, V.G. Drinfeld, Yu.I. Manin, Phys. Lett. A 65 (1978) 185; M.F. Atiyah, Geometry of Yang-Mills fields, Fermi lectures, Scuola Normale Superiore, Pisa, 1979.

[11] E. Corrigan, P. Goddard, Ann. Phys. (NY) 154 (1984) 253.

[12] S.K. Donaldson, P.B. Kronheimer, The Geometry of Four-Manifolds, Clarendon Press, Oxford, 1990.

[13] P. van Baal, Nucl. Phys. B (Proc. Suppl.) 49 (1996) 238, hep-th/9512223.

[14] K. Lee, Instantons and Magnetic Monopoles on $R^3 \times S^1$ with Arbitrary Simple Gauge Groups, hep-th/9802012.

[15] D.J. Gross, R.D. Pisarski, L.G. Yaffe, Rev. Mod. Phys. 53 (1983) 43.

[16] G.'t Hooft, Phys. Rev. D 14 (1976) 3432.

[17] C. Taubes, Commun. Math. Phys. 86 (1982) 257, 299; in: G.'t Hooft et al. (Eds.), Progress in gauge field theory, Plenum Press, New York, 1984, p. 563.

[18] P. Rossi, Nucl. Phys. B 149 (1979) 170.

[19] G.'t Hooft, as quoted in R. Jackiw, C. Nohl, C. Rebbi, Phys. Rev. D 15 (1977) 1642; R. Rajaraman, Solitons and Instantons, North-Holland, Amsterdam, 1982.

[20] H. Osborn, Ann. Phys. (NY) 135 (1981) 373.

[21] S.K. Donaldson, Commun. Math. Phys. 96 (1984) 387.

[22] T.C. Kraan, Nahm's transformation and instantons on $R^3 \times S^1$ poster presented at the 2nd DRSTP symposium, Dalfsen, The Netherlands, 5–6 June, 1997, unpublished.

[23] E.F. Corrigan, D.B. Fairlie, S. Templeton, P. Goddard, Nucl. Phys. B 140 (1978) 31.

[24] E.T. Newman, T. Unti, L. Tamburino, J. Math. Phys. 4 (1963) 915.

[25] S. Hawking, in: S. Hawking, W. Israel (Eds.), General Relativity, an Einstein centenary survey, Cambridge Univ. Press, 1979, p. 774.

[26] G. Gibbons, C. Pope, Commun. Math. Phys. 66 (1979) 267.

[27] N.J. Hitchin, A. Karlhede, U. Lindström, M. Roček, Commun. Math. Phys. 108 (1987) 535; M.F. Atiyah, N.J. Hitchin, The Geometry and Dynamics of Magnetic Monopoles, Princeton Univ. Press, 1988, ch. 9.

[28] R. Sorkin, Phys. Rev. Lett. 51 (1983) 87; D. Gross, M. Perry, Nucl. Phys. B 226 (1983) 29.

[29] K. Lee, P. Yi, Phys. Rev. D 56 (1997) 3711, hep-th/9702107.

[30] K. Lee, C. Lu, SU (2) Calorons and Magnetic Monopoles, hep-th/9802108.

ELSEVIER

Nuclear Physics B 533 (1998) 627–659

Periodic instantons with non-trivial holonomy

Thomas C. Kraan, Pierre van Baal

Instituut-Lorentz for Theoretical Physics, University of Leiden, P.O. Box 9506, NL-2300 RA Leiden, The Netherlands

Received 8 June 1998; accepted 11 August 1998

Abstract

We present the detailed derivation of the charge-1 periodic instantons – or calorons – with non-trivial holonomy for $SU(2)$. We use a suitable combination of the Nahm transformation and ADHM techniques. Our results rely on our ability to compute explicitly the relevant Green's function in terms of which the solution can be conveniently expressed. We also discuss the properties of the moduli space, $\mathbb{R}^3 \times S^1 \times$ Taub-NUT/$\mathbb{Z}_2$ and its metric, relating the holonomy to the Taub-NUT mass parameter. We comment on the monopole constituent description of these calorons, how to retrieve topological charge in the context of abelian projection and possible applications to QCD. © 1998 Elsevier Science B.V.

PACS: 11.10.Wx; 11.27.+d; 14.80.Hv
Keywords: Instantons; Monopoles; ADHM construction; Nahm transformation; Moduli spaces

1. Introduction

Instantons [1] and Bogomolny–Prasad–Sommerfield (BPS) monopoles [2] possess remarkable properties. They exist as exact solutions with arbitrary charges and with an action or energy, proportional to their integer charge. Therefore the multi-charge solutions can be seen as built from constituents of unit charge. Indeed, for BPS monopoles the absence of an interaction energy can be understood – for large separation – as a cancellation between the electro-magnetic and scalar interactions [3,4].

Instantons are self-dual solutions with finite action. For non-compact manifolds, the solutions must approach vacua in the non-compact directions. Due to the topology of the base manifold, these vacua can be non-trivial and can give rise to extra parameters for these self-dual solutions. For periodic instantons on $\mathbb{R}^3 \times S^1$, also called calorons, the vacuum label is given by the eigenvalues of the Polyakov loop (holonomy) around S^1 at spatial infinity. Equivalently, one may consider this vacuum as the background field

628   T.C. Kraan, P. van Baal/Nuclear Physics B 533 (1998) 627–659

on which the solution is superposed. In this respect, they are very similar to monopole solutions in broken gauge theories, a non-trivial vacuum generally breaking the gauge symmetry.

Calorons can be seen to have as constituents BPS monopoles [5,6] (N for $SU(N)$), as follows from Nahm's work [7]. The constituents are such that the net magnetic and electric charge of the caloron vanishes. Unlike for the ordinary multi-monopoles, the BPS constituents are hence of opposite charge, and thus have an attractive electro-magnetic interaction. Nevertheless, also here exact solutions exist with an *action* that does not depend on the parameters, though the solutions become static only for large separations. In order to have this non-trivial situation the Polyakov loop at spatial infinity has to be non-trivial, breaking the gauge invariance spontaneously. The eigenvalues of this Polyakov loop uniquely fix the masses of the constituent monopoles. Their separation – not surprisingly – is related to the scale parameter of the caloron solution.

In this paper we study periodic $SU(2)$ instantons with topological charge 1 and arbitrary holonomy [8]. Its purpose is to provide the necessary details for this construction. Central to our success in providing explicit and relatively simple new solutions is the construction of the relevant Green's function. In the context of the Nahm transformation, introduced here as the Fourier transform of the Atiyah–Drinfeld–Hitchin–Manin (ADHM) data [1], this can be reduced to a quantum mechanical problem on the circle with a piecewise constant potential and well-defined delta function singularities related to the holonomy. We find compact expressions for the gauge field and action density of the solution and investigate the properties of the caloron. The moduli space is described in terms of the constituent monopoles, which in our approach appear as explicit lumps in the action density. Furthermore, we relate the constituent monopole nature of these instantons to work by Taubes in which he showed how to make gauge configurations with non-trivial topological charge out of monopole fields [9].

Independently, the recent work in Ref. [10] has taken the constituent monopole description [5–7] as the starting point, suitably superposing two BPS monopole solutions to form a caloron solution.

Periodic instantons have been discussed first in the context of finite-temperature field theory [11,12], where the period (T) is the inverse temperature in euclidean field theory. A non-trivial value of the Polyakov loop will modify the vacuum fluctuations and thereby leads to a non-zero vacuum energy density as compared to a trivial Polyakov loop. It was on the basis of this observation that calorons with non-trivial holonomy were deemed irrelevant in the infinite-volume limit [12]. It should be emphasised though, that the semi-classical one-instanton calculation is no longer considered a reliable approximation. At finite temperature A_0 can be seen to play the role of a Higgs field and in a strongly interacting environment one could envisage regions with this Higgs field pointing predominantly in a certain direction, and nevertheless having at infinity a trivial Higgs field. Given a finite density of periodic instantons, in an infinite volume solutions with non-trivial holonomy (in some average sense) may well have a role to play.

In our construction of the charge-1 caloron with non-trivial Polyakov loop, we pick a particular gauge. In the periodic gauge, the spatial components of the vacuum connection at infinity can be gauged to zero. The A_0 component can only be gauged to a constant, e.g. $A_0 = 2\pi i \boldsymbol{\omega} \cdot \boldsymbol{\tau}/T$ (with τ_a the Pauli matrices), when the total magnetic charge is vanishing. This connection has obviously a non-trivial Polyakov loop at infinity

$$\mathcal{P}(\boldsymbol{x}) = P \exp\left(\int_0^T A_0(\boldsymbol{x}, x_0) dx_0 \right) \to e^{2\pi i \boldsymbol{\omega} \cdot \boldsymbol{\tau}} \tag{1}$$

(P stands for path ordering). Alternatively, connections on $\mathbb{R}^3 \times S^1$ can be formulated by embedding them in $\mathbb{R}^4$ and demanding periodicity modulo gauge transformations. Gauging with a non-periodic gauge transformation $g(\boldsymbol{x}, x_0) = e^{2\pi i x_0 \boldsymbol{\omega} \cdot \boldsymbol{\tau}/T}$, starting from the periodic gauge with zero A_i and constant $A_0 = 2\pi i \boldsymbol{\omega} \cdot \boldsymbol{\tau}/T$ at infinity, sets *all* gauge fields to zero at infinity. In that case we have

$$A_\mu(\boldsymbol{x}, x_0 + T) = e^{2\pi i \boldsymbol{\omega} \cdot \boldsymbol{\tau}} A_\mu(\boldsymbol{x}, x_0) e^{-2\pi i \boldsymbol{\omega} \cdot \boldsymbol{\tau}}, \tag{2}$$

with T the period in the imaginary time direction, i.e. the inverse temperature. Clearly, $e^{2\pi i \boldsymbol{\omega} \cdot \boldsymbol{\tau}} \equiv g_0(\boldsymbol{x})$ is the transition function or cocycle. Using the proper expression for the Polyakov loop along a path traversing the boundary between coordinate patches [13],

$$\mathcal{P}(\boldsymbol{x}) = P \exp\left(\int_0^T A_0(\boldsymbol{x}, x_0) dx_0 \right) g_0(\boldsymbol{x}), \tag{3}$$

we find the same value for the holonomy in this gauge, the holonomy at infinity now solely being carried by the cocycle $g_0(\boldsymbol{x})$. It is in this so-called "algebraic" gauge that we will calculate the generalised caloron solutions.

The charge-k instantons on $\mathbb{R}^4$ are given by the ADHM construction [1]. The Nahm transformation forms a modification of this approach, initially introduced by Nahm to study BPS monopoles [14,15]. Later developments culminated in the Nahm duality transformation on generalised tori, which forms a powerful tool for studying self-dual connections. Also caloron solutions can be treated along these lines [7]. The Nahm transformation is outlined in Section 2. We summarise in Section 3 the details of the ADHM formalism necessary for our construction of the caloron in Section 4. We make it evident how the two approaches are related by Fourier transformation. By relying on the ADHM construction we profit from the vast knowledge on multi-instanton calculus within this formalism. Section 4 forms the calculational core of this paper, in which we derive the gauge potential and a particularly simple expression for the action density. The various properties and symmetries of the caloron are unraveled in Section 5. In Section 6 we describe the moduli space of the caloron. In Section 7 we discuss the relation to Taubes' work, abelian projection [16] and possible applications to QCD.

For the benefit of the reader let us point out what is new in this paper with respect to the Letter [8], where we announced our results. Section 4 contains the result for the Green's function, crucial for the construction of the new caloron solution; Section 5 contains a more in-depth discussion of the properties; the details of the computation for the metric in Section 6, useful in their own right, make precise how the Nahm transformation preserves the metric properties of the moduli space. Finally in Section 7 a completely new application of the new caloron solutions in the context of QCD is presented (we have been able to establish similar results for $SU(3)$ as well [61], even extending to arbitrary $SU(n)$, where the caloron has n BPS constituents).

2. The Nahm transformation

We will consider a $U(n)$ bundle E with self-dual gauge connection A_μ on a four-manifold $M = \mathbb{R}^4/H$, with instanton number k. Here H is a subgroup of translation symmetries under which the physics is invariant. When H is a four-dimensional lattice, M will be the four-torus [17]. Other four-manifolds are obtained by taking appropriate limits [18]. We demand the gauge potential to be invariant modulo gauge transformations under the action of H.

An essential ingredient in Nahm's construction [15] is to add a curvature free abelian connection, $-2\pi i z_\mu dx_\mu$, to the gauge field and to study the Weyl operator

$$D_z(A) = \sigma_\mu D_z^\mu(A), \quad D_z^\dagger(A) = -\bar\sigma_\mu D_z^\mu(A), \quad D_z^\mu(A) = \partial_\mu + A_\mu - 2\pi i z_\mu. \quad (4)$$

$\sigma_\mu = (1_2, i\tau)$ and $\bar\sigma_\mu = \sigma_\mu^\dagger = (1_2, -i\tau)$ are unit quaternions. As compared to usual conventions [17,18], we replaced z by $-z$ to facilitate matching with the ADHM construction. When A is without flat factors (WFF, meaning that the vector bundle E does not split in $E' \oplus L$ for any flat line bundle L), then $D_z(A)$ will have a trivial kernel [19]. For such gauge fields $G_z(x, y) = (D_z^\dagger(A)D_z(A))^{-1}$ is well defined. The index theorem [20,21] shows that there are k normalisable zero-modes of the Weyl operator $D_z^\dagger(A)$, for each value of $z \in \hat{M} = \mathbb{R}^4/\hat{H}$, $\hat{H} = \{z \in \mathbb{R}^4 | z \cdot y \in \mathbb{Z}, \forall y \in H\}$, cf. Ref. [18]. We can therefore define a $U(k)$ connection on the space $\hat{M}$,

$$\hat{A}_\mu^{ij}(z) = \int_M dx \, \psi_z^{i\dagger}(x) \frac{\partial}{\partial z_\mu} \psi_z^j(x), \quad (5)$$

where $\psi_z^i(x)$, $i = 1, \ldots, k$ form an orthonormal basis for the Nahm bundle $\hat{E}$ of fermionic zero-modes. This is called the Nahm transformed connection.

The Weitzenböck identity [19] states

$$D_z^\dagger(A)D_z(A) = -(D_z^\mu(A)D_z^\mu(A) + \tfrac{1}{2}\bar\eta_{\mu\nu}F^{\mu\nu}(x)), \quad (6)$$

where $\bar\eta_{\mu\nu} = \bar\eta_{\mu\nu}^a \sigma_a = \bar\sigma_{[\mu}\sigma_{\nu]}$ is the anti-selfdual and $\eta_{\mu\nu} = \eta_{\mu\nu}^a \sigma_a = \sigma_{[\mu}\bar\sigma_{\nu]}$ the self-dual 't Hooft tensor [22] (note that in our conventions time is labelled by x_0 rather than x_4, and to conform with anti-selfduality of $\bar\eta$ we define $\epsilon_{1230} = 1$). As F is self-dual, the second term will vanish, and hence $D_z^\dagger(A)D_z(A)$ and $G_z(x, y)$ commute with the quaternions. This has profound consequences for the curvature associated to the Nahm connection. One finds [18]

T.C. Kraan, P. van Baal/Nuclear Physics B 533 (1998) 627–659   631

$$\hat{F}_{\mu\nu}(z) = 8\pi^2 \int\limits_{M \times M} dx\, dy\, \psi_z^\dagger(x) G_z(x,y) \eta_{\mu\nu} \psi_z(y)$$

$$+ 4\pi i \int\limits_{\partial M \times M} dS_\lambda(x)\, dy\, \frac{\partial}{\partial z_{[\mu}} \psi_z^\dagger(x) \sigma_\lambda \bar{\sigma}_{\nu]} G_z(x,y) \psi_z(y), \tag{7}$$

where ψ_z denotes the matrix with the zero-modes ψ_z^i as columns. Here we used that G commutes with the quaternions. The first term is clearly self-dual. The boundary term shows possible deviations from self-duality, which occur at the points z for which the zero-modes do not decay exponentially in the non-compact directions. In these directions the connection necessarily approaches a vacuum for the action to be finite. These vacua are labelled by the eigenvalues of the Polyakov loops $\mathcal{P}_i = P \exp \int_{C_i} A_\mu dx_\mu$ along the circles C_i corresponding to the compact directions. In the case that $e^{2\pi i z}$ becomes equal to one of these eigenvalues, the component of $A_\mu - 2\pi i z_\mu$ along C_i in Eq. (4) will develop a zero eigenvalue when approaching infinity. This gives rise to a surviving boundary term in Eq. (7) and as a result a deviation from self-duality, precisely for these specific points. As the deviations occur in single points, they will be expressible in delta functions. Hence, $\hat{F}_{\mu\nu}$ is self-dual almost everywhere. For the non-compact directions μ, the z_μ dependence of $\psi_z(x)$ is a plane wave factor, and hence $\hat{A}$ is z_μ independent. Note that the $U(k)$ symmetry in the space of zero-modes associated to A is mapped onto a gauge symmetry for $\hat{A}$. On the other hand, gauge transformations on A leave $\hat{A}$ unchanged.

For the four-torus T^4 the boundary terms are absent and instantons are mapped onto instantons. It can be shown, using the family index theorem, that under this Nahm transformation a $U(n)$ connection with topological charge k is mapped onto a $U(k)$ connection with topological charge n. The Nahm transformation on T^4 squares to the identity [17]. More explicitly, the dual Weyl operator $\hat{D}_x^\dagger(\hat{A}) = -\bar{\sigma}_\mu(\hat{\partial}_\mu + \hat{A}_\mu - 2\pi i x_\mu)$ has n zero-modes,

$$\hat{D}_x^\dagger(\hat{A})\hat{\psi}_x^i(z) = 0, \quad i = 1,\dots,n, \tag{8}$$

in terms of which the original connection $A_\mu(x)$ is reconstructed

$$A_\mu^{ij}(x) = \int\limits_{\hat{M}} dz\, \hat{\psi}_x^{\dagger i}(z) \frac{\partial}{\partial x_\mu} \hat{\psi}_x^j(z). \tag{9}$$

This suggests to use the Nahm transformation in the construction of self-dual connections on modified tori, in situations when one can explicitly find the dual connection $\hat{A}$. Generally one expects this to be feasible when the Nahm transformed bundle is simpler than the original, in particular when $\hat{M}$ is of lower dimension than M. Another simplification arises for the case of topological charge $k = 1$, since in that case the Nahm connection $\hat{A}$ is abelian. Because of the boundary terms, the second Nahm transformation will have to be modified by properly handling the singularities. The extreme case is $M = \mathbb{R}^4$, $H = 0$, where the dual manifold $\hat{M}$ is just a point and the pair formed by the dual Weyl operator and singularities reduce to matrices which precisely give the ADHM data [7,18,19,23]. The Nahm transformation on $\mathbb{R}^4/H$ encompasses the ADHM construction.

We will now consider the Nahm transformation for calorons (using the classical scale invariance of the self-duality equations, we can choose $\mathcal{T} = 1$ such that $H = \mathbb{Z}$) and monopoles in the BPS limit ($H = \mathbb{R}$). For the latter, A_0 is interpreted as the Higgs field. Thus we can unify these two cases by considering them as connections on $\mathbb{R}^3 \times S^1$. The connections for $\mathbb{R}^3 \times S^1$ are topologically classified according to their behaviour at the boundary, $S^2 \times S^1$. We give a short summary of the classification presented in Ref. [12].

For the action to be finite, it is necessary that the connections go to a vacuum at spatial infinity. Generally, gauge vacua are labelled by the conjugacy classes of representations of maps of the first homotopy group to the gauge group [19]. For $S^2 \times S^1$, we can characterise each vacuum by a gauge equivalence class of an element of the gauge group. Using a gauge transformation, this element can be chosen diagonal. The vacuum at infinity is related to the holonomy along the S^1 (or Polyakov loop), Eq. (1) in the periodic gauge. The difference of a closed Wilson loop evaluated along two curves C and C' is related with the flux through the surface swept out by the curves interpolating between C and C'. Hence, at spatial infinity where the curvature vanishes, a small deformation of the path C around which the holonomy is measured does not influence $\mathcal{P}(x)$. Only the homotopy of C is important, and the holonomy at spatial infinity is a topological invariant. Therefore, at spatial infinity the eigenvalues of $\mathcal{P}(x)$ become constants,

$$\mathcal{P}(x) \to V(\hat{x}) \exp[2\pi i \operatorname{diag}(\mu_1,\dots,\mu_n)] V^{-1}(\hat{x}), \quad \sum_i \mu_i = 0, \tag{10}$$

and we have, up to an $\hat{x}$-dependent gauge transformation V,

$$A_0 = 2\pi i \operatorname{diag}(\mu_1,\dots,\mu_n) - i \operatorname{diag}(k_1,\dots,k_n)/2r + \mathcal{O}(r^{-2}), \quad \sum_i k_i = 0. \tag{11}$$

The gauge transformation V induces a map from S^2 to the factor group G/H_∞, with H_∞ the isotropy group of $\exp[2\pi i \operatorname{diag}(\mu_1,\dots,\mu_n)]$. For $SU(n)$ these maps $V(\hat{x}) \to SU(n)/H_\infty$ are classified according to the fundamental group of H_∞. Generically, H_∞ consists of several $U(1)$ and $SU(N)$, $N > 1$ subgroups. Each $U(1)$ gives rise to a monopole winding number, related to the integers k_i.

The other topological quantum number is related to the homotopy class of the map $\partial M = S^2 \times S^1 \to SU(n)$ which occurs in the gauge transformation connecting the behaviour near the origin to that at infinity, which is classified by the instanton number $k \in \pi_3(SU(n)) = \mathbb{Z}$. Gauge connections on $\mathbb{R}^3 \times S^1$ are therefore classified by μ_i, k_i and k. We will consider the situation where the net magnetic charges of the solution vanish, $k_i = 0$. For non-zero k_i the situation is as for the (static) BPS monopoles, where the dimension of the space of fermionic zero-modes depends on z. The jumps occur exactly where $z = \mu_i$, according to the Callias–Bott–Seeley theorem [21] (the situation of non-maximal symmetry breaking, where two or more μ_i coincide, is more involved).

T.C. Kraan, P. van Baal / Nuclear Physics B 533 (1998) 627–659

For a periodic instanton with no net magnetic charges the fermionic zero-modes are associated to the instanton winding number k in the usual way. Hence, the rank of the Nahm transformed gauge potential is k and as the z dependence of the fermionic zero-modes is that of a plane wave, $\psi_{z_0,z}(x) = e^{2\pi i x \cdot z}\psi_{z_0,0}(x)$, one finds $\hat{A}$ to be z independent. The dual space $\hat{M}$ is an interval on the real line, with coordinate $z_0 \equiv z$. For monopoles, this interval is $[\mu_1, \mu_n]$ (when we order $\mu_i \leqslant \mu_{i+1}$). For calorons z is the coordinate on the dual circle and $\hat{A}$ is periodic. The Nahm transformed curvature reduces to

$$\hat{F}_{0i}(z) = \frac{d}{dz}\hat{A}_i + [\hat{A}_0, \hat{A}_i], \qquad \hat{F}_{ij}(z) = [\hat{A}_i, \hat{A}_j]. \tag{12}$$

Using the self-duality of the first term in Eq. (7), and the fact that the second term of this equation is zero almost everywhere, except for possible delta function singularities at $z = \mu_i$ one finds

$$\frac{d}{dz}\hat{A}_i + [\hat{A}_0, \hat{A}_i] + \tfrac{1}{2}\epsilon_{ijk}[\hat{A}_j, \hat{A}_k] = \sum_p \alpha_p^i \delta(z - \mu_p). \tag{13}$$

These are the celebrated Nahm equations. Any $\hat{A}_\mu(z)$ obeying the Nahm equation of the right dimensionalities and singularity structure gives rise to a BPS monopole or caloron. For monopoles this is generally proven using twistor methods [15,24], but for $SU(2)$ monopoles there exist direct proofs of the equivalence, without an intermediate twistor step [25]. For calorons the construction was formulated in Ref. [7] and the twistor method for these periodic instantons was given in Ref. [5], but a relation with existence theorems or a full circle reciprocity proof as it exists for monopoles and for instantons on $\mathbb{R}^4$ and T^4 seems not to be present.

For the $k = 1$ $SU(2)$ caloron, $\hat{A}_\mu$ is a $U(1)$ connection on the circle with two singularities corresponding to the holonomy. Note that here $\mu_2 = -\mu_1 = \omega$, with $\omega \equiv |\omega| \in [0, \tfrac{1}{2}]$. The magnetic components of $\hat{F}$ vanish and hence non-zero values and singularities are only assumed by the electric components $\hat{F}_{0i}$. By the Nahm equations $\hat{A}$ is forced to be piecewise constant. We will give the explicit ADHM construction for these calorons and show among other things that all aspects suggested in Ref. [7] arise automatically.

3. The ADHM construction

We first summarise the ADHM construction [1,26] for charge-k $SU(2)$ instantons on $\mathbb{R}^4$. Generalisation to higher groups is well known but will detract one from the simplicity of our construction. The ADHM data consist of a quaternionic row vector $\lambda = (\lambda_1, \ldots, \lambda_k)$ and a quaternionic, symmetric $k \times k$ matrix B ($\lambda_m \equiv \lambda_m^\mu \sigma_\mu$ and $B_{m,n} \equiv B_{m,n}^\mu \sigma_\mu$, with $\lambda_m^\mu \in \mathbb{R}$ and $B_{m,n}^\mu = B_{n,m}^\mu \in \mathbb{R}$). These objects are comprised in

$$\Delta(x) = \begin{pmatrix} \lambda \\ B - x \end{pmatrix}. \tag{14}$$

Here, the quaternion $x \equiv x_\mu \sigma_\mu$ denotes the position variable and a $k \times k$ unit matrix is implicit in our notation. For later use, we define $\Delta \equiv \Delta(x = 0)$. The ADHM gauge potential is given by

$$A_\mu(x) = v^\dagger(x)\partial_\mu v(x), \tag{15}$$

where $v(x)$ is a $(k+1)$-dimensional quaternionic vector, the normalised solution to

$$\Delta^\dagger(x)v(x) = 0. \tag{16}$$

For this construction to give a self-dual potential A_μ, $\Delta(x)$ has to satisfy the ADHM constraints. These demand that $\Delta^\dagger(x)\Delta(x)$ be real quaternionic (i.e. commutes with the quaternions) and invertible. It is sufficient for this to hold at $x = 0$, i.e. $\Delta^\dagger \Delta = B^\dagger B + \lambda^\dagger \lambda$ must be real quaternionic and invertible. Furthermore, $(B - x)$ should have a trivial kernel, except for k values of x, where $v(x)$ and, as a consequence, $A_\mu(x)$ are singular. This can be shown to be a gauge singularity and implements the non-triviality of the bundle and reflects the topological nature of these solutions.

This construction gives all instantons on $\mathbb{R}^4$. The following transformations

$$\lambda \to \lambda T^{-1}, \quad B \to TBT^{-1}, \quad T \in O(k),$$
$$\lambda \to g\lambda, \quad g \in SU(2), \tag{17}$$

both leave the quadratic ADHM constraint untouched. The first does not change $A_\mu(x)$, whereas the second induces a global gauge transformation. Local gauge transformations arise from the $U(2)$ symmetry $v(x) \to v(x)g(x)$ in the solution space of Eq. (16). One must divide out these symmetries in order to obtain all gauge inequivalent solutions. This reduces the dimension of the space of gauge inequivalent solutions to $8k(-3)$, depending on whether or not the global gauge degrees of freedom are included as moduli. Considering the g as moduli, the moduli space is an $8k$-dimensional hyper-Kähler manifold [19].

Many aspects featuring in the construction above have their counterparts in the Nahm transformation. The reality constraint is similar to the vanishing of the imaginary quaternions in the Weitzenböck formula, which leads to the self-duality of the Nahm connection. The symmetries in the ADHM construction can be traced back to the triviality of the gauge action in the Nahm transformation and the unitary symmetry of the fermionic zero-modes. We define two matrix inverses, the analogues of the Green's functions,

$$f_x = (\Delta(x)^\dagger \Delta(x))^{-1} \in \mathbb{R}^{k \times k}, \quad G_x = ((B - x)^\dagger (B - x))^{-1} \in \mathbb{H}^{k \times k}, \tag{18}$$

and a scalar function

$$\phi(x) = 1 + \lambda G_x \lambda^\dagger. \tag{19}$$

These Green's functions f_x and G_x are related, as can be seen from the expansion of f_x in terms of G_x,

$$f_x = (G_x^{-1} + \lambda^\dagger \lambda)^{-1} = G_x - G_x \lambda^\dagger \sum_{n=0}^{\infty} (-\lambda G_x \lambda^\dagger)^n \lambda G_x = G_x - \phi^{-1}(x) G_x \lambda^\dagger \lambda G_x .$$

$$(20)$$

Acting on Eq. (20) with $\lambda^\dagger$ on the right and/or with λ on the left, yields

$$G_x \lambda^\dagger = \phi(x) f_x \lambda^\dagger, \qquad \phi(x) = (1 - \lambda f_x \lambda^\dagger)^{-1}.$$

$$(21)$$

To solve for $v(x)$ in Eq. (16), we introduce (a matrix of spinors) $u(x)$, and obtain

$$v(x) = \phi^{-1/2}(x) \begin{pmatrix} -1 \\ u(x) \end{pmatrix}, \qquad u(x) = (B^\dagger - x^\dagger)^{-1} \lambda^\dagger,$$

$$(22)$$

where $\phi(x) = 1 + \lambda G_x \lambda^\dagger = 1 + u^\dagger(x) u(x)$ accounts for the normalisation of $v(x)$. In terms of these quantities, the gauge potential reads

$$A_\mu(x) = \tfrac{1}{2} \phi^{-1}(x)(u^\dagger(x) \partial_\mu u(x) - \partial_\mu u^\dagger(x) u(x)).$$

$$(23)$$

To show that the connection is indeed self-dual is best seen from using $F = dA + A \wedge A$, where $A = A_\mu dx_\mu$ and $F = \tfrac{1}{2} F_{\mu\nu} dx_\mu \wedge dx_\nu$. With $A = v^\dagger(x) dv(x)$ one finds

$$\begin{aligned} F &= dv^\dagger(x) \wedge dv(x) - dv^\dagger(x) v(x) \wedge v^\dagger(x) dv(x) \\ &= dv^\dagger(x)(1 - v(x) \otimes v^\dagger(x)) dv(x). \end{aligned}$$

$$(24)$$

As $1 - v(x) \otimes v^\dagger(x)$ is the projection on the orthogonal complement of the kernel of $\Delta^\dagger(x)$ (since $\Delta^\dagger(x) v(x) = 0$), we can use that $1 - v(x) \otimes v^\dagger(x) = \Delta(x) f_x \Delta^\dagger(x)$. Substituting this in the expression for F and using that

$$\Delta^\dagger(x) dv(x) = -dx_\mu \frac{\partial \Delta^\dagger(x)}{\partial x_\mu} v(x) = dx^\dagger(b^\dagger v(x))$$

$(dx \equiv \sigma_\mu dx_\mu, \ b^\dagger v(x) \equiv \phi^{-1/2}(x) u(x))$, we find [26]

$$F = (v^\dagger(x) b) dx \wedge f_x dx^\dagger (b^\dagger v(x)).$$

$$(25)$$

The crucial observation is now that the quadratic ADHM constraint implies that f_x commutes with the quaternions and that $dx \wedge dx^\dagger = \eta_{\mu\nu} dx_\mu \wedge dx_\nu$, much like in the Nahm transformation where the Weitzenböck identity, Eq. (6), guarantees that $D_z^\dagger(A) D_z(A)$ commutes with the quaternions. We thus find [23]

$$F_{\mu\nu}(x) = 2\phi^{-1}(x) u^\dagger(x) \eta_{\mu\nu} f_x u(x),$$

$$(26)$$

which is self-dual due to the self-duality of $\eta_{\mu\nu} = \sigma_{[\mu} \bar{\sigma}_{\nu]}$.

In the case at hand, Eq. (23) is of little practical use as it stands. We therefore rearrange it such that we can express A_μ in terms of evaluations of the Green's function f_x. Using $\partial_\mu (B^\dagger - x^\dagger)^{-1} = (B^\dagger - x^\dagger)^{-1} \bar{\sigma}_\mu (B^\dagger - x^\dagger)^{-1} = (B - x) G_x \bar{\sigma}_\mu (B - x) G_x$, we get

$$A_\mu(x) = \phi^{-1}(x)\, \lambda G_x \bar{\eta}_{\mu\nu}(B-x)_\nu G_x \lambda^\dagger. \tag{27}$$

We substitute $G_x \lambda^\dagger = \phi(x) f_x \lambda^\dagger$, Eq. (21), and noting that

$$\partial_\nu f_x^{-1} = \partial_\nu G_x^{-1} = -2(B-x)_\nu, \qquad (\partial_\mu f_x^{-1})f_x = -f_x^{-1}\partial_\mu f_x, \tag{28}$$

we arrive at the following compact result for the gauge potential (see also Ref. [27]):

$$A_\mu(x) = \tfrac{1}{2}\phi(x)\partial_\nu\left(\lambda\bar{\eta}_{\mu\nu}f_x\lambda^\dagger\right), \tag{29}$$

using once again that f_x commutes with the quaternions. When $\bar{\eta}_{\mu\nu}$ is moved through λ, one finds an expression for A_μ in terms of (derivatives of) "expectation values" of the Green's function f_x,

$$A_\mu(x) = \tfrac{1}{2}\phi(x)\sigma_\alpha\bar{\eta}_{\mu\nu}\bar{\sigma}_\beta\partial_\nu\phi_{\alpha\beta}, \tag{30}$$

where

$$\phi_{\alpha\beta}(x) = \phi_{\beta\alpha}(x) = (\lambda_\alpha f_x \lambda'_\beta). \tag{31}$$

At this point we can make contact with the well-known 't Hooft ansatz [28]. This forms a subclass of the ADHM construction with λ real ($\lambda_m = \sigma_0\rho_m$) and $B_{m,n} = \delta_{m,n}y_m$ diagonal, corresponding to k instantons with scales ρ_m at positions y_m. This $5k$-dimensional family trivially satisfies the ADHM constraints. In this simpler situation the gauge potential can be written even in terms of a *single* scalar potential $\phi(x) = 1 + \sum_k \rho_k^2/|x - y_k|^2$ as $A_\mu(x) = \tfrac{1}{2}\bar{\eta}_{\mu\nu}\partial_\nu \log\phi(x)$, since $\phi_{00}(x) = 1 - \phi^{-1}(x)$.

For the action density, the following expression can be found in the literature [29,23]

$$\operatorname{tr} F^2_{\mu\nu}(x) = -\partial_\mu^2\partial_\nu^2 \log\det f_x. \tag{32}$$

This expression is regular everywhere. We can rewrite Eq. (32) using Eqs. (20) and (28) as

$$\operatorname{tr} F^2_{\mu\nu}(x) = -\tfrac{1}{2}\partial_\mu^2\partial_\nu^2 \log\det G_x + \partial_\mu^2\partial_\nu^2 \log\phi(x). \tag{33}$$

The factor $\tfrac{1}{2}$ is due to G_x being considered as a quaternionic and f_x as a real $k \times k$ matrix. For the 't Hooft ansatz, $\partial_\mu^2 \log\det G_x$ vanishes, except for delta functions at $x = y_k$, and we retrieve the known result [28], $\operatorname{tr} F^2_{\mu\nu} = \partial_\mu^2\partial_\nu^2 \log\phi(x)$, which is indeed singular at these points (inadvertently in Section 2 of Ref. [8], $\operatorname{tr} F^2_{\mu\nu}$ was given with the wrong sign).

4. The construction of the caloron

In this section we describe the ADHM construction of caloron solutions with non-trivial holonomy. This will be a two-step process. Crucial will be the interpretation of the ADHM data as the Fourier coefficients of the Weyl operator in the Nahm transformation. In our strategy, we build the caloron as an infinite, periodic (gauge-twisted) chain of

T.C. Kraan, P. van Baal/Nuclear Physics B 533 (1998) 627–659   637

instantons. It will be shown how we can realise this within the ADHM construction, by solving the quadratic constraint on the ADHM data. To find $A_\mu(x)$ we use again a Fourier transform to construct $\hat{f}_x(z, z')$, the Green's function of an ordinary second-order differential equation, which allows for the determination of $\phi_{\alpha\beta}(x) = (\lambda_\alpha f_x \lambda'_\beta)$, see Eq. (30).

The boundary conditions $A_\mu(x + \mathcal{T}) = e^{2\pi i\omega\cdot\tau} A_\mu(x) e^{-2\pi i\omega\cdot\tau}$ are satisfied when

$$u_k(x + \mathcal{T}) = u_{k-1}(x) \exp(-2\pi i\omega \cdot \tau), \tag{34}$$

as is seen from Eq. (23). This is implemented by the periodicity constraints

$$\lambda_{k+1} = e^{2\pi i\omega\cdot\tau}\lambda_k, \qquad B(x + \mathcal{T})_{m,n} = B(x)_{m-1,n-1}, \tag{35}$$

where $B(x) = B - x$. It now follows that

$$B_{m+1,n+1} = B_{m,n} + \mathcal{T}\delta_{m,n}, \tag{36}$$

the inhomogeneous part of which is solved by having $\ldots, -2\mathcal{T}, -\mathcal{T}, 0, \mathcal{T}, 2\mathcal{T}, \ldots$ on the diagonal of B. We still have to determine the remainder of B, called $\hat{A}$ (anticipating its interpretation as Nahm connection), that contains its off-diagonal entries. In order to satisfy Eq. (35), $\hat{A}$ has to be of a convolutive type $\hat{A}_{m,n} = \hat{A}_{m-n}$, such that

$$\lambda_k = e^{2\pi ik\omega\cdot\tau}\zeta, \qquad \zeta = \rho q, \qquad B_{m,n} = \mathcal{T}m\delta_{m,n} + \hat{A}_{m-n}. \tag{37}$$

Here ζ is an arbitrary quaternion. Its length $\rho = |\zeta|$ is the scale parameter of the caloron. The $SU(2)$ element $q = \zeta/\rho$ describes its combined spatial and gauge orientation. The diagonal of $\hat{A}_{m,n}$ is necessarily constant, $\hat{A}_{m,n}^{\text{diag}} \equiv \xi$, and plays the role of the position of the caloron. The ADHM data can now be readily interpreted as describing a periodic array of instantons, with temporal spacing $\mathcal{T}$ and relative gauge orientation $e^{2\pi i\omega\cdot\tau}$, with off-diagonal terms to account for the non-linear constraints. To simplify notations, we use the scale invariance of the self-duality equations to set $\mathcal{T} = 1$. On dimensional grounds one can easily reinstate the proper $\mathcal{T}$ dependence when required.

When we perform the Fourier transformation, B will be transformed into a Weyl operator, λ and $\lambda^\dagger\lambda$ into delta function singularities and $u(x)$ into a spinor (to be more precise a 2×2 matrix with as columns $\hat{\psi}_x^i$, cf. Section 2):

$$\sum_{m,n} B_{m,n}(x) e^{2\pi i(mz - nz')} = \frac{\delta(z - z')}{2\pi i}\hat{D}_x(z'),$$

$$\hat{D}_x(z) = \sigma_\mu D_x^\mu(z) = \frac{d}{dz} + \hat{A}(z) - 2\pi ix,$$

$$\sum_m \lambda_m e^{2\pi imz} = \hat{\lambda}(z), \qquad \hat{\lambda}(z) = (P_+\delta(z - \omega) + P_-\delta(z + \omega))\zeta,$$

$$P_\pm = \tfrac{1}{2}(1 \pm \hat{\omega} \cdot \tau),$$

$$\sum_{m,n} \lambda_m^\dagger \lambda_n e^{2\pi i(mz - nz')} = \hat{\lambda}^\dagger(z')\hat{\lambda}(z) = \delta(z - z')\hat{\Lambda}(z), \qquad \hat{\Lambda}(z) = \bar{\zeta}\hat{\lambda}(z) = \hat{\lambda}^\dagger(z)\zeta.$$

$$\sum_m u_m(x)e^{2\pi imz} = \hat{\psi}_x(z).$$ (38)

All these objects are defined on S^1, or more appropriately from the Nahm perspective, on $\mathbb{R}^4/\hat{H} = \mathbb{R}^4/(\mathbb{R}^3 \times \mathbb{Z})$. Note that $\hat{A}(z) = \sigma_\mu \hat{A}_\mu(z) = 2\pi i \sum_m \exp(2\pi imz)\hat{A}_m$, such that from the symmetry of $\hat{A}_{m,n}$ (implying $\hat{A}_m = \hat{A}_{-m}$) it follows that $\hat{A}_\mu(z)$ is imaginary such that the differential operator $\hat{D}_x(z)$ is exactly the dual Weyl operator $\hat{D}_x(\hat{A})$ introduced in Section 2 (to agree with the notation there we have – unlike in Ref. [8] – included a factor $2\pi i$ in our definitions). Combining these features, we can interpret the Fourier transform of the ADHM construction as the inverse (or second) Nahm transformation.

The symmetries in the ADHM construction for periodic instantons lead to a $U(1)$ gauge symmetry for $\hat{A}_\mu(z)$. In order for Eq. (17) to preserve the periodicity constraint (35), T has to be of a convolutive type, $T_{m,n} = T_{m-n}$. Defining the periodic function $\hat{g}(z) = \sum_m \exp(2\pi imz)T_m$ and using the fact that T is orthogonal ($T_{m,n}^{-1} = T_{n-m}$ and $\sum_k T_{k+n}T_k = \delta_{n,0}$), one concludes that $\hat{g}(z) \in U(1)$. A gauge transformation with $\hat{g}(z)$ leaves $\hat{A}_i(z)$ invariant and transforms $\hat{A}_0(z)$ to $\hat{A}_0(z) - d\log\hat{g}(z)/dz$. Note that $\hat{A}_0(z)$ can be gauged away, apart from a constant (as the Polyakov loop is gauge invariant). Hence we may choose $\hat{A}_0(z) = 2\pi i\xi_0$.

To implement the quadratic ADHM constraint after Fourier transformation we note that any complex 2×2 matrix W can be decomposed as $W = W_\mu \sigma_\mu$ and that $[W, \sigma_\mu] = 0$ can be implemented by requiring $\text{Im}\, W = 0$, provided we define $\text{Im}\, W \equiv \frac{1}{2}[W - \tau_2 W' \tau_2]$. Also note that $\text{Re}\, W \equiv \frac{1}{2}[W + \tau_2 W' \tau_2] = \frac{1}{2}\text{tr}\, W = W_0$. Thus, the quadratic ADHM constraint can be formulated as

$$\text{Im}(\hat{D}_x^\dagger(z)\hat{D}_x(z) + 4\pi^2 \hat{\Lambda}(z)) = 0.$$ (39)

From the Weitzenböck formula $\hat{D}_x^\dagger(z)\hat{D}_x(z) = -(\hat{D}_x^\mu(z)\hat{D}_x^\mu(z) + \frac{1}{2}\bar{\eta}_{\mu\nu}\hat{F}_{\mu\nu})$ (compare with Eq. (6)) we find

$$\frac{d}{dz}\hat{A}(z) = \frac{1}{2}\bar{\eta}_{\mu\nu}\hat{F}_{\mu\nu}(z) = -\text{Im}\,\hat{D}_x^\dagger(z)\hat{D}_x(z) = 4\pi^2\text{Im}\,\hat{\Lambda}(z)$$
$$= 2\pi^2(\bar{\zeta}\hat{\omega}\cdot\tau\zeta)(\delta(z-\omega) - \delta(z+\omega)),$$ (40)

using Eq. (39) and $\hat{F}_{ij}(z) = 0$. This leads to [30]

$$\hat{A}(z) = \sigma_\mu\hat{A}_\mu(z) = 2\pi i[\xi + \pi(\bar{\zeta}\hat{\omega}\cdot\boldsymbol{\sigma}\zeta)\Theta_\omega(z)],$$ (41)

where (see Fig. 1)

$$\Theta_\omega(z) = (\chi_{[-\omega,\omega]}(z) - 2\omega).$$ (42)

Here $\chi_{[a,b]}(z) = 1$ if $z \in [a,b]$ and 0 elsewhere, properly defined on the circle. We have arranged $2\pi i\xi = \int dz\,\hat{A}(z)$, such that $B(x)$ has a single zero-mode for $x = \xi$, to agree with the interpretation of ξ as the position (centre of mass) of the caloron. Fourier transforming back, we retrieve the matrix representation of $B(x)$

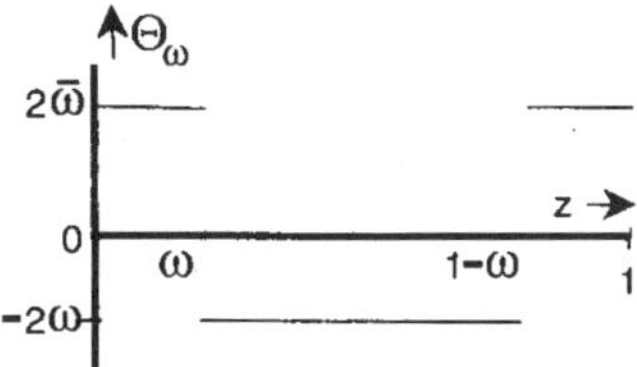

Fig. 1. The function $\Theta_\omega(z)$.

$$B_{m,n}(x) = (m + \xi - x)\delta_{m,n} + \bar\zeta\hat\omega \cdot \sigma\zeta \frac{\sin(2\pi(m-n)\omega)}{m-n}(1 - \delta_{m,n}). \tag{43}$$

The moduli space is thus parametrised by the caloron position ξ and by its scale and orientation $\zeta = \rho q$, with $\xi_0 \in S^1$, $\xi \in \mathbb{R}^3$, $\rho \in \mathbb{R}^+$ and $q \in SU(2)$.

We end this first step in the construction by noting that the delta function singularities arise precisely as predicted by the general properties of the Nahm transformation, discussed in Section 2 and that $\hat{A}(z)$ constructed in this section solves the Nahm equations, whereby we have found all self dual solutions for $\mathbb{R}^3 \times S^1$.

For the second step we have to find the Green's function f_x. For further notational simplification we absorb ξ by a translation (we have already used the scale invariance to fix $\mathcal{T} = 1$) such that after a Fourier transformation the definition of f_x, Eq. (18), can be cast into a differential equation for $\hat{f}_x(z, z') \equiv \sum_{n,m} f_x^{m,n} e^{2\pi i(mz - nz')}$,

$$\left\{ \left(\frac{1}{2\pi i}\frac{d}{dz} - x_0\right)^2 + s^2\chi_{|-\omega,\omega|}(z) + r^2\chi_{|\omega,1-\omega|}(z) \right.$$

$$\left. + \frac{\rho^2}{2}(\delta(z - \omega) + \delta(z + \omega))\right\}\hat{f}_x(z, z') = \delta(z - z'). \tag{44}$$

Here, the radii r and s are given by

$$r^2 = \tfrac{1}{2}\mathrm{tr}(x \cdot \tau + 2\pi\omega\rho^2\bar{q}\hat\omega \cdot \tau q)^2,$$

$$s^2 = \tfrac{1}{2}\mathrm{tr}(x \cdot \tau - 2\pi\bar\omega\rho^2\bar{q}\hat\omega \cdot \tau q)^2, \quad (\bar\omega = \tfrac{1}{2} - \omega), \tag{45}$$

and can be interpreted as the respective centre of mass radii of the two constituent monopoles of the caloron. Note that $\bar{q}\hat\omega \cdot \tau q$ shows how global gauge rotations are to be correlated to spatial rotations so as to keep the holonomy unchanged. We will come back to this in the next section. The symmetries of $\Delta^\dagger(x)\Delta(x)$ imply for the Green's function $\hat{f}_x(z, z')$ the following relations:

$$\hat{f}_x(z, z') = \hat{f}_x(-z, -z')^* = \hat{f}_x(-z', -z) = \hat{f}_x(z', z)^*. \tag{46}$$

In particular we have $\hat{f}_x(\omega, \omega) = \hat{f}_x(-\omega, -\omega) \in \mathbb{R}$ and $\hat{f}_x(\omega, -\omega) = \hat{f}_x(-\omega, \omega)^*$. The Green's function is that of ordinary quantum mechanics on a circle with a piecewise constant potential and delta function singularities of strength $\tfrac{1}{2}\rho^2$ at the jumping points $z = \pm\omega$. It can thus be constructed in the usual straightforward (but tedious) method of matching the value of the Green's function and its derivative (up to the appropriate jumps) at $z = \pm\omega$ and $z = z'$. The result reads

$$\hat{f}_x(z,z') = \chi_{[-\omega,\omega]}(z')\left(\chi_{[-\omega,\omega]}(z)\hat{f}_x^d(z,z',r,s,\omega)\right.$$

$$+\chi_{[\omega,1-\omega]}(z)\hat{f}_x^o(z,z',r,s,\omega))$$

$$+\chi_{[\omega,1-\omega]}(z')\left(\chi_{[\omega,1-\omega]}(z)\hat{f}_x^d(z-\tfrac{1}{2},z'-\tfrac{1}{2},s,r,\bar{\omega})\right.$$

$$+\chi_{[-\omega,\omega]}(z)\hat{f}_x^o(z',z,r,s,\omega)^*\Big). \tag{47}$$

In the following the diagonal component $\hat{f}_x^d(z,z')$ is only defined strictly for $z,z' \in [-\omega,\omega]$ and the off-diagonal component $\hat{f}_x^o(z,z')$ only for $z \in [\omega, 1-\omega]$ and $z' \in [-\omega,\omega]$. For z or z' outside of these intervals, one first has to map back to the interval $[-\omega, 1-\omega]$, using periodicity.

$$\hat{f}_x^d(z,z',r,s,\omega) = e^{2\pi i x_0(z-z')}\pi(rs\psi)^{-1}\left\{e^{-2\pi i x_0 \mathrm{sign}(z-z')}r\sinh(2\pi s|z-z'|)\right.$$

$$-s^{-1}\cosh(2\pi s(z+z'))$$

$$\times\left[\pi\rho^2 r\cosh(4\pi r\bar{\omega}) + \tfrac{1}{2}(r^2 - s^2 + \pi^2\rho^4)\sinh(4\pi r\bar{\omega})\right]$$

$$+s^{-1}\cosh(2\pi s(2\omega - |z-z'|))$$

$$\times\left[\pi\rho^2 r\cosh(4\pi r\bar{\omega}) + \tfrac{1}{2}(r^2 + s^2 + \pi^2\rho^4)\sinh(4\pi r\bar{\omega})\right]$$

$$+ \sinh(2\pi s(2\omega - |z-z'|))$$

$$\times\left[r\cosh(4\pi r\bar{\omega}) + \pi\rho^2\sinh(4\pi r\bar{\omega})\right]\Big\},$$

$$\hat{f}_x^o(z,z',r,s,\omega) = e^{2\pi i x_0(z-z')}\pi(rs\psi)^{-1}$$

$$\times\left\{\pi\rho^2\sinh(2\pi r(1-z-\omega))\sinh(2\pi s(z'+\omega))\right.$$

$$+r\cosh(2\pi r(z-1+\omega))\sinh(2\pi s(z'+\omega))$$

$$-s\sinh(2\pi r(z-1+\omega))\cosh(2\pi s(z'+\omega))$$

$$+e^{-2\pi i x_0}\left[s\sinh(2\pi r(z-\omega))\cosh(2\pi s(z'-\omega))\right.$$

$$-r\cosh(2\pi r(z-\omega))\sinh(2\pi s(z'-\omega))$$

$$-\pi\rho^2\sinh(2\pi r(z-\omega))\sinh(2\pi s(z'-\omega))\Big]\Big\}, \tag{48}$$

where we have introduced the scalar function

$$\psi = -\cos(2\pi x_0) + \cosh(4\pi r\bar{\omega})\cosh(4\pi s\omega)$$

$$+\frac{(r^2 + s^2 + \pi^2\rho^4)}{2rs}\sinh(4\pi r\bar{\omega})\sinh(4\pi s\omega)$$

$$+\pi\rho^2\left(s^{-1}\sinh(4\pi s\omega)\cosh(4\pi r\bar{\omega}) + r^{-1}\sinh(4\pi r\bar{\omega})\cosh(4\pi s\omega)\right). \tag{49}$$

In particular,

$$\hat{f}_x(\omega,-\omega) = \pi(rs\psi)^{-1}e^{4\pi i x_0\omega}\left\{e^{-2\pi i x_0}r\sinh(4\pi s\omega) + s\sinh(4\pi r\bar{\omega})\right\},$$

$$\hat{f}_x(\omega,\omega) = \pi(rs\psi)^{-1}\left\{s\sinh(4\pi r\bar{\omega})\cosh(4\pi s\omega) + r\sinh(4\pi s\omega)\cosh(4\pi r\bar{\omega})\right.$$

$$+\pi\rho^2\sinh(4\pi r\bar{\omega})\sinh(4\pi s\omega)\Big\} \tag{50}$$

T.C. Kraan, P. van Baal/Nuclear Physics B 533 (1998) 627–659

and (see Eq. (21))

$$\phi(x) = (1 - \lambda f_x \lambda^\dagger)^{-1} = (1 - \rho^2 \hat{f}_x(\omega, \omega))^{-1} \equiv \psi/\hat{\psi}, \tag{51}$$

where

$$\hat{\psi} = -\cos(2\pi x_0) + \cosh(4\pi r\bar{\omega})\cosh(4\pi s\omega)$$
$$+ \frac{(r^2 + s^2 - \pi^2 \rho^4)}{2rs}\sinh(4\pi r\bar{\omega})\sinh(4\pi s\omega). \tag{52}$$

We now use Eqs. (30), (31) to determine A_μ. The scalar functions $\phi_{\alpha\beta}$ are all defined in terms of $\hat{f}_x(\omega, \pm\omega)$. To get compact expressions, we introduce the complex function

$$\chi = \rho^2 \hat{f}_x(\omega, -\omega) = e^{4\pi i x_0 \omega}\frac{\pi\rho^2}{\psi}\left\{ e^{-2\pi i x_0}s^{-1}\sinh(4\pi s\omega) + r^{-1}\sinh(4\pi r\bar{\omega})\right\}. \tag{53}$$

We choose $\boldsymbol{\omega}$ in the z-direction and $q = 1$, which can always be achieved by performing a suitable gauge and spatial rotation. We find for those functions $\phi_{\alpha\beta}$ that are non-zero

$$\phi_{00} = \tfrac{1}{2}(1 - \phi^{-1} + \mathrm{Re}\,\chi), \quad \phi_{33} = \tfrac{1}{2}(1 - \phi^{-1} - \mathrm{Re}\,\chi), \quad \phi_{03} = \phi_{30} = \tfrac{1}{2}\mathrm{Im}\,\chi, \tag{54}$$

such that with the use of Eq. (30) (for a matrix W, $\mathrm{Re}\,W \equiv \tfrac{1}{2}(W + W^\dagger)$)

$$A_\mu = \frac{i}{2}\bar{\eta}^3_{\mu\nu}\tau_3\partial_\nu \log\phi + \frac{i}{2}\phi\mathrm{Re}\left((\bar{\eta}^1_{\mu\nu} - i\bar{\eta}^2_{\mu\nu})(\tau_1 + i\tau_2)\partial_\nu\chi\right). \tag{55}$$

For $\omega = 0$ this reduces to the Harrington–Shepard solution of the caloron with trivial holonomy, since in that case $\chi = 1 - \phi^{-1}$.

The self-duality of Eq. (55) follows from Eq. (26), but has also been checked numerically. In the asymptotic regime of large distances $|x|$,

$$\hat{f}_x(z, z') \approx \frac{\prime\prime}{|x|}e^{-2\pi|x||z-z'|+2\pi i x_0(z-z')}, \tag{56}$$

(with $|z - z'|$ the obvious distance function on the circle) from which it follows that A_μ tends to zero at spatial infinity. The holonomy at spatial infinity is then fully carried by the cocycle, and equals $e^{2\pi i \boldsymbol{\omega}\cdot\boldsymbol{\tau}}$ as required. The non-trivial holonomy becomes even more transparent in the periodic gauge by performing a gauge transformation $g(x) = e^{-2\pi i x_0 \boldsymbol{\omega}\cdot\boldsymbol{\tau}}$. This yields

$$A_\mu^{\mathrm{per}} = \frac{i}{2}\bar{\eta}^3_{\mu\nu}\tau_3\partial_\nu \log\phi + \frac{i}{2}\phi\mathrm{Re}\left((\bar{\eta}^1_{\mu\nu} - i\bar{\eta}^2_{\mu\nu})(\tau_1 + i\tau_2)(\partial_\nu + 4\pi i\omega\delta_{\nu,0})\tilde{\chi}\right)$$
$$+\delta_{\mu,0}2\pi i\omega\tau_3, \tag{57}$$

where

$$\tilde{\chi} \equiv e^{-4\pi i x_0 \omega} \chi = \frac{\pi \rho^2}{\psi} \left\{ e^{-2\pi i x_0} s^{-1} \sinh(4\pi s\omega) + r^{-1} \sinh(4\pi r\bar{\omega}) \right\}. \tag{58}$$

In this gauge we immediately read off the constant background field at spatial infinity, $A_\mu^{\mathrm{per}} = 2\pi i \omega \cdot \tau \delta_{\mu,0}$, responsible for the holonomy in the periodic gauge. This concludes the construction of the caloron solution.

To use Eq. (32) for the action density we have to regularise the determinant, which for calorons diverges. However, $\partial_\mu \log \det f_x$ turns out to be finite. With the help of Eq. (28) we find

$$\partial_\mu \log \det f_x = \partial_\mu \mathrm{Tr} \log f_x = \frac{1}{\pi i} \mathrm{Tr}\, \hat{D}_x^\mu f_x = \frac{1}{\pi i} \int_{S^1} dz \lim_{z' \to z} \hat{D}_x^\mu(z) \hat{f}_x(z, z'), \tag{59}$$

where Tr denotes the Hilbert space trace. We use point-splitting to define

$$\lim_{z' \to z} \frac{d}{dz} \hat{f}_x(z, z') \equiv \tfrac{1}{2} \left(\lim_{\epsilon \downarrow 0} \frac{d}{dz} \hat{f}_x(z + \epsilon, z') + \lim_{\epsilon \downarrow 0} \frac{d}{dz} \hat{f}_x(z - \epsilon, z') \right) \bigg|_{z' = z}, \tag{60}$$

in accordance with the Fejér theorem for the convergence of Fourier series, see e.g. Ref. [31]. Careful inspection shows that

$$\mathrm{Tr}\, \hat{D}_x^\mu f_x = -\pi i \partial_\mu \log \psi, \tag{61}$$

leading to the following miraculously simple result:

$$-\mathrm{tr}\, F_{\mu\nu}^2(x) = \partial_\mu^2 \partial_\nu^2 \log \det f_x = -\partial_\mu^2 \partial_\nu^2 \log \psi. \tag{62}$$

Note that ψ is positive definite and smooth, despite its appearance. The same applies for Eq. (62). The action density $-\tfrac{1}{2}\mathrm{tr}\, F_{\mu\nu}^2(x)$ takes its maximal value at $x_0 = 0$. We have verified Eq. (62) numerically, using Eq. (55). Since the action density is a total derivative, one can express the total action in terms of a surface integral at spatial infinity. Using that $\partial_\mu^2 \log \psi = 4\pi/|x| + \mathcal{O}(|x|^{-4})$, one easily verifies that for the topological charge

$$k = -\frac{1}{8\pi^2} \int \mathrm{tr}\, F \wedge F = -\frac{1}{16\pi^2} \int d_4 x\, \mathrm{tr}\, F_{\mu\nu}^2(x) = 1. \tag{63}$$

In Appendix A we give the expression for the Green's function G_x, from which it follows that

$$\partial_\mu^2 \partial_\nu^2 \log \det G_x = -\tfrac{1}{2} \partial_\mu^2 \partial_\nu^2 \log \hat{\psi}, \tag{64}$$

in accordance with Eq. (33) and Eq. (51).

5. Properties of the caloron solution

We first discuss the issue of orientations in colour and real space. In general only the centre $\mathbb{Z}_2$ of the group of global gauge transformations will leave the gauge potential invariant. The framing – embedding of the solution in colour space – is in general not invariant under global gauge rotations. For non-trivial holonomy ($\omega\bar{\omega} \neq 0$, or $\mathcal{P} \neq \pm 1$ at infinity), also the holonomy is not invariant under such global gauge rotations, except for a $U(1)$ subgroup generated by $\hat{\omega} \cdot \boldsymbol{\tau}$, which in the monopole terminology generates the unbroken gauge symmetry. For each choice of the holonomy – which can not change under continuous deformations – we have a separate caloron parameter space. We note that the spatial orientation is given by the preferred axis that appears in the formula for the action density and in the definition of the two radii r and s, Eq. (45). The action density has an axial symmetry around $\hat{a}$ defined through

$$\hat{a} \cdot \boldsymbol{\tau} = \bar{q}\hat{\omega} \cdot \boldsymbol{\tau}q = \bar{\zeta}\hat{\omega} \cdot \boldsymbol{\tau}\zeta/\rho^2. \tag{65}$$

When q is part of the $U(1)$ subgroup generated by $\hat{\omega}\cdot\boldsymbol{\tau}$, it does not affect the orientation of the solution, and indeed can be pulled through in Eq. (37), to be identified with the global gauge invariance of Eq. (17) associated to this residual $U(1)$. The dimension of the moduli space of *gauge inequivalent* solutions at fixed holonomy, including the position of the caloron described by ξ_μ, is thus 7 for non-trivial and 5 for trivial holonomy. A global residual $U(1)$ gauge transformation (or a global $SU(2)$ gauge transformation in case of trivial holonomy), however, does change the framing of the solutions and the moduli space of framed calorons is eight-dimensional. Including these global gauge degrees of freedom will reveal the hyper-Kähler structure of the moduli space, to be discussed in the next section.

Since a global gauge transformation leaves $\hat{a}$ invariant, we can best describe the parameters of the solutions for the choice where $\hat{\omega} = \hat{e}_3$, i.e. $\hat{\omega}$ is pointing in the positive x_3-direction. Due to the residual gauge group, to any point on the two sphere defined by the symmetry axes of the caloron solution, a full $U(1)$ can be associated. This gives the Hopf fibration of $S^3 = SU(2)$ over S^2, the fiber being $U(1)$. Using Euler angles we may choose the parametrisation

$$q = e^{-i\Upsilon\frac{\tau_3}{2}} e^{i(\frac{\pi}{2}-\theta)\frac{\tau_2}{2}} e^{-i\varphi\frac{\tau_3}{2}}, \quad 0 \leqslant \Upsilon \leqslant 4\pi, \quad 0 \leqslant \varphi \leqslant 2\pi, \quad 0 \leqslant \theta \leqslant \pi, \tag{66}$$

which leads, for $\hat{\omega} = \hat{e}_3$, to the axis of axial symmetry $\hat{a} = (\cos\theta, \sin\theta\sin\varphi, \sin\theta\cos\varphi)$. The variable Υ describes the residual $U(1)$ gauge group generated by $\tau_3(= \hat{\omega} \cdot \boldsymbol{\tau})$. With $q = q_\mu\sigma_\mu$ ($|q| = 1$) we can introduce the Maurer–Cartan one-forms

$$\Sigma_i = 2\eta^i_{\mu\nu}q_\mu dq_\nu, \quad d\Sigma_i = \tfrac{1}{2}\epsilon_{ijk}\Sigma_j \wedge \Sigma_k. \tag{67}$$

In terms of the Euler angles these read

$$\Sigma_1 = \cos\Upsilon\sin\theta d\varphi - \sin\Upsilon d\theta,$$

$$\Sigma_2 = \sin\Upsilon\sin\theta d\varphi + \cos\Upsilon d\theta,$$

$$\Sigma_3 = d\Upsilon + \cos\theta d\varphi. \tag{68}$$

In order to visualise the caloron solution, we use Eq. (62). A time-slice of the caloron shows that it generically consists of two lumps. In Fig. 2 we study the time

 T.C. Kraan, P. van Baal/Nuclear Physics B 533 (1998) 627–659

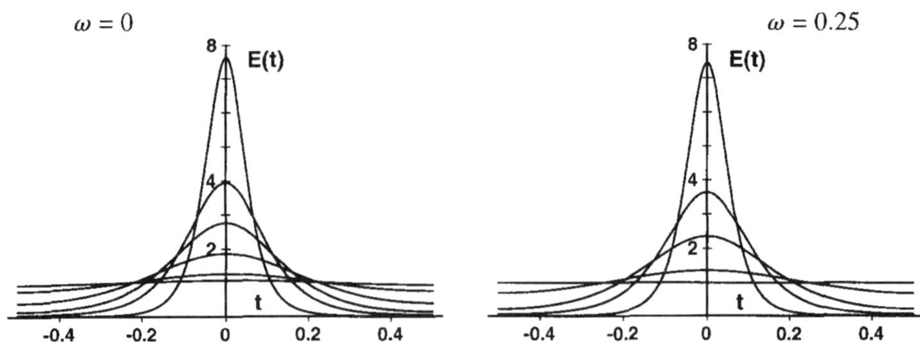

Fig. 2. Time evolution of the caloron solution. During one period ($T = 1$), we plot the "energy" as a function of time, $E(t) \equiv -(1/16\pi^2) \int_{\mathbb{R}^3} \mathrm{tr}\, F^2$, for $\rho = 0.1, 0.2, 0.3, 0.5, 1.0, 2.0$. For small values of ρ, the caloron is short-lived and instanton-like, whereas for large values, $\rho > 1$, the profile flattens and the caloron becomes static and monopole-like.

dependence for various values of ρ. For small ρ the caloron approaches the ordinary single instanton solution, with no dependence on ω, as $\rho \to 0$ is equivalent to $T \to \infty$. Finite size effects set in when the size of the instanton becomes of the order of the compactification length T, i.e. when the caloron bites in its own tail. This occurs at roughly $\rho = \frac{1}{2}T$. At this point, for $\omega\bar{\omega} \neq 0$ (i.e. the holonomy $\mathcal{P} \neq \pm 1$), two lumps are formed, whose separation grows as $\pi\rho^2/T$ (cf. Eq. (45)). At large ρ the solution spreads out over the entire circle in the euclidean time direction and becomes static in the limit $\rho \to \infty$. So for large ρ the lumps are well separated, see Fig. 3. When far apart they become spherically symmetric. As they are static and self-dual they are necessarily BPS monopoles. One can show in this limit that they have unit, but opposite, magnetic charges and that the two lumps have spatial scales proportional to respectively $1/\bar{\omega}$ and $1/\omega$ (see Section 7). Their action densities (or energy densities in this static limit) scale with $\bar{\omega}^4$ and ω^4. After integration, this results in monopole masses of respectively $16\pi^2\bar{\omega}/T$ and $16\pi^2\omega/T$ for the two lumps, their mass ratio is therefore $\bar{\omega}/\omega$. The total energy, simply obtained by addition, indeed conforms with the unit topological charge of the solution. For $\omega = 0$ or $\omega = \frac{1}{2}$, the second lump is absent and the solution is spherically symmetric. For generic ω, the solution has only an axial symmetry around the axis $\hat{a}$ connecting the two lumps. For $\omega = \frac{1}{4}$, the lumps are equally sized, and the solution has a mirror symmetry in the plane perpendicular to $\hat{a}$, see Fig. 4.

These aspects can be readily retrieved by inspecting Eq. (55), and in particular (62), for the limit of large ρ and realising that r and s are the centre off mass radii of the constituent monopoles. If ρ is small, the caloron is best described in terms of the instanton picture, whereas for large ρ the two-monopole picture is more appropriate. For the constituent picture of oppositely charged BPS monopoles to be correct, the field has to behave like a magnetic (and electric) dipole at large distances. Indeed one easily finds that the field strength decays as $1/|\boldsymbol{x}|^3$, for distances much larger than $\pi\rho^2/T$. Note that for $\omega = 0$ we have the standard Harrington–Shepard caloron [11], which in the limit of large ρ was already shown by Rossi [32] to become the standard BPS monopole (after a singular gauge transformation).

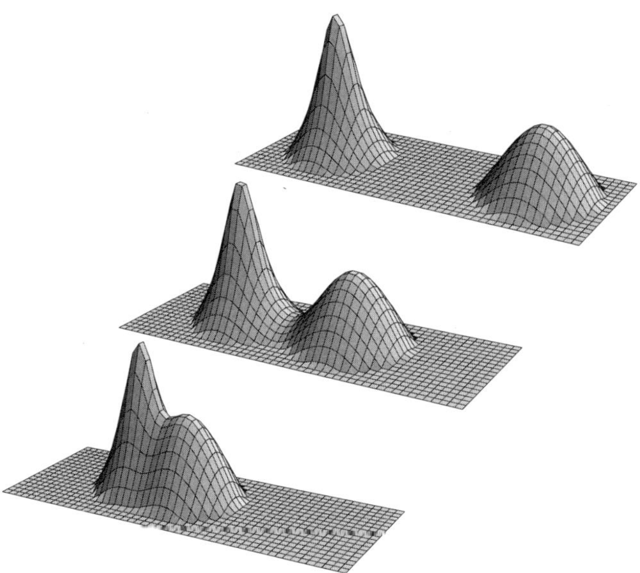

Fig. 3. Shown are caloron profiles for $\omega = 0.125$ ($\mathcal{T} = 1$), with $\rho = 0.8, 1.2, 1.6$ (from bottom to top). This illustrates the growing separation of the two lumps with ρ. Once the constituents are separated, the lumps are spherically symmetric and do not change their shape upon further separation. Vertically is plotted the action density at $x_0 = 0$, on equal logarithmic scales for all profiles. They were cut off at an action density below $1/(2e^2)$.

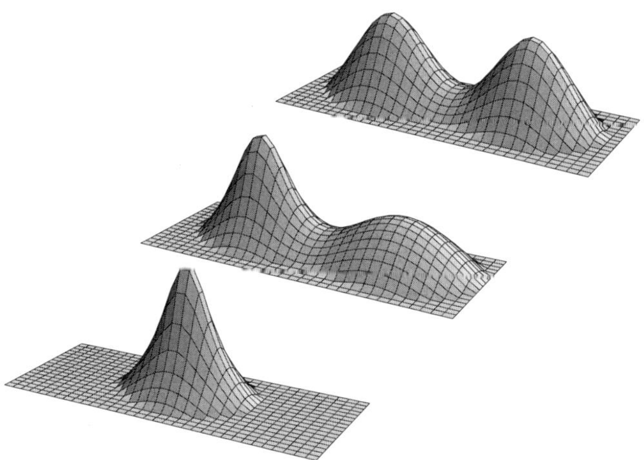

Fig. 4. Profiles for calorons at $\omega = 0, 0.125, 0.25$ (from left to right) with $\rho = \mathcal{T} = 1$. The axis connecting the lumps, separated by a distance π (for $\omega \neq 0$), corresponds to the direction of $\hat{a}$. The other direction indicates the distance to this axis, making use of the axial symmetry of the solutions. The mass ratio of the two lumps is approximately $\omega/\bar{\omega}$, i.e. zero (no second lump), a third and one (equal masses), for the respective values of ω. Vertically is plotted the action density at $x_0 = 0$, on equal logarithmic scales for all profiles. They were cut off at an action density below $1/(2e)$.

646   *T.C. Kraan, P. van Baal/Nuclear Physics B 533 (1998) 627–659*

Interpreting the Nahm data (Eq. (41)) as the juxtaposition of two sub-intervals of lengths 2ω and $2\bar{\omega}$ respectively, with constant Nahm connections $\hat{A}(z)$, leads to a more indirect way of understanding the composite nature of the caloron. Indeed, $\hat{A} = 0$ gives the standard BPS monopole, adding a constant merely translates the solution in space. Thus, each interval gives rise to a BPS monopole on $\mathbb{R}^3 \times S^1$, and we can in a good approximation add the two connections corresponding to the two sub-intervals. The ρ^2 dependence of $\hat{A}(z)$ explains the large separation of the constituent lumps for large ρ. As the lengths of the intervals are given by the asymptotic Higgs vacuum expectation value of the corresponding monopoles, the mass ratio $\bar{\omega}/\omega$ of the lumps is easily explained by noting that the mass of a BPS monopole is proportional to the Higgs vacuum expectation value. The above interpretation underlies the approach taken in Ref. [10]. The expression for the gauge potential given there is precisely the sum alluded to above, plus gauge-like terms and gauge transformations required for gluing them together.

6. The moduli space

The moduli space $\mathcal{M}$ of the caloron solutions has as its coordinates ξ and ζ. We should include the global gauge degrees of freedom ($SU(2)$ for trivial and $U(1)$ for non-trivial holonomies), so as to make the solution space hyper-Kähler. The moduli space of these so-called framed calorons is a product of the base manifold $\mathbb{R}^3 \times S^1$, parametrised by ξ and the (Taub-NUT) space parametrised by ζ, forming the non-trivial part of the moduli space and describing the relative coordinates of the two constituent monopoles, quite similar to that for the two-monopole moduli space [33]. It should be noted that $\zeta \rightarrow -\zeta$, corresponding to the centre of the $SU(2)$, leaves $A_\mu(x)$ invariant, such that we have to mod-out this symmetry to obtain the space of framed calorons.

The metric on this space is given by the Riemannian metric on the gauge theory configuration space, restricted to the space of solutions. The metric is then given by ($g_M^{\mu\nu} = \delta^{\mu\nu}$ being the flat metric on $M = \mathbb{R}^3 \times S^1$)

$$g_{\mathcal{M}}(Z, Z') = \int_M g_M^{\mu\nu} \mathrm{tr}\left(Z_\mu^\dagger(x) Z_\nu'(x)\right). \tag{69}$$

with Z_μ and Z_μ' two vectors tangent to the space of caloron solutions. The gauge structure requires them to be transverse to gauge deformations, thereby satisfying the background (or Coulomb) gauge condition

$$D^\mu(A) Z_\mu = 0. \tag{70}$$

The requirement that $A_\mu + Z_\mu$ is self-dual leads to the so-called deformation equations

$$D_{[\mu} Z_{\nu]} = \tfrac{1}{2} \epsilon_{\mu\nu\alpha\beta} D_{[\alpha} Z_{\beta]}, \tag{71}$$

and in the algebraic gauge we have to require in addition (see Eq. (2))

$$Z_\mu(x+1) = e^{2\pi i\omega\cdot\tau} Z_\mu(x) e^{-2\pi i\omega\cdot\tau}. \tag{72}$$

The tangent vectors (zero-modes) can be found by varying the caloron solution with respect to the coordinates ξ and ζ, which will automatically satisfy the deformation equation (71) and periodicity (72), but generally one has to apply an infinitesimal gauge transformation Φ, compatible with Eq. (72), to transform to the Coulomb gauge, Eq. (70). Hence,

$$Z_\mu^r = \delta_r A_\mu + D_\mu \Phi^r, \tag{73}$$

where the label r indicates the parameters (or coordinates) of the moduli space. For the metric (Eq. (69)) to exist, zero-modes should of course be normalisable.

For instantons on $\mathbb{R}^4$, the zero-modes can be determined within the ADHM formalism [29]. Thus one can calculate the metric in terms of the ADHM data. This reflects the fact that the Nahm transformation (and the ADHM construction) is a hyper-Kähler isometry [19,17]. We have

$$\delta\Delta = \begin{pmatrix} \delta\lambda \\ \delta B \end{pmatrix} \equiv \begin{pmatrix} c \\ Y \end{pmatrix} = C, \tag{74}$$

and for the calorons, in addition periodicity (Eq. (72)) requires

$$Y_{m,n} = Y_{m-1,n-1}, \qquad c_{m+1} = e^{2\pi i\omega\cdot\tau} c_m. \tag{75}$$

In terms of the deformation C of the ADHM data, the zero-mode reads [29]

$$Z_\mu = v^\dagger(x) C \bar\sigma_\mu f_x u(x) \phi^{-1/2}(x) - \phi^{-1/2}(x) u^\dagger(x) f_x \sigma_\mu C^\dagger v(x) \tag{76}$$

and

$$D^\mu Z_\mu(x) = \phi^{-1} u^\dagger(x) f_x \sigma_\mu \left(C^\dagger \Delta(x) - \Delta^\dagger(x) C \right) \bar\sigma_\mu f_x u(x). \tag{77}$$

(Note that for $W = W_\mu \sigma_\mu$, $\sigma_\mu W \bar\sigma_\mu = 4W_0 = 2\mathrm{tr}\,W$.) Combining the deformation of the quadratic ADHM constraint (as the imaginary part) with the Coulomb gauge condition (as the real part), imposes

$$(\Delta^\dagger(x)C) = (\Delta^\dagger(x)C)^t. \tag{78}$$

To satisfy the Coulomb gauge condition, the real part of this equation being equivalent to

$$\mathrm{tr}(\Delta^\dagger(x)C - C^\dagger\Delta(x)) = 0, \tag{79}$$

one has only the T invariance of Eq. (17) available (i.e. the $U(1)$ gauge invariance of $\hat A$), since a global gauge rotation would distort the framing. We write $T \in O(k)$ as $T = \exp(-\delta X)$, with $\delta X^t = -\delta X$. Like T, also δX has to satisfy $\delta X_{m,n} = \delta X_{m-n}$, and can be interpreted as the Fourier coefficients of a gauge function $\delta\hat g(z)$ on the circle. We now write the zero-modes C in ADHM language as a variation of Δ plus a compensating gauge transformation,

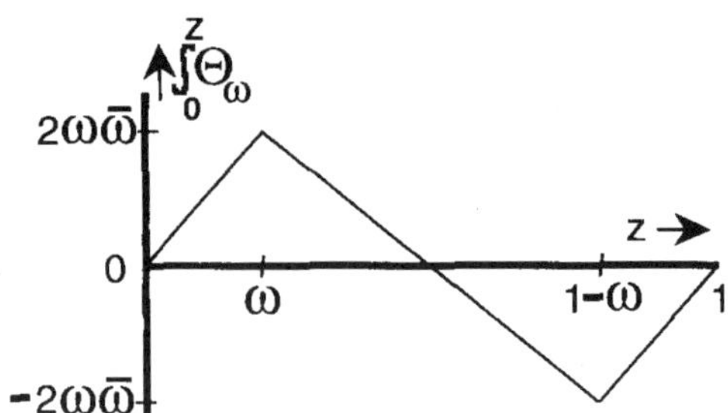

Fig. 5. The function $\int_0^z \Theta_\omega(z')dz'$.

$$C = \delta\Delta + \delta_X\Delta, \quad \delta_X\Delta = \begin{pmatrix} \lambda\delta X \\ [B,\delta X] \end{pmatrix}. \tag{80}$$

Inserting this into Eq. (79) we find

$$\mathrm{tr}\left((\delta\Delta^\dagger)\Delta - \Delta^\dagger\delta\Delta - 2\delta X\Lambda + \left[B^\dagger,\delta X\right]B - B^\dagger\left[B,\delta X\right]\right) = 0, \tag{81}$$

where we used that $[\Lambda,\delta X] = 0$, since the gauge symmetry (described by δX) is abelian. After Fourier transformation, with $\delta(z - z')\delta\hat{X}(z) = \sum_{m,n} X_{m,n}e^{2\pi i(mz-nz')}$, this equation reads

$$-\frac{1}{4\pi^2}\frac{d^2\delta\hat{X}(z)}{dz^2} + |\zeta|^2(\delta(z-\omega) + \delta(z+\omega))\delta\hat{X}(z)$$

$$= \frac{i}{2}\rho^2(\delta(z-\omega) - \delta(z+\omega))\hat{\omega}\cdot\Sigma, \tag{82}$$

using that with the help of the Maurer–Cartan one-forms, Eq. (67), we can write

$$\mathrm{tr}\left((\delta\zeta\bar{\zeta} - \zeta\delta\bar{\zeta})\hat{\omega}\cdot\boldsymbol{\sigma}\right) = 4\hat{\omega}^a\eta_{\mu\nu}^a\zeta_\mu\delta\zeta_\nu = 2\rho^2\hat{\omega}\cdot\Sigma. \tag{83}$$

The solution to the differential equation for $\delta\hat{X}(z)$ gives the infinitesimal gauge transformation needed to go to Coulomb gauge. One finds

$$\delta\hat{X}(z) = -2\pi^2 i\rho^2\hat{\omega}\cdot\Sigma(1 + 8\pi^2\omega\bar{\omega}\rho^2)^{-1}\int_0^z \Theta_\omega(z')dz', \tag{84}$$

which is a zig-zag wave (see Fig. 5), vanishing at $2z \in \mathbb{Z}$ and taking its extremal values at $z = \pm\omega$,

$$\delta\hat{X}_\omega \equiv \pm\delta\hat{X}(\pm\omega) = -4i\pi^2\omega\bar{\omega}\rho^2\hat{\omega}\cdot\Sigma(1 + 8\pi^2\omega\bar{\omega}\rho^2)^{-1}. \tag{85}$$

Note that for the variations with respect to the caloron position, ξ, no compensating gauge transformation is needed.

In order to evaluate the metric, Eq. (69), it is sufficient to compute $g_{\mathcal{M}}(Z,Z)$ for Z related to an arbitrary deformation of the moduli parameters, as determined by C in Eqs. (80) and (84). For this we employ the following relation due to Osborn [29]:

T.C. Kraan, P. van Baal/Nuclear Physics B 533 (1998) 627–659

$$\text{tr}(Z_\mu^\dagger(x)Z_\mu(x)) = -\tfrac{1}{2}\partial^2\,\text{tr}\,\text{Tr}\left(C^\dagger(2 - \Delta(x)f_x\Delta^\dagger(x))Cf_x\right) \tag{86}$$

$$= -\tfrac{1}{2}\partial^2\,\text{tr}\,\text{Tr}\left(2(c^\dagger c + Y^\dagger Y)f_x\right.$$

$$\left. -(c^\dagger\lambda + Y^\dagger B(x))f_x(\lambda^\dagger c + B^\dagger(x)Y)f_x\right),$$

which is derived from Eq. (76), making use of Eq. (78). We introduce the Fourier transforms $\hat c(z) \equiv \sum_m e^{2\pi i m z}c_m$ and $\hat Y(z)$, with $\delta(z - z')\hat Y(z) = \sum_{m,n} e^{2\pi i(mz - nz')}Y_{m,n}$, such that

$$\hat c(z) = \delta\hat\lambda(z) + \hat\lambda(z)\delta\hat X(z) = P_+\delta(z - \omega)(\delta\zeta + \zeta\delta\hat X_\omega)$$

$$+ P_-\delta(z + \omega)(\delta\zeta - \zeta\delta\hat X_\omega),$$

$$\hat Y(z) = \frac{1}{2\pi i}\left(\delta\hat A(z) + \frac{d}{dz}\delta\hat X(z)\right)$$

$$= \delta\xi + \pi\left(\delta\bar\zeta\hat\omega\cdot\boldsymbol\sigma\zeta + \zeta\hat\omega\cdot\boldsymbol\sigma\delta\zeta - \rho^2\hat\omega\cdot\boldsymbol\Sigma(1 + 8\pi^2\omega\bar\omega|\zeta|^2)^{-1}\right)\Theta_\omega(z).$$

$$\tag{87}$$

We may use the special structure of $\hat c(z)$ and $\hat\lambda(z)$ (Eq. (38)) – formed out of the combinations $\delta(z \mp \omega)P_\pm$, with $P_\pm^2 = P_\pm$ and $P_\pm P_\mp = 0$ – to deduce

$$\hat c^\dagger(z)\hat c(z') = \delta(z - z')\hat c^\dagger(z)\langle\hat c\rangle,$$

$$\hat c^\dagger(z)\hat\lambda(z') = \delta(z - z')\hat c^\dagger(z)\langle\hat\lambda\rangle, \quad \langle h\rangle \equiv \int_{S^1} h(z)\,dz. \tag{88}$$

For calorons, this allows us to turn Osborn's equation into

$$\text{tr}(Z_\mu^\dagger(x)Z_\mu(x)) = -\partial^2\int_{S_1} dz\ \text{tr}\left([\hat Y^\dagger(z)\hat Y(z) + \hat c^\dagger(z)\langle\hat c\rangle]\,\hat f_x(z,z)\right)$$

$$+ \tfrac{1}{2}\partial^2\int_{S^1} dz\,dz'\,\text{tr}\left([\hat C(z) + \hat{\mathcal{Y}}_x(z)]\hat f_x(z,z')[\hat{\mathcal{Y}}_x^\dagger(z') + \hat C^\dagger(z')]\hat f_x(z',z)\right),$$

$$\tag{89}$$

with

$$\hat{\mathcal{Y}}_x(z) = (2\pi i)^{-1}\hat Y^\dagger(z)\hat D_x(z), \qquad \hat C(z) = \hat c^\dagger(z)\langle\hat\lambda\rangle. \tag{90}$$

When integrating over space-time, the ∂_0^2 part does not contribute due to periodicity. The integral is therefore reduced to a boundary term at spatial infinity, $|x| \to \infty$. In this limit ρ^2 can be neglected and the two radii r and s in Eq. (45) become equal. In particular in this limit the potential in Eq. (44) equals $|x|^2$, independent of z. From this one concludes that asymptotically $\hat f_x(z,z')$ becomes a function of $z - z'$ and therefore that $\hat D_x(z)\hat f_x(z,z')\hat D_x^\dagger(z') = 4\pi^2\delta(z - z') + \mathcal{O}(r^{-1})$. This can also be deduced from the asymptotic form of $\hat f_x(z,z')$ in Eq. (56), which implies $\hat f_x(z,z) = \pi/r + \mathcal{O}(r^{-2})$. Thus

$$\int_{S^1} dz' \, \mathrm{tr}[\hat{\mathcal{Y}}_x(z)\hat{f}_x(z,z')\hat{\mathcal{Y}}_x^\dagger(z')\hat{f}_x(z',z)] = \mathrm{tr}[\hat{Y}^\dagger(z)\hat{Y}(z)\hat{f}_x(z,z)],$$

to be combined with the first term in Eq. (89). Using Eq. (88), we also have

$$\int_{S^1} dz' \, \mathrm{tr}[\hat{\mathcal{C}}(z)\hat{f}_x(z,z')\hat{\mathcal{C}}^\dagger(z')\hat{f}_x(z',z)] = \mathrm{tr}[\langle\hat{c}^\dagger\rangle\langle\hat{\lambda}\rangle\langle\hat{\lambda}^\dagger\rangle\hat{c}(z)\hat{f}_x^2(z,z)] = \mathcal{O}(r^{-2}),$$

and

$$2\pi i \int_{S^1} dz' \, \mathrm{tr}[\hat{\mathcal{Y}}(z)\hat{f}_x(z,z')\hat{\mathcal{C}}^\dagger(z')\hat{f}_x(z',z)]$$

$$= \sum_{s=\pm} \mathrm{tr}[\hat{Y}^\dagger(z)D_x(z)f_x(z,s\omega)\langle\hat{\lambda}^\dagger\rangle\langle\hat{c}\rangle P_s f_x(s\omega,z)],$$

which after integration over z is $\mathcal{O}(r^{-2})$. Only those terms that are $\mathcal{O}(r^{-1})$ will contribute and we obtain the following remarkably simple result:

$$g_{\mathcal{M}}(Z,Z) = 2\pi^2 \mathrm{tr}(\langle\hat{Y}^\dagger\hat{Y}\rangle + 2\langle\hat{c}^\dagger\rangle\langle\hat{c}\rangle). \tag{91}$$

Inserting Eq. (87) gives the metric (we put $\hat{\omega} = \hat{e}_3$)

$$ds^2 = 2\pi^2 \left\{ 2d\xi_\mu d\xi^\mu + (1 + 8\pi^2\omega\bar{\omega}\rho^2)\left(4d\rho^2 + \rho^2\left(\Sigma_1^2 + \Sigma_2^2\right)\right) \right.$$
$$\left. + \rho^2(1 + 8\pi^2\omega\bar{\omega}\rho^2)^{-1}\Sigma_3^2 \right\}. \tag{92}$$

The first part describes the flat metric of the base manifold $\mathbb{R}^3 \times S^1$, the remainder forms the non-trivial part of the metric. They separate because $\int \Theta_\omega(z)dz = 0$, see Eq. (42).

We introduce a "radial" coordinate X and "mass" parameter M

$$X^2 = 8\pi^2\rho^2, \qquad M^{-2} = 16\omega\bar{\omega}, \tag{93}$$

and rewrite the non-trivial part of the moduli space metric as

$$ds^2 = \left(1 + \frac{X^2}{16M^2}\right)\left(dX^2 + \tfrac{1}{4}X^2(\Sigma_1^2 + \Sigma_2^2)\right) + \tfrac{1}{4}X^2\left(1 + \frac{X^2}{16M^2}\right)^{-1}\Sigma_3^2. \tag{94}$$

This metric is the Taub-NUT metric [34] as given in Ref. [35]. It is a self-dual Einstein manifold [36,37] and is hyper-Kähler [33,38]. The latter property is inherited from the hyper-Kähler structure of the base manifold $\mathbb{R}^3 \times S^1$, preserved by the self-duality equations [19]. Therefore, the $SU(2)$ moduli space for calorons becomes

$$\mathcal{M}_{\text{framed}} = (\mathbb{R}^3 \times S^1) \times \text{Taub-NUT}/\mathbb{Z}_2. \tag{95}$$

Note that $\mathbb{Z}_2$ corresponds with $\zeta = q = \pm 1$, i.e. the centre of the SU(2) gauge group. With $\zeta \to -\zeta$ leaving the gauge fields unchanged, this gives rise to an orbifold singularity (at $\zeta = 0$) and $(\mathbb{R}^3 \times S^1) \times \text{Taub-NUT}$ is a double cover of $\mathcal{M}_{\text{framed}}$.

For small ρ or ω, when $X^2/M^2 \to 0$, the metric becomes that of $\mathbb{R}^4$, since $\tfrac{1}{4}(\Sigma_1^2 + \Sigma_2^2 + \Sigma_3^2)$ (see Eq. (67)) is the metric on the unit three-sphere. With $\rho \to 0$ corresponding to $T \to \infty$, this describes the moduli space of a charge-1 instanton on $\mathbb{R}^4$, whereas for $\omega = 0$ we have the standard Harrington–Shepard caloron moduli space. In both cases this is parametrised by the scale and $SU(2)$ group orientation (to make the moduli space hyper-Kähler) and $\mathcal{M}_{\text{framed}} = \mathbb{R}^4 \times \mathbb{R}^4/\mathbb{Z}_2$, see Refs. [19,39].

For large ρ (i.e. large X), or equivalently for $T \to 0$, Taub-NUT space is a squashed S^3, that is $S^2 \times S^1$, with S^1 a non-trivial (Hopf) fibration over S^2. This is best studied by introducing a radial coordinate R through $X^2 = 8MR$, which brings the Taub-NUT metric to the form [36]

$$
ds^2 = \left(1 + \frac{2M}{R} \right) (dR^2 + R^2 (d\theta^2 + \sin^2 \theta d\phi^2))
$$
$$
+ 4M^2 \left(1 + \frac{2M}{R} \right)^{-1} (dT + \cos \theta d\phi)^2, \tag{96}
$$

also familiar from the (spatial part of the) Kaluza–Klein monopole solution [40]. Since $T \in [0, 4\pi]$ we read off from the asymptotic form of the metric that the compactification radius equals $4M$. At large ρ, $\mathcal{M}_{\text{framed}}$ is therefore of the form $(\mathbb{R}^3 \times S^1) \times (\mathbb{R}^3 \times' S^1)/\mathbb{Z}_2$. It is natural to view this as the product space of two single BPS monopole moduli spaces. The first $\mathbb{R}^3$ represents the centre of mass and the second the relative coordinates. The first S^1 corresponds to x_0 and can be seen as a global $U(1)$. The other gives the relative $U(1)$ orientation on which the $\mathbb{Z}_2$ acts. This $\mathbb{Z}_2$ does *not* act on the positions, as the monopoles have in general different masses and are hence not identical objects. A similar description is valid for the $SU(2)$ two-monopole moduli space, see Refs. [41,33]. At *large* separations its metric is precisely of the Taub-NUT form, but with a *negative* mass parameter M, as was shown by Manton [42], using the asymptotic form of the interactions. The complete metric was constructed by Atiyah and Hitchin (see Ref. [33]) in terms of elliptic integrals, using its symmetries and hyper-Kähler structure.

Finally it is interesting to note that the moduli space of an $SU(3)$ monopole with maximal symmetry breaking to $U(1) \times U(1)$ and charges $(1, 1)$ (see Refs. [43,44]) is Taub-NUT with a positive mass parameter, as for the caloron. This is not surprising, as its Nahm data are precisely those of the caloron. For an $SU(3)$ monopole with maximal symmetry breaking there are two sub-intervals $[\mu_1, \mu_2]$ and $[\mu_2, \mu_3]$. On each sub-interval, $\hat{A}$ has dimension one and is therefore constant, due to the Nahm equations. The matching at μ_2, implied by the delta function singularity, is of the correct form (defining $\frac{1}{2}\rho^2$), and implies we can identify μ_1 with μ_3 (defining T) with correct matching as well. As the metric can be formulated in terms of the Nahm data, we would indeed expect the metric to be the same.

7. Discussion

The interpretation of the ADHM data for a periodic instanton as the Fourier co-efficients of the Nahm-transformed Weyl operator extends naturally to higher charges and other gauge groups [45]. For charge-1 calorons in $SU(N)$ the determination of

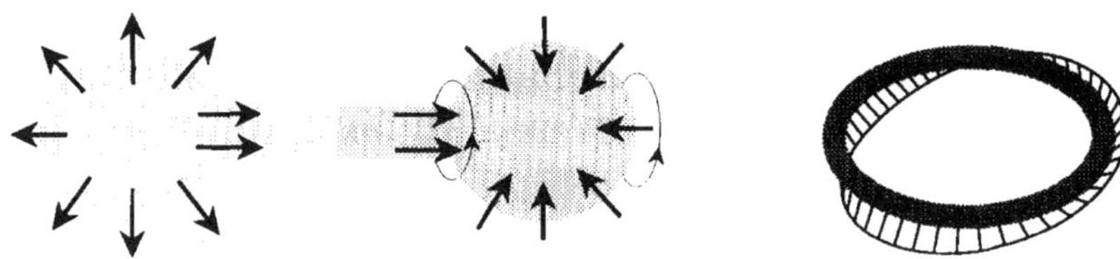

Fig. 6. The non-contractible loop is constructed from two oppositely charged monopoles by rotating one of them, as indicated on the left. On the right is a closed monopole line, rotating its frame when completing the circle.

the Green's function remains a problem of quantum mechanics on the circle with a piecewise constant potential [61] (on N sub-intervals, separated by delta functions). Also our formalism to compute the metric on the moduli space can be generalised. Due to the relation with the ADHM approach, one may wonder [6] whether there is some advantage in obtaining monopole solutions from the calorons by sending certain scales to infinity [62] – in the limit of which the solution becomes static and constituents separate. For higher charges the Nahm bundle is no longer abelian and the construction is more complicated. For generalisations to further compactifications, e.g. $\mathbb{R}^2 \times T^2$ and $\mathbb{R} \times T^3$ (see Ref. [18]), note that the 't Hooft ansatz [28] diverges when summing over more than one direction. This will correspond to all holonomies trivial and one may well have no solutions in that case. A dramatic particular example of a non-existence proof for charge-1 instantons is T^4, see Ref. [17], a situation where indeed an existence proof of Taubes [46] does not hold.

We now recall briefly Taubes' arguments for building gauge fields with topological charge 1 out of monopole fields [9,47]. Although his construction was within the standard model with a genuine Higgs field, the same argument applies to the caloron case, using A_0 as the Higgs field. As we saw in Section 2 (Eqs. (10), (11)), non-trivial $SU(2)$ monopole fields can be classified by the winding number $k_1 = -k_2$ of maps from S^2 to $SU(2)/U(1) \sim S^2$. We consider at this point configurations at a *fixed* time t, $\Psi = \{A_\mu(\mathbf{x})\}$. In the sector where the net winding vanishes we study a one-parameter family of configurations, $\Psi_t = \{A_\mu(\mathbf{x}, t)\}$ (the parameter can, but need not, be seen as the time t). When this configuration is made out of monopoles with opposite charges, in a suitable gauge the isospin orientations behave as shown in Fig. 6, sufficiently far from the core of both monopoles. We note that the arrows match in the "throat" of the configuration. This remains true if we rotate *only one* of the monopoles around the axis of the throat. Clearly, the net magnetic winding remains zero, but the fields of two monopoles will no longer cancel when brought together, despite the fact that the long-range abelian components do cancel. The non-contractible loop is now constructed by letting t affect a *full* rotation.

Taubes describes this by creating a monopole–anti-monopole pair, bringing them far apart, rotating one of them over a full rotation and finally bringing them together to annihilate. The four-dimensional configuration constructed this way is topologically non-trivial. Since an anti-monopole travelling forward in time is a monopole travelling backwards in time, we can describe the same as a closed monopole line (or loop).

It represents a topologically non-trivial configuration when the monopole makes a full rotation while moving along the closed monopole line (see Fig. 6). The non-trivial topology discussed by Taubes [9] ($\pi_1(M_0(S^2; S^2)) = \mathbb{Z}$) is just the Hopf fibration, except that now it is more natural to see S^1 as the base manifold and S^2 as the fibre, which rotates (twists) while moving along the circle formed by the closed monopole line. The only topological invariant available to characterise this homotopy type is precisely the Pontryagin index.

We mentioned that the short-range components of the fields can not be fully canceled due to the non-trivial "twist" along the monopole line, so they have to be responsible for the Pontryagin index. Indeed, in the computation of the total topological charge of our configurations as the integral over $\partial_\mu^2 \partial_\nu^2 \log \psi$ (Eq. (63)), the massive component of the field gives rise to

$$\psi = 2e^{4\pi(r\bar\omega + s\omega)}(1 + \mathcal{O}(|x|^{-1})) = 2e^{2\pi|x|}(1 + \mathcal{O}(|x|^{-1})),$$

$$\partial_\mu^2 \log \psi = 4\pi/|x| + \mathcal{O}(|x|^{-4}), \tag{97}$$

and thus yields the surviving boundary term, but at the same time does not contribute to the action density, since $\partial_\mu^2 |x|^{-1} = 0$.

We now inspect more closely the monopole content of our calorons. For this we choose ρ/T large, such that the monopoles are well separated and static. We have two world lines of monopoles running in opposite directions (due to their opposite charges), closed due to the periodic boundary conditions. At smaller separations the solutions are far from static, with the attractive force driving the constituents together, after which they annihilate. In that case the world lines form a single closed monopole line, as mentioned above. It should be noted, though, that for small ρ/T the constituents become rather extended. Nevertheless, such closed monopole lines are characterised by rotation of the local monopole field over precisely one full rotation when completing the circle, since our solution has unit topological charge. It prevents the field from decaying to the trivial configuration. It is this "twist" that provides the closed monopole line its stability.

Before continuing, we observe that the calorons are given in a singular gauge, as is usual for the ADHM construction. The function $\hat\psi$ (Eq. (52)) has an isolated zero at $x = 0$. This can be traced to the zero-mode in $B - x$, responsible for the non-trivial topology of the solution. This singularity is easily seen to be removed by a gauge transformation, locally of the form $x/|x|$ (viewing x as a quaternion). We now assume that $\omega\bar\omega \neq 0$ and consider the region outside the core of both monopoles, i.e. $r\bar\omega > 1$ and $s\omega > 1$. In this region

$$\phi = \frac{r + s + \pi\rho^2}{r + s - \pi\rho^2} + \mathcal{O}(e^{-8\pi\min(s\omega, r\bar\omega)}),$$

$$\tilde\chi = \frac{4\pi\rho^2}{(r + s + \pi\rho^2)^2}\left\{re^{-4\pi r\bar\omega}e^{-2\pi i x_0} + se^{-4\pi s\omega}\right\}\left(1 + \mathcal{O}(e^{-8\pi\min(s\omega, r\bar\omega)})\right). \tag{98}$$

Substituting this in Eq. (57) we find the solution to be time independent and abelian, up to exponential correction

$$A_0 = \frac{i}{2}\tau_3\partial_3 \log\phi + 2\pi i\omega\tau_3, \qquad A_k = \frac{i}{2}\tau_3\epsilon_{kj3}\partial_j \log\phi,$$

$$E_k = F_{k0} = \frac{i}{2}\tau_3\partial_k\partial_3 \log\phi, \qquad B_k = \epsilon_{kij}\partial_i A_j = \frac{i}{2}\tau_3\left(\partial_k\partial_3 \log\phi - \delta_{k3}\partial_j^2 \log\phi\right).$$

$$(99)$$

For convenience we rotated $\hat{a}$ to $\hat{e}_3$. Self-duality, $E = B$, requires $\log\phi$ to be harmonic. We first note that when *neglecting* the exponential corrections, ϕ^{-1} vanishes on the interval $-2\pi\rho^2\omega \leqslant x_3 \leqslant 2\pi\rho^2\bar{\omega}$ at $x_1 = x_2 = 0$ (we denote the characteristic function on this interval by $\chi_\omega(x_3)$). A careful analysis reveals $\partial_j^2 \log\phi = -4\pi\delta(x_1)\delta(x_2)\chi_\omega(x_3)$ (that it vanishes away from the zeros of ϕ^{-1} follows by a direct computation). The term $-\frac{i}{2}\tau_3\delta_{k3}\partial_j^2 \log\phi$ in the expression for the magnetic field corresponds precisely to the Dirac string singularity, carrying the return flux. One finds that $\partial_k E_k = \partial_k B_k = \frac{i}{2}\tau_3\partial_3\partial_j^2 \log\phi = 2\pi i\tau_3(\delta_3(s) - \delta_3(r))$, when ignoring this return flux, which in the full theory is absent [48] (indeed as noted before ϕ^{-1} has only an isolated zero at $x = 0$, corresponding to a gauge singularity).

Finally, to confirm our expectations it remains to identify the rotation of one of the monopoles so as to guarantee the topologically non-trivial nature of the configuration. Inspecting the behaviour in the core region of the monopoles, described by $\tilde{\chi}$ in Eq. (98), gives the following factorisation:

$$\tilde{\chi} = \chi^{(1)}(r) + e^{-2\pi i x_0}\chi^{(2)}(s).$$

$$(100)$$

While one of the monopoles has a static core, the other has a time-dependent phase rotation – equivalent to a (gauge) rotation – precisely of the type required to form a non-contractible loop, as the phase makes a full rotation when closing by the periodic boundary conditions in the time direction.

Although interpreting A_0 as the Higgs field allows one to introduce monopoles in pure gauge theory, there are some subtle differences. In the static limit the BPS equations imply $F_{i0} = D_i A_0$ and we would be tempted to call the solution a dyon. In the Higgs model dyons are constructed by taking A_0 proportional to the Higgs field Φ [2,49]. By a time-dependent gauge transformation A_0 can be gauged to zero. This gauge transformation is generated by A_0, precisely the unbroken generator, as A_0 is proportional to the Higgs field. The resulting electric field is now given by $\partial_0 A_i$ and is *not* quantised. In the Higgs model $B_i = D_i\Phi$ and $E_i = -\partial_0 A_i$. In pure gauge theory it makes, however, no sense to separate $D_i\Phi = D_i A_0$ from $\partial_0 A_i$. Gauge invariance requires that they occur in the combination $F_{i0} = D_i A_0 - \partial_0 A_i$. The electric field is necessarily fixed and quantised as soon as we interpret A_0 as the Higgs field. Nevertheless, we *can* consider dyons also in pure gauge theories, but for this we have to add a term proportional to $\theta F\tilde{F}$ to the lagrangian [50]. The electric charge is now proportional to θ, so no longer quantised. But, unlike in the Higgs model, it is the same for all monopoles.

We note that in the Higgs model the construction of the non-contractible loop generates an electric charge due to the (gauge) rotation along the closed monopole line, when interpreting the loop parameter as time. The electric charge is proportional to the rate

T.C. Kraan, P. van Baal / Nuclear Physics B 533 (1998) 627–659   655

of rotation and can vary along the monopole line. However, integrated along a closed monopole line the charge is fixed and proportional to the number of rotations, which hence plays the role of a winding number. In pure gauge theory this winding can not be read off from the long-range field components, but for both cases the fields in the core are responsible for the Pontryagin number (an abelian field can not contribute to this topological charge).

There is a natural context in which the analogy with the Higgs model is more precise. For this we have to add time as a fifth dimension, such that four-dimensional space is compactified on a circle. In the limit of zero compactification radius ($\mathcal{T} \rightarrow 0$) our solutions become genuine monopoles and can obtain dyonic charges in the sense of Julia and Zee [49]. It is in this context that Taub-NUT metric describes the scattering of oppositely charged monopoles on $\mathbb{R}^3 \times S^1$, in exactly the same way as the Atiyah–Hitchin metric describes the scattering of like-charged monopoles on $\mathbb{R}^3$ [51,33].

Monopoles appear also in the context of 't Hooft's abelian projection [16] as (gauge) singularities. The lesson we have learned from the above analysis is that in order to include the non-trivial topological charge, important for fermion zero-modes, breaking of the axial $U(1)$ symmetry [22] and presumably for chiral symmetry breaking, one needs to keep some information on the behaviour near the core of these monopoles. This allows one to combine the attractiveness of the dual superconductor picture of confinement [52] in terms of monopole degrees of freedom, with the success of the instanton liquid model [53]. There have been many attempts to make an effective monopole model for the long-range confining properties of QCD, see e.g. Ref. [54]. Also there have been many studies in lattice gauge theory, using the idea of abelian projection implemented by the so-called maximal abelian gauge [55], in order to extract the monopole content of the theory. It was observed that the string tension is saturated by the monopole fields [56]. More recently it was found that after abelian projection instantons contain closed monopole lines [57]. As emphasised in Ref. [58], in the light of Taubes' construction this was to be expected. Here we have shown in more detail how one can make fields with non-zero topological charge out of monopole degrees of freedom, with as example the well defined setting of calorons with non-trivial Polyakov loop. What is *minimally* required is a frame associated to each monopole, whose rotation is a topological invariant for closed monopole lines. Such closed monopole lines can shrink, but one will be left over with what represents an instanton. It would be interesting to build a hybrid model based on the instanton liquid and monopoles, and see how successful it is in capturing the appropriate phenomenology.

To conclude, it is sensible to take the monopole content of instantons serious in the broader context sketched here. Our gauge-invariant method of investigating the monopoles inside an instanton is somewhat destructive (but reversible). First we heat the instanton just a little. Then we add a non-trivial value of the Polyakov loop at infinity, without disturbing the instanton significantly (true for $\mathcal{T}$ sufficiently large). Now we have to squeeze (or heat) it hard. Out come the two constituent monopoles, in a direction determined by the choice we have made for the Polyakov loop at infinity (which does not change under heating). The new caloron solutions can be studied on the lattice by taking all links in the time direction, at the spatial boundary of the lattice, equal to $U_0 = \exp(2\pi i \boldsymbol{\omega} \cdot \boldsymbol{\tau}/N_0)$ (in lattice units $\mathcal{T}$ equals N_0). One can look for solutions using improved cooling [59] (to prevent calorons to disappear due to scaling violations). When interested in seeing the monopole constituents one may just as well take the time direction to be one lattice spacing ($N_0 = 1$). The lattice study of Ref. [60] hints at regular monopole solutions in pure gauge theory, although with a vanishing electric component, not what we would expect for our constituents.

Acknowledgements

We thank Kimyeong Lee for useful correspondence and Dmitri Diakonov, Simon Hands and Bernd Schroers for discussions. T.C.K. was supported by a grant from the FOM/SWON Association for Mathematical Physics.

Appendix A

In this appendix we present the result for the Green's function $G_x = (B^\dagger(x)B(x))^{-1}$, which after Fourier transformation satisfies the equation

$$\left\{ \left(\frac{1}{2\pi i}\frac{d}{dz} - x_0 \right)^2 + s^2 \chi_{[-\omega,\omega]}(z) + r^2 \chi_{[\omega,1-\omega]}(z) \right.$$
$$\left. -\frac{\rho^2}{2}\hat{\omega}\cdot\boldsymbol{\tau}(\delta(z-\omega)-\delta(z+\omega)) \right\}\hat{G}_x(z,z') = \delta(z-z'), \tag{A.1}$$

with r and s given as in Eq. (45). Its solution is given by

$$\hat{G}_x(z,z') = \chi_{[-\omega,\omega]}(z')\left(\chi_{[-\omega,\omega]}(z)\hat{G}_x^d(z,z',r,s,\omega)\right.$$
$$+\chi_{[\omega,1-\omega]}(z)\hat{G}_x^o(z,z',r,s,\omega))$$
$$+\chi_{[\omega,1-\omega]}(z')\left(\chi_{[\omega,1-\omega]}(z)\hat{G}_x^d(\tfrac{1}{2}-z,\tfrac{1}{2}-z',s,r,\bar{\omega})^*\right.$$
$$\left. +\chi_{[-\omega,\omega]}(z)\hat{G}_x^o(z',z,r,s,\omega)^*\right). \tag{A.2}$$

Like for $\hat{f}_x$ the diagonal component $\hat{G}_x^d(z,z')$ is only defined strictly for $z,z' \in [-\omega,\omega]$ and the off-diagonal component $\hat{G}_x^o(z,z')$ only for $z \in [\omega,1-\omega]$ and $z' \in [-\omega,\omega]$. For z or z' outside of these intervals, one first has to map back to the interval $[-\omega,1-\omega]$, using periodicity.

$$\hat{G}_x^d(z,z',r,s,\omega) = e^{2\pi i x_0(z'-z)}\pi(rs\hat{\psi})^{-1}\left\{e^{-2\pi i x_0\mathrm{sign}(z-z')}r\sinh(2\pi s|z-z'|)\right.$$
$$+s^{-1}\sinh(4\pi r\bar{\omega})[\pi s\rho^2\hat{\omega}\cdot\boldsymbol{\tau}\sinh(2\pi s(z+z'))$$
$$+\tfrac{1}{2}(s^2-r^2+\pi^2\rho^4)\cosh(2\pi s(z+z'))$$
$$+\tfrac{1}{2}(s^2+r^2-\pi^2\rho^4)\cosh(2\pi s(2\omega-|z-z'|))]$$
$$\left. +r\cosh(4\pi r\bar{\omega})\sinh(2\pi s(2\omega-|z-z'|))\right\},$$

$$\hat{G}^o_x(z, z', r, s, \omega) = e^{2\pi i x_0(z-z')}\pi(rs\hat{\psi})^{-1}$$

$$\times\{\pi\rho^2\hat{\omega}\cdot\boldsymbol{\tau}\sinh(2\pi r(1-z-\omega))\sinh(2\pi s(z'+\omega))$$

$$+r\cosh(2\pi r(z-1+\omega))\sinh(2\pi s(z'+\omega))$$

$$-s\sinh(2\pi r(z-1+\omega))\cosh(2\pi s(z'+\omega))$$

$$\times e^{-2\pi i x_0}[\,s\sinh(2\pi r(z-\omega))\cosh(2\pi s(z'-\omega))$$

$$-r\cosh(2\pi r(z-\omega))\sinh(2\pi s(z'-\omega))$$

$$+\pi\rho^2\hat{\omega}\cdot\boldsymbol{\tau}\sinh(2\pi r(z-\omega))\sinh(2\pi s(z'-\omega))]\}. \qquad (A.3)$$

In particular,

$$\hat{G}_x(\omega, -\omega) = \pi(rs\hat{\psi})^{-1}e^{4\pi i x_0\omega}\left\{e^{-2\pi i x_0}r\sinh(4\pi s\omega) + s\sinh(4\pi r\bar{\omega})\right\}, \qquad (A.4)$$

$$\hat{G}_x(\pm\omega, \pm\omega) = \pi(rs\hat{\psi})^{-1}\Big\{s\sinh(4\pi r\bar{\omega})\cosh(4\pi s\omega)$$

$$+r\sinh(4\pi s\omega)\cosh(4\pi r\bar{\omega}) \pm \pi\rho^2\hat{\omega}\cdot\boldsymbol{\tau}\sinh(4\pi r\bar{\omega})\sinh(4\pi s\omega)\Big\}.$$

which can be used to verify Eq. (64) and Eq. (51), as derived from Eq. (19).

References

[1] M.F. Atiyah, N.J. Hitchin, V.G. Drinfeld, Yu.I. Manin, Phys. Lett. 65 A (1978) 185.

[2] E.B. Bogomol'ny, Yad. Fiz. 24 (1976) 861; Sov. J. Nucl. Phys. 24 (1976) 449; M.K. Prasad and C.M. Sommerfield, Phys. Rev. Lett. 35 (1975) 760.

[3] N.S. Manton, Nucl. Phys. B 126 (1977) 525.

[4] C. Montonen and D. Olive, Phys. Lett. B 72 (1977) 117.

[5] H. Garland and M.K. Murray, Commun. Math. Phys. 120 (1988) 335.

[6] K. Lee and P. Yi, Phys. Rev. D 56 (1997) 3711 (hep-th/9702107).

[7] W. Nahm, Self-dual monopoles and calorons, *in* Lecture Notes in Physics, Vol. 201, ed. G. Denardo et al. (1984) p. 189.

[8] T.C. Kraan and P. van Baal, Exact T-duality between calorons and Taub-NUT spaces, Phys. Lett. B 428 (1998) 268 (hep-th/9802049).

[9] C. Taubes, Morse theory and monopoles: topology in long-range forces, *in* Progress in Gauge Field Theory, ed. G. 't Hooft et al. (Plenum Press, New York, 1984) p. 563.

[10] K. Lee, Instantons and magnetic monopoles on $R^3 \times S^1$ with arbitrary simple gauge groups, Phys. Lett. B 426 (1998) 323 (hep-th/9802012); K. Lee and C. Lu, SU(2) calorons and magnetic monopoles, Phys. Rev. D 58 (1998) 25011 (hep-th/9802108).

[11] B.J. Harrington and H.K. Shepard, Phys. Rev. D 17 (1978) 2122; D 18 (1978) 2990.

[12] D.J. Gross, R.D. Pisarski and L.G. Yaffe, Rev. Mod. Phys. 53 (1983) 43.

[13] P. van Baal, Twisted boundary conditions: a non-perturbative probe for pure non-abelian gauge theories, Ph.D. thesis, Utrecht, July 1984.

[14] W. Nahm, Phys. Lett. B 90 (1980) 413.

[15] W. Nahm, All self-dual multimonopoles for arbitrary gauge groups, CERN preprint TH-3172 (1981), published in Freiburg ASI 301 (1981); The construction of all self-dual multimonopoles by the ADHM method, *in* Monopoles in Quantum Field Theory, ed. N. Craigie et al. (World Scientific, Singapore, 1982) p. 87.

[16] G. 't Hooft, Nucl. Phys. B 190 [FS3] (1981) 455; Physica Scripta 25 (1982) 133.

[17] P.J. Braam and P. van Baal, Commun. Math. Phys. 122 (1989) 267.

[18] P. van Baal, Nucl. Phys. B (Proc. Suppl.) 49 (1996) 238 (hep-th/9512223).

[19] S.K. Donaldson and P.B. Kronheimer, The Geometry of Four-Manifolds (Clarendon Press, Oxford, 1990).

[20] M.F. Atiyah and I.M. Singer, Ann. Math. 93 (1971) 119.

[21] C. Callias, Commun. Math. Phys. 62 (1978) 213;
R. Bott and R. Seeley, Commun. Math. Phys. 62 (1978) 235.

[22] G. 't Hooft, Phys. Rev. D 14 (1976) 3432.

[23] E. Corrigan and P. Goddard, Ann. Phys. (N.Y.) 154 (1984) 253.

[24] N.J. Hitchin, Commun. Math. Phys. 89 (1983) 145;
J. Hurtubise and M.K. Murray, Commun. Math. Phys. 122 (1989) 35.

[25] H. Nakajima, Monopoles and Nahm's Equations, *in* Einstein Metrics and Yang–Mills Connections, Sanda, 1990, ed. T. Mabuchi and S. Mukai (Marcel Dekker, New York, 1993).

[26] M.F. Atiyah, Geometry of Yang–Mills fields, Fermi Lectures (Scuola Normale Superiore, Pisa, 1979).

[27] E.F. Corrigan, D.B. Fairlie, S. Templeton and P. Goddard, Nucl. Phys. B 140 (1978) 31.

[28] G. 't Hooft, unpublished, quoted in R. Jackiw, C. Nohl, C. Rebbi, Phys. Rev. D 15 (1977) 1642;
R. Rajaraman, Solitons and Instantons (North-Holland, Amsterdam, 1982).

[29] H. Osborn, Ann. Phys. (N.Y.) 135 (1981) 373; Nucl. Phys. B 159 (1979) 497.

[30] T.C. Kraan, Nahm's transformation and instantons on $R^3 \times S^1$, poster presented at the 2nd DRSTP symposium, Dalfsen, The Netherlands, 5-6 June, 1997 (unpublished).

[31] E.T. Whitakker and G.N. Watson, A Course in Modern Analysis (Cambridge Univ. Press, Cambridge, 1927) p. 169.

[32] P. Rossi, Nucl. Phys. B 149 (1979) 170.

[33] M.F. Atiyah and N.J. Hitchin, The Geometry and Dynamics of Magnetic Monopoles (Princeton Univ. Press, Princeton, 1988).

[34] E.T. Newman, T. Unti and L. Tamburino, J. Math. Phys. 4 (1963) 915.

[35] S. Hawking, *in* General Relativity, an Einstein Centenary Survey, ed. S. Hawking and W. Israel (Cambridge Univ. Press, Cambridge, 1979) p. 774.

[36] S. Hawking, Phys. Lett. A 60 (1977) 81.

[37] T. Eguchi and A.T. Hanson, Phys. Lett. B 74 (1978) 249;
G.W. Gibbons and C.N. Pope, Commun. Math. Phys. 66 (1979) 267.

[38] N.J. Hitchin, A. Karlhede, U. Lindström and M. Roček, Commun. Math. Phys. 108 (1987) 535.

[39] J.P. Gauntlett and J.A. Harvey, *S*-duality and the spectrum of magnetic monopoles in heterotic string theory, hep-th/9407111.

[40] R. Sorkin, Phys. Rev. Lett. 51 (1983) 87;
D. Gross and M. Perry, Nucl. Phys. B 226 (1983) 29.

[41] G.W. Gibbons and N.S. Manton, Nucl. Phys. B 274 (1986) 183.

[42] N.S. Manton, Phys. Lett. B 154 (1985) 397, with erratum B 157 (1985) 475.

[43] S.A. Connell, The dynamics of the SU(3) charge (1,1) magnetic monopole (1991), to be found at: ftp: //maths.adelaide.edu.au/pure/murray/oneone.tex, unpublished preprint.

[44] J.P. Gauntlett and D.A. Lowe, Nucl. Phys. B 472 (1996) 194 (hep-th/9601085);
K. Lee, E.J. Weinberg and P. Yi, Phys. Lett. B 376 (1996) 97 (hep-th/9601097).

[45] K. Lee and P. Yi, Dyons in $N = 4$ supersymmetric theories and three-pronged strings, hep-th/9804174.

[46] C.H. Taubes, J. Diff. Geom. 19 (1984) 517.

[47] C.H. Taubes, Commun. Math. Phys. 86 (1982) 257, 299.

[48] G. 't Hooft, Nucl. Phys. B 79 (1974) 276;
A.M. Polyakov, JETP Lett. 20 (1974) 194.

[49] B. Julia and A. Zee, Phys. Rev. D 11 (1975) 2227.

[50] E. Witten, Phys. Lett. B 86 (1979) 283.

[51] N.S. Manton, Phys. Lett. B 110 (1982) 54.

[52] G. 't Hooft, *in* High Energy Physics, EPS conference, Palermo, 1975, ed. A. Zichichi (Editrice Compositori, Bologna, 1976);
S. Mandelstam, Phys. Rep. C 23 (1976) 245.

[53] E. Shuryak, Phys. Rep. 264 (1996) 357;
T. Schäfer and E. Shuryak, Rev. Mod. Phys. 70 (1998) 323 (hep-ph/9610451).

[54] J. Smit and A. van der Sijs, Nucl. Phys. B 355 (1991) 603.

[55] A.S. Kronfeld, G. Schierholz and U.J. Wiese, Nucl. Phys. B 293 (1987) 461.

[56] T. Suzuki and I. Yotsuyanagi, Phys. Rev. D 42 (1990) 4257;
     M. Polikarpov, Nucl. Phys. B (Proc. Suppl.) 53 (1997) 134 (hep-lat/9609020), and references therein.
[57] M.N. Chernodub and F.V. Gubarev, JETP Lett. 62 (1995) 100 (hep-th/9506026);
     A. Hart and M. Teper, Phys. Lett. B 372 (1996) 261 (hep-lat/9511016);
     V. Bornyakov and G. Schierholz, Phys. Lett. B 384 (1996) 190 (hep-lat/9605019);
     R.C. Brower, K.N. Orginos and Ch.-I. Tan, Phys. Rev. D 55 (1997) 6313 (hep-th/9610101).
[58] P. van Baal, Nucl. Phys. B (Proc. Suppl.) 63A–C (1998) 126 (hep-lat/9709066).
[59] M. García Pérez, A. González-Arroyo, J. Snippe and P. van Baal, Nucl. Phys. B 413 (1994) 535. (hep-lat/9309009).
[60] M.L. Laursen and G. Schierholz, Z. Phys. C 38 (1988) 501.
[61] T.C. Kraan and P. van Baal, Monopole constituents inside SU(n) calorons, Phys. Lett. B 435 (1998) 389 (hep-th/9806034).
[62] T.C. Kraan and P. van Baal, Constituent monopoles without gauge fixing, hep-lat/9808015.

ELSEVIER

10 September 1998

Physics Letters B 435 (1998) 389–395

PHYSICS LETTERS B

Monopole constituents inside $SU(n)$ calorons

Thomas C. Kraan, Pierre van Baal

Instituut-Lorentz for Theoretical Physics, University of Leiden, P.O. Box 9506, NL-2300 RA Leiden, The Netherlands

Received 8 June 1998
Editor: P.V. Landshoff

Abstract

We present a simple result for the action density of the $SU(n)$ charge one periodic instantons – or calorons – with arbitrary non-trivial Polyakov loop $\mathscr{P}_\infty$ at spatial infinity. It is shown explicitly that there are n lumps inside the caloron, each of which represents a BPS monopole, their masses being related to the eigenvalues of $\mathscr{P}_\infty$. A suitable combination of the ADHM construction and the Nahm transformation is used to obtain this result. © 1998 Elsevier Science B.V. All rights reserved.

1. Introduction

Instantons and BPS monopoles are self-dual finite action solutions to the Yang-Mills equations of motion. Their origin is topological and related to windings in the gauge transformations describing their behaviour at infinity, where these solutions approach vacua, necessary for the action to be finite. The action of these solutions is proportional to these winding numbers or charges, which suggests that they are composed out of elements with unit charge. We will find by computing the action density that even charge one periodic instantons (calorons) for the gauge group $SU(n)$ are further composed out of constituents. The constituents are n basic BPS monopoles [1], whose magnetic charges cancel exactly.

Periodic instantons were first discussed in the context of finite temperature field theory [2,3]. Motivated by issues of T-duality [4] and D-brane constructions [5] in string theory, periodic instantons with a non-trivial Polyakov loop were recently constructed. The ingredients of the Nahm transformation for $SU(n)$ calorons [6], can be naturally presented in terms of n monopole constituents [7]. This has been the basis of the approach followed by Lee and co-workers [8,9]. Our work relies on a more direct approach, combining the ADHM construction of multi-instanton solutions [10] with the Nahm construction [6], related by Fourier transformation. The constituent monopoles appear as explicit lumps in the action density [11].

The composite nature of the periodic instantons can also be appreciated by an argument due to Taubes [12] on how to build out of monopoles, configurations with non-trivial topological charge. It has far reaching consequences, which go beyond the existence of exact caloron solutions [11]. Its relevance for QCD has motivated us to extend our work to $SU(3)$.

Finite action solutions on $\mathbb{R}^3 \times S^1$ should approach vacua at spatial infinity. Due to the topology of this base manifold, these vacua can be non-trivial. This endows the caloron with extra parameters – labeling these vacua

390　　　　　　　　　　*T.C. Kraan, P. van Baal / Physics Letters B 435 (1998) 389–395*

– which are studied in terms of the eigenvalues of the Polyakov loop around S^1 at spatial infinity. The Polyakov loop is defined in the periodic gauge, $A_\mu(x, x_0 + \mathcal{T}) = A_\mu(x, x_0)$, as

$$\mathcal{P}(x) = P\exp\left(\int_0^{\mathcal{T}} A_0(x, x_0)\, dx_0\right), \tag{1}$$

where $\mathcal{T}$ is the period and P denotes path-ordering. At infinity the value of the Polyakov loop does not change under continuous deformations of the loop and its eigenvalues are topological invariants.

In extending our work to the $SU(3)$ case it turned out to be natural to generalise to $SU(n)$ (see also the appendices of Ref. [9]). We will present the formula for the action density in Section 2. The derivation is outlined in Section 3. A detailed description will be given elsewhere. In Section 4 we discuss the properties of the solution.

2. The result

We consider the calorons with no net magnetic charge, in which case the Polyakov loop (holonomy) at spatial infinity becomes constant. Its eigenvalues $e^{2\pi i \mu_m}$ will play an important role in the construction,

$$\lim_{|x| \to \infty} \mathcal{P}(x) = \mathcal{P}_\infty = V\mathcal{P}_\infty^0 V^{-1}, \quad \mathcal{P}_\infty^0 = \exp\left[2\pi i\,\mathrm{diag}(\mu_1, \ldots, \mu_n)\right]. \tag{2}$$

Making use of the gauge symmetry, we can choose the eigenvalues such that

$$\mu_1 < \ldots < \mu_n < \mu_{n+1} \equiv \mu_1 + 1, \quad \sum_{m=1}^{n} \mu_m = 0, \tag{3}$$

assuming maximal symmetry breaking for the moment. We define $\nu_m = \mu_{m+1} - \mu_m$, related to the mass of the mth constituent monopole. Standard arguments, summarised below, gives $4n$ instanton parameters for fixed $\mathcal{P}_\infty$, including the global gauge transformations that do not change $\mathcal{P}_\infty$. We will see that $3n$ parameters can be interpreted as the positions (y_m) of the constituents. The remaining parameters in this interpretation are the $n-1$ phases related to the unbroken gauge group $U(1)^{n-1}$, on which the action density does not depend and the position of the caloron in time, which we fix to be 0 by translational invariance. Also we will use the scale invariance to set $\mathcal{T} = 1$. Where needed, the proper $\mathcal{T}$ dependence can be reinstated on dimensional grounds. We find the following surprisingly simple formula:

$$-\tfrac{1}{2}\mathrm{tr}\, F_{\mu\nu}^2 = -\tfrac{1}{2}\partial_\mu^2\partial_\nu^2 \log \psi. \tag{4}$$

where the positive scalar potential ψ is defined as

$$\psi(x) = \tfrac{1}{2}\mathrm{tr}\prod_{m=1}^{n}\left\{\begin{pmatrix} r_m & |y_m - y_{m+1}| \\ 0 & r_{m+1} \end{pmatrix}\begin{pmatrix} \cosh(2\pi\nu_m r_m) & \sinh(2\pi\nu_m r_m) \\ \sinh(2\pi\nu_m r_m) & \cosh(2\pi\nu_m r_m) \end{pmatrix}\frac{1}{r_m}\right\} - \cos(2\pi x_0). \tag{5}$$

Here $r_m = |x - y_m|$ denotes the center of mass radius of the mth monopole. The order of matrix multiplication is crucial, $\prod_{m=1}^{n} A_m \equiv A_n \ldots A_1$.

3. The construction

In our description of the caloron with non-trivial Polyakov loop, we pick the so-called algebraic gauge,

$$A_\mu(x, x_0 + \mathcal{T}) = \mathcal{P}_\infty A_\mu(x, x_0)\mathcal{P}_\infty^{-1}, \tag{6}$$

which is related to the periodic gauge by the non-periodic gauge transformation $g(x, x_0) = V\exp[2\pi i x_0 \mathrm{diag}(\mu_1, \ldots, \mu_n)/\mathcal{T}]V^{-1}$. In the algebraic gauge *all* gauge field components approach zero at infinity. The technique we use is to interpret the ADHM data as the Fourier coefficients of the functions that appear in the Nahm transformation. This is to solve the quadratic ADHM constraint, which is non-trivial for a periodic array of instantons, twisted in colour space going from one time slice to the next.

We summarise the ADHM formalism for $SU(n)$ charge k instantons [10], to fix our notation. It employs a k dimensional vector $\lambda = (\lambda_1, \ldots, \lambda_k)$, where $\lambda_i^\dagger$ is a two-component spinor in the $\bar{n}$ representation of $SU(n)$. Alternatively, λ can be seen as an $n \times 2k$ complex matrix. In addition one has four complex hermitian $k \times k$ matrices B_μ, combined into a $2k \times 2k$ complex matrix $B = \sigma_\mu \otimes B_\mu$, using the unit quaternions $\sigma_\mu = (1_2, i\tau)$ and $\bar{\sigma}_\mu = (1_2, -i\tau)$, where τ_i are the Pauli matrices. With abuse of notation, we often write $B = \sigma_\mu B_\mu$. Together λ and B constitute the $(n + 2k) \times 2k$ dimensional matrix $\Delta(x)$, to which is associated a complex $(n + 2k) \times n$ dimensional normalised zero mode vector $v(x)$,

$$\Delta(x) = \begin{pmatrix} \lambda \\ B(x) \end{pmatrix}, \quad B(x) = B - x, \quad \Delta^\dagger(x)v(x) = 0, \quad v^\dagger(x)v(x) = 1_n. \tag{7}$$

Here the quaternion $x = x_\mu \sigma_\mu$ denotes the position (a $k \times k$ unit matrix is implicit) and $v(x)$ can be solved explicitly in terms of the ADHM data by

$$v(x) = \begin{pmatrix} -1_n \\ u(x) \end{pmatrix}\phi^{-\frac{1}{2}}, \quad u(x) = (B^\dagger - x^\dagger)^{-1}\lambda^\dagger, \quad \phi(x) = 1_n + u^\dagger(x)u(x). \tag{8}$$

As $\phi(x)$ is an $n \times n$ positive hermitian matrix, its square root $\phi^{\frac{1}{2}}(x)$ is well-defined. The gauge field is given by

$$A_\mu(x) = v^\dagger(x)\partial_\mu v(x) = \phi^{-\frac{1}{2}}(x)\big(u^\dagger(x)\partial_\mu u(x)\big)\phi^{-\frac{1}{2}}(x) + \phi^{\frac{1}{2}}(x)\partial_\mu \phi^{-\frac{1}{2}}(x). \tag{9}$$

For $A_\mu(x)$ to be a self-dual connection, $\Delta(x)$ has to satisfy the quadratic ADHM constraint, which states that $\Delta^\dagger(x)\Delta(x) = B^\dagger(x)B(x) + \lambda^\dagger\lambda$ (considered as $k \times k$ complex quaternionic matrix) has to commute with the quaternions, or equivalently

$$\Delta^\dagger(x)\Delta(x) = \sigma_0 \otimes f_x^{-1}, \tag{10}$$

defining f_x as a hermitian $k \times k$ Green's function. The self-duality follows by computing the curvature:

$$F_{\mu\nu} = 2\phi^{-\frac{1}{2}}(x)u^\dagger(x)\eta_{\mu\nu}f_x u(x)\phi^{-\frac{1}{2}}(x), \tag{11}$$

making essential use of the fact that f_x commutes with the quaternions, and $\eta_{\mu\nu} \equiv \sigma_{[\mu}\bar{\sigma}_{\nu]}$ being self-dual ($\bar{\eta}_{\mu\nu} \equiv \bar{\sigma}_{[\mu}\sigma_{\nu]}$ is anti-self-dual). The quadratic constraint can be formulated as $\Im m(\Delta^\dagger(x)\Delta(x)) = 0$, where $\Im m W \equiv \frac{1}{2}[W - \tau_2 W^t \tau_2]$, and one obtains

$$\bar{\eta}_{\mu\nu} \otimes B_\mu B_\nu + \frac{1}{2}\tau_a \otimes \mathrm{tr}_2(\tau_a \lambda^\dagger\lambda) = 0, \tag{12}$$

where tr_2 is the spinorial trace. Note that this implies that $\mathrm{tr}_2(\tau_a \lambda^\dagger\lambda)$ is traceless for $a = 1,2,3$. To count the number of instanton parameters we observe that the transformation $\lambda \to \lambda T^\dagger$, $B_\mu \to TB_\mu T^\dagger$, with $T \in U(k)$ leaves the gauge field and the ADHM constraint untouched. Taking this symmetry into account, we find the dimension of the instanton moduli space to be $4kn$ dimensional. Global gauge transformations are realised by $\lambda \to g\lambda$, with $g \in SU(n)$. Those that leave $\mathscr{P}_x$ invariant reduce the dimension of the gauge invariant parameter space (by $n - 1$ for maximal symmetry breaking). Finally, we quote an elegant result [13] for the action density in terms of f_x

$$\mathrm{tr}F_{\mu\nu}^2(x) = -\partial_\mu^2\partial_\nu^2 \log\det f_x. \tag{13}$$

392 *T.C. Kraan, P. van Baal / Physics Letters B 435 (1998) 389–395*

The charge one caloron with Polyakov loop $\mathscr{P}_\infty$ at infinity is built out of a periodic array of instantons, twisted by $\mathscr{P}_\infty$. This is implemented in the ADHM formalism by requiring (suppressing colour and spinor indices)

$$u_{p+1}(x+1) = u_p(x)\mathscr{P}_\infty^{-1} \tag{14}$$

with $p \in \mathbb{Z}$. Using that $\phi^{\pm\frac{1}{2}}(x+1) = \mathscr{P}_\infty \phi^{\pm\frac{1}{2}}(x)\mathscr{P}_\infty^{-1}$, Eq. (9) leads to the required result, Eq. (6). Demanding

$$\lambda_{p+1} = \mathscr{P}_\infty \lambda_p, \quad B_{p,q} = B_{p-1,q-1} + \sigma_0 \delta_{p,q}, \tag{15}$$

suitably implements Eq. (14) and is partially solved by imposing

$$\lambda_p = \mathscr{P}_\infty^p \zeta, \quad B_{p,q} = p\sigma_0 \delta_{p,q} + \hat{A}_{p-q}, \tag{16}$$

with $\hat{A}$ still to be determined to account for Eq. (12), which also constrains the spinor $\zeta^\dagger$ in the $\bar{n}$ representation of $SU(n)$ to

$$\mathrm{tr}_2\left(\tau_a \zeta^\dagger \zeta\right) = 0, \quad a = 1,2,3. \tag{17}$$

It is useful to introduce the n projectors $P_m = VP_m^{(0)}V^{-1}$, with $(P_m^{(0)})_{a,b} = \delta_{m,a}\delta_{m,b}$ and $P_m P_{m'} = \delta_{m,m'}P_m$, such that $\mathscr{P}_\infty = \sum_m \exp(2\pi i \mu_m)P_m$ and $\lambda_p = \sum_m \exp(2\pi i p\mu_m)P_m\zeta$.

We now perform the Fourier transformation to the Nahm setting [6], which casts B into a Weyl operator and $\lambda^\dagger\lambda$ into a singularity structure on S^1,

$$\sum_{p,q} B_{p,q}(x)e^{2\pi i(pz-qz')} = \frac{\delta(z-z')}{2\pi i}\hat{D}_x(z'), \quad \hat{D}_x(z) = \sigma_\mu \hat{D}_x^\mu(z) = \frac{d}{dz} + \hat{A}(z) - 2\pi i x,$$

$$\hat{A}(z) = \sigma_\mu \hat{A}^\mu(z), \quad \hat{A}^\mu(z) = 2\pi i \sum_p e^{2\pi i pz}\hat{A}_p^\mu, \tag{18}$$

$$\sum_p e^{-2\pi i pz}\lambda_p = \sum_p e^{2\pi i p(\mu_m - z)}P_m\zeta = \hat{\lambda}(z), \quad \hat{\lambda}(z) = \sum_m \delta(z-\mu_m)P_m\zeta,$$

$$\sum_{p,q} \lambda_p^\dagger e^{2\pi i(pz-qz')}\lambda_q = \delta(z-z')\hat{\Lambda}(z), \quad \hat{\Lambda}(z) = \sum_m \delta(z-\mu_m)\zeta^\dagger P_m\zeta = \zeta^\dagger\hat{\lambda}(z).$$

Introducing the vector $\zeta_{(m)} = P_m\zeta$, it is standard to show

$$\zeta^\dagger P_m\zeta = \zeta_{(m)}^\dagger \zeta_{(m)} = \frac{1}{2\pi}\left(|\boldsymbol{\rho}_m| - \boldsymbol{\tau}\cdot\boldsymbol{\rho}_m\right), \tag{19}$$

and the quadratic ADHM constraint, which takes the form

$$\frac{1}{2}\left[\hat{D}_\mu(z),\hat{D}_\nu(z)\right]\bar{\eta}_{\mu\nu} = 4\pi^2 \Im m\,\hat{\Lambda}(z), \tag{20}$$

leads to the (for $k = 1$ abelian) Nahm equation

$$\frac{d}{dz}\hat{A}_j(z) = 2\pi i \sum_m \delta(z-\mu_m)\rho_m^j. \tag{21}$$

The T symmetry in the ADHM construction translates into a $U(1)$ gauge symmetry on S^1, which leaves $\hat{A}_j$ invariant and allows one to set $\hat{A}_0 = 2\pi i \xi_0$, ξ_0 being the position in time, which we absorb in x_0. Since $\boldsymbol{\rho}_m = -\pi\,\mathrm{tr}_2(\boldsymbol{\tau}\zeta^\dagger P_m\zeta)$, it follows that $\sum_m \boldsymbol{\rho}_m = -\pi\,\mathrm{tr}_2(\boldsymbol{\tau}\zeta^\dagger\zeta) = 0$, see Eq. (17), such that we may introduce y_m ($y_0 \equiv y_n$), with $\boldsymbol{\rho}_m = y_m - y_{m-1}$. In terms of these we find

$$\hat{A}_j(z) = 2\pi i \sum_m \chi_{[\mu_m,\mu_{m+1}]}(z)y_m^j, \tag{22}$$

where $\chi_{[\mu_m,\mu_{m+1}]}(z) = 1$ for $z \in [\mu_m,\mu_{m+1}]$ and 0 elsewhere, taking into account z has period 1. Note that y_m is only fixed up to a constant ξ, related to the freedom of adding a constant to the solution of the Nahm equation, Eq. (21). The 4-vector ξ_μ describes the position of the caloron. Also note that the ρ_m are independent of the phases of $P_m\zeta$, affected by the residual gauge symmetry, of which $n-1$ are independent due to the gauge group being $SU(n)$ rather than $U(n)$.

The vector y_m is interpreted as the position of the mth constituent. On each sub-interval $[\mu_m,\mu_{m+1}]$, $\hat{A}_j(z) = 2\pi i y_m^j$ is constant, which is precisely the Nahm datum for a single BPS monopole located at y_m [14]. The length of the Nahm interval for the single BPS monopole corresponds to its asymptotic Higgs value, and thereby to its mass. Thus, the mth subinterval $[\mu_m,\mu_{m+1}]$ corresponds to a BPS monopole at y_m, with mass proportional to $\nu_m = \mu_{m+1} - \mu_m$. Together, the n BPS monopoles form the $SU(n)$ caloron.

The Green's function f_x, central in the ADHM construction, is found after a Fourier transformation of Eq. (10), introducing $\hat{f}_x(z,z') \equiv \sum_{p,q} f_x^{p,q} e^{2\pi i(pz - qz')}$ as the solution to the differential equation

$$\left\{ \left(\frac{1}{2\pi i}\frac{d}{dz} - x_0 \right)^2 + \sum_m \chi_{[\mu_m,\mu_{m-1}]}(z)\, r_m^2 + \frac{1}{2\pi}\sum_m \delta(z - \mu_m)|y_m - y_{m-1}| \right\} \hat{f}_x(z,z') = \delta(z - z'), \quad (23)$$

where the radii are given by $r_m = |x - y_m|$, to be interpreted as the center of mass radii of the constituent monopoles. The solution of a quantum-mechanical problem on the circle with a piecewise constant potential and delta function impurities is obtained by solving it on each sub-interval, where $\hat{f}_x(z,z')$ is of simple exponential form. Starting at $z = z'$ and matching properly at $z = \mu_m$ so as to account for the scattering by the impurity, we can go full circle to return at $z = z'$ where one last matching accounts for the delta function at the rhs. of Eq. (23).

With the solution for $f_x(z,z')$ available, we can compute $\mathrm{Tr}\hat{D}_x^\mu f_x$ and show it to be equal to $-\pi i \partial_\mu \log\psi$, ψ being the positive scalar function defined in Eq. (5). Using Eq. (13), this leads to Eq. (4). One retrieves our $SU(2)$ results of Refs. [4,11] by putting $\mu_1 = -\omega$, $\mu_2 = \omega$ and $\mu_3 = 1 - \omega$, such that $\nu_1 = \mu_2 - \mu_1 = 2\omega$ and $\nu_2 = \mu_3 - \mu_2 = 1 - 2\omega \equiv 2\bar{\omega}$. Furthermore we identify $r_1 = s$, $r_2 = r$ and $|y_1 - y_0| = |y_2 - y_1| = \pi\rho^2$.

4. Discussion

We briefly discuss the properties of the $SU(n)$ charge one caloron. From Eq. (5) we see that the mth constituent monopole can be located at arbitrary y_m, with arbitrary mass $8\pi^2\nu_m/\mathcal{T}$, subject only to the constraint $\sum_m \nu_m = 1$, choosing $\mathscr{P}_\infty$, ξ and ζ appropriately. As the action density, Eq. (4), is expressed as a total derivative, the action is easily found by partial integration, with the expected result of $8\pi^2$. The size of the instanton is related to the differences in position of the constituent monopoles. As we work in units of $\mathcal{T}$, the situation of a small scale (nearby constituents), corresponds to large $\mathcal{T}$, i.e. to an instanton on $\mathbb{R}^4$. At the other extreme one has well separated lumps for small $\mathcal{T}$, i.e. in the static limit. In Fig. 1 we present a typical $SU(3)$ caloron for decreasing values of $\mathcal{T}$ using Eq. (4). We will resist the temptation of showing results for other n, as Eq. (4) can be readily implemented.

When the lumps are far apart, they do not deform each other and become spherically symmetric. Since the solution is self-dual, the constituents have to be basic BPS monopoles. This can be proven by carefully analysing Eq. (4) for the limit where $r_m \ll r_l$ for all $l \neq m$, in which case the action density approaches $-\frac{1}{2}\partial_\mu^2\partial_\nu^2 \log[\sinh(2\pi\nu_m r_m)/r_m]$. This is precisely the behaviour of the BPS monopole [15]. The other constituents need not be well-separated from each other for the above argument to hold. In particular sending the mth constituent to infinity (i.e. $|y_m| \to \infty$) suffices to make the caloron static. What remains are $n-1$ monopole constituents with a combined magnetic charge opposite to the magnetic charge of the mth constituent monopole that has been removed. As the solution is static in this limit one is left with an $SU(n)$ BPS monopole. Indeed, for $|y_m| \to \infty$ we see from the solution of the Nahm equation, Eq. (21), that $\hat{A}_j(z)$ lives on an interval, rather

 T.C. Kraan, P. van Baal / Physics Letters B 435 (1998) 389–395

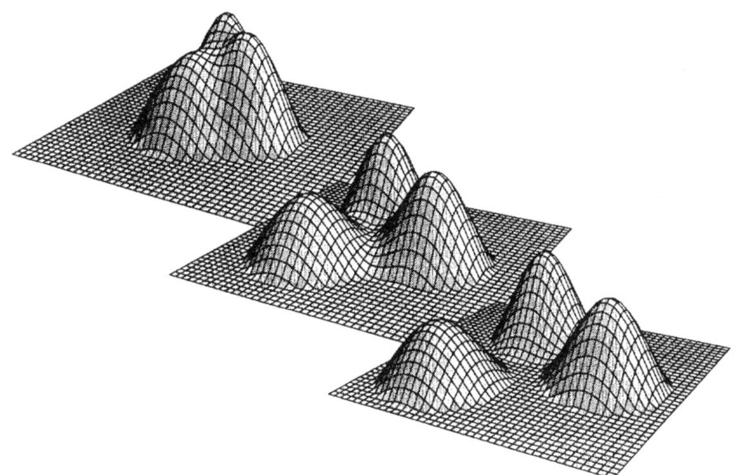

Fig. 1. Action densities for the $SU(3)$ caloron at $x_0 = 0$ in the plane defined by the centers of the three constituents for $1/\mathcal{T} = 1.5, 3$ and 4 (increasing temperature from top to bottom). We choose mass parameters $(\nu_1, \nu_2, \nu_3) = (0.4, 0.35, 0.25)$, implemented by $(\mu_1, \mu_2, \mu_3) = (-17/60, -2/60, 19/60)$. The constituents are located at $y_1 = (-\frac{1}{2}, \frac{1}{2}, 0)$, $y_2 = (0, \frac{1}{2}, 0)$ and $y_3 = (\frac{1}{2}, -\frac{1}{4}, 0)$, in units of $\mathcal{T}$. The profiles are given on equal logarithmic scales, cut-off at an action density below $1/e$.

than on the circle, as is appropriate for the $SU(n)$ monopole [16]. One readily obtains the energy density of this monopole by taking the limit $|y_m| \to \infty$ in Eq. (5), verifying that it decays as $1/|x|^4$, as opposed to $1/|x|^6$ without removing the mth constituent.

Our results have been derived for the case of maximal symmetry breaking, $\mu_m \neq \mu_{m+1}$. The situation of non-maximal symmetry breaking corresponds to a constituent obtaining zero mass, $\nu_m = 0$. In that case its center of mass radius *drops out* of Eq. (5), using

$$\begin{pmatrix} r_m & |y_m - y_{m+1}| \\ 0 & r_{m+1} \end{pmatrix} \frac{1}{r_m} \begin{pmatrix} r_{m-1} & |y_m - y_{m-1}| \\ 0 & r_m \end{pmatrix} = \begin{pmatrix} r_{m-1} & |y_m - y_{m+1}| + |y_m - y_{m-1}| \\ 0 & r_{m+1} \end{pmatrix}. \tag{24}$$

This was also observed for $SU(2)$, in which case non-maximal symmetry breaking corresponds to a trivial Polyakov loop, $\mathcal{P}_\infty = \pm 1$, and the solution becomes that of Harrington and Shepard [2]. Hence our formula for the action density should also be valid for non-maximal symmetry breaking.

Although our formalism can be extended easily to higher topological charges [9], the appropriate Nahm equation (i.e. solving the quadratic ADHM constraint) becomes a non-abelian problem, and finding solutions requires more powerful tools. Nevertheless, it is interesting to note that it is natural to conjecture that k instantons (i.e. an instanton of charge k) can be built from kn monopoles, since each instanton can be considered as being built from n BPS monopoles. The monopole constituents are only well separated when $\mathcal{T}$ is small, where the $4kn$ instanton parameters can be interpreted as $3kn$ positions and kn phases (including $\exp(2\pi i \xi_0 / \mathcal{T})$). We will not speculate further on these matters here, but want to emphasise that the monopole constituent picture has some interesting phenomenological implications for the description of the long distance properties of QCD, discussed in detail in Ref. [11], and which will be the subject of further investigations.

Acknowledgements

T.C.K. was supported by a grant from the FOM/SWON Association for Mathematical Physics.

References

[1] E.B. Bogomol'ny, Yad. Fiz. 24 (1976) 861; Sov. J. Nucl. 24 (1976) 449; M.K. Prasad, C.M. Sommerfield, Phys. Rev. Lett. 35 (1975) 760.

[2] B.J. Harrington, H.K. Shepard, Phys. Rev. D 17 (1978) 2122; D 18 (1978) 2990.

[3] D.J. Gross, R.D. Pisarski, L.G. Yaffe, Rev. Mod. Phys. 53 (1983) 43.

[4] T.C. Kraan, P. van Baal, Phys. Lett. B 428 (1998) 268, hep-th/9802049.

[5] K. Lee, P. Yi, Phys. Rev. D 56 (1997) 3711, hep-th/9702107.

[6] W. Nahm, Self-dual monopoles and calorons, in: G. Denardo, e.a. (Eds.), Lect. Notes in Physics. 201, 1984, p. 189.

[7] H. Garland, M.K. Murray, Commun. Math. Phys. 120 (1988) 335.

[8] K. Lee, Instantons and magnetic monopoles on $R^3 \times S^1$ with arbitrary simple gauge groups, hep-th/9802012; K. Lee, C. Lu, SU (2) calorons and magnetic monopoles, hep-th/9802108.

[9] K. Lee, P. Yi, Dyons in $N = 4$ supersymmetric theories and three-pronged strings, hep-th/9804174.

[10] M.F. Atiyah, N.J. Hitchin, V.G. Drinfeld, Yu.I. Manin, Phys. Lett. 65 A (1978) 185; M.F. Atiyah, Geometry of Yang-Mills fields Fermi lectures, Scuola Normale Superiore, Pisa, 1979.

[11] T.C. Kraan, P. van Baal, Periodic Instantons with non-trivial Holonomy, hep-th/9805168; New Instanton Solutions at Finite Temperature, hep-th/9805201.

[12] C. Taubes, Morse theory and monopoles: topology in long range forces, in: G.'t Hooft et al. (Eds.), Progress in gauge field theory Plenum Press, New York, 1984, p. 563.

[13] H. Osborn, Nucl. Phys. B 159 (1979) 497.

[14] W. Nahm, Phys. Lett. B 90 (1980) 413.

[15] P. Rossi, Nucl. Phys. B 149 (1979) 170.

[16] W. Nahm, All self-dual multimonopoles for arbitrary gauge groups CERN preprint TH-3172, 1981, published in Freiburg ASI 301, 1981; The construction of all self-dual multimonopoles by the ADHM method, in: N. Craigie, e.a. (Eds.), Monopoles in quantum field theory, World Scientific, Singapore, 1982, p. 87.

RECEIVED: *March 16, 1999*, ACCEPTED: *June 2, 1999*

Calorons on the lattice - a new perspective

Margarita García Pérez, Antonio González-Arroyo*and Alvaro Montero

Departamento de Física Teórica C-XI,
Universidad Autónoma de Madrid,
28049 Madrid, Spain.
E-mail: marga@martin.ft.uam.es, tony@martin.ft.uam.es,
montero@martin.ft.uam.es

Pierre van Baal

Instituut-Lorentz for Theoretical Physics,
University of Leiden, PO Box 9506,
NL-2300 RA Leiden, The Netherlands.
E-mail: vanbaal@lorentz.leidenuniv.nl

ABSTRACT: We discuss the manifestation of instanton and monopole solutions on a periodic lattice at finite temperature and their relation to the infinite volume analytic caloron solutions with asymptotic non-trivial Polyakov loops. As a tool we use improved cooling and twisted boundary conditions. Typically we find $2Q$ lumps for topological charge Q. These lumps are BPS monopoles.

KEYWORDS: Nonperturbative Effects, Solitons Monopoles and Instantons, Lattice Gauge Field Theories.

---

*Also Instituto de Física Teórica C-XVI, Universidad Autónoma de Madrid, 28049 Madrid, Spain.

Contents

1. Introduction

Calorons are characterised by their holonomy, defined by the value of the Polyakov loop at spatial infinity. When non-trivial, it resolves the fact that a caloron is built from constituent monopoles, their mass ratios directly determined by the holonomy [1, 2]. These solutions differ from the (deformed) instantons described by the Harrington-Shepard solution [3], for which the holonomy is trivial. What we find by (improved [4]) cooling on a finite lattice, to relatively high accuracy, is SU(2) configurations that fit these infinite volume caloron solutions for arbitrary constituent monopole mass ratios. Twist [5] in the time direction constrains the masses of the two constituent monopoles to be equal.

The constituent nature becomes evident when the instanton scale parameter ρ is larger than the time extent β (inverse temperature) of the system. The masses of the monopoles are for SU(2) proportional [1] to ω and $1/2 - \omega$, where ω ($0 \leq \omega \leq 1/2$) follows from the trace of the holonomy: $2\cos(2\pi\omega)$. The distance between the monopole constituents is given by $\pi\rho^2/\beta$. At $\rho/\beta \ll 1$ the constituents therefore hide deep inside the core of the instanton and the non-trivial holonomy plays no discernible role. But for $\rho/\beta \gg 1$ the situation is opposite; the instanton becomes static and will dissolve in two BPS monopoles [6, 7]. The transition occurs [8, 9] for $\beta/2 < \rho < \beta$. When, however, the holonomy is trivial one of the monopoles is massless and will hide in the background.

JHEP06(1999)001

Charge one SU(N) calorons have N constituent monopoles [10] for non-trivial holonomy. These have the same location in time, but the spatial position of each constituent monopole can be arbitrary. There are (at fixed holonomy) $N-1$ phases associated to the residual U(1)$^{N-1}$ gauge symmetry that leaves the holonomy invariant. The total number of parameters describing these calorons is therefore $4N$. One may speculate that the $N-1$ phases are replaced in a finite volume by the holonomy itself, indeed described by $N-1$ eigenvalues taking values in U(1) ($\exp(2\pi i\omega)$ for SU(2)). Also it is likely that, in general, a charge Q caloron is characterised by NQ constituent monopoles, which we confirm for a $Q=2$ caloron solution obtained from cooling. At zero temperature it is tempting to explain the $4NQ$ parameters of an SU(N) charge Q instanton in terms of the positions of NQ objects [11]. Indeed, there are charge $Q=1/N$ instanton solutions on a torus with twisted boundary conditions, whose four parameters specify its position [12]. Subdividing a given finite volume in boxes with the appropriate twisted boundary conditions, such that each cell supports a $Q=1/N$ instanton, provides an exact solution that has NQ lumps. In ref. [11] it is suggested that a *typical* self-dual configuration would appear as an ensemble of N randomly placed lumps of charge $Q=1/N$, whose locations would account for the $4NQ$ parameters. The results at finite temperature presented here suggest that the assignment of $Q=1/N$ charge to each lump might only hold on the average.

Our results point to the usefulness of studying the dynamical role of these configurations. A first attempt in that direction is hampered by the fact that at high temperatures where the constituent monopoles should be well separated, the fluctuations are so large that on average topological charge cannot be supported over large enough domains of space-time to capture the configurations with cooling [13].

On a semiclassical basis one is tempted to argue against non-trivial holonomy. It polarises the vacuum at infinity and raises the energy density above the one with a trivial holonomy [14]. But now we have seen that these BPS bound states can be supported in a finite volume, it is time to acknowledge this as an irrelevant objection, given the non-perturbative and non-trivial nature of the QCD vacuum. As a consequence, constituent monopoles, at least at high temperatures, are tangible objects that do not depend on a choice of Abelian projection [15], which till now has been used to address the monopole content of the theory [16]. In extracting the non-trivial topological content of the theory constituent monopoles introduce an extra parameter: their mass, $16\pi^2\omega/\beta$. Up to now only the *maximal* mass, $8\pi^2/\beta$, of such a BPS monopole was considered. It arises in terms of the caloron with trivial holonomy, described by the Harrington-Shepard solution [3]. Rossi [17] showed that at high temperature, equivalent to a large scale parameter, this solution indeed becomes a BPS monopole [7].

In section 2 we discuss the numerical procedure of constructing the configurations. Apart from cooling with improved actions, twisted boundary conditions are used as a tool for biasing the cooling towards non-trivial holonomy. The twist can then be removed, while preserving the non-trivial holonomy and the constituent monopole nature of the configuration, although it should be pointed out that no exact charge one instanton solutions can exist on T^4, which *remains* true at finite temperature. Interesting in this respect is that the well-established $Q = 1/2$ instanton solutions that occur with suitable combinations of spatial and temporal twists (so-called non-orthogonal twist), can be argued to become a *single* static BPS monopole in the infinite volume limit at finite temperature. This is discussed in section 3. Configurations of higher charge are discussed in section 4 and we conclude with some speculations and possible applications. An appendix summarises the formulae for the SU(2) analytic caloron solutions.

2. Non-trivial holonomy from time-twist

For finite temperature ($T = 1/\beta$) and volume (L^3, $L \gg \beta$) caloron configurations with non-trivial holonomy were discovered on lattices with twisted boundary conditions. Starting with a random configuration and after applying a standard cooling algorithm one frequently reaches $Q = 1$ self-dual configurations which are stable under many cooling steps. These configurations are later analysed. An automatic peak-searching routine identified one or (actually more frequently) two lumps in them. These are our candidate caloron configurations on the lattice. Below, we will show how twist in the time direction can help in bringing about non-trivial holonomy.

To appreciate the ease with which twist can be implemented on the lattice, and because twist has been a very useful tool [4, 18, 19], neglected by large parts of the lattice community, we think it is useful to review the notion of twist-carrying plaquettes that introduce twist by modifying the lattice action [20], but not the measure. In the initial formulation of 't Hooft [5], SU(N) twisted boundary conditions were implemented by defining gauge functions $\Omega_\mu(x)$ (which are assumed independent of x_μ), such that with a^μ the periods of the torus in the four directions ($a^\nu_\mu = L_\mu \delta_{\mu\nu}$)

$$U_\nu(x + a^\mu) = \Omega_\mu(x)\, U_\nu(x)\, \Omega_\mu^\dagger(x + \hat{\nu})\,, \tag{2.1}$$

here re-formulated for a lattice of size $\prod_\mu N_\mu$. Calculating $U_\nu(x + a^\mu + a^\lambda)$ in two ways shows that for all x one should have

$$\Omega_\mu(x + a^\lambda)\, \Omega_\lambda(x) = Z_{\mu\lambda}\, \Omega_\lambda(x + a^\mu)\, \Omega_\mu(x)\,, \tag{2.2}$$

with $Z_{\mu\lambda} = \exp(2\pi i n_{\mu\lambda}/N)$ an element of the center of the gauge group. (We define $k_i = n_{0i}$ and $m_i = \frac{1}{2}\varepsilon_{ijk} n_{jk}$ to distinguish the twist in the time and space directions respectively). The center freedom arises because $U_\mu(x)$ is invariant under constant center gauge transformations (i.e. the gauge field is in the adjoint representation). In the presence of site variables (fields in the fundamental representation) one is required to put all $Z_{\mu\nu}$ equal to 1. We now perform the following change of variables [20]

$$U'_\mu(x) = U_\mu(x)\,\Omega_\mu(x)\,, \qquad \text{for} \quad x_\mu = N_\mu - 1\,. \tag{2.3}$$

As a consequence, the plaquettes at $x_\lambda = N_\lambda - 1$ and $x_\mu = N_\mu - 1$ (for any value of the other two components of x) can be shown to have acquired an additional factor $Z_{\lambda\mu}$. These corner plaquettes are called twist-carrying and the change of variables has absorbed the twist in the action, by multiplying these plaquettes by the appropriate center element (the action involves the real part of the plaquette variables *after* this multiplication). The location of the twist-carrying plaquette is arbitrary, as one is free to choose the boundary of the box used for defining the torus. Alternatively, the twist-carrying plaquette can be moved around by a *periodic* gauge transformation. It corresponds to the non-Abelian analogue of a Dirac string, and is at the heart of 't Hooft's definition of magnetic flux for non-Abelian gauge theories [21, 5]. Thus, twist is introduced by the trivial modification of the weights of the plaquettes in terms of multiplication with appropriate center elements and causes no computational overhead.

Note that we have just shown that if $Z_{\mu\nu} = 1$ for all μ and ν, in a suitable gauge the links can be chosen periodic without changing the weights of the plaquettes. In the continuum, however, there remains an obstruction in making the gauge field periodic when the topological charge of the configuration is non-trivial [22, 18]. This shows that on the lattice, only the center charges are unambiguously defined. Interestingly this includes configurations [12] that in the continuum would be assigned a non-trivial fractional Pontryagin index [5] (so-called twisted instantons).

To understand what the effect of the twist is on the holonomy, we use the observation that the presence of the Z_N flux can be measured by taking a Polyakov loop in the a^λ direction, which when translated over a period in the a^μ direction picks up a factor $Z_{\mu\lambda}$.

$$P_\lambda(x) = \frac{1}{N}\mathrm{Tr}\, P \exp\left(\int_0^1 A_\lambda(x + sa^\lambda)ds\right)\Omega_\lambda(x)\,, \tag{2.4}$$

$$P_\lambda(x + a^\mu) = Z_{\lambda\mu}P_\lambda(x)\,. \tag{2.5}$$

There are various ways to see this [23, 4], but becomes most evident when 'pulling' the loop over the twist-carrying plaquette. For SU(2) this means that the Polyakov loop is anti-periodic in case the twist is non-trivial. In particular for $Z_{0i} = -1$, $P_0(\vec{x})$ is anti-periodic in the x_i-direction. As we increase the size of the spatial torus it is natural to expect that the self-dual configuration would approach a caloron solution. Then $P_0(\vec{x})$ would approach a constant at spatial infinity. This is only compatible

JHEP06(1999)001

with the anti-periodicity implied by the non-trivial time-twist when $P_0(\vec{x}) \to 0$ for $|\vec{x}| \to \infty$, forcing $\omega = 1/4$ and thus non-trivial holonomy. This therefore provides a sure way of obtaining caloron solutions with non-trivial holonomy on the lattice, which at high temperature gives rise to two constituent monopoles, albeit in this case of equal mass.

Since the twist in the time direction forces the constituent monopoles to have equal mass, the lattice corrections to the value of the action (which depend on the shape of the configuration [4]) are affected only by the separation of the two constituents (in the next section we will encounter the situation where the mass ratio is affected by the cooling). This allows to manipulate the positions of the two lumps by using the tool of cooling with modified actions. This can be implemented [4] by using a lattice action that combines the traces of the 1×1 and 2×2 plaquettes. The two couplings are fixed in terms of the parameters multiplying the leading (continuum) and next to leading (a^2) terms in the expansion of the lattice action in powers of the lattice spacing a. The a^2 term is given by a unique dimension six operator, and its coefficient is called ε (it is trivial to incorporate the twist-carrying plaquettes also in these modified actions). Wilson's action corresponds to $\varepsilon = 1$. The choice $\varepsilon < 1$ is known as over-improvement, whereas improved cooling is performed by choosing $\varepsilon = 0$. In this last case the lattice and continuum action differ only by corrections of order a^4. For that reason, we will choose $\varepsilon = 0$ whenever we compare with the analytic infinite volume continuum caloron solution. However, unlike for the continuum action, the value of the a^2 operator depends on the position of the constituent monopoles, and therefore we can use other values of ε to alter these positions. Cooling with the Wilson action has the effect of driving the constituent monopoles together, since the Wilson action is decreased with respect to the continuum when the field strength has a larger gradient [4]. Once the two lumps merge, and can no longer be distinguished from an instanton (at which point the solution will no longer be static), it follows the usual fate of an instanton under prolonged cooling with the Wilson action: At some point it falls through the lattice [24]. (For cooling histories see fig. 3).

Over-improved cooling has the effect of pushing the two constituent monopoles apart. One can speed-up the rate at which monopoles separate by decreasing ε. Apriori it is not clear if, when the lumps are maximally apart, the solution will not be affected significantly by the boundary conditions. This will partly depend on the ratio L/β, but we find for $L = 4\beta$ that these effects are rather small.

In figure 1 we give an example of a caloron configuration with well separated constituents on a $16^3 \times 4$ lattice with $\vec{k} = (1,1,1)$, initially generated by cooling with the Wilson action, switching to improved cooling to reduce lattice artifacts. Shown is the action density s. We see that the agreement with the infinite volume analytic result is very good, with the action peaks for the lattice result somewhat

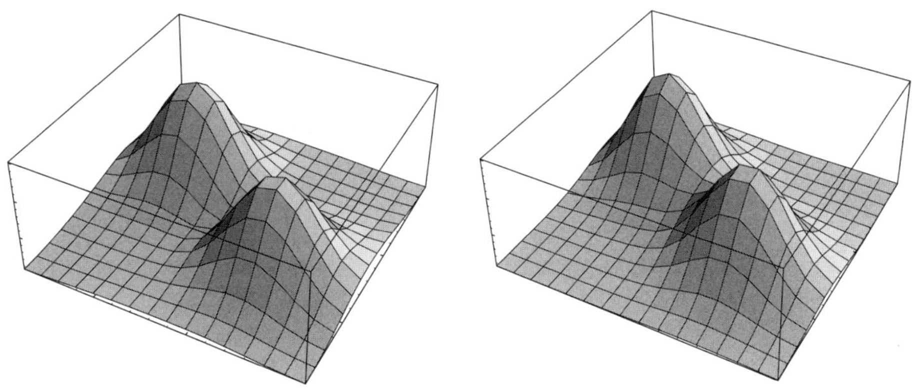

Figure 1: Lattice caloron profiles (left) on a $16^3 \times 4$ lattice for $\vec{k} = (1,1,1)$, created with improved cooling ($\varepsilon = 0$). The total action is $1.000185 \times 8\pi^2$. Vertically is plotted $\log(1+s/3)$, with s the action density at the lattice site (after clover averaging). The profile fits well to the analytic caloron solution (shown on the right at $y=t=0$) with $\omega = 1/4$ and constituents at $\vec{y}_1 = (2.50, 0.12, 0.95)$ and $\vec{y}_2 = (1.38, \quad 0.24, 2.67)$, in units where $\beta = 1$ (or $a = 1/4$) and the left most lattice point corresponding to $x = z = 0$.

lower (this feature is somewhat suppressed by plotting $\log(1 + s/3)$, rather than s). The total action of this static lattice configuration is very close to the required continuum value $8\pi^2$. An example of a non-static configuration with overlapping constituents will be presented below (see fig. 5). There seems no doubt that a continuum solution with this constituent monopole structure should exist on the time-twisted torus.

3. The case of space-twist

When both space and time twists are non-trivial and $\vec{k} \cdot \vec{m} \neq 0$ mod N (called non-orthogonal twist), the minimum of the action corresponds to a so-called twisted instanton with fractional charge. Unlike the integer charge instantons, these twisted instantons can not fall through the lattice. Their scale is fixed, only their position is a free parameter. This was used in the past [12] to find accurate lattice results using ordinary cooling ($\varepsilon = 1$). At high temperatures such a twisted instanton becomes static and represents a single BPS monopole on T^3. The twist allows for non-zero charge in the box. As discussed in the previous section it also gives rise to a holonomy characterised by $\omega = 1/4$. Indeed, we were able to fit the finite temperature twisted instanton (in a sufficiently large volume) to one of the constituent monopoles of the caloron at $\omega = 1/4$ (when placing the other constituent at a sufficiently large separation). In the appropriate limits both become ordinary BPS monopoles with mass $4\pi^2/\beta$.

JHEP06(1999)001

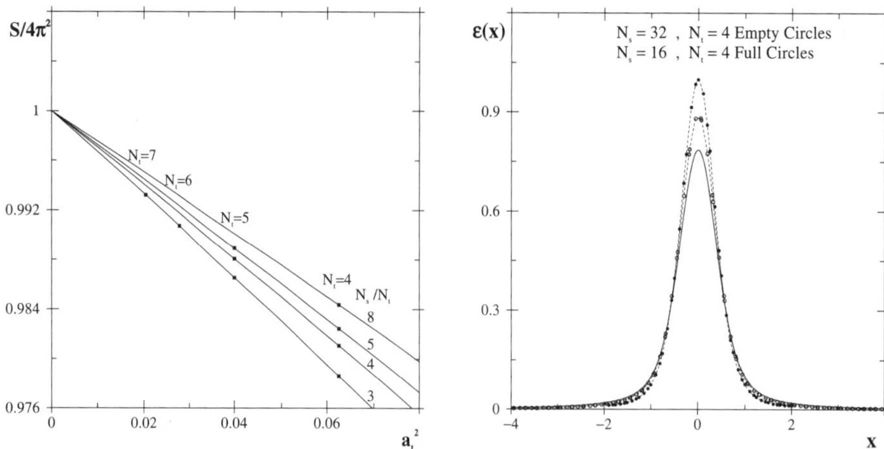

Figure 2: Lattice minimum action for various sizes (left), including the best fit to eq. (3.1) with: $S_0 = 0.999975(4)$, $b = 0.19950(7)$, $c = 0.3107(3)$ and $d = 0.0844(5)$. Also shown is the lattice action profile $\mathcal{E}(x)$, obtained by summing the action density over all but one of the spatial coordinates, as compared to the infinite volume analytic BPS result (right).

Now we will show the type and size of finite volume and lattice artifact effects. In fig. 2 (left) we display a plot of the minimum lattice (Wilson) action for lattices with different space and time extensions, $N_s^3 \times N_t$ and twist $\vec{k} = \vec{m} = (1, 1, 1)$. In the given range, $N_t = 4$—7 and $N_s = 16$—32, deviations from the continuum result are of the order of a few percent. However, the pattern of deviations from the continuum value is well understood. If we set $a_t = 1/N_t$ and $a_s = 1/N_s$, the value of the lattice action can be fitted with great accuracy to a formula:

$$\frac{S}{4\pi^2} = S_0 - ba_t^2 - ba_s^2 - ca_t a_s - d(a_t + a_s)^4 \,. \tag{3.1}$$

The extrapolated value of the continuum action matches $4\pi^2$ to a precision of a few parts in 10^5. Notice also that the extrapolation shows the existence of a self-dual continuum solution for any value of the ratio $L/\beta = N_s/N_t$. Furthermore, the lattice correction to the action decreases in absolute value with the ratio N_s/N_t, consistent with the statement made before that Wilson's action decreases with decreasing separation of the monopoles, since in this case N_s plays the role of the separation between lumps (the periodic mirrors).

To measure finite volume corrections, we performed improved cooling ($\varepsilon = 0$, to minimise lattice corrections) for $N_s = 16$ and 32. In this case, the values of the minimum lattice action attained are of the order $S/4\pi^2 = 1.0001(1)$. In fig. 2 (right) we compare the x-profiles $\mathcal{E}(x)$ obtained from the lattice minimum action configuration with the corresponding one for the BPS monopole. The x-profile is the integral of the action density over all but the x coordinate. This quantity has smaller errors and is less sensitive to the lattice discretisation than the action density itself. From the figure we see how the lattice profiles approach the infinite volume BPS monopole profile. The slow convergence is due to the powerlike Abelian tail of the BPS monopole (in contrast to the exponential tail found for other cases [25]).

Interestingly, an exact caloron solution with equal-size constituents ($\omega = 1/4$) on the twisted torus can be constructed by gluing two twisted instantons together, starting from the $Q = 1/2$ solution defined by $\vec{k} = \vec{m} = (1, 0, 0)$. Gluing two boxes in the y- or z-directions preserves $\vec{k}$, but reduces $\vec{m}$ to the trivial value (since $n_{\mu\nu}$ is defined modulo 2 for SU(2)). This exact solution corresponds to the situation studied in the previous section. Instead, gluing two boxes in the x-direction removes the time-twist, but preserves the space-twist. The same twist results when gluing the two boxes in the time-direction. In the first case we have an exact solution on a space-twisted torus with equal size constituents (corresponding to $\omega = 1/4$) at maximal separation in the direction of the twist, whereas in the second case the static nature of the finite temperature solution simply leads to doubling the mass of the monopole. Therefore this solution corresponds to an exact caloron solution on a space-twisted torus with trivial holonomy (the other constituent monopole is massless).

We have also performed lattice studies on a space-twisted torus, with $\vec{k} = \vec{0}$, which allowed us to probe the constituent monopole mass ratios, by a subtle use of the cooling procedure. It can be proven that without twist there are *no regular* charge one instanton solutions on a torus [26], but for any non-trivial twist an 8 dimensional space of regular solutions exists [27, 28]. Part of this parameter space comes about by gluing a localised instanton to the unique curvature free background supported by the twist. The eight parameters are given by scale, space-time position and so-called attachment parameters, that describe the gauge orientation of the localised instanton relative to the fixed curvature free background. For $\vec{m} \neq \vec{0}$ we find that the magnetically charged constituent monopoles, superposed on this non-Abelian magnetic flux background, experience an additional force that repels them as far as the finite volume allows. The presence of this force is evident from the fact that under prolonged cooling in all cases, $\varepsilon = 1, 0, -1$, the separation between the two constituent monopoles was increasing and that their centers lined up with the direction of $\vec{m}$. Once the constituent monopoles are placed at their maximal separation, further cooling with the Wilson action ($\varepsilon = 1$) leads to action shifting from one to the other peak, driving the constituent monopole mass ratios away from equal masses. Once one of the masses has decreased to zero, the scale parameter of the remaining (deformed) instanton configuration can shrink, resulting in the usual fate of falling through the lattice under prolonged cooling with the Wilson action. For over-improvement the effect is opposite, and the masses are pushed to equal values. The 'force' — due to lattice artifacts — changing the value of ω can be neglected for $\varepsilon = 0$ cooling. We summarise the behaviour under cooling in

JHEP06(1999)001

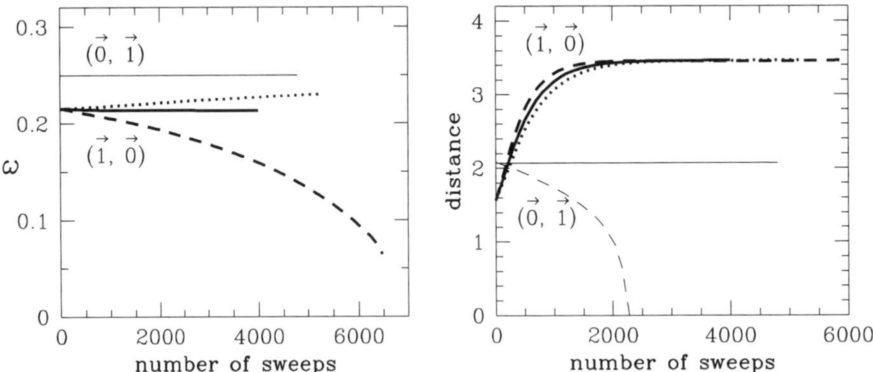

Figure 3: Cooling histories for $(\vec{m}, \vec{k}) = (\vec{0}, \vec{1})$ and $(\vec{1}, \vec{0})$ (resp. thin and fat curves), where $\vec{1} \equiv (1, 1, 1)$, on lattices of size $16^3 \times 4$. Solid, dashed and dotted curves are for resp. $\varepsilon = 0, 1, -1$ cooling. For $(\vec{0}, \vec{1})$, where $\omega \equiv 1/4$, the two ω–curves cannot be distinguished.

fig. 3, by showing the distance between the peak locations and ω (estimated by equating $(1/2-\omega)^4/\omega^4$ to the ratio of the peak heights) as a function of the number of cooling sweeps. Shown are the histories for $\vec{m} = (1, 1, 1)$ at $\varepsilon = -1, 0, 1$ and for $\vec{k} = (1, 1, 1)$ at $\varepsilon = 0, 1$.

That we can have solutions that are characterised by arbitrary mass ratios of the constituent monopoles is also illustrated in figure 4, which represents two values for the parameter ω, comparing the finite volume configurations obtained from improved cooling to the analytic infinite volume caloron solutions with non-trivial holonomy. We see again that the agreement is very good (and will improve for increasing L/β), with the peaks for the lattice result now somewhat higher as compared to the infinite volume results.

Next, we discuss the comparison with configurations that are not static. Here the constituents are close together and therefore there is considerable overlap. This is illustrated in figure 5, both in the case of twist in time as in the case of no twist. As mentioned previously, the presence of twist in time ($\vec{k} \neq 0$) forces $\omega = 1/4$, while in the absence of twist ω can be arbitrary. Obtaining configurations with no twist requires some care. By cooling random configurations one ends up quickly in the trivial vacuum configuration. Hence, it is useful to start with a $Q = 1$ configuration having $\vec{m} \neq \vec{0}$ obtained by cooling. Then twist is eliminated from this configuration by setting the weights of the twist-carrying plaquettes to their standard (untwisted) value. Additional improved cooling steps were applied to the configuration, leading to a new solution still having a non-trivial ω value. We recall that there are no exactly self-dual $Q = 1$ solutions on the torus without twist [26]. However, for solutions well-localised inside the torus the configuration is very approximately self-dual. Notice, nonetheless, that this reflects itself in higher values of the minimum lattice action. For these configurations with periodic boundary conditions performing

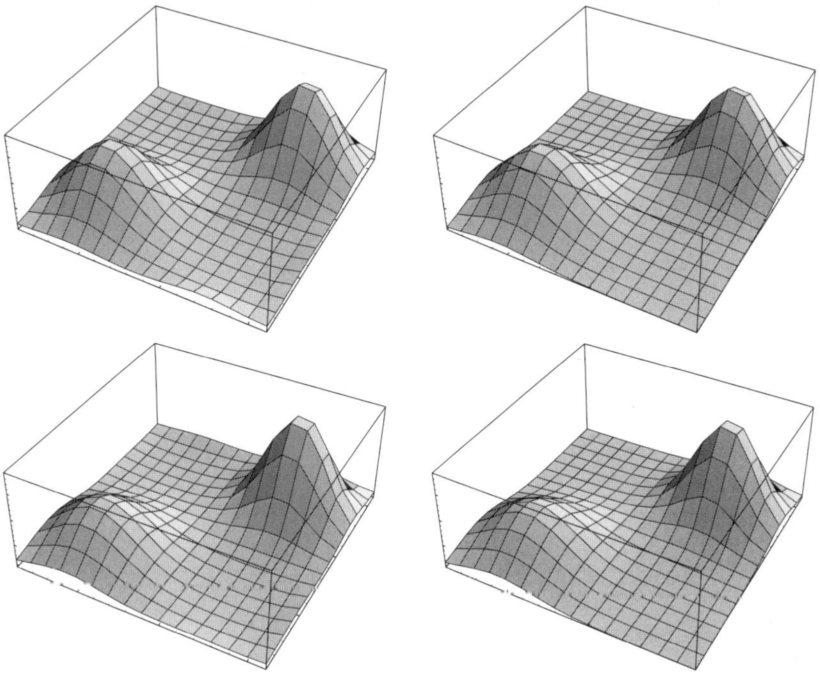

Figure 4: Lattice caloron profiles (left) for two configurations on a $16^3 \times 4$ lattice for $\vec{m} = (1,1,0)$, created with improved cooling ($\varepsilon = 0$) - after manipulating with $\varepsilon = \pm 1$ cooling to obtain the desired mass ratios. The total actions are $1.000155 \times 8\pi^2$ (top) and $1.000001 \times 8\pi^2$ (bottom). Vertically is plotted $\log(1 + s/3)$, with s the action density at the lattice site (after clover averaging). The profiles fit well to the analytic caloron solutions (shown on the right at $y = t = 0$) with top: $\omega = 0.210$ and constituents at $\vec{y}_1 = (1.04, -0.08, 0.86)$ and $\vec{y}_2 = (3.05, -0.09, 2.85)$, and bottom: $\omega = 0.175$ with constituents at $\vec{y}_1 = (0.85, -0.06, 0.85)$ and $\vec{y}_2 = (2.85, \ 0.06, 2.85)$, all in units where $\beta = 1$ (or $a = 1/4$) and the left most lattice point corresponding to $x = z = 0$.

further cooling steps with positive or zero ε will bring the constituents together and leads to the standard fate of instantons on the lattice. This can be stabilised by $\varepsilon < 0$, and the better the solution is contained within the box, the closer one can take $\varepsilon = 0$ to have a stable lattice solution [4].

The differences with the analytic infinite volume caloron solutions only show themselves by small differences in peak heights (at $t = 0$) and would not be clearly visible on the scale of figure 5. Instead, in figure 6, we display the analytic action density profile in the $z - t$ plane, where z is the axis connecting the two constituent monopole centers. The values of ω and the distance of the constituent monopoles is as in figure 5. It is clear that a two-lump structure is still visible. As a function of z the constituent monopoles are best seen for t values where the density is minimal ($t = 1/2$). The logarithmic scale enhances the regions of low action densities in favour of those with large densities, and brings out more clearly the constituents.

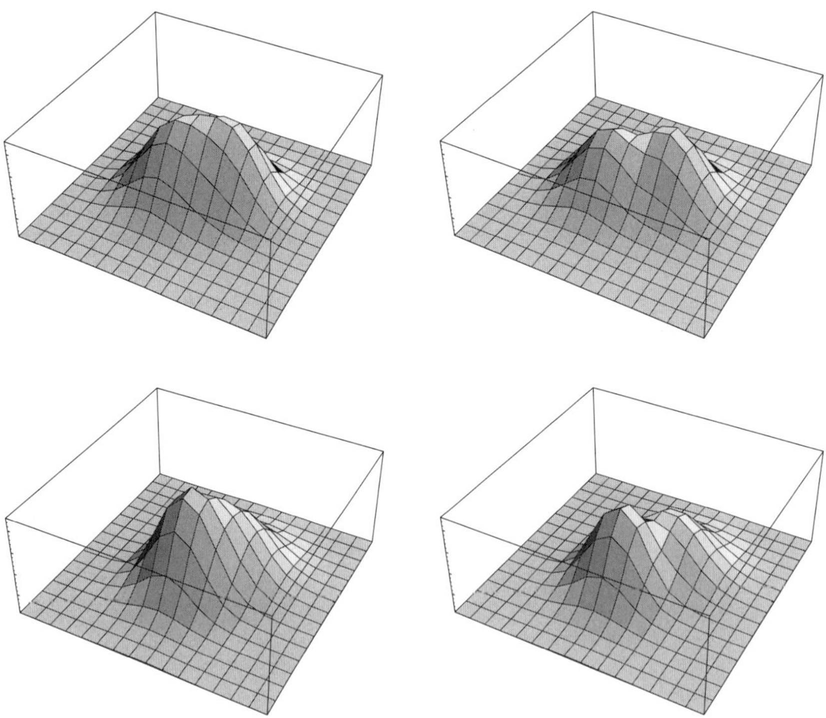

Figure 5: Caloron profile for two configurations (top and bottom) on a $16^3 \times 4$ lattice for $\vec{m} = \vec{0}$ obtained with improved ($\varepsilon = 0$) cooling. Vertically is plotted $\log(1+s/3)$, with s the action density at the lattice site (after clover averaging). The left plots correspond to the plane $y = t = 0$ and the right ones to $y = 0$, $t = 1/2$. Top: $\vec{k} = (1,1,1)$ with total action $1.000016 \times 8\pi^2$, $\omega = 1/4$, $t_0 = -0.125$, $\vec{y}_1 = (1.75, -0.15, 1.43)$ and $\vec{y}_2 = (2.20, 0.00, 2.20)$. Bottom: $\vec{k} = (0,0,0)$ with total action $1.010951 \times 8\pi^2$, $\omega = 0.185$, $t_0 = -0.09$, $\vec{y}_1 = (2.29, 0.07, 2.15)$ and $\vec{y}_2 = (1.57, 0.08, 1.73)$. All coordinates obtained from fitting to the infinite volume analytic solutions (not shown) are in units where $\beta = 1$ (or $a = 1/4$) and the left most lattice point corresponding to $x = z = 0$.

For large L/β the difference between the finite volume solutions with respect to the infinite volume calorons is mostly due to the contribution of the Coulombic tails of the periodic copies of the monopole constituents. That this depends on the nature of the twist is to be expected. For twist in time the charges change sign when shifting over a period of the torus. For twist in space there is no change in sign. This behaviour of the charges is correlated to the zeros of A_0 (which plays the role of the Higgs field) at the core of the constituent monopoles as illustrated by the behaviour of P_0 which is anti-periodic with time-twist and periodic with space-twist. It can be shown from the analytic solution (see the appendix A) that $P_0 = 1$ (corresponding to $A_0 = 0$) near one of the constituent centers, and $P_0 = -1$ near the other (related to $A_0 = 0$ by a gauge transformation that is anti-periodic in time -

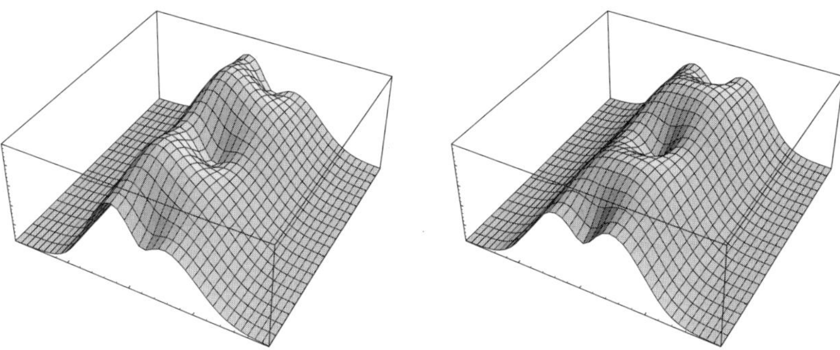

Figure 6: Space-time profile for the calorons of figure 5, using the infinite volume analytic result. Vertically is plotted $\log(1 + s/3)$, with s the action density. Horizontally is plotted time ranging over two periods $\beta = 1$ and space along the line connecting the two constituent monopole positions. Right corresponds to $\omega = 1/4$ and left to $\omega = 0.185$.

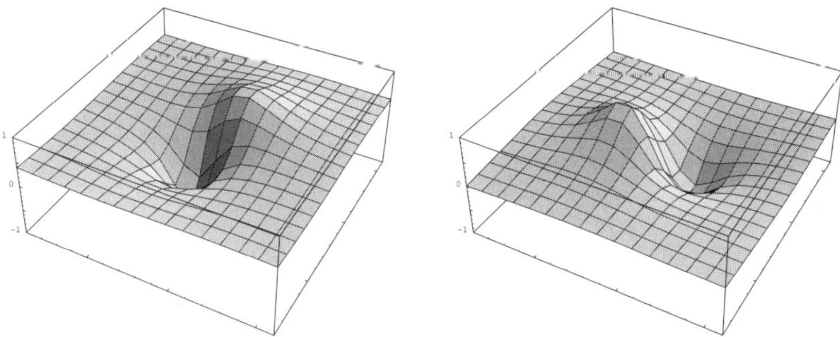

Figure 7: Profile of the Polyakov loop $P_0(\vec{x})$ for the calorons of fig. 5. Right corresponds to $\vec{k} = (1,1,1)$ and $\omega = 1/4$, left is for $\vec{k} = (0,0,0)$ with $\omega = 0.185$. Plotted is the plane $y = 0$, for other details see the caption of fig. 5.

the gauge transformation that changes ω to $1/2 - \omega$). This vanishing, i.e. $P_0^2 = 1$, of the Higgs field near to the constituent monopole centers is reproduced by the lattice data as is illustrated in figure 7.

4. Higher charge calorons

In this section we discuss our finding for higher topological charge. Analytic results in infinite volumes for higher charge calorons with non-trivial holonomy are not yet available.

Due to the local (lumpy) character of the caloron solutions one would expect that higher charge configurations can be obtained by "gluing" together lower charged solutions. Indeed for configurations on the torus, considering more than one period in any direction is a sure way of producing solutions with higher topological charge. On the basis of this it is to be expected that in the case of SU(2), for example, configurations would have $2Q$ action density lumps.

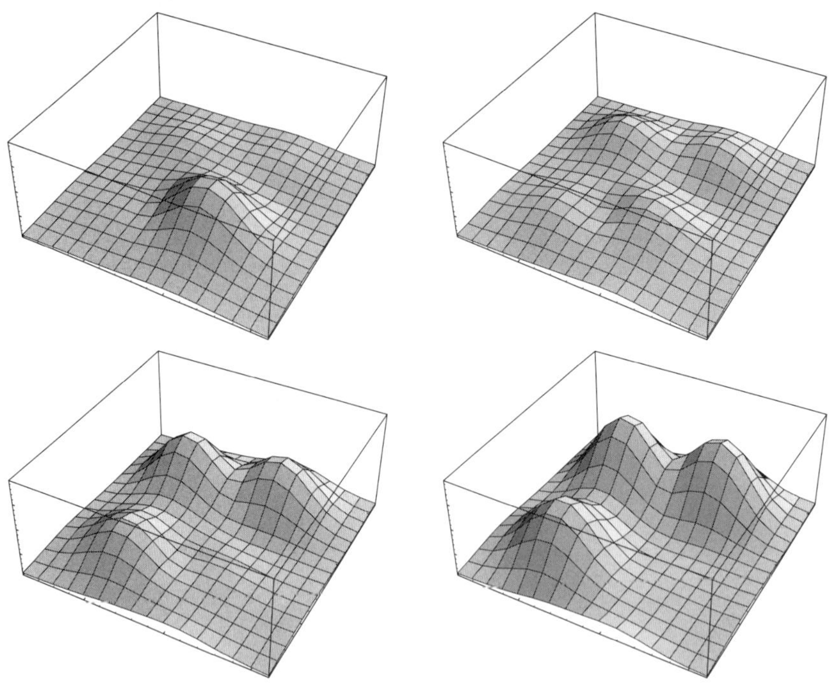

Figure 8: Lattice action density profiles of a (static) charge 2 configuration found with ordinary cooling on a $16^3 \times 4$ lattice with $\vec{k} = (1,1,1)$, as a function of x and z for four consecutive slices in the y direction.

Producing high charge configurations with our method is simple. It is sufficient to monitor the value of the lattice action during cooling. Typically, this quantity shows plateaus at integer multiples of $8\pi^2$. The cooling process can be interrupted at the desired value of the lattice action. We used ordinary ($\varepsilon = 1$) cooling; resulting configurations can subsequently be studied in more detail with other values of ε.

Figure 8 shows a configuration of charge 2, generated with ordinary cooling and twist $\vec{k} = (1,1,1)$. Indeed we find four lumps. We have been able to fit these to two $Q = 1$, $\omega = 1/4$, calorons by just adding the action densities together. Other charge 2 configurations have been obtained as well. This includes a configuration with 3 lumps, one of which seems describable as a $Q = 1$ object.

With similar techniques one can generate configurations with topological charge higher than 2. This process led us to study the whole cooling histories that go from randomly generated configurations to low action ones. On lattices $N_s^3 \times 4$, with $N_s = 16$, 20 and 24, we computed every 10 ($\varepsilon = 1$) cooling steps the total action S of the configuration and used our peak-searching algorithm to locate action density maxima. The information was recorded whenever the density of peaks, $N_{peak}/(N_s^3 \times 4)$, found by the algorithm was smaller than $50/(24^3 \times 4)$ (for higher densities the results are too sensitive to the details of the peak searching algorithm to be considered reliable).

For all recorded data the quotient $S/(4\pi^2 N_{peak})$ was found to lie between 0.8 and 2, and peaked around 1. This means that *on average* every peak is associated to an action of $4\pi^2$, a property shared with the exact $Q = 1$ caloron solution with *non-trivial* holonomy. The same follows for configurations that are aggregates of $Q = 1$ calorons, which each have either one or (more often) two lumps (the constituent monopoles). Our result shows that this pattern extends to higher densities, where a detailed analysis of individual peaks is hard to do. Furthermore, the sign of the topological charge of these lumps is not always the same, thereby pushing the picture of a constituent monopole ensemble beyond the case of self-dual configurations. Our result resembles the findings of ref. [29], where a similar behaviour was reported for Monte Carlo generated configurations at zero temperature. In our case, we have the additional advantage of having an analytic control for $Q = 1$ self-dual configurations. This allows us to conclude that the lumps correspond to constituent monopoles and hence, at least in this finite temperature case, not all lumps carry integer or half-integer topological charge. We hope these results will help to motivate other authors to investigate this point further.

5. Discussion

In this paper we have shown that $Q = 1$ self-dual solutions can be obtained profusely on asymmetric lattices $L^3 \times \beta$ with $L \gg \beta$ by using twisted boundary conditions. These configurations match quite well the analytic caloron solutions on $R^3 \times S_1$ [1]. The main change induced by the finite spatial volume is due to the contribution of the Coulombic tails of the periodic mirrors of the caloron solutions. We have shown that with judicious use of the twist values and of the parameter ε appearing in the cooling method of ref. [4], one can produce caloron solutions with different values of ρ and ω. In comparing to the continuum expressions, the choice $\varepsilon = 0$ (improved cooling) reduces considerably the size of lattice corrections.

In our analysis we have attempted to disentangle the finite size effects from the lattice artifacts, by making use of the ε engineering. We have also explored self-dual configurations with higher values of the topological charge. Our results show that these configurations look very much like ensembles of $Q = 1$ calorons with trivial or non-trivial holonomy. The conclusion, sustained by our results, is that typically a configuration with topological charge Q has $2Q$ lumps (constituent monopoles).

Given their local nature and the non-perturbative nature of the QCD (Yang-Mills) vacuum we vindicate that these configurations ought to play a role in the dynamics of the theory. It is to be emphasised that for high charges, the existence of these solutions does not rely on the use of any particular boundary conditions (twisted or not). Twist however plays a role in stabilising these solutions under cooling and this lies at the heart of the success of our method. This is most probably related to the fact that there are no exactly self-dual $Q = 1$ solutions on the torus

JHEP06(1999)001

in the absence of twist [26]. This does not happen for non-zero twist [12, 28]. Thus, lattice studies involving cooling methods could introduce distortions for low values of the topological charge [30]. We stress again that due to its simple implementation and zero computational overhead, the use of twisted boundary conditions is an ideal tool for non-perturbative investigations of non-Abelian gauge theories and QCD.

A. Analytic results

Here we summarise the infinite volume analytic solutions for the $SU(n)$ calorons with non-trivial holonomy. After a constant gauge transformation, the holonomy H is characterised by $(\sum_{m=1}^{n} \mu_m = 0)$

$$H = \exp[2\pi i \, \mathrm{diag}(\mu_1, \ldots, \mu_n)], \qquad \mu_1 < \cdots < \mu_n < \mu_{n+1} \equiv \mu_1 + 1. \tag{A.1}$$

Note that $\mathrm{tr}(H)/n = \lim_{|\vec{x}| \to \infty} P_0(\vec{x})$. Using the classical scale invariance to put $\beta = 1$, one has

$$s(x) = -\frac{1}{2}\mathrm{Tr}F_{\mu\nu}^2(x) = -\frac{1}{2}\partial_\mu^2\partial_\nu^2 \log \psi(x),$$
$$\psi(x) = \Psi(\vec{x}) - \cos(2\pi t),$$
$$\Psi(\vec{x}) = \frac{1}{2}\mathrm{tr}(A_n \cdots A_1), \tag{A.2}$$

where

$$A_m \equiv \frac{1}{r_m}\begin{pmatrix} r_m & |\vec{y}_m - \vec{y}_{m+1}| \\ 0 & r_{m+1} \end{pmatrix}\begin{pmatrix} c_m & s_m \\ s_m & c_m \end{pmatrix}. \tag{A.3}$$

Noting that $r_{n+1} \equiv r_1$ and $\vec{y}_{n+1} \equiv \vec{y}_1$ we defined $r_m = |\vec{x} - \vec{y}_m|$, with $\vec{y}_m$ the position of the mth constituent monopole, which can be assigned a mass $8\pi^2\nu_m$, where $\nu_m \equiv \mu_{m+1} - \mu_m$. Furthermore, $c_m \equiv \cosh(2\pi\nu_m r_m)$ and $s_m \equiv \sinh(2\pi\nu_m r_m)$.

Restricting to the gauge group of $SU(2)$, choosing $H = \exp(2\pi i\omega\tau_3)$ and defining $\pi\rho^2 = |\vec{y}_2 - \vec{y}_1|$, we can place the constituents at $\vec{y}_1 = (0, 0, \nu_2\pi\rho^2)$ and $\vec{y}_2 = (0, 0, -\nu_1\pi\rho^2)$ by a suitable combination of a constant gauge transformation, spatial rotation and translation. For this case the gauge field reads

$$A_\mu(x) = \frac{i}{2}\bar{\eta}_{\mu\nu}^3\tau_3\partial_\nu \log \phi(x) + \frac{i}{2}\phi(x)\mathrm{Re}\left((\bar{\eta}_{\mu\nu}^1 - i\bar{\eta}_{\mu\nu}^2)(\tau_1 + i\tau_2)\partial_\nu\chi(x)\right), \tag{A.4}$$

where the anti-selfdual 't Hooft tensor $\bar{\eta}$ is defined by $\bar{\eta}_{0j}^i = -\bar{\eta}_{j0}^i = \delta_{ij}$ and $\bar{\eta}_{jk}^i = \varepsilon_{ijk}$ (with our conventions of $t = x_0$, $\varepsilon_{0123} = -1$) and τ_a are the Pauli matrices. Furthermore,

$$\phi^{-1}(x) = 1 - \frac{\pi\rho^2}{\psi(x)}\left(\frac{s_1c_2}{r_1} + \frac{s_2c_1}{r_2} + \frac{\pi\rho^2 s_1 s_2}{r_1 r_2}\right) \tag{A.5}$$

and

$$\chi(x) = \frac{\pi\rho^2}{\psi(x)}\left(e^{-2\pi it}\frac{s_1}{r_1} + \frac{s_2}{r_2}\right)e^{2\pi i\nu_1 t},$$

JHEP06(1999)001

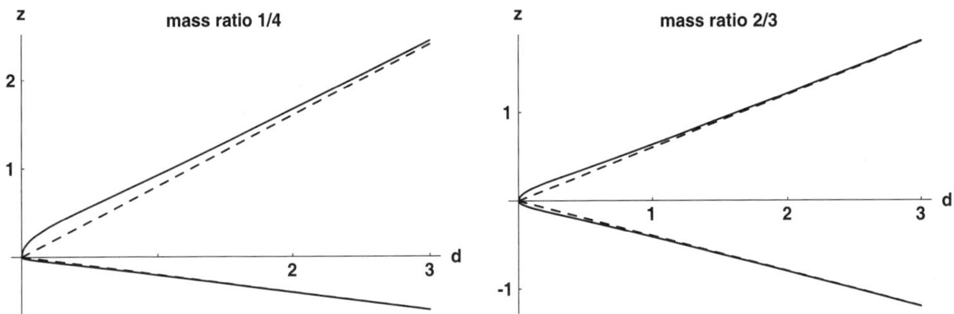

Figure 9: Shift of the locations where $P_0^2(\vec{x}) = 1$ as compared to the location of the constituent monopole centers $\vec{y}_i$. Horizontally is plotted the distance $d = \pi\rho^2$ between the constituents and vertically the position of $z_1 = (1 - 2w)d$ and $z_2 = -2wd$ and the locations where $P_0(z) = 1$ $(z > 0)$ and $P_0(z) = -1$ $(z < 0)$. Left is for $w = 0.1$ and right for $w = 0.2$.

with $\nu_1 = 2w$ and $\nu_2 = 1 - 2w$. The solution is presented in the "algebraic" gauge,

$$A_\mu(t + 1, \vec{x}) = \exp(2\pi i w \tau_3) A_\mu(t, \vec{x}) \exp(-2\pi i w \tau_3).$$

Since the radii r_i are even functions of x and y, derivatives in these two directions vanish on the z-axis. Hence, along the line connecting the two constituents A_0 is Abelian, allowing for a simple result for P_0 along this axis

$$P_0(z) = \cos(\pi\nu_1 + \Phi(z)), \qquad \Phi(z) = \frac{1}{2} \int_0^1 dt \, \partial_z \log \phi(t, z). \qquad (A.6)$$

Since $\psi(x)$ and $\phi(x)$ are even functions of r_i we may substitute $r_i = z - z_i$ (with $z_1 = \nu_2 \pi\rho^2$ and $z_2 = -\nu_1 \pi\rho^2$) to find

$$\phi(t, z) = \frac{\Psi(z) - \cos(2\pi t)}{\cosh(2\pi z) - \cos(2\pi t)},$$

with

$$\Psi(z) = \cosh(2\pi z) + \pi\rho^2 \left(\frac{s_1 c_2}{r_1} + \frac{s_2 c_1}{r_2} + \frac{\pi\rho^2 s_1 s_2}{r_1 r_2} \right) > 1$$

a smooth function of z. The pole of $\phi(x)$ at $x = 0$ represents the usual gauge singularity. It leads to a jump of 2π in $\Phi(z)$, to which the gauge invariant observable $P_0(z)$ is insensitive. The integration over time can be performed explicitly and one finds

$$P_0(z) = -\cos\left[\nu_1\pi + \frac{1}{2}\partial_z \mathrm{acosh}(\Psi(z))\right]. \qquad (A.7)$$

From this it is easily shown that each of the values $P_0(z) = \pm 1$ is taken only once. Only for large ρ one finds $P_0(z_1) = 1$ and $P_0(z_2) = -1$. When associating the constituent monopole locations to the zeros of the Higgs field (i.e. to $P_0^2(\vec{x}) = 1$), we find these are shifted *outward* from $\vec{y}_i$. This is illustrated in figure 9. For the cases we studied in this paper these shifts are small, but they tend to become large for the constituent monopoles with a small mass (ω approaching either 0 or 1/2). We should also note that the maxima of the energy density (at $t = 0$) are shifted *inward* due to overlap of the energy profiles of each constituent.

The numerical evaluation of the action density $s(x)$ and of the Polyakov loop $P_0(\vec{x})$ are straightforward, but tedious. For the action density it involves taking 4 derivatives, which is most conveniently achieved by using the fact that $\Psi(\vec{x})$ depends on $\vec{x}$ through the radii r_i. The C-programmes written for this purpose are available [31].

Acknowledgments

We are grateful to Conor Houghton, Thomas Kraan and Carlos Pena for useful discussions. This work was supported in part by a grant from "Stichting Nationale Computer Faciliteiten (NCF)" for use of the Cray Y-MP C90 at SARA. A. Gonzalez-Arroyo and A. Montero acknowledge financial support by CICYT under grant AEN97-1678. M. García Pérez acknowledges financial support by CICYT and warm hospitality at the Instituut Lorentz while part of this work was developed.

References

[1] T.C. Kraan and P. van Baal, *Phys. Lett.* **B 428** (1998) 268 [hep-th/9802049]; *Nucl. Phys.* **B 533** (1998) 627 [hep-th/9805168].

[2] K. Lee, *Phys. Lett.* **B 426** (1998) 323 [hep-th/9802012];
K. Lee and C. Lu, *Phys. Rev.* **D 58** (1998) 025011 [hep-th/9802108].

[3] B.J. Harrington and H.K. Shepard, *Phys. Rev.* **D 17** (1978) 2122; *Phys. Rev.* **D 18** (1978) 2990.

[4] M. García Pérez, A. González-Arroyo, J. Snippe and P. van Baal, *Nucl. Phys.* **B 413** (1994) 535 [hep-lat/9309009].

[5] G. 't Hooft, *Nucl. Phys.* **B 153** (1979) 141; *Acta Physica Aust. Suppl.* **XXII** (1980) 531.

[6] G. 't Hooft, *Nucl. Phys.* **B 79** (1974) 276;
A.M. Polyakov, *Sov. Phys. JETP Lett.* **20** (1974) 194.

[7] E.B. Bogomol'ny, *Sov. J. Nucl. Phys.* **24** (1976) 449;
M.K. Prasad and C.M. Sommerfield, *Phys. Rev. Lett.* **35** (1975) 760.

[8] T.C. Kraan and P. van Baal, *Nucl. Phys.* **73** *(Proc. Suppl.)* (1999) 554 [hep-lat/9808015].

[9] R.C. Brower, D. Chen, J. Negele, K. Orginos and C-I. Tan, *Nucl. Phys.* **73** *(Proc. Suppl.)* (1999) 557 [hep-lat/9810009].

[10] T.C. Kraan and P. van Baal, *Phys. Lett.* **B 435** (1998) 389 [hep-th/9806034].

[11] M. García Pérez, A. González-Arroyo and P. Martínez, *Nucl. Phys.* **34** *(Proc. Suppl.)* (1994) 228 [hep-lat/9312066];
A. González-Arroyo and P. Martinez, *Nucl. Phys.* **B 459** (1996) 337 [hep-lat/9507001].

[12] M. García Pérez, A. González-Arroyo and B. Söderberg, *Phys. Lett.* **B 235** (1990) 117;
M. García Pérez and A. González-Arroyo, *J. Phys.* **A26** (1993) 2667;
M. García Pérez, A. González-Arroyo, A. Montero and C. Pena, *Nucl. Phys.* **63** *(Proc. Suppl.)* (1998) 501 [hep-lat/9709107].

[13] P. de Forcrand et al., *Nucl. Phys.* **73** *(Proc. Suppl.)* (1999) 578 [hep-lat/9810033].

[14] D.J. Gross, R.D. Pisarski and L.G. Yaffe, *Rev. Mod. Phys.* **53** (1983) 43.

[15] G. 't Hooft, *Nucl. Phys.* **B 190** (1981) 455; *Physica Scripta* **25** (1982) 133.

[16] For a review see the contributions of G. 't Hooft, M. Polikarpov, A. Di Giacomo and T. Suzuki in: *Confinement, Duality and Non-perturbative Aspects of QCD*, ed. P. van Baal, NATO ASI Series B: Vol. 368, Plenum Press, 1998.

[17] P. Rossi, *Nucl. Phys.* **B 149** (1979) 170.

[18] A. González-Arroyo, *Yang-Mills fields on the four-dimensional torus. Part 1.: Classical theory*, hep-th/9807108, in: *Advanced school for nonperturbative quantum field physics*, eds. M. Asorey and A. Dobado, World Scientific 1998, p. 57.

[19] P. van Baal, *Nucl. Phys.* **49** *(Proc. Suppl.)* (1996) 238 [hep-th/9512223].

[20] J. Groeneveld, J. Jurkiewicz and C.P. Korthals Altes, *Physica Scripta* **23** (1981) 1022.

[21] G. 't Hooft, *Nucl. Phys.* **B 138** (1981) 1.

[22] P. van Baal, *Comm. Math. Phys.* **85** (1982) 529.

[23] P. van Baal, *Twisted boundary conditions: a non-perturbative probe for pure non-abelian gauge theories*, PhD-Thesis, Utrecht, July 1984.

[24] M. Teper, *Phys. Lett.* **B 162** (1985) 357; *Phys. Lett.* **B 171** (1986) 81.

[25] A. González-Arroyo and A. Montero, *Phys. Lett.* **B 442** (1998) 273 [hep-th/9809037].

[26] P. Braam and P. van Baal, *Comm. Math. Phys.* **122** (1989) 267.

[27] C.H. Taubes, *J. Diff. Geom.* **19** (1984) 517.

[28] P. Braam, A. Maciocia and A. Todorov, *Inv. Math.* 108 (1992) 419.

[29] A. González-Arroyo, P. Martinez and A. Montero, *Phys. Lett.* **B 359** (1995) 159 [hep-lat/9507006];
A. González-Arroyo and A. Montero, *Phys. Lett.* **B 387** (1996) 823 [hep-th/9604017].

[30] P. de Forcrand, M. García Pérez and I. O. Stamatescu, *Nucl. Phys.* **B 499** (1997) 409 [hep-lat/9701012]; *Nucl. Phys.* **53** *(Proc. Suppl.)* (1997) 557 [hep-lat/9608032].

[31] The C-programmes and instructions for their use can be found at http://www-lorentz.leidenuniv.nl/vanbaal/Caloron.html.

PHYSICAL REVIEW D, VOLUME 60, 031901

Weyl-Dirac zero mode for calorons

Margarita García Pérez,[1] Antonio González-Arroyo,[1,2] Carlos Pena,[1] and Pierre van Baal[3]

[1]*Departamento de Física Teórica C-XI, Universidad Autónoma de Madrid, 28049 Madrid, Spain*
[2]*Instituto de Física Teórica C-XVI, Universidad Autónoma de Madrid, 28049 Madrid, Spain*
[3]*Instituut-Lorentz for Theoretical Physics, University of Leiden, P.O. Box 9506, NL-2300 RA Leiden, The Netherlands*
(Received 4 May 1999; published 18 June 1999)

We give the analytic result for the fermion zero mode of the $SU(2)$ calorons *with a nontrivial holonomy*. It is shown that the zero mode is supported on *only one* of the constituent monopoles. We discuss some of its implications. [S0556-2821(99)50513-4]

PACS number(s): 11.10.Wx, 14.80.Hv

I. INTRODUCTION

In this paper, we give the exact expression for the $SU(2)$ fermion zero mode in the field of the infinite volume caloron with a nontrivial holonomy and unit charge. Study of the gauge field configurations had, somewhat surprisingly, revealed that at nontrivial holonomy calorons have two Bogomol'nyi-Prasad-Sommerfield (BPS) monopoles [N for $SU(N)$] as their constituents [1,2,3]. For the Harrington-Shepard [4] solution with trivial holonomy, this is hidden because one of the constituents is massless (it can be removed by a singular gauge transformation to show that the caloron for a large scale parameter becomes a single BPS monopole [5]). We find that, for calorons with well-separated constituents, the fermion zero mode is entirely supported on one of them. In itself it is not surprising that the zero mode is correlated to the monopole constituents. Independently, this observation was recently also made for gluino zero modes in the context of supersymmetric gauge theories [6]. Gluinos are in the adjoint representation of the gauge group, such that there are four zero modes that can be split in pairs associated with each of the two constituents [6]. However, for the Dirac fermion, there is only one zero mode. To understand the "affinity" of the zero mode to only one of the two monopoles, we will analyze in some detail what distinguishes them.

Calorons are characterized by the (fixed) holonomy [1,7]. In the gauge in which $A_\mu(x)$ is periodic, this holonomy is given by

$$\mathcal{P}_\infty = \lim_{|\vec{x}|\to\infty} \mathrm{P}\exp\left(\int_0^\beta A_0(t,\vec{x})dt\right) \equiv \exp(2\pi i\vec{\omega}\cdot\vec{\tau}). \quad (1)$$

Solutions are simplest in the so-called "algebraic" gauge, for which

$$A_\mu(t+\beta,\vec{x}) = \mathcal{P}_\infty A_\mu(t,\vec{x})\mathcal{P}_\infty^{-1},$$

$$\Psi_z^\pm(t+\beta,\vec{x}) = \pm\mathcal{P}_\infty\Psi_z^\pm(t,\vec{x}). \quad (2)$$

We will generalize the problem of finding the fermion zero mode in the field of the caloron with nontrivial holonomy, by adding a curvature free Abelian field, which

forms the basis for the Nahm transformation [8]. Why this is useful will be evident from the construction. The equation to be solved is

$$\overline{D}_z\Psi_z(x) \equiv \overline{\sigma}_\mu[\partial_\mu + A_\mu(x) - 2\pi i z_\mu]\Psi_z(x) = 0, \quad (3)$$

with $\overline{\sigma}_\mu = \sigma_\mu^\dagger = (1, -i\vec{\tau})$ and τ_a the Pauli matrices. For calorons, defined on $R^3\times S^1$, i.e., at finite temperature $1/\beta$, one can choose $z_1 = z_2 = z_3 = 0$ [the plane-wave factor $\exp(2\pi i\vec{z}\cdot\vec{x})$ does not affect the boundary conditions or the normalization of the zero mode and can be used to remove the $\vec{z}$ dependence]. But $z = z_0$ will be arbitrary (it has a dual period of $1/\beta$). The zero modes are represented as two-component spinors in the (chiral) Weyl decomposition for massless Dirac fermions.

II. ADHM CONSTRUCTION

The construction of the zero mode is best done in the Atiyah-Drinfeld-Hitchin-Manin (ADHM) formalism [9]. We will be brief in reviewing this formalism, further details can be found in Refs [1,10,11,12]. In general the ADHM construction involves an operator $\Delta(x)$ [the "dual" of Eq. (3)], whose normalized zero mode, $\Delta^\dagger(x)v(x) = 0$, gives the gauge field as $A_\mu(x) = v^\dagger(x)\partial_\mu v(x)$. For $SU(2)$ instantons of charge k, one has $\Delta^\dagger = (\lambda^\dagger, B^\dagger - x^\dagger)$, with $x^\dagger = x_\mu\overline{\sigma}_\mu$, λ a k dimensional row vector and B a $k\times k$ symmetric matrix, all with values in the quaternions (λ_{il}^l, $B_{IJ}^{l,m} = B_{IJ}^{m,l}$, with $l,m = 1,\ldots,k$ "charge" and $I,J = 1,2$ spinor indices, whereas $i = 1,2$ is a color index). Introducing the row vector $u^\dagger(x) \equiv \lambda(B-x)^{-1}$ and the scalar (real quaternion) $\phi(x) = 1 + u^\dagger(x)u(x)$, it can be shown [10,11,12] that the instanton gauge field and the k zero modes are given by

$$A_\mu(x) = \phi^{-1}(x)\mathrm{Im}(u^\dagger(x)\partial_\mu u(x)),$$

$$\Psi_{iJ}^l(x) = \pi^{-1}\phi^{-1/2}(x)(u^\dagger(x)f_x)_{il}^l\varepsilon_{IJ}, \quad (4)$$

where f_x is the matrix inverse, or Green's function,

$$f_x = ((B-x)^\dagger(B-x) + \lambda^\dagger\lambda)^{-1}. \quad (5)$$

Essential in the ADHM construction is that λ and B satisfy a quadratic constraint, which is equivalent to f_x being a symmetric matrix whose imaginary quaternion components van-

RAPID COMMUNICATIONS

MARGARITA GARCÍA PÉREZ *et al.*

PHYSICAL REVIEW D **60** 031901

ish, $(f_x)_{IJ}^{l,m} \equiv f_x^{l,m}\delta_{IJ}$. From this alone, it can be proven that the gauge field is self-dual and that the $\Psi^l(x)$ are zero modes, $\overline{D}\Psi^l(x) \equiv \overline{\sigma}_\mu(\partial_\mu + A_\mu(x))\Psi^l(x) = 0$. Its proper normalization and the topological charge are read off from the remarkable results [10,11,12]

$$\Psi^l(x)^\dagger \Psi^m(x) = -(2\pi)^{-2}\partial_\mu^2 f_x^{l,m},$$

$$\mathrm{Tr}\,F_{\mu\nu}^2(x) = -\partial_\mu^2\partial_\nu^2 \log\det f_x, \qquad (6)$$

using $\lim_{|x|\to\infty} f_x^{l,m} = \delta^{l,m}|x|^{-2}$. Before addressing the explicit form of these expressions for the caloron with nontrivial holonomy, we perform one further simplification (for the details follow Eqs. (21)–(29) in Ref. [1], see also Ref. [10]),

$$A_\mu(x) = \tfrac{1}{2}\phi(x)\partial_\nu(\lambda\,\overline{\eta}_{\mu\nu}f_x\lambda^\dagger),$$

$$\Psi_{iJ}^l(x) = (2\pi)^{-1}\phi^{1/2}(x)\partial_\mu(\lambda f_x\overline{\sigma}_\mu)_{iI}^l\varepsilon_{IJ}, \qquad (7)$$

where the anti-self-dual 't Hooft tensor $\overline{\eta}$ is defined by $\overline{\eta}_{0j}^i = -\overline{\eta}_{j0}^i = \delta_{ij}$ and $\overline{\eta}_{jk}^i = \varepsilon_{ijk}$ (furthermore, $\varepsilon_{12} = 1$ and with our conventions of $t = x_0$, $\varepsilon_{0123} = -1$).

The caloron with nontrivial holonomy is found by imposing boundary conditions to compactify time to a circle $u^{l+1}(t+\beta,\vec{x}) = u^l(t,\vec{x})\mathcal{P}_\infty^\dagger$, which is easily seen to give the correct boundary conditions for the gauge field. For the general form of λ and B which respect this symmetry, see Ref. [1]. Note that now the index l runs over the set of all integers; the R^4 configuration with these boundary conditions has infinite topological charge (unit topological charge per time period). To obtain the zero mode with the appropriate boundary condition, we note that with Eq. (4)

$$\hat{\Psi}_z(x) \equiv \sum_l e^{-2\pi i l\beta z}\Psi^l(x), \qquad (8)$$

satisfies the boundary condition $\hat{\Psi}_z(t+\beta,\vec{x}) = e^{-2\pi i\beta z}\mathcal{P}_\infty\hat{\Psi}_z(t,\vec{x})$ and satisfies $\overline{D}\hat{\Psi}_z(x) = 0$ for all z. The general solution of the Weyl equation, Eq. (3), with both periodic and antiperiodic boundary conditions is now easily found (for simplicity, we put $\beta = 1$)

$$\Psi_z^+(x) = e^{2\pi i z t}\hat{\Psi}_z(x), \quad \Psi_z^-(x) = e^{2\pi i z t}\hat{\Psi}_{z+1/2}(x). \qquad (9)$$

In particular, $\Psi^-(x) = \Psi_0^-(x) = \hat{\Psi}_{1/2}(x)$ is the, for finite temperature, physically relevant chiral zero mode in the background of a caloron, whereas $\Psi^+(x) = \Psi_0^+(x) = \hat{\Psi}_0(x)$ is relevant for compactifications.

III. NAHM-FOURIER TRANSFORMATION

The interpretation of the ''charge'' index as a Fourier index, as suggested by the construction of the caloron zero mode, has been essential for solving the quadratic ADHM constraint in the presence of nontrivial holonomy. It maps the ADHM construction to the Nahm formalism, in which, furthermore, f_x is solved in terms of a quantum mechanics problem on the circle ($z \in [0,1]$) with a piecewise constant

potential and delta function singularities determined by the holonomy [1]. The relevant quantities involved are

$$\hat{f}_x(z,z') = \sum_{l,m} f_x^{l,m} e^{2\pi i(lz-mz')}, \quad \hat{\lambda}(z) = \sum_l \lambda^l e^{-2\pi i l z}, \qquad (10)$$

where matrix multiplication is replaced by convolution in the usual sense. The solution of the ADHM constraint implies that $\hat{\lambda}(z)$ is the sum of two delta functions. Together with the explicit expression for the Green's function $\hat{f}_x(z,z')$ as given in Eqs. (47)–(49) of Ref. [1], the zero mode reads

$$\hat{\Psi}_z(x) = (2\pi)^{-1}\phi^{1/2}(x)\partial_\mu\left(\int_0^1 dz'\hat{\lambda}(z')\hat{f}_x(z',z)\overline{\sigma}_\mu\right)_{iI}\varepsilon_{IJ}, \qquad (11)$$

compare Eq. (4). Whereas Eq. (6) yields

$$\hat{\Psi}_{z'}^\dagger(x)\hat{\Psi}_z(x) = -(2\pi)^{-2}\partial_\mu^2\hat{f}_x(z',z). \qquad (12)$$

IV. EXPLICIT EXPRESSIONS

Using the classical scale invariance to put $\beta = 1$, one has [1]

$$s(x) = -\tfrac{1}{2}\mathrm{Tr}F_{\mu\nu}^2(x) = -\tfrac{1}{2}\partial_\mu^2\partial_\nu^2\log\psi(x),$$

$$\psi(x) = \hat{\psi}(\vec{x}) - \cos(2\pi t), \quad \hat{\psi}(\vec{x}) = \tfrac{1}{2}\mathrm{tr}(\mathcal{A}_2\mathcal{A}_1), \qquad (13)$$

where

$$\mathcal{A}_m \equiv \frac{1}{r_m}\begin{pmatrix} r_m & |\vec{y}_m - \vec{y}_{m+1}| \\ 0 & r_{m+1} \end{pmatrix}\begin{pmatrix} c_m & s_m \\ s_m & c_m \end{pmatrix}. \qquad (14)$$

Noting that $r_3 \equiv r_1$ and $\vec{y}_3 \equiv \vec{y}_1$, we defined $r_m = |\vec{x} - \vec{y}_m|$, with $\vec{y}_m$ the position of the mth constituent monopole, which can be assigned a mass $8\pi^2\nu_m$, where $\nu_1 = 2\omega$ and $\nu_2 = 2\bar{\omega} \equiv 1 - 2\omega$. Furthermore, $c_m \equiv \cosh(2\pi\nu_m r_m)$, $s_m \equiv \sinh(2\pi\nu_m r_m)$, and $\pi\rho^2 = |\vec{y}_2 - \vec{y}_1|$.

New is the result for the zero-mode density

$$|\Psi^-(x)|^2 = -(2\pi)^{-2}\partial_\mu^2\hat{f}_x(\tfrac{1}{2},\tfrac{1}{2}),$$

$$|\Psi^+(x)|^2 = -(2\pi)^{-2}\partial_\mu^2\hat{f}_x(0,0), \qquad (15)$$

defined by [1]

$$\hat{f}_x(\tfrac{1}{2},\tfrac{1}{2}) = \frac{\pi}{r_1 r_2\psi(x)}\Bigg(s_2[r_1 c_1 + \pi\rho^2 s_1] + r_2 s_1 + \frac{c_2 - 1}{r_2}[\pi\rho^2 r_1 c_1 + \tfrac{1}{2}(r_1^2 + r_2^2 + \pi^2\rho^4)s_1]\Bigg),$$

RAPID COMMUNICATIONS

WEYL-DIRAC ZERO MODE FOR CALORONS

PHYSICAL REVIEW D **60** 031901

$$\hat{f}_x(0,0) = \frac{\pi}{r_1 r_2 \psi(x)} \Bigg(s_1[r_2 c_2 + \pi \rho^2 s_2] + r_1 s_2$$

$$+ \frac{c_1 - 1}{r_1}[\pi \rho^2 r_2 c_2 + \tfrac{1}{2}(r_1^2 + r_2^2 + \pi^2 \rho^4) s_2] \Bigg).$$

$$(16)$$

By a suitable combination of a constant gauge transformation, spatial rotation, and translation, we can arrange both $\vec{\omega} = (0,0,\omega)$ and the constituents at $\vec{y}_1 = (0,0,\nu_2 \pi \rho^2)$ and $\vec{y}_2 = (0,0,-\nu_1 \pi \rho^2)$. For this choice, we find

$$A_\mu(x) = \frac{i}{2}\overline{\eta}^3_{\mu\nu}\tau_3 \partial_\nu \log \phi(x)$$

$$+ \frac{i}{2}\phi(x)\mathrm{Re}((\overline{\eta}^1_{\mu\nu} - i\overline{\eta}^2_{\mu\nu})(\tau_1 + i\tau_2)\partial_\nu \chi(x)),$$

$$(17)$$

$$\Psi^-_{1I}(x) = (2\pi)^{-1}\rho \phi^{1/2}(x)\begin{pmatrix} \partial_2 + i\partial_1 \\ \partial_0 - i\partial_3 \end{pmatrix}\hat{f}_x(\omega, \tfrac{1}{2}),$$

$$\Psi^-_{2I}(x) = \Psi^-_{1J}(x)^* \varepsilon_{JI},$$

$$\Psi^+_{1I}(x) = (2\pi)^{-1}\rho \phi^{1/2}(x)\begin{pmatrix} \partial_2 + i\partial_1 \\ \partial_0 - i\partial_3 \end{pmatrix}\hat{f}_x(\omega,0),$$

$$\Psi^+_{2I}(x) = \Psi^+_{1J}(x)^* \varepsilon_{JI}, \qquad (18)$$

with $\phi^{-1}(x) = 1 - [\pi \rho^2/\psi(x)](s_1 c_2/r_1 + s_2 c_1/r_2 + \pi \rho^2 s_1 s_2/r_1 r_2)$ and $\chi(x) = [\pi \rho^2/\psi(x)](e^{-2\pi i t}(s_1/r_1) + s_2/r_2)e^{2\pi i \nu_1 t}$, and

$$\hat{f}_x(\omega, \tfrac{1}{2}) = \frac{\pi e^{\pi i \nu_1 t}}{r_1 r_2 \psi(x)}\{(e^{\pi i t}r_1$$

$$+ e^{-\pi i t}[\pi \rho^2 s_1 + r_1 c_1])\sinh(\pi r_2 \nu_2)$$

$$+ e^{-\pi i t}r_2 s_1 \cosh(\pi r_2 \nu_2)\},$$

$$\hat{f}_x(\omega, 0) = \frac{\pi e^{-\pi i \nu_2 t}}{r_1 r_2 \psi(x)}\{(e^{-\pi i t}r_2$$

$$+ e^{\pi i t}[\pi \rho^2 s_2 + r_2 c_2])\sinh(\pi r_1 \nu_1)$$

$$+ e^{\pi i t}r_1 s_2 \cosh(\pi r_1 \nu_1)\}. \qquad (19)$$

V. PROPERTIES OF THE ZERO MODE

The gauge field has a symmetry under the antiperiodic gauge transformation $g(x) = \exp(\pi i t \hat{\omega} \cdot \vec{\tau})$, which changes the sign of the holonomy, $\mathcal{P}_\infty \to -\mathcal{P}_\infty$, or $\omega \leftrightarrow \overline{\omega} = \tfrac{1}{2} - \omega$. An antiperiodic gauge transformation does, however, not leave fermions invariant, and indeed it interchanges $\Psi^+_z(x)$ and $\Psi^-_z(x)$. To preserve the special choice of parametrization presented above, the change of sign in the holonomy, which interchanges ν_1 and ν_2, is also accompanied by an interchange of the constituent locations. This indeed leaves the action density invariant. That the zero mode clearly distin-

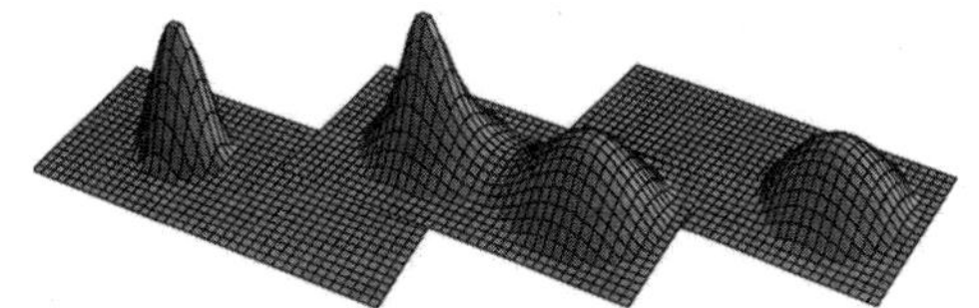

FIG. 1. For the two figures on the sides, we plot on the same scale the logarithm of the zero mode densities (cutoff below $1/e^5$) for $\omega = 1/8$ (left Ψ^-/right Ψ^+) and $\omega = 3/8$ (right Ψ^-/left Ψ^+), with $\beta = 1$ and $\rho = 1.2$. In the middle figure, we show for the same parameters (both choices of ω give the same action density), the logarithm of the action density (cutoff below $1/2e^2$).

guishes between the two cases becomes evident in the static limit, $\rho \to \infty$ (or equivalently $\beta \to 0$), in which case the zero mode is completely localized on one of the constituent monopoles, as follows from [compare Eq. (16)]

$$\lim_{\rho \to \infty}|\Psi^-(x)|^2 = -\partial_\mu^2\left(\frac{\tanh(\pi r_2 \nu_2)}{4\pi r_2}\right),$$

$$\lim_{\rho \to \infty}|\Psi^+(x)|^2 = -\partial_\mu^2\left(\frac{\tanh(\pi r_1 \nu_1)}{4\pi r_1}\right). \qquad (20)$$

Under the antiperiodic gauge transformation, the antiperiodic zero mode becomes periodic. The *new* antiperiodic zero mode is now completely localized on the *other* constituent monopole [this is consistent with the fact that $\hat{f}_x(0,0)$ can be obtained from $\hat{f}_x(\tfrac{1}{2},\tfrac{1}{2})$ by interchanging r_1 and ν_1 with r_2 and ν_2]. Figure 1 illustrates these issues [13].

In the gauge where $A_\mu(x)$ is periodic for large ρ, one of the constituent monopole fields is completely time independent, whereas the other one has a time dependence that would result from a full rotation along the axis connecting the two constituents [1]. This is read off from

$$A_\mu^{\mathrm{per}} = \frac{i}{2}\overline{\eta}^3_{\mu\nu}\tau_3 \partial_\nu \log \phi$$

$$+ \frac{i}{2}\phi \,\mathrm{Re}[(\overline{\eta}^1_{\mu\nu} - i\overline{\eta}^2_{\mu\nu})(\tau_1 + i\tau_2)(\partial_\nu + 4\pi i\omega\delta_{\nu,0})\tilde{\chi}]$$

$$+ \delta_{\mu,0}2\pi i\omega\tau_3, \qquad (21)$$

$$\tilde{\chi} = e^{-4\pi i t \omega}\chi$$

$$= \frac{4\pi\rho^2}{(r_2 + r_1 + \pi\rho^2)^2}(r_2 e^{-2\pi r_2 \nu_2}e^{-2\pi i t} + r_1 e^{-2\pi r_1 \nu_1})$$

$$\times [1 + \mathcal{O}(e^{-4\pi \min(r_1\nu_1, r_2\nu_2)})].$$

[Note that for large ρ, $\phi(x)$ becomes time independent.] This full rotation—which we will call the *Taubes-winding*—is responsible for the topological charge of the otherwise time independent monopole pair [14]. Under the antiperiodic gauge transformation that changes the sign of the holonomy, the Taubes-winding is supported by the other constituent. It has not gone unnoticed that the antiperiodic fermion zero

PHYSICAL REVIEW D **60** 031901

mode is precisely localized on the monopole constituent that carries the Taubes-winding. Another way to distinguish the two constituent monopoles is by inspecting the Polyakov loop values at their centers. One finds -1 for the monopole line with the Taubes-winding and $+1$ for the other monopole line (this is correlated to the vanishing of the would-be Higgs field), see the Appendix of Ref. [15]. For trivial holonomy, $\omega = 0$, the Polyakov loop is indeed -1 at the center of the Harrington-Shepard [4] caloron. Its zero mode, constructed before in Refs. [7,16], agrees with the results found here.

The association of the zero mode with the constituent that carries the Taubes-winding lends considerable support for the role of the monopole loops with Taubes-winding in QCD for chiral dynamics [1]. The precise embedding of these straight finite temperature monopole loops as curved monopole loops in flat space remains a nontrivial and challenging problem. Although it may seem contradictory to expect the zero mode with antiperiodic boundary conditions to be the relevant one, one should not forget that for a curved monopole loop, the spin frame makes also one full rotation due to the *bending* of the loop, thereby most likely providing the compensating sign flip.

ACKNOWLEDGMENTS

We are grateful to Maxim Chernodub, Arjan Keurentjes, Valya Khoze, and Tamas Kovács for useful discussions and correspondence. A. Gonzalez-Arroyo and C. Pena acknowledge financial support by CICYT under Grant No. AEN97-1678. M. García Pérez acknowledges financial support by CICYT.

[1] T. C. Kraan and P. van Baal, Nucl. Phys. **B533**, 627 (1998).

[2] T. C. Kraan and P. van Baal, Phys. Lett. B **428**, 268 (1998); **435**, 389 (1998).

[3] K. Lee and P. Yi, Phys. Rev. D **56**, 3711 (1997); K. Lee, Phys. Lett. B **426**, 323 (1998); K. Lee and C. Lu, Phys. Rev. D **58**, 025011 (1998).

[4] B. J. Harrington and H. K. Shepard, Phys. Rev. D **17**, 2122 (1978); **18**, 2990 (1978).

[5] P. Rossi, Nucl. Phys. **B149**, 170 (1979).

[6] T. J. Hollowood, V. V. Khoze, W. Lee, and M. P. Mattis, "Breakdown of cluster decomposition in instanton calculations of the gluino condensate," hep-th/9904116; N. M. Davies, T. J. Hollowood, V. V. Khoze, and M. P. Mattis, "Gluino condensate from magnetic monopoles in SUSY gluodynamics," hep-th/9905015.

[7] D. J. Gross, R. D. Pisarski, and L. G. Yaffe, Rev. Mod. Phys. **53**, 43 (1983).

[8] W. Nahm, Phys. Lett. **90B**, 413 (1980); *Self-dual monopoles and calorons*, Lecture Notes in Physics Vol. 201, edited by G. Denardo (Springer, Berlin, 1984), p. 189.

[9] M. F. Atiyah, N. J. Hitchin, V. G. Drinfeld, and Yu. I. Manin, Phys. Lett. **65A**, 185 (1978).

[10] E. F. Corrigan, D. B. Fairlie, S. Templeton, and P. Goddard, Nucl. Phys. **B140**, 31 (1978).

[11] H. Osborn, Nucl. Phys. **B140**, 45 (1978); Ann. Phys. (N.Y.) **135**, 373 (1981).

[12] E. Corrigan and P. Goddard, Ann. Phys. (N.Y.) **154**, 253 (1984).

[13] C-programmes for action/zero-mode densities and Polyakov loops can be found at http://www-lorentz.leidenuniv.nl/vanbaal/Caloron.html.

[14] C. Taubes, in *Progress in Gauge Field Theory*, edited by G. 't Hooft *et al.* (Plenum, New York, 1984), p. 563.

[15] M. García Pérez, A. González-Arroyo, A. Montero, and P. van Baal, "Calorons on the lattice—a new perspective," hep-lat/9903022.

[16] N. Bilić, Phys. Lett. **97B**, 107 (1980); A. González-Arroyo and Yu. A. Simonov, Nucl. Phys. **B460**, 429 (1996).

ELSEVIER

Nuclear Physics B (Proc. Suppl.) 83–84 (2000) 556–558

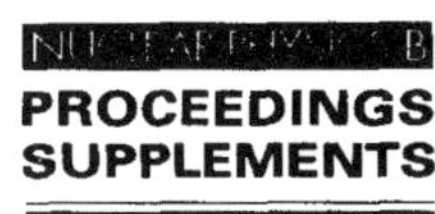

PROCEEDINGS
SUPPLEMENTS

www.elsevier.nl/locate/npe

Exact fermion zero-mode for the new calorons

M.N. Chernodub[a], Thomas C. Kraan[b] and Pierre van Baal[b]*

[a]Institute of Theoretical and Experimental Physics,
B.Cheremushkinskaya 25, Moscow, 117259, Russia

[b]Instituut-Lorentz for Theoretical Physics, University of Leiden,
PO Box 9506, NL-2300 RA Leiden, The Netherlands.

We construct the fermion zero-mode for arbitrary charge one $SU(n)$ calorons with non-trivial holonomy, both in the finite temperature context (anti-periodic boundary conditions in time) and in the Kaluza-Klein compactification context (periodic boundary conditions in time). The zero-mode is localised on one of the constituent monopoles and we discuss a relation to the Callias index theorem.

1. Introduction

The $SU(n)$ instantons at finite temperature (or calorons) can be seen as bound states of n constituent monopoles, evident only when the Polyakov loop at spatial infinity is non-trivial. In the periodic gauge, $A_\mu(t+\beta,\vec{x})=A_\mu(t,\vec{x})$,

$$\mathcal{P}_\infty = \lim_{|\vec{x}|\to\infty} P \exp(\int_0^\beta A_0(\vec{x},t)dt). \qquad (1)$$

After a suitable constant gauge transformation, it can be characterised by $\sum_{m-1}^n \mu_m = 0$ and

$$\mathcal{P}_\infty^0 = \exp[2\pi i \operatorname{diag}(\mu_1,\ldots,\mu_n)], \qquad (2)$$

$$\mu_1 \leq \ldots \leq \mu_n \leq \mu_{n+1} \equiv 1+\mu_1.$$

Using the classical scale invariance we can always arrange $\beta = 1$, as will be assumed throughout. A remarkably simple formula for the $SU(n)$ action density exists [1,2].

$$\operatorname{Tr}F_{\mu\nu}^2(x) = \partial_\mu^2\partial_\nu^2 \log\psi(x), \qquad (3)$$

$$\psi(x) = \tfrac{1}{2}\operatorname{tr}(\mathcal{A}_n\cdots\mathcal{A}_1) - \cos(2\pi t),$$

$$\mathcal{A}_m \equiv \frac{1}{r_m}\begin{pmatrix} r_m & |\vec{y}_m-\vec{y}_{m+1}| \\ 0 & r_{m+1} \end{pmatrix}\begin{pmatrix} c_m & s_m \\ s_m & c_m \end{pmatrix},$$

with $r_m = |\vec{x}-\vec{y}_m|$ the center of mass radius of the mth constituent monopole, which can be assigned a mass $8\pi^2\nu_m$, where $\nu_m \equiv \mu_{m+1}-\mu_m$. Furthermore, $c_m \equiv \cosh(2\pi\nu_m r_m)$, $s_m \equiv \sinh(2\pi\nu_m r_m)$, $r_{n+1} \equiv r_1$ and $\vec{y}_{n+1} \equiv \vec{y}_1$.

---

*Talk presented by the last author.

2. Monopole constituents

These generalised caloron solutions can be found [3] using a combination of the Nahm transformation [4] and the Atiyah-Drinfeld-Hitchin-Manin (ADHM) construction [5]. The latter is mainly needed to resolve the delta function singularities that arise in the Nahm transformation, although other methods were developed as well [6].

The Nahm equation for these charge one instantons reduces to an abelian problem on the circle, parametrised by $z \bmod 1$,

$$\frac{d}{dz}\hat{A}_j(z) = 2\pi i \sum_m \left(y_m^j - y_{m-1}^j\right)\delta(z-\mu_m), \qquad (4)$$

giving $\hat{A}_j(z) = 2\pi i y_m^j$, for $z \in [\mu_m,\mu_{m+1}]$. In the monopole literature $\hat{A}_j(z)$ is usually denoted by $T_j(z)$. Taking one interval in isolation, applying the Nahm transformation [4] gives a single static Bogomol'nyi-Prasad-Sommerfeld (BPS) monopole with mass proportional to the length (ν_m) of the interval. Taking $|\vec{y}_n| \to \infty$ leaves the interval $[\mu_1,\mu_n]$, allowing for the interpretation of an $SU(n)$ monopole with μ_i specifying the eigenvalues of the Higgs field at infinity, for which it is crucial they add to zero. Indeed, in the periodic gauge A_0 tends to a constant at spatial infinity.

Note that we have to order $\exp(2\pi i\mu_m)$ along the circle to ensure that the ν_i add to 1, an ordering inherited by μ_m when extended to the real line by insisting $\mu_{kn+m} = k + \mu_m$, for any

M.N. Chernodub et al. / Nuclear Physics B (Proc. Suppl.) 83–84 (2000) 556–558

integer k. Let us pick one to be labelled by μ_1. All we can guarantee at this point is that $\sum_{m=1}^{n}\mu_m = \ell$, an integer. With $\mu_{kn+m} = k+\mu_m$, we find $\sum_{m=1}^{n}\mu_{m+p} = \ell + p$, for *any* integer p. A cyclic shift of the labels by $p = -\ell$ proves that there is a *unique* choice of the μ_m that satisfy eq. (2). It demonstrates why $\vec{y}_n$ does play a special role, and in the limit $|\vec{y}_n| \to \infty$ one therefore has a *static* monopole solution [4], which can be seen as the composite of $n-1$ BPS monopoles of mass ν_m, located at $\vec{y}_m$, for $m = 1, \cdots, n-1$. From the general formalism it is clear these $n-1$ monopole constituents are time independent, as was verified explicitly for $SU(2)$ [2,3]. Note that our argument demonstrates that for $|\vec{y}_m| \to \infty$ with $m \neq n$, one is left with a gauge field that cannot be time independent, even though the resulting action density is [2].

The significance of one constituent carrying a time dependent field lies in the fact that the n constituent monopoles form an instanton, and the topological charge can be associated to the so-called Taubes-winding [7], described by a time dependent (gauge) rotation, going full circle when t progresses over one period. For $SU(2)$ this can be read-off from the explicit expression for the gauge field [2,3]. We thus conclude that the constituent located at $\vec{y}_n$ is the one that carries this Taubes-winding, even though its action density is time independent for well-separated constituents. This conclusion can also be drawn from the formalism developed in ref. [6], see also ref. [8].

3. Fermion zero-mode

The basic ingredient in the construction of caloron solutions is a Greens function defined by

$$\left(D_z^2 + r^2(x;z) + \sum_m \delta_m(z)\right)\hat{f}_x(z,z') = \delta(z-z'), \quad (5)$$

where $D_z = (2\pi i)^{-1}\partial_z - t$, $r^2(x;z) = r_m^2(x)$ for $z \in [\mu_m, \mu_{m+1}]$ and $\delta_m(z) = \delta(z-\mu_m)|\vec{y}_m - \vec{y}_{m-1}|/2\pi$. A similarity with the impurity scattering problem allows for a straightforward solution [1], which we present here for the case that $\mu_m \leq z' \leq z \leq \mu_{m+1}$ (extended to $z < z'$ by $\hat{f}_x(z',z) = \hat{f}_x^*(z,z')$)

$$\hat{f}_z(z,z') = \frac{\pi e^{2\pi i t(z-z')}}{r_m \psi}\left(e^{-2\pi i t}\sinh\left(2\pi(z-z')r_m\right)+\right.$$

$$\left. < v_m(z')|\mathcal{A}_{m-1}\cdots\mathcal{A}_1\mathcal{A}_n\cdots\mathcal{A}_m|w_m(z)>\right), \quad (6)$$

where the spinors v_m and w_m are defined by

$$v_m^1(z) = -w_m^2(z) = \sinh\left(2\pi(z-\mu_m)r_m\right),$$
$$v_m^2(z) = w_m^1(z) = \cosh\left(2\pi(z-\mu_m)r_m\right). \quad (7)$$

For the zero-mode densities we find

$$|\Psi_z(x)|^2 = -(2\pi)^{-2}\partial_\mu^2 \hat{f}_x(z,z), \quad (8)$$

derived exactly as for $SU(2)$ [9], not repeated here. With the gauge field in the periodic gauge one has, $\Psi_z(t+1,\vec{x}) = \exp(2\pi i z)\Psi_z(t,\vec{x})$. To obtain the finite temperature fermion zero-mode one puts $z = \frac{1}{2}$, whereas for the fermion zero-mode with periodic boundary conditions, relevant in supersymmetric applications, one takes $z = 0$.

In figure 1 we show a typical $SU(3)$ caloron, illustrating that also for $n > 2$ the fermion zero-modes are localised on one of the constituents. This localisation can be established easily in the

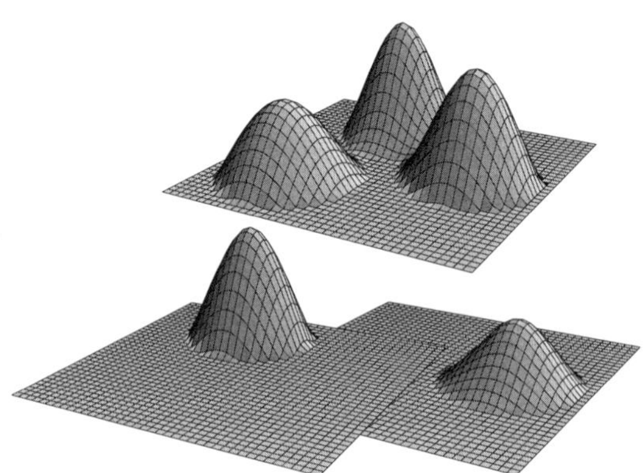

Figure 1. The action densities (top) for the $SU(3)$ caloron, cut off at $1/(2e)$, on a logarithmic scale, with $(\mu_1,\mu_2,\mu_3) = (-17,-2,19)/60$ for $t=0$ in the plane defined by $\vec{y}_1 = (-2,-2,0)$, $\vec{y}_2 = (0,2,0)$ and $\vec{y}_3 = (2,-1,0)$, for $\beta = 1$, with masses $8\pi^2\nu_i$, $(\nu_1,\nu_2,\nu_3) = (0.25,0.35,0.4)$. On the bottom-left is shown the zero-mode density for fermions with anti-periodic boundary conditions in time and on the bottom-right for periodic boundary conditions, at equal logarithmic scales, cut off below $1/e^5$.

limit of large $|\vec{y}_i - \vec{y}_{i+1}|$ for all i, in which case one finds, when $z \in [\mu_m, \mu_{m+1}]$,

$$\hat{f}_x(z,z) = \frac{\sinh[2\pi(z-\mu_m)r_m]\sinh[2\pi(\mu_{m+1}-z)r_m]}{r_m \sinh[2\pi\nu_m r_m]/2\pi}$$

making explicit that the location of the zero-mode is determined by the interval that contains the appropriate value of z. From eq. (2) it follows that $\mu_1 \leq 0 \leq \mu_n$, such that the periodic zero-mode is associated to the *static* constituent at $\vec{y}_m$, with $\mu_m \leq 0 \leq \mu_{m+1}$. This is precisely the condition for the existence of a zero-mode given by the Callias index theorem [10] (see also the appendix of ref. [11]). Due to the static background (for well-separated constituents), time dependence of the zero-mode would be of the form $\exp(2\pi i k t)$ for k integer, shifting $z = 0$ by k, out of the interval that allows for a zero-mode.

Allowing for $k = \pm\frac{1}{2}$, for which $\exp(2\pi i k t)$ turns the periodic zero-mode anti-periodic, we can have situations where this anti-periodic zero-mode is associated to one of the static monopole constituents. A specific example for $SU(3)$ where this occurs is $(\mu_1, \mu_2, \mu_3) = (-0.48, -0.03, 0.51)$, yielding $(\nu_1, \nu_2, \nu_3) = (0.45, 0.54, 0.01)$. *Both* the periodic and the anti-periodic zero-mode are associated to the 2nd constituent. We note that, apart from the fact that the 3rd constituent is nearly massless, both zero-modes are very broad since $\min(z-\mu_2, \mu_3 - z) = 0.03$ for $z = 0$ and 0.01 for $z = \frac{1}{2}$. For $SU(2)$ $z = 0$ is always midway between μ_1 and μ_2 and $z = \frac{1}{2}$ midway between μ_2 and $\mu_3 = 1+\mu_1$). When z coincides with μ_i, the zero-mode is no longer normalisable, which is the origin of the delta function singularities in the Nahm transformation.

4. Conclusions

In conclusion, for well-separated constituents the fermion zero-mode is localised to a single constituent. For $SU(2)$ the anti-periodic zero-mode is always associated to the constituent that carries Taubes-winding [9]. For $SU(n > 2)$ this is also typically true (see fig. 1), in particular when well localised, something that may be significant for developing a model for the QCD vacuum that combines monopoles and instantons [2,3,9]. However, exceptions exist where both the periodic and anti-periodic zero-mode are associated to (possibly the same) static constituent(s), although this tends to be accompanied by nearly massless constituents, and rather delocalised zero-modes.

Acknowledgements

We thank, Jan de Boer for pointing us to the connection with the Callias theorem, Tamás Kovács and Mikhail Voloshin for discussions, Diego Bellisai for correspondence and Margarita García Pérez for collaboration in the early stages. MNCh thanks the Theoretical Physics Institute at the University of Minnesota and the Institute-Lorentz for hospitality. This work was supported in part by INTAS grant RFBR 95-0681. MNCh was supported by the INTAS fellowship YSF 98-95 and TCK by a grant from FOM/SWON. PvB thanks the Aspen Center of Physics, where part of this work was done, and workshop organisers Urs Heller and Rajamani Narayanan, and participants, for creating a stimulating atmosphere.

REFERENCES

1. T.C. Kraan and P. van Baal, Phys. Lett. B435 (1998) 389.
2. T.C. Kraan and P. van Baal, Nucl. Phys. B(Proc.Suppl.) 73 (1999) 554.
3. T.C. Kraan and P. van Baal, Phys. Lett. B428 (1998) 268; Nucl. Phys. B533 (1998) 627.
4. W. Nahm, *Self-dual monopoles and calorons*, in: Lecture Notes in Physics 201 (1984) 189.
5. M.F. Atiyah, N.J. Hitchin, V. Drinfeld and Yu.I. Manin, Phys. Lett. 65A (1978) 185
6. K.Lee and P. Yi, Phys. Rev. D56(1997)3711; K.Lee and C.Lu, Phys.Rev.D58(1998)025011.
7. C. Taubes, in: *Progress in gauge field theory*, eds. G.'t Hooft e.a., (Plenum Press, New York, 1984) p.563
8. N.M. Davies, T.J. Hollowood, V.V. Khoze and M.P. Mattis, hep-th/9905015.
9. M.García Pérez, A.González-Arroyo, C.Pena and P. van Baal, Phys.Rev. D60(1999)031901.
10. C.J. Callias, Comm.Math.Phys. 62(1978)213.
11. J. de Boer, K. Hori and Y. Oz, Nucl. Phys. B500 (1997) 163.

QCD IN A FINITE VOLUME

PIERRE VAN BAAL

Instituut-Lorentz for Theoretical Physics, University of Leiden,
P.O. Box 9506, Nl-2300 RA Leiden, The Netherlands

We will review our understanding of non-abelian gauge theories in finite physical volumes. It allows one in a reliable way to trace some of the non-perturbative dynamics. The role of gauge fixing ambiguities related to large field fluctuations is an important lesson that can be learned. The hamiltonian formalism is the main tool, partly because semiclassical techniques are simply inadequate once the coupling becomes strong. Using periodic boundary conditions, continuum results can be compared to those on the lattice. Results in a spherical finite volume will be discussed as well.

Contents

1 Introduction

We have decided to take this opportunity of contributing to a handbook of QCD, in honor of Boris Lazarevich Ioffe, to bring together results and methods that were developed in solving for the low-lying spectrum in a finite volume. The emphasis will be on the dynamical aspects of the classically scale invariant theory for non-abelian gauge theories in 3+1 dimensions.[1] The challenge is to understand how the mass gap is generated. Due to the need for regularization, at the quantum level breaking the scale invariance, a running coupling appears. The difficulty lies in the fact that this coupling increases at low-energies and long-distance scales, beyond the point where one has control over the field fluctuations, which become so large that they probe the essential non-linearities of the theory. A finite volume explicitly breaks the scale invariance. However, it does so in a rather mild way. Classically one can use the scale transformation to go from one physical volume to another, and only the running of the coupling constant prevents us from taking the infinite volume limit. As the longest distance scale, i.e. the lowest energy scale, is set by the volume, one can keep in check the growth of the running coupling constant.

We will describe the results mainly in the context of a hamiltonian picture[2] with wave functionals on configuration space. Although rather cumbersome from a perturbative point of view, where the covariant path integral approach of Feynman is vastly superior, it provides more intuition on how to deal with non-perturbative contributions in situations where semi-classical techniques can no longer be used, like for observables that do not vanish in perturbation theory. The high energy modes can be well-approximated by a harmonic oscillator contribution to the wave functional. In the direction of these field modes the potential energy rises steeply. Their contribution, including regulating the ultraviolet behavior, is treated perturbatively, giving in particular rise to the running of the coupling. The finite volume allows us to have a well-defined mode expansion. Due to the classical scale invariance, the hamiltonian can be formulated in terms of dimensionless fields. This can be extended to the quantum theory, as Ward identities allow for a field definition without anomalous scaling. Apart from the overall scaling dimension of the hamiltonian, only the running coupling introduces a non-trivial volume dependence.

Due to the non-abelian nature of the theory the physical configuration space, formed by the set of gauge orbits $\mathcal{A}/\mathcal{G}$ ($\mathcal{A}$ is the collection of connections, $\mathcal{G}$ the group of local gauge transformations) is non-trivial.[3] Most frequently, coordinates of this orbit space are chosen by picking a representative gauge field on the orbit in a smooth and preferably unique way. Linear gauge conditions like the Landau or Coulomb gauge suffer from what is known as Gribov ambiguities.[4] The region of field space that contains no further gauge copies is called a fundamental domain for non-abelian gauge theories.[5]

Having arranged the low-energy field modes (and all those modes not affected by the cutoff) to be scale invariant, the spreading of the wave functional is completely caused by an increasing coupling. This is what leads to non-perturbative effects. Asymptotic freedom on the other hand guarantees that in small volumes the running coupling is small and it thus keeps the wave functional localized near the classical vacuum manifold. In a periodic geometry, this perturbative analysis was pioneered by Bjorken[6] and further developed by Lüscher[7]. The essential ingredient we have added to address non-perturbative effects is boundary conditions in *field space*, at the boundary of the fundamental domain, with gauge invariance implemented properly at all stages.

The non-perturbative features due to spreading of the wave functional can be followed out to a physical volume of about one cubic fermi (setting the scale by the string tension or lowest glueball state). This is, on the basis of a comparison with lattice Monte Carlo data, the point where in the pure gauge theory the confining string is being formed, with no significant finite volume dependence beyond this volume. What has become clear is that the transition from the finite to the infinite volume is driven by field fluctuations that cross the barrier which is associated with tunneling between different classical vacua. This is natural, since this barrier (the finite volume sphaleron), will be the direction beyond which the wave functional can first spread most significantly, as it provides the lowest mountain pass in the energy landscape.

In Sec. 2 we describe the process of complete gauge fixing, and the fact that the boundary of the fundamental domain, unlike its interior, has gauge copies that implement the non-trivial topology of configuration space. This is applied in Sec. 3 to formulating non-abelian gauge theories on a torus, both for SU(2) and in Sec. 3.1 for SU(3). The influence of massless quarks on the small volume vacuum structure is discussed in Sec. 3.2, whereas Sec. 3.3 gives a short review of the non-perturbative evaluation of the running renormalized coupling, making explicit use of the finite volume geometry.[8] In Sec. 3.4 we consider the case of twisted boundary conditions,[9] and in Sec. 3.5 we discuss supersymmetric Yang-Mills theory in a finite volume in the light of some recent developments concerning the Witten index.[10] Sec. 4 outlines what is known about instantons on the torus, reviewing the Nahm transformation[11] in Sec. 4.1. Sec. 5 analyzes the situation for a spherical geometry, particularly well adapted to study the role of instantons in the low-lying glueball spectrum. Finally, we shall consider the behavior in large finite volumes. We discuss how the previous results fit together and what this implies. In Secs. 6.1 and 6.2 we review the volume dependence whenever the polarization clouds are well-contained in the finite volume.[12] In Sec. 6.3 we briefly mention the situation in the absence of a mass gap, particularly relevant for QCD with its light up and down quarks, for which chiral perturbation theory applies.[13] We conclude with a review of 't Hooft's electric-magnetic duality on the torus[9] in Sec. 6.4.

2 Complete Gauge Fixing

An (almost) unique representative of the gauge orbit is found by minimizing the L^2 norm of the vector potential along the gauge orbit[5,14]

$$G_A(h) \equiv \|[h]A\|^2 = -\int_M \operatorname{tr}\left(\left(h^{-1}(\mathbf{x})A_i(\mathbf{x})h(\mathbf{x}) + h^{-1}(\mathbf{x})\partial_i h(\mathbf{x})\right)^2\right), \quad (1)$$

where the vector potential is taken anti-hermitian, the integral over the finite spatial volume M is with the appropriate canonical volume form and $h(\mathbf{x})$ is a Lie-group element, with $[h]A$ a short-hand notation for the associated gauge transformation. Note that in these conventions the field strength is given by

$$F_{\mu\nu}(\mathbf{x},t) = \partial_\mu A_\nu(\mathbf{x},t) - \partial_\nu A_\mu(\mathbf{x},t) + [A_\mu(\mathbf{x},t), A_\nu(\mathbf{x},t)] \quad (2)$$

and the action by

$$L(t) = \tfrac{1}{2}\int_M g^{-2}\operatorname{tr}\left(F_{\mu\nu}^2(\mathbf{x},t)\right). \quad (3)$$

Expanding around the minimum of Eq. (1), writing $h(\mathbf{x}) = \exp(X(\mathbf{x}))$ ($X(\mathbf{x})$ is, like the gauge field $A_i(\mathbf{x})$, an element of the Lie-algebra) one easily finds

$$\|[h]A\|^2 = \|A\|^2 + 2\int_M \operatorname{tr}(X\partial_i A_i) + \int_M \operatorname{tr}(X^\dagger FP(A)X) \quad (4)$$

$$+\frac{1}{3}\int_M \operatorname{tr}(X[[A_i,X],\partial_i X]) + \frac{1}{12}\int_M \operatorname{tr}([\mathcal{D}_i X,X][\partial_i X,X]) + \mathcal{O}(X^5),$$

where $FP(A)$ is the Faddeev-Popov operator $(\operatorname{ad}(A)X \equiv [A,X])$

$$FP(A) = -\partial_i \mathcal{D}_i(A) \equiv -\partial_i(\partial_i + \operatorname{ad}(A_i)). \quad (5)$$

At a local minimum the vector potential is therefore transverse, $\partial_i A_i = 0$, and $FP(A)$ is a positive operator. The set of all these vector potentials is by definition the Gribov region Ω. Using the fact that $FP(A)$ is linear in A, Ω is seen to be a convex subspace of the set of transverse connections Γ. Its boundary $\partial\Omega$ is called the Gribov horizon. At the Gribov horizon, the *lowest* non-trivial eigenvalue of the Faddeev-Popov operator vanishes, and points on $\partial\Omega$ are associated with coordinate singularities. Any point on $\partial\Omega$ can be seen to have a finite distance to the origin of field space and in some cases even uniform bounds can be derived.[15,16]

The Gribov region is the set of *local* minima of the norm functional, Eq. (1), and needs to be further restricted to the *absolute* minima to form a fundamental domain,[5] which will be denoted by Λ. The fundamental domain is clearly contained within the Gribov region. To show that also Λ is convex, note that

$$\| [h]A \|^2 - \|A\|^2 = \int \mathrm{tr}\left(A_i^2\right) - \int \mathrm{tr}\left(\left(h^{-1}A_i h + h^{-1}\partial_i h\right)^2\right) \tag{6}$$

$$= \int \mathrm{tr}\left(h^\dagger FP_f(A)\,h\right) \equiv \langle h, FP_f(A)\,h\rangle, \quad FP_f(A) \equiv -\partial_i D_i(A),$$

where $FP_f(A)$ acts on the fundamental representation and is similar to the Faddeev-Popov operator. Both $FP(A)$ and $FP_f(A)$ are hermitian operators when A is a critical points of the norm functional, i.e. for A transverse. We can define Λ in terms of the absolute minima over $h \in \mathcal{G}$ of $\langle h, FP_f(A)\,h\rangle$

$$\Lambda = \{A \in \Gamma \,|\, \min_{h \in \mathcal{G}} \langle h, FP_f(A)\,h\rangle = 0\}. \tag{7}$$

Using that $FP_f(A)$ is linear in A, the convexity of Λ is automatic: A line connecting two points in Λ lies within Λ.

If we would not specify anything further, as a convex space is contractible, the fundamental region could never reproduce the non-trivial topology of the configuration space. This means that Λ should have a boundary.[17] Indeed, as Λ is contained in Ω, this means Λ is also bounded in each direction. Clearly $A = 0$ is in the interior of Λ, which allows us to consider a ray extending out from the origin into a given direction, where it will have to cross the boundary of Λ and Ω. For any point along this ray in Λ, the norm functional is at its absolute minimum as a function of the gauge orbit. However, for points in Ω that are not also in Λ, the norm functional is necessarily at a relative minimum. The absolute minimum for this orbit is an element of Λ, but in general not along the ray. Continuity therefore tells us that at some point along the ray, this absolute minimum has to pass the local minimum. At the point they are exactly degenerate, there are two gauge equivalent vector potentials with the same norm, both at the absolute minimum. As in the interior the norm functional has a unique minimum, again by continuity, these two degenerate configurations have to both lie on the boundary of Λ.

When L denotes the linear size of M, we may express the gauge fields in the dimensionless combination of LA (in our conventions the fields have no anomalous scale dependence), and *the shape and geometry of the Gribov and fundamental regions are scale independent*. We should note that the norm functional is degenerate along the constant gauge transformations and indeed the Coulomb gauge does not fix these gauge degrees of freedom. We simply demand that the wave functional is in the singlet representation under the constant gauge transformations and we have $\Lambda/G = \mathcal{A}/\mathcal{G}$. Here Λ is assumed to include the non-trivial boundary identifications that restore the non-trivial topology of $\mathcal{A}/\mathcal{G}$.

If a degeneracy at the boundary is continuous, other than by constant gauge transformations, one necessarily has at least one non-trivial zero eigenvalue for $FP(A)$ and the Gribov horizon will touch the boundary of the fundamental domain at these so-called singular boundary points. We sketch the general situation in Fig. 1. In principle, by choosing a different gauge fixing in the neighborhood of these points one could resolve the singularity. If singular boundary points would not exist, all that would have been required is to complement the hamiltonian in the Coulomb gauge with the appropriate boundary conditions in field space. Since the boundary identifications are by gauge transformations the boundary condition on the wave functionals is simply that they are identical under the boundary identifications, possibly up to a phase in case the gauge transformation is homotopically non-trivial.

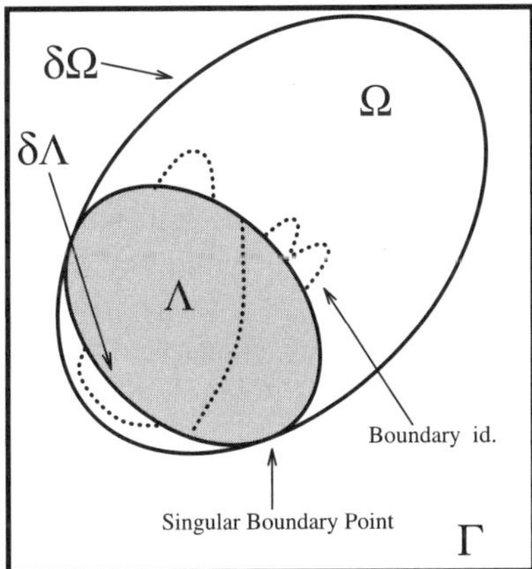

Figure 1: Sketch of the fundamental (shaded) and Gribov regions. The dotted lines indicate the boundary identifications.

Unfortunately, one can argue that singular boundary points are to be expected.[17] Generically, at singular boundary points the norm functional undergoes a bifurcation moving from inside to outside the fundamental (and Gribov) region. The absolute minimum turns into a saddle point and two local minima appear, as indicated in Fig. 2. These are necessarily gauge copies of each other. The gauge transformation is homotopically trivial as it reduces to the identity at the bifurcation point, evolving continuously from there on.

Also Gribov's original arguments for the existence of gauge copies[4] (showing that points just outside the horizon are gauge copies of points just inside) can be easily understood from the perspective of bifurcations in the norm functional. It describes the generic case where the zero-mode of the Faddeev-Popov operator arises because of the coalescence of a local minimum with a saddle point with only one unstable direction. At the Gribov horizon the norm functional locally behaves in that case as X^3, with X the relevant zero eigenfunction of the Faddeev-Popov operator. The situation sketched in Fig. 2 corresponds to the case where the leading behavior is like X^4.

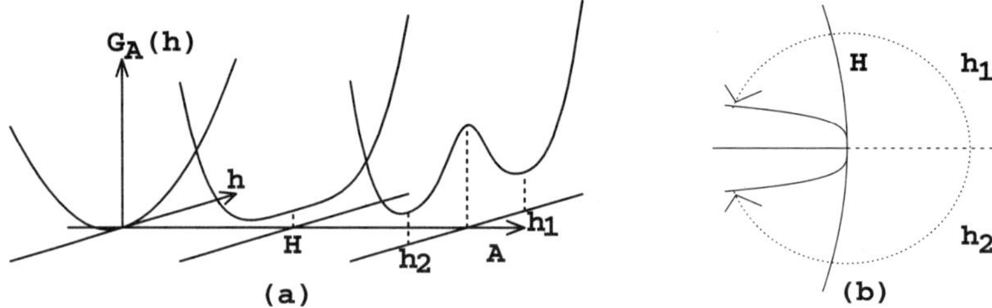

Figure 2: Sketch of a singular boundary point due to a bifurcation of the norm functional. It can be used to show that there are homotopically trivial gauge copies inside the Gribov horizon (H).

The necessity to restrict to the fundamental domain, a subset of the transverse gauge fields, introduces a non-local procedure in configuration space. This cannot be avoided since it reflects the non-trivial topology of this space. We stress again that its topology and geometry is scale independent. Homotopical non-trivial gauge transformations are in one to one correspondence with non-contractible loops in configuration space, which give rise to conserved quantum numbers. The quantum numbers are like the Bloch momenta in a periodic potential and label representations of the homotopy group of gauge transformations. On the fundamental domain the non-contractible loops arise through identifications of boundary points. Although slightly more hidden, the fundamental domain will therefore contain all the information relevant for the topological quantum numbers. Sufficient knowledge of the boundary identifications will allow for an efficient and natural projection on the various superselection sectors. Typically we integrate out the high-energy modes, being left with the low-energy modes whose dynamics is determined by an effective hamiltonian defined on the fundamental domain (restricted to these low-energy modes). In this it is assumed that the contributions of the high-energy modes can be dealt with perturbatively, generating the running coupling and the effective interactions of the low-energy modes. We will in detail discuss the results for finite volumes with a torus and sphere geometry.

3 Gauge Fields on the three-Torus

Probably the most simple example to illustrate the relevance of the fundamental domain is provided by gauge fields on the torus in the abelian zero-momentum sector. Let us first take $G=\mathrm{SU}(2)$ and $A_j = i\frac{C_j}{2L}\tau_3$ (L is the size of the torus, τ_j are the Pauli matrices). These modes are dynamically motivated as they form the set of gauge fields on which the classical potential vanishes. It is called the vacuum valley (sometimes also referred to as toron valley) and one can attempt to perform a Born-Oppenheimer-like approximation for deriving an effective hamiltonian in terms of these "slow" degrees of freedom. To find the Gribov horizon, one easily verifies that the part of the spectrum for $FP(A)$ that depends on $\mathbf{C}$, is given by $\lambda_{\mathbf{n}}^{gh}(\mathbf{C}) = 2\pi\mathbf{n} \cdot (2\pi\mathbf{n} \pm \mathbf{C})/L^2$, with $\mathbf{n} \neq \mathbf{0}$ an integer vector. The lowest eigenvalue vanishes if $C_j = \pm 2\pi$. The Gribov region is therefore a cube with sides of length 4π, centered at the origin, specified by $|C_j| \leq 2\pi$ for all j, see Fig. 3.

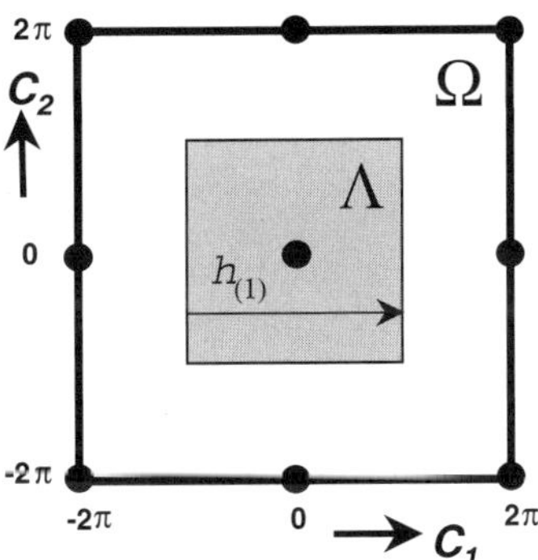

Figure 3: A two dimensional slice of the vacuum valley along the (C_1, C_2) plane. The fat square give the Gribov horizon, the grey square is the fundamental domain. The dots at the Gribov horizon are Gribov copies of the origin.

The gauge transformation $h_{(j)} = \exp(\pi i x_j \tau_3/L)$ maps C_j to $C_j + 2\pi$, leaving the other components $C_{i \neq j}$ untouched. As $h_{(j)}$ is anti-periodic it is homotopically non-trivial (they are 't Hooft's twisted gauge transformations.[9]) We thus see explicitly that gauge copies occur inside Ω, but furthermore the naive vacuum $A = 0$ has (many) gauge copies under these shifts of 2π that lie on the Gribov horizon. It can actually be shown in the Coulomb gauge that for any three-manifold, any Gribov copy by a homotopically non-trivial gauge transformation of $A = 0$ will have a non-trivial zero eigenvalue for the Faddeev-Popov operator.[17] Taking the symmetry under homotopically non-trivial gauge transformations properly into account is crucial for describing the non-perturbative dynamics and one sees that the singularity of the hamiltonian at Gribov copies of $A = 0$, where the wave functionals are in a sense maximal, could form a severe obstacle in obtaining reliable results.

To find the boundary of the fundamental domain we note that the gauge copies $\mathbf{C} = (\pi, C_2, C_3)$ and $\mathbf{C} = (-\pi, C_2, C_3)$ have equal norm. The boundary of the fundamental domain, restricted to the vacuum valley formed by the abelian zero-momentum gauge fields therefore occurs where $|C_j| = \pi$, well inside the Gribov region, see Fig. 3. The boundary identifications are by the homotopically non-trivial gauge transformations $h_{(j)}$. The fundamental domain, described by $|C_j| \leq \pi$ with all boundary points regular, has the topology of a torus. To be more precise, as the remnant of the constant gauge transformations (the Weyl group) changes $\mathbf{C}$ to $-\mathbf{C}$, the fundamental domain Λ/G restricted to the abelian constant modes is the orbifold T^3/Z_2. As we will see, the orbifold singularities are associated with problems in using the harmonic approximation in perturbation theory for modes orthogonal to those of vacuum valley.

Formulating the hamiltonian on Λ, with the boundary identifications implied by the gauge transformations $h_{(j)}$, avoids the singularities at the Gribov copies of $A = 0$. "Bloch momenta" associated with the 2π shift, implemented by the non-trivial homotopy of $h_{(j)}$, label 't Hooft's electric flux quantum numbers[9] $\Psi(C_j = -\pi) = \exp(\pi i e_j)\Psi(C_j = \pi)$. Note that the phase factor is not arbitrary, but ± 1. This is because $h_{(j)}^2$ is homotopically trivial. In other words, the homotopy group of these anti-periodic gauge transformations is Z_2^3. Considering a slice of Λ can obscure some of the topological features. A loop that winds around the slice twice is contractible in Λ as soon as it is allowed to leave the slice. Indeed including the lowest modes transverse to this slice will make the Z_2 nature of the relevant homotopy group evident.[18] It should be mentioned that for the torus in the presence of fields in the fundamental representation (quarks), only periodic gauge transformations are allowed.[19] This will be discussed in Sec. 3.2.

In weak coupling Lüscher showed unambiguously that the wave functionals are localized around $A = 0$, that they are normalizable and that the spectrum is discrete.[6,7] In this limit the spectrum is insensitive to the boundary identifications (giving rise to a degeneracy in the topological quantum numbers). This is manifested by a vanishing electric flux energy, defined by the difference in energy of a state with $|\mathbf{e}| = 1$ and the vacuum state with $\mathbf{e} = \mathbf{0}$. Although there is no classical potential barrier to achieve this suppression, it comes about by a quantum induced barrier, in strength down by two powers of the coupling constant.

$$V(\mathbf{C}) = 2L^{-1} \sum_{\mathbf{p} \in \mathbb{Z}^3} (|2\pi\mathbf{p} + \mathbf{C}| - |2\pi\mathbf{p}|) = \frac{4}{L\pi^2} \sum_{\mathbf{n} \neq 0} \frac{\sin^2(\frac{1}{2}\mathbf{n} \cdot \mathbf{C})}{(\mathbf{n} \cdot \mathbf{n})^2}. \tag{8}$$

The second identity is obtained by Poisson resummation,[7] and the periodicity reflects the gauge symmetry discussed above. This quantum induced barrier leads to a suppression[20] with a factor $\exp(-S/g)$ instead of the usual factor of $\exp(-8\pi^2/g^2)$ for instantons.[21,22] The associated tunneling configuration was called a *pinchon*.[20] At stronger coupling the wave functional spreads out over the vacuum valley and drastically changes the spectrum.[23] At this point the energy of electric flux suddenly switches on.

Integrating out the non-zero momentum modes, for which Bloch degenerate perturbation theory[24] provides a rigorous framework,[7] one finds an effective hamiltonian. Near $A = 0$, due to the quartic nature of the potential energy $V(A) = -\frac{1}{2} \int d^3x \, \mathrm{tr}\,(F_{ij})$ for the zero-momentum modes (the derivatives vanish, such that the field strength is quadratic in the field), there is no separation in time scales between the abelian and non-abelian modes. Away from $A = 0$ one could further reduce the dynamics to one along the vacuum valley, but near the origin the adiabatic approximation breaks down. This gives rise to a conic singularity, $V(\mathbf{C}) = 2|\mathbf{C}|/L + \mathcal{O}\left(C^2\right)$, due to fluctuations of the non-abelian zero-momentum modes (contributing the $\mathbf{p} = \mathbf{0}$ term in Eq. (8)).

The effective hamiltonian is expressed in terms of the coordinates c_j^a, where $i = \{1, 2, 3\}$ is the spatial index ($c_0^a = 0$) and $a = \{1, 2, 3\}$ the SU(2)-color index. It will be convenient to introduce $f_{jk}^a \equiv -\epsilon_{abd} c_j^b c_k^d$ and $r_j \equiv \sqrt{\sum_a c_j^a c_j^a}$. The latter are gauge-invariant "radial" coordinates, that will play a crucial role in specifying the boundary conditions. The zero-momentum gauge field and field strength read $A_j(\mathbf{x}) = i c_j^a \frac{\tau_a}{2L}$ and $F_{jk} = i f_{jk}^a \frac{\tau_a}{2L^2}$. For dimensional reasons the effective hamiltonian is proportional to $1/L$. It will furthermore depend on L through the renormalized coupling constant ($g(L)$) at the scale $\mu = 1/L$. To one loop order (for small L) $g^2(L) = 12\pi^2/[-11\ln(\Lambda_{MS}L)]$. One expresses the masses and the size of the finite volume in dimensionless quantities, like mass-ratios and the parameter $z = mL$. In this way, the explicit dependence of g on L is irrelevant. This is also the preferred way of comparing results obtained within different regularization schemes (e.g. dimensional and lattice regularization). The effective hamiltonian is now given by

$$
\begin{aligned}
LH_{\mathrm{eff}}(c) \;=\; & \frac{-g^2}{2(1 + \alpha_1 g^2)} \sum_{i,a} \frac{\partial^2}{(\partial c_i^a)^2} + \frac{1}{4}\left(\frac{1}{g^2} + \alpha_2\right) \sum_{ij,a} (f_{ij}^a)^2 + \gamma_1 \sum_i r_i^2 \\
& + \gamma_2 \sum_i r_i^4 + \gamma_3 \sum_{i>j} r_i^2 r_j^2 + \gamma_4 \sum_i r_i^6 + \gamma_5 \sum_{i\neq j} r_i^2 r_j^4 + \gamma_6 \prod_i r_i^2 \\
& + \alpha_3 \sum_{ijk,a} r_i^2 (f_{jk}^a)^2 + \alpha_4 \sum_{ij,a} r_i^2 (f_{ij}^a)^2 + \alpha_5 \mathrm{det}^2 c \\
& + \gamma_7 \sum_i r_i^8 + \gamma_8 \sum_{i\neq j} r_i^6 r_j^2 + \gamma_9 \sum_{i>j} r_i^4 r_j^4 + \gamma_{10} \sum_i r_i^2 (r_1^2 r_2^2 r_3^2).
\end{aligned}
\tag{9}
$$

We have organized the terms according to the importance of their contributions. The first two terms give (when ignoring $\alpha_{1,2}$) the lowest order effective hamiltonian,[6] whose energy eigenvalues are $\mathcal{O}\left(g^{2/3}\right)$, as can be seen by rescaling c with $g^{2/3}$. Thus, in a perturbative expansion $c = \mathcal{O}\left(g^{2/3}\right)$. Lüscher's computation[7] to $\mathcal{O}\left(g^{8/3}\right)$, determined by $\alpha_1, \alpha_2, \gamma_1, \gamma_2$ and γ_3, was extended to $\mathcal{O}\left(g^4\right)$, to ensure that the terms parametrized by γ_i included the vacuum-valley effective potential (i.e. the part that does not vanish on the set of abelian configurations) to sufficient numerical accuracy.[18] The order r^8 terms were just included to check stability of the results. The vacuum-valley effective potential could actually be computed at two loops to all orders in the field, up to an irrelevant constant

$$V_{2-\text{loop}}(\mathbf{C}) = \frac{g^2 L}{32}(\Delta V(\mathbf{C}) + \Delta V(\mathbf{0}))^2, \quad \Delta V(\mathbf{C}) = \sum_i \frac{\partial^2 V(\mathbf{C})}{\partial C_i \partial C_i}, \qquad (10)$$

but more cumbersome to compute were terms of the form $g^4 c^2 \partial_c^2$, also of two loop order.[25] Their effect on the spectrum can, however, be neglected and we have not listed these terms. The coefficients appearing in H_{eff} have the following numerical values

$$\gamma_1 = -3.0104661 \cdot 10^{-1} - 3.0104661 \cdot 10^{-1}(g/2\pi)^2, \quad \alpha_1 = +2.1810429 \cdot 10^{-2},$$
$$\gamma_2 = -1.4488847 \cdot 10^{-3} - 9.9096768 \cdot 10^{-3}(g/2\pi)^2, \quad \alpha_2 = +7.5714590 \cdot 10^{-3},$$
$$\gamma_3 = +1.2790086 \cdot 10^{-2} + 3.6765224 \cdot 10^{-2}(g/2\pi)^2, \quad \alpha_3 = +1.1130266 \cdot 10^{-4},$$
$$\gamma_4 = +4.9676959 \cdot 10^{-5} + 5.2925358 \cdot 10^{-5}(g/2\pi)^2, \quad \alpha_4 = -2.1475176 \cdot 10^{-4},$$
$$\gamma_5 = -5.5172502 \cdot 10^{-5} + 1.8496841 \cdot 10^{-4}(g/2\pi)^2, \quad \alpha_5 = -1.2775652 \cdot 10^{-3},$$
$$\gamma_6 = -1.2423581 \cdot 10^{-3} - 5.7110724 \cdot 10^{-3}(g/2\pi)^2,$$
$$\gamma_7 = -9.8738947 \cdot 10^{-7} - 5.1311245 \cdot 10^{-6}(g/2\pi)^2,$$
$$\gamma_8 = +9.1911536 \cdot 10^{-6} + 9.1452409 \cdot 10^{-5}(g/2\pi)^2,$$
$$\gamma_9 = -2.7911565 \cdot 10^{-5} - 2.5203366 \cdot 10^{-5}(g/2\pi)^2,$$
$$\gamma_{10} = +1.8208802 \cdot 10^{-5} + 6.0939067 \cdot 10^{-5}(g/2\pi)^2. \qquad (11)$$

A challenge was to rigorously determine the semiclassical expansion for the energy of electric flux due to the tunneling through a *quantum induced* potential barrier. The so-called Path Decomposition Expansion,[26] was an important tool that could be tailored to this situation.[27,28] One required matching the perturbative tail of the wave function derived from Lüscher's effective hamiltonian to the semiclassical contribution in the classically forbidden region. The resulting asymptotic expansion for the energy of one-unit ($|\mathbf{e}| = 1$) of electric flux, $\Delta E = E_0(\mathbf{e}) - E_0(\mathbf{0})$ was found to be[29,30]

$$\Delta E = 2L^{-1}\lambda B^2 g^{5/3}(L) \exp\left[-S/g(L) + \varepsilon_1 T/g^{2/3}(L)\right]\{1 + f(g(L))\}. \qquad (12)$$

Here $S = 12.4637$ is the tunneling action and $T = 3.9186$ is related to the tunneling time, and comes from the classical turning points of the quantum induced potential $V(\mathbf{C})$, which also determines $\lambda = 0.6997$, due to its transverse fluctuations along the tunneling path. It was shown[18] that $f(g) = \mathcal{O}(g)$. The constant $\varepsilon_1 = 4.116719735$ determines to lowest order the mass of the scalar glueball, $m_0 = L^{-1}\varepsilon_1 g^{2/3}(L)$, which was already computed by Lüscher and Münster.[31] Finally, the quantity $B = 0.206$ is taken from the asymptotic expansion of the lowest order perturbative wave function (in the direction of the vacuum valley). To compute ε_1 and B it suffices to consider H_{eff} in lowest non-trivial order, by putting all constants α_i and γ_i equal to zero.

Of more practical importance is to go to the domain where the typical energies are above those of the quantum induced potential barrier, when the wave functional has spread out over the full fundamental domain, see Fig. 3. In this case one will become sensitive to the boundary identifications on the fundamental domain. The influence of the boundary conditions on the low-lying glueball states is felt as soon as $z = m_0 L > 0.9$. We summarize below the ingredients that enter the calculation.

The choice of boundary conditions, associated with each of the irreducible representations of the cubic group $O(3, \mathbb{Z})$ and the electric flux quantum numbers,[9] is best described by observing that the cubic group is the semidirect product of the group of coordinate permutations S_3 and the group of coordinate reflections Z_2^3. We denote the parity under the coordinate reflection $c_i^a \to -c_i^a$ ($\forall a$) by $p_i = \pm 1$. The electric flux quantum number for the same direction will be denoted by $q_i = \pm 1$. This is related to the more usual additive (mod 2) quantum number e_j by $q_j = \exp(i\pi e_j)$. Note that for SU(2) electric flux is invariant under coordinate reflections. If not all of the electric fluxes are identical, the cubic group is broken to $S_2 \times Z_2^3$, where $S_2 (\cong Z_2)$ corresponds to interchanging the two directions with identical electric flux (unequal to the other electric flux). If all the electric fluxes are equal, the wave functions are irreducible representations of the cubic group. These are the four singlets $A_{1(2)}^{\pm}$, which are completely (anti-)symmetric with respect to S_3 and have $p_1 = p_2 = p_3 = \pm 1$. Then there are two doublets $E^{\pm}$, also with $p_1 = p_2 = p_3 = \pm 1$ and finally one has four triplets $T_{1(2)}^{\pm}$. Each of these triplet states can be decomposed into eigenstates of the coordinate reflections. Explicitly,[18,32] for $T_{1(2)}^{\pm}$ we have one state that is (anti-)symmetric under interchanging the two- and three-directions, with $p_1 = -p_2 = -p_3 = \pm 1$. The other two states are obtained through cyclic permutation of the coordinates. Thus, any eigenfunction of the effective hamiltonian with specific electric flux quantum numbers q_i can be chosen to be an eigenstate of the parity operators p_i. The boundary conditions of these eigenfunctions $\Psi_{\mathbf{q},\mathbf{p}}(c)$ are given by

$$\Psi_{\mathbf{q},\mathbf{p}}(c)\big|_{r_i=\pi} = 0 \quad , \qquad \text{if} \quad p_i q_i = -1$$

$$\frac{\partial}{\partial r_i}(r_i \Psi_{\mathbf{q},\mathbf{p}}(c))\big|_{r_i=\pi} = 0 \quad , \qquad \text{if} \quad p_i q_i = +1 \tag{13}$$

and one easily shows that with these boundary conditions the hamiltonian is hermitian with respect to the inner product $< \Psi, \Psi' > = \int_{r_i \leq \pi} d^9 c\, \Psi^*(c) \Psi'(c)$.

In Sec. 3.1, when discussing the case of SU(3), we show in more detail how one arrives at these boundary conditions. For negative parity states ($\prod_i p_i = -1$) this description is, however, not accurate[32] as parity restricted to the vacuum valley is equivalent to a Weyl reflection (a remnant of the invariance under constant gauge transformations). One can use as a basis for $\Psi_{\mathbf{p},\mathbf{q}}(c)$

$$< \mathbf{c}|\mathbf{l},\mathbf{n};\mathbf{e}> = \sum_{\mathbf{m}} W(\mathbf{l},\mathbf{m}) \prod_{i=1}^{3} r_i^{-1} \chi_{n_i,l_i}^{(e_i)}(r_i) Y_{l_i,m_i}(\theta_i,\phi_i), \qquad (14)$$

with $Y_{l,m}$ spherical harmonics, and $W(\mathbf{l},\mathbf{m})$ the Wigner coefficients for adding three angular momenta $\sum_i \mathbf{L}_i = \mathbf{0}$ (ensuring gauge invariance). The angular quantum numbers l_i are restricted to be even or odd, for resp. $p_i = 1$ or $p_i = -1$, and the radial wave function $\chi_{n,l}^{(e)}(r)$ either vanishes, or has vanishing derivative at $r = \pi$ (for resp. $l+e$ even or odd, see Eq. (13)). One computes the matrix elements of the effective hamiltonian (Eq. (9)) and solves for the low-lying spectrum by Rayleigh-Ritz (providing *also* lower bounds from the second moment of the hamiltonian). The first two lines in Eq. (9) are sufficient to obtain the mass-ratios to an accuracy of better than 5%.

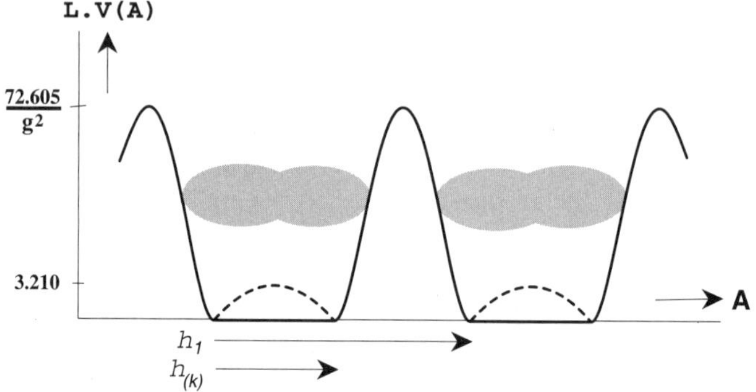

Figure 4: Sketch of the potential for the torus. Shown are two vacuum valleys, related to each other by a gauge transformation h_1, with winding number $\nu(h_1) = 1$. The induced one-loop effective potential, of height $3.210/L$, has degenerate minima related to each other by the anti-periodic gauge transformations $h_{(k)}$. The classical barrier, separating the two valleys, has a height $72.605/Lg^2$.

At larger volumes extra degrees of freedom will behave non-perturbatively. We know from the existence of gauge transformations with non-trivial winding number

$$\nu(h) = \frac{1}{24\pi^2} \int_M \mathrm{tr}\,((h^{-1}dh)^3), \qquad (15)$$

that there exist gauge equivalent vacuum valleys, that can be reached only by crossing a saddle point (called the finite volume sphaleron), see Fig. 4. Typically this saddle point lies on the tunneling path (the instanton), with the euclidean time of the instanton solution playing the role of the path parameter. We expect, as will be shown for the three-sphere, that the boundary of the fundamental domain along this path in field space across the barrier occurs at the saddle point in between the two minima. The degrees of freedom along this tunneling path go outside of the space of zero-momentum gauge fields and if the energy of a state flows over the barrier, its wave functional will no longer be exponentially suppressed below the barrier and will in particular be non-negligible at the boundary of the fundamental domain. Boundary identifications in this direction of field space now become dynamically important too. The relevant "Bloch momentum" is in this case the θ parameter. Wave functionals pick up a phase factor $e^{i\theta}$ under a gauge transformation with winding number one.

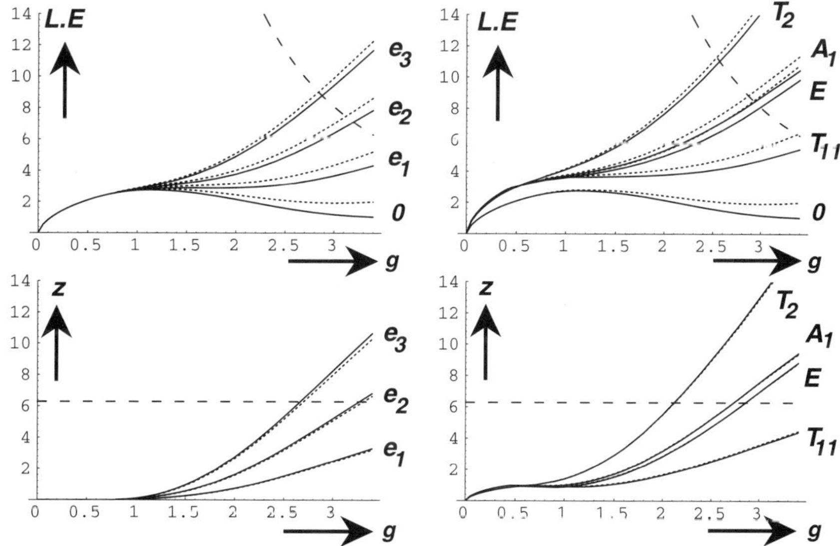

Figure 5: The top figures show $LE(g)$ for the relevant (positive parity) representations. The dotted lines are without the two loop correction included. The dashed curve denotes the barrier height $LE_{sph}(g)$. The bottom figures show $z(g) = Lm(g) = L(E(g) - E_0(g))$, and the dashed line is at $z = 2\pi$.

To estimate for which volumes the extra degrees of freedom start to contribute non-perturbatively, the minimal barrier height that separates two vacuum valleys was found to be $E_{sph}(g) = 72.605/Lg^2$, using the lattice approximation and carefully taking the continuum limit.[33] As long as the states under consideration have energies below this value, the transitions over this barrier can be neglected (or treated semi-classically if there is no perturbative contribution) and the zero-momentum effective hamiltonian provides an accurate description. One can now find for which volume the energy starts to be of the order of this barrier height. For this purpose we have collect the energy levels as a function of g in Fig. 5. In the top two figures we have plotted $LE(g)$ for the various representations of interest (0 stands for the vacuum state, which is in the A_1^+ representation of the cubic group). The dotted curves are *without* the two loop corrections included. In the same figures the dashed curve shows the height of the instanton barrier. In the lower two figures we show $z = Lm \equiv L(E - E_0)$. The dashed line is at $z = 2\pi$. We see that the sensitivity to the two loop corrections almost entirely drops out for the energy differences, at the scale of this figure. We now read off that instantons only become important for $z_0 = 5$ to 6, i.e. for L roughly 5 to 6 times the correlation length set by the scalar glueball mass. Certain states may actually be less sensitive to instantons, in case the representation is such that the wave functional in the direction of the barrier is suppressed. This may for example be the case for the T_2^+ state. It should also be noted that $z = 2\pi$ is associated with the energy scale $2\pi/L$ of the non-zero momentum field modes that have been integrated out. Therefore, z should also not be much bigger than 2π, although also here the particular representation may matter.

On the three-torus we have therefore achieved a self-contained picture of the low-lying glueball spectrum in intermediate volumes from first principles with *no free parameters*, apart from the overall scale. For very small volumes the energy of electric flux vanish and there is an accidental rotational symmetry, only split by the $\mathcal{O}\left(g^{8/3}\right)$ terms in Eq. (9). The 2^+ tensor glueball splits in a nearly degenerate doublet E^+ and a triplet T_2^+. Both are lighter than the scalar 0^+, also denoted by A_1^+ as the scalar singlet representation of the cubic group. Going to larger volumes, $L > 0.1$ fermi, the energy of electric flux per unit length, which in an infinite volume would be the string tension K, is surprisingly constant in intermediate volumes, whereas the splitting of the tensor states becomes quite large. In intermediate volumes the doublet E^+ and triplet T_2^+ have respectively masses of roughly 0.9 and 1.7 times the scalar mass. This doublet E^+, *lighter* than the scalar, was first observed in the lattice studies of Berg and Billoire,[34] and caused some stir at the time.

The lattice data for the triplet[35] were obtained only after our continuum results first appeared[23,18] and did not confirm the predictions for this state. This was resolved by Vohwinkel[32] by observing that the state we had initially identified as T_2^+ (taking $p_1 = p_2 = p_3 = 1$ instead of $p_1 = -p_2 = -p_3 = 1$) actually carried two units of electric flux (cmp. Eq. (13)), making it even more of a surprise that it was found to be even lighter than the doublet E^+ in intermediate volumes. In the infinite volume limit it is pushed out of the low-lying spectrum. Around the same time this state (named T_{11}) was as such measured on the lattice,[36] confirming the proper interpretation of these states.

Electric flux energies (for the trivial representation) are labelled by $e = e_1$ for $\mathbf{e} = (1, 0, 0)$, e_2 for $(1, 1, 0)$ and e_3 for $(1, 1, 1)$. For e_n we speak of n units of electric flux, sometimes called "torelon" energies. As was argued by 't Hooft,[9] if a confining string would have formed, $E_e = KL$ (or $z_e = LE_e = KL^2$), it would be energetically favorable to run along the direction of $\mathbf{e}$, giving $R_n \equiv E(e_n)/E(e_1) = \sqrt{n}$, instead of splitting in separate strings, each winding in the direction for which $e_j = 1$, which would give $R_n = n$. In intermediate volumes it is the latter behavior that we found,[23] confirmed by Monte Carlo results.[37] One has to go to quite large volumes to start to observe the expected $\sqrt{n}$ behavior.[38] The same is found for SU(3), in a study where a different lattice Monte Carlo method was used to get the electric flux energies.[39]

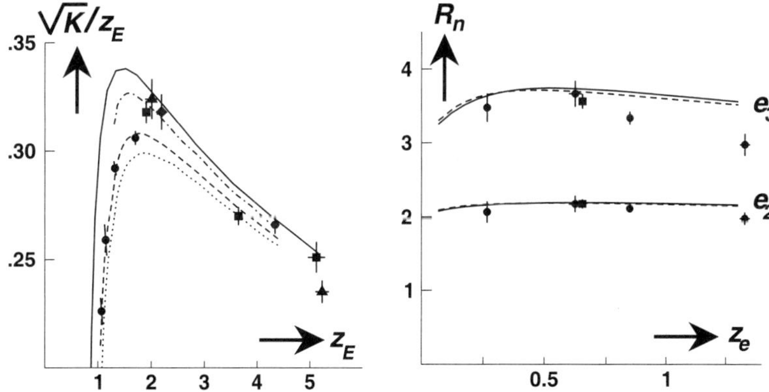

Figure 6: Left: The ratio $\sqrt{K(L)}/m_E$ as a function of $z_E = m_E(L)L$. Full (continuum), dashed-dotted (6^3 lattice) and dashed (4^3 lattice) curves give the hamiltonian results. Rayleigh-Ritz diagonalization errors are smaller than the thickness of the lines. The sensitivity to the two loop correction can be read off from the dotted curve (4^3 lattice), where this correction was **not** included. Note the blown-up scale. Right: electric flux ratios R_n as a function of $z_e = E_e(L)L$. Monte Carlo data[37,40] obtained with the Wilson action on a $N^3 \times N_t$ lattice ($N_t > 6N$), for $N = 4$ (dots), 6 (squares), 8 (triangles) and 10 (diamond).

The lattice Monte Carlo data of Berg and Billoire[37,40] are compared with the hamiltonian results in Fig. 6. On the left we show the ratio $\sqrt{K(L)}/z_E$ as a function of z_E. Their methods had considerable difficulty in dealing with the scalar glueball state, since it has the same quantum numbers as the vacuum. To push the results to low values of z_E, they used lattices $N^3 \times N_t$ with $N_t = 256$ for $N = 4$. Please note that $z = m_E L$ around 0.95 is nearly a constant (even not single valued) function of $g(L)$, see Fig. 5. This overemphasizes the steep behavior where the wave function starts to spread over the whole vacuum valley, leading to non-zero energies of electric flux. We took the Monte Carlo data from Table V of their paper,[40] but removed those entries with $N_t/N \leq 6$. On the right in Fig. 6 we compare with the lattice Monte

Carlo data of Berg[37] for the electric flux ratios, but used where available, the more accurate results[40] for $z_e = LE_e$. The data at $\beta = 4/g_0^2 = 2.7$ from both papers[37,40] were left out because they were not consistent with each other.

Considerable progress was achieved by the so-called variational method, which allowed much more accurate results, not only for the scalar glueball, but also for arbitrary other representations.[35] In Fig. 7 we present a comparison with the Monte Carlo results of Michael,[41] obtained for a lattice of spatial size 4^3, confirmed in a study of improved lattice actions.[42] The hamiltonian results below $z = 0.95$ are due to Lüscher and Münster,[31] which is where the spectrum is insensitive to *any* identifications at the boundary of Λ. The electric flux energy ratios R_2 and R_3 are shown on the right, cmp. Fig. 6.

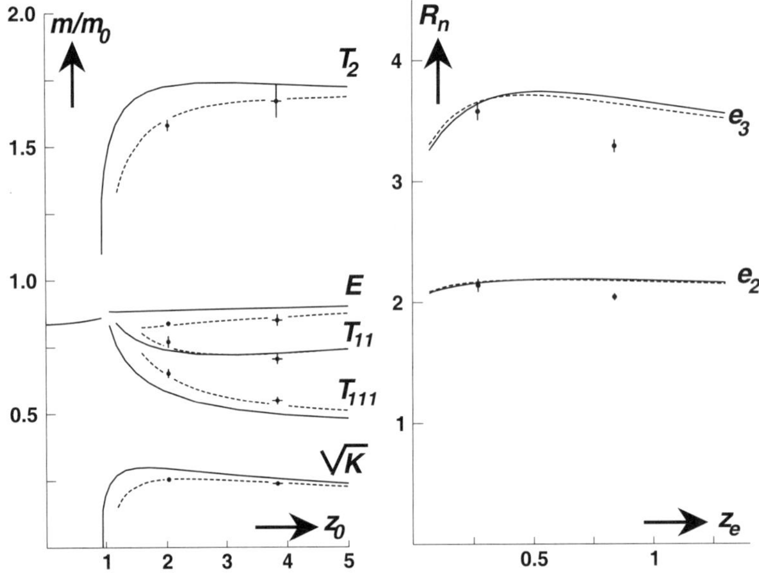

Figure 7: Left: mass ratios m/m_0 as a function of $z = m_0(L)L$, with m_0 the scalar glueball mass. Full (continuum) and dashed (4^3 lattice) curves give the hamiltonian results. Shown are string tension $\sqrt{K(L)}$, E^+ and T_2^+ tensor masses and the exotic states T_{11} with two, and $T_{111} = T_2(111)$ with three units of electric flux. Right: electric flux ratios R_n as a function of z_e. Monte Carlo data[41] obtained with Wilson action on $4^3 \times N_t$ lattice ($N_t = 32, \beta = 2.4$ and $N_t = 99, \beta = 3.0$).

We note that the solid curves that represent the continuum results, which were reproduced by Berg and Vohwinkel,[43,44] deviate significantly from the lattice data. Initially the lattice data were not accurate enough to show this deviation. Even though one should not expect a 4^3 lattice to be an accurate approximation for the continuum, it was cause for some doubt that the approximations made in the continuum studies were not under control as well.[40]

To settle this issue we redid the complete derivation of the effective hamiltonian starting from the lattice theory, *without taking the continuum limit*.[45,25] The hamiltonian is basically of the same form as in Eq. (9), except that the coefficients in Eq. (11) depend on the lattice spacing (some extra corrections appear, e.g. to correct for the discrete time evolution). One can follow the renormalization group flow of the hamiltonian to its continuum fixed point in this formalism in all detail, see Fig. 8. Using the same analysis as in the continuum leads for a finite lattice to the dashed curves in Figs. 6,7. The lattice

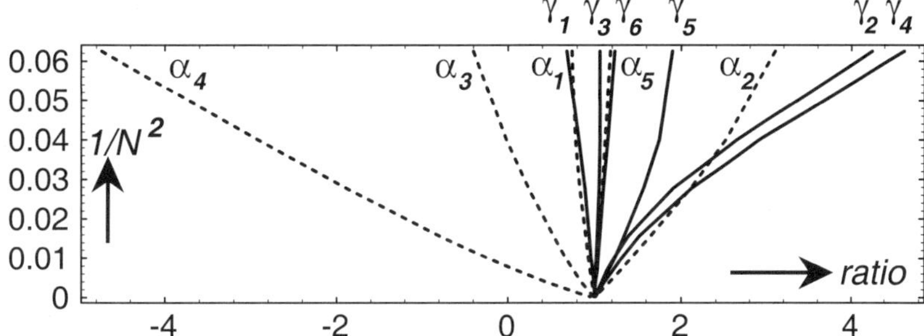

Figure 8: Flow of the lattice hamiltonian coefficients to their continuum ($N \to \infty$) values. Shown are $\alpha_i(N)/\alpha_i$ (dashed lines) and $\gamma_i(N)/\gamma_i$ (full lines) as a function of the square of the lattice spacing $(a/L)^2 = 1/N^2$.

data now agree perfectly, up to a volume of about 0.75 fermi, the regime for which we have shown that the effective hamiltonian in the zero-momentum modes should provide a good approximation. The deviations for R_2 and in particular R_3 are in accordance with the fact that these are of relatively high energy, and therefore expected to be sensitive to other non-perturbative effects that invalidate integrating out all the non-zero momentum modes, see Fig. 5.

3.1 General Gauge Group

The question of extending the previous results to SU(3) is a natural one in the light of QCD. The perturbative expansion,[7,46] vacuum-valley effective potential,[7,30] and the semi-classical evaluation of the energy of electric flux[47] (due to the tunneling through the quantum induced vacuum-valley effective potential) are more or less straightforward. The non-trivial problem of formulating the appropriate boundary conditions on the boundary of the fundamental domain for SU(3) was solved by Vohwinkel and qualitative agreement with the lattice data was found.[48,44] With the results for SU(3) in hand generalization to SU(N) was achieved.[49] The formalism sketched in Sec. 2 was not yet developed in those early days. What is now called the fundamental domain was then called the unit cell. Although the argumentation was more cumbersome, the underlying principles were the same.

A simple approximation for Lüscher's effective hamiltonian in terms of the zero-momentum gauge fields $A_j(\mathbf{x}) = i c_j^a T_a / L$ (T_a the hermitian generators of $\mathrm{SU}(N)$ and $f_{ij} \equiv L^2 F_{ij}(\mathbf{x})$ the dimensionless field strength) is given by

$$H_{\mathrm{eff}}(c) = \frac{-g^2}{2L(1 + \alpha_1 g^2)} \sum_{i,a} \frac{\partial^2}{\partial c_i^{a\,2}} - \frac{1}{2L}\left(\frac{1}{g^2} + \alpha_2\right) \sum_{i,j} \mathrm{tr}\,(f_{ij}^2) + V_1(c), \quad (16)$$

with

$$V_1(c) = \frac{1}{L} \sum_{\mathbf{p} \neq 0} \left(\mathrm{Tr}_{\mathrm{ad}} \sqrt{(2\pi\mathbf{p} + \mathbf{c}^a \mathrm{ad}\, T_a)^2} - \mathrm{Tr}_{\mathrm{ad}} \sqrt{(2\pi\mathbf{p})^2} \right), \quad (17)$$

where g is the renormalized coupling constant at the scale $\mu = 1/L$. In a perturbative expansion,[7,46] this gives the correct result up to $\mathcal{O}\left(g^{8/3}\right)$. Along the vacuum valley, parametrized by $A_j(\mathbf{x}) = i \sum_{b=1}^{N-1} C_j^b T_b / L$, where the first $N - 1$ generators are assumed to commute (forming a basis of the Cartan subalgebra H_G), the effective potential $V_1(c)$ is exact to one loop order. A more explicit result can be found in Sec. 3.2, Eq. (25). As for SU(2), one does not include the $\mathbf{p} = \mathbf{0}$ term, since only the non-zero momentum modes are to be integrated out.

The effective potential $V_1(c)$ only depends on the Casimir invariants

$$r_i^2 = 2\,\mathrm{tr}\,[(c_i^a T_a)^2], \quad s_i = 4\,\mathrm{tr}\,[(c_i^a T_a)^3], \quad \cdots, \quad (18)$$

where the sum over *all* color indices is implicit. There are as many independent Casimir invariants as the rank of the gauge group. For SU(2) only r_i will be non-trivial. These coordinates are uniquely related to those obtained by restricting the zero-momentum gauge field to be abelian. For SU(2), $c_i^a T_a = C_i \tau_3 / 2$ yields $r_i^2 = C_i^2$, whereas for SU(3), in terms of the Gell-Mann matrices, $c_i^a T_a = (C_i^1 \lambda_8 + C_i^2 \lambda_3)/2$ gives $s_i = C_i^1 \sqrt{3}((C_i^2)^2 - (C_i^1)^2/3)$ and $r_i^2 = (C_i^1)^2 + (C_i^2)^2$. In this manner the effective potential on the set of abelian zero-momentum modes can indeed be minimally extended in a unique way to all constant gauge fields.

By adding the zero-momentum ($\mathbf{p} = \mathbf{0}$) contribution to the one-loop effective potential,

$$V_{\mathrm{eff}}(c) = V_1(c) + L^{-1} \mathrm{Tr}_{\mathrm{ad}} \sqrt{(\mathbf{c}^a T_a)^2}, \quad (19)$$

restricted to the vacuum valley gives the appropriate effective potential when integrating out all the field modes orthogonal to the vacuum valley. Its symmetries are a consequence of gauge invariance, which can be divided in two classes. The constant gauge transformations, that leave H_G invariant. This is represented by the Weyl group $\mathcal{W}$ acting on $\mathbf{C}^b$ (for SU(2) $\mathbf{C} \to -\mathbf{C}$). The other class of gauge transformations that leaves the set $\mathbf{C}^b$ invariant are of the form $h_{\Theta}(\mathbf{x}) = \exp(2\pi i \mathbf{x} \cdot \mathbf{\Theta}/L)$, where $\Theta_i \in H_G$ such that h does not affect the periodic boundary conditions on the gauge fields. These lead to shift symmetries on $\mathbf{C}^b$. The associated lattice of Θ_j corresponds to the dual root lattice $\tilde{\Lambda}_r$, as follows from the condition that $\exp(2\pi i \Theta_j) = 1$. The vacuum valley, i.e. the moduli space of flat connections ($F_{ij} = 0$), corresponds to the orbifold $[H_G/2\pi\tilde{\Lambda}_r]^3/\mathcal{W}$. Recently it has become clear that for orthogonal and exceptional Lie-groups there are other, disconnected, components in the moduli space of flat connections, which we will briefly discuss in Sec. 3.5.

A crucial role is played by the twisted gauge transformations, only periodic up to an element of the center Z_N of the gauge group.[9] These can be realized with h_{Θ}, but now Θ_j belongs to the dual weight lattice $\tilde{\Lambda}_w$, which follows from the requirement that $\exp(2\pi i \Theta_j)$ is an element of the center Z_G of the gauge group. Indeed, $\tilde{\Lambda}_w/\tilde{\Lambda}_r \cong Z_G$. These twisted gauge transformations generate a shift symmetry associated with $\tilde{\Lambda}_w$. Their homotopy type is specified by $\mathbf{\Theta} \in (\tilde{\Lambda}_w/\tilde{\Lambda}_r)^3 \cong Z_G^3$. A particular representative for SU(N) is given by $h_{\mathbf{k}}(\mathbf{x}) = \exp(2\pi i \mathbf{k} \cdot \mathbf{x}\Theta_0/L)$, where $\Theta_0 \in H_G$ is a generator for the center, $\exp(2\pi i \Theta_0) = \exp(2\pi i/N)$. For SU(2) we can choose $\Theta_0 = \tau_3/2$ and for SU(3) there are two independent choices (that span the dual weight lattice $\tilde{\Lambda}_w$) $\Theta_0 = \lambda_8/\sqrt{3}$ and $\Theta_0' = \frac{1}{2}(\lambda_3 - \lambda_8/\sqrt{3})$. The non-trivial homotopy is labelled by $\mathbf{k} \in Z_N^3$ and is thus in general non-trivially represented on the wave functionals. Any gauge transformation can be decomposed[50] in $h = h_{\mathbf{k}}h_1^{\nu}(h)h_0$, where h_1 is a particular strictly periodic gauge transformation with unit winding number, $\nu(h_1) = 1$. What is left is a homotopically trivial gauge function h_0, under which the wave functional is invariant. We therefore have

$$\Psi([h]A) = \Psi([h_{\mathbf{k}}h_1^{\nu}]A) = \exp(2\pi i \mathbf{e} \cdot \mathbf{k}/N)\exp(i\theta\nu)\Psi(A), \qquad (20)$$

where θ is the usual vacuum parameter associated with instantons and $\mathbf{e}$ (defined modulo N) is the gauge invariant definition of electric flux.[9] As for SU(2) the electric flux quantum number can be implemented within the zero-momentum effective hamiltonian by imposing suitable boundary conditions on the boundary of the fundamental domain. For this we study the action of $h_{\mathbf{k}}$ on $\mathbf{C}^b$, corresponding to a shift. For SU(2) $\mathbf{C} \to \mathbf{C} + 2\pi\mathbf{k}$ and for SU(3) $(\mathbf{C}^1, \mathbf{C}^2) \to (\mathbf{C}^1 + 2\pi(2\mathbf{n}-\mathbf{l})/\sqrt{3}, \mathbf{C}^2 + 2\pi\mathbf{l})$, with $\mathbf{k} = \mathbf{n} + \mathbf{l} \pmod 3$ specifying the homotopy type of the gauge transformation.

Fig. 9 gives for SU(3) the equipotential lines of the effective potential V_{eff}, in a plane specified by putting $C_2^b = C_3^b = 0$, the fundamental domain Λ (bounded by the dashed hexagon) and the Gribov region Ω (bounded by the fat hexagon). The triangular lattice structure is due to the invariance under twisted gauge transformations and reflects the dual weight lattice of SU(3). Indeed, the gauge transformations $\exp(2\pi i \Theta_0 x_j/L)$ and $\exp(2\pi i \Theta_0' x_j/L)$ all map points on the boundary of the fundamental domain to the same boundary.

They also map $\mathbf{C}^b = 0$ to the Gribov horizon. The Weyl transformations are generated by the reflections in the three principle axes of the dual weight lattice.

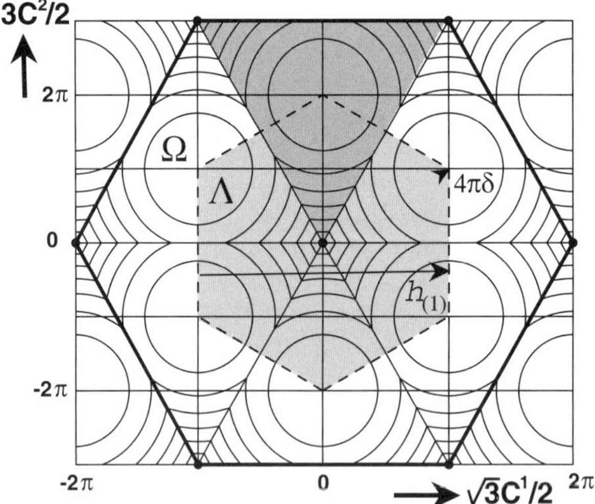

Figure 9: Two dimensional cross-section of V_{eff}, for $C_1^b T_b = C_1^1 \lambda_8/2 + C_1^2 \lambda_3/2$ and $C_{2,3}^b = 0$. The circles indicate the equipotential lines (in increments of $V_{\text{eff}}(4\pi\delta)/7$). The light shaded area represents the fundamental region, with the dashed hexagon as its boundary, the fat hexagon indicates the Gribov horizon, with gauge copies of $A = 0$ indicated by the dots. The darker shaded area is the so-called fundamental Weyl chamber.

The near spherical behavior of the equipotential lines within each triangle is due to a remarkably accurate approximation (cmp. Fig. 10) that holds for any SU(N) when restricting C_i^b to one dimension, say $C_1'^b = C^b$ (normalizing as usual $\operatorname{tr}(T_a T_b) = \frac{1}{2}\delta_{ab}$)

$$V_{\text{eff}}(C) \approx N\left((4\pi\delta)^2 - (C - 4\pi\delta)^2\right)/2\pi L, \tag{21}$$

for C in the Weyl chamber restricted to the fundamental domain, centered around $4\pi\delta$, where δ is the highest weight. For SU(2) $\delta = 1/4$ and for SU(3) $\delta = (\delta^1, \delta^2) = (\sqrt{3}, 1)/6$. It is also instructive from Fig. 9 to identify the orbifold singularities, along the axes of the dual weight lattice (edges of the Weyl chambers), which are fixed under the Weyl reflections. This is where the $U(1)^2$ symmetry in a generic point of the vacuum valley is restored to SU(2), whereas at the corners (all gauge copies of $A = 0$) the full SU(3) symmetry is restored. At these locations also some of the directions in which the classical potential is quadratic turn quartic. It is this that gives rise to the conic singularities in V_{eff}. The effective potential V_1 in Eq. (17), when restricted to the fundamental domain, is free from any of these singularities, since they are caused by fluctuations in the non-abelian constant modes, not integrated out in Eq. (17). The conic singularities of V_{eff}, Eq. (19), in Λ are given by $L^{-1}\operatorname{Tr}_{\text{ad}}\sqrt{(\mathbf{C}^b T_b)^2}$ (where Eq. (21) is valid, equal to $4NL^{-1}\delta_b C^b$).

To implement the symmetries on the wave functional we perform a change of coordinates $c_i^a \rightarrow (C_i^b, \Omega_i)$, where Ω_i stands for the collection of SU(N)-angular coordinates parametrizing $SU(N)/U(1)^{N-1}$ and (C_i^a) are restricted to a fundamental Weyl chamber, for SU(3) the triangular shaded region in Fig. 9, where the relation between (C^1, C^2) and (r, s) is one to one. For SU(2) these new coordinates are the spherical coordinates (r_i, Ω_i), with $\Omega_i = (\theta_i, \phi_i)$ coordinates on $S^2 = SU(2)/U(1)$. The square root of the Jacobian of this transformation, is given by $J = \prod_i J_i$, with $J_i = C_i$ for SU(2) and $J_i = C_i^2((C_i^2)^2 - 3(C_i^1)^2)$ for SU(3). The wave functions can be decomposed as

$$\Psi(c) = \prod_i J_i^{-1}(C_i^b)\chi_i(C_i^b)\mathcal{Y}_i(\Omega_i). \tag{22}$$

The "radial" part χ will be antisymmetric with respect to Weyl reflections, so as to cancel the zero's of the Jacobian. For SU(2) the angular wave functions $\mathcal{Y}(\Omega)$ are nothing but the spherical harmonics, see Eq. (14), whereas in general they are irreducible representations of the gauge group. Helpful in making suitable choices for χ is that the vacuum valley kinetic term after the rescaling with the square root J of the Jacobian becomes $-\frac{1}{2}g^2 \sum_{i,a}(\partial/\partial C_i^a)^2$, compatible with the canonical flat metric on the orbifold. Furthermore, as is familiar from the radial reduction in three dimensions, this rescaling does not create an additional potential term, since J is in all cases harmonic, $\sum_{i,a}\partial^2 J/(\partial C_i^a)^2 = 0$.

We note that suitable combinations of a shift and Weyl reflection leave invariant the lines that constitute the polygon at the boundary of the fundamental domain, see Figs. 3 and 9. This implies that alternatively the properties of χ can be described in terms of boundary conditions at the boundary of the fundamental domain. E.g., for SU(2) this leads to two possible choices of boundary conditions at $r_j = \pi$, for each j, see Eq. (13). One can use $\chi(C) = \sin(nC/2)$, although spherical Bessel functions provide a more effi cient choice, keeping the hamiltonian more sparse.[18] Carefully working out the consequences of the symmetries one can show[48,49] that a complete basis for χ in the case of SU(3) is given by

$$\begin{aligned}
\chi(C^1, C^2) &= \sin\left(mC^2/2\right)\exp\left(inC^1\sqrt{3}/2\right) \\
&+ \sin\left(m(\sqrt{3}C^1 - C^2)/4\right)\exp\left(in\sqrt{3}(\sqrt{3}C^1 + C^2)/2\right) \\
&- \sin\left(m(\sqrt{3}C^1 + C^2)/4\right)\exp\left(in\sqrt{3}(\sqrt{3}C^1 - C^2)/2\right),
\end{aligned} \tag{23}$$

for each of the three coordinate directions. The quantum numbers n and m will be restricted by the electric flux and irreducible representations of the Weyl and cubic groups, which is the part that requires most of the care. Finally if one restricts $\prod_i \mathcal{Y}_i$ to transform as a singlet under $SU(N)$ one obtains a complete and gauge invariant basis for the effective hamiltonian, that through the boundary conditions carries the information of electric flux. Needless to say that for $SU(3)$ the computation of all the relevant matrix elements[48] is rather more cumbersome than for $SU(2)$. However, once the matrix of the hamiltonian for a particular basis is computed one performs a simple Rayleigh-Ritz analysis to determine the spectrum. The region where the wave functional spreads out over the whole vacuum valley, with the energy of electric flux no longer suppressed by the quantum induced barrier of V_{eff}, is well described by this Rayleigh-Ritz analysis, based on Eq. (23). Of course, also for $SU(3)$ the results are valid as long as the classical barrier is sufficiently high as compared to the energy-levels studied, cmp. Fig. 4. Like for $SU(2)$ the approximations break down for L large than 5 to 6 times the correlation length of the scalar glueball. That beyond this volume, $L \sim 0.75$ fermi, instanton effects set in can be seen from the rather sudden onset of the topological susceptibility[51]

3.2 Including Massless Quarks

Quarks fields are not invariant under the center of the gauge groups. This means that on the space of zero-momentum abelian gauge fields, which still form the classical vacuum valley, the twisted gauge transformations no longer represent a symmetry. This complicates matters since without the equivalence under twisted gauge transformations the fundamental domain extends to the Gribov horizon. Some of these complications may be avoided by introducing on the original fundamental domain additional quark fields, obtained by applying a twisted gauge transformation. The boundary identification on the boundary of the fundamental domain are include an operation that permutes these fermion field components. This has not been worked out so far.

Nevertheless, interesting statements can be made about the vacuum structure in small enough volumes, for which the wave functional is sufficiently localized around the vacuum configuration. One simply adds in one loop order the quantum effects of the quark field fluctuations. The resulting effective potential will no longer respect the center symmetry, but it still properly reflects invariance under constant and periodic gauge transformations. The quark fields can satisfy either periodic or anti-periodic spatial boundary conditions. Actually, for $SU(2)$ (with -1 a non-trivial element of Z_2) these are equivalent by a twisted gauge transformation with homotopy type $\mathbf{k} = (1,1,1)$. Under this gauge transformation the gauge field is shifted and shows it is not a priori clear that $A = 0$ will represent the proper classical vacuum to expand around. As we will show, it will be the correct one *only* with anti-periodic boundary conditions for the quark fields, both for $SU(2)$ and $SU(3)$. In that case, due to the anti-periodicity, there will be no zero-energy modes for the quark fields, and chiral symmetry is *unbroken* in the finite volume. For SU(3) no gauge equivalence of periodic to anti-periodic boundary conditions holds, and the vacuum structure with periodic quark fields actually leads to spontaneous breaking of some discrete symmetries. Yet, no zero-energy quark modes appear, and chiral symmetry remains unbroken. It also means that in a small volume, with quark momenta of order π/L and glueball masses of order $g^{2/3}(L)/L$, that glueballs cannot decay in mesons. The quark degrees of freedom can be integrated out.

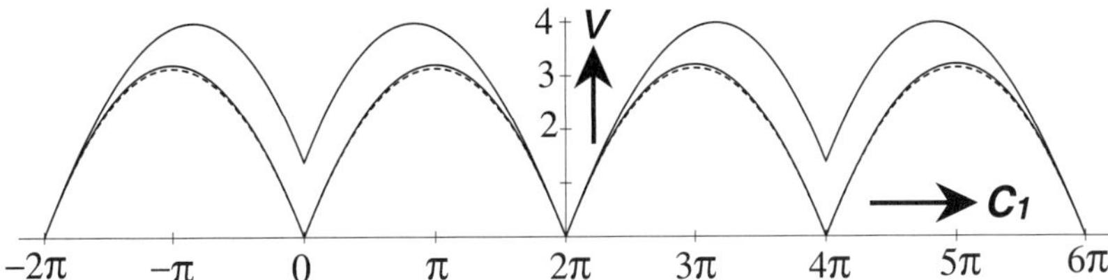

Figure 10: Vacuum-valley effective potential at $C_2 = C_3 = 2\pi$ for SU(2) and two flavors of periodic massless quarks (Eq. (25) with $\mathbf{k} = \mathbf{0}$), normalized to 0 at its minimum, as well as the effective potential for zero flavors (Eq. (8)). The dashed curve represents Eq. (21).

To be more specific let us first generalize the computation of the vacuum-valley effective potential to include the quark field fluctuations. The most efficient way to represent the result is to introduce the weight vectors $\mu^{(i)}$, determined by the eigenvalues of the abelian generators,

$$T_b = \mathrm{diag}(\mu_b^{(1)}, \mu_b^{(2)}, \cdots, \mu_b^{(N)}), \quad T_b \in H_G. \tag{24}$$

For SU(2) one finds $\mu^{(1)} = \frac{1}{2}(1, -1)$, whereas for SU(3) $\mu^{(1)} = (1, 1, -2)/\sqrt{12}$ and $\mu^{(2)} = \frac{1}{2}(1, -1, 0)$ (the conventions used in this paper are $T_1 = \frac{1}{2}\lambda_8$ and $T_2 = \frac{1}{2}\lambda_3$). With n_f flavors of massless quark fields we find[19] (see Fig. 10)

$$V_{\mathrm{eff}}^{\mathbf{k}}(\mathbf{C}^b) = \sum_{i>j} V(\mathbf{C}^b[\mu_b^{(i)} - \mu_b^{(j)}]) - n_f \sum_i V(\mathbf{C}^b \mu_b^{(i)} + \pi\mathbf{k}), \tag{25}$$

with $\mathbf{k} = \mathbf{0}$ or $\mathbf{k} = (1, 1, 1)$, for resp. periodic or anti-periodic boundary conditions on the quark fields. The function $V(\mathbf{C})$ is the SU(2) one-loop effective potential for $n_f = 0$, Eq. (8). The correct quantum vacuum is to be found at the minimum of this effective potential.

Observe that the gauge symmetry should not be spontaneously broken, which implies that the Polyakov loop observables

$$P_j(\mathbf{x}) \equiv \frac{1}{N} \mathrm{tr} \left(P \exp\left(\int_0^L ds\, A_j(\mathbf{x} + s e^{(j)}) \right) \right), \tag{26}$$

should be a constant center element at the vacuum configurations, or

$$P_j = N^{-1}\,\mathrm{tr}\left(\exp(iC_j^b T_b)\right) = N^{-1}\sum_n \exp(i\mu_b^{(n)}C_j^b) = \exp(2\pi i l_j/N). \quad (27)$$

This implies that $\mu_b^{(n)}\mathbf{C}^b = 2\pi\mathbf{l}/N \pmod{2\pi}$, independent of n, and gives $V_{\mathrm{eff}}^{\mathbf{k}} = -n_f N V(2\pi\mathbf{l}/N + \pi\mathbf{k})$. In the case of anti-periodic boundary conditions, $\mathbf{k} = (1,1,1)$, this is minimal only when $\mathbf{l} = \mathbf{0} \pmod{2\pi}$. This means the quantum vacuum in this case is the naive one, $A = 0$ ($P_j = 1$). In the case of periodic boundary conditions, $\mathbf{k} = \mathbf{0}$, the above candidate vacua have $\mathbf{l} \neq \mathbf{0}$, that is P_j correspond to non-trivial center elements. Both for SU(2) and SU(3) this means $l_j = \pm 1$. For SU(2) the vacuum with $P_j = -1$ is unique, as also follows from the gauge equivalence argument given above. The only difference is that one now needs to expand around $\mathbf{C} = 2\pi(1,1,1)$. For SU(3), however, there are now 8 possible choices $P_j = \exp(\pm 2\pi i/3)$, related by the three coordinate reflections. As this is a symmetry of the full hamiltonian, each is indeed equivalent. But it does mean that in a small volume parity (P) and charge conjugation (C) are spontaneously broken, although CP is still a good symmetry.[19] A consequence of the spontaneous breaking of parity is that the mass gap, the lowest excitation above the vacuum, is exponentially small in a small volume. All these intricate effects would make this an ideal testing ground for dynamical fermion algorithms in lattice gauge theory.

The minima of the effective potential are obtained from $A = 0$ by the twisted gauge transformation $h_{\mathbf{l}}(\mathbf{x})$. As there are no zero-energy quark field modes, also the effective hamiltonian can be expressed as in the bosonic case in terms of the zero-momentum gauge fields, after taking this shift into account. In the case of SU(3) with periodic boundary conditions, because of the spontaneous break down of parity and charge conjugation invariance, extra terms appear in Lüscher's effective hamiltonian. To the order in which this hamiltonian was worked out the new interaction $\mathrm{tr}\left(\{A_i, A_j\}A_k\right)$ appears. In addition couplings that were equal because they were related by the discrete symmetries, now spontaneously broken, will split.[19] All these corrections are linear in the number (n_f) of quark flavors. The flavor dependence of the effective hamiltonian for the case of anti-periodic boundary is simpler, since no new couplings appear. For SU(2) one simply replaces α_i and γ_i in Eq. (9) by $\alpha_i - 2n_f\alpha_i'$ and $\gamma_i - 2n_f\gamma_i'$, with

$$\begin{aligned}
\gamma_1' &= -2.1272012 \cdot 10^{-2}, & \alpha_1' &= +3.098211 \cdot 10^{-5}, \\
\gamma_2' &= +4.2255250 \cdot 10^{-4}, & \alpha_2' &= +1.7211922 \cdot 10^{-3}, \\
\gamma_3' &= -7.3994300 \cdot 10^{-4}, & \alpha_3' &= +3.0178786 \cdot 10^{-5}, \\
\gamma_4' &= -2.8659656 \cdot 10^{-6}, & \alpha_4' &= +3.2156523 \cdot 10^{-5}, \\
\gamma_5' &= +1.1578663 \cdot 10^{-5}, & \alpha_5' &= -3.2271736 \cdot 10^{-5}, \\
\gamma_6' &= -7.9447492 \cdot 10^{-5}.
\end{aligned} \quad (28)$$

Independently Kripfganz and Michael calculated for SU(N) to $\mathcal{O}\left(g^{8/3}\right)$ the change in the coefficients of Lüscher's effective hamiltonian, due to quarks with anti-periodic boundary conditions only.[52] They confirmed the values in Eq. (28) and also introduced for SU(2) a lagrangian formulation of the effective hamiltonian in terms of compact group variables, that incorporates in a simple way the proper boundary conditions in field space.[36] After full equivalence was established,[53,54] the Monte Carlo analysis of this effective lagrangian model continued to suffer from a technical difficulty in efficiently implementing the kinetic term,[41] only fully understood by Vohwinkel a number of years later.[55] This has hampered using the lagrangian formulation as a reliable alternative[56] for the hamiltonian Rayleigh-Ritz analysis.

There is another choice of boundary conditions that strongly reduces the center symmetry. This is called C periodic boundary conditions, where the field is periodic up to a charge conjugation. The boundary conditions can be used to avoid the system to be charge neutral, as is the case for periodic boundary conditions, both for magnetic and electric charge. This was the original motivation to introduce these boundary conditions for the abelian theory,[57] which were subsequently studied for non-abelian gauge theories.[58] For SU(2), which is pseudo real, one retrieves the periodic case, but for SU(3) the center symmetry is completely broken and a number of the features we saw when including quark fields, appear here as well. The vacuum valley for SU(3) is reduced from six to three dimensional in terms of the real gauge field $\mathbf{A} = \frac{1}{2}i\mathbf{C}\lambda_2/L$, with[59]

$$V_{\text{eff}}(\mathbf{C}) = V(\tfrac{1}{2}\mathbf{C}) + V(\mathbf{C} - \pi\mathbf{l}) + V(\tfrac{1}{2}\mathbf{C} - \pi\mathbf{l}), \tag{29}$$

where $V(\mathbf{C})$ is the SU(2) effective potential, see Eq. (8), and $\mathbf{l} = (1,1,1)$. The minimum of this effective potential occurs at the four points $\mathbf{C} = (\pi, \pi, \pi)$, $(-\pi, -\pi, \pi)$, $(\pi, -\pi, -\pi)$, and $(-\pi, \pi, -\pi)$, which correspond to orbifold singularities with quartic modes (associated with the three real generators $i\lambda_A/2$, for $A = 2, 5$ and 7, forming an SU(2) subalgebra). The effective hamiltonian is again of the Lüscher type, at $\mathcal{O}\left(g^{2/3}\right)$ identical to it, at higher order additional couplings appear because of the spontaneous breaking of parity, i.e. the cubic group is broken down to the permutation group. No attempts have been made to study the fundamental domain for this theory, and we may expect similar difficulties as in the presence of quark fields.

3.3 The Renormalized Coupling

We have assumed there is a renormalized coupling in terms of which perturbation theory in the field modes that are integrated out is well-behaved. By expressing quantities in dimensionless combinations, lattice Monte Carlo results and continuum (or lattice) hamiltonian results can be compared without being sensitive to any problems in expressing the renormalized coupling in terms of the bare coupling. Determining the renormalized strong coupling constant non-perturbatively in a reliable way is, however, an important problem. The integration constant of the beta-function, the so-called Lambda parameter, ideally should be fixed in terms of the infrared quantities of the theory, like the mass gap and string tension, or other observables in the low-lying spectrum of the theory. The running of the coupling allows one to compute unambiguously the strong coupling constant at, say the Z-boson mass. The most accurate such method is based on a finite volume study, proposed by Lüscher,[60] long before it was feasible to be implemented.[61] It makes use of a discrete version of the beta-function, the so-called step scaling function. The scale at which the renormalized coupling is defined is fixed by the volume. The volume is subsequently changed by an integer factor (usually, but not necessarily) of 2. So instead of an infinitesimal scale transformation it considers a finite one. The change in the coupling can of course be obtained by intergrating the beta-function, but this function is not available non-perturbatively. Instead, one picks a suitable definition of the renormalized coupling constant at the scale L, set by a given physical volume, than doubles L and calculates the value of the coupling at this new scale $2L$. This can all be performed on the lattice (hence the integer scaling factor) using Monte Carlo calculations, at each step carefully extrapolating to the continuum limit. Also (euclidean) time is taken finite, $T = L$. Small volumes go together with high temperatures in such geometries. The naive strategy of taking L large, and defining an observable set by a variable scale *within the same* finite volume, fails because the lattice spacing gives a limit to the shortest distance one can probe in a given volume, due to computer limitations.

Many definitions of the renormalized coupling could in principle be used, but technical requirements have led to a particular one that is related to the effective action in the background of a constant chromo-electric field, based on the so-called Schrödinger functional[62] (SF). In the spatial directions the boundary conditions are periodic, but in the (finite) time direction one prescribes an initial and a final configuration of gauge fields taking values in the vacuum valley.[62,8]

$$\text{SU(2)}: \quad C_j(t=0) = 2\pi\delta + 2\eta, \quad C_j(t=T) = 2\pi - C_j(t=0),$$

$$\text{SU(3)}: \quad (C_j^1, C_j^2)(t=0) = 2\pi\delta + (-\sqrt{3}\eta, 0), \tag{30}$$

$$(C_j^1, C_j^2)(t=T) = (4\pi/\sqrt{3}, 0) - (C_j^1, C_j^2)(t=0), \quad \forall j,$$

with δ the highest weight, defined below Eq. (21). The classical equations of motion lead to a linear interpolation in time, giving rise to a constant chromo-electric background. The euclidean quantum effective action, $\Gamma(\eta)$, describes the reaction of the system to this background. The renormalized coupling is defined by $g^2(L) = \frac{d\Gamma(\eta)}{d\eta} / \frac{d\Gamma_0(\eta)}{d\eta}$ evaluated at $\eta = 0$, where $\Gamma_0(\eta)$ is the bare effective action. This background has been chosen to stay well away from the orbifold singularities along the vacuum valley for the entire classical path that interpolates between the initial and final configuration, to simplify the perturbative analysis (in part for estimating the lattice artifacts). This particular choice of coupling fits *extremely well* to the perturbative running of the coupling constant up to the largest volume probed,[8] with L up to $\sim$0.35 fermi. At large volumes the non-perturbative running is bound to deviate.

An alternative definition for the non-perturbatively defined running coupling for SU(2) has been based on ratios of the expectation values of suitable Polyakov loop operators, using twisted boundary conditions,[9] the so-called twisted Polyakov (TP) scheme.[63] We will see next that twisted boundary conditions remove the zero-momentum modes, making perturbation theory well behaved. Without these twisted boundary conditions, computing expectation values in the four dimensional euclidean finite volume is difficult to control.[64] The TP coupling also agrees well with perturbation theory and after matching of the scales, with the SF coupling.[65] Only the largest volume result ($L = 0.28$ fermi) probed by the TP coupling lies slightly, but significantly, below the perturbative result. The near perturbative behavior seems to support the fact that non-zero momentum modes do behave perturbatively in intermediate volumes.

3.4 Twisted Boundary Conditions

With twisted boundary conditions, in the absence of zero-momentum modes, the classical vacuum at $A = 0$ is isolated and the small volume behavior for the glueball masses is described by a perturbative series in $g^2(L)$, as opposed to $g^{2/3}(L)$ in absence of twist. The volume dependence will be quite different, this is in particular true for electric flux energies.[66] Nevertheless, in large volumes the results should not depend on boundary conditions. Therefore, comparing the different boundary conditions gives valuable information about the transition to large volumes. In the hamiltonian formulation the twisted boundary conditions are most easily implemented in a gauge where the so-called twist matrices $\Omega_j \in \mathrm{SU}(N)$ are constant

$$\mathbf{A}(\mathbf{x} + L\mathbf{e}^{(j)}) = \Omega_j^\dagger \mathbf{A}(\mathbf{x})\Omega_j. \tag{31}$$

They satisfy 't Hooft's consistency condition,[9] which also gives the relation to the magnetic flux $\mathbf{m} \in \mathbb{Z}_N^3$,

$$\Omega_k^\dagger \Omega_\ell^\dagger \Omega_k \Omega_\ell = \exp(2\pi i \epsilon_{k\ell j} m_j / N). \tag{32}$$

These generate a so-called Heisenberg group (the group commutator is central, i.e. commutes with all group elements). The finite group theory allows one to construct in an elegant way the most general set $\Omega_k \in SU(N)$ for any given $\mathbf{m}$, and its generalizations to higher dimensions.[67,68]

It seems that twisted boundary conditions spontaneously break the gauge invariance, due to the explicit choice of Ω_k. However, this is of course similar to the case of periodic boundary conditions, which also represents a gauge choice in formulating the boundary conditions. Once one has specified the gauge choice for the boundary conditions, Gauss's law tells us that local gauge transformations have to satisfy

$$h(\mathbf{x} + L\mathbf{n}) = \prod_j \mathrm{Ad}^{n_j}(\Omega_j)h(\mathbf{x}), \quad \mathrm{Ad}(\Omega)(h) \equiv \Omega^\dagger h \Omega, \tag{33}$$

for $\mathbf{n} \in \mathbb{Z}^3$, indicating the shifts over multiple periods. Note that in the adjoint representation the twist matrices commute. Thus, in a sense L is (typically) a $1/N$ period (intimately related to the notion of color momentum, underlying the principle of the Twisted-Eguchi-Kawai one-point lattice model.[69]) However, for finite size effects in large volumes, where the degrees of freedom that propagate "around" the boundary are colorless, this has no consequence (see Sec. 6.1).

The candidate classical vacuum configuration satisfying the twisted boundary conditions is $A = 0$, of zero classical energy despite the presence of magnetic flux.[70,71] As we argued in Sec. 3.2, the Polyakov loop (before taking the trace), evaluated at the classical vacuum configuration should be invariant under gauge transformations. In the case of periodic boundary conditions, this implies it should be in the center of the gauge group, uniquely specified by $P_j \in Z_N$. To address the same question with twisted boundary conditions, the proper definition of the Polyakov loop has to be used[72]

$$P_j(\mathbf{x}) \equiv \frac{1}{N} \, \mathrm{tr} \left[\left(P \exp\left(\int_0^L ds \, A_j(\mathbf{x} + s\mathbf{e}^{(j)}) \right) \right) \Omega_j^\dagger \right], \tag{34}$$

(path ordering from left to right) which indeed is invariant under the gauge transformations, Eq. (33). For $A = 0$ this is even so before taking the trace, such that gauge invariance is indeed not spontaneously broken. Thus, $P_j(A = 0) = \mathrm{tr}\,(\Omega_j^\dagger)/N$ and using Eq. (32) one finds that $P_j = 0$, whenever $m_j \neq 0$ (mod N). This is closely related to invariance under constant gauge transformations that are compatible with the allowed twisted gauge transformations,

$$h_\mathbf{k}(\mathbf{x} + L\mathbf{n}) = \exp(2\pi i \mathbf{k} \cdot \mathbf{n}/N) \prod_j \mathrm{Ad}^{n_j}(\Omega_j) h_\mathbf{k}(\mathbf{x}). \tag{35}$$

As for the periodic case, $\mathbf{k} \in Z_N^3$ specifies the homotopy type of the gauge transformation. These gauge transformations multiply P_j with $\exp(2\pi i k_j/N)$.

We note that $A = 0$ is left unaffected by constant gauge transformations, $h_{\mathbf{k}}(\mathbf{x}) \equiv \Omega_0(\mathbf{k})$, which from Eq. (35) have to satisfy

$$\Omega_0^\dagger(\mathbf{k})\Omega_\ell^\dagger\Omega_0(\mathbf{k})\Omega_\ell = \exp(2\pi i k_\ell/N), \tag{36}$$

extending Eq. (32) to four dimensions. This equation is solved for example by Ω_j, with $k_\ell = m_i\epsilon_{ij\ell}$. In general solutions exist if and only if[50,67] $\mathbf{k} \cdot \mathbf{m} = 0$ (mod N). When $\mathbf{k} \cdot \mathbf{m} \neq 0$ (mod N), the gauge transformation $h_{\mathbf{k}}(\mathbf{x})$ does *not* leave $A = 0$ invariant, but maps to a vacuum state with fractional Chern-Simons number, see Eq. (49) (equal to $\nu(h_{\mathbf{k}})$ as defined in Eq. (15)). It is separated from $A = 0$ by a classical potential barrier related to the instanton with *fractional* topological charge for twisted gauge fields on the torus,[73,50] to be discussed in more detail in Sec. 4. In Fig. 11 we illustrate these features. Since electric flux quantum numbers are associated with the representations

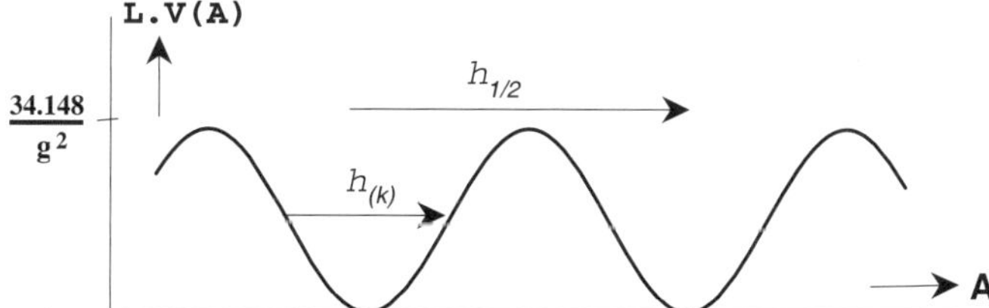

Figure 11: Sketch of the SU(2) potential energy with twisted boundary conditions. Shown are two isolated vacua, separated by a barrier with height $34.148/Lg^2$. These vacua are related by twisted gauge transformations $h_{1/2} = h_{\mathbf{p}}$ with $\mathbf{p} \cdot \mathbf{m} = 1$ (mod 2), having winding number $\nu(h_{1/2}) = \frac{1}{2}$. Gauge transformations $h_{(k)} = h_{\mathbf{k}}$ with $\mathbf{k} \cdot \mathbf{m} = 0$ (mod 2), leave the vacuum invariant.

of the Z_N^3 homotopy, this means[66] that some of the electric flux states will have energies that do not vanish in perturbation theory, whereas the electric flux energies associated with tunneling through the classical barrier will be suppressed by $\exp(-8\pi^2/Ng^2)$. Both differ from the behavior we observed when $\mathbf{m} = \mathbf{0}$. For example, for SU(2) and $\mathbf{m} = (0,0,1)$, with $\Omega_1 = i\tau_1$, $\Omega_2 = i\tau_2$ and $\Omega_3 = 1$, one finds $\Omega_0(1,0,0) = \pm i\tau_2$, $\Omega_0(0,1,0) = \pm i\tau_1$ and $\Omega_0(1,1,0) = \pm i\tau_3$ as the only non-trivial constant gauge transformations that leave $A = 0$ unchanged. Therefore, energies of electric flux with e_1 and/or e_2 non-trivial are perturbatively lifted, whereas states that differ only by the e_3 quantum number are degenerate.

For SU(2), the boundary conditions are solved by a Fourier expansion[66,74]

$$\mathbf{A}^a(\mathbf{x}) = \sum_{\mathbf{k}\in\mathbb{Z}^3} \mathbf{A}^a(\mathbf{k} + \mathbf{r}^{(a)})\exp(2\pi i(\mathbf{k} + \mathbf{r}^{(a)}) \cdot \mathbf{x}/L), \tag{37}$$

with the *color dependent* "fractional" momentum

$$\mathbf{r}^{(a)} = (\mathbf{e}^{(a)} \times \mathbf{m})/2. \tag{38}$$

Momentum conservation ensures that gauge invariance is not broken by these color dependent momenta. Using the Coulomb gauge $\partial_i A_i = 0$, one checks the classical vacuum $\mathbf{A} = 0$ is isolated with all fluctuations quadratic. To illustrate the perturbative analysis we consider the simpler case with $\mathbf{m} = (1,1,1)$, realized by $\Omega_k = i\tau_k$, because this choice of $\mathbf{m}$ does not break the invariance under the cubic group. Glueball states can thus be classified as for the periodic case. Generalizations to $SU(3)$ (or arbitrary $SU(N)$ and magnetic flux) are easy to obtain. The allowed non-trivial constant gauge transformations are now as follows: $\Omega_0(0,1,1) = \pm i\tau_1$, $\Omega_0(1,0,1) = \pm i\tau_2$ and $\Omega_0(1,1,0) = \pm i\tau_3$. The one-particle state associated with the Fourier mode $\mathbf{A}_\pm^a(\mathbf{p})$, with creation operator $b^\dagger(a,\mathbf{p},\pm)$, has non-zero electric flux. They are further characterized by the momentum $2\pi\mathbf{p}/L$, energy $2\pi|\mathbf{p}|/L$ and the polarization $\pm$ of the gauge field, satisfying $\mathbf{A}_\pm^a(\mathbf{p}) \cdot \mathbf{p} = 0$. To be precise, the electric flux vector belonging to this state is $\mathbf{e} = \mathbf{e}^{(a)}$. This motivates interpreting[74] $2\pi\mathbf{r}^{(a)}/L$ as a Poynting vector. It also plays an interesting role in how the wave functional behaves under translations over L in the three coordinate directions.[72] For any $|\Psi\rangle_{\mathbf{m},\mathbf{e}}$ with magnetic flux $\mathbf{m}$ and electric flux $\mathbf{e}$

$$|\Psi(\mathbf{x} + L\mathbf{n})\rangle_{\mathbf{m},\mathbf{e}} = \exp(i\mathbf{n} \cdot \mathbf{P}L)|\Psi(\mathbf{x})\rangle_{\mathbf{m},\mathbf{e}}, \quad \mathbf{P} = \pi(\mathbf{e} \times \mathbf{m})/L. \tag{39}$$

The electric flux $\mathbf{e} = \mathbf{e}^{(a)}$ is created with the gauge invariant Polyakov loop operator P_a, see Eq. (34). This contains the one-particle state $b^\dagger(a,\mathbf{p},\pm)|0\rangle$, the energy of one unit of electric flux is therefore given by the length of the Pointing vector, the minimal value $2\pi|\mathbf{p}|/L$ can take,

$$E_e = \pi|\mathbf{e}^{(a)} \times \mathbf{m}|/L = \pi\sqrt{2}/L. \tag{40}$$

The energy of two units of electric flux (e.g. $\mathbf{e} = (0,1,1)$) is perturbatively degenerate with this. At higher order one has to take into account that the two one-gluon transverse polarizations are now no longer degenerate. Properly creating electric flux with the gauge invariant Polyakov loop operator picks out the polarization in the direction of the loop; for the symmetric torus along $\mathbf{e}$. This causes a perturbative splitting between the energy of one and two units of electric flux.[75] The energy of three units of electric flux, $\mathbf{e} = (1,1,1)$, is entirely due to instanton effects,[66,76] see Fig. 11. Therefore, in lowest order $R_2 = 1 + \mathcal{O}\left(g^2\right)$ and $R_3 = 0 + \mathcal{O}\left(\exp(-4\pi^2/g^2)\right)$, quite distinct from what one finds with periodic boundary conditions in small and intermediate volumes.

To find the mass gap in the zero electric flux sector, one needs two-particle states, built from states with opposite (which for $SU(2)$ is equivalent with identical) electric flux. They are of the form

$$b^\dagger(a,\mathbf{p},\alpha)\,b^\dagger(a,\mathbf{q},\beta)|0\rangle, \tag{41}$$

Table 1: One loop coefficients[74,75] for SU(2) with $\mathbf{m} = (1, 1, 1)$.

irrep r	γ_r	irrep r	γ_r
A_1^+	-92.08	T_1^+	14.74
A_1^-	-91.93	A_1^+	25.56
T_2^+	-41.26	E^-	36.07
E^+	-22.90	E^+	36.38
T_2^+	-7.39		
T_2^-	-6.59	e	-5.43
T_2^+	7.53	e_2	-1.16

with $\alpha, \beta = \pm$ the polarizations of the one-particle states. These states have total momentum $\mathbf{Q}$ and energy E satisfying

$$\mathbf{Q} = 2\pi(\mathbf{p} + \mathbf{q})/L \,, \quad E = 2\pi(|\mathbf{p}| + |\mathbf{q}|)/L. \tag{42}$$

The minimal zero-momentum state gives the mass gap, in lowest order

$$m_{gap} = 2\pi|\mathbf{e}^{(a)} \times \mathbf{m}|/L = 2\pi\sqrt{2}/L, \tag{43}$$

which is twice the length of the Poynting vector. Counting the number of ways one can form these two-particle colorless zero-momentum states from the one-particle states, one finds a 24 fold degeneracy. This degeneracy will be lifted in one loop order, arranged in irreducible representations r of the cubic group,[74] for which lattice discretization effects were reported as well,[75] but no details have been published. In the continuum, the $\mathcal{O}\left(g^2\right)$ mass and energy shifts are parametrized by the constants γ_r,

$$z_r = m_r L = 2\sqrt{2}\pi + \frac{g^2(L)\gamma_r}{16\pi^2}, \quad z_{e_n} = E_{e_n} L = \sqrt{2}\pi + \frac{g^2(L)\gamma_{e_n}}{16\pi^2} \ (n \neq 3), \tag{44}$$

which are listed in Table 1.

In comparison to the case of periodic boundary conditions we note that the E^+ tensor state is now heavier than the scalar A_1^+, but with the T_2^+ in between. Also, z_0 is here a decreasing function of the volume, which has to turn around at some point, when the mass stabilizes and z_0 becomes linear in L. A clear finite volume artifact is also the near degeneracy of the oddball (A_1^-) with the glueball (A_1^+). Both of these features we will also find for the sphere, see Sec. 5. There it will be demonstrated that taking the non-perturbative effects of instantons into account will lead to an appreciable splitting between the oddball and glueball. Also here, like for the case of periodic boundary conditions, we can estimate at which volume instantons become important, by equating the energy of the scalar glueball state with the height of the barrier between two vacua, set by the sphaleron energy,[33] $34.148/Lg^2(L)$. Again one finds the critical value of L to be of the order of 6 times the scalar glueball correlation length. First lattice Monte Carlo results with twisted boundary conditions were obtained by Stephenson and Teper.[77,78] They find in very small volumes[78] ($\beta = 4/g_0^2 = 4.7$, on a $4^3 \times 96$ lattice) that the $A_1^\pm$, $E^\pm$, $T_2^\pm$ and T_1^+ glueball masses indeed all become degenerate and equal to $2E_e$, with $R_2 = 1$ and $R_3 = 0$. The $A_2^\pm$ and T_1^- states are appropriately heavier. Because the shifts in the masses are so small, a detailed comparison with the predictions for the $\mathcal{O}\left(g^2\right)$ shifts is inconclusive. At larger volumes, $z_0 > 6$, the Monte Carlo results show that the differences between twisted and periodic boundary conditions disappear.[77,78]

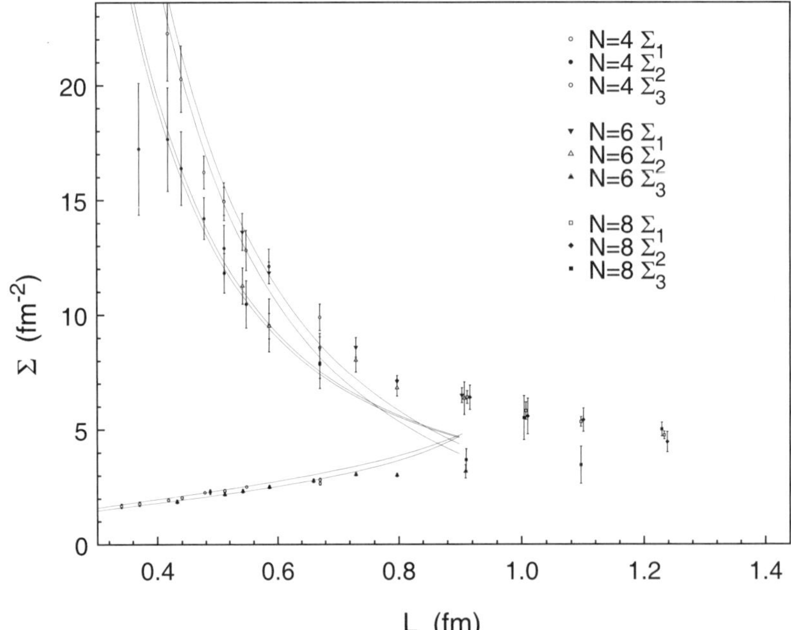

Figure 12: Numerical data[80] for the SU(2) electric flux energies, $E_{e_n} = L\Sigma_n\sqrt{n}$, with twisted boundary conditions, $\mathbf{m} = (1,1,1)$, on lattices of size $N^3 \times N_t$, with $N = 4, 6, 8$ and resp. $N_t = 128, 128, 64$. The scale is set by $L = 400N \exp(-3.38\beta)$ fm and the curves show the perturbative predictions combined with a fit to the expected tunneling contributions.

Also an extensive study was made of the electric flux energies.[79,80] Here lattices of size $8^3 \times 64$ and $N^3 \times 128$ with $N = 4, 6$ were used, with $\beta = 4/g_0^2$ ranging from 2.25 to 2.6, reaching volumes between 0.3 and 1.3 fermi. In Fig. 12 $\Sigma_n = E_{e_n}/(L\sqrt{n})$ is plotted as a function of L. This definition is such that if, as predicted by 't Hooft,[9] $R_n \to \sqrt{n}$ in the infinite volume limit,

Σ_n gives the infinite volume string tension K independent of n. The curves test the small volume expansion. They contain the perturbative contribution, corrected for the lattice artifacts[75] (the data is not accurate enough to test the $\mathcal{O}\left(g_0^2\right)$ corrections), together with const.$\times g_0^{-4} \exp(-S/g_0^2)$, the expected shift due to tunneling (cmp. Fig. 11 and Sec. 6.4) mediated by the charge $\frac{1}{2}$ instanton with classical action S ($4\pi^2$ in the continuum, see Sec. 4). The constant is fitted separately for $N = 4$ and 6 ($N = 8$ fits not shown). One finds fair agreement with the predicted behavior, with confirmation that the transition to the large volume starts around 0.75 fermi.

3.5 Supersymmetry and the Witten Index

The case of supersymmetric Yang-Mills theories in a finite volume has been considered in the context of the Witten index[10] in some detail. The torus geometry is crucial to preserve the supersymmetry. The Witten index involves counting the number of quantum states (fermionic states with a negative sign). At non-zero energy, because of the supersymmetry, this number cancels between the bosonic and fermionic states, such that the counting can be reduced to the vacuum sector. Here the number can be non-zero, because the supersymmetry generator can annihilate vacuum states. A zero Witten index is a sign of spontaneous breakdown of supersymmetry, where the vacuum energy is non-zero, explaining the physical significance of this index. A zero vacuum energy is a direct consequence of unbroken supersymmetry, where the hamiltonian is given by $H = \frac{1}{2}\{Q, Q^\dagger\}$ with Q, $Q^\dagger$ the supersymmetry generators, that annihilate the vacuum state.

In perturbation theory bosonic loops are typically cancelled by fermionic loops, e.g. in the vacuum energy. Applied to the problem of non-abelian gauge theories in a finite volume this implies that the vacuum-valley effective potential vanishes. The cancellation is caused by the contribution from the gluino fluctuations, which are the superpartners of the gluons. They are Weyl fermions in the adjoint representation of the gauge group, denoted by λ_α^a, with α a two-component spinor index. This means that the wave function is no longer localized to $A = 0$ or any of the other orbifold singularities of the vacuum valley (the moduli space of flat connections). It has been shown in the context of the supermembrane, when taking the supersymmetric Yang-Mills hamiltonian restricted to the zero-momentum modes only, that the spectrum is continuous, down to zero-energy. One can construct trial wave functions with a support arbitrarily far from $A = 0$ that nevertheless have finite energy.[93] The compactness of the vacuum valley is crucial to obtain a discrete spectrum.

The counting of quantum vacuum states was based on the assumption that for all gauge groups the moduli space of flat connections is given in terms of the Cartan subalgebra, as we discussed for SU(N). The gluonic part of the groundstate wave function $|0\rangle$ is assumed to be constant over the vacuum valley. In the reduction to the vacuum valley there are r gluinos (associated with the generators in the Cartan subalgebra), each with two helicities. They are constant and carry no energy, which is the source of the vacuum degeneracy. These gluinos have to be combined in Weyl invariant combinations, respecting Fermi-Dirac statistics. There are r independent invariants, made from

$$U = \delta_{ab}\epsilon^{\alpha\beta}\lambda_\alpha^a\lambda_\beta^b \tag{45}$$

and its powers. So one has $U^n|0\rangle$, $n = 0, 1, \cdots, r$, as the $r+1$ bosonic vacuum states, with no invariant fermionic vacuum states. Thus, one finds an index equal to the rank of gauge group plus one, $r+1$.

Because of possible problems with the adiabatic approximation near the orbifold singularities, as we encountered in the previous sections, Witten considered the alternative of twisted boundary conditions.[10] For $SU(N)$ the same result, $r+1 = N$, for the index follows. Other groups in general do not admit the type of twisted boundary conditions that completely remove the continuous vacuum degeneracy with its orbifold singularities. So it was natural to attempt to find the exact zero-energy ground state for the supersymmetric generalization of Lüscher's zero-momentum effective hamiltonian,[94] or even ignoring the time dependence, studying the path integral in the ultra local limit.[95] None of these studies took the compactness of the vacuum valley into account and therefore fail to address the proper situation.[93]

The problem of the adiabatic approximation remained an urgent one because of a discrepancy between the finite volume calculation of the Witten index, and the one based on the infinite volume determination of the gluino condensate through instanton contributions.[96,97,99] One relies on the fact that an index can not change under smooth perturbations, like increasing the volume. Also the infinite volume calculation has its problems. It uses the semiclassical approximation for a strongly interacting theory. This could, however, be circumvented by first adding matter fields to introduce an external mass scale to control the instanton calculation and then rely on the index being constant under a smooth deformation (through holomorphy), that decouples the extra matter sector.[97] This resulted nevertheless in a discrepancy, $\sqrt{5/4}$ for $SU(2)$, between the so-called strong and weak coupling calculations. Both calculations rely on the cluster decomposition property, since the instantons have more than two gluino zero modes, which seems to make the condensate $\langle\lambda\lambda\rangle$ vanish. The instanton calculation instead computes the appropriate power of the gluino condensate, $\langle(\lambda\lambda)^h\rangle$, that saturates the $2h$ gluino zero modes, where h is the so-called dual Coxeter number of the gauge group, $h = N$ for $SU(N)$. It is this power that gives the number of vacuum states,

$$\langle\lambda\lambda\rangle \equiv e^{2\pi in/h}\left(|\langle(\lambda\lambda)^h\rangle|\right)^{1/h}, \quad n = 1, 2, \cdots, h. \tag{46}$$

These arguments seem reasonable, but are not rigorous.[99] See for further details a review by Shifman.[98] Recently, use has been made of the constituent nature of periodic instantons (or calorons),[100,101] in the context of a Kaluza-Klein reduction with periodic gluinos, as opposed to a high temperature reduction with anti-periodic gluinos, which would break the supersymmetry. The constituent monopoles have exactly two zero-modes and saturate the condensate, $\langle\lambda\lambda\rangle$. The strong coupling calculation now agrees with the weak coupling result.[102] The period can of course be used to control the coupling constant, but it is assumed the index does not change, going from a small to a large period.

The mismatch in the Witten index between small and infinite volumes occurs for $SO(N > 6)$ and the exceptional groups. There has, however, been a recent revision in counting the number of vacuum states in a finite volume. In a study of D-brane orientifolds in string theory, Witten[103] constructed for $SO(7)$ an extra disconnected component on the moduli space of flat gauge connections, which can be embedded easily in $SO(N > 7)$. For $SO(7)$ and $SO(8)$ this gives an isolated component of the moduli space, contributing only one extra vacuum state. For $SO(N > 8)$ the extra component in the moduli space behaves like the trivial component for $SO(N\text{-}7)$. Adding $r+1$ coming from the $SO(N)$ and $SO(N-7)$ moduli space components gives the dual Coxeter number of $SO(N)$, thereby giving the same number of vacuum states as obtained in the infinite volume.

Witten's construction based on orientifolds does not work for the exceptional groups. This naturally led to a derivation of the extra vacuum states in a field theoretic context,[104] trivially extended to the exceptional group G_2, as a subgroup of $SO(7)$. Three different groups have independently managed to solve the problem for other exceptional groups with periodic boundary conditions[105,106] and for any group with twisted boundary conditions.[107] As we remarked before, twisted boundary conditions usually do not remove all the vacuum degeneracies, but it is important that the number of vacuum states is independent of the twist for all gauge groups that have a non-trivial center. The origin of the extra moduli space components is actually not too hard to understand.[105] Large gauge groups can have subgroups that are products of unitary groups, which each would allow for twisted boundary conditions. By choosing twists from all subgroups to cancel one obtains periodic flat connections that can not be deformed to the Cartan subalgebra, which supports the trivial component of flat connections. Of course one need not cancel these twists completely.[107] Needless to say, the group theory involved to sort out all the constraints and count the number of vacuum components is rather involved.

Supersymmetry does not play a role in establishing the existence of these extra vacuum components. Supersymmetry is, however, crucial for these extra components to lead to extra quantum vacua. As soon as the perturbative quantum fluctuations in the vacuum energy do not cancel, this will in general be different for different vacuum components. In a small volume, i.e. at weak coupling, the wave functional will localize around the one with the lowest vacuum energy. Within such a connected component it will localize around the minimum of the effective potential, as we discussed in the previous sections. It is likely, since the trivial vacuum component is the widest, that this is the one where the wave functional localizes. But interestingly, one now has potential energy barriers between vacuum components that are not related by a homotopically non-trivial gauge transformation (since the different vacuum components are not isomorphic). Still these vacua can be characterized by fractional Chern-Simons numbers and tunneling between them would be described by new types of instanton solutions with fractional topological charge.[107,105] It considerably adds to the richness of non-abelian gauge theories.

Although these new results for counting the number of vacuum states in a finite volume remove the urgency of addressing the problem with the adiabatic approximation, it does remain a sore point in the finite volume analysis, as also stressed recently by Witten.[108] One immediate problem we encounter is that Eq. (22) seems to imply that the vacuum-valley wave function has to vanish at the orbifold singularities, which seems inconsistent with it being constant sufficiently far from the orbifold singularities. However, one should take into account the behavior under Weyl reflections of the occupied negative energy gluino states in the Dirac sea near the orbifold singularities in the zero-momentum hamiltonian. Attempting to incorporate this in the formalism developed in the previous sections, we run into the problem that spin $\frac{1}{2}$ fields do not decompose in three components, which can be associated with each of the coordinate directions. In the bosonic sector we could conveniently ignore the transversality, with gauge invariance restored at the end by restricting to the invariant wave functions. Thereby each of the coordinate directions separately allowed a polar decomposition, not compatible with the nature of the gluino fields. For SU(2) there is, however, an elegant polar decomposition using the spherical and gauge symmetry of the zero-momentum hamiltonian,[109] which even holds for H_{eff} in Eq. (9) to $\mathcal{O}\left(g^{4/3}\right)$. In the supersymmetric case one would require the wave function to become constant after reduction to the vacuum valley, sufficiently far away from $A = 0$, to match to the expected behavior away from each of the orbifold singularities. Such a matching can in principle be controlled sufficiently rigorously,[28] as was done in the semiclassical calculation of the energy of electric flux,[29,30] but here we do not have the benefit of a well localized zero-momentum wave function at the orbifold singularities,[93] and judging the incomplete results in the literature,[94] this seems not the way to go. It is in the light of the robustness of supersymmetry somewhat surprising (and frustrating) this problem remains so technically demanding.

4 Instantons and Sphalerons on the Torus

Instantons are associated with the tunneling through barriers that separate different vacuum components. We are interested in these barriers to establish the directions in field space in which the spreading of the wave functional has to be taken into account, when energies of the low-lying states no longer are well below the barriers. The barrier is a saddle point of the energy functional with one unstable mode, a so-called sphaleron.[81] They exist here by virtue of the finite volume. Remarkably, on any compact four manifold instantons exist,[82,83] still with an arbitrary scale parameter, only limited by the size of the manifold. The finite volume sphaleron is typically associated with the instanton that achieves this maximal scale.

For low topological charge, sometimes smooth instantons can *not* be proven to exist.[82] An example of non-existence actually occurs[84] on T^4. In general one takes a small localized instanton from $\mathbb{R}^4$ that is matched smoothly to the flat connection. When the moduli space of flat connections has a continuous degeneracy, as is the case for the torus, there can be obstructions against this procedure. One can get arbitrarily close to satisfying the self-duality equations, $F_{\mu\nu} = \pm\tilde{F}_{\mu\nu} \equiv \pm\frac{1}{2}\epsilon_{\mu\nu\alpha\beta}F^{\alpha\beta}$, that saturate the bound for the *euclidean* action

$$S = -\tfrac{1}{2}\int \mathrm{tr}\,(F_{\mu\nu}^2) = -\tfrac{1}{4}\int \mathrm{tr}\,(F_{\mu\nu}\pm\tilde{F}_{\mu\nu})^2 \pm \tfrac{1}{2}\int \mathrm{tr}\,(F_{\mu\nu}\tilde{F}^{\mu\nu}) \leq 8\pi^2|\nu|, \quad (47)$$

with ν the topological charge. However, an exact self-dual solution is only achieved in the limit where the scale parameter is forced to zero or the volume is taken to infinity. For twisted boundary conditions[9], existence of charge one instantons[85] can be understood from the fact that twist removes the continuous degeneracy in the moduli space of flat connections.

From the hamiltonian point of view we are interested in instantons on $T^3 \times \mathbb{R}$, where time is not constrained to be finite. This can be seen as a limit of T^4 where one of the periods, called T, is taken to infinity. In the $A_0 = 0$ gauge the field is periodic up to a gauge transformation h in the time direction, $A(\mathbf{x}, T) = [h]A(\mathbf{x}, 0)$, and

$$-\frac{1}{16\pi^2}\int \mathrm{tr}\,(F_{\mu\nu}\tilde{F}^{\mu\nu}) = \int_0^T CS(A) = CS([h]A) - CS(A) = \nu(h), \quad (48)$$

with $\nu(h)$ defined in Eq. (15). Using that $\mathrm{tr}\,(F_{\mu\nu}\tilde{F}^{\mu\nu}) = \partial_\mu K_\mu(A)$, we find for the Chern-Simons functional

$$CS(A) = -\frac{1}{16\pi^2}\int d^3x\, K_0(A) = -\frac{1}{8\pi^2}\int d^3x\, \mathrm{tr}\,\left(A_i(\partial_j A_k + \tfrac{2}{3}A_j A_k)\right)\epsilon_{ijk}. \quad (49)$$

When T^3 has periodic boundary conditions there is an ambiguity *which* point in the vacuum valley to tunnel from and to. These need not be the same points, as is already evident with tunneling related by non-trivial twisted gauge transfomations (see Eq. (20) and (35)), under which the Polyakov loop is periodic in time up to a non-trivial Z_N factor. In that case charge one instanton so-

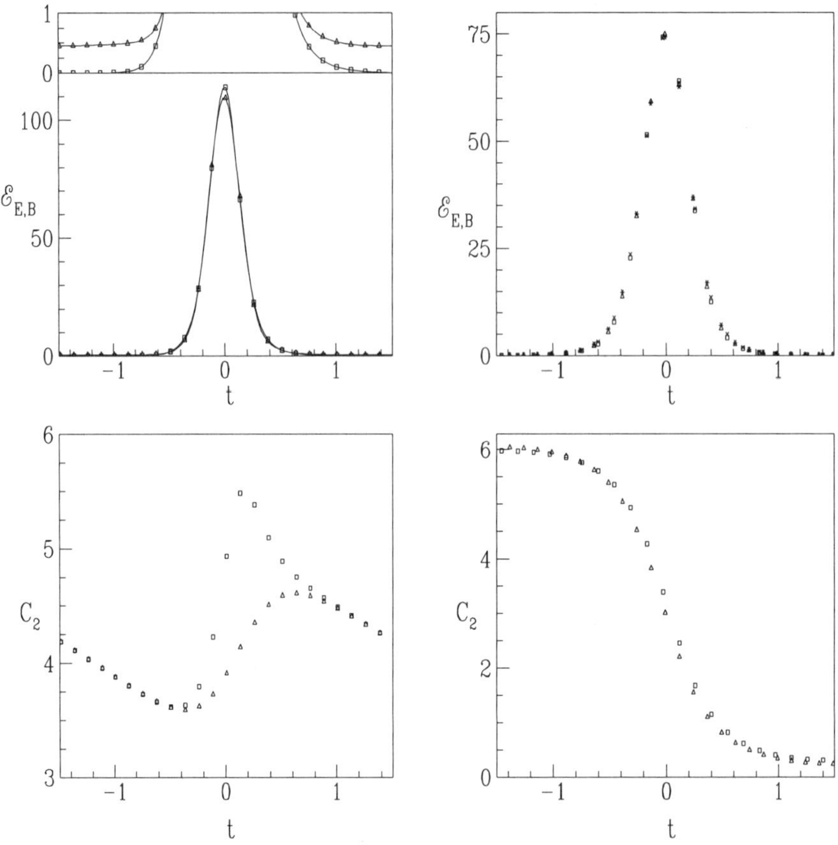

Figure 13: Numerical results[87] obtained with cooling on $N^3 \times 3N$ lattices (scaled to $L = 1$). Left is with $\mathbf{k} = \mathbf{0}$, $N = 8$ and right with $\mathbf{k} = (1, 1, 1)$, $N = 7$ and 8 ($\mathbf{k}$ indicating the twist in time). The top figures show electric ($\mathcal{E}_E(t)$ triangles) and magnetic ($\mathcal{E}_B(t)$ squares) energies ($g = 1$), the inset shows the tails at an enlarged scale. In the lower figures we plot $C_2(t)$ through two distinct (on the right one) point(s).

lutions can be shown to exist.[85] The obstruction for the periodic four-torus is reflected in the fact that the instanton does not want to tunnel between the *same* points in the vacuum valley. This is illustrated in Fig. 13, obtained by a numerical procedure called cooling[86] that minimizes the lattice action

Table 2: The lowest eigenvalues[33] for the Hessian of the energy functional evaluated at the twisted and periodic sphaleron on a N^3 lattice for $N = 4$, 6 and 8.

m=(1,1,1) sphaleron			**m=0 sphaleron**		
N=4	N=6	N=8	N=4	N=6	N=8
-11.1280	-10.9003	-10.8174	-13.1960	-13.3935	-13.4432
0.0074	0.0001	0.0000	0.0073	0.0006	0.0000
0.0091	0.0001	0.0000	0.0124	0.0010	0.0000
0.0101	0.0001	0.0000	0.0139	0.0011	0.0000
13.1667	13.3371	13.3934	1.7041	1.4151	1.3009
13.1667	13.3371	13.3934	1.7070	1.4303	1.3012
13.1667	13.3372	13.3934	7.9677	8.0984	8.1260
16.0475	16.8663	17.1597	7.9682	8.1008	8.1261
16.0478	16.8663	17.1597	7.9684	8.1279	8.1266

in a given topological sector, carefully dealing with the lattice discretization problems.[87] With periodic boundary conditions the configuration first relaxes to a solution which is not self-dual, although it has reached the vacuum valleys at either end in euclidean time, judging the zero magnetic energy. However, it moves in accordance with the equations of motion along this vacuum valley with constant speed (i.e. electric field) to match the forced periodicity, as is also evident from the linear (space independent) slope for $C_2(t)$. When one would continue to lower the action, in order to reach a self-dual solution, the instanton is forced to shrink to zero size. With a twist in the time direction the configuration immediately relaxes to a self-dual one.

Shown on the right in Fig. 13 is an instanton that is close to being maximal in its size, thereby going through the lowest barrier, to maintain the same action. This lowest barrier is assumed to be a finite volume sphaleron. This was verified,[33] by explicitly looking for saddle-point solutions on T^3, minimizing the square of equations of motion using a similar numerical procedure of cooling (cmp. Fig. 14). This saddle point is not invariant under translations, giving rise to zero-modes, which in addition to the unstable (i.e. tunneling direction) will become the degrees of freedom that need to be treated non-perturbatively, by imposing the proper boundary conditions in field space. In addition it was found that there are two directions *not* associated with zero-modes, in which the saddle point is nearly flat (see Table 2). Such modes would have to be treated non-perturbatively as well. Without an analytic solution, it is hard to judge if such an analysis is feasible. This was one of the motivations to look at the situation where the torus geometry is replaced by that of a sphere. In this case the finite volume sphaleron has a constant energy density and no (almost) flat directions, see Sec. 5.

We already mentioned in the previous section that with non-zero magnetic flux, there exist instanton solutions with fractional topological charge. When first proposed by 't Hooft,[73] this was met with some disbelieve. The mathematical classification of so-called fiber bundles, to describe gauge fields on compact manifolds seemed to require integer topological charge, specified by the second Chern number (see Eq. (54)). But due to the twist, the fiber bundle has to be considered real and the appropriate classification is by the first Pontryagin class. With the proper normalizations, this makes the would-be fractional second Chern class an integer first Pontryagin class.[50] It is the difference between a unitary and orthogonal fiber bundle, e.g. $SU(2)$ versus $SO(3)$. This makes also precise the intimate connection between the topological charge and twist, as first discovered by 't Hooft.[73] As we have seen in Secs. 3.1 and 3.4 (see Eq. (35), which also applies for $\mathbf{m} = \mathbf{0}$, in which case $\Omega_j = 1$), there are homotopically non-trivial gauge transformations $h_\mathbf{k}$, "periodic" up to an element of the center of the gauge group, that do not affect the boundary conditions of the gauge fields. In particular $h_\mathbf{k}^{-1} dh_\mathbf{k}$ has the right periodicity, and $\nu(h_\mathbf{k})$ in Eq. (15) remains well-defined (in the gauge where the twist matrices Ω_j are constant and $A_0 = 0$). When $\mathbf{m} = \mathbf{0}$ we have seen that we can always choose $h_\mathbf{k}$ abelian, and $\nu(h_\mathbf{k}) = 0$. In this case there is no interplay between the twist and the topological charge ν, which is here integer. When $\mathbf{m}$ is non-trivial we stated in Sec. 3.4 that when $\mathbf{k} \cdot \mathbf{m} = 0 \ (\mathrm{mod}\ N)$, $h_\mathbf{k}$ can be chosen constant, $h_\mathbf{k} = \Omega_0(\mathbf{k})$, as can be checked for the simple case of $SU(2)$ by inspection, but can be proven from first principles.[67] However, when $\mathbf{k} \cdot \mathbf{m} \neq 0 \ \mathrm{mod}\ N$ it may be that $\nu(h_\mathbf{k}) \neq 0$. The additive nature of the topological invariant $\nu(h)$,

$$\nu(h'h) = \nu(h') + \nu(h), \tag{50}$$

implies that $\nu(h_\mathbf{k}^N) = N\nu(h_\mathbf{k})$. This has to be an integer, since $h_\mathbf{k}^N$ is "periodic" and the winding number of a periodic function $h : T^3 \to G$ is an integer. Also one can duplicate T^3, under which the topological charge is additive, as many times in the space directions as is necessary to get a new torus for which $\mathbf{m} = \mathbf{0}$, and hence with integer topological charge. Therefore,[73] $\nu(h)$ is proportional to N^{-1} and linear in $\mathbf{k}$ and $\mathbf{m}$, such that $\nu + \mathbf{m} \cdot \mathbf{k}/N \in \mathbb{Z}$, as was proven rigorously in the context of fiber bundles.[50] Conversely, using this result one easily sees that the existence of constant twist matrices in four dimensions, i.e. solutions to both Eq. (32) and Eq. (36), require $\mathbf{k} \cdot \mathbf{m} = 0 \ \mathrm{mod}\ N$. This is so because $A = 0$, with $\nu = 0$, satisfies the boundary conditions. For further details on these twist-related issues one can also consult a recent review.[88]

The above implies that the lowest non-trivial topological charge that can be reached is $1/N$. With Eq. (47), the associated euclidean action of this instanton solution is found to be $8\pi^2/N$, which can indeed be attained as one can prove existence of self-dual solutions in the appropriate sector.[89] Like for topological charge one, despite much effort up to date no analytic solutions are known. What can be deduced from general principles[90] is that the solutions with charge $1/N$ have only the four translational degrees of freedom. In particular they have no free scale parameter, their size is fixed with respect to the torus. This fixed scale is a great benefit in studying these instantons on the lattice.[91,76,92] For this fractionally charged instanton the finite volume sphaleron was constructed as well, by finding for SU(2) with twist $\mathbf{m} = (1, 1, 1)$ on T^3 the appropriate saddle point. Like for $\mathbf{m} = \mathbf{0}$ it is not invariant under translations, giving three flat directions at the saddle point, but in this case no other near zero-modes appear,[33] see Table. 2.

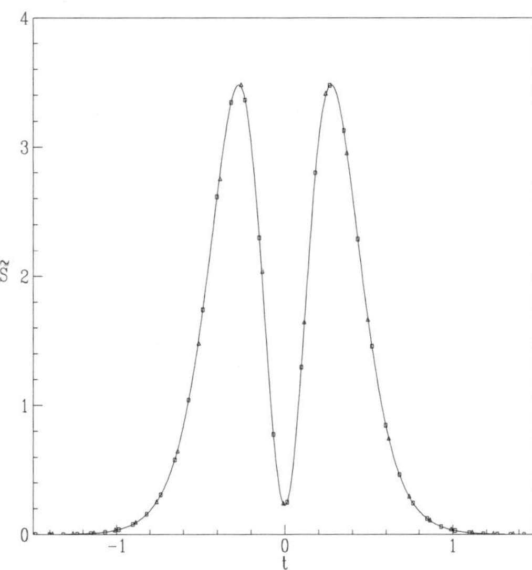

Figure 14: Numerical results[33] for $\tilde{S}(t)$, the square of the equations of motion, along the tunneling path for the SU(2), charge $\frac{1}{2}$ twisted instanton solution for $\mathbf{m} = \mathbf{k} = (1, 1, 1)$, evaluated on a $8^3 \times 24$ (triangles) and a $12^3 \times 36$ (squares) lattice (scaled to $L = 1$). The dip at $t = 0$ occurs at the finite volume sphaleron.

In Fig. 14 we illustrate the fact that along the tunneling path there is a saddle point, by plotting the square of the equations of motion,

$$\tilde{S}(t) \equiv -\frac{1}{32} \int_{T^3} d^3x \, \mathrm{tr} \left(\partial_i F_{ij}(\mathbf{x}) + [A_i(\mathbf{x}), F_{ij}(\mathbf{x})] \right)^2, \qquad (51)$$

as a function of euclidean time. The fact that $\tilde{S}$ does not exactly vanish in the middle, at the top of the barrier, is due to a limited resolution on the finite lattices employed.

4.1 T-duality for Instantons on the Torus

Although no explicit solutions of the basic instantons of charge $1/N$ and charge 1 are known, progress has been made in better understanding instantons on the torus. We will first discuss the case of T^4, before addressing $T^3 \times \mathbb{R}$. For T^4 the only explicitly known solutions are the ones constructed by 't Hooft with constant field strength.[110,111,112] These are essentially abelian and are self-dual *only* when the periods of the torus match the field strength (which is quantized due to quantization of flux). Recently these solutions to the self-duality equations were followed in perturbation theory by deforming away from the special geometry,[113] using as a starting point the exact fluctuation spectrum[111] in the background of the non-selfdual constant curvature solutions.

The essential ingredient for gaining further understanding has been the so-called Nahm or Mukai transformation.[11,114] Nahm introduced it to find magnetic monopole solutions, getting inspiration from the algebraic Atiyah-Drinfeld-Hitchin-Manin[115] (ADHM) construction of instantons in $\mathbb{R}^4$. The formalism can as a matter of fact be used to provide a *simple* proof of the ADHM construction.[83,116] It will be convenient in the following to see T^4 as $\mathbb{R}^4/\Xi$, where Ξ is a four dimensional lattice spanned by the four periods $a^{(\mu)}$, and to introduce the connection one-form $\omega(x) = A^\mu(x)dx_\mu$, which is invariant up to a gauge transformation under translation over a period. These gauge transformations, called cocycles, satisfy cocycle conditions[9] (twisted boundary conditions will be discussed later), to assure the gauge invariant quantities are periodic, and such that one has an appropriate fiber bundle over the torus,[50]

$$\begin{aligned}
\omega(x+a) &= h_a^{-1}(x)(\omega(x)+d)h_a(x), & (52) \\
h_{a+b}(x) &= h_b(x)h_a(x+b) = h_a(x)h_b(x+a), \quad a,b \in \Xi.
\end{aligned}$$

It will be convenient to take $\mathrm{U}(N)$ as gauge group, allowing for a $\mathrm{U}(1)$ factor. The associated vector bundle in the fundamental representation of $\mathrm{U}(N)$ is denoted by E. The curvature of this bundle E, given by

$$F = d\omega + \omega \wedge \omega = \tfrac{1}{2}F^{\mu\nu}dx_\mu \wedge dx_\nu, \qquad (53)$$

leads to a topological charge $\nu = \int_{T^4}(c_2(E) - \tfrac{1}{2}c_1(E) \wedge c_1(E))$ as an integral over Chern classes (cmp. Eq. (48) and note that $c_1(E) = 0$ for $\mathrm{SU}(N)$ because of the vanishing trace), where

$$c_1(E) = \mathrm{tr}\left(\frac{F}{2\pi i}\right), \quad c_2(E) = \tfrac{1}{2}\,\mathrm{tr}\left(\frac{F}{2\pi i} \wedge \frac{F}{2\pi i}\right). \qquad (54)$$

The starting point of the Nahm transformation is that gauge fields with topological charge ν, have ν positive chirality zero-modes for the massless Dirac equation,[117] which is most conveniently written in the Weyl representation. For this we introduce the unit quaternions σ_μ and their conjugates $\bar{\sigma}_\mu \equiv \sigma_\mu^\dagger$

$$\sigma_\mu = (1_2, i\tau_j), \quad \bar{\sigma}_\mu = (1_2, -i\tau_j). \tag{55}$$

They satisfy the multiplication rules

$$\sigma_\mu \bar{\sigma}_\nu = \eta^\alpha_{\mu\nu} \sigma_\alpha, \quad \bar{\sigma}_\mu \sigma_\nu = \bar{\eta}^\alpha_{\mu\nu} \sigma_\alpha, \tag{56}$$

where we used the 't Hooft η symbols,[22] generalized slightly to include $\eta^0_{\mu\nu} = \bar{\eta}^0_{\mu\nu} = \delta_{\mu\nu}$, also useful in Sec. 5. The Weyl-Dirac operator is given by

$$D \equiv \sigma_\mu D_\mu(A) = \sigma_\mu(\partial_\mu + A_\mu). \tag{57}$$

Hence, in the background of a charge ν gauge field there are ν independent solutions to $D\Psi = 0$. For $\Psi(x)$ to be defined as a two-spinor on the torus one requires $\Psi(x + a) = h_a(x)\Psi(x)$. Strictly speaking, the index theorem[117] only states that the number of zero-modes of positive chirality minus those of negative chirality (for which $D^\dagger\Psi = 0$) is equal to the topological charge, but it will be assumed that there are no negative chirality zero-modes. For self-dual gauge fields we will discuss this assumption later. One now adds a spectral parameter $z \in \mathbb{R}^4$ in the form of a flat abelian connection[11]

$$\omega_z = \omega + 2\pi i z^\mu dx_\mu, \tag{58}$$

(the unit generator of U(1) is implicit) which leaves the curvature unchanged, $F_z = F$. Hence there is a smooth family of ν normalized fermionic zero-modes

$$D_z \Psi_z^{(i)}(x) = \sigma_\mu(\partial_\mu + A_\mu + 2\pi i z_\mu)\Psi_z^{(i)}(x) = 0, \quad \int_{T^4} d^4x \, \Psi_z^{(i)}(x)^\dagger \Psi_z^{(j)}(x) = \delta_{ij}. \tag{59}$$

From this family one constructs the connection $\hat{\omega} \equiv \hat{A}_\mu(z)dz_\mu$ by

$$\hat{A}^{ij}_\mu(z) = \int_{T^4} d^4x \, \Psi_z^i(x)^\dagger \frac{\partial}{\partial z_\mu} \Psi_z^{(j)}(x). \tag{60}$$

Using that $D_{z+\hat{a}}\left(e^{-2\pi i x \cdot \hat{a}}\Psi_z^{(i)}(x)\right) = 0$ and the fact that $\{\Psi_{z+\hat{a}}^{(i)}(x)\}$ forms a complete orthogonal set of solutions, we find that

$$\Psi_{z+\hat{a}}^{(i)}(x) = e^{-2\pi i x \cdot \hat{a}}\Psi_z^{(j)}(x)\hat{h}_{\hat{a}}^{ji}(z), \tag{61}$$

with $\hat{h}_{\hat{a}}(z)$ a unitary $\nu \times \nu$ matrix, which defines the cocycle for $\hat{\omega}$

$$\hat{\omega}(z + \hat{a}) = \hat{h}_{\hat{a}}^{-1}(z)(\hat{\omega}(z) + \hat{d})\hat{h}_{\hat{a}}(z), \quad \hat{d} \equiv dz_\mu \partial_{z_\mu}, \tag{62}$$

as a U(ν) connection on the dual torus $\hat{T}^4 = \mathbb{R}^4/\tilde{\Xi}$, where

$$\tilde{\Xi} = \left\{ \hat{a} \in \mathbb{R}^4 |\ < \hat{a}, a > \in \mathbb{Z},\ \forall a \in \Xi \right\} \tag{63}$$

is the lattice dual to Ξ. For example, if Ξ is generated by $a^{(\mu)} = Le^{(\mu)}$, giving a hypercube with sides of length L, the dual lattice $\tilde{\Xi}$ is generated by $\hat{a}^{(\mu)} = L^{-1}e^{(\mu)}$, a hypercube with sides of length $1/L$. So the Nahm transformation is a T-duality.

In the absence of negative chirality zero-modes ($\mathrm{coker}D_z = \ker D_z^\dagger = 0$) $\hat{E}_z \equiv \ker D_z$ depends smoothly on z. It was realized by Braam,[118,119] that one can use the Atiyah-Singer *family* index theorem[117] to compute the Chern character of the bundle $\hat{E}$ with connection $\hat{\omega}$, as the integral over T^4 of the Chern character of $E \otimes \mathcal{P}$, which is the vector bundle relevant for ω_z, as a vector bundle over $T^4 \times \hat{T}^4$. Here $\mathcal{P}$ is the so-called[84] Poincaré line bundle with the connection $2\pi i z^\mu dx_\mu$. It was crucial this line bundle had no curvature as a bundle over T^4, but extending it to $T^4 \times \hat{T}^4$, this is no longer the case and $F_\mathcal{P} = 2\pi i dx_\mu \wedge dz^\mu$, with a first Chern class, $c_1(\mathcal{P}) = dx_\mu \wedge dz^\mu$ (for a line bundle all other Chern classes vanish). The Chern character satisfies $ch(E \otimes E') = ch(E) \wedge ch(E')$ and is a formal polynomial in the Chern classes,

$$ch(E) = c_0(E) + c_1(E) + \tfrac{1}{2}c_1(E) \wedge c_1(E) - c_2(E), \quad ch(\mathcal{P}) = \exp(c_1(\mathcal{P})). \tag{64}$$

Perhaps somewhat confusingly $c_0(E) = rk(E)$ is called the rank of the vector bundle E, here the dimension of the fundamental representation (so $rk(E) = N$ for a U(N) fiber bundle). It now follows that

$$\mathrm{ch}(\hat{E}) = \int_{T^4} \mathrm{ch}(E) \wedge \mathrm{ch}(\mathcal{P}). \tag{65}$$

The integral over T^4 only gives a non-zero result if the combined form is of degree four in dx_μ, i.e. a top-form. This means that

- The zero-form $c_0(E) = N$, becomes a volume form on $\hat{T}^4$, whose integral over $\hat{T}^4$ gives the topological charge of the dual gauge field.

- The two-form $c_1(E)$ whose cohomology class can be specified by the *abelian* fluxes $c_1(E) = \tfrac{1}{2}n^{\mu\nu}dx_\mu \wedge dx_\nu$, goes over in $c_1(\hat{E}) = \tfrac{1}{2}\tilde{n}^{\mu\nu}dz_\mu \wedge dz_\nu$.

- The four-form whose integral is the topological charge ν of the original gauge field becomes a zero-form, $c_0(\hat{E}) = \nu$.

Thus we see that the Nahm transformation interchanges the rank N of the bundle with the topological charge ν, and maps the first Chern class to its dual. In particular, an SU(N) bundle has vanishing first Chern class and is thus mapped under the Nahm transformation to $U(\nu)$ with vanishing first Chern class. It is well-known that in the latter case the bundle can be gauged to an $SU(\nu)$ bundle (its $U(1)$ part is trivial and can be gauged away).

The family index theorem only provides topological information on the dual gauge field. We will demonstrate the wondrous result[11], that if ω is a self-dual connection, than also $\hat\omega$ is self-dual. A crucial ingredient is formed by

$$D_z D_z^\dagger = -D_\mu^2(A_z) - \tfrac{1}{2}\sigma_{[\mu}\bar\sigma_{\nu]} F^{\mu\nu}, \tag{66}$$

using $\sigma_\mu\bar\sigma_\nu = \delta_{\mu\nu} + \sigma_{[\mu}\bar\sigma_{\nu]}$. Since $\sigma_{[\mu}\bar\sigma_{\nu]} \equiv \sigma_i\eta^i_{\mu\nu}$ is an *anti-selfdual* tensor,[a] we see that $D_z D_z^\dagger = -D_\mu^2(A_z)$. We first deal as promised with the condition that there are no opposite chirality zero-modes. We have seen that if ω is self-dual, and $D_z^\dagger\Psi = 0$ that $D_\mu^2(A_z)\Psi = D_z D_z^\dagger\Psi = 0$, but this would imply that the connection ω has a flat factor,[83] meaning that the bundle of rank N splits in *the direct sum* of a bundle of rank $N-1$ and a flat line bundle, $E = E' \oplus L_\xi$, where L_ξ has the connection $2\pi\xi^\mu dx_\mu$. Such a flat factor would exist for any z (simply shifting ξ proportionally), and so a sufficient condition to impose the absence of opposite chirality zero-modes is to require ω to be without flat factors (WFF, sometimes also called 1-irreducible).

A direct corollary is now that regular $SU(N)$ charge one instantons cannot exist on T^4. Suppose they would exist. The Nahm transformation gives rise to a $U(1)$ bundle of charge N, which is impossible as the first Chern class vanishes and $U(1)$ bundles have always vanishing second Chern class. Suppose flat factors prevent us from making the argument. In that case we can remove the flat factor and go down in rank by one (which keeps the first Chern class zero). This can be repeated until we are left with an $SU(N-k)$ charge one instanton without flat factors, or completely reduce to flat factors which, however, can not support topological charge.

Assuming no flat factors for the self-dual connection ω, the kernel of $D_z D_z^\dagger$ is trivial and the well-defined Greens function $G_z = (D_z D_z^\dagger)^{-1}$ commutes with the quaternions σ_μ. It is now straightforward to show that $\hat\omega$ is self-dual as well

$$\hat F^{ij}(z) = \hat d\hat\omega^{ij} + \hat\omega^{ik} \wedge \hat\omega^{kj} = <\hat d\Psi_z^{(i)}|1 - P_z|\hat d\Psi_z^{(j)}>, \tag{67}$$

where P_z is the projection on $\ker D_z$,

$$P_z \equiv \sum_{k=1}^{\prime\prime} |\Psi_z^{(k)}><\Psi_z^{(k)}| = 1 - D_z^\dagger G_z D_z. \tag{68}$$

The equality to the right-hand side is trivially true for a zero-mode, whereas the orthogonal complement of $\ker D_z$ is described by the image of D_z, on

---

[a]Note that 't Hooft had introduced x_4, instead of x_0, as the imaginary time. Using his symbols[22] replacing the index 4 by 0, flips the dualities around, since $\epsilon_{0123} = -\epsilon_{4123}$.

which both sides act as the identity. Next we use $D_z \hat{d}\Psi_z^{(i)} = [D_z, \hat{d}]\Psi_z^{(i)} = -2\pi i dz_\mu \sigma^\mu \Psi_z^{(i)}$ to find

$$\hat{F}_{\mu\nu}^{ij}(z) = 8\pi^2 < \Psi_z^{(i)}|\bar{\sigma}_{[\mu}\sigma_{\nu]}G_z|\Psi_z^{(j)} > . \tag{69}$$

Self-duality immediately follows from the fact that $\bar{\sigma}_{[\mu}\sigma_{\nu]} = \sigma_i \bar{\eta}_{\mu\nu}^i$ is self-dual.

For T^4, applying the Nahm transformation the second time, it can be shown that ω is retrieved identically, and the explicit form of $\hat{\Psi}_x(z)$ in terms of $\Psi_z(x)$ allows one to show that metric and hyperKähler structures of the moduli spaces are preserved under the Nahm transformation. In other words, if we denote by $\mathcal{M}_{N,\nu}$ the moduli space of SU(N) charge ν instantons, the Nahm transformation induces a map between moduli spaces, $\mathcal{N} : \mathcal{M}_{N,\nu} \to \mathcal{M}_{\nu,N}$, which is an involution that preserves the natural metric and hyperKähler structure of the moduli space.[84] The dimension of the moduli space, $4N\nu$, is indeed symmetric under interchanging ν and N.

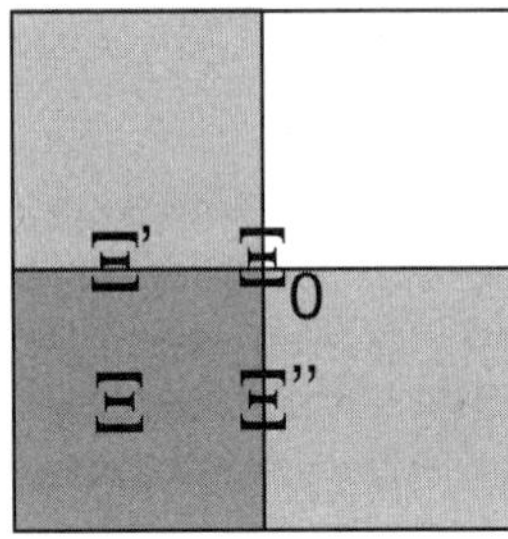
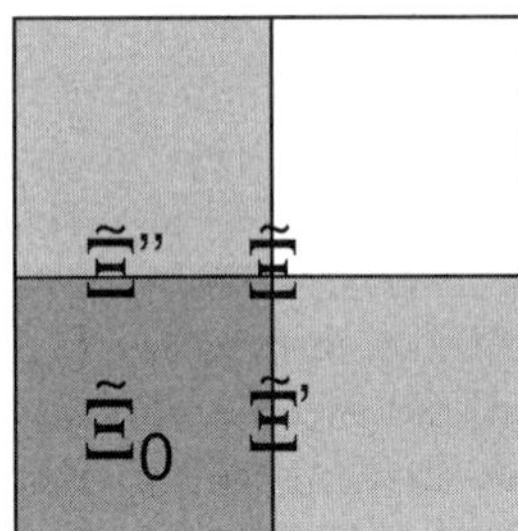

Figure 15: Shown is a cross section through the 1-2 plane of the various unit cells that appear in the Nahm transformation for SU(2) with twisted boundary conditions $n_{12} = -n_{21} = 1$ on R^4/Ξ. We have chosen here $||a^{(1)}|| = ||a^{(2)}|| = \frac{1}{2}\sqrt{2}$. Duplicated unit cells without twist are Ξ' and Ξ'', their union is denoted by Ξ_0. Its dual, $\tilde{\Xi}_0$, gives the torus on which the Nahm transformed gauge field lives, with the same gauge group and twist as for the gauge field we started with. The unit cell $\tilde{\Xi}_0$ is also obtained by intersecting $\tilde{\Xi}'$ and $\tilde{\Xi}''$, the duals of the duplicated unit cells. When a symbol overlaps with different cells it belongs to all of those cells as a whole.

The Nahm transformation on T^4 can be defined to include twisted boundary conditions, as González-Arroyo has shown recently.[120] The basic idea is simple: One duplicates the unit cell, also denoted by Ξ, in the various directions as many times as is necessary to remove the twist. Let us assume for SU(2) $n_{12} = -n_{21} = m_3 = 1$ and the rest of the twist factors trivial. We can duplicate Ξ either in the 1- or in the 2-direction in order to remove the twist, giving Ξ' and Ξ''. In both cases we can apply the Nahm to the enlarged unit cells Ξ' or Ξ'', see Fig. 15. In each case its dual $\tilde{\Xi}'$ or $\tilde{\Xi}''$ is shorter by a factor 2 in the direction we originally duplicated. These two choices of unit cell intersect in a cell we call $\tilde{\Xi}_0$ that is twice shorter in each of the two directions.

Due to the intersection it contains all the gauge invariant information of $\hat{\omega}'$ and $\hat{\omega}''$. The gauge field therefore has *half-periods* associated with twisted boundary conditions on $\mathbb{R}^4/\tilde{\Xi}_0$. In this particular example the resulting twist is the same as we started with. The argument can be generalized, either along the above lines, or using so-called flavor multiplication,[120,121] to arbitrary $SU(N)$ and twist. The example shown in Fig. 15 (with identical behavior in the 0-3 plane), maps the charge $\frac{1}{2}$ instanton to the *same* solution. Another example of self-similarity occurs for $SU(2)$ in the charge 2 sector. When applied to the special solutions of 't Hooft,[110] this can be used as an example where the Nahm transformation can be worked through analytically. The resulting Nahm transformed connection is again of the special form.[122]

To have the Nahm transformation help us finding explicit instanton solutions, simplifications have to be arranged for. These happen to be perfectly geared to finding charge one instanton solution on $T^3 \times \mathbb{R}$. For charge one the Nahm transformed gauge field is abelian and with one period infinite, the dual period collapses to zero. The decompactification limit $T \to \infty$ is associated with dimensional reduction. This reaches a dramatic height for the case that all periods are send to infinity, relevant for instantons on $\mathbb{R}^4$. The dual is now a single point, explaining why the ADHM construction[115] is in essence algebraic. However, under the decompactification limit, a partial integration is required in going from Eq. (67) to Eq. (69),

$$\hat{F}^{ij}_{\mu\nu}(z) = 4\pi i \bar{\eta}^k_{\lambda\,[\mu} \oint d^3_\lambda x \, \frac{\partial}{\partial z_{\nu]}} \left(\Psi^{(i)}_z\right)^{\dagger}(x) \, \sigma_k \left(G_z \Psi^{(j)}_z\right)(x)$$
$$+ 8\pi^2 \bar{\eta}^k_{\mu\nu} < \Psi^{(i)}_z | \sigma_k G_z | \Psi^{(j)}_z > . \tag{70}$$

These boundary terms destroy self-duality of $\hat{F}$, which can be repaired[11,116] for instantons on $\mathbb{R}^4$ and for calorons (periodic instantons) on $\mathbb{R}^3 \times S_1$. Except for $\mathbb{R}^4$, for which there is no freedom, the location of the singularities are fixed by the asymptotically flat connections at $t \to \pm\infty$, which are required to occur to keep the action finite. The Weyl-Dirac hamiltonian in the background of a flat connection has generically a mass gap. The mass gap vanishes, however, in particular when the flat connection is pure gauge. The flat connections are characterized by $\mathbf{A}_z = 2\pi i \, \mathrm{diag}\left(\mathbf{w}^{(1)} + \mathbf{z}, \cdots, \mathbf{w}^{(N)} + \mathbf{z}\right)$, and for the mass gap to vanish it is sufficient that $||\mathbf{w}^{(k)} + \mathbf{z}|| = 0$ for any of the N values of k. With two asymptotically flat connections, their are $2N$ corresponding values of $\mathbf{z}$ (which are defined modulo $\tilde{\Xi}$) where the mass gap vanishes. Only here the boundary terms in Eq. (70) can be non-vanishing. A simple steepest descent analysis in their neighborhood shows that they act as point sources with *abelian* electric *and* magnetic charges $\pm\pi$, enforced by charge quantization (positive charges associated with the flat connection at $t = \infty$). This ensures the magnetic sources are Dirac monopoles with unobservable Dirac strings.[123]

Outside of the singularities the *abelian* field is self-dual and, noting the z_0 independence, $\hat{B}_j(\mathbf{z}) = \hat{E}_j(\mathbf{z}) = -\partial\hat{A}_0(\mathbf{z})/\partial z_j$, where

$$\hat{A}_0(\mathbf{z}) = \frac{i}{2}\sum_{j=1}^{N}\sum_{\mathbf{n}\in\tilde{\Xi}}\left(||\mathbf{w}_+^{(j)} + \mathbf{z} + \mathbf{n}||^{-1} - ||\mathbf{w}_-^{(j)} + \mathbf{z} + \mathbf{n}||^{-1}\right). \quad (71)$$

For the usual normalization of abelian fields one multiplies this with $-i$. The sum over the periods $\tilde{\Xi}$ on $\hat{T}^3$, although formally divergent, can be resummed in a rapidly converging series and one has an exact result[123] for $\hat{A}_0(\mathbf{z})$ (see Fig. 16). One may argue from this that indeed time-periodic instanton so-

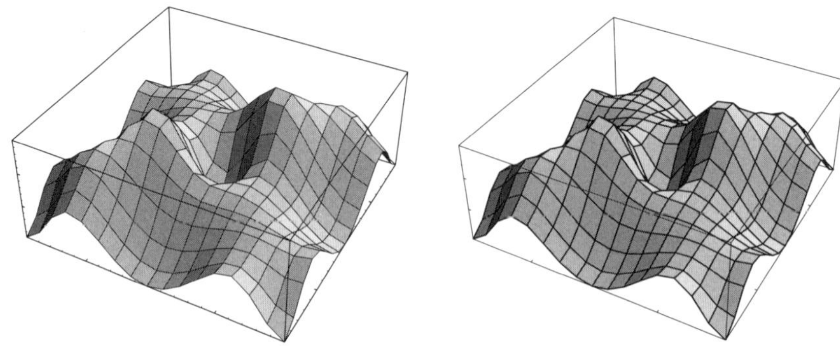

Figure 16: Comparison[121] of $\hat{\mathbf{E}}^2$ in the plane $z_3 = 0.5$ for the Nahm transformation of an SU(2) instanton on an $8^3 \times 40$ lattice (scaled to $L = 1$) and twist $\mathbf{k} = (1,1,1)$ in the time direction (left), with the analytic result on $T^3 \times \mathbb{R}$ for $P_j^- = (0.86, 0.76, 0.08) = -P_j^+$ (right).

lutions do not exist, as for $\mathbf{w}_+^{(j)} = \mathbf{w}_-^{(j)}$ we clearly find $\hat{A}_0(\mathbf{z}) = 0$, which under the Nahm transformation leads to a trivial connection on $T^3 \times \mathbb{R}$. With twisted boundary conditions in the time direction one has $\mathbf{w}_+^{(j)} = \mathbf{w}_-^{(j)} + \mathbf{k}/N$. Although it is not straightforward to solve the Weyl-Dirac equation on $\hat{T}^3$ in the background of an abelian self-dual gauge field generated by point charges, one may hope explicit solutions of charge one instantons on the torus can be found this way.

The situation with the twisted instantons of fractional charge is, perhaps somewhat paradoxically, a bit more complicated. However, much understanding has been gained through the numerical implementation of the Nahm transformation on a lattice.[124] It has been found[121] that, in staying away from the exact decompactification limit, the Dirac monopole singularities are resolved into the fully non-abelian constituent monopoles that appear in the calorons,[100] with the only difference that these calorons are now modified by the finite volume,[125] for which no exact analytic solutions are available. They nevertheless fit very well to the infinite volume solutions, whereas away from the non-abelian cores of the constituent monopoles there is a nearly perfect fit to the abelian field of the singular Dirac monopoles. This aspect is illustrated in Fig. 16, comparing the analytic result in the decompactification limit with the numerical result on a lattice. For further details one should consult the literature.[121] It is natural to conjecture, in the light of these results that the fractionally charged instanton is mapped to a single point charge on a torus with (abelian) C periodic boundary conditions[57] in the decompactification limit.

5 Gauge Fields on the three-Sphere

The most important reason to study gauge fields on the three-sphere is that conformal equivalence of $S^3 \times \mathbb{R}$ to $\mathbb{R}^4$ gives a very simple and explicit construction for the instantons.[126,127] This allows one to formulate in a precise way how the θ dependence can be encoded in the boundary conditions on the fundamental domain.[126,128] Due to the curvature of the sphere, at large volumes the corrections to the glueball masses are in powers of the inverse radius, as opposed to an exponential approach for the torus (to be discussed in Sec. 6.1). But ultimately, any geometry can be scaled-up to an infinite volume, and should in this limit give the same results. Therefore, by comparing different geometries we may indirectly get some useful information.

We embed S^3 in $\mathbb{R}^4$ by considering the unit sphere (the radius R can be reinstated on dimensional grounds where required), parametrized by a unit vector n_μ. Alternative formulations, useful for diagonalizing the Faddeev-Popov and fluctuation operators, were developed by Cutkosky.[129] We can use the 't Hooft tensors η and $\bar{\eta}$ to define orthonormal framings[130] of S^3, which were motivated by the particularly simple form of the instanton vector potentials in these framings. The framing for S^3 is obtained from the framing of $\mathbb{R}^4$ by restricting in the following equation the four-index α to a three-index j (for $\alpha = 0$ one obtains the normal on S^3, see Eq. (56)),

$$e_\mu^\alpha = \eta_{\mu\nu}^\alpha n_\nu, \quad \bar{e}_\mu^\alpha = \bar{\eta}_{\mu\nu}^\alpha n_\nu. \tag{72}$$

Note that e and $\bar{e}$ have opposite orientations. Each framing defines a differential operator with associated (mutually commuting) angular momentum operators $\mathbf{L}_1$ and $\mathbf{L}_2$:

$$\partial^i = e_\mu^i \frac{\partial}{\partial x^\mu}, \quad L_2^i = \frac{i}{2} \, \partial^i \quad , \quad \bar{\partial}^i = \bar{e}_\mu^i \frac{\partial}{\partial x^\mu}, \quad L_1^i = \frac{i}{2} \, \bar{\partial}^i. \tag{73}$$

It is easily seen that $\mathbf{L}_1^2 = \mathbf{L}_2^2$, which has eigenvalues $l(l+1)$, with $l = 0, \frac{1}{2}, 1, \cdots$.

By identifying the logarithm of the radius in $\mathbb{R}^4$ as time in the geometry $S^3 \times \mathbb{R}$, the (anti-)instantons are easily obtained from those[21] on $\mathbb{R}^4$

$$A_0 = \frac{\varepsilon^\mu \cdot \sigma_\mu}{2(1 + \varepsilon^\mu n_\mu)}, \quad A_i = \frac{\epsilon_{ijk}\sigma^j \varepsilon^k - (v + \varepsilon^\mu n_\mu)\sigma_i}{2(1 + \varepsilon^\mu n_\mu)}. \tag{74}$$

Here ε^k, σ_j and A_j are defined with respect to the framing $\bar{e}^j_\mu$ for instantons and with respect to the framing e^j_μ for anti-instantons, and we introduced

$$u = \frac{2s^2}{1 + b^2 + s^2}, \quad \varepsilon^\mu = \frac{2sb^\mu}{1 + b^2 + s^2}, \quad s = \lambda \exp(t). \tag{75}$$

The unit quaternions σ_μ were given in Eq. (55). The instanton describes tunneling from $A = 0$ at $t = -\infty$ to $A_j = -\sigma_j$ at $t = \infty$, over a potential barrier at $t = 0$ that is lowest when $b \equiv 0$. This configuration corresponds to a sphaleron,[81] i.e. the vector potential $A_j = -\frac{1}{2}\sigma_j$ is a saddle point of the energy functional with one unstable mode, corresponding to the direction (u) of tunneling. At $t = \infty$, $A_j = -\sigma_j$ has zero energy and is a gauge copy of $A_j = 0$ by a gauge transformation $h = n_\mu \sigma^\mu$ with winding number one.

We will be concentrating our attention to the 18 modes that are degenerate in energy to lowest order with the modes that describe tunneling through the sphaleron and "anti-sphaleron". The latter is a gauge copy by a gauge transformation $h' = n_\mu \bar{\sigma}^\mu$ with winding number -1 of the sphaleron. The two dimensional space containing the tunneling paths through these sphalerons is consequently parametrized by u and v through

$$A_\mu(u, v) = \left(-u\bar{e}^a_\mu - ve^a_\mu\right)\frac{\sigma_a}{2}. \tag{76}$$

The gauge transformation h with winding number one is easily seen to map $(u, v) = (w, 0)$ into $(u, v) = (0, 2 - w)$. The 18 dimensional space is defined by

$$A_\mu(c, d) = \left(c^a_j \bar{e}^j_\mu + d^a_j e^j_\mu\right)\frac{\sigma_a}{2} = A_j(c, d)\bar{e}^j_\mu. \tag{77}$$

The c and d modes are mutually orthogonal and satisfy the Coulomb gauge condition, $\partial_j A_j(c, d) = 0$. This space contains in it the (u, v) plane through $c^a_j = -u\delta^a_j$ and $d^a_j = -v\delta^a_j$. The significance of this 18 dimensional space is that the energy functional[126]

$$\mathcal{V}(c, d) \equiv -\int_{S^3} \frac{1}{2}\operatorname{tr}\left(F^2_{ij}\right) = \mathcal{V}(c) + \mathcal{V}(d) + \frac{2\pi^2}{3}\left\{(c^a_i)^2(d^b_j)^2 - (c^a_i d^a_j)^2\right\},$$

$$\mathcal{V}(c) = 2\pi^2\left\{2(c^a_i)^2 + 6\det c + \frac{1}{4}[(c^a_i c^a_i)^2 - (c^a_i c^a_j)^2]\right\}, \tag{78}$$

is degenerate to second order in c and d. Indeed, the quadratic fluctuation operator $\mathcal{M}$ in the Coulomb gauge, defined by

$$-\int_{S^3} \frac{1}{2}\,\mathrm{tr}\,(F_{ij}^2) \;=\; \int_{S^3} \mathrm{tr}\,(A_i \mathcal{M}_{ij} A_j) + \mathcal{O}\left(A^3\right),$$

$$\mathcal{M}_{ij} \;=\; 2\mathbf{L}_1^2 \delta_{ij} + 2\left(\mathbf{L}_1 + \mathbf{S}\right)_{ij}^2, \quad S_{ij}^a = -i\epsilon_{aij}, \tag{79}$$

has $A(c,d)$ as its eigenspace for the (lowest) eigenvalue 4. These modes are consequently the equivalent of the zero-momentum modes on the torus, with the difference that their zero-point frequency does not vanish.

To find the fundamental region by minimizing the norm functional, Eq. (1), we can use $FP_f(A)$ in Eq. (6) as a hermitian operator acting on the vector space $\mathcal{L}$ of functions h over S^3 with values in the space of the quaternions $\mathbb{H} = \{q^\mu \sigma_\mu | q \in \mathbb{R}^4\}$. The gauge group $\mathcal{G}$ is contained in $\mathcal{L}$ by restricting to the unit quaternions: $\mathcal{G} = \{h \in \mathcal{L} | h = h^\mu \sigma_\mu, h \in \mathbb{R}^4, h_\mu h^\mu = 1\}$. This allows us to extract detailed information about the Gribov and fundamental region within the 18 dimensional space (c,d). When minimizing the norm functional over $\mathcal{L}$, instead of $\mathcal{G}$, one obviously should find a smaller space $\tilde{\Lambda} \subset \Lambda$. With $\mathcal{L}$ a linear space, $\tilde{\Lambda}$ can now also be defined by the condition that $FP_f(A)$ be positive,

$$\tilde{\Lambda} = \{A \in \Gamma | \langle h, FP_f(A)\, h\rangle \geq 0, \ \forall h \in \mathcal{L}\}. \tag{80}$$

Similar to the Gribov horizon, the boundary of $\tilde{\Lambda}$ is now determined by the location where the lowest non-trivial eigenvalue of $FP_f(A)$ vanishes. For the (c,d) space it can be shown[128] that the boundary $\partial\tilde{\Lambda}$ will touch the Gribov horizon $\partial\Omega$. It establishes the existence of singular points on the boundary of the fundamental domain due to the inclusion $\tilde{\Lambda} \subset \Lambda \subset \Omega$. (By showing that the fourth order term in Eq. (5) is positive,[128] this is seen to correspond to the situation as sketched in Fig. 2.)

In order to simplify the notation, write $FP_{\frac{1}{2}} \equiv FP_f$ and $FP_1 \equiv FP$, with the indices related to the isospin. The associated generators are

$$\mathbf{T}_{\frac{1}{2}} = \tfrac{1}{2}\vec{\tau} \qquad \text{and} \qquad \mathbf{T}_1 = \tfrac{1}{2}\mathrm{ad}\vec{\tau}. \tag{81}$$

We can now make use of the $\mathrm{SU}(2){\times}\mathrm{SU}(2){\times}\mathrm{SU}(2)$ symmetry generated by $\mathbf{L}_1$, $\mathbf{L}_2$ and $\mathbf{T}_t$ to calculate explicitly the spectrum of $FP_t(A)$. One has

$$FP_t(A(c,d)) = 4\mathbf{L}_1^2 - \frac{2}{t}c_i^a T_t^a L_1^i - \frac{2}{t}d_i^a T_t^a L_2^i, \tag{82}$$

which commutes with $\mathbf{L}_1^2 = \mathbf{L}_2^2$, but for arbitrary (c,d) there are in general no other commuting operators (except for a charge conjugation symmetry when $t = \tfrac{1}{2}$). Restricting to the (u,v) plane one finds that

$$FP_t(A(u,v)) = 4\mathbf{L}_1^2 + \frac{2}{t}u\mathbf{L}_1 \cdot \mathbf{T}_t + \frac{2}{t}v\mathbf{L}_2 \cdot \mathbf{T}_t, \tag{83}$$

which in addition commutes with $\mathbf{J}_t = \mathbf{L}_1 + \mathbf{L}_2 + \mathbf{T}_t$, the total angular momentum, and is easily diagonalized. Fig. 17 summarizes the results for this (u, v) plane and also shows the equipotential lines, as well as exhibiting the multiple vacua and the sphalerons. As it is easily seen that the two sphalerons are gauge copies (by a unit winding number gauge transformation) with equal norm, they lie on $\partial \Lambda$, which can be extended by perturbing around these sphalerons.[131]

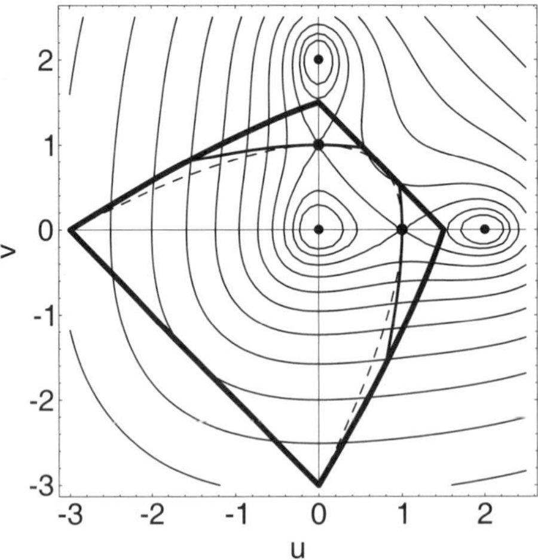

Figure 17: Location of the sphalerons (large dots), classical vacua (smaller dots), the Gribov horizon (fat sections), the boundary of $\tilde{\Lambda}$ (dashed parabola) and part of the boundary of the fundamental domain (full curves, through the two sphalerons). Also indicated are the lines of equal potential in units of 2^n times the sphaleron energy.

To obtain the result for general (c, d) one can use the invariance under rotations generated by $\mathbf{L}_1$ and $\mathbf{L}_2$ and under constant gauge transformations generated by $\mathbf{T}_t$, to bring c and d to a standard form, or express $\det\left(FP_t(A(c,d))|_{l=\frac{1}{2}}\right)$, which determines the locations of $\partial \Omega$ and $\partial \tilde{\Lambda}$, in terms of invariants. We define the matrices X and Y by $X^a_b = c^a_j c^b_j$ and $Y^a_b = d^a_j d^b_j$, in terms of which

$$\det\left(FP_{\frac{1}{2}}(A(c,d))|_{l=\frac{1}{2}}\right) = \left[81 - 18\operatorname{Tr}(X + Y) + 24(\det c + \det d)\right.$$
$$\left. - (\operatorname{Tr}(X - Y))^2 + 2\operatorname{Tr}((X - Y)^2)\right]^2. \quad (84)$$

A two-fold multiplicity (the square) is due to charge conjugation symmetry. The expression for $t = 1$, that determines the location of the Gribov horizon in the (c, d) space,[128] is somewhat more complicated. If we restrict to $d = 0$ the result simplifies considerably. In that case one can bring c to a diagonal form $c^a_i = x_i \delta^a_i$. Rotations and gauge transformations reduce to permutations of the x_i and simultaneous changes of the sign of two of the x_i. One now easily finds the invariant expression ($\mathrm{Tr}(X) = \sum_i x_i^2$ and $\det c = \prod_i x_i$)

$$\det\left(FP_1(A(c,0))|_{l=\frac{1}{2}}\right) = (2\det c - 3\,\mathrm{Tr}(X) + 27)^4. \tag{85}$$

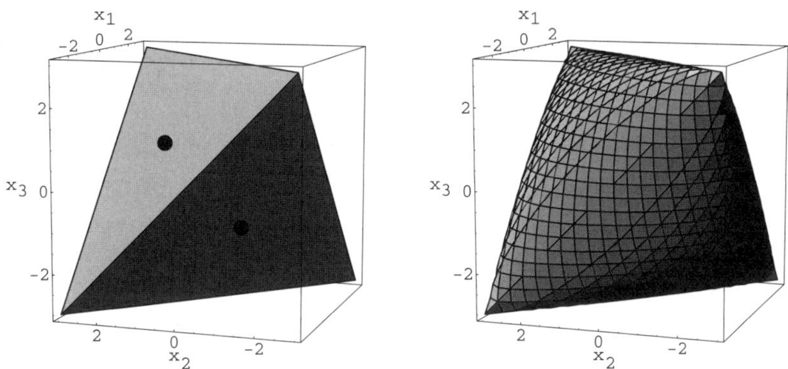

Figure 18: The fundamental domain (left) for constant gauge fields on S^3, with respect to the framing $\bar{e}^j_\mu$, in the diagonal representation $A_j = x_j\sigma_j$ (no sum over j). The dots on the faces indicate the sphalerons. On the right we show the Gribov horizon, which encloses the fundamental domain, coinciding with it at the singular boundary points along the edges of the tetrahedron.

In Fig. 18 we present the results for Λ and Ω. In this particular case, where $d = 0$, Λ coincides with $\tilde{\Lambda}$, a consequence of the convexity and the fact that both the sphalerons (indicated by the dots) and the edges of the tetrahedron lie on $\partial\Lambda$, the latter also lying on $\partial\Omega$. It is essential that the sphalerons do not lie on the Gribov horizon and that the potential energy near $\partial\Omega$ is relatively high. This is why we can take the boundary identifications near the sphalerons into account without having to worry about singular boundary points, as long as the energies of the low-lying states will be not much higher than the energy of the sphaleron. It allows one to study the glueball spectrum as a function of the CP violating angle θ, but more importantly it incorporates for $\theta = 0$ the noticeable influence of the barrier crossings, i.e. of the instantons.

An effective hamiltonian for the c and d modes is derived from the one-loop effective action.[132] To lowest order it is given by

$$H = -\frac{g^2(R)}{4\pi^2 R}\left(\left(\frac{\partial}{\partial c_i^a}\right)^2 + \left(\frac{\partial}{\partial d_i^a}\right)^2\right) + \frac{1}{g^2(R)R}\mathcal{V}(c,d) + \frac{1}{R}\mathcal{V}^{(1)}_{\mathrm{eff}}(c,d), \tag{86}$$

where $g(R)$ is the running coupling constant (related to the MS running coupling by a finite renormalization, such that the kinetic term in Eq. (86) has no corrections). The one loop correction to the effective potential is given by

$$
\begin{aligned}
\mathcal{V}^{(1)}_{\text{eff}}(c,d) &= \mathcal{V}^{(1)}_{\text{eff}}(c) + \mathcal{V}^{(1)}_{\text{eff}}(d) + \kappa_7 (c_i^a)^2 (d_j^b)^2 + \kappa_8 (c_i^a d_j^a)^2, \\
\mathcal{V}^{(1)}_{\text{eff}}(c) &= \kappa_1 (c_i^a)^2 + \kappa_2 \det c + \kappa_3 (c_i^a c_i^a)^2 + \kappa_4 (c_i^a c_j^a)^2 \\
&\quad + \kappa_5 (c_i^a)^2 \det c + \kappa_6 (c_i^a c_i^a)^3.
\end{aligned}
\tag{87}
$$

with the following numerical values for the coefficients

$$
\begin{aligned}
\kappa_1 &= -0.24534599851796, & \kappa_5 &= -0.84996541224534, \\
\kappa_2 &= +3.66869179814223, & \kappa_6 &= -0.06550330854836, \\
\kappa_3 &= +0.50070320309661, & \kappa_7 &= -0.36171221599671, \\
\kappa_4 &= -0.83935963341300, & \kappa_8 &= -2.29535686135471.
\end{aligned}
\tag{88}
$$

Along the tunneling path (e.g. $c_i^a = -u\delta_i^a$ and $d = 0$) the effective potential can be calculated with simpler methods, providing an important check.

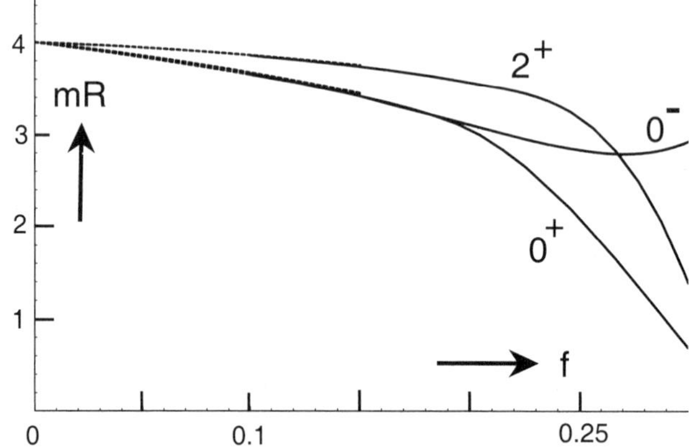

Figure 19: The full one loop results for the masses of scalar, tensor and odd glueballs on S^3 as a function of $f = g^2(R)/2\pi^2$ for $\theta = 0$. The dashed lines correspond to the perturbative result.

Unlike for the torus, where in lowest order all excitations in the zero-momentum modes are degenerate and Bloch perturbation theory[24] provides a rigorous definition of the effective hamiltonian, one has to rely here on an adiabatic approximation not controlled by the coupling constant. The low lying excitations in the c and d modes are well below the excitations of the modes that were integrated out, justifying the adiabatic approximation.[126,132] It provides enough room to achieve a satisfactory understanding of the non-perturbative dynamics due to spreading of the wave functional to the boundary of the fundamental domain. The boundary conditions are chosen such that the gauge and (left and right) rotational invariances are preserved and that they coincide with the appropriate boundary conditions near the sphalerons. Projection on the irreducible representations of these symmetries is essential to reduce the size of the matrices to be diagonalized in a Rayleigh-Ritz analysis. All this could be implemented in a tractable way,[132] see Fig. 19.

In perturbation theory the 0^- "oddball" is slightly *lighter* than the 0^+ glueball state. In lowest order the 0^+, 0^- and 2^+ glueball states are all degenerate, lifted in the next order in perturbation theory, by an amount independent of the one loop corrections

$$
\begin{aligned}
(m_{0^-} - m_{0^+})R &= -g^2(R)/4\pi^2 + \mathcal{O}\left(g^4\right), \\
(m_{2^+} - m_{0^+})R &= 5g^2(R)/6\pi^2 + \mathcal{O}\left(g^4\right),
\end{aligned}
\tag{89}
$$

whereas $m_0^+ R = 4 - (1.25 - \kappa_1)g^2(R)/2\pi^2 + \mathcal{O}\left(g^4\right)$. This perturbative result was also found by Diekmann.[133] (His one loop corrections, however, do not survive the test of reproducing the effective potential along the tunneling path.) Clearly, the 0^- glueball being the lightest state is an artifact of the finite volume, and it is an important test to see if non-perturbative effects that set in when going to larger volumes are able to correct for this unwanted feature. Indeed, when including the effects of the boundary of the fundamental domain, the $0^-/0^+$ mass ratio rapidly increases from slightly below 1 to above. At the same time the slow rise of the $2^+/0^+$ mass ratio becomes more rapid. This behavior is observed both with and without the one loop corrections included. In the latter case the non-perturbative effects set in at $g(R) \sim 2$. Beyond $g(R) \sim 2.35$ it can be shown that the wave functionals start to feel parts of the boundary of the fundamental domain which the present calculation is not representing properly. This value of $g(R)$ corresponds to a circumference of roughly 1.3 fm, when setting the scale as for the torus, assuming the scalar glueball mass in both geometries at this intermediate volume to coincide. That instanton effects are largely responsible for the spin splittings is further confirmed by studies in the instanton liquid model,[134] which concludes that instantons lead to an attractive, neutral and repulsive force in respectively the scalar, tensor and pseudo scalar glueball channels.[135]

6 Large Volume Results

We divide $\mathcal{A}$ by the set of *all* gauge transformations $\mathcal{G}$, including those that are homotopically non-trivial, to get the physical configuration space. All the non-trivial topology is then retrieved by the identifications of points on the boundary of the fundamental domain. This becomes important when the wave functional spreads out in configuration space, which happens at large volumes, whereas at very small volumes the wave functional is localized around $A = 0$ and one need not worry about these non-perturbative effects. That these effects can be dramatic, even at relatively small volumes (above a tenth of a fermi across), was demonstrated for the case of the torus. It has historically been very important that the torus with periodic boundary conditions provides a relatively wide window, from a tenth of a fermi to three quarters of a fermi, where these non-perturbative effects can *reliably* be taken in to account, and were carefully verified in lattice Monte Carlo studies. It shows the intricate dynamics due to the intrinsic non-linearities of the theory and the fact that the issue of gauge copies can not be ignored and is important for the infrared dynamics of the theory.[4]

The hierarchy of boundary effects is effectively described by the various saddle points, that typically lie at the boundary of the fundamental domain. An important lesson can be learned from comparing different geometries, between which the structure of the fundamental domain deviate considerably. As we stressed before, the shape of the fundamental domain is independent of L if the gauge field is expressed in units of $1/L$. It is only the strength of the coupling constant that controls the spreading of the wave functional. Suppose that the coupling constant will grow without bound. This would make the potential irrelevant and makes the wave functional spread out over the whole of field space (which could be seen as a strong coupling expansion). If the kinetic term would have been trivial, the wave functionals would be "plane waves" on a space with complicated boundary conditions. In that case it seems unavoidable that the infinite volume limit would depend on the geometry (like T^3 or S^3) that is scaled-up to infinity. Due to the non-triviality of the kinetic term this conclusion cannot be readily made and our present understanding only allows comparison in volumes around one cubic fermi. However, one way to avoid this undesirable dependence on the geometry is that the vacuum is unstable against domain formation. As periodic subdivisions are space filling on a torus, this seems to be the preferred geometry to study domain formation. It is hard to formulate this in a precise way, let alone find an order parameter for this domain formation. But let us assume domain formation does occur.

Since the ratio of the square root of the string tension to the scalar glueball mass shows no structure around $L = 0.75$ fermi, we assume that within a domain both reach their large volume value. The color electric string now arises from the fact that flux that enters a domain has to leave it at the opposite side. Flux conservation with these building blocks automatically leads to a string picture, with a string tension as computed within a single domain and a transverse size of the string equal to the average size of a domain. The tensor glueball in an intermediate volume is heavily split between the doublet (E^+) and triplet (T_2^+) representations of the cubic group, with resp. 0.9 and 1.7 times the scalar glueball mass. This implies that the tensor glueball is at least as large as the average size of a domain. Rotational invariance in a domain-like vacuum comes about by averaging over all orientations of the domains. This is expected to lead to a mass which is the multiplicity weighted average of the doublet and triplet, yielding a mass of 1.4 times the scalar glueball mass. In the four dimensional euclidean context, $O(4)$ invariance makes us assume that domain formation extends in all four directions. The deconfining temperature would again be set by the average domain size. As is implied by averaging over orientations, domains will not neatly stack. There will be dislocations which most naturally are gauge dislocations. A point-like gauge dislocation in four dimensions is an instanton, lines give rise to monopoles and surfaces to vortices. In the latter two cases most naturally of the Z_N type. We estimate the density of these objects to be one per average domain size. We thus predict an instanton density of $3.2\,\mathrm{fm}^{-4}$, with an average size of $1/3$ fermi. For monopoles we predict a density of $2.4\,\mathrm{fm}^{-3}$. The SU(2) spectra in volumes around the domain size, for the sphere and the torus, are compatible with $m(2^+)/m(0^+) \sim 1.5$ and $m(0^-)/m(0^+) \sim 1.7$. In this picture, we would further predict $\sqrt{K}/m_{0+} \sim 0.24$. All these results seem to give the right order of magnitude, close to the values one measures in a large volume,

$$m(2^+)/m(0^+) = 1.46 \pm 0.09, \quad \sqrt{K}/m_{0+} = 0.267 \pm 0.009,$$
$$m(0^-)/m(0^+) = 1.78 \pm 0.24, \tag{90}$$

taken from a recent review by Teper.[136] It would be useful to have *high-precision* lattice Monte Carlo data around $L = 1$ fermi. In this intermediate volume range, the nature of the states changes from being dominated by zero-momentum fields to genuine particle states. It is important to take this into account in choosing the proper variational basis for the operators in the analysis of the Monte Carlo data. It is not sufficiently appreciated that it is exactly this change of the nature of the states that can provide fundamental physical insight in the *formation* of the mass gap and the confining string.

The studies reported so far in small and intermediate volumes were from first principles. The volume plays the role of the control parameter to keep the strength of the interactions in check. The challenge is to find which are the effective degrees of freedom that drive the formation of the mass gap. The lesson we learned is that large field fluctuations are essential. They enter as sphaleron configurations in the hamiltonian formulation, and are thus associated with instantons. Nevertheless, it seems inescapable that instantons, monopoles and vortices *all* have to be part of the equation,[137] in one way or another.

6.1   Volume Dependence of Stable Particle Masses

Once a mass gap is being formed, this gives of course the relevant degrees of freedom to describe the large distance behavior. For the torus, isolating the volume dependence to come from the propagator winding around the boundary of the box, has led Lüscher to derive the exponential approach of stable particle masses.[12] The method makes cunning use of the fact that the finite volume propagator is given by $\Delta_L(x) = \sum_{n \in \mathbb{Z}^4} \Delta(x + nL)$, where $\Delta(x)$ is the propagator in the infinite volume. With a mass gap the sum over periods will converge, since $\Delta(x) \approx e^{-m|x|}$. Physically $\Delta(x + nL)$ can be identified with propagation from 0 to x, going in addition n times around the box. Going around only once gives the leading finite volume correction of the order e^{-ML}. The value of M depends on the mass gap, and which type of virtual corrections contribute to the self-energy of the particle under consideration. Obviously, without interactions their would be no volume dependence.

Applying these ideas to QCD, one assumes that the long distance behavior is described by an effective theory. In the presence of light quarks this is of course the chiral effective lagrangian. For the calculation of the volume dependence of the glueball, it is assumed to be an effective scalar theory. Confinement implies that propagation of a quark around the boundary, unlike in a small volume, is not an option for large L due to the confining string it has to stretch. Similarly, gluons are confined and thus cannot propagate around the volume. It is only the physical colorless states that can propagate around the box at large L. Therefore, the fractional periods that appear in the perturbative expansion with twisted boundary conditions, see Sec. 3.4, have no effect on the finite size corrections in a large volume. Indeed twisted boundary conditions are devised such that all gauge invariant quantities are strictly periodic.[9]

Vertex functions are not affected by the finite volume, and the expression of a Feynman diagram in a finite volume is exactly of the same form as in an infinite volume, except that integrating over the vertex positions is restricted to the finite volume. The exponential suppression is caused by the finite volume modification in the propagator, specific for periodic boundary conditions. For the sphere, due to the curvature of the manifold, there will be algebraic corrections, and the volume corrections are much harder to establish. To keep track of the volume dependence universal properties of one-particle irreducible vertex functions are studied. A general Feynman diagram $\mathcal{D}$ will correspond to the amplitude

$$\mathcal{J}_L(\mathcal{D}) = \prod_{v \in \mathcal{V} \setminus v_0} \prod_{\mu=0}^{d} \int_0^{L_\mu} \mathrm{d}x_\mu(v) \exp(i \sum_e p(e) \cdot x(e)) \mathbf{V} \prod_{\ell \in \mathcal{L}} \Delta_L(x(f(\ell)) - x(i(\ell))), \tag{91}$$

where $p(e)$ are external momenta. The set of vertices v (v_0 those that are connected to an external line) and propagators ℓ are denoted by $\mathcal{V}$ and $\mathcal{L}$. The product of all vertex factors (coupling constants) is given by $\mathbf{V}$.

To single out those contributions that correspond to propagating n times around the box, one has to take into account the invariance

$$x(v) \rightarrow x(v) + m(v)L, \quad n(\ell) \rightarrow n(\ell) + m(f(\ell)) - m(i(\ell)), \qquad (92)$$

where $m_\mu(v) \in \mathbb{Z}$ and $n(\ell)$ occurs in the expansion of $\Delta_L(x)$. With graph theory Lüscher[12] showed that the sum over $n(\ell)$ splits in a sum over the orbits $[n]$ and a sum over the integers $m(v)$. Obviously one has $\sum_m \int_0^L dx = \int_{-\infty}^\infty dx$ and only the sum over the orbits remains, $\mathcal{J}_L(\mathcal{D}) = \sum_{[n]} \mathcal{J}_L(\mathcal{D}, [n])$,

$$\mathcal{J}_L(\mathcal{D}, [n]) = \prod_{v \in \mathcal{V} \backslash v_0} \prod_{\mu=0}^{d} \int_{-\infty}^{\infty} dx_\mu(v) \exp(i \sum_e p(e) \cdot x(e)) \mathbf{V}$$

$$\times \prod_{\ell \in \mathcal{L}} \Delta(x(f(\ell)) - x(i(\ell)) + n(\ell)L). \qquad (93)$$

The advantage of this decomposition is that for the orbit $[n] = [0]$, this is precisely the infinite volume expression, i.e. $\mathcal{J}_L(\mathcal{D}, [0]) = \mathcal{J}_\infty(\mathcal{D})$. When $[n] \neq [0]$ there is net winding around the boundary, the length of the extra winding given by $W([n])$, thereby leading to an exponential suppression,

$$\mathcal{J}_L(\mathcal{D}, [n]) \approx \exp\left[-mL\kappa(\{p(e)\})W([n])\right]. \qquad (94)$$

The constant $\kappa(\{p(e)\})$ measures, in units of the mass gap m, the lowest momentum flowing through the propagators, constrained by the fixed external momenta. For a self-energy graph $\kappa(\{p(e)\}) = \frac{1}{2}\sqrt{3}$, with the external momenta on-shell $p(e) = (im, \mathbf{0})$, the only case needed here. Furthermore, $W([1]) = 1$ for a "simple" orbit $[1]$ (where $|n(\ell)| = 1$, for one ling and 0 for all others) and $W([n]) \geq \sqrt{2}$ for all other cases.[12]

To determine the finite volume corrections for the lightest stable one-particle state, one takes $L_0 = \infty$ and $L_j - L$ (this simply implies we always have $n_0 = 0$). The mass is measured using $\langle \phi(t)\phi(0) \rangle_L = \exp(-M(L)t)$, where $\phi(t)$ is an interpolating field for this state. This definition of the mass, also used in lattice calculations, coincides with the pole in the finite volume two-point function, $G_L^{-1}(iM(L), 0) = 0$. The two-point function $G_L(p)$ is defined by $G_L^{-1}(p) = p^2 + m^2 - \Sigma_L(p)$, where the normalization of the self energy Σ is such that $\Sigma_\infty(im, \mathbf{0}) = \frac{\partial^2}{\partial p^2}\Sigma_\infty(im, \mathbf{0}) = 0$, which implies that in leading order $\delta M(L) = -\frac{1}{2m}(\Sigma_L - \Sigma_\infty)(im, \mathbf{0}) = -\frac{1}{4m}\Gamma^{(2)}((im, \mathbf{0}), [1])$. This can be expressed in terms of the full four-point Greens function by

$$\Gamma^{(2)}(p, [1]) = \int \frac{d^4 q}{(2\pi)^3} \left[6e^{iq_1 L}G(q)\right] G^{(4)}(p, q, -p, -q), \qquad (95)$$

where $\left[6e^{iq_1 L}G(q)\right]$ comes from summing $\sum_{[n]=[1]} \mathcal{J}_L(\mathcal{D}_2,[1])$ over the Feynman diagrams $\mathcal{D}_2$ that contribute to the full two-point function, and making use of the cubic invariance, to equate the contributions with $n_j = \pm 1$ to $n_1 = 1$.

The full four-point function can be related with the help of Schwinger-Dyson equations to the one-particle irreducible three- and four-point vertex functions. By shifts of integration contours, taking the analyticity domains of the vertex functions properly into account, one finds that the dominating contribution comes from the on-shell three-point vertex functions, denoted by λ, a physical three particle coupling constant. With $p = (im, \mathbf{0})$, and both q and $p + q$ "on-shell", $q_1 = i\sqrt{q_\perp^2 + 3m^2/4}$ and the integral over q, assuming the three point coupling does not vanish, can be shown to give[12]

$$\delta M(L) = -\frac{3\lambda^2}{16\pi m^2 L}\exp(-\tfrac{1}{2}\sqrt{3}mL) + \mathcal{O}\left(e^{-mL}\right). \tag{96}$$

The subleading term, of order e^{-mL}, is due to the contribution coming from the four-point function $\Gamma^{(4)}$.

The derivation has been given for the volume dependence of the mass gap, e.g. for the scalar glueball mass in pure gauge theory. The results also apply for the cases of a stable boundstate particle, with mass $m_B = 2m - E_B$, and for a nucleon that couples to a pion. One only needs to adjust the "on-shell" condition, $q_1 = i\sqrt{q_\perp^2 + \mu_B^2}$, for the different kinematical situation. For the bound state one finds $\mu_B^2 \equiv m^2 - m_B^2/4$, which leads to $\delta M_B(L) = -3\lambda^2 \exp(-\mu_B L)/(16\pi m_B^2 L)$, whereas for the nucleon $\mu_N^2 \equiv m_\pi^2 - m_\pi^4/4m_N^2$, giving $\delta M_N(L) = -9m_\pi^2 g_{N\pi}^2 \exp(-\mu_N L)/(16\pi m_N^2 L)$, where λ has been expressed in the appropriate dimensionless pion-nucleon coupling constant $g_{N\pi}$. For the finite volume correction to the pion mass in QCD, the pion three-point vertex vanishes and one gets a slightly more complicated result of order $\exp(-m_\pi L)$, involving an integral over the forward scattering amplitude.[12]

6.2 Volume Dependence from Scattering Phase Shifts

Also it was shown by Lüscher how to extract scattering phase shifts from volume dependence.[138,139] A less rigorous method, based on the notion of pseudo potentials, allows for a transparent way of understanding this all order result.[140] Consider the reduced hamiltonian $H = -m^{-1}\partial_\mathbf{x}^2 + V(\mathbf{x})$ for two interacting particles in the center of mass in ordinary quantum mechanics with a (Bose) symmetric reduced wave functions, $\psi(\mathbf{x}) = \psi(-\mathbf{x})$. When the potential has a finite range λ ($V(\mathbf{x}) = 0$ for $|\mathbf{x}| > \lambda$), the wave function is a plane-wave outside of the interaction region. Scattering theory tells us there is a unique relation between the incoming wave $e^{-ip|x|}$ and the outgoing wave $e^{ik|x|}$ in terms of the scattering phase shift $\delta(k)$

$$\psi(x) = e^{-ik|x|} + e^{2i\delta(k)}e^{ik|x|}, \quad |x| > \lambda. \tag{97}$$

These states form a complete basis of scattering states. In a finite volume the periodic boundary condition, $\psi(x + L/2) = \psi(x - L/2)$, implies the following implicit equation for the momenta

$$e^{2i\delta(k)}e^{ikL} = 1, \tag{98}$$

which holds as long as $L > 2\lambda$, showing how the volume dependence of two-particle states are related to the phase shift. For a vanishing potential the phase shift vanishes, $\delta(k) = 0$, and one recovers the standard discretization of the momenta, $k = 2\pi n/L$.

In more than one dimension a complete set of scattering wave functions is given by

$$\psi_{\ell m}^{(\mathbf{k})}(\mathbf{x}) = [j_\ell(|\mathbf{k}|r) - \tan(\delta_\ell(\mathbf{k}))n_\ell(|\mathbf{k}|r)]Y_{\ell m}(\hat{\mathbf{k}}), \tag{99}$$

where $\hat{\mathbf{k}} = \mathbf{k}/|\mathbf{k}|$, $r = |\mathbf{x}|$ and j_ℓ, n_ℓ are the spherical Bessel-functions. The energy is given by $E = \mathbf{k}^2/m$. It appears that the spherical nature of the scattered waves makes it impossible to impose periodic boundary conditions. One can, however, introduce the notion of a pseudo potential as was introduced for the hard-sphere bose gas by Huang and Yang.[141] The essential idea is equally simple as powerful. Replace $V(\mathbf{x})$ by the simple pseudo potential $V_\delta(\mathbf{x})$ such that for $|\mathbf{x}| > \lambda$, Eq. (99) is still an exact solution of the relevant Schrödinger equation. Subsequently one solves the Schrödinger equation with V_δ as its (energy dependent) potential, using periodic boundary conditions. We illustrate this by making the simplified assumption that all phase shifts vanish, except for δ_0. In that case one easily verifies that the pseudopotential is given by

$$V_\delta(\mathbf{x}) = -\frac{4\pi}{mk}\tan(\delta_0(\mathbf{k}))\hat{\delta}_3(\mathbf{x}). \tag{100}$$

Since we have to allow for wave functions that are singular at the origin, we can extend our class of functions to those that, when averaged over the angles, have a Laurent expansion $\sum_{n=-N}^{\infty} c_n r^n$. We define $\hat{\delta}_3(\mathbf{x})\psi(\mathbf{x}) = c_0\delta_3(\mathbf{x})$. Alternatively, $\hat{\delta}_3(\mathbf{x})\psi(\mathbf{x}) = \frac{\partial}{\partial r}(r\psi(\mathbf{x}))\delta_3(\mathbf{x})$. We expand the wave function in plane waves, suitable for implementing periodic boundary conditions, $\psi(\mathbf{x}) = L^{-3}\sum b_{\mathbf{n}}\exp(2\pi i\mathbf{x}\cdot\mathbf{n}/L)$. Substituting this in the relevant Schrödinger equation

$$\left(-\frac{1}{m}\frac{\partial^2}{\partial\mathbf{x}^2} + V_\delta(\mathbf{x})\right)\psi(\mathbf{x}) = \frac{\mathbf{k}^2}{m}\psi(\mathbf{x}), \tag{101}$$

is easily seen to give

$$b_{\mathbf{n}} = -\frac{4\pi\tan(\delta_0(\mathbf{k}))}{|\mathbf{k}|(\mathbf{k}^2 - (2\pi\mathbf{n}/L)^2)}c_0. \tag{102}$$

With $c_0 = \sum b_{\mathbf{n}}$ we thus find a relation between the momentum in the center of mass and the scattering phase shift *at this* momentum

$$\frac{\tan(\delta_0(\mathbf{k}))\mathcal{Z}_{00}(1;\mathbf{q})}{2\pi^2|\mathbf{q}|} = 1, \quad \mathbf{q} = \frac{\mathbf{k}L}{2\pi}. \tag{103}$$

The zeta-function $\mathcal{Z}_{00}(s;\mathbf{q}) \equiv \sum_{\mathbf{n}\in\mathbb{Z}^3}(\mathbf{n}^2 - \mathbf{q}^2)^{-s}$ is defined through analytic continuation in s.

To obtain this result c_0 should be non-vanishing. There can be "singular" solutions[139] for which $c_0 = 0$ at momenta $\mathbf{k} = 2\pi\mathbf{n}/L$, if there exists $\mathbf{n}' \in \mathbb{Z}^3$, such that $|\mathbf{n}| = |\mathbf{n}'|$. In that case $\psi(\mathbf{x}) = \exp(2\pi i\mathbf{n}\cdot\mathbf{x}/L) - \exp(2\pi i\mathbf{n}'\cdot\mathbf{x}/L)$ is regular and vanishes in the origin. If ψ is in the scalar representation (A_1) of the cubic group, this singular behavior only occurs when $\mathbf{n}'$ is not related to $\mathbf{n}$ by a cubic transformation, which will make the momentum where this occurs quite large. Furthermore, restricting to the A_1 sector, Eq. (103) will also be valid if the phase shifts only vanish for angular momenta $\ell \geq 3$. This is because a spin 2 wave function decomposes in the E and T_2 representations of the cubic group and hence does not couple to the scalar sector (note that due to the Bose symmetry all odd angular momentum phase shifts will vanish).

In field theory it can be shown that the reduced hamiltonian is replaced by an effective Schrödinger equation, that can be derived from the Bethe-Salpeter equation for the four-point function.[138] Its energy-dependent "potential" is proportional to the so-called Bethe-Salpeter kernel, with the range λ determined by the polarization cloud (which is why we restrict the analysis to field theories with a mass gap). The fully relativistic two-particle energy W is given by $W = 2\sqrt{\mathbf{k}^2 + m^2} = 2\sqrt{m(m+E)}$, where $\mathbf{k}$ is the center of mass momentum of the scattering pair. Lüscher's[139] all order analysis is based on studying the solutions of the Helmholtz equation $(\partial^2/\partial\mathbf{x}^2 + \mathbf{k}^2)\psi(\mathbf{x}) = 0$ in a finite volume, allowing for power-like singularities at the origin. He demonstrates that truncating to a finite number of phase shifts δ_ℓ, ($\ell < \ell_{max}$) in general converges rapidly with ℓ_{max}. For the simplified case that only δ_0 is non-vanishing, the result is as in Eq. (103).

One of the applications is computing the energy of the two-particle states as a function of L. Using that $\tan(\delta_0(\mathbf{k})) = a_0|\mathbf{k}| + \mathcal{O}\left(k^3\right)$, with a_0 the so-called scattering length and introducing $Z_{00}(s;\mathbf{q}) = \mathcal{Z}_{00}(s;\mathbf{q}) - (-\mathbf{q}^2)^{-s}$, which satisfies

$$\mathcal{Z}_{00}(1;\mathbf{q}) = -\frac{1}{\mathbf{q}^2} + Z_{00}(1;\mathbf{0}) + \mathbf{q}^2 Z_{00}(2;\mathbf{0}) + \mathcal{O}\left(q^4\right), \tag{104}$$

one can now straightforwardly iterate Eq. (103) and find

$$\mathbf{k}^2 = -\frac{4\pi a_0}{L^3}\left(1 + \frac{c_1 a_0}{L} + \frac{c_2 a_0^2}{L^2}\right) + \mathcal{O}\left(L^{-6}\right), \tag{105}$$

$$c_1 = \frac{Z_{00}(1;\mathbf{0})}{\pi}, \quad c_2 = \frac{Z_{00}^2(1;\mathbf{0}) - Z_{00}(2;\mathbf{0})}{\pi^2}.$$

The numerical values[138,141] are $c_1 = -2.837297$ and $c_2 = 6.375183$. When the two-particle energies come close to the mass of a resonance, like the ρ or K meson coupling to two-particle pion states, the phase shifts get large and it becomes imperative to use the all order results.[139]

Many of the aspects of the large volume expansions for the one-particle and two-particle masses have been verified successfully for the two dimensional non-linear sigma[142] and Ising model[143], and for the four dimensional Ising[144] and $O(4)$ ϕ^4 model in the symmetric phase.[145] Recently the method has been revisited by Lellouch and Lüscher[146] for studying on the lattice non-perturbatively the decay $K \to \pi\pi$, important for understanding direct CP violation. By studying the decay in a finite volume, the transition amplitude of the kaon to the discrete set of finite volume two-pion final states can be related to the infinite volume decay rate. With the advance in computer power this calculation will be feasible in the future.

6.3 Goldstone Modes and Chiral Perturbation Theory

The finite size corrections to the string tension $K(L)$ also rely on an effective description, namely that of the bosonic relativistic string. The finite size behavior is described in terms of the force by

$$F(L) = \frac{dE(L)}{dL} = \frac{dLK(L)}{dL} = K + \frac{\pi}{3}L^{-2} + \mathcal{O}\left(L^{-3}\right), \qquad (106)$$

with K the infinite volume string tension. The so-called Lüscher term[147] with the $1/L^2$ correction is universal. A simple way to understand this correction is in terms of the Casimir energy in one space dimension (the string) on a periodic interval of length L (the length of the string) for two massless scalar fields (the two independent transverse fluctuations of the string), $2 \cdot \frac{1}{2} \sum_k 2\pi|k|/L = 4\pi\zeta(-1)/L$ (in dimensional regularization, with ζ the Riemann zeta function). Lüscher's original derivation was for a string with fixed end-points, for which the Casimir energy is a quarter of the one with periodic boundary conditions. The transverse fluctuations are examples of Goldstone bosons, that here arise due to the spontaneous breakdown of the transverse translational invariance in the background of a string.[147]

Goldstone bosons do of course also play a very important role in the low-energy description of QCD in the form of pions. They occur with massless quarks due to the spontaneous breakdown of chiral symmetry. With an explicit mass term for the quarks, the pions acquire a mass and the results of Secs. 6.1 and 6.2 apply as long as $m_\pi L \gg 1$. For massless pions this can of course not be realized, and instead chiral perturbation theory becomes an effective tool. It also applies for $m_\pi \neq 0$ provided L is in the so-called mesoscopic range, which is $m_\pi^{-1} \gg L \gg \Lambda_{QCD}^{-1}$. The principle behind chiral perturbation theory is that the symmetries of the low-energy theory strongly constrain the effective lagrangian. The resulting sigma model has its symmetry determined by the number of light quark flavors and the gauge group. For SU(3) with two light flavors, the up and the down quark, this is the O(4) sigma model spontaneously broken down to O(3) by the chiral condensate. Finite size effects occur in powers of $1/L$, but are determined by infinite volume quantities, computed essentially through replacing the continuous by discrete momenta.[148]

From the practical point of view this gives access to these infinite volume quantities through finite size effects. An important example is the chiral condensate, which can be related through the Leutwyler-Smilga sum rules[149] to eigenvalue distributions of the Dirac operator (averaging over the gauge fields). It is the average spacing of these eigenvalues near zero that through the Banks-Casher formula[150] gives access to the chiral condensate $\langle \Delta\lambda \rangle = \pi(L^3 T | \langle \overline{\psi}\psi \rangle |)^{-1}$. The roughness of the gauge field causes $\langle \Delta\lambda \rangle$ to be inversely proportional to the volume $(L^3 T)$ of space-time. At weak coupling, like in a small volume, the typical eigenvalue spacing would be $1/L$, but also we saw in Sec. 3.2 there are no near-zero eigenvalues and chiral symmetry is manifest.

Remarkably, in the mesoscopic domain the low-lying eigenvalue distribution $\langle \rho(\lambda) \rangle$ is related to a universal function called the microscopic spectral density,[151] $\rho_S(u) = \lim_{V \to \infty}(V\Sigma)^{-1}\langle \rho\left((V\Sigma)^{-1}u\right)\rangle$. It can be calculated in Random Matrix Theory, leaving the chiral condensate $\Sigma = |\langle \overline{\psi}\psi \rangle|$ as a single free parameter. We refer to a review by Verbaarschot[152] for details.

6.4 Electric-Magnetic Duality

Much of the finite volume work in gauge theories has started with 't Hooft's paper on the torus with twisted boundary conditions.[9] We come back to it one more time to show how it leads to an electric-magnetic duality, which can be used to derive finite size results for the energy of magnetic flux in terms of the string tension, assuming one has a confining string.

Consider the gauge group SU(N) and fix the magnetic flux $\mathbf{m}$ by twisted boundary conditions. In the $A_0 = 0$ gauge this can be specified as in Eq. (31). Given these boundary conditions, the remaining gauge freedom is specified by the twisted gauge transformations $h_{\mathbf{k}}$, defined in Eq. (35), where $\mathbf{k}$ classifies the element of the homotopy group $H_2(T^3, Z_N) \sim Z_N^3$. Two twisted gauge transformations with the same value of $\mathbf{k} \in Z_N^3$ differ by a gauge transformation with integer winding number. A particular representative of $h_{\mathbf{k}}$ can be chosen such that[50] $\nu(h_{\mathbf{k}}) = -\mathbf{k}\cdot\mathbf{m}/N$, as we argued below Eq. (50). The generalization of Eq. (20) to $\mathbf{m} \neq \mathbf{0}$ can now be written as

$$\Psi([h_{\mathbf{k}} h_1^\nu]\, A) = e^{i\theta\nu} \exp\left(\frac{2\pi i \mathbf{e}_\theta \cdot \mathbf{k}}{N}\right) \Psi(A), \quad \mathbf{e}_\theta \equiv \mathbf{e} - \frac{\theta}{2\pi}\mathbf{m}. \qquad (107)$$

The non-trivial θ dependence implies that magnetic flux will carry in general a fractional electric flux. This result is intimately related to the observation of Witten[153] that the θ dependence can be implemented at the hamiltonian level, by replacing the canonical momentum belonging to the gauge field, i.e. the electric field $E_i(\mathbf{x})$, by $E_i(\mathbf{x}) - \frac{\theta}{4\pi^2}B_i$, with $B_i(\mathbf{x}) \equiv \frac{1}{2}\epsilon_{ijk}F_{jk}(\mathbf{x})$ the magnetic field. This behavior, of magnetic charge obtaining a θ dependent electric charge, also plays a role in 't Hooft's abelian projection and oblique confinement.[154]

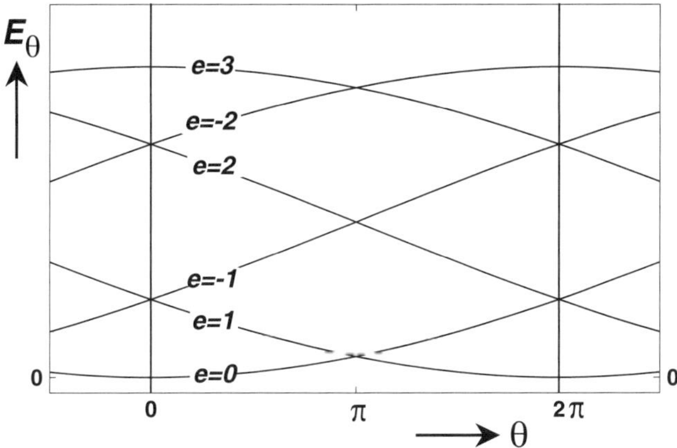

Figure 20: The SU(6) spectral flow[72] of E_θ in a small volume, Eq. (108), representing the energy of electric flux for $\mathbf{m} = (0,0,1)$ and $e_1 = e_2 = 0$. The spectrum is periodic with a period 2π, the flow only with a period of 12π.

It seems that the 2π periodicity in θ needs to be replaced by $2\pi N$. Nevertheless, the spectrum itself is periodic with period 2π. There is, however, a non-trivial spectral flow, which in the infinite volume may be non-analytic and give rise to oblique confinement.[154] The spectral flow can neatly be illustrated in a finite volume, where energies are required to be analytic in θ. For this we consider $\mathbf{m} = (0,0,1)$ and recall from Sec. 3.4 that the energy splittings for $\mathbf{e} = (0,0,e_3)$ are due to tunneling with fractionally charged instantons. One easily computes the resulting energies through standard instanton calculations,[155,156] using the fact that the twisted instanton has only four zero-modes due to translations (cmp. Sec. 4),

$$E_\theta(\mathbf{e} = e\mathbf{m} = e e^{(3)}) = \mathcal{C}\sin^2(\pi|\mathbf{e}_\theta|/N)\exp\left[-8\pi^2/Ng^2(L)\right]/Lg^4(L). \quad (108)$$

The behavior of E_θ is illustrated, for $N = 6$, in Fig. 20 from which it is immediately clear that the spectrum is periodic with a period of 2π, but that following the energy levels adiabatically, one only returns to the same situation after $N = 6$ periods.

We now consider the euclidean partition function at finite temperature, specified by $1/\beta$, where β is the period in the euclidean time direction. For simplicity we will also assume the period lattice Ξ is orthogonal, $a^{(\mu)} = L_\mu e^{(\mu)}$ with $L_0 \equiv \beta$. The free energy $F(\mathbf{e}, \mathbf{m}, L_\mu)$ is defined for the different superselection sectors as

$$\exp\left[-\beta F_\theta(\mathbf{e}, \mathbf{m}, L_\mu)\right] = \mathrm{Tr}_{\,\theta,\mathbf{e}}\left[\exp(-\beta H_\mathbf{m})\right], \qquad (109)$$

where the trace is here over the Hilbert space with definite θ, electric and magnetic quantum numbers. To relate the finite temperature partition function to the euclidean path integral, θ and the electric flux have to be implemented as a sum over the homotopy classes ($\nu \in \mathbb{Z}$ and $\mathbf{k} \in Z_N^3$) of the gauge transformations. In the $A_0 = 0$ gauge, the gauge field at $t = 0$ is related to that at $t = \beta$ by $A(\mathbf{x}, \beta) = [h_\mathbf{k} h_1^\nu] A(\mathbf{x}, 0)$. For a given choice of ν and $\mathbf{k}$ we denote the euclidean path integral by $W_\nu(\mathbf{k}, \mathbf{m}, L_\mu) = W_\nu(n, \Xi)$, where n denotes the antisymmetric integer twist tensor, with $n_{ij} = \epsilon_{ijl} m_l$ and $n_{0i} = k_i$ specifying the twisted boundary conditions, which can now be formulated in any gauge. The relation between the free energy and the euclidean path integral reads

$$\exp[-\beta F_\theta(\mathbf{c}, \mathbf{m}, L_\mu)] = N^{-3} \sum_{\mathbf{k} \in Z_N^3} \sum_{\nu \in \mathbb{Z}} e^{i\nu\theta} \mathrm{z}_N^{-\mathbf{k}\cdot\mathbf{e}_\theta} W_\nu(\mathbf{k}, \mathbf{m}, L_\mu), \qquad (110)$$

where we introduced the N-th root of unity, $\mathrm{z}_N \equiv e^{2\pi i/N}$.

The alternate notation, $W_\nu(n_{\mu\nu}, \Xi)$, makes manifest that from the point of view of the euclidean path integral there is no essential distinction between the twisted boundary conditions in the space and the time directions. As observed by 't Hooft, this path integral is invariant under a joint SO(4) rotation of the period lattice Ξ and the twist tensor n. This nearly trivial fact, however, results in a duality between the electric and magnetic sectors.[9] A simultaneous rotation in the 1-2 and the 3-0 planes over 90 degrees, interchanges (k_1, k_2) with (m_1, m_2), leaves k_3 and m_3 unchanged and maps L_μ to $\tilde{L}_\mu = (L_3, L_2, L_1, \beta)$. Substituting this in Eq. (110) leads to the *exact* duality[9]

$$\exp[-\beta F_\theta(\mathbf{e}, \mathbf{m}, L_\mu)] = N^{-2} \sum_{\tilde{\mathbf{m}}, \tilde{\mathbf{e}} \in Z_N^3} \delta_{\tilde{m}_3 m_3} \delta_{\tilde{e}_3 e_3}\, \mathrm{z}_N^{\tilde{\mathbf{e}}\cdot\mathbf{m} - \tilde{\mathbf{m}}\cdot\mathbf{e}} \exp[-L_3 F_\theta(\tilde{\mathbf{e}}, \tilde{\mathbf{m}}, \tilde{L}_\mu)].$$

$$(111)$$

It is because $\tilde{\mathbf{e}}_\theta \cdot \mathbf{m} - \tilde{\mathbf{m}} \cdot \mathbf{e}_\theta$, is independent of θ, that θ does not explicitly appear in this equation.[73] Hence, we will in the following ignore the θ label.

Let us normalize $F(\mathbf{0}, \mathbf{0}, L_\mu) = 0$, and call those flux states for which $L_\mu F(\mathbf{e}, \mathbf{m}, L_\mu)$ tends to zero *light fluxes*. That they cannot all be light is easily seen by summing Eq. (111) over m_1, m_2, e_1 and e_2,

$$N^{-2} \sum_{e_1, e_2, m_1, m_2} \exp[-\beta F(\mathbf{e}, \mathbf{m}, L_\mu)] = \exp[-L_3 F(e_3 \mathbf{e}^{(3)}, m_3 \mathbf{e}^{(3)}, \tilde{L}_\mu)]. \qquad (112)$$

If $(e_3\mathbf{e}^{(3)}, m_3\mathbf{e}^{(3)})$ produces a light flux state, there must be $N^2 - 1$ *additional* light flux states (with the same e_3 and m_3). On the other hand, since the euclidean path integral is positive, irrespective the twist, once the flux $(\mathbf{e}, \mathbf{m})$ is light, so must be the flux $(\mathbf{0}, \mathbf{m})$ (see Eq. (110)). For SU(2) and SU(3) it can thus be concluded[9] that, in the confined phase, the magnetic fluxes have a vanishing free energy in the infinite volume limit.

It will be assumed that for $\beta \gg L_i$ the free energy factorizes in a magnetic and electric component, $F(\mathbf{e}, \mathbf{m}, L_\mu) = F_m(\mathbf{m}, L_\mu) + F_e(\mathbf{e}, L_\mu)$. This can be justified by the fact that the electric flux is squeezed into a string with a finite width (to give rise to a linear potential), which at high β takes negligible space, whereas magnetic flux spreads out over the whole volume (for the free energy to vanish). We now show how to compute the volume dependence of the magnetic energy, $E_m(\mathbf{m}, L_i) \equiv \lim_{\beta \to \infty}(F_m(\mathbf{m}, L_\mu) - F_m(\mathbf{0}, L_\mu))$, for the case where $m_3 = 0$. Let us choose $e_3 = 0$ and (for simplicity) $L_1 = L_2 = L$, with $\beta, L_3 \gg L$. From Eq. (111) it follows that

$$\mathcal{C} \exp[-\beta F_m(\mathbf{m}_\perp, L_\mu)] = N^{-2} \sum_{\tilde{\mathbf{e}}_\perp \in Z_N^2} z_N^{\tilde{\mathbf{e}}_\perp \cdot \mathbf{m}_\perp} \exp[-L_3 F_e(\tilde{\mathbf{e}}_\perp, \tilde{L}_\mu)]. \tag{113}$$

The normalization constant $\mathcal{C}$ (obtained by putting $\mathbf{m} = \mathbf{0}$) can be absorbed in F_m. One can compute the electric free energies from the statistical distribution of flux strings of energy KL, running in either the 1- or 2-direction, to build up the fluxes e_i from $n_i + e_i$ strings running in the i-direction and n_i in the opposite direction. For SU(2), fluxes running in opposite directions are equivalent, and one instead sums over n_i even for $e_i = 0$ and n_i odd for $e_i = 1$. Entropy allows us to ignore higher units of electric flux (including the ones running diagonally). Strings running in the *long* 3-directions can be neglected as well. The Boltzmann weight of one string equals $LL_3 \exp(-\beta KL)/\mathcal{A}$, where $\mathcal{A}$ is a constant related to the "proper area" of the string. It is now straightforward to compute the electric free energy, substitute the result in Eq. (113) and obtain[9] the magnetic energy by extracting the limiting behavior in $\beta \to \infty$,

$$E_m = 2(2 - \delta_{N,2})\big[\sin^2(\pi m_1/N) + \sin^2(\pi m_2/N)\big]\mathcal{A}^{-1}L \exp(-LL_3K). \tag{114}$$

Thus the magnetic energy falls off in leading order with the same area law as the Wilson loop expectation value.

7 Conclusions

I have tried to convince the reader of the usefulness of a finite volume as a control in studying four dimensional non-abelian gauge theories, with for the torus the practical benefit of providing a guide for lattice Monte Carlo results. There are many issues I have not addressed here. Needless to say the choices have been determined by the things I have worked on in the past, or that were close to me. I hope to have at least succeeded in developing this theme here in a logical and pedagogical fashion. It is my belief that probing the non-perturbative dynamics through spectral properties is the only reliable way to gain insight. This is not to say it is only through finite volume studies we may make progress. There is of course the large N expansion,[158,159] and the hope for a string representation of QCD.[160] It should also be said we have followed the standard paradigm, that confinement is first to be understood in the pure gauge sector, instead of taking the attitude that the light quarks provided by nature is what mostly matters.[157]

On the basis of this review we list a number of open problems that are worthwhile addressing.

1. Solve the problem with the adiabatic approximation in the calculation of the Witten index in small volume supersymmetric gauge theories.

2. Find analytic solutions for the basic instantons on the torus, with and without twist.

3. Isolate the associated sphaleron configurations and non-perturbatively important degrees of freedom.

4. Map out with *high-precision* the low-lying spectrum around 1 fermi.

5. Establish in a *reliable* way the physical nature of the finite volume cross-over near 1 fermi, and find the intermediate-distance effective degrees of freedom, that are responsible for the mass gap and string formation.

The first two items are interesting technical problems in their own, which one may hope are close to being solved. They have a wider range of applications than in the context of QCD alone. The fourth item does not have to wait for analytic predictions, and can be addressed by Monte Carlo methods with present day resources. The last item is of course what it is all about. It is unlikely this can be solved by a single mathematical equation, or with mathematical rigor.[161] Fortunately, we do have experiment to guide us here. Despite the many phenomenological models, victory can not be claimed yet. The scenario of Mandelstam and 't Hooft, for describing QCD in terms of a dual superconductor,[162,163] remains an appealing one and gets some support from supersymmetric gauge theories exhibiting Seiberg-Witten duality,[164] even though their matter content is uncomfortably far from that of QCD.

The challenge of *really* understanding QCD non-perturbatively remains wide open, even though many in the particle physics community escape to other dimensions. Some of them hope that from this higher vantage point we can look down upon this problem that was left over from the twentieth century, and solve it with twenty-first century techniques. Anything is allowed, as long as we come up with a practical and reliable solution.

Acknowledgements

I thank Misha Shifman for inviting me to contribute to this volume. I somewhat hesitatingly accepted, but while working on it started to enjoy bringing these results together in one place. I hope he is satisfied with the final result. Of course I also hope Boris Ioffe, whom I only met briefly on one of my visits to ITEP, is happy with this chapter in his honor. I have covered an aspect of QCD that has not touched on many of his deep contributions to the field. Let me just say that I have always felt a special tie to my Russian friends, for whom physics is a way of life, something Ioffe was very much a part of. May I express the hope that in a changing world, this way of life does not completely disappear. Finally, I am grateful to the many people I worked with and who have all contributed in an essential way to developing the material presented.

References

1. C.N. Yang and R.L. Mills, *Phys. Rev.* **96**, 191 (1954).
2. N.M. Christ and T.D. Lee, *Phys. Rev.* D **22**, 939 (1980).
3. O. Babelon and C. Viallet, *Comm. Math. Phys.* **81**, 515 (1981).
4. V.N. Gribov, *Nucl. Phys.* B **139**, 1 (1978).
5. M.A. Semenov-Tyan-Shanskii and V.A. Franke, *Zapiski Nauchnykh Sem. Leningradskogo Otdeleniya Mat. Inst. im. V.A. Steklov AN SSSR* **120**, 159 (1982), Translation (Plenum, New York, 1986) p 999.
6. J.D. Bjorken, "Elements of Quantum Chromodynamics", in *Slac Summer Institute on Particle Physics*, ed. A. Mosher (SLAC, Stanford, 1980).
7. M. Lüscher, *Nucl. Phys.* B **219**, 233 (1983).
8. M. Lüscher, R. Sommer, U. Wolff and P. Weisz, *Nucl. Phys.* B **389**, 247 (1993) (hep-lat/9207010); *Nucl. Phys.* B **413**, 481 (1994) (hep-lat/9309005).
9. G. 't Hooft, *Nucl. Phys.* B **153**, 141 (1979).
10. E. Witten, *Nucl. Phys.* B **202**, 253 (1982).
11. W. Nahm, *Phys. Lett.* B **90**, 413 (1980); in *Monopoles in Quantum Field Theory*, eds. N. Craigie *et al* (World Scientific, Singapore, 1982); in *Lect. Notes in Phys.* vol **201**, eds. G. Denardo *et al* (Springer, Berlin, 1985).
12. M. Lüscher in *Progress in Gauge Field Theory*, eds. G. 't Hooft *et al* (Plenum, New York, 1984) p 451; *Comm. Math. Phys.* **104**, 177 (1986).
13. H. Leutwyler, *Chiral Dynamics*, this volume (hep-ph/0008124).
14. G. Dell'Antonio and D. Zwanziger, *Comm. Math. Phys.* **138**, 291

(1991); in *Probabilistic Methods in Quantum Field Theory and Quantum Gravity*, eds. P.H. Damgaard *et al* (Plenum, New York, 1990) p 107.

15. G. Dell'Antonio and D. Zwanziger, *Nucl. Phys.* B **326**, 333 (1989).
16. D. Zwanziger, *Nucl. Phys.* B **378**, 525 (1992).
17. P. van Baal, *Nucl. Phys.* B **369**, 259 (1992).
18. J. Koller and P. van Baal, *Nucl. Phys.* B **302**, 1 (1988).
19. P. van Baal, *Nucl. Phys.* B **307**, 274 (1988) [erratum B**312**, 752 (1989)].
20. P. van Baal, *Nucl. Phys.* B **264**, 548 (1986).
21. A. Belavin, A. Polyakov, A. Schwarz and Y. Tyupkin, *Phys. Lett.* B **59**, 85 (1975).
22. G. 't Hooft, *Phys. Rev.* D **14**, 3432 (1976) [erratum D**18**, 2199 (1978)].
23. J. Koller and P. van Baal, *Phys. Rev. Lett.* **58**, 2511 (1987).
24. C. Bloch, *Nucl. Phys.* **6**, 329 (1958).
25. P. van Baal, *Nucl. Phys.* B **351**, 183 (1991).
26. A. Auerbach, S. Kivelson and D. Nicole, *Phys. Rev. Lett.* **53**, 411 (1984); A. Auerbach and S. Kivelson, *Nucl. Phys.* B **257**, 799 (1985).
27. P. van Baal and A. Auerbach, *Nucl. Phys.* B **275**, 93 (1986).
28. P. van Baal in *Lectures on Path Integration*, eds. H.A. Cerdeira *et al* (World Scientific, Singapore, 1993) p 54.
29. J. Koller and P. van Baal, *Nucl. Phys.* B **273**, 387 (1986).
30. P. van Baal and J. Koller, *Ann. Phys.* (N.Y.) **174**, 299 (1987).
31. M. Lüscher and G. Münster, *Nucl. Phys.* B **232**, 445 (1984).
32. C. Vohwinkel, *Phys. Lett.* B **213**, 54 (1988); *Nucl. Phys.* B(Proc. Suppl.) **9**, 242 (1989).
33. M. García Pérez and P. van Baal, *Nucl. Phys.* B **429**, 451 (1994) (hep-lat/9403026).
34. B.A. Berg and A.H. Billoire, *Phys. Lett.* B **166**, 203 (1986) [erratum B**185**, 446 (1987)]; B.A. Berg, A.H. Billoire and C. Vohwinkel, *Phys. Rev. Lett.* **57**, 400 (1986).
35. C. Michael, G.A. Tickle and M.J. Teper, *Phys. Lett.* B **207**, 313 (1988).
36. J. Kripfganz and C. Michael, *Nucl. Phys.* B **314**, 25 (1989).
37. B.A. Berg, *Phys. Lett.* B **206**, 97 (1988).
38. C. Michael, *Phys. Lett.* B **232**, 247 (1989).
39. P. Hasenfratz, A. Hasenfratz and F. Niedermayer, *Nucl. Phys.* B **329**, 739 (1990).
40. B.A. Berg and A.H. Billoire, *Phys. Rev.* D **40**, 550 (1989).
41. C. Michael, *Nucl. Phys.* B **329**, 225 (1990).
42. M. García Pérez, J. Snippe and P. van Baal, *Phys. Lett.* B **389**, 112 (1996) (hep-lat/9608036).
43. B.A. Berg and C. Vohwinkel, *Ann. Phys.* (N.Y.) **204**, 351 (1990).
44. C. Vohwinkel, *Calculation of the Mass Spectrum and Deconfining Temperature in Non-Abelian Gauge Theory*, PhD thesis (Tallahassee, Septem-

ber 1989).

45. P. van Baal, *Phys. Lett.* B **224**, 397 (1989);
    Nucl. Phys. B(Proc. Suppl.) **17**, 581 (1990).

46. P. Weisz and V. Ziemann, *Nucl. Phys.* B **284**, 157 (1987).

47. J. Koller and P. van Baal, *Phys. Rev. Lett.* **57**, 2783 (1986).

48. C. Vohwinkel, *Phys. Rev. Lett.* **63**, 2544 (1989).

49. P. van Baal in *Probabilistic Methods in Quantum Field Theory and Quantum Gravity*, eds. P.H. Damgaard *et al* (Plenum, New York, 1990) p 31.

50. P. van Baal, *Comm. Math. Phys.* **85**, 529 (1982).

51. J. Hoek, *Nucl. Phys.* B **332**, 530 (1990).

52. J. Kripfganz and C. Michael, *Phys. Lett.* B **209**, 77 (1988).

53. P. van Baal in *Frontiers in Nonperturbative Field Theory*, eds. Z. Horvath *et al*, (World Scientific, Singapore, 1989), p 204.

54. P. van Baal and A.S. Kronfeld,
    Nucl. Phys. B(Proc. Suppl.) **9**, 227 (1989).

55. C. Vohwinkel, *Nucl. Phys.* B **443**, 417 (1995) (hep-lat/9410010).

56. H. Tiedemann, *Phys. Rev.* D **44**, 1280 (1991).

57. L. Polley and U. Wiese, *Nucl. Phys.* B **356**, 629 (1991).

58. A. Kronfeld and U. Wiese, *Nucl. Phys.* B **357**, 521 (1991).

59. A. Kronfeld and U. Wiese, *Nucl. Phys.* B **401**, 190 (1993)
    (hep-lat/9210008).

60. M. Lüscher, *Project Proposal for the EMC^2 Collaboration*, unpublished notes, December 1983.

61. M. Lüscher, P. Weisz and U. Wolff, *Nucl. Phys.* B **359**, 221 (1991).

62. M. Lüscher, R. Narayanan, P. Weisz and U. Wolff,
    Nucl. Phys. B **384**, 168 (1992) (hep-lat/9207009).

63. G.M. de Divitiis, R. Frezzotti, M. Guagnelli and R. Petronzio,
    Nucl. Phys. B **422**, 382 (1994) (hep-lat/9312085);
    Nucl. Phys. B **433**, 390 (1995) (hep-lat/9407028).

64. A. González-Arroyo, J. Jurkiewicz and C.P. Korthals Altes, in *Proc. 11th NATO Summer Institute*, eds. J. Honerkamp *et al* (Plenum, New York, 1982); A. Coste, A. González-Arroyo, C.P. Korthals Altes, B. Söderberg and A. Tarancon, *Nucl. Phys.* B **287**, 569 (1987).

65. G.M. de Divitiis, R. Frezzotti, M. Guagnelli, M. Lüscher, R. Petronzio, R. Sommer, P. Weisz and U. Wolff, *Nucl. Phys.* B **437**, 447 (1995)
    (hep-lat/9411017).

66. T.H. Hansson, P. van Baal and I. Zahed, *Nucl. Phys.* B **289**, 628 (1987).

67. B. van Geemen and P. van Baal, *Proc. K. Ned. Akad. Wet.* B **89**, 39 (1986); P. van Baal and B. van Geemen, *J. Math. Phys.* **27**, 455 (1986).

68. D.R. Lebedev and M.I. Polikarpov, *Nucl. Phys.* B **269**, 285 (1986).

69. A. González-Arroyo and M. Okawa, *Phys. Rev.* D **27**, 2397 (1983).

70. J. Ambjørn and H. Flyvbjerg, *Phys. Lett.* B **79**, 241 (1980).

71. J. Groeneveld, J. Jurkiewicz and C.P. Korthals Altes,
Physica Scripta **23**, 1022 (1981).

72. P. van Baal, *Twisted Boundary Conditions: A Non-Perturbative Probe
for Pure Non-Abelian Gauge Theories*, PhD thesis (Utrecht, July 1984).

73. G. 't Hooft, *Acta Physica Austriaca*, Suppl. **XXII**, 531 (1980).

74. A. González-Arroyo and C. Korthals Altes, *Nucl. Phys.* B **311**, 433
(1988/89); D. Daniel, A. González-Arroyo, C. Korthals Altes and B.
Söderberg, *Phys. Lett.* B **221**, 136 (1989).

75. D. Daniel, A. González-Arroyo and C. Korthals Altes,
Phys. Lett. B **251**, 559 (1990).

76. M. García Pérez, A. González-Arroyo and B. Söderberg,
Phys. Lett. B **235**, 117 (1990).

77. P. Stephenson and M. Teper, *Nucl. Phys.* B **327**, 307 (1989).

78. P. Stephenson, *Nucl. Phys.* B **356**, 318 (1991).

79. The RTN collaboration, M. García Pérez *et al*,
Phys. Lett. B **305**, 366 (1993) (hep-lat/9302007);
M. García Pérez, A. González-Arroyo and P. Martínez,
Nucl. Phys. B(Proc. Suppl.) **34**, 228 (1994) (hep-lat/9312066).

80. A. González-Arroyo and P. Martínez, *Nucl. Phys.* B **459**, 337 (1996)
(hep-lat/9507001).

81. F. R. Klinkhamer and M. Manton, *Phys. Rev.* D **30**, 2212 (1984).

82. C. Taubes, *J. Diff. Geom.* **19**, 517 (1984).

83. S. Donaldson and P. Kronheimer, *The Geometry of Four Manifolds*
(Oxford University Press, 1990).

84. P.J. Braam and P. van Baal, *Comm. Math. Phys.* **122**, 267 (1989).

85. P.J. Braam, A. Maciocia and A. Todorov, *Inv. Math.* **108**, 419 (1992).

86. B. Berg, *Phys. Lett.* B **104**, 475 (1981); J. Hoek, M. Teper and J.
Waterhouse, *Nucl. Phys.* B **288**, 589 (1987).

87. M. García Pérez, A. González-Arroyo, J. Snippe and P. van Baal,
Nucl. Phys. B **413**, 535 (1994) (hep-lat/9309009);
Nucl. Phys. B(Proc. Suppl.) **34**, 222 (1994) (hep-lat/9311032).

88. A. González-Arroyo in *Advanced School for Non-perturbative Quantum
Field Theory*, eds. M. Asorey *et al* (World Scientific, Singapore, 1998)
p 57 (hep-th/9807108).

89. S. Sedlacek, *Comm. Math. Phys.* **86**, 515 (1982).

90. A.S. Schwarz, *Comm. Math. Phys.* **64**, 233 (1979).

91. B. Söderberg, *Topology in Twisted Lattice Gauge Theories*, LU TP 87-2
(Lund preprint, February 1987, unpublished);
I.M. Barbour and S.J. Psycharis, *Nucl. Phys.* B **334**, 302 (1990).

92. M. García Pérez and A. González-Arroyo, *J. Phys.* A **26**, 2667 (1993)
(hep-lat/9206016); M. García Pérez, A. González-Arroyo and A. Montero, *Nucl. Phys.* B(Proc. Suppl.) **63**, 501 (1998) (hep-lat/9709107);

A. Montero, *J. High Energy Phys.* **05**, 022 (2000) (hep-lat/0004009).

93. B. de Wit, M. Lüscher and H. Nicolai, *Nucl. Phys.* B **320**, 135 (1989).

94. H. Itoyama and B. Razzaghe-Ashrafi, *Nucl. Phys.* B **354**, 85 (1991).

95. A.V. Smilga, *Nucl. Phys.* B **266**, 45 (1986); *Yad. Fyz.* **43**, 215 (1986).

96. V. Novikov, M. Shifman, A. Vainshtein and V. Zakharov,
Nucl. Phys. B **229**, 407 (1983).

97. V. Novikov, M. Shifman, A. Vainshtein and V. Zakharov,
Nucl. Phys. B **260**, 157 (1985);
M.A. Shifman and A.I. Vainshtein, *Nucl. Phys.* B **296**, 445 (1988).

98. M. Shifman in *Confinement, Duality and Nonperturbative Aspects of QCD*, ed. P. van Baal (Plenum, New York, 1998) p 477.

99. D. Amati, K. Konishi, Y. Meurice, G. Rossi and G. Veneziano,
Phys. Rep. **162**, 169 (1988).

100. T.C. Kraan and P. van Baal, *Nucl. Phys.* B **533**, 627 (1998) (hep-th/9805168); *Phys. Lett.* B **435**, 389 (1998) (hep-th/9806034).

101. K. Lee and P. Yi, *Phys. Rev.* D **56**, 3711 (1997) (hep-th/9702107);
K. Lee and C. Lu, *Phys. Rev.* D **58**, 025011 (1998) (hep-th/9802108).

102. N.M. Davies, T.J. Hollowood, V.V. Khoze and M.P. Mattis,
Nucl. Phys. B **559**, 123 (1999) (hep-th/9905015).

103. E. Witten, *J. High Energy Phys.* **02**, 006 (1998) (hep-th/9712028).

104. A. Keurentjes, A. Rosly and A.V. Smilga, *Phys. Rev.* D **58**, 081701 (1998) (hep-th/9805183).

105. A. Keurentjes, *J. High Energy Phys.* **05**, 001 (1999) (hep-th/9901154); *J. High Energy Phys.* **05**, 014 (1999) (hep-th/9902186); *New Vacua for Yang-Mills Theory on a 3-Torus*, PhD thesis (Leiden, June 2000) (hep-th/0007196).

106. V.G. Kac and A.V. Smilga, *Vacuum Structure in Supersymmetric Yang-Mills Theories with any Gauge Group*, hep-th/9902029 v.3.

107. A. Borel, M. Friedman and J.W. Morgan, *Almost Commuting Elements in Compact Lie Groups*, math/9907007.

108. E. Witten, *Supersymmetric Index in Four-Dimensional Gauge Theories*, hep-th/0006010.

109. G.K. Savvidy, *Phys. Lett.* B **159**, 325 (1985).

110. G. 't Hooft, *Comm. Math. Phys.* **81**, 267 (1981).

111. P. van Baal, *Comm. Math. Phys.* **94**, 397 (1984).

112. D.R. Lebedev, M.I. Polikarpov and A.A. Rosly, *Nucl. Phys.* B **325**, 138 (1989).

113. M. García Pérez, A. González-Arroyo and C. Pena, *Perturbative Construction of Self-dual Solutions on the Torus*, hep-th/0007113.

114. S. Mukai, *Nagoya Math. J.* **81**,153 (1981).

115. M. Atiyah, N. Hitchin, V. Drinfeld and Yu. Manin,
Phys. Lett. A **65**, 185 (1978).

116. E. Corrigan and P. Goddard, *Ann. Phys.* (N.Y.) **154**, 253 (1984).

117. M.F. Atiyah and I.M. Singer, *Ann. Math.* **93**, 119 (1971);
Proc. Natl. Acad. Sci. USA, Vol. **81**, 2597 (1984).

118. P. van Baal, *Complex Structures in Gauge Theories*, Graduate Course
Lectures, unpublished, (Stony Brook, spring 1986).

119. H. Schenk, *Comm. Math. Phys.* **116**, 177 (1988).

120. A. González-Arroyo, *Nucl. Phys.* B **548**, 626 (1999) (hep-th/9811041).

121. M. García Pérez, A. González-Arroyo, C. Pena and P. van Baal, *Nucl.
Phys.* B **564**, 159 (2000).

122. P. van Baal, *Nucl. Phys.* B(Proc. Suppl.) **49**, 238 (1996)
(hep-th/9512223).

123. P. van Baal, *Phys. Lett.* B **448**, 26 (1999) (hep-th/9811112).

124. A. González-Arroyo and C. Pena, *J. High Energy Phys.* **09**, 013 (1998)
(hep-th/9807172).

125. M. García Pérez, A. González-Arroyo, A. Montero and P. van Baal,
J. High Energy Phys. **06**, 001 (1999) (hep-lat/9903022).

126. P. van Baal and N. D. Hari Dass, *Nucl. Phys.* B **385**, 185 (1992).

127. Y. Hosotani, *Phys. Lett.* B **147**, 44 (1984).

128. P. van Baal and B. van den Heuvel, *Nucl. Phys.* B **417**, 215 (1994)
(hep-lat/9310005).

129. R.E. Cutkosky, *J. Math. Phys.* **25**, 939 (1984);
R.E. Cutkosky and K. Wang, *Phys. Rev.* D **37**, 3024 (1988);
R.E. Cutkosky, *Czech. J. Phys.* **40**, 252 (1990).

130. M. Lüscher, *Phys. Lett.* B **70**, 321 (1977).

131. P. van Baal and R.E. Cutkosky, *Int. J. Mod. Phys.* A(Proc. Suppl.)
3A, 323 (1993) (hep-lat/9208027).

132. B. van den Heuvel, *Phys. Lett.* B **368**, 124 (1996) (hep-lat/9509019);
Phys. Lett. B **386**, 233 (1996) (hep-lat/9604017); *Nucl. Phys.* B **488**,
282 (1997) (hep-lat/9608101); *Non-perturbative Phenomena in Gauge
Theory on* S^3, PhD thesis (Leiden, September 1996).

133. B. Diekmann, *Eine Modellraum-Methode zur Berechnung des Glueball-
spektrums im kompaktifizierten Minkowskiraum*, PhD thesis [TK-95-26]
(Bonn, October 1995).

134. E.V. Shuryak, *Nucl. Phys.* B **302**, 559,574,599,621 (1988); T. Schäfer
and E.V. Shuryak, *Rev. Mod. Phys.* **70**, 323 (1998) (hep-ph/9610451);
E. Shuryak in *Confinement, Duality and Nonperturbative Aspects of
QCD*, ed. P. van Baal (Plenum, New York, 1998) p 307.

135. T. Schäfer and E.V. Shuryak, *Phys. Rev. Lett.* **75**, 1707 (1995)
(hep-ph/9410372).

136. M. Teper, *Glueball Masses and other Physical Properties of SU(N) Gauge
Theories in D=(3+1): A Review of Lattice Results for Theorists*, Oxford
preprint OUTP-98-88-P, hep-lat/9812187.

137. P. van Baal, *Nucl. Phys.* B(Proc. Suppl.) **63**, 126 (1998).

138. M. Lüscher, *Comm. Math. Phys.* **105**, 153 (1986).

139. M. Lüscher, *Nucl. Phys.* B **354**, 531 (1991);
Nucl. Phys. B **364**, 237 (1991).

140. P. van Baal, *Nucl. Phys.* B(Proc. Suppl.) **20**, 3 (1991).

141. K. Huang and C.N. Yang, *Phys. Rev.* **105**, 767 (1957);
C.N. Yang, *Chinese J. Phys.* **25**,80 (1987).

142. M. Lüscher and U. Wolff, *Nucl. Phys.* B **339**, 222 (1990).

143. C.R. Gattringer and C.B. Lang, *Phys. Lett.* B **274**, 95 (1992);
Nucl. Phys. B **391**, 463 (1993) (hep-lat/9206004).

144. I. Montvay and P. Weisz, *Nucl. Phys.* B **290**, 327 (1987);

145. Ch. Frick, K. Jansen, J. Jersák, I. Montvay, P. Seuferling and G. Münster, *Nucl. Phys.* B **331**, 515 (1990).

146. L. Lellouch and M. Lüscher, *Weak Transition Matrix Elements from Finite Volume Correlation Functions*, hep-lat/0003023.

147. M. Lüscher, *Nucl. Phys.* B **180**, 317 (1981).

148. J. Gasser and H. Leutwyler, *Phys. Lett.* B **188**, 477 (1987);
P. Hasenfratz and H. Leutwyler, *Nucl. Phys.* B **343**, 241 (1990);
H.C. Hansen and H. Leutwyler, *Nucl. Phys.* B **350**, 201 (1991).

149. H. Leutwyler and A.V. Smilga, *Phys. Rev.* D **46**, 5607 (1992).

150. T. Banks and A. Casher, *Nucl. Phys.* B **169**, 103 (1980).

151. E.V. Shuryak and J.J.M. Verbaarschot, *Nucl. Phys.* A **560**, 306 (1993) (hep-th/9212088).

152. J. Verbaarschot in *Confinement, Duality and Nonperturbative Aspects of QCD*, ed. P. van Baal (Plenum, New York, 1998) p 343 (hep-th/9710114).

153. E. Witten, *Phys. Lett.* B **86**, 283 (1979).

154. G. 't Hooft, *Nucl. Phys.* B **190**, 455 (1981); *Physica Scripta*, Vol. **25**, 133 (1982); in *Confinement, Duality and Nonperturbative Aspects of QCD*, ed. P. van Baal (Plenum, New York, 1998) p 379.

155. S. Coleman, "The Uses of Instantons", in *The Whys of Subnuclear Physics*, ed. A. Zichichi (Plenum Press, New York, 1979) p 805.

156. R. Rajaraman, *Solitons and Instantons* (North-Holland, Amsterdam, 1982).

157. V.N. Gribov, *Eur. Phys. J.* C **10**, 71 (1999) (hep-ph/9807224);
Eur. Phys. J. C **10**, 91 (1999) (hep-ph/9902279).

158. G. 't Hooft, *Nucl. Phys.* B **72**, 461 (1974); *Comm. Math. Phys.* **86**, 449 (1982); *Comm. Math. Phys.* **88**, 1 (1983); in *Progress in Gauge Field Theory*, eds. G. 't Hooft *et al* (Plenum, New York, 1984) p 271.

159. Yu. Makeenko, *Large-N Gauge Theories*, hep-th/0001047.

160. A.M. Polyakov, *Gauge Fields and Strings* (Harwood, Chur, 1987);
Nucl. Phys. B **486**, 23 (1997) (hep-th/9607049);
Nucl. Phys. B(Proc. Suppl.) **68**, 1 (1998) (hep-th/9711002);

Int. J. Mod. Phys. A**14**, 645 (1999) (hep-th/9809057).

161. See www.claymath.org/prize_problems/yang_mills.htm.
162. S. Mandelstam, *Phys. Rep.* **23**, 245 (1976).
163. G.'t Hooft in *High Energy Physics*, ed. A. Zichichi (Editrice Compositori, Bolognia, 1976); *Nucl. Phys.* B **138**, 1 (1978).
164. N. Seiberg and E. Witten, *Nucl. Phys.* B **426**, 19 (1994) [erratum B**430**, 485 (1994)] (hep-th/9407087).

ELSEVIER

Nuclear Physics B 645 (2002) 105–133

www.elsevier.com/locate/npe

Multi-caloron solutions

Falk Bruckmann, Pierre van Baal

*Instituut-Lorentz for Theoretical Physics, University of Leiden, PO Box 9506,
NL-2300 RA Leiden, The Netherlands*

Received 4 September 2002; accepted 17 September 2002

Abstract

We discuss the construction of multi-caloron solutions with non-trivial holonomy, both as approximate superpositions and exact self-dual solutions. The charge k $SU(n)$ moduli space can be described by kn constituent monopoles. Exact solutions help us to understand how these constituents can be seen as independent objects, which seems not possible with the approximate superposition. An "impurity scattering" calculation provides relatively simple expressions. Like at zero temperature an explicit parametrization requires solving a quadratic ADHM constraint, achieved here for a class of axially symmetric solutions. We will discuss the properties of these exact solutions in detail, but also demonstrate that interesting results can be obtained without explicitly solving for the constraint.
© 2002 Elsevier Science B.V. All rights reserved.

PACS: -10.10.Wx; 12.38.Lg; 14.80.Hv

1. Introduction

The last four years more understanding has been gained of the interplay between instantons and monopoles in non-Abelian gauge theories, based on the ability to construct exact caloron solutions, i.e., instantons at finite temperature for which A_0 approaches a constant at spatial infinity [1,2]. This last condition is best expressed by specifying the Polyakov loop to approach a constant value at infinity, also called the holonomy. It is the finite action that demands the field strength to go to zero at infinity, and guarantees the Polyakov loop to be independent of the direction in which we approach infinity. One can parametrize this Polyakov loop in terms of its eigenvalues $\exp(2\pi i \mu_j)$ and a gauge rotation

---

E-mail address: vanbaal@lorentz.leidenuniv.nl (P. van Baal).

 F. Bruckmann, P. van Baal / Nuclear Physics B 645 (2002) 105–133

g, such that

$$\mathcal{P}_\infty = \lim_{x \to \infty} P \exp\left(\int_0^\beta A_0(\vec{x}, t)\,dt \right) = g^\dagger \exp\big(2\pi i \,\mathrm{diag}(\mu_1, \mu_2, \ldots, \mu_n)\big)g, \qquad (1)$$

which can be arranged such that $\sum_{i=1}^n \mu_i = 0$, and $\mu_1 \leqslant \mu_2 \leqslant \cdots \leqslant \mu_n \leqslant \mu_{n+1}$, with $\mu_{n+k} \equiv 1 + \mu_k$. This is in the gauge where the gauge fields, assumed to be anti-hermitian matrices taking values in the algebra of $SU(n)$, are periodic in the time direction $A_\mu(\vec{x}, t) = A_\mu(\vec{x}, t + \beta)$. In our conventions the field strength is given by $F_{\mu\nu}(x) = \partial_\mu A_\nu(x) - \partial_\nu A_\mu(x) + [A_\mu(x), A_\nu(x)]$.

We will find it convenient to construct the multi-caloron solutions from the ADHM-Nahm Fourier construction [1–4] based on taking instantons in R^4, which periodically repeat in the time direction (see also Ref. [5] which uses directly the Nahm transformation [4, 6] as the starting point for the charge 1 construction). To allow for non-trivial holonomy, the periodicity is only up to a constant gauge rotation (which is the holonomy). In this so-called algebraic gauge, all gauge field components vanish at spatial infinity, and we may approximately superpose these calorons by simply adding the gauge fields. When each gauge field is periodic up to the same constant gauge transformation, the sum satisfies the same property. It should be noted that we are not allowed to add gauge fields with different holonomy; in an *infinite* volume the holonomy is fixed by the boundary condition. As to the topological charge, we recall that in the Atiyah–Drinfeld–Hitchin–Manin (ADHM) construction [3] it is supported by gauge singularities (the algebraic gauge is for this reason also called the singular gauge), as opposed to at infinity being a pure gauge with the gauge function having the appropriate winding number. To deal with the gauge singularity of one instanton, when adding the field of the others, one has to smoothly deform the gauge field of the latter to vanish near the gauge singularity. As long as the singularities are not too close, this can be done without a significant increase in the action. On the lattice this problem does not occur, when hiding singularities between the meshes of the lattice.

Calorons, however, have an additional feature. When squeezed in the imaginary time direction (the size ρ becoming bigger than the period β), they split in constituent monopoles with masses $8\pi^2 \nu_j / \beta$ ($\nu_j \equiv \mu_{j+1} - \mu_j$). Outside the cores of these monopoles the gauge field becomes abelian. Ignoring the charged components, which decay exponentially outside the cores of the constituent monopoles, the field is described in terms of self-dual Dirac monopoles. The singularity of a Dirac string can only be avoided by *not* neglecting the contributions coming from the charged components, even when far away from the constituents. For a single caloron we would not care, since the Dirac string is not seen in gauge invariant quantities. It does, however, involve a rather subtle interplay between the charged and neutral components of the gauge field in the vicinity of the would-be Dirac string [1]. It is this subtle interplay that is disturbed when we add gauge fields of various calorons together. Unlike for the gauge singularity, the combined field will not diverge, but it shows a narrow and steep enhancement at the location of the would-be Dirac string as illustrated in Fig. 1, where we added two $SU(2)$ calorons. Also here one may shield the Dirac strings from these tails. But one always pays the price that the Dirac string no longer can be hidden, and carries energy. Let us stress again that these are genuine

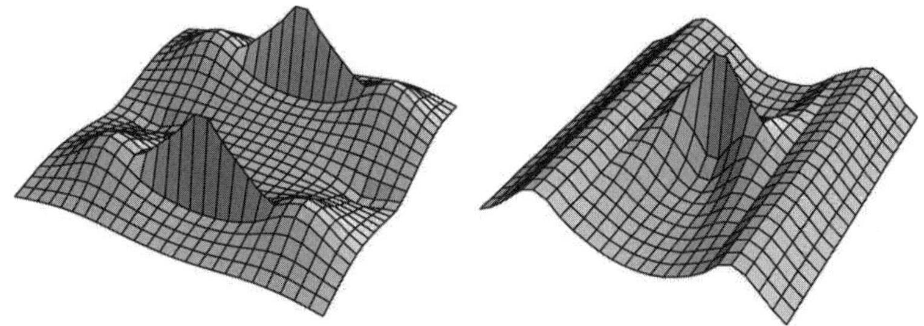

Fig. 1. Approximate superposition of two $SU(2)$ charge 1 calorons with its pairs of equal mass constituents at $\vec{x} = (2,0,2)$, $(2,0,8)$ and $\vec{x} = (8,0,2)$, $(8,0,8)$. The logarithm of the action density is plotted as a function of x and z. The plot on the right shows one of the would-be Dirac strings, zooming in by a factor 40 on the transverse direction.

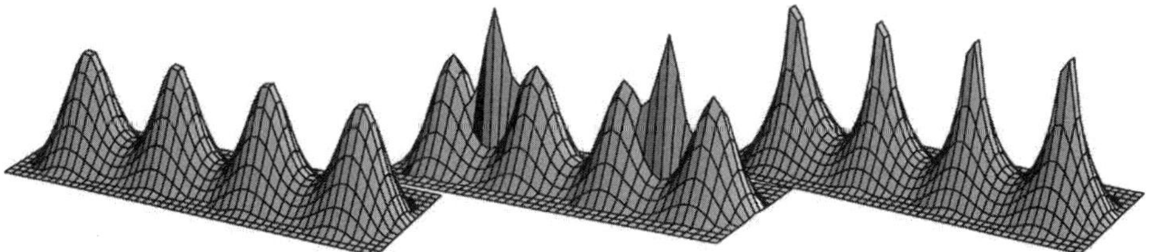

Fig. 2. Comparing the logarithm of the action density (cutoff for $\log(S)$ below -3 and above 7) as a function of x and z for the exact $SU(2)$ solution ($\mu_2 = \frac{1}{4}$) with charge 2 (left) with the approximate superposition of two charge one calorons (middle) and the Abelian solution based on Dirac monopoles (right), all on the same scale. The two pairs of constituent monopoles are located at $\vec{x} = (0,0,6.031)$, $(0,0,2.031)$ and $\vec{x} = (0,0,-2.031)$, $(0,0,-6.031)$.

gauge invariant non-singular features in the configuration, even though they are of course a consequence of our particular way of constructing a superposition.

A visible would-be Dirac string presents a formidable obstacle to considering the constituents as independent objects; they remember to which caloron they belong. Insisting the Abelian field far from the constituents to be exactly additive under the approximate superposition leaves us little room for considering other possible superpositions, apart from carefully fine-tuning the charged components of the configuration. Solving for the exact self-dual caloron solutions of higher charge we wish to show that a visible Dirac string is an artifact of the particular procedure to construct approximate caloron solutions. For the exact charge k caloron solutions we expect kn constituents and from the point of view of the parameter space these can be expected to be independent (as long the constituents do not get too close together).

In this paper we develop the formalism and give exact solutions for a class of axially symmetric solutions. For $SU(2)$ an example of such an exact solution with charge 2 is presented in Fig. 2. We compare the exact solution (left), with the one obtained by adding two charge 1 calorons, showing the would-be Dirac strings (middle) and with the exact Abelian solution determined from (self-dual) Dirac monopoles placed at the location of the constituents (right). Indeed, the would-be Dirac strings are no longer visible for the exact non-Abelian solution. This gives us good reasons to expect that moving away from the

108   *F. Bruckmann, P. van Baal / Nuclear Physics B 645 (2002) 105–133*

requirement of axial symmetry, the exact solutions will exhibit the constituent monopoles as independent objects. This is an important prerequisite for attempting to formulate the long distance features of QCD in terms of these monopoles. Non-linearities will remain important when constituents overlap, like for instantons at zero-temperature, as will be illustrated through suitable examples.

Our results can also be used to extract information on multi-monopole solutions. In the charge 1 case it has been shown [1] that sending one of the constituents to infinity one is left with exact monopole solutions. Perhaps somewhat surprisingly, it has been notoriously difficult to construct approximate superpositions of magnetic monopoles in such a way the interaction energy decreases with their separation. Within our approach this is due to the difficulty (particularly when sending some constituents to infinity) in keeping Dirac strings hidden. It would therefore still be important to find approximate superpositions that achieve this. Nevertheless, it shows how subtle it is to consider Abelian fields as embedded in non-Abelian ones. There is much more to 't Hooft's Abelian projection [7] than meets the eye at first glance.

The rest of the paper will be organized as follows. In Section 2 we formulate the Fourier analysis of the ADHM data for calorons of higher charge. This allows us in Section 2.1 to relate to the Nahm equations, providing the connection between the ADHM and Nahm data. The Nahm equation for charge k calorons [4] expresses self-duality of dual $SU(k)$ gauge fields on a circle, but with singularities. In Section 2.2 we introduce the "master" Green's function in terms of which the gauge field (Section 2.3) and the action density (Section 2.4) can be explicitly computed, relying heavily on some beautiful results [8,9] derived in the context of the ADHM construction.

This can all be achieved without explicitly solving the Nahm equation. In Section 3 we will, however, solve it for some special cases with axial symmetry. There we also illustrate some interesting features that appear when constituents overlap, similar to what was observed at zero temperature [10].

Section 4 is devoted to the far-field limit (related to the high-temperature limit), where up to exponential corrections only the Abelian component of the field survives. Again, expressions can be derived without explicitly solving the Nahm equation. We apply the formalism to the exact axially symmetric solutions in Section 4.3.1, where we prove that in the far-field region they are in a precise sense described by (self-dual) Dirac monopoles, which is conjectured to be the case for any exact solution.

We end in Section 5 with a discussion of lattice results that have been obtained by various groups, some open questions and suggestions for future studies.

2. Construction

In constructing the higher charge caloron solutions similar steps are followed as for charge 1. We distinguish two constructions, $Sp(n)$ based on quaternions, and $SU(n)$ based on complex spinors. With $SU(2) = Sp(1)$, this implies that for $SU(2)$ two constructions are possible. For $k = 1$ it was easily shown these are identical, but for $k > 1$ this is more complicated due to the quadratic ADHM constraint [3] which cannot be solved in all generality. Since the essential part of the construction that employs the impurity scattering technique is identical in both approaches, we will give a unified presentation.

The $SU(n)$ ADHM formalism for charge k instantons [3] employs a k-dimensional vector $\lambda = (\lambda_1, \ldots, \lambda_k)$, where $\lambda_i^\dagger$ is a two-component spinor in the $\bar{n}$ representation of $SU(n)$. Alternatively, λ can be seen as a $n \times 2k$ complex matrix. In addition one has four complex Hermitian $k \times k$ matrices B_μ, combined into a $2k \times 2k$ complex matrix $B = \sigma_\mu \otimes B_\mu$, using the unit quaternions $\sigma_\mu = (\mathbb{1}_2, i\vec{\tau})$ and $\bar{\sigma}_\mu = (\mathbb{1}_2, -i\vec{\tau})$, where τ_i are the Pauli matrices. With some abuse of notation, we often write $B = \sigma_\mu B_\mu$. Together λ and B constitute the $(n + 2k) \times 2k$ dimensional matrix $\Delta(x)$, to which is associated a complex $((n + 2k) \times n)$-dimensional normalized zero mode vector $v(x)$,

$$\Delta(x) = \begin{pmatrix} \lambda \\ B(x) \end{pmatrix}, \qquad B(x) = B - x, \qquad \Delta^\dagger(x)v(x) = 0, \qquad v^\dagger(x)v(x) = \mathbb{1}_n.$$

(2)

Here the quaternion $x = x_\mu \sigma_\mu$ denotes the position (a $k \times k$ unit matrix is implicit) and $v(x)$ can be solved explicitly in terms of the ADHM data by

$$v(x) = \begin{pmatrix} -\mathbb{1}_n \\ u(x) \end{pmatrix} \phi^{-\frac{1}{2}}, \qquad u(x) = (B^\dagger - x^\dagger)^{-1}\lambda^\dagger, \qquad \phi(x) = \mathbb{1}_n + u^\dagger(x)u(x).$$

(3)

As $\phi(x)$ is an $n \times n$ positive Hermitian matrix, its square root $\phi^{\frac{1}{2}}(x)$ is well-defined. The gauge field is given by

$$A_\mu(x) = v^\dagger(x)\partial_\mu v(x) = \phi^{-\frac{1}{2}}(x)\big(u^\dagger(x)\partial_\mu u(x)\big)\phi^{-\frac{1}{2}}(x) + \phi^{\frac{1}{2}}(x)\partial_\mu \phi^{-\frac{1}{2}}(x). \qquad (4)$$

The $Sp(1)$ ADHM formalism for charge k instantons [3] employs also a k-dimensional vector $\lambda = (\lambda_1, \ldots, \lambda_k)$, where now λ_i is a quaternion (4 real parameters). Again, λ can be seen as an $2 \times 2k$ complex matrix (but with $4k$ real, as opposed to complex, parameters). Now the four $k \times k$ matrices B_μ are required to be real and symmetric, still to be combined into a $2k \times 2k$ complex matrix $B = \sigma_\mu B_\mu$. The $(2 + 2k) \times 2k$ dimensional matrix $\Delta(x)$ is constructed as before. It is immediately obvious that now $\phi(x)$ is proportional to σ_0, simplifying the expression for the gauge field to

$$A_\mu(x) = v^\dagger(x)\partial_\mu v(x) = \big(u^\dagger(x)\partial_\mu u(x)\big)/\phi(x). \qquad (5)$$

For $A_\mu(x)$ to be a self-dual connection, $\Delta(x)$ has to satisfy the quadratic ADHM constraint, which states that $\Delta^\dagger(x)\Delta(x) = B^\dagger(x)B(x) + \lambda^\dagger\lambda$ (considered as a $k \times k$ complex quaternionic matrix) has to commute with the quaternions, or equivalently

$$\Delta^\dagger(x)\Delta(x) = \sigma_0 \otimes f_x^{-1}, \qquad (6)$$

defining f_x as a Hermitian (resp. symmetric) $k \times k$ Green's function. The self-duality follows by computing the curvature

$$F_{\mu\nu} = 2\phi^{-\frac{1}{2}}(x)u^\dagger(x)\eta_{\mu\nu}f_x u(x)\phi^{-\frac{1}{2}}(x), \qquad (7)$$

making essential use of the fact that f_x commutes with the quaternions, and $\eta_{\mu\nu} \equiv \sigma_{[\mu}\bar{\sigma}_{\nu]}$ being self-dual ($\bar{\eta}_{\mu\nu} \equiv \bar{\sigma}_{[\mu}\sigma_{\nu]}$ is anti-self dual). The quadratic constraint can be formulated as $\Im m(\Delta^\dagger(x)\Delta(x)) = 0$, where $\Im m W \equiv W - \frac{1}{2}\sigma_0 \mathrm{tr}_2 W$, and one obtains

$$\bar\eta_{\mu\nu} \otimes B_\mu B_\nu + \frac{1}{2}\tau_a \otimes \mathrm{tr}_2\big(\tau_a \lambda^\dagger \lambda\big) = 0, \tag{8}$$

where tr_2 is the spinorial trace. Note that this implies that $\mathrm{tr}_2(\tau_a \lambda^\dagger \lambda)$ is traceless for $a = 1, 2, 3$.

To count the number of instanton parameters we observe that the transformation $\lambda \to \lambda T^\dagger$, $B_\mu \to T B_\mu T^\dagger$, with $T \in U(k)$ (respectively, $T \in O(k)$) leaves the gauge field and the ADHM constraint untouched. Taking this symmetry into account, one checks the dimension of the instanton moduli space to be $4kn$ dimensional. We have $4kn$ and $4k^2$ real parameters from λ and B_μ. The $U(k)$ symmetry removes k^2 real parameters and finally the quadratic ADHM constraint gives $3k^2$ real equations. On the other hand, for the quaternionic construction there are $4k$ and $4 \cdot \frac{1}{2}k(k+1)$ parameters from λ and B_μ, of which $\frac{1}{2}k(k-1)$ are removed by the $O(k)$ symmetry, and the quadratic ADHM constraint gives $3 \cdot \frac{1}{2}k(k-1)$ equations. Global gauge transformations are realized by $\lambda \to g\lambda$, with $g \in SU(n)$ and are here included in the parameter count. For calorons with non-trivial holonomy the dimension of the gauge invariant parameter space is minimally reduced by $n - 1$ (maximal symmetry breaking) and maximally by $n^2 - 1$ (trivial holonomy).

We note that for $SU(2)$ the quadratic ADHM constraint and symmetry of the ADHM data for high charge differ considerably. For the caloron we have to deal in a sense with infinite topological charge, but finite within each (imaginary) time interval of length β. This infinity is resolved by Fourier transformation, relating it to the Nahm formalism [4], but the difference between the $U(k)$ and $O(k)$ symmetries (the infinity is moved to making these gauge symmetries local) remains, as well as of course the nature of the Nahm data (the Fourier transformation of the ADHM data). Henceforth we put $\beta = 1$, which can always be achieved by a rescaling.

Like for charge one [1,2] the caloron with Polyakov loop $\mathcal{P}_\infty$ at infinity is built out of a periodic array of instantons, twisted by $\mathcal{P}_\infty$. This is implemented in the ADHM formalism by requiring (suppressing color and spinor indices, respectively, quaternion indices)

$$u_{pk+k+a}(x+1) = u_{pk+a}(x)\mathcal{P}_\infty^{-1} \tag{9}$$

with $p \in \mathbb{Z}$ (the Fourier index), and $a = 1, \dots, k$ (associated to the non-Abelian nature of the Nahm data). Using that $\phi^{\pm\frac{1}{2}}(x+1) = \mathcal{P}_\infty \phi^{\pm\frac{1}{2}}(x)\mathcal{P}_\infty^{-1}$, Eq. (4) leads to the required periodicity. Demanding

$$\lambda_{pk+k+a} = \mathcal{P}_\infty \lambda_{pk+a}, \qquad B_{pk+a,qk+b} = B_{pk-k+a,qk-k+b} + \sigma_0 \delta_{pq}\delta_{ab}, \tag{10}$$

suitably implements Eq. (9) and is partially solved by imposing

$$\lambda_{pk+a} = \mathcal{P}_\infty^p \zeta_a, \qquad B_{pk+a,qk+b} = p\sigma_0 \delta_{pq}\delta_{ab} + \hat{A}_{p-q}^{ab}, \tag{11}$$

with $\hat{A}$ still to be determined to account for Eq. (8). It is useful to introduce the n projectors P_m on the mth eigenvalue of $\mathcal{P}_\infty$, such that $\mathcal{P}_\infty = \sum_m e^{2\pi i \mu_m} P_m$ and $\lambda_{pk+a} = \sum_m e^{2\pi i p \mu_m} P_m \zeta_a$.

2.1. Nahm setting

We now perform the Fourier transformation to the Nahm setting [4], which casts B into a Weyl operator and $\lambda^\dagger \lambda$ into a singularity structure on S^1,

$$
\sum_{p,q} B_{pk+a,qk+b}(x)e^{2\pi i(pz-qz')} = \frac{\delta(z-z')}{2\pi i}\widehat{D}^{ab}(z'),
$$

$$
2\pi i \sum_{p} e^{2\pi ipz}\hat{A}^{ab}_p = \hat{A}^{ab}(z),
$$

$$
\sum_{p,q}\lambda^\dagger_{pk+a}e^{2\pi i(pz-qz')}\lambda_{qk+b} = \delta(z-z')\hat{\Lambda}_{ab}(z),
\qquad
\sum_{p}e^{-2\pi ipz}\lambda_{pk+a} = \hat{\lambda}_a(z).
$$

$$(12)$$

With $B = \sigma_\mu B^\mu$, we can write $\widehat{D}(z) = \sigma_\mu \widehat{D}^\mu(z)$ and $\hat{A}(z) = \sigma^\mu \hat{A}_\mu(z)$, where

$$
\widehat{D}^{ab}_x(z) \equiv \widehat{D}^{ab}(z) - 2\pi ix\delta^{ab} = \delta^{ab}\left(\sigma_0\frac{d}{dz} - 2\pi ix\right) + \hat{A}^{ab}(z),
$$

$$
\hat{\Lambda}_{ab}(z) = \sum_{m}\delta(z-\mu_m)\zeta^\dagger_a P_m \zeta_b,
\qquad
\hat{\lambda}_a(z) = \sum_{m}\delta(z-\mu_m)P_m\zeta_a.
$$

$$(13)$$

It should be noted that B_μ Hermitian, implies that $\hat{A}_\mu(z)$ is Hermitian (as a $k \times k$ matrix), whereas a real symmetric B_μ in addition implies $\hat{A}^t_\mu(z) = \hat{A}_\mu(-z)$.

The 2×2 matrix $\hat{\Lambda}^{ab}$ can always be decomposed as

$$
\zeta^\dagger_a P_m \zeta_b = \frac{1}{2\pi}\left(\sigma_0 \widehat{S}^{ab}_m - \vec{\tau}\cdot\vec{\rho}^{\,ab}_m\right).
$$

$$(14)$$

On the diagonal one can show, as for $k = 1$, that $\widehat{S}^{aa}_m = |\vec{\rho}^{\,aa}_m|$, but in general the relation between $\vec{\rho}_m$ and $\widehat{S}_m$ is more complicated,

$$
\widehat{S}^{ab}_m \widehat{S}^{cd}_m + \widehat{S}^{ad}_m \widehat{S}^{cb}_m = \vec{\rho}^{\,ab}_m \cdot \vec{\rho}^{\,cd}_m + \vec{\rho}^{\,ad}_m \cdot \vec{\rho}^{\,cb}_m.
$$

$$(15)$$

Furthermore, $\mathrm{tr}_2(\vec{\tau}\lambda^\dagger\lambda)$ is traceless implies that $\sum_{a=1}^{k}\sum_{m=1}^{n}\vec{\rho}^{\,aa}_m = \vec{0}$. Both these conditions are equally valid for the $Sp(1)$ construction. But in the latter case, since through the Nahm equation $\vec{\rho}_m$ determines the discontinuities in $\hat{A}(z)$, compatibility with $\hat{A}^t_\mu(z) = \hat{A}_\mu(-z)$ requires $\vec{\rho}_1 + \vec{\rho}^{\,t}_2 = \vec{0}$, which will be verified below Eq. (19). The Nahm equation is obtained by Fourier transforming the quadratic ADHM constraint

$$
\frac{1}{2}[\widehat{D}_\mu(z), \widehat{D}_\nu(z)]\bar{\eta}_{\mu\nu} = 4\pi^2 \Im m\,\hat{\Lambda}(z),
$$

$$(16)$$

or in more familiar form (as a $k \times k$ matrix equation)

$$
\frac{d}{dz}\hat{A}_j(z) + [\hat{A}_0(z), \hat{A}_j(z)] + \frac{1}{2}\varepsilon_{jk\ell}[\hat{A}_k(z), \hat{A}_\ell(z)] = 2\pi i\sum_{m}\delta(z-\mu_m)\rho^j_m,
$$

$$
\hat{A}_j : \quad
\begin{array}{ccccccccc}
\rho^j_1 & \rho^j_2 & \rho^j_3 & & \rho^j_{n-2} & \rho^j_{n-1} & \rho^j_n & \rho^j_1 \\
\hline
z=\mu_1 & \mu_2 & \mu_3 & - - - - - - & \mu_{n-2} & \mu_{n-1} & \mu_n & \mu_{n+1}\equiv\mu_1+1
\end{array}
$$

$$(17)$$

where the figure illustrates the jumps of $\hat{A}_j$ at each of the singularities (an overall factor of $2\pi i$ is not shown). The T symmetry in the ADHM construction translates into a $U(k)$ gauge symmetry on S^1, which allows one to set $\hat{A}_0 = 2\pi i \xi_0$ to be constant, where ξ_0 is a Hermitian matrix which can be made diagonal by a constant gauge transformation. Its trace part can be absorbed in x_0.

2.2. Green's function

Central to the ADHM construction is the Green's function f_x, which when Fourier transformed to $\hat{f}_x^{ab}(z, z') \equiv \sum_{p,q} f_x^{pk+a,qk+b} e^{2\pi i(pz - qz')}$ is a solution to the differential equation

$$\left\{ \left(\frac{\widehat{D}_x^{\mu}(z)}{2\pi i} \right)^2 + \frac{1}{2\pi} \sum_m \delta(z - \mu_m) \widehat{S}_m \right\} \hat{f}_x(z, z') = \mathbb{1}_k \delta(z - z'),$$

$$\frac{d\hat{f}_x}{dz} : \begin{array}{ccccccccc} & \widehat{S}_1 & \widehat{S}_2 & \widehat{S}_3 & 2\pi\mathbb{1}_k & \widehat{S}_{n-2} & \widehat{S}_{n-1} & \widehat{S}_n & \widehat{S}_1 \\ \text{---} & \text{---} & \text{---} & \text{---} & \text{---} & \text{---} & \text{---} & \text{---} & \text{---} \\ z = \mu_1 & \mu_2 & \mu_3 & & z' & & \mu_{n-2} & \mu_{n-1} & \mu_n & \mu_{n+1} \equiv \mu_1 + 1 \end{array}, \tag{18}$$

where the figure illustrates the jumps for the *derivative* of $\hat{f}_x$ (itself being continuous) at each of the singularities, which now includes $z = z'$ (an overall factor of 2π is omitted). With help of the impurity scattering formalism we will be able to express the solution in a simple form, without assuming explicit knowledge of $\hat{A}(z)$. As a bonus this also gives a more transparent derivation for $k = 1$. The equation for the Green's function is valid for the $SU(n)$ and $Sp(1)$ formulations alike. In general the matrix f_x is complex Hermitian, but for $Sp(1)$ it is real symmetric, which implies that

$$\hat{f}_x^{ab}(z, z') = \hat{f}_x^{ba}(z', z)^*, \quad \text{and for } Sp(1) \text{ only} \quad \hat{f}_x^{ab}(z, z') = \hat{f}_x^{ba}(-z', -z). \tag{19}$$

For $Sp(1)$ consistency requires, $\widehat{S}_1 - \widehat{S}_2^t = 0$, which is to be compared with the condition we found in the previous section, $\vec{\rho}_1 + \vec{\rho}_2^{\,t} = 0$. To demonstrate the validity of these relations we make use of the fact that $\zeta_a = \zeta_a^{\mu} \sigma_{\mu}$, with ζ_a^{μ} real (such that $\zeta_a^{\dagger} = \bar{\zeta}_a$). Furthermore, we note that with $\mathcal{P}_{\infty} \equiv \exp(2\pi i \vec{\omega} \cdot \vec{\tau})$, $\mu_2 = -\mu_1 = \omega \equiv |\vec{\omega}|$, whereas $P_1 = \frac{1}{2}(\mathbb{1}_2 - \hat{\omega} \cdot \vec{\tau})$ and $P_2 = \frac{1}{2}(\mathbb{1}_2 + \hat{\omega} \cdot \vec{\tau})$. Introducing $\Omega \equiv \hat{\omega} \cdot \vec{\sigma} = i\hat{\omega} \cdot \vec{\tau}$, which is an imaginary unit quaternion (i.e., $\bar{\Omega} = -\Omega$, $\bar{\Omega}\Omega = \sigma_0$), we find with the help of Eq. (14)

$$\sigma_0 \widehat{S}_m^{ab} - \vec{\tau} \cdot \vec{\rho}_m^{\,ab}$$
$$= \pi \zeta_a^{\mu} \bar{\sigma}_{\mu} \big(\sigma_0 - i(-1)^m \Omega \big) \zeta_b^{\nu} \sigma_{\nu}$$
$$= \pi \zeta_a^{\mu} \zeta_b^{\nu} \big([\bar{\sigma}_{\{\mu} \sigma_{\nu\}} - i(-1)^m \bar{\sigma}_{[\mu} \Omega \sigma_{\nu]}] + [\bar{\sigma}_{[\mu} \sigma_{\nu]} - i(-1)^m \bar{\sigma}_{\{\mu} \Omega \sigma_{\nu\}}] \big)$$
$$= \sigma_0 \pi \big[\zeta_a^{\mu} \zeta_b^{\mu} - i(-1)^m \hat{\omega} \cdot \vec{\eta}_{\mu\nu} \zeta_a^{\mu} \zeta_b^{\nu} \big] + \pi \big[\bar{\eta}_{\mu\nu} \zeta_a^{\mu} \zeta_b^{\nu} + (-1)^m \bar{\zeta}_{\{a} \hat{\omega} \cdot \vec{\tau} \zeta_{b\}} \big], \tag{20}$$

for which we used that $\bar{\sigma}_{[\mu} \Omega \sigma_{\nu]}$ is a real quaternion, therefore, given by

$$\frac{1}{2} \sigma_0 \mathrm{tr}_2 \big(\bar{\sigma}_{[\mu} \Omega \sigma_{\nu]} \big) = \frac{1}{2} \sigma_0 \, \mathrm{tr}_2 (\Omega \eta_{\nu\mu}) = \sigma_0 \hat{\omega} \cdot \vec{\eta}_{\mu\nu}.$$

We directly read off $\widehat{S}_m$ and $\vec{\rho}_m$ and verify that indeed $\widehat{S}_1 = \widehat{S}_2^t$ and $\vec{\rho}_1 = -\vec{\rho}_2^t$. We wrote the result so as to easily make contact with the $k = 1$ results [1] (for which the 2nd term in $\widehat{S}_m$ and the 1st term in $\vec{\rho}_m$ do not contribute).

In formulating the impurity scattering we note that the $U(k)$ gauge transformation

$$\hat{g}(z) = \exp(2\pi i (\xi_0 - x_0 \mathbb{1}_k) z) \tag{21}$$

turns $\widehat{D}_x^0(z)$ into the ordinary derivative and conjugates all other objects in Eq. (18), such that

$$\left\{ -\frac{d^2}{dz^2} + V(z; \vec{x}) \right\} f_x(z, z') = 4\pi^2 \mathbb{1}_k \delta(z - z'), \tag{22}$$

with $f_x(z, z')$ and $V(z; \vec{x})$ given by

$$f_x(z, z') \equiv \hat{g}(z) \hat{f}_x(z, z') \hat{g}^\dagger(z'),$$
$$V(z; \vec{x}) \equiv 4\pi^2 \vec{R}^2(z; \vec{x}) + 2\pi \sum_m \delta(z - \mu_m) S_m, \tag{23}$$

where $\vec{R}(z; \vec{x})$ and S_m are defined by

$$R_j(z; \vec{x}) \equiv x_j \mathbb{1}_k - \frac{1}{2\pi i} \hat{g}(z) \hat{A}_j(z) \hat{g}^\dagger(z), \qquad S_m \equiv \hat{g}(\mu_m) \widehat{S}_m \hat{g}^\dagger(\mu_m). \tag{24}$$

Periodicity is now only up to a gauge transformation. Note that $R_j(z; \vec{x})$ is a Hermitian $k \times k$ matrix, and that $V(z; \vec{x})$ does *not* depend on x_0. By combining f_x and its derivative into a vector,

$$\vec{f}_x(z, z') \equiv \begin{pmatrix} f_x(z, z') \\ \frac{d}{dz} f_x(z, z') \end{pmatrix}, \tag{25}$$

we can turn the second order equation in to a first order equation

$$\left\{ \frac{d}{dz} - \begin{pmatrix} 0 & \mathbb{1}_k \\ V(z; \vec{x}) & 0 \end{pmatrix} \right\} \vec{f}_x(z, z') = \vec{C} \delta(z - z'), \qquad \vec{C} \equiv -4\pi^2 \begin{pmatrix} 0 \\ \mathbb{1}_k \end{pmatrix}. \tag{26}$$

Its solution is

$$\vec{f}_x(z, z') = W(z) [\vec{c}(z') - \theta(z' - z) W^{-1}(z') \vec{C}],$$
$$\frac{d}{dz} W(z) = \begin{pmatrix} 0 & \mathbb{1}_k \\ V(z; \vec{x}) & 0 \end{pmatrix} W(z) \tag{27}$$

where W is a 2×2 matrix, whereas $\vec{c}$ is a two-component vector (like $\vec{C}$), all components being Hermitian $k \times k$ matrices. The theta function takes care of the inhomogeneous part of the equation. However, we have to restrict to $z' \in [z - 1, z + 1]$ since the delta function is periodic. The solution can be extended beyond this range using the periodicity, which when imposed, as we will see, also determines $\vec{c}(z')$.

The solution for $W(z)$ can be (formally) written as a path ordered exponential integral in the usual way. To show that this, indeed, makes sense, we should specify how to deal with the delta functions in V. From the definition of the path ordered exponential integral we find $W(\mu_m + 0) = T_m W(\mu_m - 0)$ with

$$T_m \equiv P \exp\left[\int_{\mu_m-0}^{\mu_m+0}\begin{pmatrix}0 & \mathbb{1}_k \\ V(z;\vec{x}) & 0\end{pmatrix}dz\right] = \exp\begin{pmatrix}0 & 0 \\ 2\pi S_m & 0\end{pmatrix} = \begin{pmatrix}\mathbb{1}_k & 0 \\ 2\pi S_m & \mathbb{1}_k\end{pmatrix},$$

(28)

which correctly reflects the matching conditions due to the "impurity" at $z = \mu_m$, since $f_x(z, z')$ is continuous across $z = \mu_m$, whereas the derivative jumps with $2\pi S_m f_x(\mu_m, z')$ and both conditions can be summarized by $\vec{f}_x(\mu_m + 0, z') = T_m \vec{f}_x(\mu_m - 0, z')$. Note that $\vec{f}_x$ evolves with z as $\vec{f}_x(z_2, z') = W(z_2)W^{-1}(z_1)\vec{f}_x(z_1, z')$, with

$$W(z_2)W^{-1}(z_1) = P \exp\left[\int_{z_1}^{z_2}\begin{pmatrix}0 & \mathbb{1}_k \\ V(z;\vec{x}) & 0\end{pmatrix}dz\right] \equiv W(z_2, z_1).$$

(29)

In particular, when $z_1 = \mu_m$ and $z_2 = \mu_{m+1}$ this gives the "propagation" between two neighboring impurities and we can write

$$H_m \equiv W(\mu_{m+1} - 0, \mu_m + 0) = P \exp\left[\int_{\mu_m}^{\mu_{m+1}}\begin{pmatrix}0 & \mathbb{1}_k \\ 4\pi^2 \vec{R}^2(z;\vec{x}) & 0\end{pmatrix}dz\right].$$

(30)

Neither T_m nor H_m require us to specify a boundary condition for $W(z)$. A change in boundary condition, however, affects $\vec{c}(z')$. To avoid such ambiguities, we define $\vec{c}_{z_0}(z') \equiv W(z_0)\vec{c}(z')$ such that

$$\vec{f}_x(z, z') = W(z, z_0)\left[\vec{c}_{z_0}(z') - \theta(z' - z)W^{-1}(z', z_0)\vec{C}\right].$$

(31)

We determine $\vec{c}_{z_0}(z')$ by scattering "around" the circle determined from the boundary conditions of $\vec{f}_x$. Using Eq. (23), the fact that $\hat{f}_x(z, z')$ is strictly periodic and that $\hat{g}(z + 1) = \hat{g}(1)\hat{g}(z)$, one finds $\vec{f}_x(z, z' + 1) = \vec{f}_x(z, z')\hat{g}^\dagger(1)$ and $\vec{f}_x(z + 1, z') = \hat{g}(1)\vec{f}_x(z, z')$ (where the order of the *matrix* multiplication, defined componentwise, is important). The latter condition can be used to fix $\vec{c}_{z_0}(z')$ over the range of one period, $z' \in [z, 1 + z]$ (the use of $\hat{f}_x(z, z')$ requires $z' \in [z - 1, z + 1]$, that of $\hat{f}_x(z + 1, z')$ further restricts the range for z' to $z' > z$). With the help of the periodicity properties of W it gives $\vec{c}_{z_0}(z') = (\mathbb{1}_{2k} - \hat{g}^\dagger(1)W(z_0 + 1, z_0))^{-1}W^{-1}(z', z_0)\vec{C}$. We thus find for *arbitrary* z_0

$$\vec{f}_x(z, z') = W(z, z_0)\left\{(\mathbb{1}_{2k} - \mathcal{F}_{z_0})^{-1} - \theta(z' - z)\mathbb{1}_{2k}\right\}W^{-1}(z', z_0)\vec{C},$$

(32)

where we introduced the "holonomy" $\mathcal{F}_{z_0}$

$$\mathcal{F}_{z_0} \equiv \hat{g}^\dagger(1)W(z_0 + 1, z_0) = g^\dagger(1)P \exp\left[\int_{z_0}^{1+z_0}\begin{pmatrix}0 & \mathbb{1}_k \\ V(z;\vec{x}) & 0\end{pmatrix}dz\right].$$

(33)

The equation for $\vec{f}_x$ is valid for $z' \in [z, z + 1]$, but can be extended with the appropriate periodicity specified above. We note that $\hat{g}^\dagger(1)$ plays the role of a cocycle, in the gauge where $\hat{A}_0(z) - 2\pi i x_0 \mathbb{1}_k$ is transformed to 0, with $\mathcal{F}_{z_0}$ the full circle "scattering" matrix.

The z_0-independence of $\vec{f}_x$ follows from

$$\mathcal{F}_{z_0} = g^\dagger(1)W(z_0+1, z_0'+1)W(z_0'+1, z_0) = W(z_0, z_0')g^\dagger(1)W(z_0'+1, z_0)$$
$$= W(z_0, z_0')\mathcal{F}_{z_0'}W^{-1}(z_0, z_0'). \tag{34}$$

Putting things together we, therefore, find

$$\hat{f}_x(z, z') = \hat{g}^\dagger(z)(\mathbb{1}_k, 0) \cdot \vec{f}_x(z, z')\hat{g}(z'),$$
$$\widehat{D}_x^0(z)\hat{f}_x(z, z') = \hat{g}^\dagger(z)(0, \mathbb{1}_k) \cdot \vec{f}_x(z, z')\hat{g}(z'), \tag{35}$$

satisfying all required conditions as can be checked explicitly.

Interestingly, Eq. (34) implies $\mathcal{F}_{z_0+1} = \hat{g}(1)\mathcal{F}_{z_0}\hat{g}^\dagger(1)$ as should of course be the case. In the light of this we also note that (choosing $z_0 = \mu_m + 0$)

$$\mathcal{F}_{\mu_m} = \hat{g}^\dagger(1)T_{m+n}H_{m+n-1}T_{m+n-1}H_{m+n-2}\cdots T_{m+1}H_m$$
$$= T_m H_{m-1}\cdots T_2 H_1 T_1 \hat{g}^\dagger(1)H_n T_n H_{n-1}\cdots T_{m+1}H_m,$$

$$\mathcal{F}_{\mu_m}: \quad
\begin{array}{ccccccccccc}
 & T_m & & T_{m-1} & & T_1\hat{g}^\dagger(1) & & T_n & & T_{m+1} & \\
 & & H_{m-1} & & & & H_n & & & & H_m \\
\hline
z=1+\mu_m & & 1+\mu_{m-1} & & \cdots & & 1+\mu_1 & & \mu_n & \cdots & \mu_{m+1} & \mu_m
\end{array}
\tag{36}$$

using for the second identity that $T_{m+n} = \hat{g}(1)T_m\hat{g}^\dagger(1)$ and $H_{m+n} = \hat{g}(1)H_m\hat{g}^\dagger(1)$. We will use these ingredients further on to relate to the earlier results for $k = 1$, where the positioning of $\hat{g}^\dagger(1)$ is of course irrelevant.

2.3. Gauge field

The central role of the Green's function $\hat{f}_x(z, z')$ becomes clear when one appeals to the fact that it can be used to find the gauge field (working out Eq. (4)), whereas its determinant gives a simple expression for the action density. This follows from the general ADHM construction [8,9], and can be directly taken over for the caloron [1,2,11]. For the gauge field one finds

$$A_\mu(x) = \frac{1}{2}\phi^{1/2}(x)\lambda\bar{\eta}_{\mu\nu}\partial_\nu f_x\lambda^\dagger\phi^{1/2}(x) + \frac{1}{2}[\phi^{-1/2}(x), \partial_\mu\phi^{1/2}(x)]$$
$$= \frac{1}{2}\phi^{1/2}(x)\bar{\eta}_{\mu\nu}^j\partial_\nu\phi_j(x)\phi^{1/2}(x) + \frac{1}{2}[\phi^{-1/2}(x), \partial_\mu\phi^{1/2}(x)], \tag{37}$$

where $\phi(x)$ and $\phi_j(x)$ are $n \times n$ matrices defined by

$$\phi(x) \equiv (\mathbb{1}_n - \lambda f_x\lambda^\dagger)^{-1}, \qquad \phi_j \equiv \lambda\sigma_j f_x\lambda^\dagger. \tag{38}$$

To apply this to the caloron all we have to do is perform the Fourier transformation,

$$\phi(x)^{-1} = \mathbb{1}_n - \sum_{m,m'} P_m\zeta_a \hat{f}_x^{ab}(\mu_m, \mu_{m'})\zeta_b^\dagger P_{m'},$$
$$\phi_j = \sum_{m,m'} P_m\zeta_a\sigma_j \hat{f}_x^{ab}(\mu_m, \mu_{m'})\zeta_b^\dagger P_{m'}. \tag{39}$$

For the $Sp(1)$ construction $\phi(x)$ is a real quaternion, and hence a multiple of σ_0, after which the gauge field simplifies to

116   F. Bruckmann, P. van Baal / Nuclear Physics B 645 (2002) 105–133

$$A_\mu(x) = \frac{1}{2}\phi(x)\bar{\eta}^j_{\mu\nu}\partial_\nu\phi_j(x). \tag{40}$$

To simplify $\phi(x)$ and $\phi_j(x)$ we use Eq. (19), together with $\mu_1 = -\mu_2$, such that $\hat{f}^{ab}_x(\mu_1,\mu_1) = \hat{f}^{ba}_x(\mu_2,\mu_2)$ and $\hat{f}^{ab}_x(\mu_1,\mu_2) = \hat{f}^{ba}_x(\mu_1,\mu_2) = \hat{f}^{ba}_x(\mu_2,\mu_1)^*$. We note that $\phi(x)$ and $\phi_j(x)$ involve the combinations $\zeta_a\sigma_\mu\zeta^\dagger_b = \zeta_a\sigma_\mu\bar{\zeta}_b$, which can be split in symmetric and anti-symmetric combinations (cf. Eq. (20))

$$\zeta_a\sigma_0\bar{\zeta}_b = \sigma_0\zeta^\mu_a\zeta^\mu_b + \eta_{\mu\nu}\zeta^\mu_a\zeta^\nu_b, \qquad \zeta_a\sigma_j\bar{\zeta}_b = \sigma_0\bar{\eta}^j_{\mu\nu}\zeta^\mu_a\zeta^\nu_b + \zeta_{\{a}\sigma_j\bar{\zeta}_{b\}}. \tag{41}$$

No contributions from $\hat{f}^{ab}_x(\mu_1,\mu_2)$ and $\hat{f}^{ab}_x(\mu_2,\mu_1)$ can appear in $\phi(x)$ since these are symmetric in a and b, selecting from $\zeta_a\sigma_0\zeta_b$ the term proportional to σ_0, but $P_1\sigma_0 P_2$ and $P_2\sigma_0 P_1$ vanish. Therefore (cf. Eq. (20))

$$\begin{aligned}
\phi(x)^{-1} &= \sigma_0 - \hat{f}^{ab}_x(\mu_2,\mu_2)\sum_{m=1}^{2}\left(P_m\zeta^\mu_a\zeta^\mu_b P_m + (-1)^m P_m\eta_{\mu\nu}\zeta^\mu_a\zeta^\nu_b P_m\right) \\
&= \sigma_0\left[1 - \hat{f}^{ab}_x(\mu_2,\mu_2)\left(\zeta^\mu_a\zeta^\mu_b + i\hat{\omega}\cdot\bar{\eta}_{\mu\nu}\zeta^\mu_a\zeta^\nu_b\right)\right] \\
&= \sigma_0\left[1 - \pi^{-1}\mathrm{Tr}_k(\hat{f}_x(\mu_2,\mu_2)\widehat{S}_2)\right].
\end{aligned} \tag{42}$$

Similarly we can simplify the expression for $\phi_j(x)$, which we split in a charged component and Abelian, or neutral, component $\phi_j(x) = \phi^{\text{ch}}_j(x) + \phi^{\text{abel}}_j(x)$ with

$$\phi^{\text{ch}}_j = \hat{f}^{ab}_x(\mu_1,\mu_2)P_1\zeta_{\{a}\sigma_j\bar{\zeta}_{b\}}P_2 - \text{h.c.} \tag{43}$$

and

$$\begin{aligned}
\phi^{\text{abel}}_j(x) &= \hat{f}^{ab}_x(\mu_2,\mu_2)\sum_{m=1}^{2}\left(P_m\zeta_{\{a}\sigma_j\bar{\zeta}_{b\}}P_m + (-1)^m P_m\bar{\eta}^j_{\mu\nu}\zeta^\mu_a\zeta^\nu_b P_m\right) \\
&= i\hat{\omega}\cdot\vec{\tau}\,\hat{f}^{ab}_x(\mu_2,\mu_2)\left[\frac{1}{2}\mathrm{tr}_2(\hat{\omega}\cdot\vec{\tau}\zeta_{\{a}\tau_j\bar{\zeta}_{b\}}) - i\bar{\eta}^j_{\mu\nu}\zeta^\mu_a\zeta^\nu_b\right] \\
&= -i\hat{\omega}\cdot\vec{\tau}\pi^{-1}\mathrm{Tr}_k(\hat{f}_x(\mu_2,\mu_2)\rho^j_2),
\end{aligned} \tag{44}$$

where we used that $\frac{1}{2}\mathrm{tr}_2(\hat{\omega}\cdot\vec{\tau}\zeta_{\{a}\tau_j\bar{\zeta}_{b\}}) = \frac{1}{2}\mathrm{tr}_2(\bar{\zeta}_{\{b}\hat{\omega}\cdot\vec{\tau}\zeta_{a\}}\tau_j)$ to correctly identify $\bar{\rho}^{ba}_2$, see Eq. (20). We may of course express $\phi(x)$ and $\phi^{\text{abel}}_j(x)$ also in terms of the first impurity, $\phi^{-1}(x) = \sigma_0[1 - \pi^{-1}\mathrm{Tr}_k(\hat{f}_x(\mu_1,\mu_1)\widehat{S}_1)]$ and $\phi^{\text{abel}}_j(x) = i\hat{\omega}\cdot\vec{\tau}\pi^{-1}\mathrm{Tr}_k(\hat{f}_x(\mu_1,\mu_1)\rho^j_1)$, as is easily verified.

Like for $k = 1$ we will show further on that $\hat{f}^{ab}_x(\mu_1,\mu_2)$ decays exponentially, away from the cores of the constituent monopoles, where only the Abelian component survives,

$$\begin{aligned}
A^{\text{abel}}_\mu(x) = &-\frac{i}{2}\hat{\omega}\cdot\vec{\tau}\left[1 - \pi^{-1}\mathrm{Tr}_k(\hat{f}_x(\mu_2,\mu_2)\widehat{S}_2)\right]^{-1} \\
&\times \bar{\eta}^j_{\mu\nu}\partial_\nu\left[\pi^{-1}\mathrm{Tr}_k(\hat{f}_x(\mu_2,\mu_2)\rho^j_2)\right].
\end{aligned} \tag{45}$$

Note that for $k = 1$ we are able to write $A^{\text{abel}}_\mu(x) = -\frac{i}{2}\hat{\omega}\cdot\vec{\tau}e_j\bar{\eta}^j_{\mu\nu}\partial_\nu\log\phi(x)$, with $\vec{e} = \vec{\rho}_2/|\vec{\rho}_2|$, making use of the fact that $\widehat{S}_m = |\vec{\rho}_m|$, but that this is no longer true for higher charge, despite the fact that on the diagonal we still have $\widehat{S}^{aa}_m = |\vec{\rho}^{aa}_m|$. This seems to allow

F. Bruckmann, P. van Baal / Nuclear Physics B 645 (2002) 105–133

for the dipoles of k well-separated calorons to point in different directions. We will discuss this issue further when studying the limiting behavior far from the cores of the constituents, where the field becomes algebraic and constituent locations are readily identified.

2.4. Action density

Within the ADHM formalism the action density is given by [9],

$$-\frac{1}{2}\mathrm{Tr}_n\, F_{\mu\nu}^2(x) = -\frac{1}{2}\partial_\mu^2\partial_\nu^2 \log\psi(x), \tag{46}$$

where $\psi(x)$ equals $1/\det(f_x)$ after regularization to extract an irrelevant overall and for calorons divergent constant. We will be able to find a simple expression for $\psi(x)$ at any k, generalizing the result for charge 1 calorons. We use that $\partial_\nu\log\det(f_x) = -\mathrm{Tr}(\{\partial_\nu f_x^{-1}\}f_x) = \frac{1}{\pi i}\widehat{\mathrm{Tr}}(\widehat{D}_x^\nu \hat{f}_x)$, where in the last step we performed the Fourier transformation, and $\widehat{\mathrm{Tr}}$ includes an integration with respect to z. The case $\nu = 0$ is treated separately, due to the discontinuity in the z derivative of $\hat{f}_x(z, z')$ at $z = z'$, which we regularize using point-splitting:

$$\frac{1}{\pi i}\widehat{\mathrm{Tr}}\big(\widehat{D}_x^0 \hat{f}_x\big)$$

$$= \lim_{\varepsilon\to 0}\frac{1}{2\pi i}\int_0^1 dz\,\mathrm{Tr}_k\left(\frac{d}{dz}f_x(z+\varepsilon, z') + \frac{d}{dz}f_x(z-\varepsilon, z')\right)_{z'=z}$$

$$= 4\pi i\int_0^1 dz\,\mathrm{Tr}\left(W^{-1}(z, z_0)\begin{pmatrix} 0 & 0 \\ 0 & \mathbb{1}_k \end{pmatrix} W(z, z_0)\left[(\mathbb{1}_{2k} - \mathcal{F}_{z_0})^{-1} - \frac{1}{2}\mathbb{1}_{2k}\right]\right)$$

$$= 2\pi i\int_0^1 dz\left[\mathrm{Tr}\left((\mathbb{1}_{2k} - \mathcal{F}_{z_0})^{-1} - \frac{1}{2}\mathbb{1}_{2k}\right) - \frac{1}{4\pi^2}\frac{d}{dz}\mathrm{Tr}_k\big(f_x(z, z)\big)\right]. \tag{47}$$

The Tr without an index or hat indicates the full trace over the $2k \times 2k$ matrix involved. To see how the total derivative term appears (not contributing to the integral due to the periodicity of $\mathrm{Tr}_k[f_x(z, z)]$) we use that

$$\frac{d}{dz}\mathrm{Tr}_k\big(\hat{f}_x(z, z)\big)$$

$$= -4\pi^2\frac{d}{dz}\mathrm{Tr}\left(W^{-1}(z, z_0)\begin{pmatrix} 0 & 0 \\ \mathbb{1}_k & 0 \end{pmatrix} W(z, z_0)\left[(\mathbb{1}_{2k} - \mathcal{F}_{z_0})^{-1} - s\mathbb{1}_{2k}\right]\right)$$

$$= -4\pi^2\,\mathrm{Tr}\left(W^{-1}(z, z_0)\left[\begin{pmatrix} 0 & 0 \\ \mathbb{1}_k & 0 \end{pmatrix}, \begin{pmatrix} 0 & \mathbb{1}_k \\ V(z;\vec{x}) & 0 \end{pmatrix}\right]\right.$$

$$\left. \times W(z, z_0)\left[(\mathbb{1}_{2k} - \mathcal{F}_{z_0})^{-1} - s\mathbb{1}_{2k}\right]\right)$$

$$= -4\pi^2\,\mathrm{Tr}\left(W^{-1}(z, z_0)\begin{pmatrix} -\mathbb{1}_k & 0 \\ 0 & \mathbb{1}_k \end{pmatrix} W(z, z_0)\left[(\mathbb{1}_{2k} - \mathcal{F}_{z_0})^{-1} - s\mathbb{1}_{2k}\right]\right), \tag{48}$$

with $s = 1/2$ (as for point-spitting, although one checks that the s dependent term actually vanishes). We note that $\mathcal{F}_{z_0}$ depends on x_0 only through $\hat{g}(1)$, and that $\partial_0 \hat{g}(1) = 2\pi i \hat{g}(1)$, such that $\partial_0 \mathcal{F}_{z_0} = 2\pi i \mathcal{F}_{z_0}$. With this we find

$$\partial_0 \log \det(\hat{f}_x) = -\partial_0 \log \psi, \qquad \psi \equiv \det\left(i e^{-\pi i x_0}(\mathbb{1}_{2k} - \mathcal{F}_{z_0})/\sqrt{2}\right), \tag{49}$$

which is *independent* of z_0. The factor $i/\sqrt{2}$ in the argument of the determinant was inserted just so ψ agrees with the definition introduced earlier for $k = 1$.

Next we compute $\partial_j \log \det(\hat{f}_x)$,

$$\frac{1}{\pi i} \int_0^1 dz \, \mathrm{Tr}_k\left(\widehat{D}_x^j \hat{f}_x(z, z)\right)$$

$$= -2 \int_0^1 dz \, \mathrm{Tr}_k\left(R_j(z; \vec{x}) f_x(z, z)\right)$$

$$= 8\pi^2 \int_0^1 dz \, \mathrm{Tr}\left(W^{-1}(z, z_0)\begin{pmatrix} 0 & 0 \\ R_j(z; \vec{x}) & 0 \end{pmatrix} W(z, z_0)\left[(\mathbb{1}_{2k} - \mathcal{F}_{z_0})^{-1} - s\mathbb{1}_{2k}\right]\right)$$

$$= \mathrm{Tr}\left((\mathbb{1}_{2k} - \mathcal{F}_{z_0})^{-1}\mathcal{F}_{z_0}\int_{z_0}^{1+z_0} dz \, W^{-1}(z, z_0)\partial_j\begin{pmatrix} 0 & \mathbb{1}_k \\ V(z; \vec{x}) & 0 \end{pmatrix} W(z, z_0)\right), \tag{50}$$

where again s can take any value, but for convenience is best set to 1 here. Finally noting that $\mathcal{F}_{z_0}$ only depends on $\vec{x}$ through $\vec{R}(z; \vec{x})$ and using that

$$W^{-1}(z, z_0)\partial_j W(z, z_0) = \int_{z_0}^{1+z_0} dz \, W^{-1}(z, z_0)\partial_j\begin{pmatrix} 0 & \mathbb{1}_k \\ V(z; \vec{x}) & 0 \end{pmatrix} W(z, z_0), \tag{51}$$

we verify that $\partial_\mu \log \det(\hat{f}_x) = -\partial_\mu \log \psi$ for all μ and k. It is amusing to note that this implies the remarkable formula

$$-\frac{1}{2}\mathrm{Tr}_n F_{\mu\nu}^2(x) = -\frac{1}{2}\partial_\mu^2\partial_\nu^2 \log \det\left(\mathbb{1}_{2k} - \hat{g}^\dagger(1)P \exp\left[\int_0^1\begin{pmatrix} 0 & \mathbb{1}_k \\ V(z; \vec{x}) & 0 \end{pmatrix} dz\right]\right), \tag{52}$$

even though explicit evaluation can be quite cumbersome. Not so for some special cases, including the single caloron $k = 1$, where $\vec{R}(z; \vec{x})$ is *piecewise constant* as we will discuss next.

3. Special cases

Consider $\vec{R}(z; \vec{x})$ to be piecewise constant, for $z \in [\mu_m, \mu_{m+1}]$ defined to be $\vec{x} - \vec{Y}_m$ (cf. Eq. (24)), with $\vec{Y}_m$ constant Hermitian $k \times k$ matrices related to the Nahm potential by

$\hat{A}_j(z) = 2\pi i \,\hat{g}^\dagger(z) Y_m^j \hat{g}(z)$. In this case we can easily deal with the path ordered exponential integrals. To be specific, the "propagation" from μ_m to μ_{m+1} defined through H_m in Eq. (30), is given by

$$H_m = \begin{pmatrix} \cosh(2\pi v_m R_m) & (2\pi R_m)^{-1} \sinh(2\pi v_m R_m) \\ 2\pi R_m \sinh(2\pi v_m R_m) & \cosh(2\pi v_m R_m) \end{pmatrix}, \tag{53}$$

where $v_m = \mu_{m+1} - \mu_m$ (with $\mu_{m+n} = 1 + \mu_m$ such that $\sum_m v_m = 1$) and $R_m \equiv (\vec{R}_m \cdot \vec{R}_m)^{\frac{1}{2}}$ (a Hermitian $k \times k$ matrix). Since $\cosh(y)$ and $y^{\pm 1} \sinh(y)$ are both quadratic in y, actually no square root is involved in this expression for H_m.

3.1. The known charge 1-caloron

For charge 1 the Nahm equation (Eq. (17)) has no commutator terms and $\vec{R}(z; \vec{x})$ is always piecewise constant. It reduces to the mth constituent radius for $z \in [\mu_m, \mu_{m+1}]$, $R_m = r_m \equiv |\vec{r}_m|$, whereas the prefactor at the impurity becomes $S_m = |\vec{r}_m - \vec{r}_{m-1}| = |\vec{\rho}_m|$. With

$$A_m \equiv r_m^{-1} \begin{pmatrix} r_m & |\vec{\rho}_{m+1}| \\ 0 & r_{m+1} \end{pmatrix} \begin{pmatrix} \cosh(2\pi v_m r_m) & \sinh(2\pi v_m r_m) \\ \sinh(2\pi v_m r_m) & \cosh(2\pi v_m r_m) \end{pmatrix}, \tag{54}$$

a link to the earlier charge 1 results [2] is established by noting that

$$A_m = \begin{pmatrix} 0 & 1 \\ 2\pi r_{m+1} & 0 \end{pmatrix} T_{m+1} H_m \begin{pmatrix} 0 & 1 \\ 2\pi r_m & 0 \end{pmatrix}^{-1}. \tag{55}$$

With the placing of $\hat{g}^\dagger(1)$ irrelevant, and the possibility of absorbing ξ_0 in x_0, we therefore find

$$\mathcal{F}_{\mu_m}^{k=1} = \begin{pmatrix} 0 & 1 \\ 2\pi r_m & 0 \end{pmatrix} e^{2\pi i x_0} A_{m-1} A_{m-2} \cdots A_1 A_n \cdots A_{m+1} A_m \begin{pmatrix} 0 & 1 \\ 2\pi r_m & 0 \end{pmatrix}^{-1}, \tag{56}$$

cf. Eq. (36). In particular this shows that $\psi = -\frac{1}{2} e^{-2\pi i x_0} \det(\mathbb{1}_2 - \mathcal{F}_{\mu_m})$ agrees with the result found earlier, $\psi = \frac{1}{2} \operatorname{tr}(A_n A_{n-1} \cdots A_1) - \cos(2\pi x_0)$. The Green's functions can be shown to agree as well, using $(\mathbb{1}_2 - \mathcal{F}_{\mu_m})^{-1} = (\mathbb{1}_2 - \bar{\sigma}_2 \mathcal{F}_{\mu_m}^t \sigma_2) / \det(\mathbb{1}_2 - \mathcal{F}_{\mu_m})$.

3.2. Exact axially symmetric solution

Arranging $\vec{R}(z; \vec{x})$ to be piecewise constant when $k > 1$ requires one to fulfill some constraints. To solve the Nahm equation, Eq. (17), in terms of the $Y_m^j = \frac{1}{2\pi i} \hat{g}(z) \hat{A}_j(z) \hat{g}^\dagger(z)$, the commutator term should vanish. One way to achieve this, is by choosing $\vec{Y}_m = Y_m \vec{e}$. The Nahm equation relates the discontinuities of $\hat{A}_j(z)$ to $\vec{\rho}_m$,

$$\hat{g}^\dagger(\mu_m)(Y_m - Y_{m-1})\hat{g}(\mu_m)\vec{e} = \vec{\rho}_m, \tag{57}$$

which imposes constraints on ζ_a, see Eq. (14). To seek a solution we choose all ζ^a to be parallel in group space, $\zeta^a = \rho_a \zeta$ (ρ_a a positive real number). This reduces the problem to $k = 1$, since $\zeta_a^\dagger P_m \zeta_b = \rho_a \rho_b \zeta^\dagger P_m \zeta$ is proportional to $\zeta^\dagger P_m \zeta$. For $SU(2)$ this already solves the constraint, since for $k = 1$ one has $\vec{\rho}_1 = -\vec{\rho}_2$. For $n > 2$ it has been shown [2] that $\vec{\rho}_m$ can take any value, provided $\sum_{m=1}^{n} \vec{\rho}_m = \vec{0}$, in particular we may choose all $\vec{\rho}_m$ to be proportional to $\vec{e}$ (by properly choosing ζ). For $k = 1$ it is convenient to parametrize $\zeta^\dagger P_m \zeta = (|\vec{\rho}_m| - \vec{\tau} \cdot \vec{\rho}_m)/(2\pi)$ in terms of constituent locations, $\vec{\rho}_m = \Delta \vec{y}_m \equiv \vec{y}_m - \vec{y}_{m-1}$. As in Section 2.3 we will take $\vec{e} = \vec{\rho}_2/|\vec{\rho}_2|$.

We obtain a larger class of ζ^a for which the $\vec{\rho}_m$ are parallel, by taking ζ^a to be parallel up to a gauge rotation with an element of the unbroken subgroup $U(1)^{n-1} \subset SU(n)$ which leaves the holonomy unchanged,

$$\zeta_a = \rho_a \exp(2\pi i \alpha_a)\zeta, \qquad \alpha_a \equiv \sum_{m=1}^{n} \alpha_a^m P_m, \quad \mathrm{Tr}_n \, \alpha_a = 0. \tag{58}$$

This leads to

$$\vec{\rho}_m^{ab} = \rho^a \rho^b \exp\big(2\pi i \big(\alpha_b^m - \alpha_a^m\big)\big)\Delta \vec{y}_m, \qquad \widehat{S}_m^{ab} = \vec{\rho}_m^{ab} \cdot \Delta \vec{y}_m/|\Delta \vec{y}_m|. \tag{59}$$

Note that for $Sp(1)$ we verify that $\vec{\rho}_1 + \vec{\rho}_2^t = \vec{0}$ and $\widehat{S}_1 - \widehat{S}_2^t = 0$ (using $\alpha_a^1 = -\alpha_a^2$). With $\Delta \vec{y}_m = \Delta y_m \vec{e}$ for all m (by definition $\Delta y_2 > 0$) we may solve Eq. (57),

$$\begin{aligned}
Y_m^{ab} = {}&\big(\xi_a + \rho_a^2 y_m\big)\delta_{ab} + i(1 - \delta_{ab})\rho_a \rho_b \\
&\times \sum_{j=1}^{n} \Delta y_j \frac{\exp\big(2\pi i [\alpha_b^j - \alpha_a^j - (\mu_j + s_j^m)(\xi_0^b - \xi_0^a)]\big)}{2\sin(\pi[\xi_0^b - \xi_0^a])},
\end{aligned} \tag{60}$$

where the ξ_a are arbitrary, $m = 1, \ldots, n$ and $s_j^m = \frac{1}{2}$ for $j = 1, 2, \ldots, m$ and $s_j^m = -\frac{1}{2}$ for $j = m + 1, \ldots, n$. The eigenvalues of these Hermitian matrices determine the constituent locations, all lined-up along $\vec{e}$. It should be noted that there is no reason to expect that all the Y_m can be diagonalized simultaneously. We will come back to this in the following section.

Returning to the simplest case of parallel gauge orientations, i.e., putting $\alpha_a^j = 0$, we may take the limit $\xi_0 \to 0$ (related to vanishing time separations) to find

$$Y_m^{ab} = \big(\xi_a + y_c \rho_a^2\big)\delta_{ab} + (y_m - y_c)\rho_a \rho_b, \qquad y_c \equiv \sum_{m=1}^{n} v_m y_m. \tag{61}$$

The ($k = 1$) "center of mass" coordinate y_c can be freely chosen and the ξ_a play the role of "center of mass" of each constituent caloron. How exactly this is realized becomes clear when we diagonalize Y_m. Let us first consider Eq. (61) for $SU(2)$ and charge 2, $k = n = 2$, with $\xi_1 = -\xi_2 \equiv \xi$. Without loss of generality we choose $y_c = 0$, such that the two eigenvalues of Y_m are given by $y_m^{(j)} = \frac{1}{2} y_m \rho^2 + (-1)^j \sqrt{\xi^2 + \frac{1}{4} y_m^2 \rho^4 + y_m \xi \Delta \rho^2}$, where $\rho^2 \equiv \rho_1^2 + \rho_2^2$ and $\Delta \rho^2 \equiv \rho_1^2 - \rho_2^2$. For large and positive ξ we find $y_m^{(j)} = (-1)^j \xi + y_m \rho_j^2 + \mathcal{O}(\xi^{-1})$, representing two charge 1 calorons centered at ξ and $-\xi$, with separations between their constituents monopoles given in terms of $\Delta y_2 \rho_{1,2}^2$. We plot the constituent locations as a function of ξ in Fig. 3 for $y_2 = -y_1 = v_j = \frac{1}{2}$ and $\rho_j = 2$.

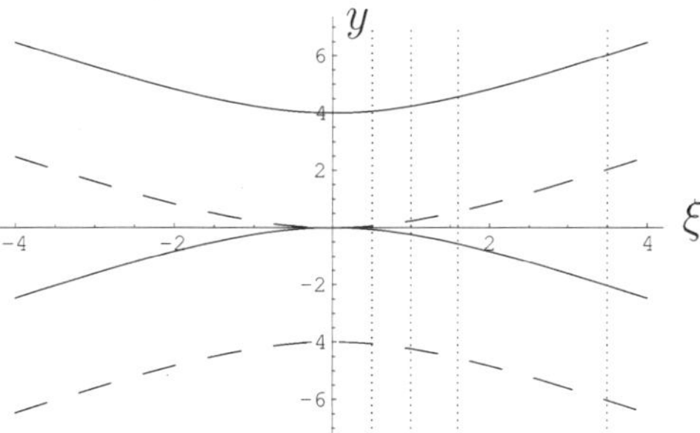

Fig. 3. Constituent locations $y_m^{(j)}$ based on Eq. (61) (i.e., $\alpha_a = 0$ and $\xi_0 \to 0$) as a function of $\xi = \xi_1 = -\xi_2$ for $y_2 = -y_1 = v_1 = v_2 = \frac{1}{2}$ and $\rho_1 = \rho_2 = 2$. Dashed versus full lines distinguish the magnetic charge of the constituents. The dotted lines represent the four cases shown in Figs. 2 and 4.

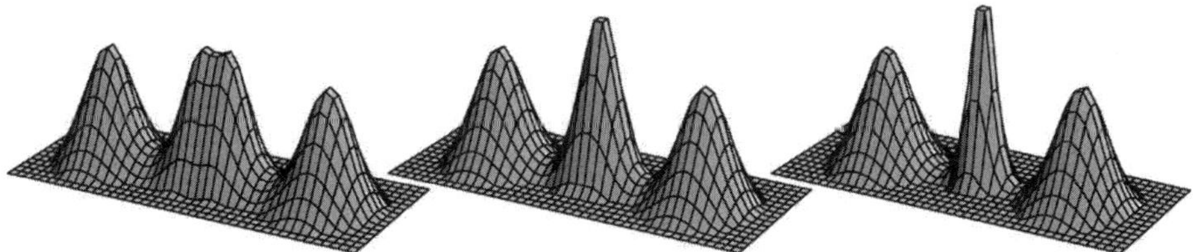

Fig. 4. The action density (cutoff for $\log(S)$ below -3) as a function of x and z for the $SU(2)$ solution with charge 2 ($\mu_2 = \frac{1}{4}$, $\alpha_a^j = \xi_0^a = 0$, $\rho_j = 2$) and increasing values of $\xi \equiv \xi_1 = -\xi_2$ (see Fig 2 (left) for $\xi = 3.5$), with $\xi = 1.6$ (left), $\xi = 1.0$ (middle) and $\xi = 0.5$ (right). Compare Fig. 3 for the corresponding constituent locations.

Action density profiles are shown in Fig. 2 (left) for $\xi = 3.5$ and in Fig. 4 for $\xi = 1.6, 1.0, 0.5$. From the dotted lines in Fig. 3 one reads off the associated constituent locations. Note that the magnetic moments of the two calorons are pointing in the same direction and that we cannot freely interchange constituent monopole locations within our axially symmetric ansatz. However, when ξ is small it is more natural to interpret the configuration as a narrow caloron (i.e., instanton) with inverted magnetic moment in the background of a large caloron. This is the proper setting to understand the non-trivial time dependence for $\xi = 0.5$ illustrated in Fig. 5 (left).

For $\xi \to 0$ a singular caloron arises due to the fusion of two constituents (with opposite magnetic charge). This singularity is avoided when $\xi_0^a \neq 0$, which can be understood by observing that the eigenvalues of ξ_0^a parametrize time-locations. If $\alpha_a^j \neq 0$, with ξ_0 and ξ made small, one will find two calorons (and their constituents) to be pushed far from each other. This can be understood as well, in terms of a short-to-long distance duality in the ADHM data for an instanton pair with non-parallel group orientation [10], but can also be read off from the eigenvalues of Y_m defined in Eq. (60). As an example we take again $SU(2)$ and charge 2, but now with $\xi_0 \equiv \xi_0^1 = -\xi_0^2$ and $\alpha_1^2 = -\alpha_2^2 = -\alpha_1^1 = \alpha_2^1 \equiv \alpha$ in general non-zero. For the case $\alpha = 1/8$ (perpendicular relative color orientations), $\rho_1 = \rho_2 \equiv \rho$ and $\mu_2 = \frac{1}{4}$ (equal mass constituents), the eigenvalues are

$$y_m^{(j)} = y_m \rho^2 + (-1)^j \sqrt{\xi^2 + \tfrac{1}{4}(\Delta y_2)^2 \rho^4 \sin^{-2}(\pi \xi_0)}, \quad \text{whereas for } \alpha = 0 \text{ one finds } y_m^{(j)} =$$

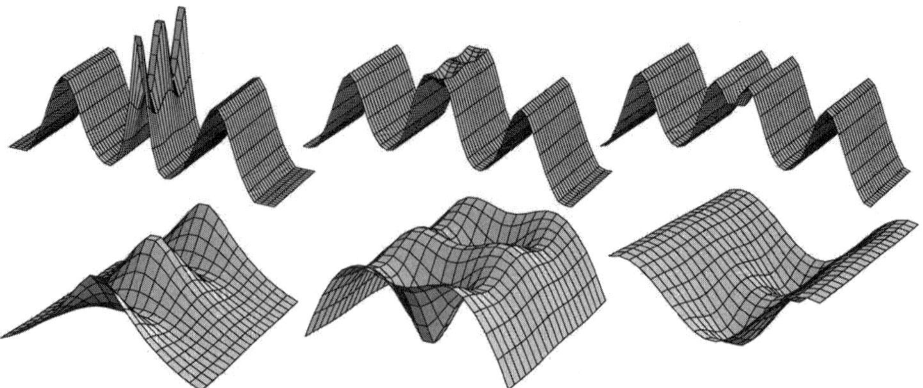

Fig. 5. The action density (cutoff for $\log(S)$ below -3) as a function of z and t (doubling the time-period) for the $SU(2)$ solution with charge 2 ($\mu_2 = \frac{1}{4}$, $\xi = \frac{1}{2}$, $\alpha_a^j = 0$, $\rho_j = 2$) and increasing values of ξ_0, with $\xi_0 = 0$ (left), $\xi_0 = 0.2$ (middle) and $\xi_0 = 0.25$ (right), for the top row all on the same scale, zooming in on the middle region on the bottom row (not to scale). See Fig. 4 (right) for the case of $\xi_0 = 0$ shown as a function of x and z.

$y_m \rho^2 + (-1)^j \sqrt{\xi^2 + \frac{1}{4}(\Delta y_2)^2 \rho^4 \cos^{-2}(\pi \xi_0)}$. In the light of this it is interesting to observe, as shown in Fig. 5, that with $\alpha = 0$ and *increasing* ξ_0 the constituents are pushed out in the z direction as well. When $\xi_0 \to 0.5$ the constituents would otherwise come close together through the periodicity in the time direction. Effectively these constituents thus have perpendicular color orientations (due to our choice of holonomy with $\mu_2 = \frac{1}{4}$). The transition from constituents separating in the time direction for ξ_0 near 0 to constituents separating in the z direction for ξ_0 near $\frac{1}{2}$ occurs for $\rho = 2$ at approximately $\xi_0 = 0.2$.

With a little imagination one detects the ring-shaped structure also observed [10] in the case of instantons at zero temperature, see Fig. 5 (middle). A more direct analogy of course occurs when two calorons (with ρ small, i.e., instantons with unresolved constituent monopoles) approach each other. We checked that for $\xi = \alpha = 0$ and $\xi_0 \to 0$ a singular caloron forms due to the overlap of two calorons with parallel gauge orientations, whereas for $\xi_0 \to \frac{1}{2}$ the two calorons are pushed away (to infinity) in the z direction as is appropriate for the non-parallel group orientation due to the non-trivial holonomy. At an intermediate value ($\xi_0 = 0.25$ for $\rho_1 = \rho_2 = 0.1$) one observes a small ring in the t-z plane. Choosing ξ large one may check that $\pm \xi_0$ indeed gives the time location for each caloron. When, however, ξ_0 approaches $\frac{1}{2}$ they can no longer keep parallel gauge orientations due to the non-trivial holonomy. As noted before, this may be described by a solution with $\alpha \neq 0$ and $\xi_0 \to 0$. Computing the eigenvalues of Y_m therefore allows one to easily predict the behavior of the exact solution.

For charge 1 it had been shown [2] that as soon as one of the constituents is far removed from the others the solution becomes static. For the "dimensional reduction" to take place at higher charge this is no longer sufficient. We have seen (generalization to $SU(n)$ is straightforward) that any magnetically neutral cluster of constituents, when small with respect to β, will behave like an instanton that is localized in time. For the special case with parallel group orientations, putting all $\xi_a = 0$ in Eq. (61) one would even be left with $k - 1$ singular instantons on top of one regular caloron, whose scale parameter is set by

$\rho^2 = \sum_{a=1}^{k} \rho_a^2$, which can be understood from the fact that the matrix $\rho_a \rho_b$ has rank 1. It does, however, give us one opportunity to go beyond the axial symmetry considered so far. When $\xi_a = 0$ we could solve the Nahm equation for parallel gauge orientations by $\vec{Y}_m^{ab} = \vec{y}_m \rho_a \rho_b$ *without* insisting all the $\vec{y}_m$ line-up. This still describes $k-1$ singular instantons on top of one regular caloron, except that now the singular instantons can have *arbitrary* locations.

Even though our ansatz to obtain exact solutions has been restrictive (as is clear from the axial symmetry), we stress that the solutions for the Nahm equation we found provide genuine multi-caloron solutions, which reveal kn isolated lumps for each of its constituent monopoles (with suitably chosen ξ^a so the constituents do not overlap). This is not only illustrated in Fig. 2, but can also be understood analytically for any charge k and $SU(n)$ as follows. We diagonalize each Y_m with (in general *different*) similarity transformations U_m, or $Y_m \equiv U_m \operatorname{diag}(y_m^{(1)}, \ldots, y_m^{(k)}) U_m^\dagger$. These bring R_m to the diagonal form $R_m^{\mathrm{diag}} = \operatorname{diag}(r_m^{(1)}, \ldots, r_m^{(k)})$, with $r_m^{(j)} \equiv |\vec{x} - y_m^{(j)} \vec{e}|$ the constituent radii, such that $\tilde{H}_m \equiv U_m^\dagger H_m U_m$ (U_m acting componentwise) simplifies to (cf. Eq. (53))

$$\tilde{H}_m \equiv \begin{pmatrix} \cosh\!\left(2\pi v_m R_m^{\mathrm{diag}}\right) & \left(2\pi R_m^{\mathrm{diag}}\right)^{-1} \sinh\!\left(2\pi v_m R_m^{\mathrm{diag}}\right) \\ 2\pi R_m^{\mathrm{diag}} \sinh\!\left(2\pi v_m R_m^{\mathrm{diag}}\right) & \cosh\!\left(2\pi v_m R_m^{\mathrm{diag}}\right) \end{pmatrix}. \tag{62}$$

The action density can now be explicitly expressed in terms of the constituent radii

$$-\frac{1}{2}\operatorname{Tr}_n F_{\mu\nu}^2(x) = -\frac{1}{2}\partial_\mu^2 \partial_\nu^2 \log \psi(x), \qquad \psi = \det\!\left(ie^{-\pi i x_0}(\mathbb{1}_{2k} - \mathcal{F})/\sqrt{2}\right),$$

$$\mathcal{F} \equiv \exp\!\left(2\pi i (x_0 \mathbb{1}_k - \xi_0)\right) U_n \tilde{H}_n \tilde{T}_n \tilde{H}_{n-1} \tilde{T}_{n-1} \cdots \tilde{H}_1 U_1^\dagger T_1 \tag{63}$$

cf. Eqs. (36), (46), (49), (53), where $\tilde{T}_m \equiv U_m^\dagger T_m U_{m-1}$ (U_m again acting componentwise). The size of the constituent monopoles is read off to be $(2\pi v_m)^{-1}$ (or $\beta(2\pi v_m)^{-1}$ when $\beta \neq 1$), and one concludes that the action density will contain kn lumps for sufficiently well separated constituents. The figures were produced by computing the action density using precisely this method.

4. Far-field limit

The non-trivial value of the Polyakov loop at spatial infinity (holonomy) leads to a spontaneous breaking of the gauge symmetry, but without the need of introducing a Higgs field. One may view A_0 as the Higgs field in the adjoint representation. This is one way to understand why constituent monopoles emerge. The best way to describe the caloron solutions, in case of well separated constituents, is by analyzing the field outside the cores of these constituents, where only the Abelian field survives. Since in our case the asymptotic Polyakov loop value defines a global direction in color space the Abelian generator in terms of which we can describe the so-called far-field configuration is fixed, giving rise to a global embedding in the full gauge group. Extrapolating the Abelian fields back to inside the core of the constituents leads to Dirac monopoles. Such an extrapolation is well defined in terms of the *high temperature limit*, which makes the core of the constituents shrink to zero size and the field to become a smooth Abelian gauge field everywhere except for the singularities of the Dirac monopoles. Thus we anticipate that in this limit the self-dual Abelian field is still described by point like constituents, despite the fact that $\vec{R}(z, \vec{x})$ is no longer piecewise constant. Any "fuzziness" of the constituent location that may result from this, would be confined to the non-Abelian core, and not visible from afar.

4.1. Green's function

Despite the somewhat formal expression for the Green's function $\hat{f}_x(z, z')$, one can extract information about the long-range fields from it. In the following we will show how to neglect the exponentially decaying fields in the cores of the monopole constituents, being left with the Abelian components of fields which decay algebraically. We only need to consider the "bulk" contributions H_m, Eq. (30), which contain all the dependence on $\vec{x}$. Our starting point is Eq. (22) restricted to the mth interval, $z \in (\mu_m, \mu_{m+1})$

$$\left\{ -\frac{d^2}{dz^2} + 4\pi^2 \vec{R}^2(z; \vec{x}) \right\} f_m(z) = 0. \tag{64}$$

To distinguish between exponentially growing and decreasing contributions for this homogeneous equation we take as a basis for $f_m(z)$ functions $f_m^{\pm}(z)$ with the following asymptotic behavior

$$|\vec{x}| \to \infty: f_m^{\pm}(z) \to \exp\big(\pm 2\pi |\vec{x}|(z - \mu_m)\mathbb{1}_k\big), \tag{65}$$

relying on the fact that $\vec{R}^2(z; \vec{x}) \to \vec{x}^2 \mathbb{1}_k$. This prompts us to introduce on each interval the matrix valued functions $R_m^{\pm}(z)$ (we suppress the dependence on $\vec{x}$) such that

$$f_m^{\pm}(z) = P \exp\left[\pm 2\pi \int_{\mu_m}^{z} R_m^{\pm}(z) dz \right] \tag{66}$$

from which it follows that $R_m^{\pm}(z)$ is a solution of the Riccati equation

$$R_m^{\pm}(z)^2 \pm \frac{1}{2\pi} \frac{d}{dz} R_m^{\pm}(z) = \vec{R}^2(z; \vec{x}). \tag{67}$$

We note that for $|\vec{x}| \to \infty$, $R_m^{\pm}(z) \to |\vec{x}|$ and that for piecewise constant $\vec{R}(z; \vec{x})$ both $R_m^+(z)$ and $R_m^-(z)$ are constant and equal to R_m, introduced in Eq. (53).

We can write for $z, z' \in (\mu_m, \mu_{m+1})$ the "propagator" $W(z, z')$ defined in Eq. (29) in terms of $f_m^{\pm}(z)$ as $W(z, z') = W_m(z) W_m^{-1}(z')$ with

$$W_m(z) \equiv \begin{pmatrix} f_m^+(z) & f_m^-(z) \\ 2\pi R_m^+(z) f_m^+(z) & -2\pi R_m^-(z) f_m^-(z) \end{pmatrix}. \tag{68}$$

Using that $H_m = W_m(\mu_{m+1}) W_m^{-1}(\mu_m)$ and $f_m^{\pm}(\mu_m) = \mathbb{1}_k$, we find by neglecting the exponentially decreasing factors $f_m^-(\mu_{m+1})$ the required limiting behavior for H_m. Paying special attention to the ordering of the $k \times k$ matrices $R_m^{\pm}$, observing that

$$W_m^{-1}(\mu_m) = \big(4\pi R_m(\mu_m)\big)^{-1} \begin{pmatrix} 2\pi R_m^-(\mu_m) & \mathbb{1}_k \\ 2\pi R_m^+(\mu_m) & -\mathbb{1}_k \end{pmatrix},$$

F. Bruckmann, P. van Baal / Nuclear Physics B 645 (2002) 105–133 125

$$R_m \equiv \frac{1}{2}\left(R_m^+(\mu_m) + R_m^-(\mu_m)\right), \tag{69}$$

we find well outside the cores of the constituents

$$H_m \to \begin{pmatrix} \mathbb{1}_k & 0 \\ 2\pi R_m^+(\mu_{m+1}) & 0 \end{pmatrix} f_m^+(\mu_{m+1})(4\pi R_m)^{-1} \begin{pmatrix} 2\pi R_m^-(\mu_m) & \mathbb{1}_k \\ 0 & 0 \end{pmatrix}. \tag{70}$$

The sparse nature of the matrices involved will be of considerable help to simplify the limiting behavior of the Green's function. A crucial ingredient is the combination

$$\begin{pmatrix} 2\pi R_m^-(\mu_m) & \mathbb{1}_k \\ 0 & 0 \end{pmatrix} \begin{pmatrix} \mathbb{1}_k & 0 \\ 2\pi S_m & \mathbb{1}_k \end{pmatrix} \begin{pmatrix} \mathbb{1}_k & 0 \\ 2\pi R_{m-1}^+(\mu_m) & 0 \end{pmatrix} = \begin{pmatrix} 2\pi \Sigma_m & 0 \\ 0 & 0 \end{pmatrix},$$

$$\Sigma_m \equiv R_{m-1}^+(\mu_m) + R_m^-(\mu_m) + S_m, \tag{71}$$

for clarity summarizing the various ingredients in the following picture

$$
\begin{array}{ccc}
 & \Sigma_m = & \\
\Sigma_{m-1} & R_{m-1}^+(\mu_m)+S_m+R_m^-(\mu_m) & \Sigma_{m+1} \\
 & R_m = & \\
R_{m-1} & (R_m^+(\mu_m)+R_m^-(\mu_m))/2 & R_{m+1} \\
\end{array}
$$

$$z = \mu_{m-1} \quad R_{m-1}^\pm(z), f_{m-1}^\pm(z) \quad \mu_m \quad R_m^\pm(z), f_m^\pm(z) \quad \mu_{m+1}$$

This leads to the following far-field approximation for $\mathcal{F}_{\mu_m}$ (cf. Eq. (36)),

$$\mathcal{F}_{\mu_m} \to \hat{g}^\dagger(1) \begin{pmatrix} \mathbb{1}_k & 0 \\ 2\pi(R_{m+n-1}^+(\mu_{m+n}) + S_{m+n}) & 0 \end{pmatrix} \mathcal{G}_m \begin{pmatrix} 2\pi R_m^-(\mu_m) & \mathbb{1}_k \\ 0 & 0 \end{pmatrix} \tag{72}$$

where $\mathcal{G}_m \equiv \mathcal{G}_{m+n,m}$ and

$$\mathcal{G}_{m',m} \equiv f_{m'-1}^+(\mu_{m'})(2R_{m'-1})^{-1} \Sigma_{m'-1} f_{m'-2}^+(\mu_{m'-1})(2R_{m'-2})^{-1}$$

$$\times \Sigma_{m'-2} \cdots \cdots f_{m+1}^+(\mu_{m+2})(2R_{m+1})^{-1} \Sigma_{m+1} f_m^+(\mu_{m+1})(4\pi R_m)^{-1}. \tag{73}$$

One might have expected a factor $2\pi \Sigma_m$ on the right, but this is contained in the remaining terms of Eq. (72). For example $\mathrm{Tr}_k(\mathcal{F}_{\mu_m}) = \mathrm{Tr}_k(\hat{g}^\dagger(1)\mathcal{G}_m(2\pi \Sigma_m))$.

As we have seen in Eqs. (37) and (39), the gauge field only requires us to know the Green's function at the impurities. Without loss of generality we may assume $\mu_{m'} > \mu_m$ and take $z_0 = \mu_m + 0$, such that (see Eqs. (23), (32), (35))

$$f_x(\mu_{m'}, \mu_m) = -4\pi^2(\mathbb{1}_k, 0) \cdot W(\mu_{m'}, \mu_m + 0)(\mathbb{1}_{2k} - \mathcal{F}_{\mu_m})^{-1} \begin{pmatrix} 0 \\ \mathbb{1}_k \end{pmatrix}. \tag{74}$$

The matrix $(\mathbb{1}_{2k} - \mathcal{F}_{\mu_m})$ has a 2×2 block structure, and one can verify that in general

$$\begin{pmatrix} a & b \\ c & d \end{pmatrix}^{-1} = \begin{pmatrix} \left(a - bd^{-1}c\right)^{-1} & \left(c - db^{-1}a\right)^{-1} \\ \left(b - ac^{-1}d\right)^{-1} & \left(d - ca^{-1}b\right)^{-1} \end{pmatrix}. \tag{75}$$

Identifying the blocks, in the high temperature limit we find

$$a \equiv (\mathbb{1}_{2k} - \mathcal{F}_{\mu_m})_{11} \to \mathbb{1}_k - 2\pi \hat{g}^\dagger(1)\mathcal{G}_m R_m^-(\mu_m),$$

$$b \equiv (\mathbb{1}_{2k} - \mathcal{F}_{\mu_m})_{12} \to -\hat{g}^\dagger(1)\mathcal{G}_m,$$

$$c \equiv (\mathbb{1}_{2k} - \mathcal{F}_{\mu_m})_{21} \to -4\pi^2(R_{m-1}^+(\mu_m) + S_m)\hat{g}^\dagger(1)\mathcal{G}_m R_m^-(\mu_m),$$

$$d \equiv (\mathbb{1}_{2k} - \mathcal{F}_{\mu_m})_{22} \to \mathbb{1}_k - 2\pi(R_{m-1}^+(\mu_m) + S_m)\hat{g}^\dagger(1)\mathcal{G}_m, \tag{76}$$

where we used that $R^{\pm}_{m+n-1}(\mu_{m+n}) + S_{m+n} = \hat{g}(1)(R^{\pm}_{m-1}(\mu_m) + S_m)\hat{g}^{\dagger}(1)$, cf. Eqs. (34), (36). Evaluating the Green's function at the *same* impurities, $\mu_{m'} = \mu_m$, is simplified by the fact that $W(\mu_m, \mu_m) = \mathbb{1}_{2k}$. This gives the following remarkably simple result in the far-field limit,

$$f_x(\mu_m, \mu_m) = -4\pi^2(\mathbb{1}_{2k} - \mathcal{F}_{\mu_m})^{-1}_{12} \to 4\pi^2\big(2\pi\,\Sigma_m - \mathcal{G}_m^{-1}\hat{g}(1)\big)^{-1} \to 2\pi(\Sigma_m)^{-1}. \tag{77}$$

For the Green's function evaluated at *different* impurities, $\mu_m \neq \mu_{m'}$, we need to determine $W(\mu_{m'} - 0, \mu_m + 0)$, for which we can follow the same method as for $\mathcal{F}_{\mu_m}$

$$W(\mu_{m'} - 0, \mu_m + 0) = \begin{pmatrix} \mathbb{1}_k & 0 \\ 2\pi R^+_{m'-1}(\mu_{m'}) & 0 \end{pmatrix} \mathcal{G}_{m',m} \begin{pmatrix} 2\pi R^-_m(\mu_m) & \mathbb{1}_k \\ 0 & 0 \end{pmatrix}, \tag{78}$$

with $\mathcal{G}_{m',m}$ as defined in Eq. (73). This leads to

$$f_x(\mu_{m'}, \mu_m) \to -4\pi^2 \mathcal{G}_{m',m}\big((\mathbb{1}_{2k} - \mathcal{F}_{\mu_m})^{-1}_{22} + 2\pi R_m(\mu_m)(\mathbb{1}_{2k} - \mathcal{F}_{\mu_m})^{-1}_{12}\big)$$

$$\to \mathcal{G}_{m',m}\mathcal{G}_m^{-1}\hat{g}(1)f_x(\mu_m, \mu_m), \tag{79}$$

which is exponentially suppressed since $\mathcal{G}_{m',m}$ grows as $\exp(2\pi|\vec{x}|(\mu_{m'} - \mu_m))$. This cannot compensate for the decay of $\mathcal{G}_m^{-1}$, provided all μ_m are unequal, i.e., all constituents have a non-zero mass. Massless constituents have a so-called non-Abelian cloud [12], which has no Abelian far-field limit.

4.2. Total action

To determine ψ in the expression for the action density, Eqs. (46), (49), we need to compute $\det(\mathbb{1}_{2k} - \mathcal{F}_{\mu_m})$. Using Eq. (76) we find

$$\det(\mathbb{1}_{2k} - F_{\mu_n}) = \det\begin{pmatrix} a & b \\ c & d \end{pmatrix} = \det\begin{pmatrix} 0 & b \\ c - ab^{-1}d & d \end{pmatrix} = \det(b)\det\big(ab^{-1}d - c\big)$$

$$\to \det\big(\hat{g}^{\dagger}(1)\mathcal{G}_m\big)\det\big(\mathcal{G}_m^{-1}\hat{g}(1) - 2\pi\,\Sigma_m\big) \to \det\big(-2\pi\,\hat{g}^{\dagger}(1)\mathcal{G}_m\,\Sigma_m\big), \tag{80}$$

such that

$$\psi \to \det(\pi\mathcal{G}_m\,\Sigma_m) = 2^{-k} \prod_{m=1}^{n} \big\{\det\big(f_m^+(\mu_{m+1})\big)\det(\Sigma_m)/\det(2R_m)\big\}. \tag{81}$$

For $|\vec{x}| \to \infty$, $f_m^+(\mu_{m+1}) \to \exp(2\pi\nu_m|\vec{x}|\mathbb{1}_k)$ (see Eq. (65)) and $\frac{1}{2}\Sigma_m$, $R_m \to |\vec{x}|\mathbb{1}_k$, which implies that $\psi \to 2^{-k} \prod_{m=1}^{n} \det\big[\exp(2\pi\nu_m|\vec{x}|\mathbb{1}_k)\big] = 2^{-k}\exp(2\pi k|\vec{x}|)$ (recall that $\sum_{m=1}^{n}\nu_m = 1$). Therefore, the action is given by $S = -\frac{1}{2}\int d^4x\,\partial_\mu^2\partial_\nu^2\log\psi(x) = 8\pi^2 k$, as should be the case for a self-dual charge k solution.

4.3. Gauge field

Without the off-diagonal components of the Green's function contributing to the far-field region, the functions ϕ and ϕ_j in Eqs. (37), (39) can be further simplified to

$$\phi(x)^{-1} \rightarrow 1 - \sum_m \hat{f}^{ab}(\mu_m, \mu_m) P_m \zeta_a \zeta_b^\dagger P_m,$$

$$\phi_j \rightarrow \sum_m \hat{f}^{ab}(\mu_m, \mu_m) P_m \zeta_a \hat{\sigma}_j \zeta_b^\dagger P_m, \tag{82}$$

and only the Abelian components of the gauge field survive. Particularly the case of $Sp(1)$ discussed in Section 2.3 is easy to deal with. Using Eqs. (23), (40), (42), (44), (77) we find

$$\phi^{-1}(x) \rightarrow \sigma_0 \left(1 - \text{Tr}_k\left[2\Sigma_2^{-1} S_2\right]\right) \equiv \sigma_0 \phi_{\text{ff}}^{-1}(x),$$

$$\vec{\phi}(x) \rightarrow \hat{\omega} \cdot \vec{\sigma} \, \text{Tr}_k\left[2\Sigma_2^{-1} \hat{g}(\mu_2) \vec{\rho}_2 \hat{g}^\dagger(\mu_2)\right],$$

$$A_\mu(x) \rightarrow \frac{i}{2} \hat{\omega} \cdot \vec{\tau} \left(1 - \text{Tr}_k\left[2\Sigma_2^{-1} S_2\right]\right)^{-1} \bar{\eta}_{\mu\nu}^j \partial_\nu \, \text{Tr}_k\left[2\Sigma_2^{-1} \hat{g}(\mu_2) \rho_2^j \hat{g}^\dagger(\mu_2)\right], \tag{83}$$

where we recall (see Eq. (71)) that $\Sigma_2 = (R_1^+(\mu_2) + R_2^-(\mu_2) + S_2)$. This is in perfect agreement with the earlier $k = 1$ results [1]. Note that the gauge rotation which relates $\hat{f}_x(z, z')$ to $f_x(z, z')$ also relates $\widehat{S}_m$ to S_m (see Eqs. (23), (24)) and therefore does not appear in the final expression for $\phi(x)$.

It is interesting to note that the dipole moment of the Abelian gauge field is particularly simple and does not require us to solve for $R_m^\pm(z)$, since $\lim_{|\vec{x}|\to\infty} \Sigma_2 = 2|\vec{x}| \mathbb{1}_k$ such that

$$\lim_{|\vec{x}|\to\infty} A_\mu(x) = \frac{i}{2} \hat{\omega} \cdot \vec{\tau} \, \bar{\eta}_{\mu\nu}^j \partial_\nu \frac{\text{Tr}_k(\rho_2^j)}{|\vec{x}|}. \tag{84}$$

Hence the dipole moment $\vec{p} \equiv \frac{1}{2} \text{Tr}_k(\vec{\rho}_2)$ only involves ζ_a, and we do except it allows for configurations with a vanishing dipole moment. For higher multipole moments, through $R_m^\pm(z)$, we need to deal with the full quadratic ADHM constraint, or equivalently with the Riccati and Nahm equations. Nevertheless, it is remarkable that in the high temperature limit the $\vec{x}$ dependence is restricted to $R_1^+(\mu_2)$ and $R_2^-(\mu_2)$ *only*. We would like to prove that each of its eigenvalues vanish at an isolated point, as one way to identify the $2k$ constituent locations. We will defer the study of this interesting issue, and its generalization to $SU(n)$, to a future publication.

4.3.1. Axially symmetric case

The far-field approximation further simplifies when considering the axially symmetric solutions discussed in Section 3.2. We restrict ourselves here to $Sp(1)$. Since $\vec{R}(z; \vec{x})$ is piecewise constant, the Riccati equation is trivial to solve,

$$R_m^\pm(z) = R_m = \sqrt{(\vec{x}\mathbb{1}_k - \vec{e}\, Y_m) \cdot (\vec{x}\mathbb{1}_k - \vec{e}\, Y_m)}. \tag{85}$$

The square root involves a positive $k \times k$ matrix, and is well-defined. Due to the fact that $S_m = \hat{g}(\mu_m) \widehat{S}_m \hat{g}^\dagger(\mu_m) = \hat{g}(\mu_m) \Delta \vec{y}_m \cdot \vec{\rho}_m / |\Delta y_m| \hat{g}^\dagger(\mu_m)$ (see Eqs. (24), (59)), the Abelian

 F. Bruckmann, P. van Baal / Nuclear Physics B 645 (2002) 105–133

component of the gauge field is of the simple form ($\vec{e} = \vec{\rho}_2/|\vec{\rho}_2|$, see Section (2.3))

$$A_\mu^{\text{abel}}(x) = -\frac{i}{2}\hat{\omega} \cdot \vec{\tau} e_j \bar{\eta}_{\mu\nu}^j \partial_\nu \log \phi(x).$$

(86)

In the far field limit $\phi(x) \to \phi_{\text{ff}}(x)$ (see Eq. (83)). Since S_2 has rank 1, the matrix $M \equiv 2\Sigma_2^{-1} S_2$ has only one non-vanishing column with respect to a suitably chosen basis, which implies that $1 - \text{Tr}_k(M) = \det(\mathbb{1}_k - M)$. This allows us to write in the far-field region

$$\phi_{\text{ff}}(x) = \frac{\det(R_1 + R_2 + S_2)}{\det(R_1 + R_2 - S_2)},$$

(87)

from which we immediately read off the result [1] for $k = 1$, in which case it is easy to show that $\phi_{\text{ff}}(x) = (r_2 + \vec{e} \cdot \vec{r}_2)/(r_1 + \vec{e} \cdot \vec{r}_1)$, revealing $A_\mu(x)$ to be a linear superposition of two oppositely charged self-dual Dirac monopoles. We would like $\phi_{\text{ff}}(x)$ to similarly factorize for $k > 1$ in $2k$ Dirac monopoles, but since R_1, R_2 and S_2 in general do not commute, some care is required in demonstrating the factorization. We will rely on the fact that $(R_m - \vec{e} \cdot \vec{R}_m)(R_m + \vec{e} \cdot \vec{R}_m) = (R_m + \vec{e} \cdot \vec{R}_m)(R_m - \vec{e} \cdot \vec{R}_m) = R_m^2 - (\vec{e} \cdot \vec{R}_m)^2 = (\vec{x} \times \vec{e})^2 \mathbb{1}_k$ and hence independent of m. Since by definition $\Delta y_2 > 0$ (see Section 3.2), one finds that

$$\phi_{\text{ff}}(x) = \frac{\det((R_1 - \vec{e} \cdot \vec{R}_1) + (R_2 + \vec{e} \cdot \vec{R}_2))}{\det((R_2 - \vec{e} \cdot \vec{R}_2) + (R_1 + \vec{e} \cdot \vec{R}_1))}.$$

(88)

This can now be reorganized according to

$$\phi_{\text{ff}}(x) = \frac{\det(((R_1 - \vec{e} \cdot \vec{R}_1) + (R_2 + \vec{e} \cdot \vec{R}_2))(R_1 + \vec{e} \cdot \vec{R}_1)) \det(R_2 + \vec{e} \cdot \vec{R}_2)}{\det((R_2 + \vec{e} \cdot \vec{R}_2)((R_2 - \vec{e} \cdot \vec{R}_2) + (R_1 + \vec{e} \cdot \vec{R}_1))) \det(R_1 + \vec{e} \cdot \vec{R}_1)}$$

$$= \frac{\det(R_2 + \vec{e} \cdot \vec{R}_2)}{\det(R_1 + \vec{e} \cdot \vec{R}_1)},$$

(89)

after which we can separately diagonalize $\vec{R}_1$ and $\vec{R}_2$ to find

$$\phi_{\text{ff}}(x) = \prod_i \frac{r_2^{(i)} + \vec{e} \cdot \vec{r}_2^{(i)}}{r_1^{(i)} + \vec{e} \cdot \vec{r}_1^{(i)}} = \prod_i \frac{r_1^{(i)} + r_2^{(i)} + |y_1^{(i)} - y_2^{(i)}|}{r_1^{(i)} + r_2^{(i)} - |y_1^{(i)} - y_2^{(i)}|},$$

(90)

where $y_m^{(i)}\vec{e}$ give the locations of the constituent monopoles, with $y_m^{(i)}$ the eigenvalues of Y_m and $\vec{r}_m^{(i)} = \vec{x} - y_m^{(i)}\vec{e}$ (the index m distinguishing their charge). The second expression for the factorized version of $\phi_{\text{ff}}(x)$ uses the fact that the constituent locations can be ordered according to

$$y_1^{(1)} < y_2^{(1)} < y_1^{(2)} < y_2^{(2)} < \cdots < y_1^{(k)} < y_2^{(k)}.$$

(91)

It should be noted that this prevents the constituents to pass each other while varying the parameters for the axially symmetric configuration, see also Fig. 3 and the discussion in Section 3.2. We need to go beyond this simple axially symmetric configuration to allow for the constituents to rearrange themselves more freely.

F. Bruckmann, P. van Baal / Nuclear Physics B 645 (2002) 105–133

5. Discussion

We have presented the general formalism for finding exact instanton solutions at finite temperature (calorons) with any non-trivial holonomy and topological charge. In an infinite volume holonomy and charge are fixed. The solution is described by $4kn$ parameters of which $3kn$ give the spatial locations of the kn constituent monopoles. The remaining parameters are given by k time locations and $(n-1)k$ phases associated to gauge rotations in the subgroup that leaves the holonomy unchanged. Of these, $n-1$ can be considered as global gauge rotations. The dimension of the moduli space, i.e., the number of gauge invariant parameters, is therefore equal to $4kn - (n-1)$. Our subset of axially symmetric solutions has $2nk + 4$ parameters of which there are $n-1$ global gauge rotations, or $2nk - n + 5$ gauge invariant parameters.

Explicit solutions were found for the case of axial symmetry, an important limitation being the difficulty of solving the Nahm equation, or equivalently the quadratic ADHM constraint. We certainly expect more progress can be made on this in the near future. Nevertheless, we already found a rich structure that bodes well for being able to consider the constituent monopoles as independent objects. This is surprisingly subtle, as we have illustrated by the fact that an approximate superposition of charge 1 calorons tends to give rise to a *visible* Dirac string. This is also related to the difficulty of finding approximate multi-monopole solutions, which can be obtained from the caloron solutions by sending a subset of the constituent monopoles to infinity, as has been well studied in the charge 1 case [1,13].

An important tool has been our study of the far-field limit, describing the Abelian gauge field far removed from any of the constituent monopoles. This allows for a description of the long distance properties in terms of (self-dual) Dirac monopoles. Much could be extracted concerning its properties without the need to explicitly solve the Nahm equation. We conjecture in general to be able to identify the Dirac monopole location, but some work remains to be done here.

It may seem that all these results are somewhat academic since until recently none of these constituent monopoles were found in dynamical lattice configurations. First of all one would be tempted to search for them at high temperature, but it should be noted that above the deconfining phase transition the average Polyakov loop takes on trivial values, associated to the center of the gauge group, which is not the environment in which a caloron will reveal its constituents. This would give the well-known Harrington–Shepard solution constructed long ago [14]. With our present understanding this solution can be seen as having $n-1$ massless constituents which cannot be localized. Only when sending all these to infinity one is left with a monopole [15]. Furthermore, at high temperature classical configurations will be heavily suppressed due to their Boltzmann weight. Rather, the hope is that the constituent monopoles play an important role *below* the deconfining temperature, where the average of the Polyakov loop is non-trivial, and tends to favor *equal* mass constituents. This is why in this paper our examples were for that case, see Figs. 2, 4, 5 and the discussion in Section 3.2. Nevertheless, the formalism developed here gives results for any choice of the holonomy, and a sample of unequal mass constituents is given in Fig. 6. In general the constituent monopoles can be characterized by their magnetic

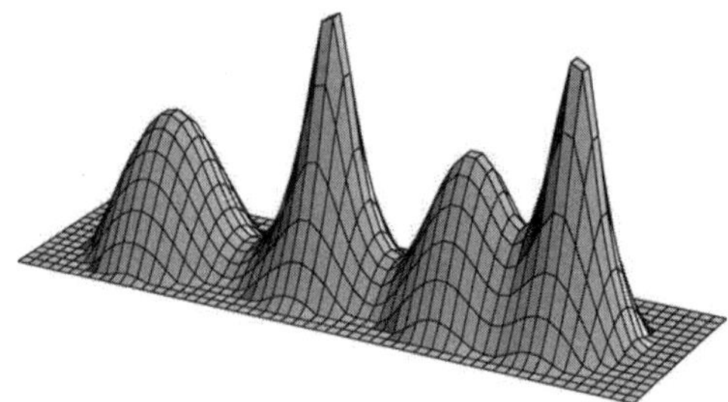

Fig. 6. The logarithm of the action density (cutoff for $\log(S)$ below -3) for an $SU(2)$ charge 2 caloron with one type of constituent three times more massive than the other ($\mu_2 = 1/8$, $\alpha_a^j = \xi_0^a = 0$, $\rho_j = 2, \xi = 3.5$).

(= electric) charge. For $SU(n)$ there are n different types of Abelian charges involved [1], and all k constituents of a given type have the same mass.

We will discuss briefly the lattice evidence for the presence of constituent monopoles that has accumulated the last few years. A first numerical study using cooling was performed with twisted boundary conditions, which implies non-trivial holonomy [16]. Good agreement was found with the infinite volume charge 1 analytic results, in particular so for the fermion zero-modes [17,18] which are more localized than the action density. A charge 2 solution was also found, shown in Fig. 8 of Ref. [16]. Fermion zero-modes played an intricate role in an extensive numerical study of Nahm dualities on the torus [19].

As suggested in Ref. [1] one may also enforce non-trivial holonomy on the lattice by putting at the spatial boundary of the box all links in the time direction to the same constant value U_0, such that $U_0^{N_t} = \mathcal{P}_\infty$ (N_t the number of lattice sites in the time direction). This has been implemented in $SU(2)$ lattice Monte Carlo studies as well, where $\mathcal{P}_\infty$ was set to the average value of the Polyakov loop, appropriate for the temperature at which the simulations were performed [20,21]. Cooling was applied to find calorons, including those at higher charge. Apart from the configurations that in the continuum would be exactly self-dual, the lattice allows one to also consider configurations in which both self-dual and anti-selfdual lumps appear. This revealed constituent monopoles that seem not directly associated to calorons, called $D\overline{D}$ (as opposed to DD). Both objects in such a $D\overline{D}$ configuration have fractional topological charged, but opposite in sign. Perhaps these arise when two near constituent monopoles, one belonging to a caloron, the other to an anti-caloron, "annihilate". Our analytic methods cannot directly address this situation due to the lack of self-duality. The same holds for configurations that seem to only carry magnetic fields, which were already seen long ago [22].

One point of criticism that applies to both methods is that the choice of boundary conditions may force the "dissociation" of instantons into constituent monopoles, particularly since volumes cannot yet be chosen so large that many instantons are contained within a given configuration. A recent study [23] has done away with the fixed boundary conditions that enforce the non-trivial holonomy. Nevertheless, still one finds in many cases that calorons "dissociate" into constituent monopoles below the deconfinement transition temperature. A particularly useful tool has turned out to be the fermionic (near) zero-modes to detect the monopole constituents when they are too close together to reveal themselves from the action density [23]. This relies on the observation that the zero-mode is localized

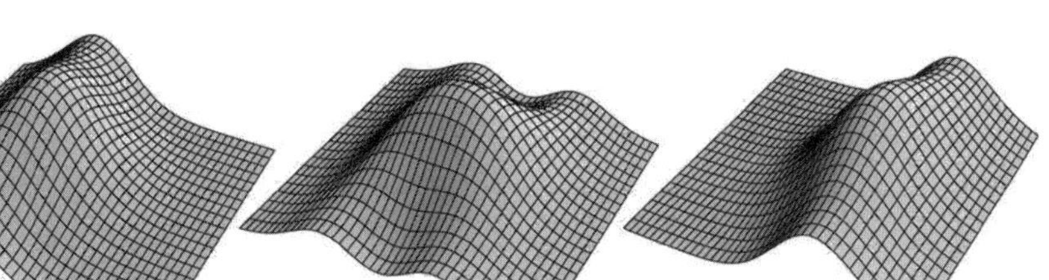

Fig. 7. Fermion zero-mode densities as a function of t and z for a charge 1 caloron ($\mu_2 = \frac{1}{4}$ and $\rho = \frac{1}{2}$) with periodic (right) and anti-periodic (left) boundary conditions, compared to the action density (middle) (see also Fig. 5 in Ref. [26]). Based on Ref. [25]; produced with Ref. [27]).

on *only* one of the constituent monopoles, determined by the boundary conditions imposed on the fermions in the time direction [17,18]. For $SU(2)$ this is particularly simple, with periodic and anti-periodic boundary conditions of the fermions making the zero-mode switch from one constituent to the other, as is illustrated for a close pair of constituents in Fig. 7. In addition one may use the Polyakov loop for diagnostic purposes [20,21,23], for $SU(2)$ taking the values $\mathbb{1}_2$ and $-\mathbb{1}_2$ near the respective constituent locations (at these points the gauge symmetry is restored, providing an alternative definition for the center of a constituent monopole).

Fermion eigenfunctions with eigenvalues near zero have also been used as an alternative to cooling, to filter out the high frequency modes and identify topological lumps. For a recent lattice study, including some discussion of calorons with non-trivial holonomy, see Ref. [24] and references therein. Using the near zero-modes as a filter, constituent monopoles have even been identified recently for $SU(3)$ well below the deconfining temperature [25], resembling Fig. 7 (see also Fig. 14 of Ref. [24]).

For the exact axially symmetric multi-caloron solutions constructed in this paper the k associated fermion zero-modes (for charge k) will be derived in the near future. We anti-cipate that one can choose a basis where each is localized on one of the constituent monopoles, the type of which is determined by the choice of fermionic boundary conditions in the time direction. Analyzing these zero-modes is particularly interesting in the light of some puzzles that were presented in a recent study [28] of the normalizable fermion zero-modes in the background of a collection of so-called bipoles, i.e., pairs of oppositely charged (but self-dual) Dirac monopoles, which are of interest in a wider context as well. This will be one of the many topics we have access to with our analytic tools. But ultimately our main aim is to develop a reliable method to describe the long distance features of non-Abelian gauge theories in terms of monopole constituents to understand both confinement and chiral symmetry breaking. The results of this paper, and in particular the recent lattice results, provide some encouragement in this direction.

Acknowledgements

We thank Conor Houghton for initial collaboration on the monopole aspects of this work and him as well as Chris Ford for extensive discussions. P.v.B. also thanks Michael

Müller-Preussker and Christof Gattringer for discussions concerning calorons with non-trivial holonomy on the lattice. Furthermore he is grateful to Leo Stodolsky and Valya Zakharov for hospitality at the MPI in Munich and to Poul Damgaard, Urs Heller and Jac Verbaarschot for inviting him to the ECT* workshop "Non-perturbative Aspects of QCD" in Trento. He thanks both institutions for their support, while some of the work presented in this paper was performed. F.B. likes to thank the organizers of the "Channel Meeting on Theoretical Particle Physics" for a well organized and stimulating meeting as well as Dimitri Diakonov, Gerald Dunne, Alexander Gorsky and Peter Orland for discussions. The research of F.B. is supported by FOM.

References

[1] T.C. Kraan, P. van Baal, Phys. Lett. B 428 (1998) 268, hep-th/9802049;
    T.C. Kraan, P. van Baal, Nucl. Phys. B 533 (1998) 627, hep-th/9805168.
[2] T.C. Kraan, P. van Baal, Phys. Lett. B 435 (1998) 389, hep-th/9806034.
[3] M.F. Atiyah, N.J. Hitchin, V.G. Drinfeld, Yu.I. Manin, Phys. Lett. A 65 (1978) 185;
    M.F. Atiyah, Geometry of Yang–Mills fields, Fermi lectures, Scuola Normale Superiore, Pisa, 1979.
[4] W. Nahm, Self-dual monopoles and calorons, in: G. Denardo (Ed.), in: Lecture Notes in Physics, Vol. 201, 1984, p. 189.
[5] K. Lee, P. Yi, Phys. Rev. D 56 (1997) 3711, hep-th/9702107;
    K. Lee, Phys. Lett. B 426 (1998) 323, hep-th/9802012;
    K. Lee, C. Lu, Phys. Rev. D 58 (1998) 025011, hep-th/9802108.
[6] W. Nahm, Phys. Lett. B 90 (1980) 413.
[7] G. 't Hooft, Nucl. Phys. B 190 (1981) 455;
    G. 't Hooft, Phys. Scr. 25 (1982) 133.
[8] E.F. Corrigan, D.B. Fairlie, S. Templeton, P. Goddard, Nucl. Phys. B 140 (1978) 31.
[9] H. Osborn, Nucl. Phys. B 159 (1979) 497.
[10] M. García Pérez, T.G. Kovács, P. van Baal, Phys. Lett. B 472 (2000) 295, hep-ph/9911485.
[11] P. van Baal, in: V. Mitrjushkin, G. Schierholz (Eds.), Lattice Fermions and Structure of the Vacuum, Kluwer Academic, Dordrecht, 2000, p. 269, hep-th/9912035.
[12] K. Lee, E.J. Weinberg, P. Yi, Phys. Lett. B 376 (1996) 97, hep-th/9601097;
    K. Lee, E.J. Weinberg, P. Yi, Phys. Rev. D 54 (1996) 6351, hep-th/9605229;
    E.J. Weinberg, Massive and massless monopoles and duality, hep-th/9908095.
[13] T.C. Kraan, Commun. Math. Phys. 212 (2000) 503, hep-th/9811179.
[14] B.J. Harrington, H.K. Shepard, Phys. Rev. D 17 (1978) 2122;
    B.J. Harrington, H.K. Shepard, Phys. Rev. D 18 (1978) 2990.
[15] P. Rossi, Nucl. Phys. B 149 (1979) 170.
[16] M. García Pérez, A. González-Arroyo, A. Montero, P. van Baal, J. High Energy Phys. 06 (1999) 001, hep-lat/9903022.
[17] M. García Pérez, A. González-Arroyo, C. Pena, P. van Baal, Phys. Rev. D 60 (1999) 031901, hep-th/9905016.
[18] M.N. Chernodub, T.C. Kraan, P. van Baal, Nucl. Phys. B (Proc. Suppl.) 83–84 (2000) 556, hep-lat/9907001.
[19] M. García Pérez, A. González-Arroyo, C. Pena, P. van Baal, Nucl. Phys. B 564 (1999) 159, hep-th/9905138.
[20] E.-M. Ilgenfritz, M. Müller-Preussker, A.I. Veselov, in: V. Mitrjushkin, G. Schierholz (Eds.), Lattice Fermions and Structure of the Vacuum, Kluwer Academic, Dordrecht, 2000, p. 345, hep-lat/0003025.
[21] E.-M. Ilgenfritz, B.V. Martemyanov, M. Müller-Preussker, A.I. Veselov, Nucl. Phys. B (Proc. Suppl.) 94 (2001) 407, hep-lat/0011051;
    E.-M. Ilgenfritz, B.V. Martemyanov, M. Müller-Preussker, A.I. Veselov, Nucl. Phys. B (Proc. Suppl.) 106 (2002) 589, hep-lat/0110212.
[22] M.L. Laursen, G. Schierholz, Z. Phys. C 38 (1988) 501.

F. Bruckmann, P. van Baal / Nuclear Physics B 645 (2002) 105–133   133

[23] E.-M. Ilgenfritz, B.V. Martemyanov, M. Müller-Preussker, S. Shcheredin, A.I. Veselov, hep-lat/0206004.
[24] C. Gattringer, M. Göckeler, P.E.L. Rakow, S. Schaefer, A. Schäfer, Nucl. Phys. B 618 (2001) 205, hep-lat/0105023.
[25] Christof Gattringer, private communications. The zero-mode densities for the case of Fig. 7 were produced for the purpose of illustrating the behavior observed by C. Gattringer and co-workers in $SU(3)$ lattice gauge theory.
[26] T.C. Kraan, P. van Baal, Nucl. Phys. B (Proc. Suppl.) 73 (1999) 554, hep-lat/9808015.
[27] http://www.lorentz.leidenuniv.nl/vanbaal/Caloron.html.
[28] P. van Baal, Chiral zero-modes for Abelian BPS dipoles, in: J. Greensite, S. Olejnik (Eds.), Confinement, Topology, and other Non-Perturbative Aspects of QCD, Kluwer Academic, in press, hep-th/0202182.

Available online at www.sciencedirect.com

SCIENCE @ DIRECT®

Nuclear Physics B 698 (2004) 233–254

www.elsevier.com/locate/npe

Higher charge calorons with non-trivial holonomy

Falk Bruckmann, Dániel Nógrádi, Pierre van Baal

*Instituut-Lorentz for Theoretical Physics, University of Leiden,
P.O. Box 9506, NL-2300 RA Leiden, The Netherlands*

Received 28 April 2004; accepted 27 July 2004

Available online 25 August 2004

Abstract

The full ADHM-Nahm formalism is employed to find exact higher charge caloron solutions with non-trivial holonomy, extended beyond the axially symmetric solutions found earlier. Particularly interesting is the case where the constituent monopoles, that make up these solutions, are not necessarily well-separated. This is worked out in detail for charge 2. We resolve the structure of the extended core, which was previously localized only through the singularity structure of the zero-mode density in the far field limit. We also show that this singularity structure agrees exactly with the Abelian charge distribution as seen through the Abelian component of the gauge field. As a by-product zero-mode densities for charge 2 magnetic monopoles are found.
© 2004 Elsevier B.V. All rights reserved.

PACS: 11.10.Wx; 12.38.Lg; 14.80.Hv

1. Introduction

Calorons are instantons at finite temperature. For a long time the influence of a background Polyakov loop on the properties of these topological excitations has been neglected. Solutions were constructed long ago [1] and were studied in detail in the semi-classical approximation [2]. In all these studies the Polyakov loop at spatial infinity (also called the holonomy) was trivial, i.e., an element of the center of the gauge group. That the influence of the background Polyakov loop on the topological excitations can be dramatic is partic-

E-mail address: vanbaal@lorentz.leidenuniv.nl (P. van Baal).

234   *F. Bruckmann et al. / Nuclear Physics B 698 (2004) 233–254*

ularly clear in the confined phase, where on average its trace vanishes. Caloron solutions in such backgrounds were constructed only relatively recently [3,4] and can be seen as composed of massive monopole constituents with their magnetic charges adding to zero.

It was observed that the one-loop correction to the action for configurations with a non-trivial asymptotic value of the Polyakov loop gives rise to an infinite action barrier, which were therefore considered irrelevant [2]. However, the infinity simply arises due to the integration over the finite energy *density* induced by the perturbative fluctuations in the background of a non-trivial Polyakov loop [5]. The proper setting would therefore rather be to calculate the non-perturbative contribution of calorons (with a given asymptotic value of the Polyakov loop) to this energy density, as was first successfully implemented in supersymmetric theories [6], where the perturbative contribution vanishes. It has a minimum where the trace of the Polyakov loop vanishes, i.e., at maximal non-trivial holonomy.

In a recent study at high temperatures, where one presumably can trust the semi-classical approximation, the non-perturbative contribution of these monopole constituents (also called dyons) was computed [7]. When added to the perturbative contribution [5] with its minima at center elements, a local minimum develops where the trace of the Polyakov loop vanishes, deepening further for decreasing temperature. This gives support for a phase in which the center symmetry, broken in the high temperature phase, is restored and provides an indication that the monopole constituents are the relevant degrees of freedom for the confined phase.

Also lattice studies, both using cooling [8] and chiral fermion zero-modes [9] as filters, have now confirmed that monopole constituents do dynamically occur in the confined phase. A charge 1 caloron is seen for $SU(n)$ to consist of n constituent monopoles. In the deconfined phase, due to the fact that the average Polyakov loop becomes a center element, the caloron returns to the form known as the Harrington–Shepard solution [1]. The latter can also be interpreted as consisting of constituent monopoles, however, with $n-1$ of them being massless.

To be precise, for self-dual configurations in the background of non-trivial holonomy the masses of constituent monopoles are given by $8\pi^2 \nu_j/\beta$, with $\nu_j \equiv \mu_{j+1} - \mu_j$. The μ_l are related to the eigenvalues of the Polyakov loop at spatial infinity,

$$\mathcal{P}_\infty = \lim_{x \to \infty} \mathrm{Pexp}\left(\int_0^\beta A_0(t, \vec{x})\, dt \right) = g^\dagger \exp\big(2\pi i\, \mathrm{diag}(\mu_1, \mu_2, \ldots, \mu_n)\big) g \tag{1}$$

(this expression assumes the periodic gauge $A_\mu(t, \vec{x}) = A_\mu(t + \beta, \vec{x})$) where g is the gauge rotation used to diagonalize $\mathcal{P}_\infty$ and β is the period in the imaginary time direction, related to the inverse temperature. The eigenvalues $\exp(2\pi i \mu_j)$ are to be ordered on the circle such that $\mu_1 \leqslant \mu_2 \leqslant \cdots \leqslant \mu_n \leqslant \mu_{n+1}$, with $\mu_{n+j} \equiv 1 + \mu_j$ and $\sum_{i=1}^n \mu_i = 0$, which guarantees that the masses add up to $8\pi^2/\beta$, the instanton action per unit (imaginary) time. At higher topological charge k, the parameter space of widely separated constituent monopoles is described by kn monopole constituents, k of each of the n types of Abelian charges (with overall charge neutrality).

We established in an earlier paper [10] that well-separated constituents act as point sources for the so-called far field (that is far removed from any of the cores). When constituents of opposite charge (n constituents of different type) come together, the action

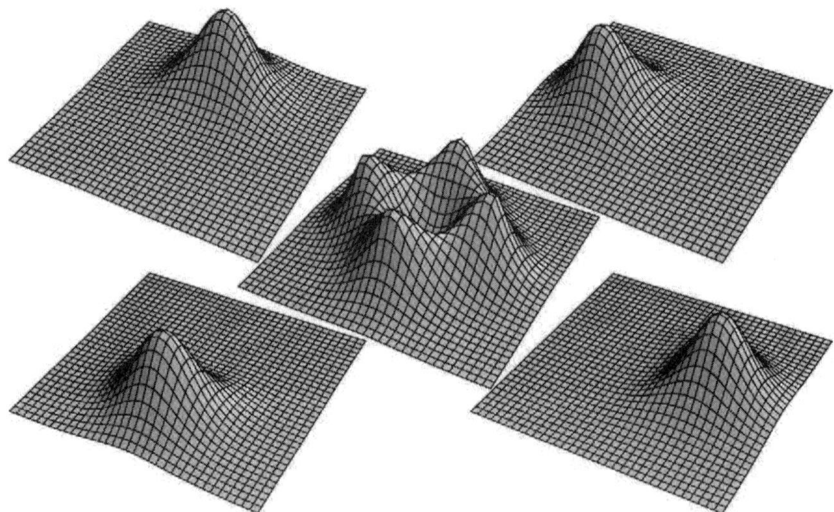

Fig. 1. In the middle is shown the action density in the plane of the constituents at $t = 0$ for an SU(2) charge 2 caloron with tr $\mathcal{P}_\infty = 0$, in the regime where constituents are not well separated. On a scale enhanced by a factor $10\pi^2$ are shown the densitities for the two zero-modes, using either periodic (left) or anti-periodic (right) boundary conditions in the time direction. This solution is for the so-called "crossed" configuration with $\mathbf{k} = 0.997$ and $D = 8.753$, see Section 4 for more details.

density no longer deviates significantly from that of a standard instanton. Its scale parameter ρ is related to the constituent separation d through $\rho^2/\beta \approx d$. Yet, the gauge field is vastly different, as is seen from the fact that within the confines of the peak there are n locations where two of the eigenvalues of the Polyakov loop coincide [11,12], thus in some sense varying over the maximal range available (e.g., for SU(2) from $\mathbb{1}_2$ to $-\mathbb{1}_2$), whereas for trivial holonomy only one such location is found.

On the other hand, when constituents of equal charge come together typically an extended core structure is found. This was deduced, in particular for the case of charge 2 calorons, from our ability to analytically determine the zero-mode density (summed over the two zero-modes implied by the index theorem) in the far field limit, neglecting exponential contributions.[1] In this limit it forms a singular distribution on a disc bounded by an ellipse, but approaches two delta functions for well-separated like-charge constituents. This zero-mode density only sees constituents of a given charge, depending on the boundary condition for the fermions in the imaginary time direction, which can be chosen to be a U(1) phase (containing the physically relevant choice of anti-periodic boundary conditions for thermal field theory). We will show for SU(2) that their difference for periodic and anti-periodic boundary conditions coincides *exactly* with the (Abelian) charge distribution extracted from the gauge field in the same limit, making contact with an old result due to Hurtubise [13] for the asymptotic behaviour of the monopole Higgs field.

---

[1] This is in some sense equivalent to the high temperature limit, with constituent masses given by $8\pi^2 v_m/\beta$, such that the range of the exponential contributions shrinks inversely proportional with the temperature.

We found two particular parametrizations within the SU(2) charge 2 moduli space that exhibited these extended charge distributions. The first includes as a limit arbitrary charge 2 monopoles. The second of these parametrizations contains as a limiting case the axially symmetric configurations constructed for arbitrary charge in Ref. [14]. Deforming away from the axial configuration the two discs overlap. Describing the intricate behaviour for the non-Abelian core of these configurations in this region of the parameter space requires one to find exact solutions, which are presented here. We rely on early work of Nahm [15] and Panagopoulos [16] for charge 2 magnetic monopoles, which is simplified to some extent by our formalism that uses the relation between the Fourier transformation of the ADHM construction (as relevant for the finite temperature case) and the Nahm transformation, a crucial ingredient for our success to find explicit solutions [3]. Fig. 1 gives a particular example for the action and zero-mode densities of a charge 2 caloron solution. The two-dimensional zero-mode basis is chosen such that each zero-mode only localizes on one of the constituents of a given charge, showing both the zero-modes with periodic and anti-periodic boundary conditions in the imaginary time direction.

This paper is organized as follows. In Section 2 we will outline the construction, introduce the Green's function that is computed through the analogy of an impurity scattering problem, and summarize the various limits that can be formulated before explicitly solving for the Green's function. In Section 3 we present the method that allows one to find the exact solution for the Green's function, first for the general case and then applied in more detail to that of topological charge 2 calorons. Readers only interested in the results could skip Sections 2.2 and 3. In Section 4 we discuss the two classes of configurations in the moduli space of the charge 2 calorons and provide plots of the various quantities to illustrate the properties of the exact results. In Section 5 we discuss the relation between the algebraic tail of the gauge field and the zero-mode density. We end with some discussions. An Appendix A presents a new result for the limiting behaviour of the action density.

2. Outline of the construction

There are two steps in the construction of charge k caloron solutions. The first step involves finding a $U(k)$ gauge field $\hat{A}_\mu(z)$, which satisfies the self-duality equation, i.e., the Nahm equation [15], on a circle parametrized by z, with z introduced through replacing the original SU(n) gauge field $A_\mu(x)$ by $A_\mu(x) - 2\pi i z \delta_{0\mu} \mathbb{1}_n$. Although not affecting the field strength, this changes the holonomy to $\exp(-2\pi i z \beta)\mathcal{P}_\infty$, revealing that z has period β^{-1}. The index theorem guarantees the existence of k zero-modes $\Psi(x; z)$ which satisfy the Dirac equation, or in the two-component Weyl form

$$\bar{D}_z \Psi(x; z) \equiv \bar{\sigma}_\mu D_z^\mu \Psi(x; z) \equiv \bar{\sigma}_\mu \big(\partial_\mu + A_\mu(x) - 2\pi i z \delta_{0\mu} \mathbb{1}_n\big) \Psi(x; z) = 0 \qquad (2)$$

with $\bar{\sigma}_\mu = \sigma_\mu^\dagger = (\mathbb{1}_2, -i\vec{\tau})$ (τ_i are the usual Pauli matrices). We may remove z from the gauge field $A_\mu(x)$ by transforming the zero-mode to $\hat{\Psi}_z(x) \equiv \exp(-2\pi i t z)\Psi(x; z)$, which is at the expense of making the zero-mode only periodic up to a phase factor, $\hat{\Psi}_z(t + \beta, \vec{x}) = \exp(-2\pi i z \beta)\hat{\Psi}_z(t, \vec{x})$. In a similar way we could introduce $\vec{z}$ through $\Psi(x; z, \vec{z}) \equiv \exp(2\pi i \vec{z} \cdot \vec{x})\Psi(x; z)$, which replaces in Eq. (2) $A_\mu(x)$ by $A_\mu(x) - 2\pi i z_\mu \mathbb{1}_n$, where $z_0 \equiv z$. Assuming the k zero-modes $\Psi^{(a)}(x; z, \vec{z})$ to be orthonormal one has

$$\hat{A}_\mu^{ab}(z) = \int \Psi^{(a)}(x; z, \vec{z})^\dagger \frac{\partial}{\partial z_\mu} \Psi^{(b)}(x; z, \vec{z})\, d^4x, \tag{3}$$

or equivalently (demonstrating as well that $\hat{A}$ does not depend on $\vec{z}$)

$$\hat{A}_0^{ab}(z) = \int \Psi^{(a)}(x; z)^\dagger \frac{\partial}{\partial z} \Psi^{(b)}(x; z)\, d^4x,$$

$$\hat{A}_k^{ab}(z) = 2\pi i \int \Psi^{(a)}(x; z)^\dagger x_k \Psi^{(b)}(x; z)\, d^4x. \tag{4}$$

We have shown how $\hat{A}_\mu(z)$ can alternatively be related to a Fourier transformation of the ADHM construction of instantons [17], that periodically repeat (up to a gauge rotation with $\mathcal{P}_\infty$) in the imaginary time direction, so as to turn an infinite charge instanton in R^4 to one of finite charge and finite temperature. The derivation of this relation will not be repeated here, see Refs. [3,14] for the details.

The connection to the ADHM construction has been useful to simplify the second step in the construction of the caloron solutions, namely how to reconstruct the original gauge field when given a solution to the Nahm equation [15] (which is equivalent to the quadratic ADHM constraint),

$$\frac{d}{dz}\hat{A}_j(z) + \left[\hat{A}_0(z), \hat{A}_j(z)\right] + \frac{1}{2}\varepsilon_{jk\ell}\left[\hat{A}_k(z), \hat{A}_\ell(z)\right] = 2\pi i \sum_m \delta(z - \mu_m)\rho_m^j. \tag{5}$$

For convenience of notation we will henceforth use the classical scale invariance to set $\beta = 1$. The singularities in the Nahm equation appear precisely for those values where $e^{-2\pi i z}\mathcal{P}_\infty$ has one of its eigenvalues equal to 1, at $z = \mu_i$. This is where some fermion field components become massless, i.e., the zero-mode becomes delocalized, whereas for generic values of z it is exponentially localized, which has turned out to be a useful tool to pinpoint the constituent monopoles.

One could apply the Nahm transformation again, introducing $\hat{A}_\mu(z) - 2\pi i x_\mu \mathbb{1}_k$ (with $x_0 \equiv t$) and find the n chiral fermion zero-modes in the background of this gauge field. The construction is somewhat complicated due to the presence of the singularities, whose structure is determined from the matrices $\vec{\rho}_m$ which appear in the Nahm equation. Not all $\vec{\rho}_m$ are independent, e.g., integrating and tracing the Nahm equation yields the conditions, $\sum_{m=1}^n \text{Tr}\, \vec{\rho}_m = \vec{0}$. Further constraints are implied [14] by the fact that one may introduce (as is most easily seen in relation to the ADHM construction) k two-component spinors $\zeta_a^\dagger$ in the $\bar{n}$ representation of SU(n), such that

$$2\pi \zeta_a^\dagger P_m \zeta_b = \sigma_0 \hat{S}_m^{ab} - \vec{\tau} \cdot \vec{\rho}_m^{ab}, \tag{6}$$

where P_m are projections defined through $\mathcal{P}_\infty = \sum_{m=1}^n e^{2\pi i \mu_m} P_m$ ($\hat{S}_m$ will appear in Eq. (8)).

2.1. The Green's function

However, with reference to the ADHM construction, there is great benefit in first finding the solution for the Green's function, $\hat{f}_x(z, z') \equiv \hat{g}^\dagger(z) f_x(z, z') \hat{g}(z')$, where

$$\left\{ -\frac{d^2}{dz^2} + V(z; \vec{x}) \right\} f_x(z, z') = 4\pi^2 \mathbb{1}_k \delta(z - z'), \tag{7}$$

with

$$V(z; \vec{x}) \equiv 4\pi^2 \vec{R}^2(z; \vec{x}) + 2\pi \sum_m \delta(z - \mu_m) S_m, \quad S_m \equiv \hat{g}(\mu_m) \hat{S}_m \hat{g}^\dagger(\mu_m),$$

$$R_j(z; \vec{x}) \equiv x_j - (2\pi i)^{-1} \hat{g}(z) \hat{A}_j(z) \hat{g}^\dagger(z), \tag{8}$$

and S_m playing the role of "impurities". This is formulated in the gauge where first we transform $\hat{A}_0(z)$ to a constant (diagonal) matrix, $2\pi i \xi_0$, as is always possible in one dimension, and then use

$$\hat{g}(z) \equiv \exp\left(2\pi i (\xi_0 - x_0 \mathbb{1}_k) z\right), \qquad \mathrm{Tr}\, \xi_0 = 0 \tag{9}$$

(when $\mathrm{Tr}\, \xi_0 \neq 0$ it is absorbed in a shift of x_0) to transform $\hat{A}_0 - 2\pi i x_0 \mathbb{1}_k$ to zero. This is at the expense of introducing periodicity up to a gauge transformation; although $\hat{f}_x(z, z')$ is periodic in z and z' with period 1 (for $\beta = 1$), $f_x(z, z')$ no longer is.[2]

Given a solution for the Green's function, there are straightforward expressions for the gauge field [14] (only involving the Green's function evaluated at the "impurity" locations) and the fermion zero-modes [10,18]. For the zero-mode density this gives

$$\hat{\Psi}_z^{(a)}(x)^\dagger \hat{\Psi}_z^{(b)}(x) = -(2\pi)^{-2} \partial_\mu^2 \hat{f}_x^{ab}(z, z). \tag{10}$$

In this paper we will only have need for the Green's function at $z' = z$, which formally can be expressed as

$$f_x(z, z) = -4\pi^2 \left((\mathbb{1}_{2k} - \mathcal{F}_z)^{-1} \right)_{12},$$

$$\mathcal{F}_z \equiv \hat{g}^\dagger(1)\, \mathrm{Pexp} \int_z^{z+1} \begin{pmatrix} 0 & \mathbb{1}_k \\ V(w; \vec{x}) & 0 \end{pmatrix} dw, \tag{11}$$

where the $(1, 2)$ component on the right-hand side of the first identity is with respect to the 2×2 block matrix structure. In particular this leads to a compact expression for the action density [3,14]

$$S(x) \equiv -\frac{1}{2} \mathrm{tr}\, F_{\mu\nu}^2(x) = -\frac{1}{2} \partial_\mu^2 \partial_\nu^2 \log \det\left(i e^{-\pi i x_0} (\mathbb{1}_{2k} - \mathcal{F}_z) \right), \tag{12}$$

which can be shown to be independent of the choice of z.

---

[2] It is in this respect interesting to note that $\hat{g}(1)$ plays the role of the holonomy associated to the dual Nahm gauge field $\hat{A}_\mu(z)$.

The formal expression for $\mathcal{F}_z$ can be made explicit by a decomposition into the "impurity" contributions T_m at $z = \mu_m$ and the "propagation" $H_m \equiv W_m(\mu_{m+1}, \mu_m)$ between μ_m and μ_{m+1}. For $z \in (\mu_m, \mu_{m+1})$ this gives

$$\mathcal{F}_z = W_m(z, \mu_m) T_m H_{m-1} \cdots T_2 H_1 T_1 \hat{g}^{\dagger}(1) H_n T_n H_{n-1} \cdots T_{m+1} H_m W_m(\mu_m, z),$$

$$(13)$$

with

$$T_m \equiv \begin{pmatrix} \mathbb{1}_k & 0 \\ 2\pi S_m & \mathbb{1}_k \end{pmatrix},$$

$$W_m(z, z') \equiv \begin{pmatrix} f_m^+(z) & f_m^-(z) \\ \frac{d}{dz} f_m^+(z) & \frac{d}{dz} f_m^-(z) \end{pmatrix} \begin{pmatrix} f_m^+(z') & f_m^-(z') \\ \frac{d}{dz} f_m^+(z') & \frac{d}{dz} f_m^-(z') \end{pmatrix}^{-1}. \qquad (14)$$

The columns of the two $k \times k$ matrices $f_m^{\pm}(z)$, defined for $z \in (\mu_m, \mu_{m+1})$, form the $2k$ solutions of the homogeneous Green's function equation,

$$\left(\frac{d^2}{dz^2} - 4\pi^2 \vec{R}^2(z; \vec{x}) \right) \hat{v}(z) = 0, \qquad (15)$$

of which those in $f_m^+(z)$ are exponentially rising and those in $f_m^-(z)$ are exponentially falling. This implies [14] $f_m^{\pm}(z) \to \exp(\pm 2\pi |\vec{x}|(z - \mu_m)\mathbb{1}_k)C_m^{\pm}$ for $|\vec{x}| \to \infty$, in which $C_m^{\pm} \equiv f_m^{\pm}(\mu_m)$ can be arbitrary non-singular (to ensure a complete set of solutions) matrices. In Ref. [10] we put $C_m^{\pm} = \mathbb{1}_k$, but here we find it convenient to leave this choice open. With the "impurity" scattering problem solved, constructing the exact solutions of the homogeneous Green's function equation is the last step in finding analytic expressions for the higher charge calorons.

2.2. Limiting cases

Nevertheless, approximate solutions can be derived, either assuming $\vec{x}$ is far removed from any core such that the gauge field has become Abelian (which we called the far field limit, denoted by a subscript "ff"), or assuming $\vec{x}$ and the constituents of type m (belonging to the mth interval) are well separated from all others, but not necessarily from each other (which was called the zero-mode limit, denoted by a subscript "zm"). This is because we found [10] that for $z \in (\mu_m, \mu_{m+1})$ the k zero-modes $\hat{\Psi}_z^{(a)}(x)$ only "see" the constituents of type m. For $\mu_m \leqslant z \leqslant \mu_{m+1}$ we have

$$f_x^{\mathrm{zm}}(z, z) = \pi \left(f_m^-(z) f_m^-(\mu_{m+1})^{-1} - f_m^+(z) f_m^+(\mu_{m+1})^{-1} Z_{m+1}^- \right)$$

$$\times \left(f_m^-(\mu_m) f_m^-(\mu_{m+1})^{-1} - Z_m^+ f_m^+(\mu_m) f_m^+(\mu_{m+1})^{-1} Z_{m+1}^- \right)^{-1}$$

$$\times \left(f_m^-(\mu_m) f_m^-(z)^{-1} - Z_m^+ f_m^+(\mu_m) f_m^+(z)^{-1} \right) R_m^{-1}(z), \qquad (16)$$

up to *exponential* corrections in the distance to the constituents of type $m' \neq m$, with

$$Z_m^- \equiv \mathbb{1}_k - 2\Sigma_m^{-1} R_{m-1}(\mu_m), \qquad Z_m^+ \equiv \mathbb{1}_k - 2\Sigma_m^{-1} R_m(\mu_m),$$

$$R_m(z) \equiv \frac{1}{2} \left(R_m^+(z) + R_m^-(z) \right), \qquad \Sigma_m \equiv R_m^-(\mu_m) + R_{m-1}^+(\mu_m) + S_m \qquad (17)$$

and

$$2\pi R_m^\pm(z) \equiv \pm\left(\frac{d}{dz} f_m^\pm(z)\right) f_m^\pm(z)^{-1}. \tag{18}$$

In the zero-mode limit Z_m^+ and Z_{m+1}^- approach $\mathbb{1}_k$ up to *algebraic* corrections. Neglecting these contributions as well, e.g., by sending the constituents of type $m' \neq m$ to infinity, leads to the so-called monopole limit (denoted by a subscript "mon"), further simplifying the expression for the Green's function to

$$f_x^{\mathrm{mon}}(z,z) = -\pi U(z,\mu_{m+1}) U_m^{-1}(\mu_m,\mu_{m+1}) U_m(\mu_m,z) R_m^{-1}(z), \tag{19}$$

with[3]

$$U_m(z,z') \equiv f_m^+(z) f_m^+(z')^{-1} - f_m^-(z) f_m^-(z')^{-1}. \tag{20}$$

In turn, from Eq. (16) one may derive the far field limit, giving up to *exponential* corrections in the distance to *all* constituents

$$f_x^{\mathrm{ff}}(\mu_m,\mu_m) = 2\pi \Sigma_m^{-1}, \tag{21}$$

as is relevant for the expression of the gauge field in this limit [14]. For the zero-mode density (Eq. (10)) we may use for $\mu_m < z < \mu_{m+1}$ (z strictly different from μ_m, μ_{m+1})

$$f_x^{\mathrm{ff}}(z,z) = \pi R_m^{-1}(z). \tag{22}$$

The discontinuity in this limit, $\pi R_m(\mu_m) \neq 2\pi \Sigma_m^{-1} \neq \pi R_{m-1}(\mu_m)$, arises due to the zero-mode developing a massless component when z approaches μ_m. It might seem that Eq. (22), combined with Eq. (10), is inconsistent with an exponential decay. However, it turns out that [10]

$$\mathcal{V}_m(\vec{x}) \equiv (4\pi)^{-1} \mathrm{Tr}\big(R_m^{-1}(z)\big) \tag{23}$$

is independent of z in the mth interval and harmonic everywhere (hence giving vanishing zero-mode density) except for some singularities in the cores of the constituents of type m. It is this feature, and our ability to compute $\mathcal{V}_m(\vec{x})$ exactly for SU(2) charge 2 calorons, that allowed us to make statements about the localization of the cores, without solving the Green's function exactly.

In Appendix A we derive the following new result for the monopole limit of the action density (Eq. (12)). In the limit where $\vec{x}$ and the constituent locations of type m are well separated from all other constituents, for which both the action and zero-mode densities are static, we find (with U_m defined in Eq. (20))

$$\mathcal{S}^{\mathrm{mon}}(\vec{x}) = -\frac{1}{2} \partial_i^2 \partial_j^2 \log \det\big[U_m(\mu_{m+1},\mu_m) R_m^{-1}(\mu_m)\big], \tag{24}$$

up to *algebraic* corrections. This is a direct generalization for the action density of a single BPS monopole, $\mathcal{S}(\vec{x}) = -\frac{1}{2}\partial_i^2 \partial_j^2 \log[\sinh(2\pi \nu_m |\vec{x}|)/|\vec{x}|]$ (located at the origin and with

---

[3] Note that $U_m(z,z')$ satisfies the Green's function equation with boundary conditions $U_m(z,z) = 0$ and $\frac{d}{dz} U_m(z,z') = -\frac{d}{dz'} U_m(z,z') = 2\pi R_m(z)$, for $z' \to z$.

mass $8\pi^2 v_m$). We emphasize that this result can be used irrespective of the distance between the constituents of the same type m. Eq. (24) therefore gives in terms of $f_m^{\pm}(z)$ a closed form expression for the multi-monopole energy density. The same holds, when using Eq. (19), for the zero-mode density (Eq. (10)). For the caloron it involves taking the limit where all constituents of type $m' \neq m$ are sent to infinity, which is why this is called the monopole limit. We recall, that in the far field one should use [14] (see also Appendix A)

$$S^{\mathrm{ff}}(\vec{x}) = -\frac{1}{2}\partial_i^2\partial_j^2 \sum_{m=1}^{n} \log \det\big(f_m^+(\mu_{m+1})f_m^+(\mu_m)^{-1}R_m^{-1}(\mu_m)\Sigma_m\big). \tag{25}$$

3. Exact results

Finding the exact homogeneous solutions of the Green's function equation, Eq. (15), closely follows Nahm's method [15] to construct the dual chiral zero-modes. The main advantage of our approach is that we need not worry about boundary conditions, as this is solved by the "impurity" scattering formalism [14]. In the following we restrict ourselves to a given interval $z \in (\mu_m, \mu_{m+1})$ and work in the gauge where $\hat{A}_0(z) - 2\pi i x_0 \mathbb{1}_k = 0$.

Using that $\hat{A}_\mu(z)$ is self-dual, a consequence of the Nahm equation, one easily shows that $\hat{D}_x^\dagger \hat{D}_x = -\frac{d^2}{dz^2} + 4\pi^2 \vec{R}^2(z;\vec{x})$, such that it is natural to consider the equation

$$\hat{D}_x\hat{\psi}(z) = \sigma_\mu \hat{D}_x^\mu \hat{\psi}(z) = \left(\frac{d}{dz} - 2\pi\vec{\tau}\cdot\vec{R}(z;\vec{x})\right)\hat{\psi}(z) = 0. \tag{26}$$

It follows that $\hat{\psi}(z)$ would be a homogeneous solution of the Green's function equation, albeit that $\hat{\psi}(z)$ is a spinor (with a chirality opposite to that for the zero-modes involved in the Nahm transformation, cf. Eq. (2)). We follow Nahm [15] and use the ansatz $\hat{\psi}(z) = (\mathbb{1}_2 + \vec{u}(\vec{x})\cdot\vec{\tau})|s\rangle \otimes \hat{v}(z)$, where $\hat{v}(z)$ is a k-dimensional complex vector, $\vec{u}(\vec{x})$ is a unit vector that does not depend on z and $|s\rangle$ is an arbitrary normalized constant spinor (as long as it is not annihilated by $\mathbb{1}_2 + \vec{u}(\vec{x})\cdot\vec{\tau}$). It then follows that $\hat{v}(z) = \langle s|(\mathbb{1}_2 - \vec{u}(\vec{x})\cdot\vec{\tau})\hat{\psi}(z)$ satisfies Eq. (15).

The unit vector $\vec{u}(\vec{x})$ is found from a complex vector $\vec{y}(x)$ which squares to 0, $\vec{y}(\vec{x})\cdot\vec{y}(\vec{x}) = 0$, implying its real and imaginary parts are perpendicular and of equal length ($\neq 0$ as long as $\vec{y}(\vec{x}) \neq 0$), such that (for ease of notation the $\vec{x}$ dependence of $\vec{y}$ and $\vec{u}$ will henceforth be left implicit)

$$\vec{u} = i\vec{y} \times \vec{y}^* / (\vec{y}\cdot\vec{y}^*) \tag{27}$$

is well defined and $\vec{u} \times \vec{y} = -i\vec{y}$, i.e., $\mathrm{Re}(\vec{y})$, $\mathrm{Im}(\vec{y})$ and $\vec{u}$ form an orthogonal set of vectors. Using the ansatz for $\hat{\psi}(z)$ and introducing

$$\hat{Y}(z) \equiv -\vec{y}\cdot\vec{R}(z;\vec{x}), \qquad \hat{U}(z) \equiv -2\pi\vec{u}\cdot\vec{R}(z;\vec{x}), \tag{28}$$

leads to the equations

$$\hat{Y}(z)\hat{v}(z) = 0, \qquad \frac{d}{dz}\hat{v}(z) = \hat{U}(z)\hat{v}(z) \tag{29}$$

for which the first one can only have a solution provided $\det\hat{Y}(z) = 0$.

242   F. Bruckmann et al. / Nuclear Physics B 698 (2004) 233–254

It is the great beauty of Nahm's formalism that $\det \hat{Y}(z)$ is a conserved quantity. That is, $\frac{d}{dz}\hat{A}_j(z) = -\frac{1}{2}\varepsilon_{jk\ell}[\hat{A}_k(z), \hat{A}_\ell(z)]$ (the Nahm equation (5) restricted to an interval and in the gauge where $\hat{A}_0 = 0$) implies $\frac{d}{dz}\det \hat{Y}(z) = 0$ for any choice of $\vec{x}$ and $\vec{y}$ on a null-cone in C^3 (or rather in CP^2 since we may rescale $\vec{y}$ with a non-zero complex factor without changing the equations). The conserved quantities are generated by the symmetric traceless monomials, $M_{i_1 i_2 \ldots i_\ell}$, built from $\mathrm{Tr}(\hat{A}_{i_1}(z)\hat{A}_{i_2}(z) \cdots \hat{A}_{i_\ell}(z))$ with ℓ arbitrary, as one readily verifies. For example, $\mathrm{Tr}\,\hat{A}_i(z)$ is constant and defines the center of mass. A natural way to project on the traceless symmetric monomials is precisely through introducing $\vec{y} \in CP^2$ on a null-cone, forming $y_{i_1} y_{i_2} \cdots y_{i_\ell} M_{i_1 i_2 \ldots i_\ell} = y_{i_1} y_{i_2} \cdots y_{i_\ell} \mathrm{Tr}(\hat{A}_{i_1}(z)\hat{A}_{i_2}(z) \cdots \hat{A}_{i_\ell}(z))$. It is interesting to note that $x_{i_1} x_{i_2} \cdots x_{i_\ell} M_{i_1 i_2 \ldots i_\ell} |\vec{x}|^{-\ell}$ is always a spherical harmonic of order ℓ, used in Ref. [10] to show through the multipole expansion of $\mathrm{Tr}(R_m^{-1}(z))$ that it is conserved and harmonic, except for singularities in the core of the constituents.

Once it is established that $\det \hat{Y}(z)$ is independent of z for any choice of $\vec{x}$, we can look for its zeros. Using the null-cone parametrization $\vec{y} = (\frac{1}{2}(1 - \zeta^2), -\frac{i}{2}(1 + \zeta^2), \zeta)$, and the fact that the matrix $\hat{Y}(z)$ is k-dimensional, we obtain a polynomial equation in ζ of order $2k$ and hence there are for generic $\vec{x}$ exactly $2k$ solutions. It is useful to note that these solutions come in complex conjugate pairs, where the symmetry $\vec{y} \to \vec{y}^*$ implies $\vec{u} \to -\vec{u}$ and $\zeta \to -1/\zeta^*$ (giving $\vec{y}^*$ up to a multiple, equivalent to $\vec{y}^*$ in CP^2).

Given a particular zero $\vec{y}$, we may conveniently write a vector in the kernel of $\hat{Y}(z)$ as [19] $\hat{v}_a(z) = \hat{\phi}(z)(\mathrm{adj}\,\hat{Y}(z))_{ac}$ for a fixed choice of c, where $\mathrm{adj}\,\hat{Y}(z)$ is the matrix formed by the minors of $\hat{Y}(z)$. The Nahm equation is easily seen to imply $\frac{d}{dz}\hat{Y}(z) = [\hat{U}(z), \hat{Y}(z)]$ for *any* choice of $\vec{y}$ on the null-cone. From this one derives that[4] $\frac{d}{dz}\,\mathrm{adj}\,\hat{Y}(z) = [\hat{U}(z), \mathrm{adj}\,\hat{Y}(z)]$. Substituting $\hat{v}_a(z) = \hat{\phi}(z)(\mathrm{adj}\,\hat{Y}(z))_{ac}$ into Eq. (29) gives

$$\frac{d\hat{\phi}(z)}{dz}\left(\mathrm{adj}\,\hat{Y}(z)\right)_{ac} = \hat{\phi}(z)\left(\mathrm{adj}\,\hat{Y}(z)\hat{U}(z)\right)_{ac}. \tag{30}$$

Using the fact that $\hat{U}(z) = -iu_j\hat{g}(z)\hat{A}_j(z)\hat{g}^\dagger(z) - 2\pi u_j x_j$, we get

$$\hat{\phi}(z) = \frac{\exp(\hat{\mu}(z) - 2\pi z\vec{u} \cdot \vec{x})}{\sqrt{(\mathrm{adj}\,\hat{Y}(z))_{ac}}},$$

$$\frac{d\hat{\mu}(z)}{dz} = -\frac{i\{\mathrm{adj}\,\hat{Y}(z), u_j\hat{g}(z)\hat{A}_j(z)\hat{g}^\dagger(z)\}_{ac}}{2(\mathrm{adj}\,\hat{Y}(z))_{ac}}, \tag{31}$$

where the equation for $\hat{\mu}(z)$ is the same for any value of a (it may depend on the value of c). This is useful for studying the asymptotic behaviour of the solution. For large $|\vec{x}|$, $\det \hat{Y}(z) = 0$ implies that $\vec{y} \cdot \vec{x} \to 0$, such that (cf. Eq. (27)) $\vec{u} \to \pm\vec{x}/|\vec{x}|$. We may use the symmetry $\vec{u} \to -\vec{u}$ to guarantee that there are k zeros $\vec{y}^{(b)}$ with the sign of $\vec{u}^{(b)} \cdot \vec{x}$ negative, leading to solutions that rise as $\exp(2\pi z|\vec{x}|)$. It then follows that the k zeros $\vec{y}^{(b+k)} = \vec{y}^{(b)*}$

---

[4] Assume first that $\vec{y}$ is such that $\det \hat{Y} \neq 0$, in which case $\mathrm{adj}\,\hat{Y} = \hat{Y}^{-1}\det \hat{Y}$ and therefore $\frac{d}{dz}\,\mathrm{adj}\,\hat{Y} = -\hat{Y}^{-1}[\hat{U}, \hat{Y}]\hat{Y}^{-1}\det \hat{Y} + \hat{Y}^{-1}\frac{d}{dz}\det \hat{Y} = [\hat{U}, \mathrm{adj}\,\hat{Y}] + \hat{Y}^{-1}\mathrm{Tr}(\hat{Y}^{-1}\frac{d}{dz}\hat{Y}) = [\hat{U}, \mathrm{adj}\,\hat{Y}]$. Observe that $\mathrm{adj}\,\hat{Y}$ is analytic in $\vec{y}$, such that the result is valid also when $\vec{y}$ leads to $\det \hat{Y} = 0$.

F. Bruckmann et al. / Nuclear Physics B 698 (2004) 233–254

give $\vec{u}^{(b+k)} = -\vec{u}^{(b)}$, leading to the solutions that decay as $\exp(-2\pi z|\vec{x}|)$ for large $|\vec{x}|$. Hence we may put $f_{ab}^{+}(z) = \hat{v}_a^{(b)}(z)$ and $f_{ab}^{-}(z) = \hat{v}_a^{(b+k)}(z)$. Defining

$$\hat{m}_{ab}^{+}(z) = -\left(\text{adj}\,\vec{y}^{(b)} \cdot \vec{R}(z;\vec{x})\right)_{ac}, \qquad \hat{m}_{ab}^{-}(z) = -\left(\text{adj}\,\vec{y}^{(b)*} \cdot \vec{R}(z;\vec{x})\right)_{ac}, \tag{32}$$

which are *algebraic* in $\hat{A}_j(z)$ and x_j, we find

$$f_{ab}^{+}(z) = \hat{m}_{ab}^{+}(z)\hat{\phi}^{(b)}(z), \qquad f_{ab}^{-}(z) = \hat{m}_{ab}^{-}(z)\hat{\phi}^{(b+k)}(z), \tag{33}$$

where $\hat{\phi}(z)$ contains the exponential dependencies, and thus seemingly all the information about the cores of the constituents. To make this more precise we compute $R^{\pm}(z)$, see Eq. (18), using that Eq. (29) implies (with $\vec{u}^{(b+k)} = -\vec{u}^{(b)}$) $\frac{d}{dz}f_{ab}^{\pm}(z) = \mp 2\pi \sum_{d=1}^{k} \vec{u}^{(b)} \cdot \vec{R}_{ad}(z;\vec{x})f_{db}^{\pm}(z)$, finding that the factors $\hat{\phi}^{(b)}(z)$ *drop out*

$$R_{ad}^{\pm}(z) = -\sum_{b,e=1}^{k} \vec{u}^{(b)} \cdot \vec{R}_{ae}(z;\vec{x})\hat{m}_{eb}^{\pm}(z)\left(\hat{m}^{\pm}(z)^{-1}\right)_{bd}. \tag{34}$$

This proves that $R_m^{\pm}(z)$ is purely algebraic in $\hat{A}_j(z)$ and x_j, as are Σ_m and $R_m(z)$, which determine the far field limit for the zero-mode density and the gauge field.

One might wonder how, given that the $\hat{\psi}^{(b)}(z)$ are of the "wrong" chirality in the context of the Nahm transformation, one could use these results to reconstruct the gauge field for magnetic monopoles where the relation to the ADHM construction is not readily available. For this one observes that the columns of $\hat{\psi}^{(b)}(z)$ can be used to form a $2k \times 2k$ matrix $w(z)$. Using that $\hat{D}_x w(z) = 0$, one finds $\hat{D}_x^{\dagger}(w^{\dagger}(z)^{-1}) = 0$. Thus the columns of $w^{\dagger}(z)^{-1}$ give $2k$ independent solutions for each interval, from which n normalizable solutions $\hat{\Psi}^{(p)}(z;\vec{x})$ should remain after imposing the appropriate boundary (cq. matching) conditions. These are then used in Nahm's original construction to compute the gauge field (cf. Eq. (3))

$$A_{\mu}^{pq}(x) = \int \hat{\Psi}^{(p)}(z;x_0,\vec{x})^{\dagger} \frac{\partial}{\partial x_{\mu}} \hat{\Psi}^{(q)}(z;x_0,\vec{x})\,dz, \tag{35}$$

where $\hat{\Psi}^{(p)}(z;x_0,\vec{x}) \equiv \hat{g}^{\dagger}(z)\hat{\Psi}^{(p)}(z;\vec{x})$.

There seems to be considerable advantage in using the Green's function (Fourier transformed ADHM) method, since it can solve the matching conditions without relying on the availability of exact solutions for the normalizable dual zero-modes. To go beyond the approximations discussed in Section 2.2 and resolve the constituent cores we need to solve for $\hat{\mu}(z)$. This cannot always be done in closed form, but it is given by an explicit integral which can be performed numerically when required. For charge 2 monopoles Panagopoulos [16] was, however, able to find the exact integral. We can use the same ingredients for the caloron case and explicitly solve for the Green's function in the case of charge 2 calorons.

3.1. Analytic expressions for charge 2

For charge 2 the number of invariants associated to the conserved quantities of the Nahm equation is 8, of which $\text{Tr}\,\hat{A}_j(z) \equiv 4\pi i a_j$ are related to the center of mass for the

     F. Bruckmann et al. / Nuclear Physics B 698 (2004) 233–254

constituents of given magnetic charge, coming from the interval under consideration. Assuming now that $\hat{A}_j(z)$ is traceless, 5 invariants remain, given in terms of the symmetric traceless tensor

$$M_{ij} \equiv -\frac{1}{2}\left(\mathrm{Tr}\big(\hat{A}_i(z)\hat{A}_j(z)\big) - \frac{1}{3}\delta_{ij}\,\mathrm{Tr}\big(\hat{A}_k^2(z)\big)\right). \tag{36}$$

Three of its parameters are associated to the rotation $\mathcal{R}$ which diagonalizes the 3×3 matrix, $M = \mathcal{R}\,\mathrm{diag}(c_1, c_2, c_3)\mathcal{R}^t$, where $\mathcal{R}$ is fixed by requiring $c_2 \leqslant c_1 \leqslant c_3$. The c_i add to zero and can be expressed in terms of the so-called scale (D) and shape ($\mathbf{k}$) parameters,

$$c_1 = D^2\frac{1 - 2\mathbf{k}^2}{12}, \qquad c_2 = D^2\frac{\mathbf{k}^2 - 2}{12}, \qquad c_3 = D^2\frac{1 + \mathbf{k}^2}{12}. \tag{37}$$

The Nahm equation for the case of charge 2 can be solved completely in terms of Jacobi elliptic functions [20,21], which was summarized in Ref. [10]

$$\hat{g}(z)\hat{A}_j(z; \vec{a}, \mathcal{R}, h, D, \mathbf{k})\hat{g}^\dagger(z) \equiv 2\pi i a_j \mathbb{1}_2 + \frac{1}{2}i D\mathcal{R}_{jb}f_b\big(D(z - z_0)\big)h^\dagger \tau_b h, \tag{38}$$

where h is a global gauge parameter and

$$f_1(z) \equiv \frac{\mathbf{k}'}{\mathrm{cn}_\mathbf{k}(z)}, \qquad f_2(z) \equiv \frac{\mathbf{k}'\,\mathrm{sn}_\mathbf{k}(z)}{\mathrm{cn}_\mathbf{k}(z)}, \qquad f_3(z) \equiv \frac{\mathrm{dn}_\mathbf{k}(z)}{\mathrm{cn}_\mathbf{k}(z)}, \quad \mathbf{k}' \equiv \sqrt{1 - \mathbf{k}^2} \tag{39}$$

with[5] $\mathrm{sn}_\mathbf{k}(z) = \sin(\varphi(z))$, $\mathrm{cn}_\mathbf{k}(z) = \cos(\varphi(z))$ and $\mathrm{dn}_\mathbf{k}(z) = \sqrt{1 - \mathbf{k}^2\,\mathrm{sn}_\mathbf{k}^2(z)}$ the standard Jacobi elliptic functions. This does not yet address the matching of $\hat{A}_j(z)$ on the different intervals, where some difference between the monopole and caloron application appears. For the caloron, apart from the axially symmetric solutions constructed in Ref. [14], we found two sets of non-trivial solutions that interpolate between overlapping and well-separated constituents. It is for these classes of solutions that we will resolve the cores, when constituents overlap and the non-linearity plays an important role.

The next step in the construction is finding the zeros $\vec{y}$ of $\det \hat{Y}(z)$. One has

$$\det \hat{Y}(z) = y_i y_j\left(x_i x_j - \frac{1}{3}\vec{x}^2\delta_{ij} - (2\pi)^{-2}M_{ij}\right), \tag{40}$$

where we used $\vec{y}^2 = 0$. Substituting a parametrization for this null-cone in $\mathbb{C}P^2$, e.g., $\vec{y} = (\frac{1}{2}(1 - \zeta^2), -\frac{i}{2}(1 + \zeta^2), \zeta)$, gives a 4th order polynomial. However, for finding the 4 solutions we find it in this case more convenient to first diagonalize the matrix $x_i x_j - \frac{1}{3}\vec{x}^2\delta_{ij} - (2\pi)^{-2}M_{ij} = \mathcal{O}_{ik}\lambda_k\mathcal{O}_{jk}$. Introducing $\vec{y}' = \mathcal{O}^t\vec{y}$, the equation for the zeros reduces to $(y_1')^2\lambda_1 + (y_2')^2\lambda_2 + (y_3')^2\lambda_3 = 0$, which in addition to the null-cone condition, $(y_1')^2 + (y_2')^2 + (y_3')^2 = 0$, is now easily solved by

$$\vec{y}^{(a)\prime} = \big(\sqrt{\lambda_2 - \lambda_3}, (-1)^{a+1}\sqrt{\lambda_3 - \lambda_1}, i\sqrt{\lambda_2 - \lambda_1}\,\big),$$
$$\vec{y}^{(a+2)\prime} = \big(\vec{y}^{(a)\prime}\big)^*, \quad a = 1, 2, \tag{41}$$

---

[5] $\varphi(z)$ implicitly defined by the elliptic integral of the first kind $z = \int_0^{\varphi(z)} 1/\sqrt{1 - \mathbf{k}^2\sin^2\theta}\,d\theta$.

where we fixed (for generic $\vec{x}$) the diagonalizing rotation $\mathcal{O}$ by ordering $\lambda_1 \leqslant \lambda_3 \leqslant \lambda_2$. Using Eq. (27) we find for $\vec{u}' \equiv \mathcal{O}^t \vec{u}$

$$\vec{u}^{(a)'} = -\left((-1)^a \frac{\sqrt{\lambda_3 - \lambda_1}}{\sqrt{\lambda_2 - \lambda_1}}, \frac{\sqrt{\lambda_2 - \lambda_3}}{\sqrt{\lambda_2 - \lambda_1}}, 0 \right), \qquad \vec{u}^{(a+2)'} = -\vec{u}^{(a)'}, \quad a = 1, 2. \quad (42)$$

One easily checks that $\vec{u}^{(1,2)}$ and $\vec{u}^{(3,4)}$ give rise to, respectively, the exponentially rising and falling solutions.

It is also instructive to give the explicit expressions for the matrices $\hat{m}^{\pm}(z)$ in Eq. (32) (choosing $c = 2$). We note that for a 2×2 matrix adj $\hat{Y} = (\text{Tr}\,\hat{Y})\mathbb{1}_2 - \hat{Y}$, and without loss in generality[6] we take $z_0 = 0$, $\vec{a} = \vec{0}$, $\mathcal{R} = \mathbb{1}_3$, $D = 1$ and $h = \mathbb{1}_2$ in Eq. (38), such that

$$\hat{m}^+(z) = -\frac{1}{4\pi} \begin{pmatrix} y_1^{(1)} f_1(z) - i y_2^{(1)} f_2(z) & y_1^{(2)} f_1(z) - i y_2^{(2)} f_2(z) \\ 4\pi \vec{x} \cdot \vec{y}^{(1)} - y_3^{(1)} f_3(z) & 4\pi \vec{x} \cdot \vec{y}^{(2)} - y_3^{(2)} f_3(z) \end{pmatrix},$$

$$\hat{m}^-(z) = -\frac{1}{4\pi} \begin{pmatrix} y_1^{(3)} f_1(z) - i y_2^{(3)} f_2(z) & y_1^{(4)} f_1(z) - i y_2^{(4)} f_2(z) \\ 4\pi \vec{x} \cdot \vec{y}^{(3)} - y_3^{(3)} f_3(z) & 4\pi \vec{x} \cdot \vec{y}^{(4)} - y_3^{(4)} f_3(z) \end{pmatrix}. \quad (43)$$

We checked that Eq. (23), $\mathcal{V}(\vec{x}) = (2\pi)^{-1}\,\text{Tr}(R^+(z) + R^-(z))^{-1}$, evaluated using Eq. (34) is independent of z and agrees with the result derived in Ref. [10].

We next address solving Eq. (31), which for charge 2 can be written as[7]

$$\frac{d\hat{\mu}(z)}{dz} = \frac{M_{ij} u_i y_j \delta^{ac} - 2\pi i (\vec{x} \cdot \vec{y}) u_i \hat{A}_i^{ac}(z)}{2\pi (\vec{x} \cdot \vec{y}) \delta^{ac} - i y_i \hat{A}_i^{ac}(z)}. \quad (44)$$

For the same choice of parameters in Eq. (38) as above, $z_0 = 0$, $\vec{a} = \vec{0}$, $\mathcal{R} = \mathbb{1}_3$, $D = 1$ and $h = \mathbb{1}_2$, this gives (with $a = 1$ and $c = 2$)

$$\frac{d\hat{\mu}(z)}{dz} = 2\pi (\vec{x} \cdot \vec{y}) \frac{u_1 f_1(z) - i u_2 f_2(z)}{y_1 f_1(z) - i y_2 f_2(z)}. \quad (45)$$

Although the dependence on $\vec{x}$ is complicated, the integral over z turns out to be manageable (as was observed before in the context of charge 2 monopoles, although we here choose not to express the solution in terms of theta functions [16]). To solve the equation we first rewrite the right-hand side of Eq. (45) using the fact that $\vec{u} \times \vec{y} = -i\vec{y}$ (cf. Eq. (27)),

$$\frac{u_1 f_1(z) - i u_2 f_2(z)}{y_1 f_1(z) - i y_2 f_2(z)} = \frac{(u_3 y_1 + i y_2) f_1(z) - i(u_3 y_2 - i y_1) f_2(z)}{y_3 (y_1 f_1(z) - i y_2 f_2(z))}$$

$$= \frac{f_1(z) f_2(z) y_3^2 + 4 i y_1 y_2 (\mathbf{k}')^2}{y_3 (16\pi^2 (\vec{x} \cdot \vec{y})^2 - y_3^2 f_3^2(z))} + \frac{u_3}{y_3}.$$

In the last identity we used that $\vec{y}^2 = 0$, $f_1^2(z) - f_2^2(z) = 1 - \mathbf{k}^2 = (\mathbf{k}')^2$ and the fact that $\det \hat{Y}(z) = 0$ implies $(y_1 f_1(z) - i y_2 f_2(z))(y_1 f_1(z) + i y_2 f_2(z)) = 16\pi^2 (\vec{x} \cdot \vec{y})^2 - y_3^2 f_3^2(z)$.

---

[6] We may change z_0, $\vec{a}$, h, $\mathcal{R}$ and D by, respectively, translations, (gauge) rotations, and suitable rescalings.

[7] Using that $\frac{1}{2}\{\hat{A}_i, \hat{A}_j\} y_i u_j = \frac{1}{2}\mathbb{1}_2 \text{Tr}(\hat{A}_i \hat{A}_j) y_i u_j = -M_{ij} y_i u_j \mathbb{1}_2$, since $\vec{u} \cdot \vec{y} = 0$.

With $\frac{d}{dz} f_3(z) = f_1(z) f_2(z)$ we can now integrate Eq. (45),

$$
\hat{\mu}(z) = 2\pi z (\vec{x} \cdot \vec{y}) \frac{u_3}{y_3} + \frac{1}{4} \log\left(\frac{4\pi \vec{x} \cdot \vec{y} + f_3(z) y_3}{4\pi \vec{x} \cdot \vec{y} - f_3(z) y_3} \right) + i \frac{\text{sign}(z)(\mathbf{k}')^2}{2\pi (\vec{x} \cdot \vec{y})} \frac{y_1 y_2}{4 y_3} I(z),
$$

$$
I(z) \equiv \Pi_{\mathbf{k}}\big(f_3^{-1}(z), n\big) - \Pi_{\mathbf{k}}(1, n) + |z|, \quad n \equiv \frac{(4\pi \vec{x} \cdot \vec{y})^2}{y_3^2}, \tag{46}
$$

up to an irrelevant constant, where $\Pi_{\mathbf{k}}(s, n)$ is the elliptic integral of the third kind[8]

$$
\Pi_{\mathbf{k}}(s, n) \equiv \int_0^s \frac{dt}{(1 - nt^2)\sqrt{(1 - \mathbf{k}^2 t^2)(1 - t^2)}}. \tag{47}
$$

We now combine these ingredients to give in terms of $\vec{y}$ the exact form for the homogeneous solution of the Green's function equation,

$$
\hat{v}_a(z) = \hat{\phi}(z)\big(\text{adj}\,\hat{Y}(z)\big)_{a2} = \exp\big(\hat{\mu}(z) - 2\pi z \vec{u} \cdot \vec{x}\big) \frac{(\text{adj}\,\hat{Y}(z))_{a2}}{\sqrt{(\text{adj}\,\hat{Y}(z))_{12}}} \tag{48}
$$

or putting in all the relevant expressions

$$
\hat{v}_a(z) = \exp\left(i \frac{z}{y_3} \left[2\pi (\vec{y} \times \vec{x})_3 + \frac{y_1 y_2 (\mathbf{k}')^2}{8\pi (\vec{x} \cdot \vec{y})|z|} \big(|z| + \Pi_{\mathbf{k}}\big(f_3^{-1}(z), n\big) - \Pi_{\mathbf{k}}(1, n)\big) \right] \right)
$$
$$
\times (4\pi)^{-1/2} \big(-y_1 f_1(z) - (-1)^a i y_2 f_2(z)\big)^{1/2} \left(\frac{4\pi \vec{x} \cdot \vec{y} - (-1)^a y_3 f_3(z)}{4\pi \vec{x} \cdot \vec{y} + (-1)^a y_3 f_3(z)} \right)^{1/4}.
$$
$$
\tag{49}
$$

Substituting $\vec{y} = \vec{y}^{(b)} = \mathcal{O}^t \vec{y}^{(b)\prime}$, with $\vec{y}^{(b)\prime}$ as defined in Eq. (41) gives $f_{ab}^{+}(z) = \hat{v}_a^{(b)}(z)$ and $f_{ab}^{-}(z) = \hat{v}_a^{(b+2)}(z)$, and thereby the Green's function, once we specify the parameters involved in the solutions to the Nahm equation.

4. Action and zero-mode density plots

The discontinuities in $\hat{A}_j(z)$ at $z = \mu_1$ and $z = \mu_2$ implied by the Nahm equation, Eq. (5), impose constraints which are (like the quadratic ADHM constraint) in general difficult to solve. Work is in progress to describe the full parameter space for SU(2) and charge 2, but in Ref. [10] we did find two non-trivial parametrizations for which we illustrate here in a number of figures how the full caloron solutions look like, using the exact Green's function as constructed in the previous section. Taking advantage of the possibility

---

[8] More commonly the elliptic integral of the third kind is defined as $\Pi(n; \varphi, \mathbf{k}) = \Pi_{\mathbf{k}}(\sin\varphi, n)$. Note that $I(z)$ can alternatively be written as $\int_1^{f_3(z)} (1 - t^2/n)^{-1}(t^2 - \mathbf{k}^2)^{-1/2}(t^2 - 1)^{-1/2} dt$, from which it follows that $\frac{d}{dz} I(z) = ((1 - n f_3^2(z))|f_1(z) f_2(z)|)^{-1} \frac{d}{dz} f_3(z) = \text{sign}(z)(1 - n f_3^2(z))^{-1}$, using $f_3^2(z) - 1 = f_2^2(z)$, $f_3^2(z) - \mathbf{k}^2 = f_1^2(z)$ and $\frac{d}{dz} f_3(z) = f_1(z) f_2(z)$.

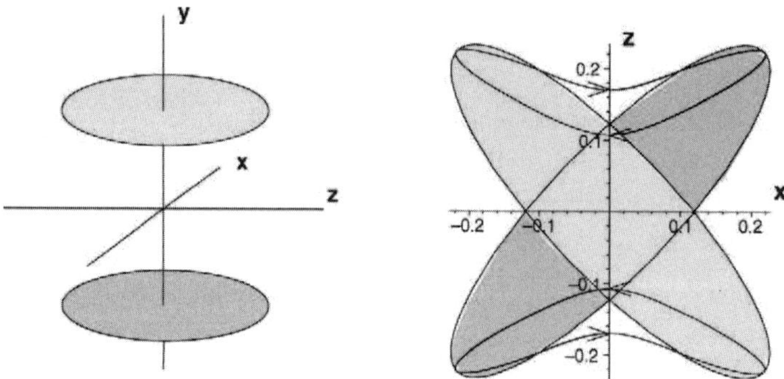

Fig. 2. An example to illustrate the location of the disc singularities (light and dark shaded according to magnetic charge) for a rectangular (left) and crossed (right) configuration. The latter is shown at $\alpha = -\pi/2$, for $d = \pi/32$ and $\theta = \pi/4$ ($\mathbf{k} = 0.962$, $D = 3.894$ and $\varphi = -\pi/4$). The curves indicate the would-be constituent locations at fixed d and θ, varying α from $-\pi$ (to which the arrows point) to 0. For $\alpha = -\pi$ and 0 the discs collapse to lines ($\mathbf{k} = 1$) with no singularities remaining, except at the endpoints.

to work with arbitrary arithmetic precision the programme Maple has been used for these calculations. The configurations are formulated in terms of the two intervals $z \in [\mu_1, \mu_2]$ and $z \in [\mu_2, 1 + \mu_1]$, each associated with two constituent monopoles of equal magnetic charge, but opposite in sign from one interval to the next. Apart from a shift and (gauge) orientation, the configuration is described by a shape ($\mathbf{k}$) and scale (D) parameter (see Section 3.1), for simplicity assumed to be the same on both intervals. We also take all constituents to be of equal mass, $\nu_{1,2} = \frac{1}{2}$ ($\mu_2 = -\mu_1 = \frac{1}{4}$), most relevant for the confined phase with $\operatorname{tr} \mathcal{P}_\infty = 0$.

The two periodic and two anti-periodic chiral fermion zero-modes each have support on oppositely charged constituents (see Fig. 1) and in the far field limit it was found that the zero-mode density (summed over the two zero-modes implied by the index theorem) is described by a disc singularity, bounded by an ellipse with semi-major axis $D/4\pi$ and eccentricity $(1 - \mathbf{k}')/(1 + \mathbf{k}')$ (where $\mathbf{k}' = \sqrt{1 - \mathbf{k}^2}$). This revealed that the core of a cluster of like-charged constituents is in general extended, unless the individual constituents are well separated. The far field only describes the (algebraic) Abelian component of the gauge field and to resolve the structure of the core we need to determine the full non-Abelian structure.

For the first parametrization (called "rectangular") the two discs are parallel and separated in height by a distance d. The configuration is charaterized by the two extremal points along the major axis of the disc, also called *would-be* constituent locations[9]

$$\vec{y}_m^{(j)} = \left(0, \frac{1}{2}(-1)^m d, (-1)^j (4\pi)^{-1} D\right), \qquad 2\pi d = D f_2 \left(\frac{1}{4} D\right), \tag{50}$$

up to an overall shift and orientation (the definition of $f_2(z)$, which involves $\mathbf{k}$, can be found in Eq. (39)). A typical example is shown in Fig. 2(left). Apart from $\mathbf{k}$ and D, the parameters

---

[9] In their immediate neighbourhood the action density is maximal; they are the constituent locations in the point-like limit, $\mathbf{k} \to 1$.

F. Bruckmann et al. / Nuclear Physics B 698 (2004) 233–254

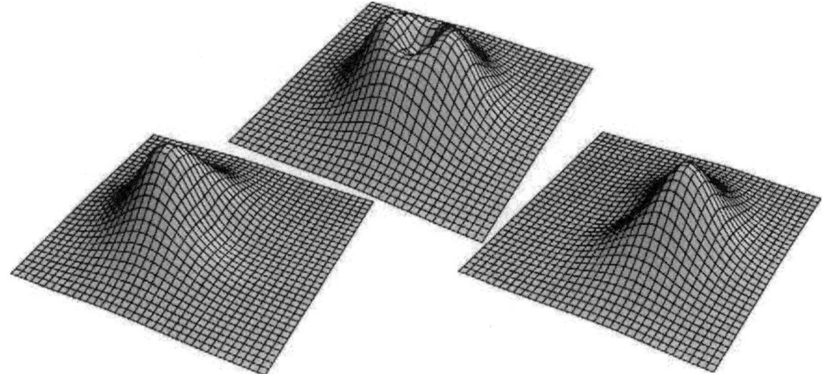

Fig. 3. Shown is the energy density (middle) for $\mu_2 = 0.25$, $\mathbf{k} = 0.570$ and $D = 6.915$, in the monopole limit $d \to \infty$ and in the x–z plane through one of the discs for the rectangular configuration, see Fig. 2(left). On a scale enhanced by a factor $4\pi^2$ are shown the densities for the two monopole zero-modes (left and right).

that enter Eq. (38) for the mth interval are $\vec{a} = (0, \frac{1}{2}(-1)^m d, 0)$, $\mathcal{R} = \mathbb{1}_3$, $h = \mathbb{1}_2$ and $z_0 = \frac{1}{4}(1 + (-1)^m)$. In all cases discussed here we have also put $\xi_0 = 0$. One verifies that the discontinuities of $\hat{A}_j(z)$ at $z = \mu_m$ are given by $2\pi i \rho_m^j$ with appropriately chosen ζ_a (cf. Eq. (6)), as discussed in detail in Ref. [10].

For the second parametrization (called "crossed") the two discs are coplanar and intersect, see Fig. 2(right) for a typical example. Their relative orientations can vary between perpendicular and coinciding (for which $\mathbf{k}$ is forced to 1). Here we choose for the parameters in Eq. (38) h and $\mathcal{R}$ to be non-trivial (isospin) rotations around the y-axis with angles $(-1)^m \theta$, respectively, $(-1)^m \varphi$, and $\vec{a} = (0, 0, -\frac{1}{2}(-1)^m d \cos \alpha)$, whereas z_0 and ξ_0 are as in the rectangular case. The would-be constituent locations are now given (up to an overall shift and orientation) by

$$\vec{y}_m^{(j)} = \left((-1)^j (4\pi)^{-1} D \sin\varphi, 0, (-1)^{m+j}(4\pi)^{-1} D \cos\varphi - \frac{1}{2}(-1)^m d \cos\alpha\right), \quad (51)$$

where $\pm\varphi$ conveniently gives the orientation of each of the two discs with respect to the z-axis. The angle α originates from the definition of ζ_a, which through Eq. (6) determines the discontinuity of $\hat{A}_j(z)$. To ensure the proper matching, the following three equations need to be satisfied [10]

$$D \sin(\theta \pm \varphi)\left[f_3\left(\frac{1}{4}D\right) \pm f_1\left(\frac{1}{4}D\right)\right] = 8\pi d(1 \pm \sin\alpha),$$

$$D f_2\left(\frac{1}{4}D\right) + 8\pi d \sin\alpha = 0, \quad (52)$$

which determine φ, $\mathbf{k}$ and D for given α, θ and d. Fig. 2 illustrates that for these crossed configurations the two discs always overlap, unlike for the rectangular case. This formed an important motivation for the present study, so as to determine in how far the overlapping discs would affect the behaviour in the core.

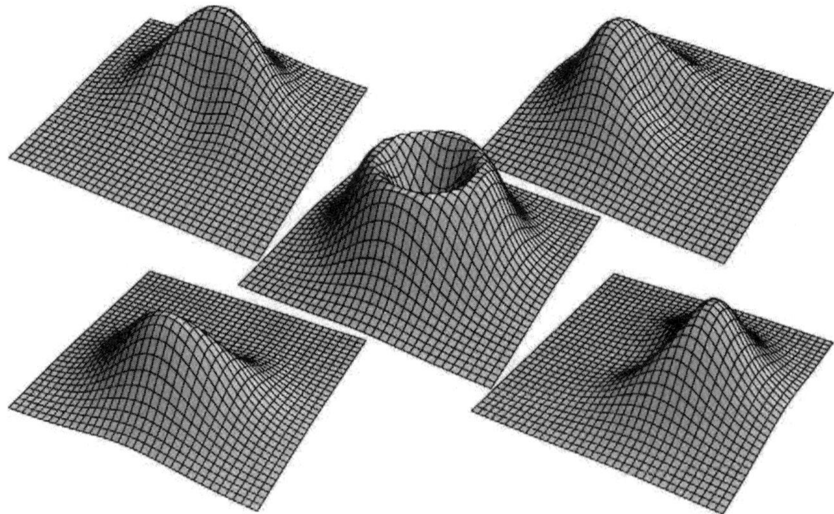

Fig. 4. In the middle is shown the action density in the plane of the constituents at $t = 0$ for an SU(2) charge 2 caloron with $\text{tr}\,\mathcal{P}_\infty = 0$ in the crossed configuration of Fig. 2, hence with $\mathbf{k} = 0.962$ and $D = 3.894$. On a scale enhanced by a factor $16\pi^2$ are shown the densities for the two zero-modes, using either periodic (left) or anti-periodic (right) boundary conditions in the time direction.

We will illustrate, using the exact formalism, how the action and zero-mode densities for these solutions behave. For the rectangular case we are mostly interested in the monopole limit, $d \to \infty$. For finite d, the density separates in two contributions, where each in the limit $d \to \infty$ is exactly the density for a charge 2 monopole with the same values of $\mathbf{k}$ and D. Note that in the limit $d \to \infty$ the matching conditions for $\hat{A}_j(z)$ "decouple" the intervals, turning into pole conditions at $z = \mu_{1,2}$, as is appropriate for multi-monopoles [20]. We do recover from this the known results for the charge 2 monopoles [22], like the doughnut structure for $\mathbf{k} = 0$, corresponding to two superimposed monopoles. The interest in the monopole limit comes from the fact that the zero-mode densities for multi-monopoles had not been studied in detail before [23]. In Fig. 3 we give the densities for $\mathbf{k} = 0.570$ and $D = 6.915$, which is intermediate between the doughnut and well-separated monopole configurations. On the other hand, for d small (compared to $\beta = 1$) the configuration will look like two non-dissociated calorons, and in particular is no longer static. When D remains much bigger than d, forcing $\mathbf{k} \to 1$, these behave as two well-separated charge 1 instantons. Otherwise, when D is comparable to d, one finds overlapping instantons [24].

An example for the crossed configuration with $\mathbf{k} = 0.997$ and $D = 8.753$ was already shown in Fig. 1. In this case D is large enough for the like-charge constituents to be separated, as is particularly clear from the zero-modes, which are essentially no longer overlapping. But two nearest neighbour (oppositely) charged constituents still show appreciable overlap. The distance between these nearest neighbours is $\frac{1}{4}D\sqrt{2}/\pi = 0.985$. As this is comparable to $\beta = 1$ we would expect the configuration to depend on time. Indeed, at the maxima of the action density its value of $1.18 \times 16\pi^2$ at $t = 0$ is reduced by almost 50% at $t = 0.5$. At the center of mass, where the action dencity is much lower, there is still a time dependence. However, far from all cores the field becomes static.[10] Increasing D further (which will push $\mathbf{k}$ closer to 1) the configuration quickly turns into well separated spherically symmetric static BPS monopoles.

More interesting is to consider the case with smaller D, like $D = 3.894$ and $\mathbf{k} = 0.962$, for which the disc singularities were illustrated in Fig. 2. The corresponding densities are shown in Fig. 4. We see that the constituents are now so close that they form a doughnut, but we stress this is different from the *static* monopole doughnut, which has $\mathbf{k} = 0$. Since the oppositely charged constituents now are as close as 0.438, which is considerably smaller than the time extent, the solution will have a strong time dependence. When D is decreased even further, it will turn into a charge 2 instanton localized in both space and time.

5. Higgs field asymptotics

In this section we make some comments on the far field limit of the gauge field. As we discussed before, the gauge field far removed from any core becomes Abelian (as well as static). The Abelian subgroup is the one that leaves the holonomy invariant, in the periodic gauge equivalent to leaving the constant asymptotic value of the adjoint Higgs field A_0 invariant. For definiteness, let us consider the case of SU(2) with $k = 2$, $\beta = 1$ and $\mathcal{P}_\infty = \exp(2\pi i \vec{\omega} \cdot \vec{\tau})$ (i.e., $\mu_2 = -\mu_1 = |\vec{\omega}|$). Up to exponential corrections we have

$$A_0^{\mathrm{ff}}(\vec{x}) = 2\pi i \vec{\omega} \cdot \vec{\tau} - \frac{1}{2} i \vec{\omega} \cdot \vec{\tau} \Phi(\vec{x}),\tag{53}$$

where we can express $\Phi(\vec{x})$ in terms of the far field limit of the Green's function at the impurities [14]

$$\Phi(\vec{x}) = \pi^{-1}\big[1 - \pi^{-1}\,\mathrm{Tr}\big(\hat{f}_x^{\mathrm{ff}}(\mu_2,\mu_2)\hat{S}_2\big)\big]^{-1}\partial_i\,\mathrm{Tr}\big(\hat{f}_x^{\mathrm{ff}}(\mu_2,\mu_2)\rho_2^i\big).\tag{54}$$

Using the twistor description of magnetic monopoles Hurtubise [13] was able to explicitly compute the asymptotic Higgs field for the SU(2) magnetic monopole long ago. The function he found for this algebraic tail amazingly agrees exactly with $\mathcal{V}_m(\vec{x})$, Eq. (23), which was introduced to describe the caloron zero-mode density (m denotes the interval and hence the type of constituents to which the corresponding zero-modes would localize [10]). As mentioned before, from our multi-caloron results one can recover the multi-monopole results by sending the constituent monopoles with the "unwanted" magnetic charge to infinity, cf. Eq. (24).

Although this tends to be cumbersome to show, $\Phi(\vec{x})$ in the far field can be written as $\Phi_1(\vec{x}) - \Phi_2(\vec{x})$, where $\Phi_m(\vec{x})$ is the contribution coming from the type m constituent monopoles and the difference in sign is due to the sign change in the magnetic charge. This is simply because the field is Abelian in the far field and linear superposition preserves the self-duality. Hence, $\Phi_m(\vec{x}) = 2\pi \mathcal{V}_m(\vec{x})$, such that for the SU(2) caloron

$$\Phi(\vec{x}) = 2\pi \mathcal{V}_1(\vec{x}) - 2\pi \mathcal{V}_2(\vec{x}).\tag{55}$$

---

[10] For any SU(n) and topological charge k the far field limit is static, *provided* all constituent monopoles have a non-vanishing mass.

We checked that this result indeed holds for the solutions discussed in Section 4, even when the two types of monopole structures are not well separated. This relation trivially holds for the axially symmetric solutions that were introduced in Ref. [14], where $\Phi(\vec{x})$ was explicitly shown to factorize in a sum of point charge contributions, compatible with what was found in Ref. [10] for the zero-mode densities. Therefore, in the far field limit (i.e., for the algebraic part) the singularity structure in the zero-mode density agrees *exactly* with the Abelian charge distribution, as given by $\partial_i^2 \Phi(\vec{x})$. Such a relation is at the heart of using chiral fermion zero-modes as a filter to isolate the underlying topological lumps from rough lattice Monte Carlo configurations [9].

6. Discussions

In this paper we have analyzed the higher charge caloron solutions and showed how to obtain exact results by suitably combining techniques developed in the context of the Nahm transformation and the ADHM formalism. The aim of these studies has been to establish that $SU(n)$ caloron solutions of charge k can be described in terms of kn monopole constituents, and that these can be viewed as independent constituents. A natural way to get an ensemble would be to consider approximate superpositions of k charge 1 calorons, but this would lead to an unwanted memory effect, with constituents remembering from which caloron they originated [14]. Our studies, within the context of self-dual configurations, have shown nevertheless that the constituents have an independent identity, with the only requirement that the net magnetic and electric charge of the configuration vanishes (each of the n types of constituents should occur with the same number). A recent lattice study [25], using the technique of over-improvement [26], fully confirms this picture.

It is therefore reasonable to consider the constituents as the independent building blocks for constructing an ensemble of monopole constituents, something that was not questioned in Ref. [7], but like for the instanton liquid [27] forms an essential assumption in a semi-classical study. Clearly the expectation is that semi-classical methods no longer work in the confined regime, at least for the part of the parameter space that corresponds to well-separated constituents, that is typically associated to instantons with a large scale parameter. It is not unlikely that the density of these constituents at low temperatures is so high that they form a coherent background and as such will no longer easily be recognized as lumps. With high quark densities leading to deconfinement, it may perhaps be that a high constituent monopole density will lead to confinement [28], although for now we have to leave this as a speculation.

Instantons that overlap get deformed and depending on the relative gauge orientation tend to "repel", i.e., inspecting the action density distribution they do not get closer than a certain distance [24]. When deconstructing instantons in monopole constituents, interestingly only like-charge constituents will show this effect, manifesting itself through the extended core structure. For unlike charges, from the point of view of the Abelian field, the configuration behaves as with linear superposition. If as a consequence of this all Abelian charge is annihilated, it disappears through forming a small instanton (localized in space and time), which in the limit of zero size describes the boundary of the moduli-space. The interaction between constituents of opposite duality is more complicated [8,29].

In conclusion, calorons with non-trivial holonomy have revealed a rich structure, incorporating traditional instanton physics, but allowing for gauge fields that inherit some essential features associated to a confining background not present in the traditional formulations. The fact that the underlying constituents are monopoles opens the way to describe the confining aspects of the theory in terms of these degrees of freedom. Much work remains to be done when it comes to understanding the dynamics, but we hope to have convinced the reader that a consistent picture is developing that holds considerable promise for the future.

Acknowledgements

We thank Conor Houghton for pointing us to Hurtubise's paper, and Chris Ford for discussions on the analytical aspects. We are grateful to Christof Gattringer, Michael Ilgenfritz, Boris Martemyanov and Michael Müller-Preussker for stimulating discussions and correspondence on lattice implementations. The research of F.B. is supported by FOM. He is grateful to Chris Ford and Conor Houghton for their hospitality while visiting Trinity college in Dublin, and thanks them, as well as Werner Nahm and Samson Shatashvili, for their interest.

Appendix A

In this appendix we derive the zero-mode limit for the action density, which assumes that the distance of $\vec{x}$ and the constituents of type m to all constituents of type $m' \neq m$ is large, but where $\vec{x}$ and the constituent locations of type m may otherwise be arbitrary.

As in Ref. [10] we take $z = \mu_m + 0$ for computing $\mathcal{F}_z$ and use that we can write $\det(\mathbb{1}_{2k} - \mathcal{F}_{\mu_m}) = \det(\mathbb{1}_{2k} - LK)$, where $K \equiv F_{m-1}\Theta_{m-1}\cdots\Theta_1\hat{g}^\dagger(1)F_n\Theta_n\cdots\Theta_{m+2}F_{m+1}$ and $L \equiv \Theta_{m+1}F_m\Theta_m$, with

$$
\Theta_m \equiv \begin{pmatrix} \mathbb{1}_k & \mathbb{1}_k \\ 2\pi R_m^+(\mu_m) & -2\pi R_m^-(\mu_m) \end{pmatrix}^{-1} T_m \begin{pmatrix} \mathbb{1}_k & \mathbb{1}_k \\ 2\pi R_{m-1}^+(\mu_m) & -2\pi R_{m-1}^-(\mu_m) \end{pmatrix},
$$
$$
F_m \equiv \begin{pmatrix} f_m^+(\mu_{m+1})f_m^+(\mu_m)^{-1} & 0 \\ 0 & f_m^-(\mu_{m+1})f_m^-(\mu_m)^{-1} \end{pmatrix} \tag{A.1}
$$

We note the K has *no* remaining dependence on the constituent locations of type m. Writing $LK \equiv \hat{L}\hat{K} + \tilde{L}\tilde{K}$, with

$$
\hat{K} \equiv \begin{pmatrix} K_{++} & K_{+-} \\ 0 & \mathbb{1}_k \end{pmatrix}, \qquad \tilde{K} \equiv \begin{pmatrix} 0 & 0 \\ K_{-+} & K_{--} \end{pmatrix},
$$
$$
\hat{L} \equiv \begin{pmatrix} L_{++} & 0 \\ L_{-+} & 0 \end{pmatrix}, \qquad \tilde{L} \equiv \begin{pmatrix} 0 & L_{+-} \\ 0 & L_{--} \end{pmatrix}, \tag{A.2}
$$

we find $\det(\mathbb{1}_{2k} - LK) = \det(\hat{K})\det(\hat{K}^{-1} - \hat{L} - \tilde{L}\tilde{K}\hat{K}^{-1})$.

We next use

$$
\hat{K}^{-1} = \begin{pmatrix} K_{++}^{-1} & -K_{++}^{-1}K_{+-} \\ 0 & \mathbb{1}_k \end{pmatrix}, \qquad \tilde{K}\hat{K}^{-1} = \begin{pmatrix} 0 & 0 \\ K_{-+}K_{++}^{-1} & (K^{-1})_{--} \end{pmatrix} \tag{A.3}
$$

and note that in the zero-mode limit K_{++}^{-1}, $K_{++}^{-1}K_{+-}$, $K_{-+}K_{++}^{-1}$ and $(K^{-1})_{--}$ are exponentially small (cf. Ref. [10, Appendix A]), such that

$$\det\big(ie^{-\pi ix_0}(\mathbb{1}_{2k} - LK)\big) = \det\big(e^{-2\pi ix_0}K_{++}\big)\det(L_{++}).\tag{A.4}$$

With the definition of L we now find

$$L_{++} = \frac{1}{4}R_{m+1}^{-1}(\mu_{m+1})\big(R_{m+1}^{-}(\mu_{m+1}) + S_{m+1}\big)\tilde{U}_m\big(R_{m-1}^{+}(\mu_m) + S_m\big),\tag{A.5}$$

where

$$\begin{aligned}
\tilde{U}_m &= \mathcal{Z}_{m+1}^{+}f_m^{+}(\mu_{m+1})f_m^{+}(\mu_m)^{-1}R_m^{-1}(\mu_m)\tilde{\mathcal{Z}}_m^{+}\\
&\quad - \mathcal{Z}_{m+1}^{-}f_m^{-}(\mu_{m+1})f_m^{-}(\mu_m)^{-1}R_m^{-1}(\mu_m)\tilde{\mathcal{Z}}_m^{-},\\
\mathcal{Z}_m^{\pm} &= \mathbb{1}_k \pm \big(R_m^{-}(\mu_m) + S_m\big)^{-1}R_{m-1}^{\pm}(\mu_m),\\
\tilde{\mathcal{Z}}_m^{\pm} &= \mathbb{1}_k \pm R_m^{\mp}(\mu_m)\big(R_{m-1}^{+}(\mu_m) + S_m\big)^{-1},
\end{aligned}\tag{A.6}$$

and $\tilde{U}_m$ contains *all* contributions due to the constituent locations of type m, up to *exponential* corrections in the distance of these, *and* of $\vec{x}$, to the other constituents. Hence $\log\det(ie^{-\pi ix_0}(\mathbb{1}_{2k} - LK))$ splits into the sum of two contributions, $\log\det(\tilde{U}_m)$ and $\log\det[\frac{1}{4}(R_{m-1}^{+}(\mu_m) + S_m)e^{2\pi ix_0}K_{++}R_{m+1}^{-1}(\mu_{m+1})(R_{m+1}^{-}(\mu_{m+1}) + S_{m+1})]$, where the last term only depends on the constituent locations of type $m' \neq m$ whose contribution will decay inversely proportional to the fourth power of their distance. Allowing for *algebraic* decay (or in the monopole limit, sending all constituents of type $m' \neq m$ to infinity) such that in addition $\mathcal{Z}_{m+1}^{\pm} = \tilde{\mathcal{Z}}_m^{\pm} = \mathbb{1}_k$, one thus finds Eq. (24).

A simple way to derive the result for the far field limit in Eq. (25) is by noting that in this case *all* $f_m^{-}(\mu_{m+1})f_m^{-}(\mu_m)^{-1}$ are exponentially small and F_m can be approximated by $\mathrm{diag}(F_m^{++}, 0)$, with $F_m^{++} = f_m^{+}(\mu_{m+1})f_m^{+}(\mu_m)^{-1}$. This therefore acts as a projection on the $++$ component and is thus seen to lead to $\det(ie^{-\pi ix_0}(\mathbb{1}_{2k} - LK)) = \det(e^{-2\pi ix_0}\hat{g}^{\dagger}(1)F_n^{++}\Theta_n^{++}\cdots F_1^{++}\Theta_1^{++})$. Using the fact that [10] $\Theta_m^{++} = \frac{1}{2}R_m^{-1}(\mu_m)\Sigma_m$ gives the required result.

References

[1] B.J. Harrington, H.K. Shepard, Phys. Rev. D 17 (1978) 2122;
    B.J. Harrington, H.K. Shepard, Phys. Rev. D 18 (1978) 2990.
[2] D.J. Gross, R.D. Pisarski, L.G. Yaffe, Rev. Mod. Phys. 53 (1981) 43.
[3] T.C. Kraan, P. van Baal, Phys. Lett. B 428 (1998) 268, hep-th/9802049;
    T.C. Kraan, P. van Baal, Nucl. Phys. B 533 (1998) 627, hep-th/9805168;
    T.C. Kraan, P. van Baal, Phys. Lett. B 435 (1998) 389, hep-th/9806034.
[4] K. Lee, P. Yi, Phys. Rev. D 56 (1997) 3711, hep-th/9702107;
    K. Lee, Phys. Lett. B 426 (1998) 323, hep-th/9802012;
    K. Lee, C. Lu, Phys. Rev. D 58 (1998) 025011, hep-th/9802108.
[5] N. Weiss, Phys. Rev. D 24 (1981) 475.
[6] N.M. Davies, T.J. Hollowood, V.V. Khoze, M.P. Mattis, Nucl. Phys. B 559 (1999) 123, hep-th/9905015.
[7] D. Diakonov, N. Gromov, V. Petrov, S. Slizovskiy, Quantum weights of dyons and of instantons with nontrivial holonomy, hep-th/0404042.

 F. Bruckmann et al. / Nuclear Physics B 698 (2004) 233–254

[8] E.-M. Ilgenfritz, M. Müller-Preussker, A.I. Veselov, in: V. Mitrjushkin, G. Schierholz (Eds.), Lattice Fermions and Structure of the Vacuum, Kluwer, Dordrecht, 2000, p. 345, hep-lat/0003025;
E.-M. Ilgenfritz, B.V. Martemyanov, M. Müller-Preussker, S. Shcheredin, A.I. Veselov, Phys. Rev. D 66 (2002) 074503, hep-lat/0206004.
[9] C. Gattringer, Phys. Rev. D 67 (2003) 034507, hep-lat/0210001;
C. Gattringer, S. Schaefer, Nucl. Phys. B 654 (2003) 30, hep-lat/0212029;
C. Gattringer, R. Pullirsch, hep-lat/0402008.
[10] F. Bruckmann, D. Nógrádi, P. van Baal, Nucl. Phys. B 666 (2003) 197, hep-th/0305063.
[11] M. García Pérez, A. González-Arroyo, A. Montero, P. van Baal, J. High Energ. Phys. 9906 (1999) 001, hep-lat/9903022;
P. van Baal, in: V. Mitrjushkin, G. Schierholz (Eds.), Lattice Fermions and Structure of the Vacuum, Kluwer, Dordrecht, 2000, p. 269, hep-th/9912035.
[12] E.M. Ilgenfritz, B.V. Martemyanov, M. Müller-Preussker, A.I. Veselov, Phys. Rev. D 69 (2004) 114505, hep-lat/0402010.
[13] J. Hurtubise, Commun. Math. Phys. 97 (1985) 381.
[14] F. Bruckmann, P. van Baal, Nucl. Phys. B 645 (2002) 105, hep-th/0209010.
[15] W. Nahm, Self-dual monopoles and calorons, in: G. Denardo, et al. (Eds.), Lecture Notes in Physics, vol. 201, 1984, p. 189.
[16] H. Panagopoulos, Phys. Rev. D 28 (1983) 380.
[17] M.F. Atiyah, N.J. Hitchin, V.G. Drinfeld, Yu.I. Manin, Phys. Lett. A 65 (1978) 185;
M.F. Atiyah, Geometry of Yang–Mills fields, in: Fermi Lectures, Scuola Normale Superiore, Pisa, 1979.
[18] M. García Pérez, A. González-Arroyo, C. Pena, P. van Baal, Phys. Rev. D 60 (1999) 031901, hep-th/9905016;
M.N. Chernodub, T.C. Kraan, P. van Baal, Nucl. Phys. B (Proc. Suppl.) 83–84 (2000) 556, hep-lat/9907001.
[19] W. Nahm, The algebraic geometry of multi-monopoles, in: M. Serdaroglu, et al. (Eds.), Group Theoretical Methods in Physics, Istanbul, 1982, p. 456.
[20] W. Nahm, Multi-monopoles in the ADHM construction, in: Z. Horvath, et al. (Eds.), Gauge Theories and Lepton Hadron Interactions, CRIP, Budapest, 1982.
[21] S.A. Brown, H. Panagopoulos, M.K. Prasad, Phys. Rev. D 26 (1982) 854;
A.S. Dancer, Commun. Math. Phys. 158 (1993) 545.
[22] P. Forgács, Z. Horváth, L. Palla, Nucl. Phys. B 192 (1981) 141;
M.F. Atiyah, N.J. Hitchin, The Geometry and Dynamics of Magnetic Monopoles, Princeton Univ. Press, Princeton, NJ, 1988.
[23] E.F. Corrigan, P. Goddard, Ann. Phys. (N.Y.) 154 (1984) 253.
[24] M. García Pérez, T.G. Kovács, P. van Baal, Phys. Lett. B 472 (2000) 295, hep-ph/9911485.
[25] F. Bruckmann, M. Ilgenfritz, B. Martemyanov, P. van Baal, hep-lat/0408004.
[26] M. García Pérez, A. González-Arroyo, J. Snippe, P. van Baal, Nucl. Phys. B 413 (1994) 535, hep-lat/9309009.
[27] T. Schäfer, E.V. Shuryak, Rev. Mod. Phys. 70 (1998) 323, hep-ph/9610451;
D. Diakonov, Prog. Part. Nucl. Phys. 51 (2003) 173, hep-ph/0212026.
[28] F. Bruckmann, D. Nógrádi, P. van Baal, Acta Phys. Pol. B 34 (2003) 5717, hep-th/0309008.
[29] E.M. Ilgenfritz, B.V. Martemyanov, M. Müller-Preussker, A.I. Veselov, Eur. Phys. J. C 34 (2004) 439, hep-lat/0310030;
E.M. Ilgenfritz, M. Müller-Preussker, B.V. Martemyanov, P. van Baal, Phys. Rev. D 69 (2004) 097901, hep-lat/0402020.

Vol. 34 (2003)   *ACTA PHYSICA POLONICA B*   No 12

INSTANTONS AND CONSTITUENT MONOPOLES*

FALK BRUCKMANN, DÁNIEL NÓGRÁDI AND PIERRE VAN BAAL

Instituut-Lorentz for Theoretical Physics, University of Leiden
P.O.Box 9506, 2300 RA Leiden, The Netherlands

(Received September 2, 2003)

Dedicated to the memory of Ian Kogan

We review how instanton solutions at finite temperature can be seen as boundstates of constituent monopoles, discuss some speculations concerning their physical relevance and the lattice evidence for their presence in a dynamical context.

PACS numbers: 11.10.Wx, 12.38.Lg, 14.80.Hv

1. Introduction

Over the last five years there has been a revived interest in studying instantons at finite temperature T, so-called calorons [1,2]. The main reason is that new explicit solutions could be obtained in the case where the Polyakov loop at spatial infinity is non-trivial, necessary to reveal more clearly the constituent nature of these calorons. This asymptotic value of the Polyakov loop is called the holonomy. In gauge theories trivial holonomy, for which the asymptotic value of the Polyakov loop takes values in the center of the gauge group, is typical for the deconfined phase. Therefore, caloron solutions with non-trivial holonomy are more expected to play a role in the confined phase, still at finite temperature, but where the average of the trace of the Polyakov loop is small. In the introduction we start with a pedagogical overview discussing monopoles, instantons, and their physical significance. The next section discusses the construction of the caloron solutions. Secs. 2.1 and 2.2 are more technical, but tutorial in nature. One could skip in particular Secs. 2.3 and 2.4, describing in Sec. 3 the properties of the solutions in less technical terms. Lattice evidence for the dynamical significance of constituent monopoles is discussed in Sec. 4.

---

* Presented by the last author at the XLIII Cracow School of Theoretical Physics, Zakopane, Poland, May 30–June 8, 2003.

        F. Bruckmann, D. Nógrádi, P. van Baal

1.1. Monopoles

That monopoles should play a role in describing the constituent nature of calorons is in itself not really a surprise, because at finite temperature A_0 plays in some sense the role of a Higgs field in the adjoint representation. However, a gauge transformation,

$$^{g}A_\mu(x) = g(x)A_\mu(x)g^{-1}(x) + g(x)\partial_\mu g^{-1}(x)\,, \tag{1}$$

shows that A_0 does not transform correctly (unless the gauge transformation is time independent), due to the inhomogeneous term. Instead, the Polyakov loop

$$P(t,\vec{x}) = \mathrm{Pexp}\left(\int_0^\beta A_0(t+s,\vec{x})ds\right)\,, \tag{2}$$

transforms as it should, $^{g}P(x) = g(x)P(x)g^{-1}(x)$. Here $\beta = 1/kT$ is the period in the imaginary time direction, under which the gauge field is assumed to be periodic. We also will consider other gauges, where $A_\mu(x)$ is periodic up to a gauge transformation, in which case the expression for P has to be modified accordingly [3]. For example, in the so-called algebraic gauge, $A_0(x)$ is transformed to zero at spatial infinity. In this case the gauge fields satisfy the boundary condition ($\mathcal{P}_\infty$ to be defined below)

$$A_\mu(t+\beta,\vec{x}) = \mathcal{P}_\infty A_\mu(t,\vec{x})\mathcal{P}_\infty^{-1}\,. \tag{3}$$

We will require that the total Euclidean action of these calorons is finite, such that the field strength[1]

$$F_{\mu\nu}(x) = \partial_\mu A_\nu(x) - \partial_\nu A_\mu(x) + [A_\mu(x), A_\nu(x)] \tag{4}$$

has to go to zero at spatial infinity. It is this that forces the Polyakov loop to become constant at spatial infinity. For $\mathrm{SU}(n)$ gauge theory this gives

$$\mathcal{P}_\infty = \exp(\beta A_0^\infty) \equiv \lim_{|\vec{x}|\to\infty} P(x), \quad A_0^\infty \equiv \frac{2\pi i}{\beta} U_0 \mathrm{diag}\,(\mu_1,\mu_2,\ldots,\mu_n)\,U_0^{-1}, \tag{5}$$

independent of the direction and time. Unlike for a Higgs field, however, $\mathcal{P}_\infty$ is unitary with determinant 1. Choosing U_0 (the constant gauge function that brings $\mathcal{P}_\infty$ to its diagonal form) appropriately, the n eigenvalues,

---

[1] Our conventions are $A_\mu(x) = eA_\mu^a(x)T_a$, where e is the coupling constant and $T_a^\dagger = -T_a$, $\mathrm{Tr}(T_a T_b) = -\frac{1}{2}\delta_{ab}$, $[T_a, T_b] = f_{abc}T_c$. For $\mathrm{SU}(2)$ $f_{abc} = \varepsilon_{abc}$ and $T_a = -i\frac{1}{2}\tau_a$ in terms of the familiar Pauli matrices.

$\exp(2\pi i \mu_m)$, are ordered. The freedom in adding an arbitrary integer to μ_m is fixed by ordering the μ_m themselves, and requiring their sum to vanish,

$$\sum_{i=1}^{n} \mu_i = 0, \quad \mu_1 \leq \mu_2 \leq \ldots \leq \mu_n \leq \mu_{n+1} \equiv 1 + \mu_1 . \tag{6}$$

For trivial holonomy $\mathcal{P}_\infty$ is an element of the center of the gauge group, $\mathcal{P}_\infty = \exp(2\pi i q/n)\mathbb{1}_n$ with q an integer between 0 and $n-1$, hence $\mu_m = q/n$.

One might now immediately object that a Higgs field that goes to a constant at infinity does not have the usual hedgehog form expected for a non-Abelian 't Hooft–Polyakov monopole [4], but this is because the caloron solutions are actually such that the total magnetic charge vanishes. The force stability of these solutions is based, as for exact multi-monopole solutions in the Bogomol'ny–Prasad–Sommerfeld (BPS) limit [5] on balancing the electromagnetic with the scalar (Higgs) force [6, 7]. For the caloron the difference is in interchanging repulsive and attractive forces. For a single caloron with topological charge one, this is because there are $n-1$ monopoles with a unit magnetic charge in the i-th U(1) subgroup, all compensated by the n-th monopole of so-called type $(1, 1, \ldots, 1)$, having a magnetic charge in each of these subgroups. This special monopole is also called a Kaluza–Klein (KK) monopole [8], as will be explained below. The well-known Harrington-Shepard solution [1] has trivial holonomy, and although all eigenvalues of $\mathcal{P}_\infty$ are equal, such that there is no spontaneous symmetry breaking, one still finds a genuine BPS monopole [9] in a suitable limit (it will be the KK monopole that survives).

In a Higgs theory, switching off the Higgs potential and splitting off a square in the energy density,

$$- \mathrm{Tr}\left((D_i \Phi)^2 + B_i^2\right) = - \mathrm{Tr}(D_i \Phi \mp B_i)^2 \mp 2\, \mathrm{Tr}\left(B_i D_i \Phi\right) , \tag{7}$$

exact monopole solutions are constructed using the BPS condition [5], which imposes the covariant derivative of the Higgs field Φ to be equal (up to a sign) to the magnetic field, $D_i \Phi \equiv \partial_i \Phi + [A_i, \Phi] = \pm B_i$, where $B_i \equiv \frac{1}{2}\varepsilon_{ijk}F_{jk}$. One is then left with a total derivative

$$\mathrm{Tr}\left(B_i D_i \Phi\right) = \mathrm{Tr}\left(D_i(B_i \Phi)\right) = \partial_i \,\mathrm{Tr}(B_i \Phi), \tag{8}$$

whose integral is proportional to the magnetic charge. The integral can also be associated to the mapping degree of the map $\hat{x} \rightarrow \Phi(r\hat{x})$ ($\hat{x} \equiv \vec{x}/|\vec{x}|$) for $r \rightarrow \infty$. With Φ taking values in the algebra, and of fixed length at infinity, this gives for SU(2) a map from S^2 to S^2. For a caloron the BPS condition is simply a consequence of the self-duality conditions characteristic of instanton solutions, $E_i = D_i A_0 - \partial_0 A_i = \pm B_i$, with $\Phi = A_0$. One might thus be

  F. Bruckmann, D. Nógrádi, P. van Baal

tempted to call the constituents dyons, rather than monopoles. In the Higgs model the Julia–Zee dyons are constructed by taking A_0 proportional to the Higgs field Φ [10]. By a time dependent gauge transformation A_0 can then be gauged to zero. The resulting electric field is now given by $E_i = -\partial_0 A_i$ and is *not* quantized (and in particular not equal to $\pm B_i$). In pure gauge theory it makes, however, no sense to separate $D_i \Phi = D_i A_0$ from $\partial_0 A_i$. Gauge invariance requires that they occur in the combination $F_{i0} = D_i A_0 - \partial_0 A_i$. The electric field is necessarily fixed and quantized as soon as we interpret A_0 as the Higgs field. As discussed in Ref. [11] adding a fifth dimension for Minkowski time, compactifying the Euclidean time direction to zero size ($\beta \to 0$) allows one to have electric charge as for the Julia–Zee dyon. A compactified Euclidean direction is what one also considers in Kaluza–Klein theories. It is in this sense that the type $(1, 1, \ldots, 1)$ monopole is called a KK monopole, because it turns out it is static up to a gauge transformation that makes one full rotation in the unbroken subgroup when going from 0 to β, thus of lowest non-trivial Kaluza–Klein momentum[2].

1.2. Instantons

We therefore consider it most appropriate to call the constituents monopoles, as is also clear from Nahm's formalism which provides one of the essential tools to find these self-dual solutions [13]. The Nahm transformation, as well as the Atiyah–Drinfeld–Hitchin–Manin (ADHM) construction [14] for instanton solutions on $\mathbb{R}^4$, form indispensable tools to find the exact caloron solutions. It should be added that from the topological point of view Taubes showed how to make out of two oppositely charged monopoles a Euclidean four dimensional gauge field that has non-zero topological charge [15]. This result is more general, since only when minimizing the action in a sector with non-trivial topological charge will one find a self-dual instanton solution. His construction is based on creating a monopole anti-monopole pair, bringing them far apart, rotating one of them over a full rotation (the so-called Taubes winding) and finally bringing them together to annihilate, see Fig. 1. The four dimensional configuration constructed this way is topologically non-trivial.

In $\mathbb{R}^4$ the topological charge is related to the winding of a gauge function $g(x)$ defined at infinity, which therefore is a mapping from S^3 to the gauge group. The way $g(x)$ enters is through requiring the Euclidean action,

$$S = -\frac{1}{2e^2} \int d^4x \ \mathrm{Tr}\left(F_{\mu\nu}^2(x)\right) , \tag{9}$$

---

[2] Nevertheless, there is a context in pure gauge theories where dyons appear, but for this we have to add a term proportional to $\theta F\tilde{F}$ to the Lagrangian [12]. The electric charge is now given by $\theta/(2\pi)$ times the magnetic charge.

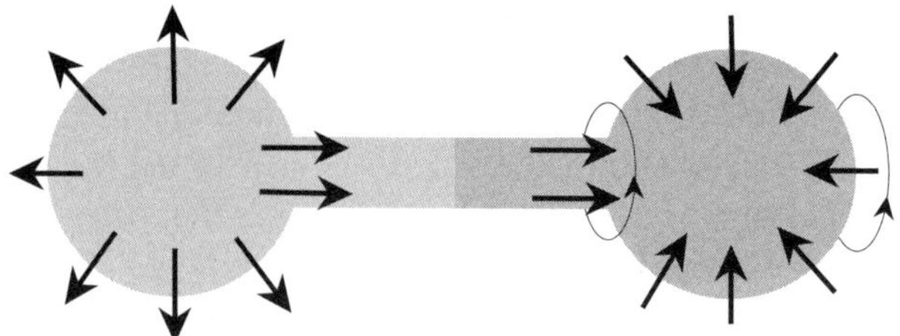

Fig. 1. The topologically non-trivial field configuration is constructed from two oppositely charged monopoles, rotating one of them over a full 2π rotation.

to remain finite. As before this implies the field strength at infinity to go to zero, where $A_\mu(x)$ can be written as a pure gauge, $A_\mu(x) = g(x)\partial_\mu g^{-1}(x)$. For SU(2) a simple parametrization as $g(x) = a_\mu(x)\sigma_\mu$ in terms of a unit vector a_μ and unit quaternions $\sigma_\mu = (\mathbb{1}_2, i\vec{\tau})$ $(\bar{\sigma}_\mu \equiv \sigma_\mu^\dagger)$ makes this winding most transparent as the mapping degree of $a(x)$ at infinity. In the simplest case of degree 1, $a(x) = x/|x|$ is a 4-dimensional hedgehog. The relation to Taubes' construction is using the fact that S^3 can be viewed as a twisted product of S^1 (the Taubes winding) and S^2 (the hedgehog formed by the Higgs field), the so-called Hopf fibration [11, 16].

In the case of periodicity in the imaginary time direction as it occurs for calorons, let us transform $A_0(x) \to 0$ everywhere. This can be done by a time dependent gauge transformation $U(x)$, which in general will *not* be periodic. Actually $P(t; \vec{x})$ itself is the gauge transformation that relates the gauge field at $t + \beta$ to that at t in the $A_0 = 0$ gauge. Since $P(x)$ goes to a constant $(\mathcal{P}_\infty)$ at spatial infinity, this provides a non-trivial mapping from S^3 (as $\mathbb{R}^3$ compactified at infinity) to the gauge group. The topological charge is precisely its winding number.

As in the case of the monopole energy density, we can rewrite the action density in terms of a square and a boundary term,

$$\mathrm{Tr}(F_{\mu\nu}^2) = \tfrac{1}{2}\,\mathrm{Tr}(F_{\mu\nu} \pm \tilde{F}_{\mu\nu})^2 \mp \mathrm{Tr}(F_{\mu\nu}\tilde{F}_{\mu\nu}), \quad \mathrm{Tr}(F_{\mu\nu}\tilde{F}_{\mu\nu}) = \partial_\mu K_\mu,$$
$$K_\mu = 2\varepsilon_{\mu\nu\alpha\beta}\,\mathrm{Tr}(A_\nu\partial_\alpha A_\beta + \tfrac{2}{3}A_\nu A_\alpha A_\beta)\,, \tag{10}$$

with $\tilde{F}_{\mu\nu} = \tfrac{1}{2}\varepsilon_{\mu\nu\alpha\beta}F_{\alpha\beta}$ the dual field strength interchanging electric and magnetic components. Hence, for self-dual solutions at finite temperature

$$S = \frac{1}{2e^2}\int (K_0(0, \vec{x}) - K_0(\beta, \vec{x}))\,d^3x = \frac{1}{3e^2}\int \mathrm{Tr}(P(\vec{x})dP^{-1}(\vec{x}))^3\,, \tag{11}$$

using that in the $A_0 = 0$ gauge $A(\beta, \vec{x}) = {}^P A(0, \vec{x})$, expressing the result in a compact differential form notation. Note that the integral is invariant under

         F. Bruckmann, D. Nógrádi, P. van Baal

small deformations of P, using $\delta(PdP^{-1})^3 = 3d\,\mathrm{Tr}(P\delta P^{-1}(PdP^{-1})^2)$, and therefore should be proportional to the winding number of P. A map of degree k can be obtained by multiplying k maps of degree 1 (the inverse gives a negative mapping degree). Each of these maps of degree 1 can be deformed at will, as the integral is invariant under continuous deformations, and in particular can be arranged to be the identity except for a small region. Choosing these regions to have no overlap, the integral is easily seen to be proportional to k. To fix for SU(2) the constant of proportionality we may take $U(\vec{x}) = ((1 - |\vec{x}|^2)\sigma_0 + 2\vec{x} \cdot \vec{\sigma})/(1 + |\vec{x}|^2)$, for the map of degree 1 related to stereographic projection from S^3 to $\mathbb{R}^3$. For SU(n) we first deform the map to lie in an SU(2) subgroup. This then gives the celebrated result that self-dual solutions with given topological charge, $k = (16\pi^2)^{-1} \int d^4x\, \mathrm{Tr}\left(F_{\mu\nu}(x)\tilde{F}_{\mu\nu}(x)\right)$, have an action equal $8\pi^2|k|/e^2$.

This is also a convenient setting to understand why in the limit of zero temperature, $\beta \to \infty$, an instanton corresponds to vacuum to vacuum tunneling. Finite action requires the field strength to go to zero at $|t| \to \infty$, but at the same time, the field at $t \to \infty$ is related to the field at $t \to -\infty$ by a topologically non-trivial gauge transformation. Strictly speaking, we need to identify gauge field configuration that are gauge equivalent. Instead of having multiple vacua we can alternatively say there is one vacuum, with the field space being non-contractible. This is quite analogous to considering a periodic quantum mechanics problem, which could be reinterpreted as quantum mechanics on a circle. Tunneling now corresponds to penetration of the wave function in the classically forbidden region, to reach back to the vacuum going around the circle in either direction. Essential is that the support of the wave function, the region where it is non-vanishing, becomes sensitive to the non-trivial topology of the configuration space [3].

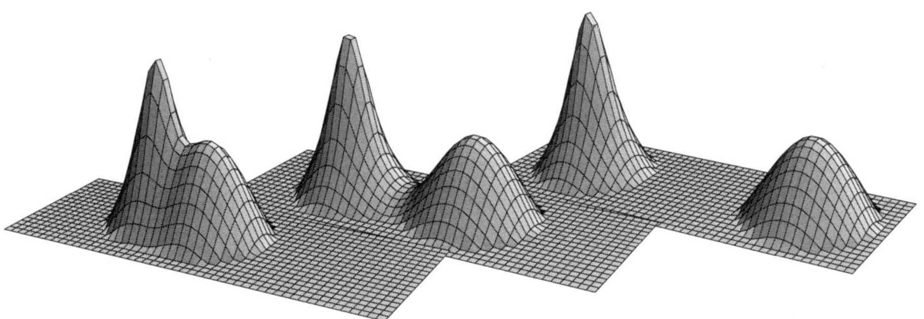

Fig. 2. Shown is a typical example for the action density (on equal logarithmic scales at the time where it is maximal) of an SU(2) caloron with non-trivial holonomy $\mu_2 = -\mu_1 = 0.125$ (for, from left to right, $\rho/\beta = 0.8, 1.2, 1.6$).

The tunneling path in non-Abelian gauge theories is of course described by the one parameter (t) family of gauge fields $A_i(t, \vec{x})$ (in the $A_0 = 0$ gauge), and a finite action means we have to cross a potential barrier when going around the non-contractible loop (or going from vacuum to vacuum). From Taubes' argument it is already clear that this intermediate configuration can be associated to a monopole-antimonopole boundstate, which is made more precise by the caloron solutions to be described. At zero-temperature this is less obvious, since there monopoles form close pairs, *cf.* Fig. 2 (left). They behave very much like virtual particles, that can only be created for a short period of time, thereby explaining the instantaneous character. At finite temperature, the monopoles can be separated much more easily due to the interactions with the thermal bath. At high temperature, however, they will be suppressed due to the Boltzmann weight, and so monopoles (like calorons) are expected to be dilute. At low temperature instantons form a more dense ensemble, possibly leading to monopoles to be dense as well, particularly when instantons overlap. One may then have some hope that the confining electric phase could be characterized by a deconfining magnetic phase, where the dual deconfinement is due to the large monopole density, in a similar spirit to high density induced quark deconfinement. It would offer a possible alternative to existing scenarios.

Instantons describe virtual processes and are quite often discussed in the context of the semiclassical approximation. For a one-dimensional particle with mass m in a positive potential V we may again split off a square,

$$L(t) = \tfrac{1}{2} m \dot{x}^2(t) + V(x(t)) = \tfrac{1}{2} m \left(\dot{x}(t) - \sqrt{2V(x(t))/m} \right)^2 + \dot{x}(t)\sqrt{2mV(x(t))},$$

$$S_{\text{cl}} = \int dt\, \dot{x}(t)\sqrt{2mV(x(t))} = \int dx\,\sqrt{2mV(x)}, \tag{12}$$

where we typically integrate between the classical turning points, related to the WKB expression for the wave function in the classically forbidden region, $\exp(-\int_{x_0}^{x} dy \sqrt{2mV(y)}/\hbar)$. In a double well this leads to a tunnel splitting proportional to $\exp(-S_{\text{cl}}/\hbar)$. One problem in the theory of strong interactions is that the effective coupling tends to become too big for large instantons. Instantons can have an arbitrary size ρ, due to the classical scale invariance of the theory. This classical scale invariance gets broken by the regularization and one is left with a scale dependent running coupling constant after renormalization. This causes a problem when integrating over the scale parameter in the one-instanton tunneling amplitude given by the celebrated result of 't Hooft [17], $dW \propto d\rho dx^4 \rho^{-5} \exp\left(-8\pi^2/e^2(\rho)\right)$. Actually, for calorons with non-trivial holonomy and $\rho > \beta$, it is more natural to associate ρ with the distance between constituent monopoles ($\pi\rho^2/\beta$ for SU(2)), and this may help alleviate the problem one encounters when deal-

                F. BRUCKMANN, D. NÓGRÁDI, P. VAN BAAL

ing with large scale instantons. This is also one way to understand why there will still be calorons with ρ arbitrary, despite the fact that β fixes the scale[3]. At zero temperature, a large ρ leads to a low energy barrier along the tunneling path, and at some point this will no longer describe a virtual process and the semiclassical approximation will break down. Nevertheless, the instanton liquid model has been very successful in describing much of the low energy phenomenology, in particular for chiral dynamics and aspects related to breaking the axial U(1) symmetry. We refer to the reviews by Schäfer and Shuryak [20] and by Diakonov [21] for more details.

1.3. Fermion zero-modes

In the instanton liquid model the interaction of instantons with fermions plays an important role. This is because instantons have a remarkable influence on the spectrum of Dirac fermions, it namely implies the presence of zero eigenvalue solutions with fixed chirality (the so-called chiral zero-modes). These chiral zero-modes will play an important role in the Nahm transformation and in studying the properties of the calorons with non-trivial holonomy, so we wish to mention some of the interesting and far reaching physical consequences. This is most easily discussed in terms of the spectral flow of the Dirac Hamiltonian along the tunneling path. Its spectrum is of course gauge invariant, and this implies that the energy levels at $t = 0$ and $t = \beta$ are identical. The gauge field provides a smooth interpolation between these, which leads to each of the energies to be a continuous function of t, the so-called spectral flow, see Fig 3. If we draw the energy level at zero, the number of crossings is clearly a conserved quantity, it cannot change under continuous deformations of the gauge field background (the background need not be a solution of the equations of motion). It is also clear, by putting instantons in a row, that the number of crossings is proportional to the topological charge, and not surprisingly it is actually equal to it. The necessity of such a crossing is related to the existence of a chiral zero-mode, whose number for fermions in the fundamental representation is equal to the topological charge, as follows from the Atiyah–Singer index theorem [22]. The argument is most simple in the case of zero-temperature, where the zero-mode at $t \to \pm\infty$ has to behave as $\exp(-tE_\pm)$, where $E_\pm$ is one of (free) fermion energies. Clearly this would only give a normalizable zero-mode if $E_+ > 0$ and $E_- < 0$, which forces a crossing! At this point the Dirac vacuum is degenerate and particles or anti-particles are created with definite chirality. Compatible with the breaking of the axial U(1) symmetry, the divergences of its current is proportional to the topological density

---

[3] Remarkably, it can be shown [18, 19] under some mild conditions, that any four dimensional manifold will have instanton solutions with arbitrary ρ.

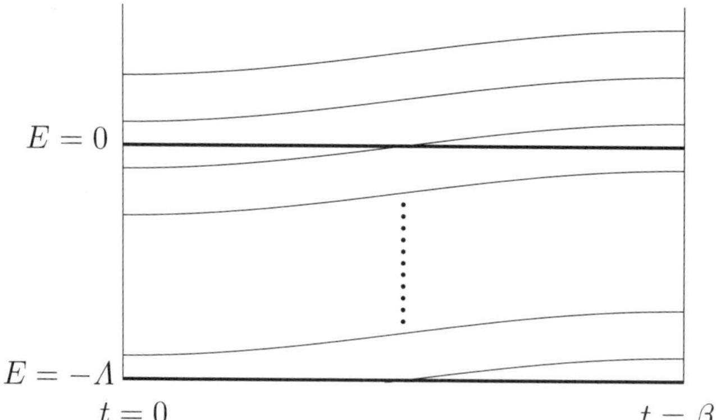

Fig. 3. Schematic representation of the spectral flow.

$\tilde{F}_{\mu\nu}F_{\mu\nu}$. It gives rise to the so-called 't Hooft interaction [17], which together with a finite density of fermion modes at zero eigenvalue (required for the Banks–Casher mechanism [23] to work) play an important role in the spontaneous breaking of chiral symmetry (or soft breaking in case of small up and down quark masses) [20]. Finally, the spectral flow also makes it easy to understand the origin of the axial anomaly, which occurs due to the need to regularize the theory. If cutting off modes with $E < -\Lambda$, where Λ is the ultraviolet cutoff, we see that the spectral flow leads to the fact that modes we had removed in the trivial vacuum reappear due to the spectral flow at the vacuum equivalent to it by a topologically non-trivial gauge transformation. That the violation in conserving the axial U(1) charge is proportional to the topological charge is in this setting simply a consequence of the number of crossings in the spectral flow. This is the celebrated infrared-ultraviolet connection, and also makes it understandable why the anomaly is robust (fully determined by the lowest order result in perturbation theory).

2. Construction of solutions

Let us start with the well know SU(2) Harrington–Shepard solution [1] for the caloron with trivial holonomy. In that case one simply takes a periodic array of instantons parallel in group space ($\mathcal{P}_\infty = \mathbb{1}_2$), placed at $a_{(p)} \equiv (a_0 + p\beta, \vec{a})$ for integer p. The solution is found in terms of the 't Hooft ansatz [24], which in its general form is given by

$$A_\mu(x) = \tfrac{1}{2}\bar{\eta}_{\mu\nu}\partial_\nu \log \phi(x). \tag{13}$$

Here $\eta_{\mu\nu} = \tfrac{1}{2}(\sigma_\mu\bar{\sigma}_\nu - \sigma_\nu\bar{\sigma}_\mu) = i\eta^a_{\mu\nu}\tau_a$ and $\bar{\eta}_{\mu\nu} = \tfrac{1}{2}(\bar{\sigma}_\mu\sigma_\nu - \bar{\sigma}_\nu\sigma_\mu) = i\bar{\eta}^a_{\mu\nu}\tau_a$, are the self-dual and anti-selfdual 't Hooft tensors. Substituting this in the

              F. Bruckmann, D. Nógrádi, P. van Baal

self-duality equation, $F_{\mu\nu} = \tilde{F}_{\mu\nu}$, one finds this is a solution if and only if $\partial_\mu^2 \phi(x) = 0$. Therefore $\phi = 1 + \sum_{i=1}^{k} \rho_i^2 |x - b_{(i)}|^2$, where $b_{(i)}$ are the four dimensional locations and ρ_i the sizes of the k instantons[4]. A singularity at $x = b_{(i)}$ can be removed by a gauge transformation. For the Harrington–Shepard solution, taking $b_{(p)} = a_{(p)}$ and $\rho_p = \rho$ one can perform the sum over $p \in \mathbb{Z}$ to find [1],

$$\phi_{\mathrm{HS}}(x) = 1 + \frac{\pi \rho^2 \sinh(2\pi r/\beta)/(\beta r)}{\cosh(2\pi r/\beta) - \cos(2\pi(t - a_0)/\beta)}, \quad r = |\vec{x} - \vec{a}|. \quad (14)$$

The case of non-trivial holonomy cannot be treated in the same way, because we will have to sum over a periodic array of instantons that has a color rotation by $\mathcal{P}_\infty$ when shifting time over β. This can only be treated within the full ADHM ansatz. The rest of this section is more technical and could be skipped, although the short tutorials on the ADHM construction and the Nahm transformation in Secs. 2.1 and 2.2 are still recommended.

2.1. ADHM formalism

The $\mathrm{SU}(n)$ ADHM construction for charge k instantons [14] starts with a k dimensional vector $\lambda = (\lambda_1, \ldots, \lambda_k)$, where $\lambda_i^\dagger$ is a two-component spinor in the $\bar{n}$ representation of $\mathrm{SU}(n)$ (i.e. λ is a $n \times 2k$ complex matrix) and a $2k \times 2k$ complex matrix $B = \sigma_\mu \otimes B_\mu$ (each B_μ is a hermitian $k \times k$ matrix). These are combined to form a $(n + 2k) \times 2k$ dimensional matrix $\Delta(x)$, which has n normalized eigenvectors with vanishing eigenvalue, combined in a complex matrix $v(x)$ of size $(n + 2k) \times n$,

$$\Delta(x) = \begin{pmatrix} \lambda \\ B(x) \end{pmatrix}, \quad B(x) = B - x\mathbb{1}_k, \quad \Delta^\dagger(x) v(x) = 0. \quad (15)$$

Here the quaternion $x = x_\mu \sigma_\mu$ (a 2×2 matrix with spinor indices) denotes the position and $v(x)$ can be solved explicitly in terms of the ADHM data,

$$v(x) = \begin{pmatrix} -\mathbb{1}_n \\ u(x) \end{pmatrix} \phi^{-1/2}, \quad u(x) = (B^\dagger - x^\dagger)^{-1} \lambda^\dagger, \quad \phi(x) = \mathbb{1}_n + u^\dagger(x) u(x). \quad (16)$$

The square root $\phi^{1/2}(x)$ is well-defined because $\phi(x)$ is a positive $n \times n$ hermitian matrix. The gauge field is now given by

$$A(x) \equiv A_\mu(x) dx_\mu = v^\dagger(x) dv(x) \quad (17)$$

---

[4] Using conformal transformations a generalization of this ansatz including non-trivial color orientations exists [25], but it will only give all possible solutions for charge 2.

and to check its self-duality it is best to use form notation. For the field strength two form $F(x) = \frac{1}{2}F_{\mu\nu}(x)dx_\mu \wedge dx_\nu = dA(x) + A(x) \wedge A(x)$ we find

$$F = d(v^\dagger dv) + v^\dagger dv \wedge v^\dagger dv = dv^\dagger \wedge (1 - v \otimes v^\dagger)dv\,. \tag{18}$$

We note that $v \otimes v^\dagger$ projects to the kernel of $\Delta^\dagger$. Assuming that Δ has no zero eigenvalues such that $\Delta^\dagger \Delta$ is invertible, we find that $1 - v \otimes v^\dagger = \Delta(\Delta^\dagger\Delta)^{-1}\Delta^\dagger$. Indeed, when acting on elements in the kernel of $\Delta^\dagger$, left- and right-hand side are equal. Any vector in the orthogonal complement of this kernel can be written as Δw, since $< v, \Delta w > = < \Delta^\dagger v, w > = 0$, such that left- and right-hand side are again equal. Hence

$$F = dv^\dagger \Delta \wedge (\Delta^\dagger\Delta)^{-1}\Delta^\dagger dv = dv^\dagger b dx \wedge (\Delta^\dagger\Delta)^{-1}dx^\dagger b^\dagger dv\,, \tag{19}$$

where we use the fact that $\Delta^\dagger dv = (d\Delta^\dagger)v = dx^\dagger b^\dagger v$, with $b^\dagger = (0 \; \mathbb{1}_2 \otimes \mathbb{1}_k)$ as a $2k \times (n + 2k)$ matrix[5]. It might seem this does not help that much, but remarkably, using $dx = dx_\mu \sigma_\mu$ we find $dx \wedge dx^\dagger = \eta_{\mu\nu}dx_\mu \wedge dx_\nu$, and since $\eta_{\mu\nu}$ is self-dual, we are done. Not quite so yet! We had to take $dx \otimes \mathbb{1}_k$ through $(\Delta^\dagger\Delta)^{-1}$ and this is only possible if

$$(\Delta^\dagger(x)\Delta(x))^{-1} = \mathbb{1}_2 \otimes f_x\,, \tag{20}$$

where f_x is a hermitian $k \times k$ matrix. This condition, stating that $\Delta^\dagger(x)\Delta(x)$ is invertible and commutes with the quaternions, is what is known as the quadratic ADHM constraint. It has reduced solving a set of non-linear partial differential equations to solving a quadratic matrix equation.

To construct a charge k caloron with non-trivial holonomy [26], we place k instantons in the time interval $[0, \beta[$, performing a color rotation with $\mathcal{P}_\infty$ for each shift of t over β, *cf.* Eq. (3). This is implemented by requiring (suppressing color and spinor indices and scaling β to 1 in this section)

$$\lambda_{pk+k+a} = \mathcal{P}_\infty\lambda_{pk+a}\,, \quad B_{pk+a,qk+b} = B_{pk-k+a,qk-k+b} + \sigma_0\delta_{pq}\delta_{ab}\,. \tag{21}$$

Solutions to these equations are parametrized by ζ_a and $\hat{A}_p^{ab}$,

$$\lambda_{pk+a} = \mathcal{P}_\infty^p\zeta_a\,, \quad B_{pk+a,qk+b} = p\sigma_0\delta_{pq}\delta_{ab} + \hat{A}_{p-q}^{ab}\,, \tag{22}$$

with $\hat{A}$ to be determined by the quadratic ADHM constraint.

---

[5] The notation $dx^\dagger b^\dagger$ may be a bit misleading, it should be read as $(dx^\dagger \otimes \mathbb{1}_k)b^\dagger$.

    F. Bruckmann, D. Nógrádi, P. van Baal

2.2. Fourier-Nahm transformation

We introduce the n projectors P_m on the mth eigenvalue of $\mathcal{P}_\infty$, such that $\mathcal{P}_\infty = \sum_m e^{2\pi i \mu_m} P_m$ and $\lambda_{pk+a} = \sum_m e^{2\pi i p \mu_m} P_m \zeta_a$. Fourier transformation now leads to

$$2\pi i \sum_p e^{2\pi i p z} \hat{A}_p^{ab} = \hat{A}^{ab}(z), \quad \sum_p e^{-2\pi i p z} \lambda_{pk+a} = \sum_m \delta(z - \mu_m) P_m \zeta_a. \quad (23)$$

Here $\zeta_a^\dagger$ is again a two-component spinor in the $\bar{n}$ representation of SU(n) and $\hat{A}^{ab}(z) = \sigma^\mu \hat{A}_\mu^{ab}(z)$, with $\hat{A}_\mu(z)$ an anti-hermitian $k \times k$ matrix. In terms of the latter

$$\sum_{p,q} B_{pk+a,qk+b}(x) e^{2\pi i(pz-qz')} = \frac{\delta(z-z')}{2\pi i} \hat{D}_x^{ab}(z'), \quad (24)$$

$$\hat{D}_x^{ab}(z) \equiv \hat{D}^{ab}(z) - 2\pi i x \delta^{ab} = \sigma_0 \delta^{ab} \frac{d}{dz} + \hat{A}^{ab}(z) - 2\pi i x \delta^{ab},$$

which is the Weyl operator (positive chirality Dirac operator) for the U(k) gauge field $\hat{A}_\mu(z) - 2\pi i x_\mu \mathbb{1}_k$ defined on the circle, $z \in [0,1]$, i.e. with periodicity 1 (β^{-1} in case $\beta \neq 1$). The quadratic ADHM constraint now reads

$$\left[\sigma_i \otimes \mathbb{1}_k, D_x^\dagger(z) D_x(z) + 4\pi^2 \sum_m \delta(z - \mu_m) \zeta_a^\dagger P_m \zeta_b \right] = 0. \quad (25)$$

Introducing

$$2\pi \zeta_a^\dagger P_m \zeta_b \equiv \mathbb{1}_2 \hat{S}_m^{ab} - \vec{\tau} \cdot \vec{\rho}_m^{ab}, \quad (26)$$

this leads precisely to the so-called Nahm equation [13],

$$\frac{d}{dz} \hat{A}_j(z) + [\hat{A}_0(z), \hat{A}_j(z)] + \tfrac{1}{2} \varepsilon_{jk\ell}[\hat{A}_k(z), \hat{A}_\ell(z)] = 2\pi i \sum_m \delta(z - \mu_m) \rho_m^j. \quad (27)$$

Note that the left-hand side is the difference between the magnetic and electric field for $\hat{A}_\mu(z)$, with the right-hand side a violation of self-duality.

Although we do not want to go into too much detail here, it is instructive to discuss the standard setting of the Nahm transformation [13,27]. One starts from an SU(n), charge k self-dual gauge field on $\mathbb{R}^4$ with periods in all four directions, some of which may be infinite or zero (in the latter case effectively leading to a reduced dimension). This extends to a family of self-dual U(n) gauge fields[6] when adding $-2\pi i z_\mu \mathbb{1}_n$ to the SU(n) gauge field $A_\mu(x)$. The index theorem now guarantees there is a family of k chiral zero-modes

---

[6] One easily checks it does not change the field strength $F_{\mu\nu}(x)$.

for the Weyl equation $D_z^\dagger \Psi(x;z) = -\bar\sigma_\mu D_\mu(x;z)\Psi(x;z) = 0$, and this allows one to construct the gauge field $\hat A_\mu^{ab}(z) = \int d^4x \Psi^a(x;z)^\dagger \partial\Psi^b(x;z)/\partial z_\mu$. This dual $U(k)$ gauge field can be shown to be self-dual with charge n, and is defined again on $\mathbb{R}^4$, but with its periods inverted. Remarkably, one can then perform the transformation again, and come back to the original gauge field [27].

The ADHM construction performs precisely this last step. For instantons on $\mathbb{R}^4$ all periods are infinite and the Nahm transformation reduces self-duality to *algebraic* equations. Singularities may appear (except when all periods are finite), as we have seen from our analysis using Fourier transformation of the ADHM data and like in $\mathbb{R}^4$ are related to introducing λ, which encodes the asymptotic behavior of the zero-modes.

One final result is crucial to appreciate the strength of this formalism,

$$D_z^\dagger D_z = -\mathbb{1}_2 \otimes D_\mu^2(x;z) - \bar\eta_{\mu\nu} \otimes [D_\mu(x;z), D_\nu(x;z)], \qquad (28)$$

which uses the fact that $\bar\sigma_\mu\sigma_\nu = \delta_{\mu\nu}\mathbb{1}_2 + \bar\eta_{\mu\nu}$. Since $[D_\mu(x;z), D_\nu(x;z)] = F_{\mu\nu}(x)$ and the contraction of an anti-self dual tensor $(\bar\eta_{\mu\nu})$ with a self-dual tensor $(F_{\mu\nu})$ vanishes, self-duality implies that $D_z^\dagger D_z = -\mathbb{1}_2 \otimes D_\mu^2(x;z)$, which therefore commutes with the quarternions. This is completely analogous to our discussion for the ADHM construction, and proves that the Nahm gauge field $\hat A_\mu(z)$ is self-dual. Doing the Nahm transformation for the second time leads us to perform the same calculation as in Eq. (28), this time for the dual Weyl operator $\hat D_x$, and we see that the quadratic ADHM constraint is fully equivalent with stating that the dual gauge field is self-dual. The only complication is the possible presence of singularities when some of the periods are infinite, in particular for the ADHM construction somewhat hiding this profound relationship to the Nahm transformation.

2.3. Some explicit formulae

One might think all of this is just pushing the same problem around, but as we noted before, there is a dramatic simplification due to the dimensional reduction. In addition, for topological charge 1 the dual gauge field is Abelian, further simplifying the Nahm equation. In that case all the commutator terms in Eq. (27) vanish, and $\hat A_\mu(z)$ is piecewise constant, only jumping at the singularities. It is this that has allowed us to make progress in finding explicit solutions, together with the "magic" formulae for gauge field, action density, fermion zero-modes and density found previously within the ADHM formalism [28, 29],

$$A_\mu(x) = \tfrac{1}{2}\phi^{1/2}(x)\bar\eta^j_{\mu\nu}\partial_\nu\phi_j(x)\phi^{1/2}(x) + \tfrac{1}{2}[\phi^{-1/2}(x),\partial_\mu\phi^{1/2}(x)]\,,$$

$$\Psi^l_{iI}(x) = (2\pi)^{-1}\left(\phi^{1/2}(x)\lambda\partial_\mu f_x\bar\sigma_\mu\varepsilon\right)^l_{iI}\,,$$

$$\Psi^l_{iI}(x)^*\Psi^m_{iI}(x) = -(2\pi)^{-2}\partial^2_\mu f^{lm}_x\,, \tag{29}$$

where $l = 1,\ldots,k$ labels the zero-modes (i, I are the gauge and spin index),

$$\phi(x) = (\mathbb{1}_n - \phi_0)^{-1}\,, \quad \phi_\mu \equiv \lambda(\sigma_\mu\otimes f_x)\lambda^\dagger\,, \quad \varepsilon \equiv \sigma_2 = i\tau_2\,. \tag{30}$$

The gauge field for SU(2) further simplifies to $A_\mu(x) = \tfrac{1}{2}\phi(x)\bar\eta^j_{\mu\nu}\partial_\nu\phi_j(x)$ (which could be viewed as a generalized 't Hooft ansatz), since in that case $\phi(x)$ as a 2×2 matrix is a multiple of $\mathbb{1}_2$. For the action density one finds [29],

$$\mathrm{Tr}\, F^2_{\mu\nu}(x) = -\partial^2_\mu\partial^2_\nu \log\det f_x\,. \tag{31}$$

It is thus very convenient to first find the matrix f_x, as defined in Eq. (20). For calorons, after Fourier transformation, it is replaced by the Green's function $\hat f_x(z, z')$, which satisfies the equation [26]

$$\left\{\left(\frac{\hat D_\mu(z;x)}{2\pi i}\right)^2 + \frac{1}{2\pi}\sum_m \delta(z-\mu_m)\hat S_m\right\}\hat f_x(z, z') = \mathbb{1}_k\delta(z-z')\,. \tag{32}$$

The gauge field, as determined through $\phi_\mu(x)$, see Eq. (29), is read off from

$$\phi_\mu(x) = \sum_{m,m'} P_m\zeta_a\sigma_\mu\hat f^{ab}_x(\mu_m,\mu_{m'})\zeta^\dagger_b P_{m'}\,, \tag{33}$$

whereas the fermion zero-modes satisfying a generalized boundary condition $\hat\Psi^a_z(t+1,\vec x) = e^{2\pi iz}\mathcal{P}_\infty\hat\Psi^a_z(t,\vec x)$, cf. Eq. (3), are given by [30, 31]

$$\hat\Psi^a_z(x) = (2\pi)^{-1}\phi^{1/2}(x)\sum_m P_m\zeta_b\bar\sigma_\mu\varepsilon\partial_\mu\hat f^{ba}_x(\mu_m, z)\,. \tag{34}$$

For $z = \tfrac{1}{2}$ this gives the usual anti-periodic boundary conditions for fermions at finite temperature, but the general z dependence will turn out to be extremely useful as a diagnostic tool. Finally, the zero-mode density reads

$$\hat\Psi^a_z(x)^\dagger\hat\Psi^b_z(x) = -(2\pi)^{-2}\partial^2_\mu\hat f^{ab}_x(z, z)\,. \tag{35}$$

To compute $\hat f_x(z, z')$ we note that we can always transform to a gauge where $\hat A_0(z) \equiv 2\pi i\xi_0$ is constant, after which $\hat g(z) \equiv \exp(2\pi i(\xi_0 - x_0\mathbb{1}_k)z)$ transforms $\hat A_0(z) - 2\pi ix_0$ to zero. This turns Eq. (32) into

$$\left\{-\frac{d^2}{dz^2} + V(z;\vec x)\right\}f_x(z, z') = 4\pi^2\mathbb{1}_k\delta(z-z')\,, \tag{36}$$

with $f_x(z, z')$ and $V(z; \vec{x})$ given by

$$f_x(z, z') \equiv \hat{g}(z) \hat{f}_x(z, z') \hat{g}^\dagger(z'), \quad V(z; \vec{x}) \equiv 4\pi^2 \vec{R}^2(z; \vec{x}) + 2\pi \sum_m \delta(z - \mu_m) S_m,$$

$$R_j(z; \vec{x}) \equiv x_j \mathbb{1}_k - (2\pi i)^{-1} \hat{g}(z) \hat{A}_j(z) \hat{g}^\dagger(z), \quad S_m \equiv \hat{g}(\mu_m) \hat{S}_m \hat{g}^\dagger(\mu_m). \quad (37)$$

Periodicity is now only up to the gauge transformation $\hat{g}(1)$. In particular when $\vec{R}^2(z; \vec{x})$ is piecewise constant explicitly computing the Green's function becomes doable. Nevertheless, in general terms one finds

$$f_x(z, z') = 4\pi^2 \left\{ W(z, z_0) \left(\theta(z' - z) \mathbb{1}_{2k} - (\mathbb{1}_{2k} - \mathcal{F}_{z_0})^{-1} \right) W^{-1}(z', z_0) \right\}_{12},$$

$$W(z_2, z_1) \equiv \mathrm{Pexp} \int_{z_1}^{z_2} \begin{pmatrix} 0 & \mathbb{1}_k \\ V(z; \vec{x}) & 0 \end{pmatrix} dz, \quad \mathcal{F}_{z_0} \equiv \hat{g}^\dagger(1) W(z_0 + 1, z_0), (38)$$

where the $(1, 2)$ component on the right-hand side for $f_x(z, z')$ is with respect to the 2×2 block matrix structure. This equation for $f_x(z, z')$ is valid for $z' \in [z, z+1]$, but can be extended with the appropriate periodicity. We can now also find an explicit result for the action density

$$\mathrm{Tr}\, F_{\mu\nu}^2(x) = \partial_\mu^2 \partial_\nu^2 \log \psi(x), \quad \psi(x) \equiv \det\left(i e^{-\pi i x_0} (\mathbb{1}_{2k} - \mathcal{F}_{z_0}) / \sqrt{2} \right).$$

$$(39)$$

Note that z_0 can be chosen at will, *e.g.* for $z_0 = \mu_m + 0$ we find

$$\mathcal{F}_{\mu_m} = T_m H_{m-1} \cdots T_2 H_1 T_1 \hat{g}^\dagger(1) H_n T_n H_{n-1} \cdots T_{m+1} H_m,$$

$$T_m \equiv \begin{pmatrix} \mathbb{1}_k & 0 \\ 2\pi S_m & \mathbb{1}_k \end{pmatrix}, \quad H_m \equiv \mathrm{Pexp} \int_{\mu_m}^{\mu_{m+1}} \begin{pmatrix} 0 & \mathbb{1}_k \\ 4\pi^2 \vec{R}^2(z; \vec{x}) & 0 \end{pmatrix} dz. (40)$$

The main pay off to have these expressions in terms of possibly unknown solutions to the Nahm equations, is that it allows us to look at the far field limit. When constituent monopoles are well separated, the charged components of the field in the core decay exponentially, being left with the Abelian fields in the far field region. This is related to the high temperature limit, in which case the cores collapse to zero size.

2.4. Limits and special cases

To extract exponential factors it turned out to be convenient to define

$$f_m^\pm(z) = \mathrm{Pexp} \left[\pm 2\pi \int_{\mu_m}^z R_m^\pm(z) dz \right], \quad R_m^\pm(z)^2 \pm \frac{1}{2\pi} \frac{d}{dz} R_m^\pm(z) = \vec{R}^2(z; \vec{x}). \quad (41)$$

                F. Bruckmann, D. Nógrádi, P. van Baal

Since $\vec{R}(z;\vec{x}) \to \vec{x}\mathbb{1}_k$ for $|\vec{x}| \to \infty$, we find in this limit that $R_m^{\pm}(z) \to |\vec{x}|\mathbb{1}_k$ and $f_m^{\pm}(z) \to \exp\left(\pm 2\pi|\vec{x}|(z - \mu_m)\mathbb{1}_k\right)$. For $z, z' \in [\mu_m, \mu_{m+1}]$ we can now write $W(z, z') = W_m(z)W_m^{-1}(z')$, see Eq. (38), with

$$W_m(z) \equiv \begin{pmatrix} f_m^+(z) & f_m^-(z) \\ 2\pi R_m^+(z)f_m^+(z) & -2\pi R_m^-(z)f_m^-(z) \end{pmatrix}. \tag{42}$$

This has allowed us to show that (ff stands for far field limit)

$$f_x^{\mathrm{ff}}(\mu_m, \mu_m) = 2\pi \sum_m{}^{-1}, \text{ and } f_x^{\mathrm{ff}}(z, z) = \pi R_m^{-1}(z) \text{ for } \mu_m < z < \mu_{m+1}, \tag{43}$$

where $\sum_m \equiv R_m^-(\mu_m) + R_{m-1}^+(\mu_m) + S_m$ and $R_m(z) \equiv \frac{1}{2}(R_m^+(z) + R_m^-(z))$. For this to hold $\vec{x}$ has to be far removed from any constituent. In addition $f^{\mathrm{ff}}(\mu_m, \mu_{m+1}) = 0$, which implies that for SU(2) and SU(3) only the Abelian components of the gauge field survive in this limit.

Useful is also the so-called zero-mode limit (zm), which assumes $\vec{x}$ to be far removed from any constituent monopole other than of type m, distinguished by their magnetic charge and mass $8\pi^2 \nu_m$, with $\nu_m \equiv \mu_{m+1} - \mu_m$, see the next section. In this limit one finds up to *exponential* corrections [31] for $\mu_m \le z' \le z \le \mu_{m+1}$ $(f_x(z', z) = f_x^\dagger(z, z')$ for $z' > z)$

$$f_x^{\mathrm{zm}}(z, z') = \pi \left(f_m^-(z)f_m^-(\mu_{m+1})^{-1} - f_m^+(z)f_m^+(\mu_{m+1})^{-1}Z_{m+1}^-\right)$$
$$\times \left(f_m^-(\mu_{m+1})^{-1} - Z_m^+ f_m^+(\mu_{m+1})^{-1}Z_{m+1}^-\right)^{-1}\left(f_m^-(z')^{-1} - Z_m^+ f_m^+(z')^{-1}\right)R_m^{-1}(z'), \tag{44}$$

with $Z_m^- \equiv \mathbb{1}_k - 2\sum_m{}^{-1} R_{m-1}(\mu_m)$ and $Z_m^+ \equiv \mathbb{1}_k - 2\sum_m{}^{-1} R_m(\mu_m)$. One concludes that the zero-modes (see Eq. (34) and Eq. (35)) are localized exponentially to the mth constituent monopoles (for z away from the interval boundaries).

Explicit solutions are found when $\vec{R}(z;\vec{x}) \equiv \vec{x} - \vec{Y}(z)$ is piecewise constant, *i.e.* $\vec{Y}(z) \equiv \vec{Y}_m$ for $z \in [\mu_m, \mu_{m+1}]$, in which case $|\vec{R}| = R_m^{\pm} = R_m$. Apart from $k = 1$, this is so for a class of axially symmetric multi-caloron solutions [26], in which case $3k$ eigenvalues of $\vec{Y}_m$ give the locations of the type m constituent monopoles. The expression for the zero-mode in that case can be used to show that in the high temperature limit the zero-mode densities reduce to delta functions localized at these constituent locations, thus establishing that the constituents in this limit are point-like monopoles. Particularly simple axially symmetric solutions are found for SU(2),

$$\zeta_{iI}^a = \rho_a \delta_{iI}, \quad \vec{Y}_m^{ab} = \vec{e}(\delta_{ab}\xi_a + y_m \rho_a \rho_b), \quad \nu_1 y_1 + \nu_2 y_2 = 0, \quad \rho_a > 0. \tag{45}$$

For charge 2, choosing $\rho_1 = \rho_2 = 2$ and equal mass constituents ($\nu_1 = \nu_2 = \frac{1}{2}$) the constituent locations y are plotted as a function of $\xi = \xi_1 = -\xi_2$ in Fig. 4. We note that opposite charges are found to alternate. The properties of these solutions will be further discussed in the next section.

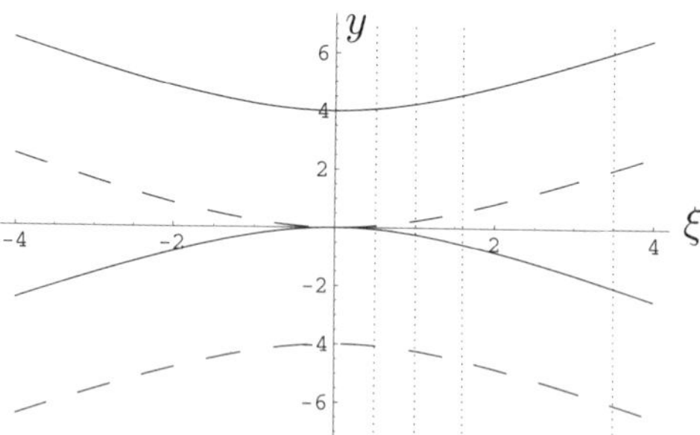

Fig. 4. Constituent locations y based on Eq. (45) as a function of $\xi = \xi_1 = -\xi_2$ for $y_2 = -y_1 = \nu_1 = \nu_2 = \frac{1}{2}$ and $\rho_1 = \rho_2 = 2$. Dashed versus full lines distinguish the magnetic charge of the constituents. The dotted lines apply to Fig. 3.

More generally, for SU(2) and charge 2 we may borrow from the study of monopoles [13,32] the general solution of the Nahm equations for $z \in [\mu_1, \mu_2]$

$$\hat{A}_j(z) \equiv 2\pi i \hat{g}^\dagger(z) h^\dagger \left(a_j \mathbb{1}_2 + \mathcal{D}\mathcal{R}_{jb} f_b(4\pi \mathcal{D}z)\tau_b \right) h\hat{g}(z), \qquad (46)$$

where $f_j(z)$ are the Jacobi elliptic functions

$$f_1(z) = \frac{k'}{cn_k(z)}, \quad f_2(z) = \frac{k' sn_k(z)}{cn_k(z)}, \quad f_3(z) = \frac{dn_k(z)}{cn_k(z)}. \qquad (47)$$

Here $\vec{a}$ is the center of mass for monopoles of a given type, $\mathcal{R}$ and h are arbitrary spatial and gauge rotations, $\mathcal{D}$ is a scale parameter, and $0 \leq k \leq 1$ ($k' \equiv \sqrt{1 - k^2}$) playing the role of a shape parameter, as will be discussed in the next section. In general all these parameters differ on the second interval where in addition z is shifted to $z - \frac{1}{2}$, but they are to be related through the discontinuities in the Nahm equation, Eq. (27). This tends to be rather restrictive, and is the point where the construction for calorons deviates from that for static monopoles.

Quite remarkably, although $\vec{R}(z; \vec{x})$ is no longer piecewise constant, the function $\operatorname{Tr} R_m^{-1}(z)$ *is independent* of $z \in [\mu_m, \mu_{m+1}]$. This is a highly non-trivial consequence of the Nahm equations (from the point of view of integrability, it gives a constant of motion). The physical significance here is that it is directly related to the zero-mode density (summed over the zero-modes) in the high temperature limit, see Eq. (35) and Eq. (43),

$$\left(\sum_a \hat{\Psi}_z^a(x)^\dagger \hat{\Psi}_z^a(x)\right)^{\text{ff}} = -\partial_i^2 \mathcal{V}_m(\vec{x}), \quad \mathcal{V}_m(\vec{x}) \equiv (4\pi)^{-1} \operatorname{Tr} R_m^{-1}(z). \quad (48)$$

Miraculously we have been able to calculate $\mathcal{V}_m(\vec{x})$ exactly [31], from which we will be able to draw conclusions on the pointlike nature of the constituents.

3. Properties of solutions

We start our tour of $SU(n)$ caloron solutions with those of charge 1. In this case the action density has a particularly simple form [33]

$$-\tfrac{1}{2}\operatorname{Tr} F_{\mu\nu}^2(x) = -\tfrac{1}{2}\partial_\mu^2\partial_\nu^2 \log \psi(x), \quad \psi(x) = \tfrac{1}{2}\operatorname{tr}(\mathcal{A}_n \cdots \mathcal{A}_1) - \cos(2\pi t/\beta),$$

$$\mathcal{A}_m \equiv \frac{1}{r_m}\begin{pmatrix} r_m & |\vec{\rho}_{m+1}| \\ 0 & r_{m+1} \end{pmatrix}\begin{pmatrix} \cosh(2\pi\nu_m r_m/\beta) & \sinh(2\pi\nu_m r_m/\beta) \\ \sinh(2\pi\nu_m r_m/\beta) & \cosh(2\pi\nu_m r_m/\beta) \end{pmatrix}. \quad (49)$$

Here $\nu_m = \mu_{m+1} - \mu_m$, $r_m = |\vec{r}_m|$, $\vec{r}_m = \vec{x} - \vec{y}_m$ and $\vec{\rho}_m = \vec{y}_m - \vec{y}_{m-1}$, where $\vec{y}_m$ ($\vec{y}_{n+m} \equiv \vec{y}_m$) are the constituent locations[7]. A few typical examples for $SU(2)$, where $|\vec{\rho}_1| = \pi\rho^2/\beta$ in terms of the instanton scale parameter ρ, are shown in Fig. 2 and for $SU(3)$ in Fig. 5 (we apologize for a somewhat awkward choice of conventions in naming the vectors $\vec{\rho}_m$ and scale parameters $\rho_a = |\zeta_a|$). From this it is already clear that there are n lumps, centered at $\vec{y}_m$ and that when well-separated they are static and spherically symmetric. The self-duality then guarantees each lump is a basic BPS monopole, which can be seen to contribute $8\pi^2\nu_m$ to the action, correctly summing to a total action $8\pi^2$. Interesting properties on the geometry of the moduli space and alternative approaches can be found in Refs. [34–36].

It is instructive to provide further evidence for the monopole nature of these constituent lumps. For this we take the far field limit (ff) of Eq. (49). In this limit $\vec{x}$ is assumed to be far from all constituent locations $\vec{y}_m$ and we find

$$\psi^{\text{ff}}(x) = \tfrac{1}{2}\prod_{m=1}^{n} \frac{(r_m + r_{m+1} + |\vec{\rho}_{m+1}|)}{2r_m} e^{2\pi\nu_m r_m/\beta}. \quad (50)$$

Note that there is no longer any time dependence and the far field limit is equivalent to the high temperature limit $\beta \to 0$, where the monopole mass becomes infinite and the non-Abelian core collapses to zero size (the range of the exponentially decaying charged fields shrinks to zero). As we will

---

[7] This can be derived from Eqs. (39),(40) using $S_m = \hat{S}_m = |\vec{\rho}_m|$ (see Eq. (26)) and $\mathcal{A}_m = B_{m+1}T_{m+1}H_m B_m^{-1}$, where $B_m^{11} = B_m^{22} = 0$, $B_m^{12} = 1$ and $B_m^{21} = 4\pi r_m$.

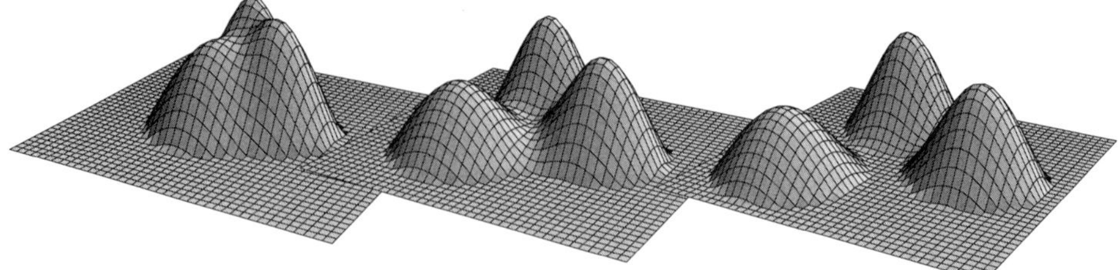

Fig. 5. Action densities for the SU(3) caloron with $(\mu_1, \mu_2, \mu_3) = (-17, -2, 19)/60$ and $(\nu_1, \nu_2, \nu_3) = (0.25, 0.35, 0.4)$ at $t = 0$ in the plane of the three constituents for $1/\beta = 1.5$, 3 and 4 (from left to right) on equal logarithmic scales.

argue, using $A^{\mathrm{D}}_\mu(\vec{x})$ as the gauge field for the basic self-dual Abelian–Dirac monopole, one finds in a suitable gauge for the Abelian far field embedded in SU(n), $A^{ab}_\mu(\vec{x}) = \delta_{ab} \tilde{A}^a_\mu(\vec{x})$, with $\tilde{A}^m_\mu(\vec{x}) = \frac{1}{2} A^{\mathrm{D}}_\mu(\vec{r}_m) - \frac{1}{2} A^{\mathrm{D}}_\mu(\vec{r}_{m-1})$, where $\vec{E}^{\mathrm{D}}(\vec{x}) = \vec{B}^{\mathrm{D}}(\vec{x}) = \vec{x}|\vec{x}|^{-3}$ are electric and magnetic fields of this self-dual Dirac monopole. One verifies[8] that indeed $-\frac{1}{2} \mathrm{Tr}\, F^2_{\mu\nu}(x) = -\frac{1}{2} \partial^2_i \partial^2_j \log \psi^{\mathrm{ff}}(x)$ when substituting this gauge field.

To give a little more insight in why the gauge field takes the above form, we remark that another way to define the location of an SU(2) monopole is to find the zeros of the Higgs field, or for SU(n) to find where two of its eigenvalues coincide. At these points the broken gauge symmetry is partially restored to include an unbroken SU(2) subgroup. In the case of the caloron we of course need to find coinciding eigenvalues of the Polyakov loop[9], $P(\vec{x})$ [37]. The gauge field $\tilde{A}^m(\vec{x})$ defined above is associated to the Abelian component of the embedded SU(2) monopole associated to the unbroken subgroup. The Polyakov loop therefore is a useful diagnostic tool particularly when monopoles are too close to be seen as separate lumps. For SU(2) the constituents are simply found where $P(\vec{x}) = \pm \mathbb{1}_2$ [38].

Important is that when coinciding eigenvalues occur at infinity, that is in $\mathcal{P}_\infty$, some constituent masses will vanish. For trivial holonomy only the nth constituent is massive and the action density shows a single lump, the usual (deformed) instanton, *cf.* Fig. 6. In the presence of massless constituents one cannot take the far field limit, due to the infinite range of the fields in these massless cores. One may, however, move massless constituents off to infinity. For SU(2) this requires one to take $\rho \to \infty$, and one is left with

---

[8] It is best to first check $-\frac{1}{2} \partial^2_i \partial^2_j \log[r_1^{-1} r_2^{-1}(r_1 + r_2 + |\vec{\rho}_2|)^2] = \left(r_1^{-3} \vec{r}_1 - r_2^{-3} \vec{r}_2\right)^2$, relevant for SU(2). The exponential terms in Eq. (50) do not contribute, except for $\sum_m \delta(\vec{r}_m) 8\pi^2 \nu_m / \beta$. However, in the core Eq. (50) should not be used since the singularity in $\psi^{\mathrm{ff}}(x)$ is simply due to approximations made outside the core.

[9] In technical terms the locations of constituents are found where $\log P(\vec{x})$ takes values on the boundary of the Weyl chamber [39].

　　　　F. Bruckmann, D. Nógrádi, P. van Baal

a BPS monopole in a singular gauge to adjust the mismatch in boundary conditions between caloron and monopole, or put differently, to compensate for the magnetic charge that is pushed past infinity [9].

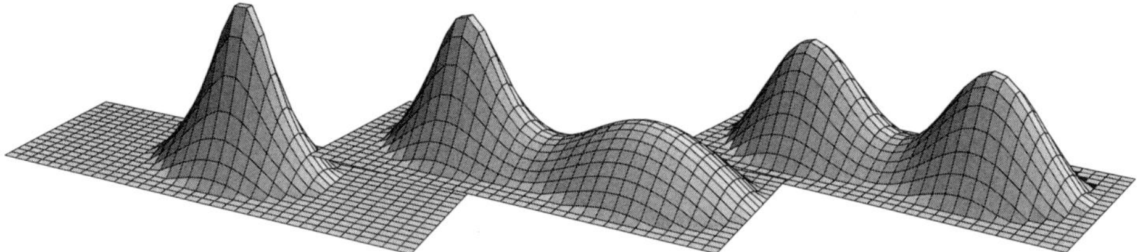

Fig. 6. The action density for SU(2) calorons with $\rho = \beta$ and (from left to right) $\mu_2 = 0$ (*i.c.* trivial holonomy), 0.125 and 0.25 (*i.e.* $\operatorname{Tr}\mathcal{P}_\infty = 0$), at $t = 0$ in a plane through the constituent locations.

As we have explained in the introduction, chiral zero-modes play an extremely important role in the physics of instantons. They will also turn out to be very useful as a diagnostic tool for studying the properties of the caloron solutions. The Atiyah index theorem has taught us there is one chiral zero-mode for a charge 1 caloron. But when the caloron has "dissociated" in n constituents the natural question is if the zero-mode is going to follow this. There is a compelling reason to expect it will opt for being localized on one constituent only. When well separated, the constituents become BPS monopoles. These are known to have zero-modes, but only for a given range of z values in terms of the Higgs expectation value, as dictated by the Callias index theorem [40]. We will show how this determines to which constituent the zero-mode will be localized. Even more useful is that we can change this by introducing an arbitrary phase in the boundary conditions for the fermions, which reads in the algebraic gauge (Eq. (3))

$$\hat{\Psi}_z(t + \beta, \vec{x}) = \exp(2\pi i z \beta)\mathcal{P}_\infty\hat{\Psi}_z(t, \vec{x})\,. \tag{51}$$

Physical fermions at finite temperature are required to be anti-periodic, $\Psi(t, \vec{x}) = \hat{\Psi}_{z=1/2\beta}(t, \vec{x})$. That z determines the location of the zero-mode is seen as follows. The gauge transformation $g(t) = \exp(2\pi i z t)\exp(tA_0^\infty)$ makes the fermions periodic at the expense of changing $A_0 = 0$ at spatial infinity to $A_0 = A_0^\infty - 2\pi i z$, *cf.* Eq. (5). This acts as an effective mass term in the Dirac equation, which differs for each of the gauge components. As discussed earlier in this section, each constituent monopole is associated with an SU(2) embedding and the two isospin components of these embeddings effectively have "masses" $2\pi(\mu_m/\beta - z)$ and $2\pi(\mu_{m+1}/\beta - z)$. This only gives a normalizable zero-mode when these are of opposite sign [40], that is for $z \in [\mu_m/\beta, \mu_{m+1}/\beta]$. In the interior of this interval the zero-mode is exponentially localized (to the constituent monopole at $\vec{y}_m$), but at

$z = \mu_m$ or $z = \mu_{m+1}$ one of the isospin components becomes massless and the zero-mode delocalizes, having in addition support on the constituent at respectively $\vec{y}_{m-1}$ or $\vec{y}_{m+1}$.

The expression for the zero-mode density can be given in a simple form for arbitrary charge 1 calorons as well [41]. With $\mu_m/\beta \le z' \le z \le \mu_{m+1}/\beta$

$$\hat{f}_x(z, z') = \frac{\pi e^{2\pi i t(z-z')}}{\beta r_m \psi} \langle v_m(z') | \mathcal{A}_{m-1} \cdots \mathcal{A}_1 \mathcal{A}_n \cdots \mathcal{A}_m - e^{-2\pi it/\beta} | \sigma_2 v_m(z) \rangle,$$

$$v_m(z) \equiv \begin{pmatrix} \sinh[2\pi(z - \mu_m/\beta)r_m] \\ \cosh[2\pi(z - \mu_m/\beta)r_m] \end{pmatrix}, \quad |\hat{\Psi}_z(x)|^2 = -\frac{1}{4\pi^2} \partial_\mu^2 \hat{f}_x(z, z), \quad (52)$$

where for later use we have introduced also the off-diagonal expression for the Green's function $\hat{f}_x(z, z')$ (for $z \le z'$ one can use the fact that $\hat{f}_x(z', z) = \hat{f}_x^*(z, z')$), which can be computed using impurity scattering calculations in a piecewise constant potential, see Sec. 2.3, the details of which need not concern us here. We illustrate the behavior of the zero-mode in Fig. 7 at various values of z for the caloron of Fig. 5. It is also interesting to consider the so-called zero-mode limit (zm) in which $\vec{x}$ is far removed from

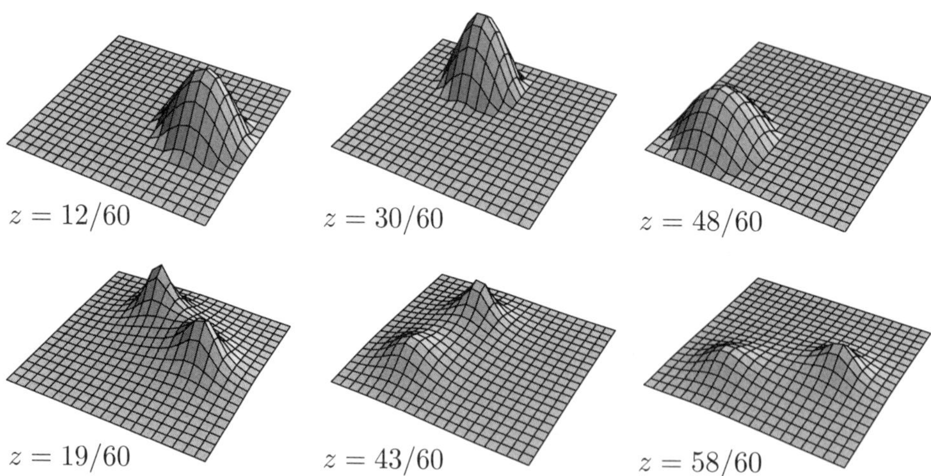

Fig. 7. Normalized zero-mode densities at $\beta = 1$, for the SU(3) caloron of Fig. 5 shown for $z = \mu_j$ on equal linear scales (bottom) and on equal logarithmic scales for three values of z roughly in the middle of each interval $z \in [\mu_j, \mu_{j+1}]$ (top). The zero-mode with anti-periodic boundary conditions is found at $z = 30/60$.

            F. Bruckmann, D. Nógrádi, P. van Baal

any constituent location, except the one at $\vec{y}_m$. This gives[10]

$$\hat{f}_x^{\mathrm{zm}}(z,z) = \frac{2\pi \sinh\left(2\pi r_m(\mu_{m+1}/\beta - z)\right) \sinh\left(2\pi r_m(z - \mu_m/\beta)\right)}{\beta r_m \sinh\left(2\pi r_m \nu_m/\beta\right)}. \tag{53}$$

At $z = \frac{1}{2}(\mu_m + \mu_{m+1})/\beta$ (*i.e.* $z = 0$ or $z = 1/2\beta$ for SU(2)) we find

$$\hat{f}_x^{\mathrm{zm}}(z,z) = \pi(r_m\beta)^{-1}\tanh(\pi r_m \nu_m/\beta), \tag{54}$$

with $-(4\pi^2)^{-1}\partial_\mu^2 \hat{f}_x^{\mathrm{zm}}(z,z)$ giving precisely the zero-mode density of a basic BPS monopole, confirming once again the nature of the constituents. In the high temperature limit, as long as $z \neq \mu_m/\beta$, the zero-modes become infinitely localized to the constituent locations $\vec{y}_m$. In this limit the zero-mode density is given by $\beta^{-1}\delta(\vec{x} - \vec{y}_m)$ for $\mu_m/\beta < z < \mu_{m+1}/\beta$. The zero-modes are therefore ideal for localizing the cores of the constituent monopoles, particularly useful for the higher charge calorons discussed below. It should be noted that in general the definitions of the constituent locations based on the peaks in the energy density, the degeneracy of two eigenvalues in the Polyakov loop and the peak in the zero-mode densities will only coincide with $\vec{y}_m$ when all constituents have a non-zero mass and are well separated (as compared to the size of the monopole cores) from each other.

Before discussing the higher charge calorons we give the simple expression for the SU(2) charge 1 gauge field in the algebraic gauge with $\mathcal{P}_\infty = \exp(\beta\omega\tau_3)$ (hence $\mu_2 = -\mu_1 = \beta\omega$),

$$A_\mu = \frac{i}{2}\bar{\eta}_{\mu\nu}^3\tau_3\partial_\nu\log\phi + \frac{i}{2}\phi\mathrm{Re}\left((\bar{\eta}_{\mu\nu}^1 - i\bar{\eta}_{\mu\nu}^2)(\tau_1 + i\tau_2)\partial_\nu\chi\right), \tag{55}$$

where $\phi^{-1} \equiv 1 - \rho^2 \hat{f}_x(\omega,\omega)$ and $\chi \equiv \rho^2 \hat{f}_x(\omega,-\omega)$, which has some resemblance to the 't Hooft ansatz in Eq. (13). It reduces to this ansatz for trivial holonomy, $\omega = 0$, for which $\chi = 1 - \phi^{-1}$, and one easily checks that this gives precisely the Harrington–Shepard solution with $\phi = \phi_{\mathrm{HS}}$, see Eq. (14). In the high temperature limit one finds for non-trivial holonomy $\chi^{\mathrm{ff}} = 0$, $\phi^{\mathrm{ff}} = (|\vec{x} - \vec{y}_1| + |\vec{x} - \vec{y}_2| + |\vec{y}_2 - \vec{y}_1|)/(|\vec{x} - \vec{y}_1| + |\vec{x} - \vec{y}_2| - |\vec{y}_2 - \vec{y}_1|)$ and that $A_\mu = \frac{i}{2}\bar{\eta}_{\mu\nu}^3\tau_3\partial_\nu\log\phi^{\mathrm{ff}}$ describes a pair of oppositely charged Dirac monopoles with the Dirac string on the line connecting them, where ϕ^{ff} diverges, but outside of which $\log\phi^{\mathrm{ff}}$ is harmonic.

3.1. Higher charge calorons

When ignoring charged components of the gauge field, outside the core the Abelian field has unavoidable Dirac strings. We can trace how the exact

---

[10] Allowing only for errors decaying *exponentially* in $r_{l\neq m}$, one needs to include $\mathcal{O}(r_{m\pm1}^{-1})$ dependent shifts in μ_m, μ_{m+1} and ν_m, see Eq. (44) and Ref. [31].

solution instead takes care of the return flux, namely through the Abelian component in the magnetic field coming from the commutator of the charged components of the non-Abelian field, as is of course well-known from the 't Hooft–Polyakov monopole. But from the numerical point of view this will require exponential fine tuning outside the core. It is notoriously difficult to find approximate superpositions for magnetic monopoles without seeing a remnant of the Dirac string, and in that sense there is "no free lunch". Like in the instanton liquid we would like to make approximate superpositions of calorons, which also allows us to mix calorons of different charges. However, for those that "dissociate", the constituents should not "remember" from which caloron they originated. Although the constituents themselves might be well separated, interference can occur in the regions between them. An example of this is shown in Fig. 8, where we added two charge 1 caloron gauge fields in the algebraic gauge, called the sum ansatz [20, 21]. This preserves the gauge condition but some care is required at the gauge singularity, which is removed by a gauge transformation with non-trivial winding number (sometimes one refers to these gauge fields as being in the singular gauge). Adding the gauge fields after such a gauge transformation is performed, destroys the proper decay at infinity. Instead, one first smoothly deforms to zero the gauge fields of all other calorons in a small neighborhood centered around the gauge singularity one wishes to remove. This only costs a small amount of action, particularly when the calorons are not too close. In principle a similar construction is possible for keeping Dirac strings hidden, but in this case the gauge singularity is due to approximating the fields by their Abelian component, far from the constituent cores. However, this would require exponential fine tuning. Nevertheless, not performing any adjustment the action density along the Dirac string (or sheet due to the additional extent in the time direction) actually stays finite[11], see Fig. 8.

It is due to these complications we felt compelled to analyze the higher charge caloron solutions. The disadvantage is that one can only consider the exact self-dual solutions this way, and not superpositions of opposite charges. On the other hand, it should guarantee absence of Dirac strings, as we indeed confirm for a set of exact axially symmetric solutions that share all the properties of the charge 1 solutions, as illustrated for the example of SU(2) charge 2 solutions in Figs. 9–11. In the high temperature limit these solutions are again described by point-like constituent monopoles [26]. Most of the memory effect has disappeared. In the light of this it is interesting to point out that when further separating the two pairs on the left from

---

[11] Only at the gauge singularity the action density of the combined field diverges, which may be removed as discussed.

     F. Bruckmann, D. Nógrádi, P. van Baal

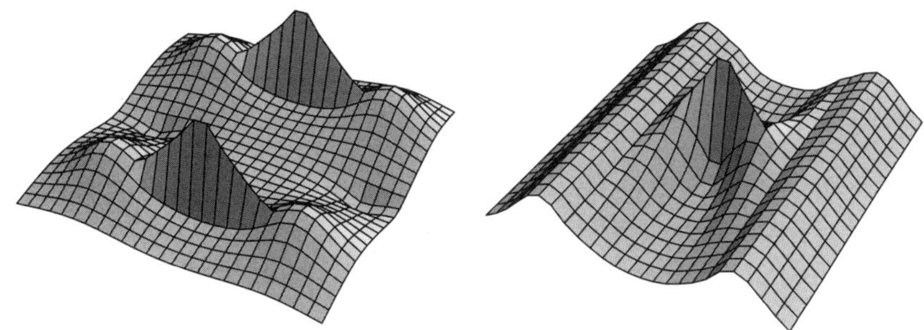

Fig. 8. Approximate superposition of two SU(2) charge 1 calorons (left). The logarithm of the action density is plotted through the plane of the constituents at $t = 0$ for $\beta = 1$ we zoom in by a factor 40 for the transverse direction on the would-be Dirac strings (right).

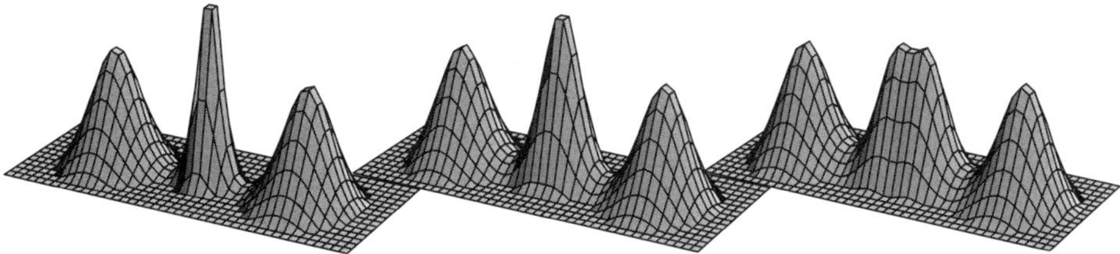

Fig. 9. Action density on equal logarithmic scales for exact axially symmetric SU(2) charge 2 solutions at $\beta = 1$ for $t = 0$, on a plane through the constituents. Locations can be read off from the dotted lines in Fig. 4, from left to right. For the last case see Fig. 10.

those on the right, each can be viewed as coming from a single caloron[12]. However, when the two middle constituents start to get closer than the size of their cores, as in Fig. 9 it is more natural to interpret the solution in terms of one well "dissociated" caloron formed by the two outer constituents, on top of which is superposed a small "non-dissociated" caloron, which is no longer static. Indeed, its peak is nearly O(4) symmetric, whereas the other two constituents are static and O(3) symmetric to a high degree, as is appropriate for a BPS monopole (see Fig. 9 (left)). A very subtle memory effect, however, remains. It can be shown that for these point-like axially symmetric solutions the magnetic charges have to alternate. In particular one cannot move one monopole through the other. It may perhaps come as a surprise, but in part there is a good reason for being cautious, since as we have just seen two oppositely charged constituent monopoles really form a

---

[12] In this context the parameters ξ_a and ρ_a appearing in Eq. (45) can be interpreted as the center of mass and size of those calorons.

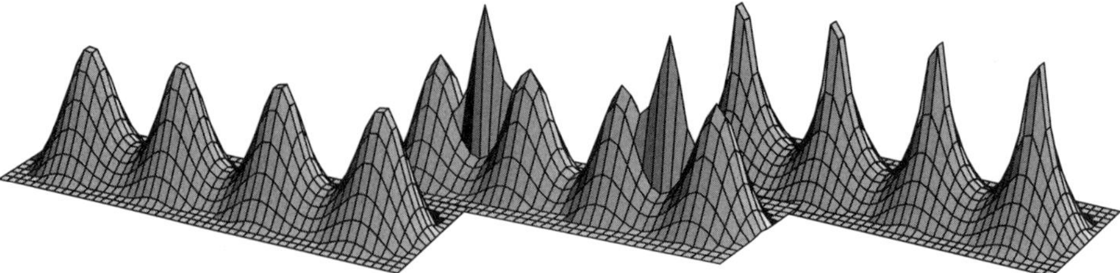

Fig. 10. Action density on equal logarithmic scales for the exact axially symmetric SU(2) charge 2 solution with all constituents well-separated (left), the approximate solution based on the sum ansatz (middle) and the high temperature limit with point-like Dirac monopoles (right). See also Fig. 9.

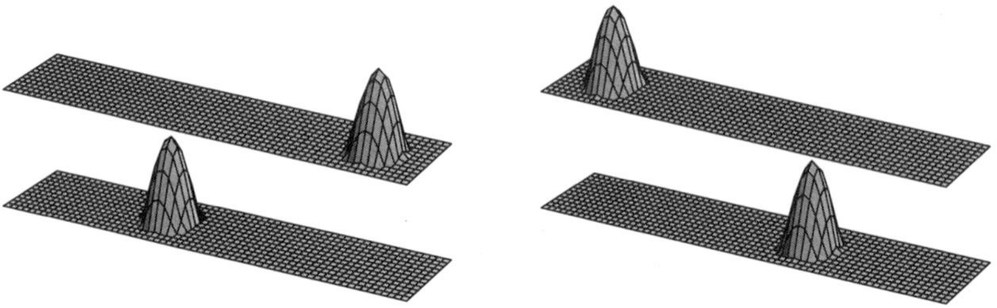

Fig. 11. Zero-mode densities on equal logarithmic scales for the caloron in Fig. 10 (same horizontal scale). On the left are shown the two periodic zero-modes ($z = 0$) and on the right the two anti-periodic zero-modes ($z = 1/2$).

small caloron. Since the distance between constituents is given by $\pi\rho^2/\beta$, this means that we have to go through a singular caloron when interchanging constituent locations on the line. A singular caloron lies on the boundary of the moduli (*i.e.* parameter) space and one cannot use continuity arguments.

The reason to expect that in general one no longer deals exclusively with point-like monopoles comes from the known multi-monopole solutions. It is well-known that when putting monopoles of equal charge on top of each other, these are deformed in for example the shape of a doughnut [42], as famously illustrated in the scattering of two magnetic monopoles [43]. At the technical level this is related to the fact that the Green's function mentioned above no longer involves a piecewise constant potential. Only for the axially symmetric solutions this was still the case. Nevertheless, for SU(2) and charge 2 we have been able to find an exact expression for the zero-mode density in the high temperature limit. As long as $z \neq \mu_j$, the fermions have an infinite mass in this limit, and the zero-modes will vanish outside the cores of the constituent monopoles, which themselves need not necessarily be isolated points. We thus are able to trace with the help of the zero-modes to which region the cores have to be localized. Here we will just present the

         F. Bruckmann, D. Nógrádi, P. van Baal

result [31] and discuss its physical significance. For $\mu_m < z < \mu_{m+1}$ and in the far field limit $\sum_a |\hat{\Psi}_z^a(x)|^2 = -\beta^{-1}\partial_i^2 \mathcal{V}_m(\vec{x})$, with

$$\mathcal{V}_m(\vec{x}) = \frac{1}{2\pi|\vec{x}|} + \frac{\mathcal{D}}{4\pi^2} \oint_{r<\mathcal{D}} r d\varphi \, \frac{\partial_r |\vec{x} - r\vec{y}(\varphi)|^{-1}}{\sqrt{\mathcal{D}^2 - r^2}}, \tag{56}$$

where $\vec{y}(\varphi) = (\sqrt{1-k^2}\cos\varphi, 0, \sin\varphi)$, up to an arbitrary coordinate shift and rotation. Here $\mathcal{D}$ is a scale and k a shape parameter to characterize *arbitrary* SU(2) charge 2 solutions. In this representation it is clear that $\mathcal{V}_m(\vec{x})$ is harmonic everywhere except on a disk bounded by an ellipse with minor axes $2\mathcal{D}\sqrt{1-k^2}$ and major axes $2\mathcal{D}$. Although not directly obvious, when $k \to 1$ the support of the singularity structure is on two points only, separated by a distance $2\mathcal{D}$. Taking an arbitrary test function $f(\vec{x})$ one can prove that [31]

$$\lim_{k\to 1} -\int f(\vec{x})\partial_i^2 \mathcal{V}_m(\vec{x}) d^3x = f(0,0,\mathcal{D}) + f(0,0,-\mathcal{D}). \tag{57}$$

For the caloron k and $\mathcal{D}$ are in general not independent, monopoles of different charges have to adjust to each other to form an exact caloron solution. As an example we illustrate in Fig. 12 the relation for the case studied in Ref. [31], which is a two parameter family of exact SU(2) charge 2 calorons solutions in terms of an instanton scale ρ and relative gauge orientation angle α. For fixed ρ it interpolates between two axially symmetric solutions. On the left is shown the relation between k and ρ for some values of α. Together with the fact that $\mathcal{D} = \pi\sqrt{2}\rho^2(1 + \mathcal{O}(1-k^2))$, this shows that $k \to 1$ for increasing $\mathcal{D}$. It can be shown this approach is exponential, *cf.* Fig. 12 (left).

The important conclusion is that, when well separated, the constituent monopoles become point-like objects. This is a necessary requirement for the constituent monopoles to be used as entities to describe the field configurations at larger distances. Much work, of course, remains to be done, but the results so far have been encouraging. In addition important lattice evidence has been accumulated by now, which we will briefly summarize in the next section.

4. Lattice evidence

There are a number of lattice studies, that clearly point to constituent monopoles to have a dynamical role to play. Two methods have been employed. One is based on so-called cooling, the other one uses fermion zero-modes. In both cases the purpose is to filter out ultraviolet noise, *i.e.* one is interested in the long distance fluctuations.

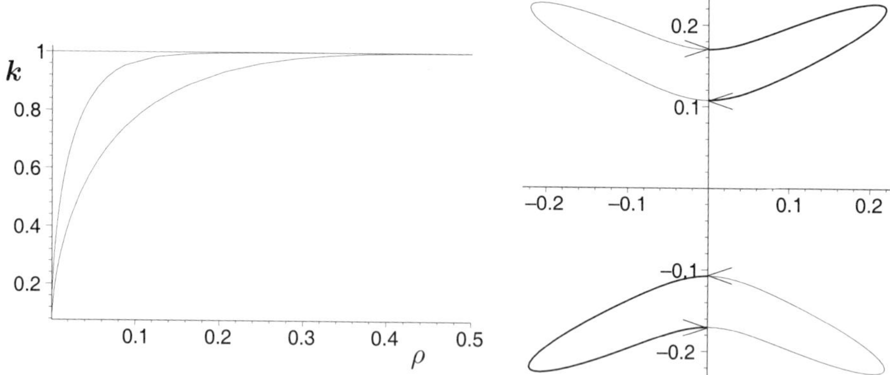

Fig. 12. On the left we plot k *versus* ρ for $\alpha = 0$ ($k \equiv 1$), $\alpha = -\pi/100$ and $\alpha = -\pi/2$. On the right are shown the locations of the constituent monopoles (fat *vs.* thin curves for opposite charges) at $\rho = 1/4$, varying α from $-\pi$ (indicated by the arrows) to 0.

4.1. Cooling

Cooling is the process by which one lowers the lattice action through local updates, replacing a link by a suitable combination of neighboring links, such that when remaining unchanged it satisfies the lattice equations of motion. There are by now many variants, using improved lattice actions (to reduce discretization errors), and criteria to stop the cooling [44]. Such a criterion is important, because when cooling too long either all the non-trivial fields are removed, or at best one reaches a self-dual solution. The latter is of course sometimes done on purpose, so as to reproduce the classical solutions on the lattice to quite some precision, or to look for solutions not known exactly. For the calorons this has been extensively studied [38], even in connection with a numerical implementation of the Nahm transformation [45], mainly in the presence of so-called twisted boundary conditions [46] to coach the system in having non-trivial holonomy.

From the dynamical point of view, one would like to find how often, and with which properties, calorons appear in the long distance fluctuations. In this case one would like to start from Monte Carlo generated configurations and stop the cooling process when reaching a plateau in the plot for the action as a function of the number of cooling steps. The plateau would be infinitely long if the configuration is a solution of the lattice equations of motion, but would still be sizeable if these are satisfied approximately. Therefore, configurations consisting of any number of well-separated instantons (Q) and anti-instantons ($\bar{Q}$) can form such plateaus, with an action

  F. Bruckmann, D. Nógrádi, P. van Baal

approximately equal to $8\pi^2(Q + \bar{Q})$, whereas the topological charge of the equivalent continuum configuration equals $Q - \bar{Q}$. An instanton can shrink to such a small size ρ (from the caloron point of view its two constituents monopoles getting too close together) that the lattice no longer supports it as a solution[13]. Also, when instantons and anti-instantons come close together they annihilate, this effect is independent of the lattice discretization. In both cases the plateau ends, and the cooling curve typically settles down at the next plateau where $Q + \bar{Q}$ has decreased by 1 or 2 unit, respectively (whereas $Q - \bar{Q}$ changes by ± 1 or 0, respectively). The expectation is that the annihilation of oppositely charged configurations is slow, at least when they are sufficiently separated, and that from the plateaus one can recover the statistical properties of instantons and calorons.

To prompt the system to have a given holonomy, one can freeze the time-like links at the spatial boundary of the box to the required holonomy [48], but in larger volumes it is expected that the confining environment itself will provide local regions with a sufficiently coherent background A_0 field associated with non-trivial holonomy. One important finding has been that after cooling the non-trivial holonomy is preserved to some degree [49]. This has been analyzed for SU(2) in terms of the so-called "asymptotic holonomy" L_∞, defined as the average of $\frac{1}{2}\operatorname{Tr}P(\vec{x})$ over all points $\vec{x}$ for which the action density, summed over t, is smaller than .0001. Histograms of L_∞ in the confined phase are shown in Fig. 13 (taken from Ref. [49]). Early on in the cooling process a clear peak at $L_\infty = 0$ is observed, which becomes a flat distribution when cooled down to the plateau associated to $Q+\bar{Q}=1$. Important is in particular that the distribution is *not* becoming peaked towards $L_\infty = \pm 1$. Many configurations with well "dissociated" calorons in this dynamical setting have been found [49], and with this method one can study the configurations in great detail. For SU(2) one can thus look for the monopole centers by finding where the Polyakov loop equals $\pm\mathbb{1}_2$ and use the fermion zero-modes for periodic and anti-periodic boundary conditions to test if they are localized on the appropriate constituents[14].

In these studies one also finds configurations that cannot be directly interpreted in terms of instantons, calorons and their possible constituent monopoles [49]. Interestingly, there are cases where two constituents appear, but of opposite fractional topological charge. These could arise from the "annihilation" of two other constituents of opposite duality [50] but where originally each of these, together with one of the surviving constituents, formed an (anti-)caloron. More recently these authors also performed cooling studies for SU(2) at considerably lower temperature than just around the

---

[13] This can be prevented by a suitable choice of improved lattice action [47].

[14] For SU(2) one profits from the fact that $z = 0$ always lies in the middle of the interval $[\mu_1, \mu_2]$, whereas $z = \frac{1}{2}$ always lies in the middle of $[\mu_2, 1 + \mu_1]$.

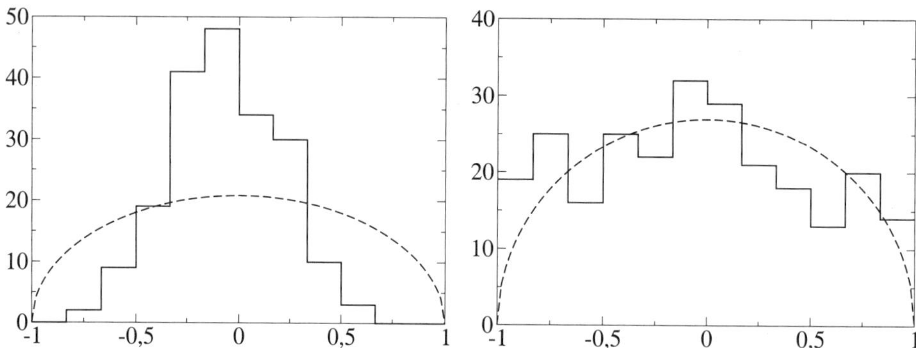

Fig. 13. The "asymptotic holonomy" L_∞ for an ensemble of $\mathcal{O}(200)$ configurations generated on a $16^3 \times 4$ lattice with periodic boundary conditions for a lattice coupling $4/e^2 = 2.2$. Left is shown the measurement on the first plateau, right on the last plateau. For comparison the Haar measure $\sqrt{1 - L^2}$ is shown by the dashed curve. Figures taken from Ref. [49].

deconfining phase transition [51]. The calorons in this case do not tend to "dissociate" in isolated lumps of action density. Nevertheless, constituents could be identified through the behavior of the Polyakov loop and were characteristic for non-trivial holonomy. This may explain why in the past constituent monopoles were never noticed in cooling studies. Finally let us mention that many results for SU(3) have now been obtained as well [51].

4.2. Zero-mode filter

The use of zero-modes as a filter relies on two observations. The first is, as we have seen, that zero-modes quite accurately trace the underlying gauge field of instantons, calorons and constituent monopoles. Secondly, a zero-mode probes the long distance features of the configuration and ignores the ultraviolet high momentum components, otherwise one could never have a zero-mode. In some sense the Dirac operator is a particular projection of the covariant momentum. For SU(3), comparing periodic and anti-periodic fermion boundary conditions [52], or cycling through all possible phases [53], Eq. (51), a significantly different behavior in the two phases was found. In the deconfined phase, the proper z-dependent behavior of the trivial holonomy configuration is seen. These are of course the old Harrington–Shepard solutions, but their zero-modes were previously only considered for anti-periodic boundary conditions. In the confined phase indications for a three lump structure is seen (see Fig. 14, taken from Ref. [53]) in a reasonable fraction of the configurations. These results are based on dynamical configurations generated with the Lüscher–Weisz [54] improved lattice action.

          F. Bruckmann, D. Nógrádi, P. van Baal

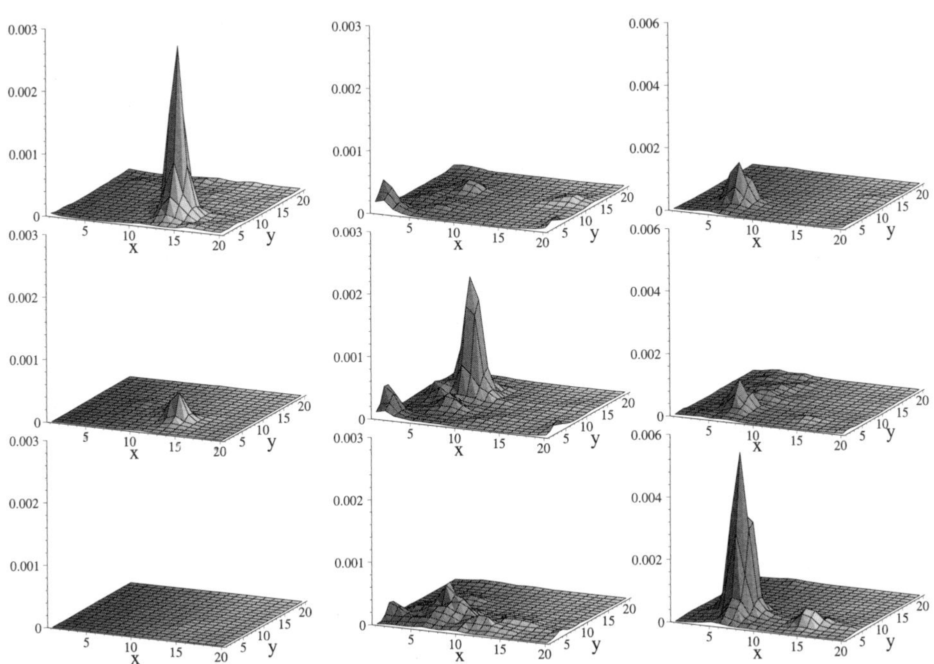

Fig. 14. The zero-mode density for a particular configuration in the confined phase on a 6×20^3 lattice with (Lüscher–Weisz) coupling $6/e^2 = 8.20$, taken from Ref. [53]. Shown are slices in (x, y) at $(t, z) = (5, 9)$ (left), $(t, z) = (2, 19)$ (center) and $(t, z) = (5, 18)$ (right), for $z = 0.05$, 0.3 and 0.65 (from top to bottom).

Configurations that had exactly one chiral zero-mode (which requires the use of a chirally improved lattice Dirac operator [55]) were singled out. By the index theorem, this implies the topological charge is equal to one. The dynamically generated configurations can, and typically will, consist of $Q + 1$ instantons and Q anti-instantons, giving rise to near zero-modes with less perfect chiral behavior. Work is in progress to analyze the near zero-modes, as well as performing cooling on the SU(3) gauge field configurations [56]. This will help to further rule out the unlikely possibility for the zero-mode to jump between instanton, rather than monopole constituents. A more theoretical study [57] has recently shown that it seems indeed impossible for the full signature of a single caloron with non-trivial holonomy to be emulated by the effect of hopping between instantons.

Even more remarkable is that a similar structure of multiple lumps for the topological charge 1 sector was found on a symmetric lattice of size 16^4 and 12^4 at low temperature, well in the confined phase [53,58]. In this case the lattice does not single out one direction as being (imaginary) time. Yet, using one of these directions to impose the z dependent boundary condition for the fermions, in almost half of the configurations one finds more than one lump. The results seem to indicate these lumps are randomly distributed over the volume. How to reconcile this with the results found with cooling, that pointed to "non-dissociated" calorons at low temperature, is at the moment not clear. But let us recall that fractionally charged instantons on the torus were long ago introduced by 't Hooft [46], and have been extensively studied on the lattice [60]. Twisted boundary conditions and, related to it, the quantization of the topological charge in units of $1/n$, make it more difficult to embed these solutions in a dynamical environment. Nevertheless, on the basis of some simple assumptions concerning the dynamical properties of these configurations, quasi realistic results have been obtained [61]. It is indeed compelling to interpret the $4kn$ dimensions of the SU(n) charge k instanton parameter space on the torus in terms of the space-time locations of kn instanton quarks, as they were called long ago [59] (even though their meaning at that time was more abstract). Results for instantons on $T^2 \times R^2$, that in some sense interpolate between the case of calorons and instantons on the torus, give further evidence [62] for this conjecture.

5. Conclusion

It is clear we can do no justice here to all the results that have been obtained in using lattice simulations. The great advantage of using zero-modes as a filter is that it minimizes the bias, since it directly uses the Monte Carlo configurations themselves. On the other hand, in the cooling studies one has access to the relation between the behavior of the Polyakov loop, which plays the role of an order parameter in SU(n) gauge theories, and the presence of constituent monopoles, which may give some insight in the underlying dynamics. Of course, we would like to make constituent monopoles into a precise tool, *e.g.* for testing the celebrated idea of a dual superconductor to describe confinement [63, 64], or as we speculated in the introduction, to describe a dense phase of monopoles to lead to deconfinement of magnetic charges, like quark deconfinment at high density.

On the more theoretical side we have also not touched upon the role these constituent monopoles play in supersymmetric gauge theories [65, 67] and how the initial motivation for reconsidering calorons with non-trivial holonomy came from the study of D-branes [8, 35, 66]. Nevertheless, we feel the applications to the dynamics of non-Abelian gauge theories is very promising. We hope to have given the reader some insight in this matter.

PvB would like to say to the organizers, and Michal Praszalowicz in particular, *dziękuję bardzo* for inviting him to Zakopane again, this time coinciding with the historic moment of the referendum to join the European Union. He also thanks Maxim Chernodub, Margarita García Pérez, Tony González-Arroyo, Thomas Kraan, Alvaro Montero and Carlos Pena for the many fruitful collaborations over the last 5 years, and Christof Gattringer, Michael Ilgenfritz, Boris Martemyanov and Michael Müller-Preussker for inspiring discussions concerning their lattice studies. The research of FB is supported by FOM.

REFERENCES

[1] B.J. Harrington, H.K. Shepard, *Phys. Rev.* **D17**, 2122 (1978); **D18**, 2990 (1978).

[2] D.J. Gross, R.D. Pisarski, L.G. Yaffe, *Rev. Mod. Phys.* **53**, 43 (1981).

[3] P. van Baal, in: *At the Frontiers of Particle Physics – Handbook of QCD*, Vol.2, ed. M. Shifman, World Scientific, Singapore, 2001, p. 683.

[4] G. 't Hooft, *Nucl. Phys.* **B79**, 276 (1974); A.M. Polyakov, *JETP Lett.* **20**, 194 (1974).

[5] E.B. Bogomol'ny, *Sov. J. Nucl. Phys.* **24**, 449 (1976); M.K. Prasad, C.M. Sommerfield, *Phys. Rev. Lett.* **35**, 760 (1975).

[6] N.S. Manton, *Nucl. Phys.* **B126**, 525 (1977).

[7] C. Montonen, D. Olive, *Phys. Lett.* **72B**, 117 (1977).

[8] K. Lee, P. Yi, *Phys. Rev.* **D56**, 3711 (1997).

[9] P. Rossi, *Nucl. Phys.* **B149**, 170 (1979).

[10] B. Julia, A. Zee, *Phys. Rev.* **D11**, 2227 (1975).

[11] T.C. Kraan, P. van Baal, *Nucl. Phys.* **B533**, 627 (1998).

[12] E. Witten, *Phys Lett.* **86B**, 283 (1979).

[13] W. Nahm, *Lect. Notes Phys.* **201**, 189 (1984).

[14] M.F. Atiyah, N.J. Hitchin, V. Drinfeld, Yu.I. Manin, *Phys. Lett.* **65A**, 185 (1978); M.F. Atiyah, *Geometry of Yang-Mills fields*, Fermi lectures, Scuola Normale Superiore, Pisa, 1979.

[15] C. Taubes, *Progress in Gauge Field Theory*, eds. G. 't Hooft *et al.*, Plenum Press, New York, 1984, p. 563.

[16] T.C. Kraan, P. van Baal, *Nucl. Phys.* **A642**, 229 (1998).

[17] G. 't Hooft, *Phys. Rev. Lett.* **37**, 8 (1976); *Phys. Rev.* **D14**, 3432 (1976); *Phys. Rep.* **142**, 357 (1986).

[18] C. Taubes, *J. Diff. Geom.* **17**, 139 (1982); **19**, 517 (1984).

[19] S.K. Donaldson, P.B. Kronheimer, *The Geometry of Four-Manifolds*, Clarendon Press, Oxford 1990.

[20] T. Schäfer, E.V. Shuryak, *Rev. Mod. Phys.* **70**, 323 (1998).

[21] D. Diakonov, `hep-ph/0212026`, to appear in *Prog. Part. Nucl. Phys.*

[22] M.F. Atiyah, I.M. Singer, *Ann. Math.* **87**, 484 (1968); **93**, 119 (1971).

[23] T. Banks, A. Casher, *Nucl. Phys.* **B169**, 103 (1980).

[24] R. Rajaraman, *Solitons and Instantons*, North-Holland, Amsterdam 1982.

[25] R. Jackiw, C. Nohl, C. Rebbi, *Phys. Rev.* **D15**, 1642 (1977).

[26] F. Bruckmann, P. van Baal, *Nucl. Phys.* **B653**, 105 (2002).

[27] P.J. Braam, P. van Baal, *Commun. Math. Phys.* **122**, 267 (1989); P. van Baal, *Nucl. Phys. B (Proc. Suppl.)* **49**, 238 (1996).

[28] E.F. Corrigan, D.B. Fairlie, S. Templeton, P. Goddard, *Nucl. Phys.* **B140**, 31 (1978).

[29] H. Osborn, *Ann. Phys. (N.Y.)* **135**, 373 (1981); *Nucl. Phys.* **B159**, 497 (1979).

[30] M. García Pérez, A. González-Arroyo, C. Pena, P. van Baal, *Phys. Rev.* **D60**, 031901 (1999).

[31] F. Bruckmann, D. Nógrádi, P. van Baal, *Nucl. Phys.* **B666**, 195 (2003).

[32] S.A. Brown, H. Panagopoulos, M.K. Prasad, *Phys. Rev.* **D26**, 854 (1982); H. Panagopoulos, *Phys. Rev.* **D28**, 380 (1983); A.S. Dancer, *Comm. Math. Phys.* **158**, 545 (1993).

[33] T.C. Kraan, P. van Baal, *Phys. Lett.* **B435**, 389 (1998).

[34] K. Lee, *Phys. Lett.* **B426**, 323 (1998); K. Lee, C. Lu, *Phys. Rev.* **D58**, 025011 (1998).

[35] T.C. Kraan, P. van Baal, *Phys. Lett.* **B428**, 268 (1998).

[36] T.C. Kraan, *Commun. Math. Phys.* **212**, 503 (2000).

[37] P. van Baal, in: *Lattice Fermions and Structure of the Vacuum*, eds. V. Mitrjushkin and G. Schierholz, Kluwer, Dordrecht 2000, p. 269.

[38] M. García Pérez, A. González-Arroyo, A. Montero, P. van Baal, *J. High Energy Phys.* **06**, 001 (1999).

[39] P. van Baal, *Nucl. Phys. B (Proc.Suppl.)* **106**, 586 (2002); *Nucl. Phys. B (Proc.Suppl.)* **108**, 3 (2002).

[40] C.J. Callias, *Commun. Math. Phys.* **62**, 213 (1978).

[41] M.N. Chernodub, T.C. Kraan, P. van Baal, *Nucl. Phys. B (Proc.Suppl.)* **83**, 556 (2000).

[42] P. Forgács, Z. Horváth, L. Palla, *Nucl. Phys.* **B192**, 141 (1981).

[43] M.F. Atiyah, N.J. Hitchin, *The Geometry and Dynamics of Magnetic Monopoles*, Princeton Univ. Press, 1988.

[44] M. García Pérez, O. Philipsen, I.O. Stamatescu, *Nucl. Phys.* **B551**, 293 (1999).

[45] A. González-Arroyo, C. Pena, *J. High Energy Phys.* **09**, 013 (1998); M. García Pérez, A. González-Arroyo, C. Pena, P. van Baal, *Nucl. Phys.* **B564**, 159 (2000).

[46] G. 't Hooft, *Nucl. Phys.* **B153**, 141 (1979).

5750   F. BRUCKMANN, D. NÓGRÁDI, P. VAN BAAL

[47] M. García Pérez, A. González-Arroyo, J. Snippe, P. van Baal, *Nucl. Phys.* **B413**, 535 (1994).

[48] E.-M. Ilgenfritz, M. Müller-Preussker, A.I. Veselov, in: *Lattice Fermions and Structure of the Vacuum*, eds. V. Mitrjushkin and G. Schierholz, Kluwer, Dordrecht 2000, p. 345; E.-M. Ilgenfritz, B. Martemyanov, M. Müller-Preussker, A.I. Veselov, *Nucl. Phys. B (Proc. Suppl.)* **94**, 407 (2001).

[49] E.-M. Ilgenfritz, B.V. Martemyanov, M. Müller-Preussker, S. Shcheredin, A.I. Veselov, *Phys. Rev.* **D66**, 074503 (2002).

[50] E.-M. Ilgenfritz, B.V. Martemyanov, M. Muller-Preussker, A.I. Veselov, *Nucl. Phys. B (Proc. Suppl.)* **106**, 589 (2002).

[51] E.-M. Ilgenfritz, talk presented at Lattice 2003, Tsukuba, July 2003.

[52] C. Gattringer, *Phys. Rev.* **D67**, 034507 (2003).

[53] C. Gattringer, S. Schaefer, *Nucl. Phys.* **B654**, 30 (2003).

[54] M. Lüscher, P. Weisz, *Commun. Math. Phys.* **97**, 59 (1985).

[55] C. Gattringer, *Phys. Rev.* **D63**, 114501 (2001); C. Gattringer, I. Hip, C.B. Lang, *Nucl. Phys.* **B597**, 451 (2001).

[56] C. Gattringer, M. Ilgenfritz, B. Martemyanov, M. Müller-Preussker, private communication.

[57] F. Bruckmann, D. Nógrádi, M. García Pérez, P. van Baal, hep-lat/0308017, to appear in the proceedings of Lattice 2003, Tsukuba, July 2003.

[58] C. Gattringer, talk presented at Lattice 2003, Tsukuba, July 2003.

[59] A.A. Belavin, V.A. Fateev, A.S. Schwarz, Y.S. Tyupkin, *Phys. Lett.* **B83**, 317 (1979).

[60] M. García Pérez, A. González-Arroyo, *J. Phys.* **A26**, 2667 (1993); M. García Pérez, PhD. Thesis, Univ. Atonoma de Madrid (1992).

[61] A. Gonzalez-Arroyo, P. Martínez, *Nucl. Phys.* **B459**, 337 (1996); A. González-Arroyo, P. Martínez, A. Montero, *Phys. Lett.* **B359**, 159 (1995).

[62] C. Ford, J. Pawlowski, *Phys. Lett.* **B540**, 153 (2002); hep-th/0302117.

[63] S. Mandelstam, *Phys. Rep.* **23**, 245 (1976).

[64] G. 't Hooft, in: *High Energy Physics*, ed. A. Zichichi, Editrice Compositori, Bolognia 1976; *Nucl. Phys.* **B138**, 1 (1978).

[65] N.M. Davies, T.J. Hollowood, V.V. Khoze, M.P. Mattis, *Nucl. Phys.* **B559**, 123 (1999).

[66] K. Lee, P. Yi, *Phys. Rev.* **D58**, 066005 (1998).

[67] D. Diakonov, V. Petrov, *Phys. Rev.* **D67**, 105007 (2003).